Student Solutions Guide

CALCULUS I WITH PRECALCULUS
A ONE-YEAR COURSE
SECOND EDITION
Larson / Hostetler / Edwards

Bruce H. Edwards
University of Florida

Dianna L. Zook
Indiana University, Purdue University at Fort Wayne

Houghton Mifflin Company   Boston   New York

Publisher: Jack Shira
Associate Sponsoring Editor: Cathy Cantin
Development Manager: Maureen Ross
Editorial Assistant: Elizabeth Kassab
Supervising Editor: Karen Carter
Senior Project Editor: Patty Bergin
Editorial Assistant: Julia Keller
Production Technology Supervisor: Gary Crespo
Executive Marketing Manager: Michael Busnach
Senior Marketing Manager: Danielle Potvin Curran
Marketing Coordinator: Nicole Mollica

Printed in the United States of America

ISBN: 0-618-56807-7

123456789-EB-09 08 07 06 05

PREFACE

This *Student Solutions Guide* is designed as a supplement to *Calculus I with Precalculus: A One-Year Course,* Second Edition, by Ron Larson, Robert P. Hostetler, and Bruce H. Edwards. All references to chapters, theorems, and exercises relate to the main text. Solutions to every odd-numbered exercise in the text are given with all essential algebraic steps included. Although this supplement is not a substitute for good study habits, it can be valuable when incorporated into a well-planned course of study. For suggestions that may assist you in the use of this text, your lecture notes, and this *Guide*, please refer to the student website for your text at *college.hmco.com*.

We have made every effort to see that the solutions are correct. However, we would appreciate hearing about any errors or other suggestions for improvement. Good luck with your study of calculus.

Bruce H. Edwards
Department of Mathematics
University of Florida

Dianna L. Zook
Indiana University
Purdue University
Fort Wayne, IN 46806

CONTENTS

CHAPTER P
Prerequisites

CHAPTER P
Prerequisites

Section P.1 Solving Equations

1. $2(x - 1) = 2x - 2$ is an *identity* by the Distributive Property. It is true for all real values of x.

3. $-6(x - 3) + 5 = -2x + 10$ is *conditional*. There are real values of x for which the equation is not true.

5. $4(x + 1) - 2x = 4x + 4 - 2x = 2x + 4 = 2(x + 2)$ is an *identity* by simplification. It is true for all real values of x.

7. $x^2 - 8x - 5 = (x - 4)^2 - 11$ is *conditional*. These are real values of x for which the equation is not true.

9. $3 + \dfrac{1}{x + 1} = \dfrac{4x}{x + 1}$ is *conditional*. There are real values of x for which the equation is not true.

11.
$$x + 11 = 15$$
$$x + 11 - 11 = 15 - 11$$
$$x = 4$$

13.
$$7 - 2x = 25$$
$$7 - 7 - 2x = 25 - 7$$
$$-2x = 18$$
$$\frac{-2x}{-2} = \frac{18}{-2}$$
$$x = -9$$

15.
$$8x - 5 = 3x + 20$$
$$8x - 3x - 5 = 3x - 3x + 20$$
$$5x - 5 = 20$$
$$5x - 5 + 5 = 20 + 5$$
$$5x = 25$$
$$\frac{5x}{5} = \frac{25}{5}$$
$$x = 5$$

17.
$$2(x + 5) - 7 = 3(x - 2)$$
$$2x + 10 - 7 = 3x - 6$$
$$2x + 3 = 3x - 6$$
$$-x = -9$$
$$x = 9$$

19.
$$x - 3(2x + 3) = 8 - 5x$$
$$x - 6x - 9 = 8 - 5x$$
$$-5x - 9 = 8 - 5x$$
$$-5x + 5x - 9 = 8 - 5x + 5x$$
$$-9 \neq 8$$

No solution

21.
$$\frac{5x}{4} + \frac{1}{2} = x - \frac{1}{2}$$
$$4\left(\frac{5x}{4}\right) + 4\left(\frac{1}{2}\right) = 4(x) - 4\left(\frac{1}{2}\right)$$
$$5x + 2 = 4x - 2$$
$$x = -4$$

23.
$$\frac{3}{2}(z + 5) - \frac{1}{4}(z + 24) = 0$$
$$4\left(\frac{3}{2}\right)(z + 5) - 4\left(\frac{1}{4}\right)(z + 24) = 4(0)$$
$$6(z + 5) - (z + 24) = 0$$
$$6z + 30 - z - 24 = 0$$
$$5z = -6$$
$$z = -\frac{6}{5}$$

25.
$$0.25x + 0.75(10 - x) = 3$$
$$4(0.25x) + 4(0.75)(10 - x) = 4(3)$$
$$x + 3(10 - x) = 12$$
$$x + 30 - 3x = 12$$
$$-2x = -18$$
$$x = 9$$

27. $x + 8 = 2(x - 2) - x$
$$x + 8 = 2x - 4 - x$$
$$x + 8 = x - 4$$
$$8 \neq -4$$
No solution

29.
$$\frac{100 - 4x}{3} = \frac{5x + 6}{4} + 6$$
$$12\left(\frac{100 - 4x}{3}\right) = 12\left(\frac{5x + 6}{4}\right) + 12(6)$$
$$4(100 - 4x) = 3(5x + 6) + 72$$
$$400 - 16x = 15x + 18 + 72$$
$$-31x = -310$$
$$x = 10$$

31.
$$\frac{5x - 4}{5x + 4} = \frac{2}{3}$$
$$3(5x - 4) = 2(5x + 4)$$
$$15x - 12 = 10x + 8$$
$$5x = 20$$
$$x = 4$$

33. $10 - \frac{13}{x} = 4 + \frac{5}{x}$
$$\frac{10x - 13}{x} = \frac{4x + 5}{x}$$
$$10x - 13 = 4x + 5$$
$$6x = 18$$
$$x = 3$$

35. $\frac{x}{x + 4} + \frac{4}{x + 4} + 2 = 0$
$$\frac{x + 4}{x + 4} + 2 = 0$$
$$1 + 2 = 0$$
$$3 \neq 0$$

Contradiction : no solution
The variable is divided out.

37. $\frac{1}{x} + \frac{2}{x - 5} = 0$ Multiply both sides by $x(x - 5)$.
$$1(x - 5) + 2x = 0$$
$$3x - 5 = 0$$
$$3x = 5$$
$$x = \frac{5}{3}$$

39. $\frac{2}{(x - 4)(x - 2)} = \frac{1}{x - 4} + \frac{2}{x - 2}$ Multiply both sides by $(x - 4)(x - 2)$.
$$2 = 1(x - 2) + 2(x - 4)$$
$$2 = x - 2 + 2x - 8$$
$$2 = 3x - 10$$
$$12 = 3x$$
$$4 = x$$

A check reveals that $x = 4$ is an extraneous solution–it makes the denominator zero. There is no real solution.

41.

$$\frac{1}{x-3} + \frac{1}{x+3} = \frac{10}{x^2-9}$$

$$\frac{(x+3)+(x-3)}{x^2-9} = \frac{10}{x^2-9}$$

$$2x = 10$$

$$x = 5$$

43. $\dfrac{3}{x^2-3x} + \dfrac{4}{x} = \dfrac{1}{x-3}$ Multiply both sides by $x(x-3)$.

$$3 + 4(x-3) = x$$

$$3 + 4x - 12 = x$$

$$3x = 9$$

$$x = 3$$

A check reveals that $x = 3$ is an extraneous solution, so there is no solution.

45.

$$(x+2)^2 + 4 = (x+3)^2$$

$$x^2 + 4x + 4 + 4 = x^2 + 6x + 9$$

$$4x + 8 = 6x + 9$$

$$-2x = 1$$

$$x = -\tfrac{1}{2}$$

47.

$$(x+2)^2 - x^2 = 4(x+1)$$

$$x^2 + 4x + 4 - x^2 = 4x + 4$$

$$4 = 4$$

The equation is an identity; every real number is a solution.

49. $2x^2 = 3 - 8x$

General form: $2x^2 + 8x - 3 = 0$

51. $(x-3)^2 = 3$

$$x^2 - 6x + 9 = 3$$

General form: $x^2 - 6x + 6 = 0$

53. $\tfrac{1}{5}(3x^2 - 10) = 18x$

$$3x^2 - 10 = 90x$$

General form: $3x^2 - 90x - 10 = 0$

55. $6x^2 + 3x = 0$

$$3x(2x+1) = 0$$

$$3x = 0 \quad \text{or} \quad 2x+1 = 0$$

$$x = 0 \quad \text{or} \qquad x = -\tfrac{1}{2}$$

57. $x^2 - 2x - 8 = 0$

$$(x-4)(x+2) = 0$$

$$x - 4 = 0 \quad \text{or} \quad x + 2 = 0$$

$$x = 4 \quad \text{or} \qquad x = -2$$

59. $x^2 + 10x + 25 = 0$

$$(x+5)(x+5) = 0$$

$$x + 5 = 0$$

$$x = -5$$

61. $3 + 5x - 2x^2 = 0$

$$(3-x)(1+2x) = 0$$

$$3 - x = 0 \quad \text{or} \quad 1 + 2x = 0$$

$$x = 3 \quad \text{or} \qquad x = -\tfrac{1}{2}$$

63. $x^2 + 4x = 12$

$$x^2 + 4x - 12 = 0$$

$$(x+6)(x-2) = 0$$

$$x + 6 = 0 \quad \text{or} \quad x - 2 = 0$$

$$x = -6 \quad \text{or} \qquad x = 2$$

65. $\tfrac{3}{4}x^2 + 8x + 20 = 0$

$$4\left(\tfrac{3}{4}x^2 + 8x + 20\right) = 4(0)$$

$$3x^2 + 32x + 80 = 0$$

$$(3x + 20)(x + 4) = 0$$

$$3x + 20 = 0 \quad \text{or} \quad x + 4 = 0$$

$$x = -\tfrac{20}{3} \quad \text{or} \qquad x = -4$$

67. $x^2 + 2ax + a^2 = 0$

$$(x+a)^2 = 0$$

$$x + a = 0$$

$$x = -a$$

69. $x^2 = 49$

$$x = \pm\sqrt{49}$$

$$= \pm 7$$

$$= \pm 7.00$$

71. $x^2 = 24$

$$x = \pm\sqrt{24}$$

$$x = \pm 2\sqrt{6}$$

$$\approx \pm 4.90$$

73. $3x^2 = 81$

$$x^2 = 27$$

$$x = \pm\sqrt{27} = \pm 3\sqrt{3}$$

$$x \approx \pm 5.20$$

75. $(x-12)^2 = 16$

$$x - 12 = \pm\sqrt{16}$$

$$x = 12 \pm 4$$

$$x = 16 \quad \text{or} \quad x = 8$$

$$x = 16.00 \quad \text{or} \quad x = 8.00$$

77. $(x+2)^2 = 14$

$$x + 2 = \pm\sqrt{14}$$

$$x = -2 \pm \sqrt{14}$$

$$x \approx 1.74 \quad \text{or} \quad x \approx -5.74$$

79. $(2x-1)^2 = 18$

$$2x - 1 = \pm\sqrt{18}$$

$$2x = 1 \pm 3\sqrt{2}$$

$$x = \frac{1 \pm 3\sqrt{2}}{2}$$

$$x \approx 2.62 \quad \text{or} \quad x \approx -1.62$$

81. $(x - 7)^2 = (x + 3)^2$

$x - 7 = \pm(x + 3)$

$x - 7 = x + 3$ or $x - 7 = -x - 3$

$-7 \neq 3$ $\qquad\qquad 2x = 4$

No solution $\qquad\qquad x = 2 = 2.00$

83. $\qquad x^2 - 2x = 0$

$x^2 - 2x + 1 = 0 + 1$

$(x - 1)^2 = 1$

$x - 1 = \pm\sqrt{1}$

$x = 1 \pm 1$

$x = 0$ or $x = 2$

85. $x^2 + 4x - 32 = 0$

$x^2 + 4x = 32$

$x^2 + 4x + 2^2 = 32 + 2^2$

$(x + 2)^2 = 36$

$x + 2 = \pm\sqrt{36}$

$x = -2 \pm 6$

$x = 4$ or $x = -8$

87. $x^2 + 6x + 2 = 0$

$x^2 + 6x = -2$

$x^2 + 6x + 3^2 = -2 + 3^2$

$(x + 3)^2 = 7$

$x + 3 = \pm\sqrt{7}$

$x = -3 \pm \sqrt{7}$

89. $9x^2 - 18x = -3$

$x^2 - 2x = -\dfrac{1}{3}$

$x^2 - 2x + 1 = -\dfrac{1}{3} + 1$

$(x - 1)^2 = \dfrac{2}{3}$

$x - 1 = \pm\sqrt{\dfrac{2}{3}}$

$x = 1 \pm \sqrt{\dfrac{6}{9}}$

$x = 1 \pm \dfrac{\sqrt{6}}{3}$

91. $\qquad 8 + 4x - x^2 = 0$

$-x^2 + 4x + 8 = 0$

$x^2 - 4x - 8 = 0$

$x^2 - 4x = 8$

$x^2 - 4x + 2^2 = 8 + 2^2$

$(x - 2)^2 = 12$

$x - 2 = \pm\sqrt{12}$

$x = 2 \pm 2\sqrt{3}$

93. $2x^2 + x - 1 = 0$

$x = \dfrac{-b \pm \sqrt{b^2 - 4ac}}{2a}$

$= \dfrac{-1 \pm \sqrt{1^2 - 4(2)(-1)}}{2(2)}$

$= \dfrac{-1 \pm 3}{4} = \dfrac{1}{2}, -1$

95. $16x^2 + 8x - 3 = 0$

$x = \dfrac{-b \pm \sqrt{b^2 - 4ac}}{2a}$

$= \dfrac{-8 \pm \sqrt{8^2 - 4(16)(-3)}}{2(16)}$

$= \dfrac{-8 \pm 16}{32} = \dfrac{1}{4}, -\dfrac{3}{4}$

97. $2 + 2x - x^2 = 0$

$x = \dfrac{-b \pm \sqrt{b^2 - 4ac}}{2a}$

$= \dfrac{-2 \pm \sqrt{2^2 - 4(-1)(2)}}{2(-1)}$

$= \dfrac{-2 \pm 2\sqrt{3}}{-2} = 1 \pm \sqrt{3}$

99. $14x = -44 - x^2$

$x = \dfrac{-b \pm \sqrt{b^2 - 4ac}}{2a}$

$= \dfrac{-14 \pm \sqrt{14^2 - 4(1)(44)}}{2(1)}$

$= \dfrac{-14 \pm 2\sqrt{5}}{2} = -7 \pm \sqrt{5}$

101. $x^2 + 8x - 4 = 0$

$x = \dfrac{-b \pm \sqrt{b^2 - 4ac}}{2a}$

$= \dfrac{-8 \pm \sqrt{8^2 - 4(1)(-4)}}{2(1)}$

$= \dfrac{-8 \pm 4\sqrt{5}}{2}$

$= -4 \pm 2\sqrt{5}$

103.
$$12x - 9x^2 = -3$$
$$-9x^2 + 12x + 3 = 0$$
$$x = \frac{-b \pm \sqrt{b^2 - 4ac}}{2a}$$
$$= \frac{-12 \pm \sqrt{12^2 - 4(-9)(3)}}{2(-9)}$$
$$= \frac{-12 \pm 6\sqrt{7}}{-18} = \frac{2}{3} \pm \frac{\sqrt{7}}{3}$$

105.
$$9x^2 + 24x + 16 = 0$$
$$x = \frac{-b \pm \sqrt{b^2 - 4ac}}{2a}$$
$$= \frac{-24 \pm \sqrt{24^2 - 4(9)(16)}}{2(9)}$$
$$= \frac{-24 \pm 0}{18}$$
$$= -\frac{4}{3}$$

107.
$$4x^2 + 4x = 7$$
$$4x^2 + 4x - 7 = 0$$
$$x = \frac{-b \pm \sqrt{b^2 - 4ac}}{2a}$$
$$= \frac{-4 \pm \sqrt{4^2 - 4(4)(-7)}}{2(4)}$$
$$= \frac{-4 \pm 8\sqrt{2}}{8} = -\frac{1}{2} \pm \sqrt{2}$$

109.
$$28x - 49x^2 = 4$$
$$-49x^2 + 28x - 4 = 0$$
$$x = \frac{-b \pm \sqrt{b^2 - 4ac}}{2a}$$
$$= \frac{-28 \pm \sqrt{28^2 - 4(-49)(-4)}}{2(-49)}$$
$$= \frac{-28 \pm 0}{-98} = \frac{2}{7}$$

111.
$$8t = 5 + 2t^2$$
$$-2t^2 + 8t - 5 = 0$$
$$t = \frac{-b \pm \sqrt{b^2 - 4ac}}{2a}$$
$$= \frac{-8 \pm \sqrt{8^2 - 4(-2)(-5)}}{2(-2)}$$
$$= \frac{-8 \pm 2\sqrt{6}}{-4} = 2 \pm \frac{\sqrt{6}}{2}$$

113.
$$(y - 5)^2 = 2y$$
$$y^2 - 12y + 25 = 0$$
$$x = \frac{-b \pm \sqrt{b^2 - 4ac}}{2a}$$
$$= \frac{-(-12) \pm \sqrt{(-12)^2 - 4(1)(25)}}{2(1)}$$
$$= \frac{12 \pm 2\sqrt{11}}{2} = 6 \pm \sqrt{11}$$

115.
$$\frac{1}{2}x^2 + \frac{3}{8}x = 2$$
$$4x^2 + 3x = 16$$
$$4x^2 + 3x - 16 = 0$$
$$x = \frac{-b \pm \sqrt{b^2 - 4ac}}{2a}$$
$$= \frac{-3 \pm \sqrt{3^2 - 4(4)(-16)}}{2(4)}$$
$$= \frac{-3 \pm \sqrt{265}}{8} = -\frac{3}{8} \pm \frac{\sqrt{265}}{8}$$

117.
$$5.1x^2 - 1.7x - 3.2 = 0$$
$$x = \frac{1.7 \pm \sqrt{(-1.7)^2 - 4(5.1)(-3.2)}}{2(5.1)}$$
$$x \approx 0.976, \, -0.643$$

119.
$$-0.067x^2 - 0.852x + 1.277 = 0$$
$$x = \frac{-(-0.852) \pm \sqrt{(-0.852)^2 - 4(-0.067)(1.277)}}{2(-0.067)}$$
$$x \approx -14.071, \, 1.355$$

121.
$$422x^2 - 506x - 347 = 0$$
$$x = \frac{506 \pm \sqrt{(-506)^2 - 4(422)(-347)}}{2(422)}$$
$$x \approx 1.687, \, -0.488$$

123. $12.67x^2 + 31.55x + 8.09 = 0$

$$x = \frac{-31.55 \pm \sqrt{(31.55)^2 - 4(12.67)(8.09)}}{2(12.67)}$$

$$x \approx -2.200, -0.290$$

125. $-16t^2 + 120t + 5 = 0$

$$16t^2 - 120t - 5 = 0$$

$$t = \frac{-(-120) \pm \sqrt{(-120)^2 - 4(16)(-5)}}{2(16)}$$

$$= \frac{120 \pm \sqrt{14,720}}{32}$$

$$t \approx 7.541 \quad \text{or} \quad t \approx -0.041$$

127. $-4.9t^2 + 46t + 1 = 0$

$$4.9t^2 - 46t - 1 = 0$$

$$t = \frac{-(-46) \pm \sqrt{(-46)^2 - 4(4.9)(-1)}}{2(4.9)}$$

$$= \frac{46 \pm \sqrt{2135.6}}{9.8}$$

$$t \approx 9.409 \quad \text{or} \quad t \approx -0.022$$

129. $x^2 - 2x - 1 = 0$

$$x^2 - 2x = 1$$

$$x^2 - 2x + 1^2 = 1 + 1^2$$

$$(x - 1)^2 = 2$$

$$x - 1 = \pm\sqrt{2}$$

$$x = 1 \pm \sqrt{2}$$

131. $(x + 3)^2 = 81$

$$x + 3 = \pm 9$$

$$x + 3 = 9 \quad \text{or} \quad x + 3 = -9$$

$$x = 6 \quad \text{or} \quad x = -12$$

133. $x^2 - x - \frac{11}{4} = 0$ Complete the Square

$$x^2 - x = \frac{11}{4}$$

$$x^2 - x + \left(\frac{1}{2}\right)^2 = \frac{11}{4} + \left(\frac{1}{2}\right)^2$$

$$\left(x - \frac{1}{2}\right)^2 = \frac{12}{4}$$

$$x - \frac{1}{2} = \pm\sqrt{\frac{12}{4}}$$

$$x = \frac{1}{2} \pm \sqrt{3}$$

135. $(x + 1)^2 = x^2$ Extract Square Roots

$$x^2 = (x + 1)^2$$

$$x = \pm(x + 1)$$

For $x = +(x + 1)$:

$$0 \neq 1 \quad \text{No solution}$$

For $x = -(x + 1)$:

$$2x = -1$$

$$x = -\frac{1}{2}$$

137. $3x + 4 = 2x^2 - 7$ Quadratic Formula

$$0 = 2x^2 - 3x - 11$$

$$x = \frac{-(-3) \pm \sqrt{(-3)^2 - 4(2)(-11)}}{2(2)}$$

$$= \frac{3 \pm \sqrt{97}}{4} = \frac{3}{4} \pm \frac{\sqrt{97}}{4}$$

139. $a^2x^2 - b^2 = 0$

$$(ax + b)(ax - b) = 0$$

$$ax + b = 0 \Rightarrow x = -\frac{b}{a}$$

$$ax - b = 0 \Rightarrow x = \frac{b}{a}$$

141. $4x^4 - 18x^2 = 0$

$$2x^2(2x^2 - 9) = 0$$

$$2x^2 = 0 \Rightarrow x = 0$$

$$2x^2 - 9 = 0 \Rightarrow x = \pm\frac{3\sqrt{2}}{2}$$

143.
$$x^4 - 81 = 0$$
$$(x^2 + 9)(x + 3)(x - 3) = 0$$
$$x^2 + 9 = 0 \implies \text{No real solutions}$$
$$x + 3 = 0 \implies x = -3$$
$$x - 3 = 0 \implies x = 3$$

145.
$$x^3 + 216 = 0$$
$$x^3 + 6^3 = 0$$
$$(x + 6)(x^2 - 6x + 36) = 0$$
$$x + 6 = 0 \implies x = -6$$
$$x^2 - 6x + 36 = 0 \implies \text{No real solutions}$$

(By completing the square)

147. $5x^3 + 30x^2 + 45x = 0$
$$5x(x^2 + 6x + 9) = 0$$
$$5x(x + 3)^2 = 0$$
$$5x = 0 \implies x = 0$$
$$x + 3 = 0 \implies x = -3$$

149.
$$x^3 - 3x^2 - x + 3 = 0$$
$$x^2(x - 3) - (x - 3) = 0$$
$$(x - 3)(x^2 - 1) = 0$$
$$(x - 3)(x + 1)(x - 1) = 0$$
$$x - 3 = 0 \implies x = 3$$
$$x + 1 = 0 \implies x = -1$$
$$x - 1 = 0 \implies x = 1$$

151.
$$x^4 - x^3 + x - 1 = 0$$
$$x^3(x - 1) + (x - 1) = 0$$
$$(x - 1)(x^3 + 1) = 0$$
$$(x - 1)(x + 1)(x^2 - x + 1) = 0$$
$$x - 1 = 0 \implies x = 1$$
$$x + 1 = 0 \implies x = -1$$
$$x^2 - x + 1 = 0 \implies \text{No real solutions} \quad \text{(By the Quadratic Formula)}$$

153.
$$x^4 - 4x^2 + 3 = 0$$
$$(x^2 - 3)(x^2 - 1) = 0$$
$$\left(x + \sqrt{3}\right)\left(x - \sqrt{3}\right)(x + 1)(x - 1) = 0$$
$$x + \sqrt{3} = 0 \implies x = -\sqrt{3}$$
$$x - \sqrt{3} = 0 \implies x = \sqrt{3}$$
$$x + 1 = 0 \implies x = -1$$
$$x - 1 = 0 \implies x = 1$$

155.
$$4x^4 - 65x^2 + 16 = 0$$
$$(4x^2 - 1)(x^2 - 16) = 0$$
$$(2x + 1)(2x - 1)(x + 4)(x - 4) = 0$$
$$2x + 1 = 0 \implies x = -\tfrac{1}{2}$$
$$2x - 1 = 0 \implies x = \tfrac{1}{2}$$
$$x + 4 = 0 \implies x = -4$$
$$x - 4 = 0 \implies x = 4$$

157.
$$x^6 + 7x^3 - 8 = 0$$
$$(x^3 + 8)(x^3 - 1) = 0$$
$$(x + 2)(x^2 - 2x + 4)(x - 1)(x^2 + x + 1) = 0$$
$$x + 2 = 0 \implies x = -2$$
$$x^2 - 2x + 4 = 0 \implies \text{No real solutions} \quad \text{(By the Quadratic Formula)}$$
$$x - 1 = 0 \implies x = 1$$
$$x^2 + x + 1 = 0 \implies \text{No real solutions} \quad \text{(By the Quadratic Formula)}$$

159. $\sqrt{2x} - 10 = 0$

$\qquad \sqrt{2x} = 10$

$\qquad 2x = 100$

$\qquad x = 50$

161. $\sqrt{x - 10} - 4 = 0$

$\qquad \sqrt{x - 10} = 4$

$\qquad x - 10 = 16$

$\qquad x = 26$

163. $\sqrt[3]{2x + 5} + 3 = 0$

$\qquad \sqrt[3]{2x + 5} = -3$

$\qquad 2x + 5 = -27$

$\qquad 2x = -32$

$\qquad x = -16$

165. $x + \sqrt{26 - 11x} = 4$

$\qquad 4 - x = \sqrt{26 - 11x}$

$\qquad 16 - 8x + x^2 = 26 - 11x$

$\qquad x^2 + 3x - 10 = 0$

$\qquad (x + 5)(x - 2) = 0$

$\qquad x + 5 = 0 \implies x = -5$

$\qquad x - 2 = 0 \implies x = 2$

167. $\sqrt{x + 1} = \sqrt{3x + 1}$

$\qquad x + 1 = 3x + 1$

$\qquad -2x = 0$

$\qquad x = 0$

169. $(x - 5)^{3/2} = 8$

$\qquad (x - 5)^3 = 8^2$

$\qquad x - 5 = 8^{2/3}$

$\qquad x = 5 + 4$

$\qquad x = 9$

171. $(x + 3)^{2/3} = 8$

$\qquad (x + 3)^2 = 8^3$

$\qquad x + 3 = \pm\sqrt{8^3}$

$\qquad x + 3 = \pm\sqrt{512}$

$\qquad x = -3 \pm 16\sqrt{2}$

173. $(x^2 - 5)^{3/2} = 27$

$\qquad x^2 - 5 = 27^{2/3}$

$\qquad x^2 = 5 + 9$

$\qquad x^2 = 14$

$\qquad x = \pm\sqrt{14}$

175. $3x(x - 1)^{1/2} + 2(x - 1)^{3/2} = 0$

$\qquad (x - 1)^{1/2}[3x + 2(x - 1)] = 0$

$\qquad (x - 1)^{1/2}(5x - 2) = 0$

$\qquad (x - 1)^{1/2} = 0 \implies x - 1 = 0 \implies x = 1$

$\qquad 5x - 2 = 0 \implies x = \frac{2}{5}$ which is extraneous.

177. $\dfrac{20 - x}{x} = x$

$\qquad 20 - x = x^2$

$\qquad 0 = x^2 + x - 20$

$\qquad 0 = (x + 5)(x - 4)$

$\qquad x + 5 = 0 \implies x = -5$

$\qquad x - 4 = 0 \implies x = 4$

179.

$$\frac{1}{x} - \frac{1}{x+1} = 3$$

$$x(x+1)\frac{1}{x} - x(x+1)\frac{1}{x+1} = x(x+1)(3)$$

$$x + 1 - x = 3x(x+1)$$

$$1 = 3x^2 + 3x$$

$$0 = 3x^2 + 3x - 1; \ a = 3, \ b = 3, \ c = -1$$

$$x = \frac{-3 \pm \sqrt{(3)^2 - 4(3)(-1)}}{2(3)} = \frac{-3 \pm \sqrt{21}}{6}$$

181.

$$x = \frac{3}{x} + \frac{1}{2}$$

$$(2x)(x) = (2x)\left(\frac{3}{x}\right) + (2x)\left(\frac{1}{2}\right)$$

$$2x^2 = 6 + x$$

$$2x^2 - x - 6 = 0$$

$$(2x + 3)(x - 2) = 0$$

$$2x + 3 = 0 \implies x = -\frac{3}{2}$$

$$x - 2 = 0 \implies x = 2$$

183.

$$\frac{4}{x+1} - \frac{3}{x+2} = 1$$

$$4(x+2) - 3(x+1) = (x+1)(x+2), x \neq -2, -1$$

$$4x + 8 - 3x - 3 = x^2 + 3x + 2$$

$$x^2 + 2x - 3 = 0$$

$$(x - 1)(x + 3) = 0$$

$$x - 1 = 0 \implies x = 1$$

$$x + 3 = 0 \implies x = -3$$

185. $|2x - 1| = 5$

$$2x - 1 = 5 \implies x = 3$$

$$-(2x - 1) = 5 \implies x = -2$$

187. $|x| = x^2 + x - 3$

$$x = x^2 + x - 3 \quad \text{OR} \qquad\qquad -x = x^2 + x - 3$$

$$x^2 - 3 = 0 \qquad\qquad\qquad x^2 + 2x - 3 = 0$$

$$x = \pm\sqrt{3} \qquad\qquad\qquad (x - 1)(x + 3) = 0$$

$$x - 1 = 0 \implies x = 1$$

$$x + 3 = 0 \implies x = -3$$

Only $x = \sqrt{3}$, and $x = -3$ are solutions to the original equation. $x = -\sqrt{3}$ and $x = 1$ are extraneous.

189. $|x + 1| = x^2 - 5$

$$x + 1 = x^2 - 5 \qquad \text{OR} \qquad -(x + 1) = x^2 - 5$$

$$x^2 - x - 6 = 0 \qquad\qquad\qquad -x - 1 = x^2 - 5$$

$$(x - 3)(x + 2) = 0 \qquad\qquad\qquad x^2 + x - 4 = 0$$

$$x - 3 = 0 \implies x = 3 \qquad\qquad\qquad x = \frac{-1 \pm \sqrt{17}}{2}$$

$$x + 2 = 0 \implies x = -2$$

Only $x = 3$ and $x = \dfrac{-1 - \sqrt{17}}{2}$ are solutions to the original equation. $x = -2$ and $x = \dfrac{-1 + \sqrt{17}}{2}$ are extraneous.

191. The student should have subtracted $15x$ from both sides so that the equation is equal to zero. By factoring out an x, there are **two** solutions.

$$x = 0 \quad \text{or} \quad x = 6$$

193. Answers will vary. For example:
Two equations are equivalent if they differ only by algebraic simplification steps and have the same solutions.

$x^2 + 3x + 4 = x + 1$ is equivalent to $x^2 + 2x + 3 = 0$.

195. $C = 0.40m + 2750$

When $C = 10,000$ we have:

$10,000 = 0.40m + 2750$

$7250 = 0.40m$

$m = 18,125$ miles

(c) $(w - 34)(w + 48) = 0$

$w = 34$ or $w = -48$ Extraneous

$l = 34 + 14 = 48$

width: 34 feet

length: 48 feet

197. (a)

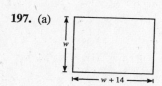

(b) $\qquad w(w + 14) = 1632$

$w^2 + 14w - 1632 = 0$

199. $\quad 16 = 0.432x - 10.44$

$26.44 = 0.432x$

$\dfrac{26.44}{0.432} = x$

$x \approx 61.2$ inches

201.

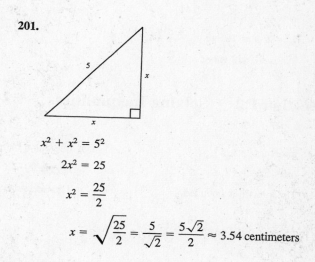

$x^2 + x^2 = 5^2$

$2x^2 = 25$

$x^2 = \dfrac{25}{2}$

$x = \sqrt{\dfrac{25}{2}} = \dfrac{5}{\sqrt{2}} = \dfrac{5\sqrt{2}}{2} \approx 3.54$ centimeters

203. Let $r = $ speed of the eastbound plane and $r + 50 = $ speed of the northbound plane. After 3 hours the eastbound plane has traveled $3r$ miles and the northbound plane has traveled $3(r + 50)$ miles.

$$(3r)^2 + [3(r + 50)]^2 = 2440^2$$

$9r^2 + 9(r^2 + 100r + 2500) = 5,953,600$

$18r^2 + 900r - 5,931,100 = 0$

By the Quadratic Formula, $r \approx 550$ (discard the negative value of r as extraneous).

Speed of the eastbound plane: 550 miles per hour

Speed of the northbound plane: 600 miles per hour

205. $x(20 - 0.0002x) = 500,000$

$20x - 0.0002x^2 = 500,000$

$0 = 0.0002x^2 - 20x + 500,000$

By the Quadratic Formula, $x = 50,000$ units.

207. $p = 0.0014t^2 + 0.004t + 0.23$ where $t = 0$ corresponds to 1940.

$0.0014t^2 + 0.004t + 0.23 > 8.00$

$0.0014t^2 + 0.004t - 7.77 > 0$

By the Quadratic Formula we have $t \approx -76$, which is extraneous, and $t \approx 73$, which would correspond to the year 2013.

209. False. $x(3 - x) = 10$ is a quadratic equation.

211. True. See Example 11. In solving the radical equation $\sqrt{2x + 7} - x = 2$, an extraneous solution of $x = -3$ was produced.

213. $x + |x - a| = b$

Solving for x we have:

First equation:

$x + x - a = b$

$$x = \frac{a + b}{2}$$

Second equation:

$x - x + a = b$

$a = b$

Thus, $x = 9$ will be the only solution if $9 = \dfrac{a + b}{2}$ and $a \neq b$ and $9 > a$.

For example,

$a = 0, b = 18$

$a = 2, b = 16$, etc.

Answers will vary.

Section P.2   Solving Inequalities

1. Interval: $[-1, 5]$

Inequality: $-1 \leq x \leq 5$

The interval is bounded.

3. Interval: $(11, \infty)$

Inequality: $11 < x < \infty$

The interval is unbounded.

5. Interval: $(-\infty, -2)$

Inequality:
$-\infty < x < -2$ or $x < -2$

The interval is unbounded.

7. $x < 3$

Matches (b)

9. $-3 < x \leq 4$

Matches (d)

11. $|x| < 3 \implies -3 < x < 3$

Matches (e)

13. (a) $x = 3$

$5(3) - 12 \overset{?}{>} 0$

$3 > 0$

Yes, $x = 3$ is a solution.

(b) $x = -3$

$5(-3) - 12 \overset{?}{>} 0$

$-27 \not> 0$

No, $x = -3$ is not a solution.

(c) $x = \frac{5}{2}$

$5\left(\frac{5}{2}\right) - 12 \overset{?}{>} 0$

$\frac{1}{2} > 0$

Yes, $x = \frac{5}{2}$ is a solution.

(d) $x = \frac{3}{2}$

$5\left(\frac{3}{2}\right) - 12 \overset{?}{>} 0$

$-\frac{9}{2} \not> 0$

No, $x = \frac{3}{2}$ is not a solution.

15. (a) $x = 4$

$0 \overset{?}{<} \dfrac{4 - 2}{4} \overset{?}{<} 2$

$0 < \dfrac{1}{2} < 2$

Yes, $x = 4$ is a solution.

(b) $x = 10$

$0 \overset{?}{<} \dfrac{10 - 2}{4} \overset{?}{<} 2$

$0 < 2 \not< 2$

No, $x = 10$ is not a solution.

(c) $x = 0$

$0 \overset{?}{<} \dfrac{0 - 2}{4} \overset{?}{<} 2$

$0 \not< -\dfrac{1}{2} < 2$

No, $x = 0$ is not a solution.

(d) $x = \dfrac{7}{2}$

$0 \overset{?}{<} \dfrac{\frac{7}{2} - 2}{4} \overset{?}{<} 2$

$0 < \dfrac{3}{8} < 2$

Yes, $x = \frac{7}{2}$ is a solution.

17. (a) $x = 13$

$|13 - 10| \overset{?}{\geq} 3$

$3 \geq 3$

Yes, $x = 13$ is a solution.

(b) $x = -1$

$|-1 - 10| \overset{?}{\geq} 3$

$11 \geq 3$

Yes, $x = -1$ is a solution.

(c) $x = 14$

$|14 - 10| \overset{?}{\geq} 3$

$4 \geq 3$

Yes, $x = 14$ is a solution.

(d) $x = 9$

$|9 - 10| \overset{?}{\geq} 3$

$1 \not\geq 3$

No, $x = 9$ is not a solution.

19. $4x < 12$

$\frac{1}{4}(4x) < \frac{1}{4}(12)$

$x < 3$

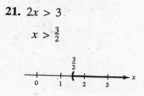

21. $2x > 3$

$x > \frac{3}{2}$

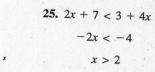

23. $x - 5 \geq 7$

$x \geq 12$

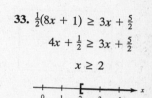

25. $2x + 7 < 3 + 4x$

$-2x < -4$

$x > 2$

27. $2x - 1 \geq 1 - 5x$

$7x \geq 2$

$x \geq \frac{2}{7}$

29. $4 - 2x < 3(3 - x)$

$4 - 2x < 9 - 3x$

$x < 5$

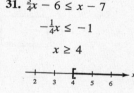

31. $\frac{3}{4}x - 6 \leq x - 7$

$-\frac{1}{4}x \leq -1$

$x \geq 4$

33. $\frac{1}{2}(8x + 1) \geq 3x + \frac{5}{2}$

$4x + \frac{1}{2} \geq 3x + \frac{5}{2}$

$x \geq 2$

35. $3.6x + 11 \geq -3.4$

$3.6x \geq -14.4$

$x \geq -4$

37. $1 < 2x + 3 < 9$

$-2 < 2x < 6$

$-1 < x < 3$

39. $-4 < \frac{2x - 3}{3} < 4$

$-12 < 2x - 3 < 12$

$-9 < 2x < 15$

$-\frac{9}{2} < x < \frac{15}{2}$

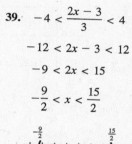

41. $\frac{3}{4} > x + 1 > \frac{1}{4}$

$-\frac{1}{4} > x > -\frac{3}{4}$

$-\frac{3}{4} < x < -\frac{1}{4}$

43. $3.2 \leq 0.4x - 1 \leq 4.4$

$4.2 \leq 0.4x \leq 5.4$

$10.5 \leq x \leq 13.5$

45. $|x| < 6$

$-6 < x < 6$

47. $\left|\frac{x}{2}\right| > 5$

$\frac{x}{2} < -5$ or $\frac{x}{2} > 5$

$x < -10 \qquad x > 10$

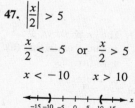

49. $|x - 5| < -1$

No solution. The absolute value of a number cannot be less than a negative number.

51. $|x - 20| \leq 6$

$-6 \leq x - 20 \leq 6$

$14 \leq x \leq 26$

53. $|3 - 4x| \geq 9$

$3 - 4x \leq -9$ or $3 - 4x \geq 9$

$-4x \leq -12 \qquad -4x \geq 6$

$x \geq 3 \qquad x \leq -\frac{3}{2}$

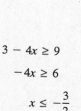

55. $\left|\frac{x - 3}{2}\right| \leq 5$

$-5 \leq \frac{x - 3}{2} \leq 5$

$-10 \leq x - 3 \leq 10$

$-7 \leq x \leq 13$

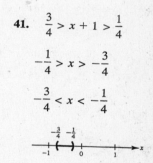

57. $|9 - 2x| - 2 < -1$

$|9 - 2x| < 1$

$-1 < 9 - 2x < 1$

$-10 < -2x < -8$

$5 > x > 4$

$4 < x < 5$

59. $2|x + 10| \geq 9$

$|x + 10| \geq \frac{9}{2}$

$x + 10 \leq -\frac{9}{2}$ or $x + 10 \geq \frac{9}{2}$

$x \leq -\frac{29}{2}$ $x \geq -\frac{11}{2}$

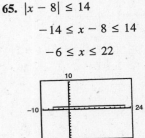

61. $6x > 12$

$x > 2$

63. $5 - 2x \geq 1$

$-2x \geq -4$

$x \leq 2$

65. $|x - 8| \leq 14$

$-14 \leq x - 8 \leq 14$

$-6 \leq x \leq 22$

67. $2|x + 7| \geq 13$

$|x + 7| \geq \frac{13}{2}$

$x + 7 \leq -\frac{13}{2}$ or $x + 7 \geq \frac{13}{2}$

$x \leq -\frac{27}{2}$ $x \geq -\frac{1}{2}$

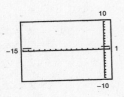

69. $x - 5 \geq 0$

$x \geq 5$

$[5, \infty)$

71. $x + 3 \geq 0$

$x \geq -3$

$[-3, \infty)$

73. $7 - 2x \geq 0$

$-2x \geq -7$

$x \leq \frac{7}{2}$

$\left(-\infty, \frac{7}{2}\right]$

75. $|x - 10| < 8$

All real numbers within 8 units of 10

77. The midpoint of the interval $[-3, 3]$ is 0. The interval represents all real numbers x no more than 3 units from 0.

$|x - 0| \leq 3$

$|x| \leq 3$

79. The graph shows all real numbers at least 3 units from 7.

$|x - 7| \geq 3$

81. All real numbers within 10 units of 12

$|x - 12| < 10$

83. All real numbers more than 5 units from -3

$|x - (-3)| > 5$

$|x + 3| > 5$

85. $x^2 - 3 < 0$

(a) $x = 3$

$(3)^2 - 3 \overset{?}{<} 0$

$6 \not< 0$

No, $x = 3$ is not a solution.

(b) $x = 0$

$(0)^2 - 3 \overset{?}{<} 0$

$-3 < 0$

Yes, $x = 0$ is a solution.

(c) $x = \frac{3}{2}$

$\left(\frac{3}{2}\right)^2 - 3 \overset{?}{<} 0$

$-\frac{3}{4} < 0$

Yes, $x = \frac{3}{2}$ is a solution.

(d) $x = -5$

$(-5)^2 - 3 \overset{?}{<} 0$

$22 \not< 0$

No, $x = -5$ is not a solution.

87. $\dfrac{x + 2}{x - 4} \geq 3$

(a) $x = 5$

$\dfrac{5 + 2}{5 - 4} \overset{?}{\geq} 3$

$7 \geq 3$

Yes, $x = 5$ is a solution.

(b) $x = 4$

$\dfrac{4 + 2}{4 - 4} \overset{?}{\geq} 3$

$\dfrac{6}{0} \not\geq 3; \dfrac{6}{0}$ is undefined.

No, $x = 4$ is not a solution.

(c) $x = -\dfrac{9}{2}$

$\dfrac{-\frac{9}{2} + 2}{-\frac{9}{2} - 4} \overset{?}{\geq} 3$

$\dfrac{5}{17} \not\geq 3$

No, $x = -\dfrac{9}{2}$ is not a solution.

(d) $x = \dfrac{9}{2}$

$\dfrac{\frac{9}{2} + 2}{\frac{9}{2} - 4} \overset{?}{\geq} 3$

$13 \geq 3$

Yes, $x = \dfrac{9}{2}$ is a solution.

89. $2x^2 - x - 6 = (2x + 3)(x - 2)$

$2x + 3 = 0 \implies x = -\dfrac{3}{2}$

$x - 2 = 0 \implies x = 2$

Critical numbers: $x = -\dfrac{3}{2}, x = 2$

91. $2 + \dfrac{3}{x - 5} = \dfrac{2(x - 5) + 3}{x - 5} = \dfrac{2x - 7}{x - 5}$

$2x - 7 = 0 \implies x = \dfrac{7}{2}$

$x - 5 = 0 \implies x = 5$

Critical numbers: $x = \dfrac{7}{2}, x = 5$

93.

$x^2 \leq 9$

$x^2 - 9 \leq 0$

$(x + 3)(x - 3) \leq 0$

Critical numbers: $x = \pm 3$

Test intervals: $(-\infty, -3), (-3, 3), (3, \infty)$

Test: Is $(x + 3)(x - 3) \leq 0$?

Interval	x-value	Value of $x^2 - 9$	Conclusion
$(-\infty, -3)$	$x = -4$	$16 - 9 = 7$	Positive
$(-3, 3)$	$x = 0$	$0 - 9 = -9$	Negative
$(3, \infty)$	$x = 4$	$16 - 9 = 7$	Positive

Solution set: $[-3, 3]$

95.

$(x + 2)^2 < 25$

$x^2 + 4x + 4 < 25$

$x^2 + 4x - 21 < 0$

$(x + 7)(x - 3) < 0$

Critical numbers: $x = -7, x = 3$

Test intervals: $(-\infty, -7), (-7, 3), (3, \infty)$

Test: Is $(x + 7)(x - 3) < 0$?

Interval	x-value	Value of $(x + 7)(x - 3)$	Conclusion
$(-\infty, -7)$	$x = -10$	$(-3)(-13) = 39$	Positive
$(-7, 3)$	$x = 0$	$(7)(-3) = -21$	Negative
$(3, \infty)$	$x = 5$	$(12)(2) = 24$	Positive

Solution set: $(-7, 3)$

97. $x^2 + 4x + 4 \geq 9$

$x^2 + 4x - 5 \geq 0$

$(x + 5)(x - 1) \geq 0$

Critical numbers: $x = -5, x = 1$

Test intervals: $(-\infty, -5), (-5, 1), (1, \infty)$

Test: Is $(x + 5)(x - 1) \geq 0$?

Interval	x-value	Value of $(x + 5)(x - 1)$	Conclusion
$(-\infty, -5)$	$x = -6$	$(-1)(-7) = 7$	Positive
$(-5, 1)$	$x = 0$	$(5)(-1) = -5$	Negative
$(1, \infty)$	$x = 2$	$(7)(1) = 7$	Positive

Solution set: $(-\infty, -5] \cup [1, \infty)$

99. $x^2 + x < 6$

$x^2 + x - 6 < 0$

$(x + 3)(x - 2) < 0$

Critical numbers: $x = -3, x = 2$

Test intervals: $(-\infty, -3), (-3, 2), (2, \infty)$

Test: Is $(x + 3)(x - 2) < 0$?

Interval	x-value	Value of $(x + 3)(x - 2)$	Conclusion
$(-\infty, -3)$	$x = -4$	$(-1)(-6) = 6$	Positive
$(-3, 2)$	$x = 0$	$(3)(-2) = -6$	Negative
$(2, \infty)$	$x = 3$	$(6)(1) = 6$	Positive

Solution set: $(-3, 2)$

101. $x^2 + 2x - 3 < 0$

$(x + 3)(x - 1) < 0$

Critical numbers: $x = -3, x = 1$

Test intervals: $(-\infty, -3), (-3, 1), (1, \infty)$

Test: Is $(x + 3)(x - 1) < 0$?

Interval	x-value	Value of $(x + 3)(x - 1)$	Conclusion
$(-\infty, -3)$	$x = -4$	$(-1)(-5) = 5$	Positive
$(-3, 1)$	$x = 0$	$(3)(-1) = -3$	Negative
$(1, \infty)$	$x = 2$	$(5)(1) = 5$	Positive

Solution set: $(-3, 1)$

103. $x^2 + 8x - 5 \geq 0$

$x^2 + 8x - 5 = 0$ Complete the Square

$x^2 + 8x + 16 = 5 + 16$

$(x + 4)^2 = 21$

$x + 4 = \pm\sqrt{21}$

$x = -4 \pm \sqrt{21}$

Critical Numbers: $x = -4 \pm \sqrt{21}$

Test Intervals: $\left(-\infty, -4 - \sqrt{21}\right), \left(-4 - \sqrt{21}, -4 + \sqrt{21}\right), \left(-4 + \sqrt{21}, \infty\right)$

Test: Is $x^2 + 8x - 5 \geq 0$?

Interval	x-value	Value of $x^2 + 8x - 5$	Conclusion
$\left(-\infty, -4 - \sqrt{21}\right)$	$x = -10$	$100 - 80 - 5 = 15$	Positive
$\left(-4 - \sqrt{21}, -4 + \sqrt{21}\right)$	$x = 0$	$0 + 0 - 5 = -5$	Negative
$\left(-4 + \sqrt{21}, \infty\right)$	$x = 2$	$4 + 16 - 5 = 15$	Positive

Solution set: $\left(-\infty, -4 - \sqrt{21}\right] \cup \left[-4 + \sqrt{21}, \infty\right)$

105. $x^3 - 3x^2 - x + 3 > 0$

$x^2(x - 3) - 1(x - 3) > 0$

$(x^2 - 1)(x - 3) > 0$

$(x + 1)(x - 1)(x - 3) > 0$

Critical Numbers: $x = \pm 1, x = 3$

Test Intervals: $(-\infty, -1), (-1, 1), (1, 3), (3, \infty)$

Test: Is $(x + 1)(x - 1)(x - 3) > 0$?

Interval	x-value	Value of $(x + 1)(x - 1)(x - 3)$	Conclusion
$(-\infty, -1)$	$x = -2$	$(-1)(-3)(-5) = -15$	Negative
$(-1, 1)$	$x = 0$	$(1)(-1)(-3) = 3$	Positive
$(1, 3)$	$x = 2$	$(3)(1)(-1) = -3$	Negative
$(3, \infty)$	$x = 4$	$(5)(3)(1) = 15$	Positive

Solution set: $(-1, 1) \cup (3, \infty)$

107. $x^3 - 2x^2 - 9x - 2 \geq -20$

$x^3 - 2x^2 - 9x + 18 \geq 0$

$x^2(x - 2) - 9(x - 2) \geq 0$

$(x - 2)(x^2 - 9) \geq 0$

$(x - 2)(x + 3)(x - 3) \geq 0$

Critical Numbers: $x = 2, x = \pm 3$

Test Intervals: $(-\infty, -3), (-3, 2), (2, 3), (3, \infty)$

Test: Is $(x - 2)(x + 3)(x - 3) \geq 0$?

Interval	x-value	Value of $(x - 2)(x + 3)(x - 3)$	Conclusion
$(-\infty, -3)$	$x = -4$	$(-6)(-1)(-7) = -42$	Negative
$(-3, 2)$	$x = 0$	$(-2)(3)(-3) = 18$	Positive
$(2, 3)$	$x = 2.5$	$(0.5)(5.5)(-0.5) = -1.375$	Negative
$(3, \infty)$	$x = 4$	$(2)(7)(1) = 14$	Positive

Solution set: $[-3, 2] \cup [3, \infty)$

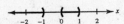

109. $4x^3 - 6x^2 < 0$

$2x^2(2x - 3) < 0$

Critical numbers: $x = 0, x = \frac{3}{2}$

Test intervals: $(-\infty, 0), \left(0, \frac{3}{2}\right), \left(\frac{3}{2}, \infty\right)$

Test: Is $2x^2(2x - 3) < 0$?

By testing an x-value in each test interval in the inequality, we see that the solution set is:

$(-\infty, 0) \cup \left(0, \frac{3}{2}\right)$

111. $x^3 - 4x \geq 0$

$x(x + 2)(x - 2) \geq 0$

Critical numbers: $x = 0, x = \pm 2$

Test intervals: $(-\infty, -2), (-2, 0), (0, 2), (2, \infty)$

Test: Is $x(x + 2)(x - 2) \geq 0$?

By testing an x-value in each test interval in the inequality, we see that the solution set is:

$[-2, 0] \cup [2, \infty)$

113. $(x - 1)^2(x + 2)^3 \geq 0$

Critical numbers: $x = 1, x = -2$

Test intervals: $(-\infty, -2), (-2, 1), (1, \infty)$

Test: Is $(x - 1)^2(x + 2)^3 \geq 0$?

By testing an x-value in each test interval in the inequality, we see that the solution set is: $[-2, \infty)$

115. $\dfrac{1}{x} - x > 0$

$\dfrac{1 - x^2}{x} > 0$

Critical numbers: $x = 0, x = \pm 1$

Test intervals: $(-\infty, -1), (-1, 0), (0, 1), (1, \infty)$

Test: Is $\dfrac{1 - x^2}{x} > 0$?

By testing an x-value in each test interval in the inequality, we see that the solution set is:

$(-\infty, -1) \cup (0, 1)$

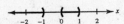

117. $\dfrac{x+6}{x+1} - 2 < 0$

$\dfrac{x+6-2(x+1)}{x+1} < 0$

$\dfrac{4-x}{x+1} < 0$

Critical numbers: $x = -1, x = 4$

Test intervals: $(-\infty, -1), (-1, 4), (4, \infty)$

Test: Is $\dfrac{4-x}{x+1} < 0$?

By testing an x-value in each test interval in the inequality, we see that the solution set is:

$(-\infty, -1) \cup (4, \infty)$

119. $\dfrac{3x-5}{x-5} > 4$

$\dfrac{3x-5}{x-5} - 4 > 0$

$\dfrac{3x-5-4(x-5)}{x-5} > 0$

$\dfrac{15-x}{x-5} > 0$

Critical numbers: $x = 5, x = 15$

Test intervals: $(-\infty, 5), (5, 15), (15, \infty)$

Test: Is $\dfrac{15-x}{x-5} > 0$?

By testing an x-value in each test interval in the inequality, we see that the solution set is: $(5, 15)$

121. $\dfrac{4}{x+5} > \dfrac{1}{2x+3}$

$\dfrac{4}{x+5} - \dfrac{1}{2x+3} > 0$

$\dfrac{4(2x+3) - (x+5)}{(x+5)(2x+3)} > 0$

$\dfrac{7x+7}{(x+5)(2x+3)} > 0$

Critical numbers: $x = -1, x = -5, x = -\dfrac{3}{2}$

Test intervals: $(-\infty, -5), \left(-5, -\dfrac{3}{2}\right),$

$\left(-\dfrac{3}{2}, -1\right), (-1, \infty)$

Test: Is $\dfrac{7(x+1)}{(x+5)(2x+3)} > 0$?

By testing an x-value in each test interval in the inequality, we see that the solution set is:

$\left(-5, -\dfrac{3}{2}\right) \cup (-1, \infty)$

123. $\dfrac{1}{x-3} \le \dfrac{9}{4x+3}$

$\dfrac{1}{x-3} - \dfrac{9}{4x+3} \le 0$

$\dfrac{4x+3-9(x-3)}{(x-3)(4x+3)} \le 0$

$\dfrac{30-5x}{(x-3)(4x+3)} \le 0$

Critical numbers: $x = 3, x = -\dfrac{3}{4}, x = 6$

Test intervals: $\left(-\infty, -\dfrac{3}{4}\right), \left(-\dfrac{3}{4}, 3\right), (3, 6), (6, \infty)$

Test: Is $\dfrac{5(6-x)}{(x-3)(4x+3)} \le 0$?

By testing an x-value in each test interval in the inequality, we see that the solution set is:

$\left(-\dfrac{3}{4}, 3\right) \cup [6, \infty)$

125. $\dfrac{x^2 + 2x}{x^2 - 9} \le 0$

$\dfrac{x(x + 2)}{(x + 3)(x - 3)} \le 0$

Critical numbers: $x = 0, x = -2, x = \pm 3$

Test intervals: $(-\infty, -3), (-3, -2), (-2, 0),$

$\qquad\qquad (0, 3), (3, \infty)$

Test: Is $\dfrac{x(x + 2)}{(x + 3)(x - 3)} \le 0$?

By testing an x-value in each test interval in the inequality, we see that the solution set is: $(-3, -2] \cup [0, 3)$

127. $\dfrac{5}{x - 1} - \dfrac{2x}{x + 1} < 1$

$\dfrac{5}{x - 1} - \dfrac{2x}{x + 1} - 1 < 0$

$\dfrac{5(x + 1) - 2x(x - 1) - (x - 1)(x + 1)}{(x - 1)(x + 1)} < 0$

$\dfrac{5x + 5 - 2x^2 + 2x - x^2 + 1}{(x - 1)(x + 1)} < 0$

$\dfrac{-3x^2 + 7x + 6}{(x - 1)(x + 1)} < 0$

$\dfrac{-(3x + 2)(x - 3)}{(x - 1)(x + 1)} < 0$

Critical Numbers: $x = -\dfrac{2}{3}, x = 3, x = \pm 1$

Test Intervals:

$\qquad (-\infty, -1), \left(-1, -\dfrac{2}{3}\right), \left(-\dfrac{2}{3}, 1\right), (1, 3), (3, \infty)$

Test: Is $\dfrac{-(3x + 2)(x - 3)}{(x - 1)(x + 1)} < 0$?

By testing an x-value in each test interval in the inequality, we see that the solution set is:

$(-\infty, -1) \cup \left(-\dfrac{2}{3}, 1\right) \cup (3, \infty)$

129. $4 - x^2 \ge 0$

$(2 + x)(2 - x) \ge 0$

Critical numbers: $x = \pm 2$

Test intervals: $(-\infty, -2), (-2, 2), (2, \infty)$

Test: Is $4 - x^2 \ge 0$?

By testing an x-value in each test interval in the inequality, we see that the domain is: $[-2, 2]$

131. $x^2 - 7x + 12 \ge 0$

$(x - 3)(x - 4) \ge 0$

Critical numbers: $x = 3, x = 4$

Test intervals: $(-\infty, 3), (3, 4), (4, \infty)$

Test: Is $(x - 3)(x - 4) \ge 0$?

By testing an x-value in each test interval in the inequality, we see that the domain is: $(-\infty, 3] \cup [4, \infty)$

133. $\dfrac{x}{x^2 - 2x - 35} \ge 0$

$\dfrac{x}{(x + 5)(x - 7)} \ge 0$

Critical Numbers: $x = 0, x = -5, x = 7$

Test Intervals: $(-\infty, -5), (-5, 0), (0, 7), (7, \infty)$

Test: Is $\dfrac{x}{(x + 5)(x - 7)} \ge 0$?

By testing an x-value in each test interval in the inequality, we see that the domain is: $(-5, 0] \cup (7, \infty)$

135. $0.4x^2 + 5.26 < 10.2$

$0.4x^2 - 4.94 < 0$

$0.4(x^2 - 12.35) < 0$

Critical numbers: $x \approx \pm 3.51$

Test intervals: $(-\infty, -3.51), (-3.51, 3.51), (3.51, \infty)$

By testing an x-value in each test interval in the inequality, we see that the solution set is: $(-3.51, 3.51)$

137. $-0.5x^2 + 12.5x + 1.6 > 0$

The zeros are $x = \dfrac{-12.5 \pm \sqrt{(12.5)^2 - 4(-0.5)(1.6)}}{2(-0.5)}$.

Critical numbers: $x \approx -0.13$, $x \approx 25.13$

Test intervals: $(-\infty, -0.13)$, $(-0.13, 25.13)$, $(25.13, \infty)$

By testing x-values in each test interval in the inequality, we see that the solution set is: $(-0.13, 25.13)$

139.
$$\frac{1}{2.3x - 5.2} > 3.4$$

$$\frac{1}{2.3x - 5.2} - 3.4 > 0$$

$$\frac{-7.82x + 18.68}{2.3x - 5.2} > 0$$

Critical numbers: $x \approx 2.39$, $x = 2.26$

Test intervals: $(-\infty, 2.26)$, $(2.26, 2.39)$, $(2.39, \infty)$

By testing x-values in each test interval in the inequality, we see that the solution set is: $(2.26, 2.39)$

141. $|x - a| \geq 2$

$x - a \leq -2$ or $x - a \geq 2$

$x \leq a - 2$ or $x \geq a + 2$

Matches graph (b)

143. $|ax - b| \leq c \Rightarrow c$ must be greater than or equal to zero.

$$-c \leq ax - b \leq c$$

$$b - c \leq ax \leq b + c$$

Let $a = 1$, then $b - c = 0$ and $b + c = 10$.
This is true when $b = c = 5$.
One set of values is: $a = 1$, $b = 5$, $c = 5$.

(Note: This solution is not unique. The following are also solutions.

$$a = 2, b = c = 10$$
$$a = 3, b = c = 15.)$$

In general, $a = k$, $b = c = 5k$, $k \geq 0$
or $a = k$, $b = 5k$, $c = -5k$, $k < 0$

145. Company A = Lemon Rental

Company B = Cars 'R Us

Company B fee > Company A fee

$$150 + 0.25x > 250$$
$$0.25x > 100$$
$$x > 400$$

If you drive more than 400 miles in a week, the rental fee for Company B is greater than the rental fee for Company A.

147. $1000(1 + r(2)) > 1062.50$

$$1 + 2r > 1.0625$$
$$2r > 0.0625$$
$$r > 0.03125$$
$$r > 3.125\% \approx 3.1\%$$

149. $\left|\dfrac{h - 68.5}{2.7}\right| \leq 1$

$$-1 \leq \frac{h - 68.5}{2.7} \leq 1$$

$$-2.7 \leq h - 68.5 \leq 2.7$$

$$65.8 \leq h \leq 71.2 \text{ inches}$$

151.
$$2L + 2W = 100 \implies W = 50 - L$$
$$LW \geq 500$$
$$L(50 - L) \geq 500$$
$$-L^2 + 50L - 500 \geq 0$$

By the Quadratic Formula we have:

Critical numbers: $L = 25 \pm 5\sqrt{5}$

Test: Is $-L^2 + 50L - 500 \geq 0$?

Solution set: $25 - 5\sqrt{5} \leq L \leq 25 + 5\sqrt{5}$

13.8 meters $\leq L \leq 36.2$ meters

153.
$$1000(1 + r)^2 > 1100$$
$$(1 + r)^2 > 1.1$$
$$1 + 2r + r^2 - 1.1 > 0$$
$$r^2 + 2r - 0.1 > 0$$

By the Quadratic Formula we have:

Critical Numbers: $r = -1 \pm \sqrt{1.1}$

Since r cannot be negative, $r = -1 + \sqrt{1.1} \approx 0.0488$
$$= 4.88\%.$$

Thus, $r > 4.88\%$.

155.
$$\frac{1}{R} = \frac{1}{R_1} + \frac{1}{2}$$
$$2R_1 = 2R + RR_1$$
$$2R_1 = R(2 + R_1)$$
$$\frac{2R_1}{2 + R_1} = R$$

Since $R \geq 1$, we have
$$\frac{2R_1}{2 + R_1} \geq 1$$
$$\frac{2R_1}{2 + R_1} - 1 \geq 0$$
$$\frac{R_1 - 2}{2 + R_1} \geq 0.$$

Since $R_1 > 0$, the only critical number is $R_1 = 2$.

The inequality is satisfied when $R_1 \geq 2$ ohms.

157. False. c has to be greater than or equal to zero for $ac \leq bc$.

159. True. $x^3 - 2x^2 - 11x + 12 = (x - 4)(x + 3)(x - 1)$.

The critical numbers are $x = 4, x = -3,$ and $x = 1$.

The test intervals are $(-\infty, -3), (-3, 1), (1, 4)$ and $(4, \infty)$.

Section P.3   Graphical Representation of Data

1.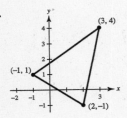

3.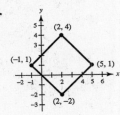

5. $A: (2, 6), \quad B: (-6, -2), \quad C: (4, -4), \quad D: (-3, 2)$

7. $(-3, 4)$

9. $(-5, -5)$

11. $x > 0$ and $y < 0$ in Quadrant IV.

13. $x > 0$ and $y > 0$ means that (x, y) is in Quadrant I.

15. $x = -4$ and $y > 0$ in Quadrant II.

17. $y < -5$ in Quadrants III and IV.

19. $(x, -y)$ is in the second Quadrant means that (x, y) is in Quadrant III.

21. $(x, y), xy > 0$ means x and y have the same signs. This occurs in Quadrants I and III.

23. The highest price of a stick of butter is 3.4 dollars. This occurred in 2001.

25.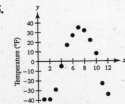

27. Percent increase:
$$\frac{2400 - 675}{675} \approx 2.56 = 256\%$$

29. $(-2 + 2, -4 + 5) = (0, 1)$
$$(2 + 2, -3 + 5) = (4, 2)$$
$$(-1 + 2, -1 + 5) = (1, 4)$$

31. $(-7 + 4, -2 + 8) = (-3, 6)$

$(-2 + 4, 2 + 8) = (2, 10)$

$(-2 + 4, -4 + 8) = (2, 4)$

$(-7 + 4, -4 + 8) = (-3, 4)$

33. $(-5 + 12, 4 - 5) = (7, -1)$

$(-3 + 12, 0 - 5) = (9, -5)$

$(-1 + 12, 2 - 5) = (11, -3)$

35. To reflect the vertices in the y-axis, negate each x-coordinate.

Original Point	Reflected Point
$(1, 5)$	$(-1, 5)$
$(5, 4)$	$(-5, 4)$
$(2, 2)$	$(-2, 2)$

37. Negate each x-coordinate.

Original Point	Reflected Point
$(2, 1)$	$(-2, 1)$
$(5, 4)$	$(-5, 4)$
$(3, 6)$	$(-3, 6)$

39. Negate each x-coordinate.

Original Point	Reflected Point
$(0, 3)$	$(0, 3)$
$(3, -2)$	$(-3, -2)$
$(6, 3)$	$(-6, 3)$
$(3, 8)$	$(-3, 8)$

41. $(-1(1), -6) = (-1, -6)$

$(-1(6), -6) = (-6, -6)$

$(-1(6), -3) = (-6, -3)$

$(-1(3), -3) = (-3, -3)$

43.

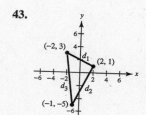

$d_1 = \sqrt{(2 - (-2))^2 + (1 - 3)^2} = \sqrt{(4)^2 + (-2)^2} = \sqrt{20} = 2\sqrt{5}$

$d_2 = \sqrt{(-1 - 2)^2 + (-5 - 1)^2} = \sqrt{(-3)^2 + (-6)^2} = \sqrt{45} = 3\sqrt{5}$

$d_3 = \sqrt{(-2 - (-1))^2 + (3 - (-5))^2} = \sqrt{(-1)^2 + (8)^2} = \sqrt{65}$

Since $d_1^2 + d_2^2 = 20 + 45 = 65 = d_3^2$, the triangle is a right triangle.

45.

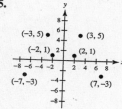

(a) The point is reflected through the y-axis.

(b) The point is reflected through the x-axis.

(c) The point is reflected through the origin.

47. Since (x_0, y_0) lies in Quadrant II, $(x_0, -y_0)$ must lie in Quadrant III. Matches (b)

49. Since (x_0, y_0) lies in Quadrant II, $(x_0, \frac{1}{2}y_0)$ must lie in Quadrant II. Matches (d)

51. The decade that shows the greatest increase in minimum wage is 1990–2000.

53. The point $(65, 83)$ represents an entrance exam score of 65.

55. $d = |5 - (-3)| = 8$

57. $d = |2 - (-3)| = 5$

59. (a) The distance between $(0, 2)$ and $(4, 2)$ is 4.

The distance between $(4, 2)$ and $(4, 5)$ is 3.

The distance between $(0, 2)$ and $(4, 5)$ is

$\sqrt{(4 - 0)^2 + (5 - 2)^2} = \sqrt{16 + 9} = \sqrt{25} = 5.$

(b) $4^2 + 3^2 = 16 + 9 = 25 = 5^2$

61. (a) The distance between $(-1, 1)$ and $(9, 1)$ is 10.

The distance between $(9, 1)$ and $(9, 4)$ is 3.

The distance between $(-1, 1)$ and $(9, 4)$ is

$\sqrt{(9 - (-1))^2 + (4 - 1)^2} = \sqrt{100 + 9} = \sqrt{109}.$

(b) $10^2 + 3^2 = 109 = \left(\sqrt{109}\right)^2$

63. (a)

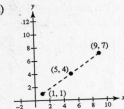

(b) $d = \sqrt{(9-1)^2 + (7-1)^2}$

$= \sqrt{64 + 36} = 10$

(c) $\left(\dfrac{9+1}{2}, \dfrac{7+1}{2}\right) = (5, 4)$

65. (a)

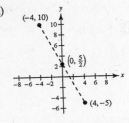

(b) $d = \sqrt{(4+4)^2 + (-5-10)^2}$

$= \sqrt{64 + 225} = 17$

(c) $\left(\dfrac{4-4}{2}, \dfrac{-5+10}{2}\right) = \left(0, \dfrac{5}{2}\right)$

67. (a)

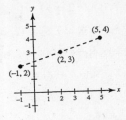

(b) $d = \sqrt{(5+1)^2 + (4-2)^2}$

$= \sqrt{36 + 4} = 2\sqrt{10}$

(c) $\left(\dfrac{-1+5}{2}, \dfrac{2+4}{2}\right) = (2, 3)$

69. (a)

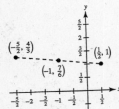

(b) $d = \sqrt{\left(\dfrac{1}{2} + \dfrac{5}{2}\right)^2 + \left(1 - \dfrac{4}{3}\right)^2}$

$d = \sqrt{9 + \dfrac{1}{9}} = \dfrac{\sqrt{82}}{3}$

(c) $\left(\dfrac{-\frac{5}{2} + \frac{1}{2}}{2}, \dfrac{\frac{4}{3} + 1}{2}\right) = \left(-1, \dfrac{7}{6}\right)$

71. (a)

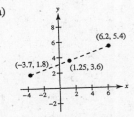

(b) $d = \sqrt{(6.2 + 3.7)^2 + (5.4 - 1.8)^2}$

$= \sqrt{98.01 + 12.96}$

$= \sqrt{110.97}$

(c) $\left(\dfrac{6.2 - 3.7}{2}, \dfrac{5.4 + 1.8}{2}\right) = (1.25, 3.6)$

73. (a)

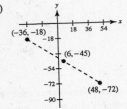

(b) $d = \sqrt{(48 + 36)^2 + (-72 + 18)^2}$

$= \sqrt{7056 + 2916}$

$= \sqrt{9972} = 6\sqrt{277}$

(c) $\left(\dfrac{-36 + 48}{2}, \dfrac{-18 - 72}{2}\right) = (6, -45)$

75. Target: $\dfrac{25.371 + 43.917}{2} = 34.644$ billion dollars

77. $d_1 = \sqrt{(4-2)^2 + (0-1)^2} = \sqrt{5}$

$d_2 = \sqrt{(4+1)^2 + (0+5)^2} = \sqrt{50}$

$d_3 = \sqrt{(2+1)^2 + (1+5)^2} = \sqrt{45}$

$\left(\sqrt{5}\right)^2 + \left(\sqrt{45}\right)^2 = \left(\sqrt{50}\right)^2$

79. $d_1 = \sqrt{(0-2)^2 + (9-5)^2} = \sqrt{4 + 16} = \sqrt{20} = 2\sqrt{5}$

$d_2 = \sqrt{(-2-0)^2 + (0-9)^2} = \sqrt{4 + 81} = \sqrt{85}$

$d_3 = \sqrt{(0-(-2))^2 + (-4-0)^2} = \sqrt{4 + 16} = \sqrt{20} = 2\sqrt{5}$

$d_4 = \sqrt{(0-2)^2 + (-4-5)^2} = \sqrt{4 + 81} = \sqrt{85}$

Opposite sides have equal lengths of $2\sqrt{5}$ and $\sqrt{85}$.

81. $d = \sqrt{(45 - 10)^2 + (40 - 15)^2} = \sqrt{35^2 + 25^2} = \sqrt{1850} = 5\sqrt{74} \approx 43$ yards

83. Polo Ralph Lauren Corp: $\dfrac{81.3 + 183.7}{2} = 132.5$ million dollars

85. False, you would have to use the Midpoint Formula 15 times.

87.

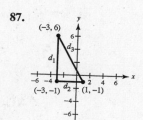

True. $d_1^2 + d_2^2 = 7^2 + 4^2 = 65 = d_3^2$

89. No. It depends on the magnitude of the quantities measured.

Section P.4   Graphs of Equations

1. $y = \sqrt{x + 4}$

  (a) $(0, 2)$: $2 \overset{?}{=} \sqrt{0 + 4}$

             $2 = 2$

    Yes, the point *is* on the graph.

  (b) $(5, 3)$: $3 \overset{?}{=} \sqrt{5 + 4}$

             $3 = \sqrt{9}$

    Yes, the point *is* on the graph.

3. $y = 4 - |x - 2|$

  (a) $(1, 5)$: $5 \overset{?}{=} 4 - |1 - 2|$

            $5 \neq 4 - 1$

    No, the point is *not* on the graph.

  (b) $(6, 0)$: $0 \overset{?}{=} 4 - |6 - 2|$

            $0 = 4 - 4$

    Yes, the point *is* on the graph.

5. $y = -2x + 5$

x	-1	0	1	2	$\frac{5}{2}$
y	7	5	3	1	0

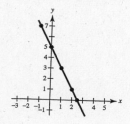

7. $y = x^2 - 3x$

x	-1	0	1	2	3
y	4	0	-2	-2	0

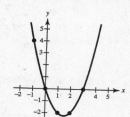

9. $y = 16 - 4x^2$

x-intercepts: $\quad 0 = 16 - 4x^2$

$\qquad\qquad 4x^2 = 16$

$\qquad\qquad x^2 = 4$

$\qquad\qquad x = \pm 2$

$\qquad\quad (-2, 0), (2, 0)$

y-intercept: $\quad y = 16 - 4(0)^2 = 16$

$\qquad\qquad (0, 16)$

11. $y = 2x^3 - 5x^2$

x-intercepts: $\; 0 = 2x^3 - 5x^2$

$\qquad\qquad 0 = x^2(2x - 5)$

$\qquad\qquad x = 0 \;$ or $\; x = \frac{5}{2}$

$\qquad\qquad (0, 0), \left(\frac{5}{2}, 0\right)$

y-intercept: $\quad y = 2(0)^3 - 5(0)^2$

$\qquad\qquad\quad = 0$

$\qquad\qquad (0, 0)$

13. $y^2 = x + 1$

x-intercept: $0 = x + 1$

$\qquad\qquad x = -1$

$(-1, 0)$

y-intercepts: $y^2 = 0 + 1$

$\qquad\qquad\; y = \pm 1$

$(0, 1), (0, -1)$

15. $x^2 + y^2 = 25$

x-intercepts: $x^2 + (0)^2 = 25$

$\qquad\qquad\qquad x^2 = 25$

$\qquad\qquad\qquad x = \pm 5$

$\qquad\qquad (\pm 5, 0)$

y-intercepts: $(0)^2 + y^2 = 25$

$\qquad\qquad\qquad y^2 = 25$

$\qquad\qquad\qquad y = \pm 5$

$\qquad\qquad (0, \pm 5)$

17. $x^2 - y = 0$

$\quad (-x)^2 - y = 0 \implies x^2 - y = 0 \implies$ y-axis symmetry

$\quad x^2 - (-y) = 0 \implies x^2 + y = 0 \implies$ No x-axis symmetry

$\quad (-x)^2 - (-y) = 0 \implies x^2 + y = 0 \implies$ No origin symmetry

19. $y = x^3$

$\quad y = (-x)^3 \implies y = -x^3 \implies$ No y-axis symmetry

$\quad -y = x^3 \implies y = -x^3 \implies$ No x-axis symmetry

$\quad -y = (-x)^3 \implies -y = -x^3 \implies y = x^3 \implies$ Origin symmetry

21. $y = \dfrac{x}{x^2 + 1}$

$\quad y = \dfrac{-x}{(-x)^2 + 1} \implies y = \dfrac{-x}{x^2 + 1} \implies$ No y-axis symmetry

$\quad -y = \dfrac{x}{x^2 + 1} \implies y = \dfrac{-x}{x^2 + 1} \implies$ No x-axis symmetry

$\quad -y = \dfrac{-x}{(-x)^2 + 1} \implies -y = \dfrac{-x}{x^2 + 1} \implies y = \dfrac{x}{x^2 + 1} \implies$ Origin symmetry

23. $xy^2 + 10 = 0$

$\quad (-x)y^2 + 10 = 0 \implies -xy^2 + 10 = 0 \implies$ No y-axis symmetry

$\quad x(-y)^2 + 10 = 0 \implies xy^2 + 10 = 0 \implies$ x-axis symmetry

$\quad (-x)(-y)^2 + 10 = 0 \implies -xy^2 + 10 = 0 \implies$ No origin symmetry

25. $y = 1 - x$ has intercepts $(1, 0)$ and $(0, 1)$.

Matches graph (c)

27. $y = x^3 - x + 1$ has a y-intercept of $(0, 1)$ and the points $(1, 1)$ and $(-2, -5)$ are on the graph.

Matches graph (b)

29. $y = -3x + 1$

x-intercept: $\left(\frac{1}{3}, 0\right)$

y-intercept: $(0, 1)$

No symmetry

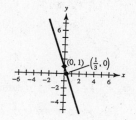

31. $y = x^2 - 2x$

Intercepts: $(0, 0)$, $(2, 0)$

No symmetry

x	-1	0	1	2	3
y	3	0	-1	0	3

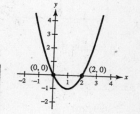

33. $y = x^3 + 3$

Intercepts: $(0, 3)$, $\left(\sqrt[3]{-3}, 0\right)$

No symmetry

x	-2	-1	0	1	2
y	-5	2	3	4	11

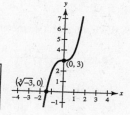

35. $y = \sqrt{x - 3}$

Domain: $[3, \infty)$

Intercept: $(3, 0)$

No symmetry

x	3	4	7	12
y	0	1	2	3

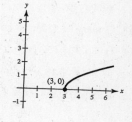

37. $y = |x - 6|$

Intercepts: $(0, 6)$, $(6, 0)$

No symmetry

x	-2	0	2	4	6	8	10
y	8	6	4	2	0	2	4

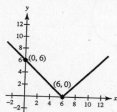
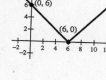
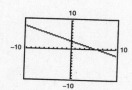

39. $x = y^2 - 1$

Intercepts: $(0, -1)$, $(0, 1)$, $(-1, 0)$

x-axis symmetry

x	-1	0	3
y	0	± 1	± 2

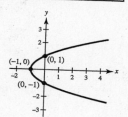

41. $y = 3 - \frac{1}{2}x$

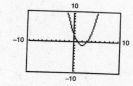

Intercepts: $(6, 0)$, $(0, 3)$

43. $y = x^2 - 4x + 3$

Intercepts: $(3, 0)$, $(1, 0)$, $(0, 3)$

45. $y = \dfrac{2x}{x - 1}$

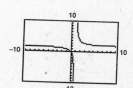

Intercept: $(0, 0)$

47. $y = \sqrt[3]{x}$

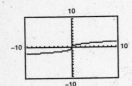

Intercept: $(0, 0)$

49. $y = x\sqrt{x + 6}$

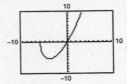

Intercepts: $(0, 0), (-6, 0)$

51. $y = |x + 3|$

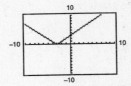

Intercepts: $(-3, 0), (0, 3)$

53. Center: $(0, 0)$; radius: 4

Standard form: $(x - 0)^2 + (y - 0)^2 = 4^2$

$$x^2 + y^2 = 16$$

55. Center: $(2, -1)$; radius: 4

Standard form: $(x - 2)^2 + (y - (-1))^2 = 4^2$

$$(x - 2)^2 + (y + 1)^2 = 16$$

57. Center: $(-1, 2)$; solution point: $(0, 0)$

$(x - (-1))^2 + (y - 2)^2 = r^2$

$(0 + 1)^2 + (0 - 2)^2 = r^2 \Longrightarrow 5 = r^2$

Standard form: $(x + 1)^2 + (y - 2)^2 = 5$

59. Endpoints of a diameter: $(0, 0), (6, 8)$

Center: $\left(\dfrac{0 + 6}{2}, \dfrac{0 + 8}{2}\right) = (3, 4)$

$(x - 3)^2 + (y - 4)^2 = r^2$

$(0 - 3)^2 + (0 - 4)^2 = r^2 \Longrightarrow 25 = r^2$

Standard form: $(x - 3)^2 + (y - 4)^2 = 25$

61. $x^2 + y^2 = 25$

Center: $(0, 0)$

Radius: 5

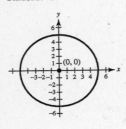

63. $(x - 1)^2 + (y + 3)^2 = 9$

Center: $(1, -3)$

Radius: 3

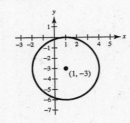

65. $\left(x - \frac{1}{2}\right)^2 + \left(y - \frac{1}{2}\right)^2 = \frac{9}{4}$

Center: $\left(\frac{1}{2}, \frac{1}{2}\right)$

Radius: $\frac{3}{2}$

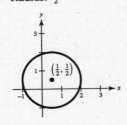

67. y-axis symmetry

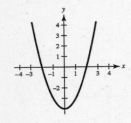

69. Origin symmetry

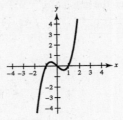

71. Answers will vary.

One possible equation with x-intercepts at $x = -2$, $x = 4$, and $x = 6$ is:

$$y = (x + 2)(x - 4)(x - 6)$$
$$= (x + 2)(x^2 - 10x + 24)$$
$$= x^3 - 10x^2 + 24x + 2x^2 - 20x + 48$$
$$= x^3 - 8x^2 + 4x + 48$$

Any non-zero multiple of the right side of this equation, $y = k(x^3 - 8x^2 + 4x + 48)$, would also have these x-intercepts.

73. $y_1 = 4 + \sqrt{25 - x^2}$

$\quad y_2 = 4 - \sqrt{25 - x^2}$

The graph represents a circle.

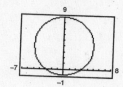

75. $y_1 = 2 + \sqrt{16 - (x - 1)^2}$

$\quad y_2 = 2 - \sqrt{16 - (x - 1)^2}$

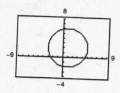

A circle is bounded by their graphs.

77. $y = 225{,}000 - 20{,}000t,\ 0 \le t \le 8$

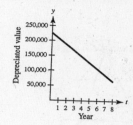

79. (a) $2x + 2w = \frac{920}{3}$

$\quad\quad x + w = \frac{460}{3}$

$\quad\quad w = \frac{460}{3} - x$

$\quad\quad A = xw = x\left(\frac{460}{3} - x\right)$

(b)

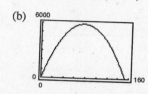

(c) The area is maximum when $x = w = 76\frac{2}{3}$ yards, or when the field is a square.

(d) A regulation football filed is 100 yards by $53\frac{1}{3}$ yards.

81. $y = -0.0026t^2 + 0.577t + 44.19$

(a) and (b)

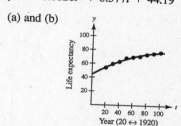

The curve seems to be a good fit for the data.

(c) For the year 2005, $t = 105$ and $y = 76.11$ years.

For the year 2010, $t = 110$ and $y = 76.2$ years.

(d) No. Because the model is quadratic, life expectancy decreases for $t > 111$ or 2011.

83. $y = \frac{10{,}770}{50^2} - 0.37 \approx 3.9$ ohms

85. False.

To find y-intercepts, let x be zero and solve the equation for y.

87. True.

A circle centered at the origin would have symmetries to the origin, the x-axis, and the y-axis.

89. False.

If $(1, -2)$ is a point on a graph that is symmetric with respect to the x-axis, then $(1, -(-2)) = (1, 2)$ is also a point on the graph.

91. True.

If $b^2 - 4ac > 0$ and $a \neq 0$, then the graph of $y = ax^2 + bx + c$

has x-intercepts at $\left(\dfrac{-b - \sqrt{b^2 - 4ac}}{2a}, 0\right)$ and $\left(\dfrac{-b + \sqrt{b^2 - 4ac}}{2a}, 0\right)$.

93. The distance between the origin and (x, y) is:

$$\sqrt{(x - 0)^2 + (y - 0)^2} = \sqrt{x^2 + y^2}$$

The distance between $(2, 0)$ and (x, y) is:

$$\sqrt{(x - 2)^2 + (y - 0)^2} = \sqrt{x^2 + y^2 - 4x + 4}$$

Since the distance between (x, y) and the origin is K times the distance between (x, y) and $(2, 0)$, we have:

$$\sqrt{x^2 + y^2} = K\sqrt{x^2 + y^2 - 4x + 4}$$

$$x^2 + y^2 = K^2(x^2 + y^2 - 4x + 4)$$

$$x^2 + y^2 - K^2x^2 - K^2y^2 + 4K^2x - 4K^2 = 0$$

$$(1 - K^2)x^2 + (1 - K^2)y^2 + 4K^2x - 4K^2 = 0$$

95. Assuming that the graph does not go beyond the vertical limits of the display, you will see the graph for the larger values of x.

Section P.5   Linear Equations in Two Variables

1.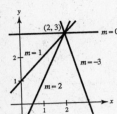

3. Two points on the line: $(0, 0)$ and $(5, 8)$

$$\text{Slope} = \frac{\text{rise}}{\text{run}} = \frac{8}{5}$$

5. Two points on the line: $(0, 3)$ and $(1, 3)$

$$\text{Slope} = \frac{\text{rise}}{\text{run}} = \frac{0}{1} = 0$$

7. Two points on the line: $(0, 8)$ and $(2, 0)$

$$\text{Slope} = \frac{\text{rise}}{\text{run}} = \frac{-8}{2} = -4$$

9.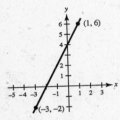

$$m = \frac{6 - (-2)}{1 - (-3)} = \frac{8}{4} = 2$$

11.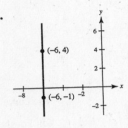

$$m = \frac{4 - (-1)}{-6 - (-6)} = \frac{5}{0}$$

m is undefined.

13.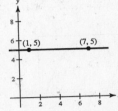

$$m = \frac{5 - 5}{7 - 1} = \frac{0}{6} = 0$$

15.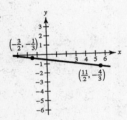

$$m = \frac{-\dfrac{1}{3} - \left(-\dfrac{4}{3}\right)}{-\dfrac{3}{2} - \dfrac{11}{2}} = -\frac{1}{7}$$

17.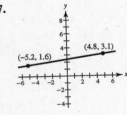

$$m = \frac{1.6 - 3.1}{-5.2 - 4.8} = \frac{-1.5}{-10}$$

$$= 0.15$$

19. Point: $(2, 1)$ Slope: $m = 0$

Since $m = 0$, y does not change.
Three points are $(0, 1)$, $(3, 1)$, and $(-1, 1)$.

21. Point: $(5, -6)$ Slope: $m = 1$

Since $m = 1$, y increases by 1 for every one unit increase in x. Three points are $(6, -5)$, $(7, -4)$, and $(8, -3)$.

23. Point: $(-8, 1)$ Slope is undefined.

Since m is undefined, x does not change.
Three points are $(-8, 0)$, $(-8, 2)$, and $(-8, 3)$.

25. Point: $(-5, 4)$ Slope: $m = 2$

Since $m = 2 = \frac{2}{1}$, y increases by 2 for every one unit increase in x. Three additional points are $(-4, 6)$, $(-3, 8)$, and $(-2, 10)$.

27. Point: $(7, -2)$ Slope: $m = \dfrac{1}{2}$

Since $m = \frac{1}{2}$, y increases by 1 unit for every two unit increase in x. Three additional points are $(9, -1)$, $(11, 0)$, and $(13, 1)$.

29. Slope of L_1: $m = \dfrac{9 + 1}{5 - 0} = 2$

Slope of L_2: $m = \dfrac{1 - 3}{4 - 0} = -\dfrac{1}{2}$

L_1 and L_2 are perpendicular.

31. Slope of L_1: $m = \dfrac{0-6}{-6-3} = \dfrac{2}{3}$

Slope of L_2: $m = \dfrac{\frac{7}{3}+1}{5-0} = \dfrac{2}{3}$

L_1 and L_2 are parallel.

33. (a) $m = \dfrac{2}{3}$. Since the slope is positive, the line rises.

Matches L_2.

(b) m is undefined. The line is vertical. Matches L_3.

(c) $m = -2$. The line falls. Matches L_1.

35. The slope is $m = -20$. This represents the decrease in the amount of the loan each week. Matches graph (b)

37. The slope is $m = 0.32$. This represents the increase in travel cost for each mile driven. Matches graph (a)

39. (a) $m = 135$. The sales are increasing 135 units per year.

(b) $m = 0$. There is no change in sales.

(c) $m = -40$. The sales are decreasing 40 units per year.

41. (a) and (b)

x	300	600	900	1200	1500	1800	2100
y	-25	-50	-75	-100	-125	-150	-175

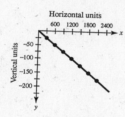

(c) $m = \dfrac{-50-(-25)}{600-300} = \dfrac{-25}{300} = -\dfrac{1}{12}$

$y-(-50) = -\dfrac{1}{12}(x-600)$

$y+50 = -\dfrac{1}{12}x + 50$

$y = -\dfrac{1}{12}x$

(d) Since $m = -\frac{1}{12}$, for every 12 horizontal measurements the vertical measurement decreases by 1.

(e) $\dfrac{1}{12} \approx 0.083 = 8.3\%$ grade

43. (a)

Years	Slope
1993–1994	0.91
1994–1995	1.47
1995–1996	0.90
1996–1997	0.72
1997–1998	1.01
1998–1999	0.84
1999–2000	0.98
2000–2001	0.63
2001–2002	0.54
2002–2003	1.29

The years 1994–1995 showed the greatest increase and the years 2001–2002 showed the smallest increase.

(b) $m = \dfrac{16.70-7.41}{13-3} = 0.929$

(c) On average, revenue increased 0.929 billion dollars each year.

45. $\dfrac{\text{rise}}{\text{run}} = \dfrac{3}{4} = \dfrac{x}{\frac{1}{2}(32)}$

$\dfrac{3}{4} = \dfrac{x}{16}$

$4x = 48$

$x = 12$

The maximum height in the attic is 12 feet.

47. $y = x - 10$

Slope: $m = 1$

y-intercept: $(0, -10)$

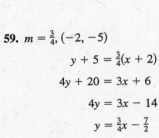

49. $3y + 5 = 0$

$3y = -5$

$y = -\dfrac{5}{3}$

Slope: $m = 0$

y-intercept: $\left(0, -\dfrac{5}{3}\right)$

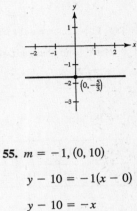

51. $2x + 3y = 9$

$3y = -2x + 9$

$y = -\dfrac{2}{3}x + 3$

Slope: $m = -\dfrac{2}{3}$

y-intercept: $(0, 3)$

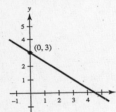

53. $-4x + 5y = -35$

$5y = 4x - 35$

$y = \dfrac{4}{5}x - 7$

Slope: $m = \dfrac{4}{5}$

y-intercept: $(0, -7)$

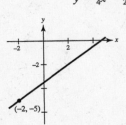

55. $m = -1, (0, 10)$

$y - 10 = -1(x - 0)$

$y - 10 = -x$

$y = -x + 10$

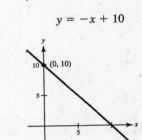

57. $m = 4, (0, 0)$

$y - 0 = 4(x - 0)$

$y = 4x$

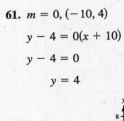

59. $m = \dfrac{3}{4}, (-2, -5)$

$y + 5 = \dfrac{3}{4}(x + 2)$

$4y + 20 = 3x + 6$

$4y = 3x - 14$

$y = \dfrac{3}{4}x - \dfrac{7}{2}$

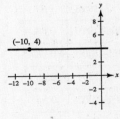

61. $m = 0, (-10, 4)$

$y - 4 = 0(x + 10)$

$y - 4 = 0$

$y = 4$

63. $m = 0, \left(-3, \frac{1}{3}\right)$

$$y - \frac{1}{3} = 0(x + 3)$$

$$y - \frac{1}{3} = 0$$

$$y = \frac{1}{3}$$

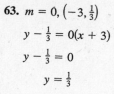

65. $m = -3, \left(-\frac{1}{2}, \frac{3}{2}\right)$

$$y - \frac{3}{2} = -3\left(x + \frac{1}{2}\right)$$

$$y - \frac{3}{2} = -3x - \frac{3}{2}$$

$$y = -3x$$

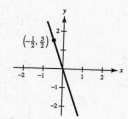

67. $m = -\frac{5}{2}, (2.3, -8.5)$

$$y - (-8.5) = -\frac{5}{2}(x - 2.3)$$

$$y + 8.5 = -2.5x + 5.75$$

$$y = -2.5x - 2.75$$

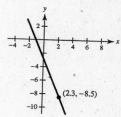

69. $(4, 3), (-4, -4)$

$$y - 3 = \frac{-4 - 3}{-4 - 4}(x - 4)$$

$$y - 3 = \frac{7}{8}(x - 4)$$

$$y - 3 = \frac{7}{8}x - \frac{7}{2}$$

$$y = \frac{7}{8}x - \frac{1}{2}$$

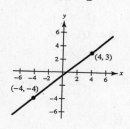

71. $(-1, 4), (6, 4)$

$$y - 4 = \frac{4 - 4}{6 - (-1)}(x + 1)$$

$$y - 4 = 0(x + 1)$$

$$y - 4 = 0$$

$$y = 4$$

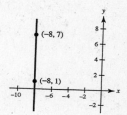

73. Since both points have $x = -8$, the slope is undefined.

$$x = -8$$

75. $(1, 1), \left(6, -\frac{2}{3}\right)$

$$y - 1 = \frac{-\frac{2}{3} - 1}{6 - 1}(x - 1)$$

$$y - 1 = -\frac{1}{3}(x - 1)$$

$$y - 1 = -\frac{1}{3}x + \frac{1}{3}$$

$$y = -\frac{1}{3}x + \frac{4}{3}$$

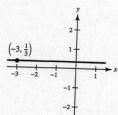

77. $\left(\frac{3}{4}, \frac{3}{2}\right), \left(-\frac{4}{3}, \frac{7}{4}\right)$

$$y - \frac{3}{2} = \frac{\frac{7}{4} - \frac{3}{2}}{-\frac{4}{3} - \frac{3}{4}}\left(x - \frac{3}{4}\right)$$

$$y - \frac{3}{2} = \frac{\frac{1}{4}}{-\frac{25}{12}}\left(x - \frac{3}{4}\right)$$

$$y - \frac{3}{2} = -\frac{3}{25}\left(x - \frac{3}{4}\right)$$

$$y - \frac{3}{2} = -\frac{3}{25}x + \frac{9}{100}$$

$$y = -\frac{3}{25}x + \frac{159}{100}$$

79. $(-8, 0.6), (2, -2.4)$

$$y - 0.6 = \frac{-2.4 - 0.6}{2 - (-8)}(x + 8)$$

$$y - 0.6 = -\frac{3}{10}(x + 8)$$

$$10y - 6 = -3(x + 8)$$

$$10y - 6 = -3x - 24$$

$$10y = -3x - 18$$

$$y = -\frac{3}{10}x - \frac{9}{5} \quad \text{or} \quad y = -0.3x - 1.8$$

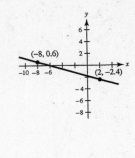

81. $(-3, 0), (0, 4)$

$$\frac{x}{-3} + \frac{y}{4} = 1$$

$$(-12)\frac{x}{-3} + (-12)\frac{y}{4} = (-12) \cdot 1$$

$$4x - 3y + 12 = 0$$

83. $\left(\frac{2}{3}, 0\right), (0, -2)$

$$\frac{x}{\frac{2}{3}} + \frac{y}{-2} = 1$$

$$\frac{3x}{2} - \frac{y}{2} = 1$$

$$3x - y - 2 = 0$$

85. $(d, 0), (0, d), (-3, 4)$

$$\frac{x}{d} + \frac{y}{d} = 1$$

$$x + y = d$$

$$-3 + 4 = d$$

$$1 = d$$

$$x + y = 1$$

$$x + y - 1 = 0$$

87. $x + y = 7$

$$y = -x + 7$$

Slope: $m = -1$

(a) $m = -1, (-3, 2)$

$$y - 2 = -1(x + 3)$$

$$y - 2 = -x - 3$$

$$y = -x - 1$$

(b) $m = 1, (-3, 2)$

$$y - 2 = 1(x + 3)$$

$$y = x + 5$$

89. $5x + 3y = 0$

$$3y = -5x$$

$$y = -\frac{5}{3}x$$

Slope: $m = -\frac{5}{3}$

(a) $m = -\frac{5}{3}, \left(\frac{7}{8}, \frac{3}{4}\right)$

$$y - \frac{3}{4} = -\frac{5}{3}\left(x - \frac{7}{8}\right)$$

$$24y - 18 = -40\left(x - \frac{7}{8}\right)$$

$$24y - 18 = -40x + 35$$

$$24y = -40x + 53$$

$$y = -\frac{5}{3}x + \frac{53}{24}$$

(b) $m = \frac{3}{5}, \left(\frac{7}{8}, \frac{3}{4}\right)$

$$y - \frac{3}{4} = \frac{3}{5}\left(x - \frac{7}{8}\right)$$

$$40y - 30 = 24\left(x - \frac{7}{8}\right)$$

$$40y - 30 = 24x - 21$$

$$40y = 24x + 9$$

$$y = \frac{3}{5}x + \frac{9}{40}$$

91. $x = 4$

m is undefined.

(a) $(2, 5), m$ is undefined.

$$x = 2$$

(b) $(2, 5), m = 0$

$$y = 5$$

93. $y = 2, m = 0$

(a) $(2, 1), m = 0$

$$y = 1$$

(b) $(2, 1), m$ is undefined.

$$x = 2$$

95. $6x + 2y = 9$

$$2y = -6x + 9$$

$$y = -3x + \frac{9}{2}$$

Slope: $m = -3$

(a) $(-3.9, -1.4), m = -3$

$$y - (-1.4) = -3(x - (-3.9))$$

$$y + 1.4 = -3x - 11.7$$

$$y = -3x - 13.1$$

(b) $(-3.9, -1.4), m = \frac{1}{3}$

$$y - (-1.4) = \frac{1}{3}(x - (-3.9))$$

$$y + 1.4 = \frac{1}{3}x + 1.3$$

$$y = \frac{1}{3}x - 0.1$$

97. (a) $y = \frac{2}{3}x$ (b) $y = -\frac{3}{2}x$ (c) $y = \frac{2}{3}x + 2$

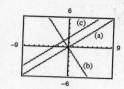

(a) is parallel to (c). (b) is perpendicular to (a) and (c).

99. (a) $y = x - 8$ (b) $y = x + 1$ (c) $y = -x + 3$

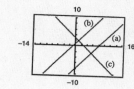

(a) is parallel to (b). (c) is perpendicular to (a) and (b).

101. $(5, 156)$, $m = 4.50$

$$V - 156 = 4.50(t - 5)$$
$$V - 156 = 4.50t - 22.50$$
$$V = 4.5t + 133.50$$

103. Set the distance between $(6, 5)$ and (x, y) equal to the distance between $(1, -8)$ and (x, y).

$$\sqrt{(x - 6)^2 + (y - 5)^2} = \sqrt{(x - 1)^2 + (y - (-8))^2}$$
$$(x - 6)^2 + (y - 5)^2 = (x - 1)^2 + (y + 8)^2$$
$$x^2 - 12x + 36 + y^2 - 10y + 25 = x^2 - 2x + 1 + y^2 + 16y + 64$$
$$x^2 + y^2 - 12x - 10y + 61 = x^2 + y^2 - 2x + 16y + 65$$
$$-12x - 10y + 61 = -2x + 16y + 65$$
$$-10x - 26y - 4 = 0$$
$$-2(5x + 13y + 2) = 0$$
$$5x + 13y + 2 = 0$$
$$13y = -5x - 2$$
$$y = -\tfrac{5}{13}x - \tfrac{2}{13}$$

105. Set the distance between $\left(-\frac{1}{2}, -4\right)$ and (x, y) equal to the distance between $\left(\frac{7}{2}, \frac{5}{4}\right)$ and (x, y).

$$\sqrt{\left(x - \left(-\tfrac{1}{2}\right)\right)^2 + (y - (-4))^2} = \sqrt{\left(x - \tfrac{7}{2}\right)^2 + \left(y - \tfrac{5}{4}\right)^2}$$
$$\left(x + \tfrac{1}{2}\right)^2 + (y + 4)^2 = \left(x - \tfrac{7}{2}\right)^2 + \left(y - \tfrac{5}{4}\right)^2$$
$$x^2 + x + \tfrac{1}{4} + y^2 + 8y + 16 = x^2 - 7x + \tfrac{49}{4} + y^2 - \tfrac{5}{2}y + \tfrac{25}{16}$$
$$x^2 + y^2 + x + 8y + \tfrac{65}{4} = x^2 + y^2 - 7x - \tfrac{5}{2}y + \tfrac{221}{16}$$
$$x + 8y + \tfrac{65}{4} = -7x - \tfrac{5}{2}y + \tfrac{221}{16}$$
$$8x + \tfrac{21}{2}y + \tfrac{39}{16} = 0$$
$$128x + 168y + 39 = 0$$
$$168y = -128x - 39$$
$$y = -\tfrac{16}{21}x - \tfrac{13}{56}$$

107. $t = 0$ represents 1999.

$(0, 0.28)$, $(5, 1.60)$

$$m = \frac{1.60 - 0.28}{5 - 0} = 0.264$$

$y = 0.264t + 0.28$

$t = 6$ represents 2005: $y = 0.264(6) + 0.28 \approx \$1.86$

$t = 11$ represents 2010: $y = 0.264(11) + 0.28 = \$3.18$

109. Using the points $(0, 32)$ and $(100, 212)$, we have

$$m = \frac{212 - 32}{100 - 0} = \frac{180}{100} = \frac{9}{5}$$

$$F - 32 = \frac{9}{5}(C - 0)$$

$$F = \frac{9}{5}C + 32.$$

111. Let $t = 0$ represent 1998.

$(0, 28{,}500), (6, 36{,}400)$

$$m = \frac{36{,}400 - 28{,}500}{6 - 0} = \frac{3950}{3}$$

$$S = \frac{3950}{3}t + 28{,}500$$

$t = 10$ represents 2008: $S = \dfrac{3950}{3}(10) + 28{,}500$

$$\approx \$41{,}666.67$$

115. Sale price $=$ List price $- \; 15\%$ of the list price

$$S = L - 0.15L$$

$$S = 0.85L$$

117. (a) $C = 36{,}500 + 5.25t + 12.50t$

$$= 17.75t + 36{,}500$$

(b) $R = 27t$

(c) $P = R - C = 27t - (17.75t + 36{,}500)$

$$= 9.25t - 36{,}500$$

(d) $P = 0$

$$9.25t - 36{,}500 = 0$$

$$t = \frac{36{,}500}{9.25} \approx 3946 \text{ hours}$$

121. $C = 120 + 0.31x$

125. (a) and (b)

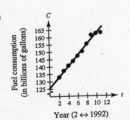

127. False. The slope with the greatest magnitude corresponds to the steepest line.

113. Using the points $(0, 875)$ and $(5, 0)$, where the first coordinate represents the year t and the second coordinate represents the value V, we have

$$m = \frac{0 - 875}{5 - 0} = -175$$

$$V = -175t + 875, \; 0 \le t \le 5.$$

119. (a)

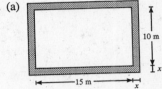

$$y = 2(15 + 2x) + 2(10 + 2x)$$

$$= 8x + 50$$

(b)

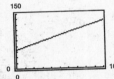

(c) Since $m = 8$, each 1 meter increase in x will increase y by 8 meters.

123. Answers will vary. One approximation is
$$y = 199.8t - 14.6.$$

(c) Answers will vary.
One approximation is $C = 3.7t + 125.5$.
(Another is $C = 3.6t + 126.3$.)

(d) $m = 3.7$. This means that fuel consumption increases 3.7 billion gallons per year.

(e) $t = 20$ represents 2010: $C = 199.5$ billion gallons

$t = 30$ represents 2020: $C = 236.5$ billion gallons

129. True. Given any two distinct points (x_1, y_1) and (x_2, y_2) on a nonvertical line, then the slope is $m = \dfrac{y_2 - y_1}{x_2 - x_1}$.

Review Exercises for Chapter P

1. $6 - (x - 2)^2 = 6 - (x^2 - 4x + 4)$

$\qquad\qquad = 2 + 4x - x^2$

The equation is an identity.

3. $\qquad -x^3 + x(7 - x) + 3 = -x^3 - x^2 + 7x + 3$

$x(-x^2 - x) + 7(x + 1) - 4 = -x^3 - x^2 + 7x + 3$

The equation is an identity.

5. $3x - 2(x + 5) = 10$

$\qquad 3x - 2x - 10 = 10$

$\qquad\qquad\qquad x = 20$

7. $4(x + 3) - 3 = 2(4 - 3x) - 4$

$\qquad 4x + 12 - 3 = 8 - 6x - 4$

$\qquad\quad 4x + 9 = -6x + 4$

$\qquad\qquad 10x = -5$

$\qquad\qquad\quad x = -\frac{1}{2}$

9. $\qquad 2x^2 - x - 28 = 0$

$\qquad (2x + 7)(x - 4) = 0$

$2x + 7 = 0 \quad$ or $\quad x - 4 = 0$

$x = -\frac{7}{2} \;$ or $\qquad\; x = 4$

11. $16x^2 = 25$

$\qquad x^2 = \dfrac{25}{16}$

$\qquad x = \pm\sqrt{\dfrac{25}{16}} = \pm\dfrac{5}{4}$

13. $(x - 8)^2 = 15$

$\qquad x - 8 = \pm\sqrt{15}$

$\qquad\quad x = 8 \pm \sqrt{15}$

15. $x^2 + 6x - 3 = 0$

$\quad a = 1, \quad b = 6, \quad c = -3$

$\quad x = \dfrac{-6 \pm \sqrt{6^2 - 4(1)(-3)}}{2(1)}$

$\qquad = \dfrac{-6 \pm \sqrt{48}}{2} = -3 \pm 2\sqrt{3}$

17. $-20 - 3x + 3x^2 = 0$

$\quad 3x^2 - 3x - 20 = 0$

$\quad a = 3, b = -3, c = -20$

$\quad x = \dfrac{-(-3) \pm \sqrt{(-3)^2 - 4(3)(-20)}}{2(3)}$

$\qquad = \dfrac{3 \pm \sqrt{249}}{6} = \dfrac{1}{2} \pm \dfrac{\sqrt{249}}{6}$

19. $4x^3 - 6x^2 = 0$

$\qquad x^2(4x - 6) = 0$

$\qquad\qquad x^2 = 0 \Longrightarrow x = 0$

$\qquad 4x - 6 = 0 \Longrightarrow x = \dfrac{3}{2}$

21. $9x^4 + 27x^3 - 4x^2 - 12x = 0$

$\quad 9x^3(x + 3) - 4x(x + 3) = 0$

$\qquad\quad (9x^3 - 4x)(x + 3) = 0$

$\qquad\quad x(9x^2 - 4)(x + 3) = 0$

$\quad x(3x + 2)(3x - 2)(x + 3) = 0$

$\qquad\qquad\qquad\qquad x = 0$

$\qquad\quad 3x + 2 = 0 \Longrightarrow x = -\dfrac{2}{3}$

$\qquad\quad 3x - 2 = 0 \Longrightarrow x = \dfrac{2}{3}$

$\qquad\qquad x + 3 = 0 \Longrightarrow x = -3$

23. $\sqrt{x - 2} - 8 = 0$

$\qquad \sqrt{x - 2} = 8$

$\qquad\quad x - 2 = 64$

$\qquad\qquad\; x = 66$

25. $\sqrt{3x-2} = 4-x$

$3x - 2 = (4-x)^2$

$3x - 2 = 16 - 8x + x^2$

$0 = 18 - 11x + x^2$

$0 = (x-9)(x-2)$

$0 = x - 9 \Rightarrow x = 9$, extraneous

$0 = x - 2 \Rightarrow x = 2$

27. $(x+2)^{3/4} = 27$

$x + 2 = 27^{4/3}$

$x + 2 = 81$

$x = 79$

29.

$8x^2(x^2-4)^{1/3} + (x^2-4)^{4/3} = 0$

$(x^2-4)^{1/3}[8x^2 + x^2 - 4] = 0$

$(x^2-4)^{1/3}(9x^2 - 4) = 0$

$(x-2)^{1/3}(x+2)^{1/3}(3x-2)(3x+2) = 0$

$x - 2 = 0 \Rightarrow x = 2$

$x + 2 = 0 \Rightarrow x = -2$

$3x - 2 = 0 \Rightarrow x = \frac{2}{3}$

$3x + 2 = 0 \Rightarrow x = -\frac{2}{3}$

31. $|2x+3| = 7$

$2x + 3 = 7$ or $2x + 3 = -7$

$2x = 4$ or $2x = -10$

$x = 2$ $x = -5$

33. $|x^2 - 6| = x$

$x^2 - 6 = x$ or

$x^2 - x - 6 = 0$

$(x-3)(x+2) = 0$

$x - 3 = 0 \Rightarrow x = 3$

$x + 2 = 0 \Rightarrow x = -2$, extraneous

$-(x^2 - 6) = x$

$x^2 + x - 6 = 0$

$(x+3)(x-2) = 0$

$x - 2 = 0 \Rightarrow x = 2$

$x + 3 = 0 \Rightarrow x = -3$, extraneous

35. Let $x =$ the number of liters of pure antifreeze.

30% of $(10 - x)$ + 100% of x = 50% of 10

$0.30(10 - x) + 1.00x = 0.50(10)$

$3 - 0.30x + 1.00x = 5$

$0.70x = 2$

$x = \dfrac{2}{0.70} = \dfrac{20}{7} = 2\dfrac{6}{7}$ liters

37. $6x - 17 > 0$

(a) $x = 3$

$6(3) - 17 \overset{?}{>} 0$

$1 > 0$

Yes, $x = 3$ is
a solution.

(b) $x = -4$

$6(-4) - 17 \overset{?}{>} 0$

$-41 \not> 0$

No, $x = -4$ is not
a solution.

39. $9x - 8 \le 7x + 16$

$2x \le 24$

$x \le 12$

$(-\infty, 12]$

41. $\dfrac{15}{2}x + 4 > 3x - 5$

$15x + 8 > 6x - 10$

$9x > -18$

$x > -2$

$(-2, \infty)$

43. $-19 < \dfrac{3x-17}{2} \le 34$

$-38 < 3x - 17 \le 68$

$-21 < 3x \le 85$

$-7 < x \le \dfrac{85}{3}$

$-7 < x \le 28\dfrac{1}{3}$ or $\left(-7, 28\dfrac{1}{3}\right]$

45. $|x + 1| \leq 5$

$-5 \leq x + 1 \leq 5$

$-6 \leq x \leq 4$ or $[-6, 4]$

47. $|x - 3| > 4$

$x - 3 > 4$ or $x - 3 < -4$

$x > 7$ $x < -1$

$(-\infty, -1) \cup (7, \infty)$

49. $R > C$

$125.33x > 92x + 1200$

$33.33x > 1200$

$x > \dfrac{1200}{33.33} \approx 36$ units

51. $x^2 - 6x - 27 < 0$

$(x + 3)(x - 9) < 0$

Critical numbers: $x = -3, x = 9$

Test intervals: $(-\infty, -3), (-3, 9), (9, \infty)$

Test: Is $(x + 3)(x - 9) < 0$?

Solution set: $(-3, 9)$

53. $6x^2 + 5x < 4$

$6x^2 + 5x - 4 < 0$

$(3x + 4)(2x - 1) < 0$

Critical numbers: $x = -\frac{4}{3}, x = \frac{1}{2}$

Test intervals: $\left(-\infty, -\frac{4}{3}\right), \left(-\frac{4}{3}, \frac{1}{2}\right), \left(\frac{1}{2}, \infty\right)$

Test: Is $(3x + 4)(2x - 1) < 0$?

Solution set: $\left(-\frac{4}{3}, \frac{1}{2}\right)$

55. $\dfrac{2}{x + 1} \leq \dfrac{3}{x - 1}$

$\dfrac{2(x - 1) - 3(x + 1)}{(x + 1)(x - 1)} \leq 0$

$\dfrac{2x - 2 - 3x - 3}{(x + 1)(x - 1)} \leq 0$

$\dfrac{-(x + 5)}{(x + 1)(x - 1)} \leq 0$

Critical numbers: $x = -5, x = \pm 1$

Test intervals: $(-\infty, -5), (-5, -1), (-1, 1), (1, \infty)$

Test: Is $\dfrac{-(x + 5)}{(x + 1)(x - 1)} \leq 0$?

Solution set: $[-5, -1) \cup (1, \infty)$

57. $\dfrac{x^2 + 7x + 12}{x} \geq 0$

$\dfrac{(x + 4)(x + 3)}{x} \geq 0$

Critical numbers: $x = -4, x = -3, x = 0$

Test intervals: $(-\infty, -4), (-4, -3), (-3, 0), (0, \infty)$

Test: Is $\dfrac{(x + 4)(x + 3)}{x} \geq 0$?

Solution set: $[-4, -3] \cup (0, \infty)$

59. $5000(1 + r)^2 > 5500$

$(1 + r)^2 > 1.1$

$1 + r > 1.0488$

$r > 0.0488$

$r > 4.9\%$

61.

$d_1 = \sqrt{(13 - 5)^2 + (11 - 22)^2} = \sqrt{64 + 121} = \sqrt{185}$

$d_2 = \sqrt{(2 - 13)^2 + (3 - 11)^2} = \sqrt{121 + 64} = \sqrt{185}$

$d_3 = \sqrt{(2 - 5)^2 + (3 - 22)^2} = \sqrt{9 + 361} = \sqrt{370}$

$d_1{}^2 + d_2{}^2 = 185 + 185 = 370 = d_3{}^2$

Thus, the triangle is a right triangle.

63. $x > 0$ and $y = -2$ in Quadrant IV.

65. $(-x, y)$ is in the third Quadrant means that (x, y) is in Quadrant IV.

67. (a)

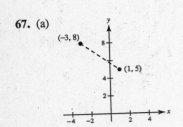

(b) $d = \sqrt{(-3-1)^2 + (8-5)^2} = \sqrt{16 + 9} = 5$

69. (a)

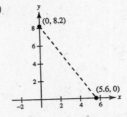

(b) $d = \sqrt{(5.6 - 0)^2 + (0 - 8.2)^2}$

$= \sqrt{31.36 + 67.24} = \sqrt{98.6} \approx 9.9$

71. (a)

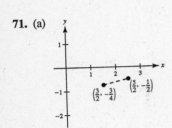

(b) $d = \sqrt{\left(\dfrac{5}{2} - \dfrac{3}{2}\right)^2 + \left[-\dfrac{1}{2} - \left(-\dfrac{3}{4}\right)\right]^2}$

$= \sqrt{(1)^2 + \left(\dfrac{1}{4}\right)^2}$

$= \sqrt{1 + \dfrac{1}{16}}$

$= \sqrt{\dfrac{17}{16}}$

$= \dfrac{\sqrt{17}}{4}$

73. (a)

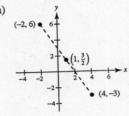

(b) Midpoint: $\left(\dfrac{-2 + 4}{2}, \dfrac{6 + (-3)}{2}\right) = \left(1, \dfrac{3}{2}\right)$

77. Change in apparent temperature: $150° \text{F} - 70° \text{F} = 80° \text{F}$

75. (a)

(b) Midpoint: $\left(\dfrac{0 + (-3.6)}{2}, \dfrac{-1.2 + 0}{2}\right) = (-1.8, -0.6)$

79. $y = 3x - 5$

x	-2	-1	0	1	2
y	-11	-8	-5	-2	1

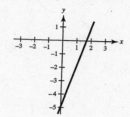

83. $y = 2x - 9$

x-intercept: $0 = 2x - 9 \implies x = \frac{9}{2}$

$\left(\frac{9}{2}, 0\right)$ is the x-intercept

y-intercept: $y = 2(0) - 9 = -9$

$(0, -9)$ is the y-intercept

87. $y = (x + 1)^2$

x-intercept: $0 = (x + 1)^2 \implies x = -1$

$(-1, 0)$ is the x-intercept.

y-intercept: $y = (0 + 1)^2 = 1$

$(0, 1)$ is the y-intercept.

91. $y = 5 - x^2$

Intercepts: $\left(\pm\sqrt{5}, 0\right), (0, 5)$

y-axis symmetry

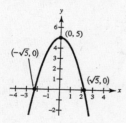

95. $y = \sqrt{x + 5}$

Domain: $[-5, \infty)$

Intercepts: $(-5, 0), \left(0, \sqrt{5}\right)$

No axis or origin symmetry

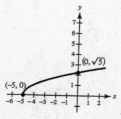

81. $y = x^2 - 3x$

x	-1	0	1	2	3	4
y	4	0	-2	-2	0	4

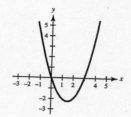

85. $y = |x - 4| - 4$

y-intercept: $y = |0 - 4| - 4 = 0$

$(0, 0)$

x-intercepts: $y = |x - 4| - 4 = 0$

$-x + 4 - 4 = 0 \qquad x - 4 - 4 = 0$

$x = 0 \qquad\qquad x = 8$

$(0, 0), (8, 0)$

89. $y = x\sqrt{9 - x^2}$

y-intercept: $y = 0\sqrt{9 - 0^2} = 0$

$(0, 0)$

x-intercepts: $y = x\sqrt{9 - x^2} = 0$

$x = 0 \qquad \sqrt{9 - x^2} = 0$

$x = \pm 3$

$(0, 0), (\pm 3, 0)$

93. $y = x^3 - 3$

Intercepts: $(0, -3), \left(\sqrt[3]{3}, 0\right) \approx (1.44, 0)$

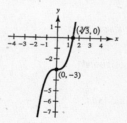

97. $y = 1 - |x|$

Intercepts: $(0, 1), (\pm 1, 0)$

Symmetry: y-axis

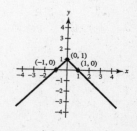

99. $x^2 + y^2 = 25$

Center: $(0, 0)$

Radius: 5

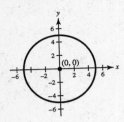

101. $(x + 2)^2 + y^2 = 16$

$(x - (-2))^2 + (y - 0)^2 = 4^2$

Center: $(-2, 0)$

Radius: 4

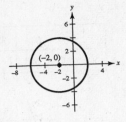

103. Endpoints of a diameter: $(0, 0)$ and $(4, -6)$

Center: $\left(\dfrac{0 + 4}{2}, \dfrac{0 + (-6)}{2} \right) = (2, -3)$

Radius: $r = \sqrt{(2 - 0)^2 + (-3 - 0)^2} = \sqrt{4 + 9} = \sqrt{13}$

Standard form: $(x - 2)^2 + (y - (-3))^2 = \left(\sqrt{13} \right)^2$

$(x - 2)^2 + (y + 3)^2 = 13$

105. $y = 0.070t + 0.37$,

$\quad t = 0$ represents 1990.

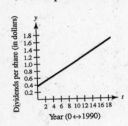
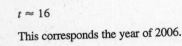

$0.070t + 0.37 = 1.50$

$0.070t = 1.13$

$t \approx 16$

This corresponds the year of 2006.

107. $y = 6$

Horizontal line

y-intercept: $(0, 6)$

Slope: $m = 0$

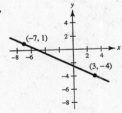

109. $y = 3x + 13$

y-intercept: $(0, 13)$

Slope: $m = 3 = \dfrac{3}{1}$

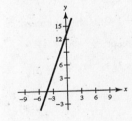

111. (a) $m = \dfrac{3}{2} > 0 \implies$ The line rises.
Matches L_2.

   (b) $m = 0 \implies$ The line is horizontal.
Matches L_3.

   (c) $m = -3 < 0 \implies$ The line falls.
Matches L_1.

   (d) $m = -\dfrac{1}{5} < 0 \implies$ The line gradually falls.
Matches L_4.

113.

Slope: $m = \dfrac{1 - (-4)}{-7 - 3} = \dfrac{5}{-10} = -\dfrac{1}{2}$

115.

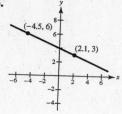

Slope: $m = \dfrac{6 - 3}{-4.5 - 2.1} = \dfrac{3}{-6.6} = -\dfrac{5}{11}$

117. $(-5, 2), (0, 10)$

$$m = \frac{10 - 2}{0 + 5} = \frac{8}{5}$$

$$y = \frac{8}{5}x + 10$$

119. $(-1, 4), (2, 0)$

Slope: $m = \dfrac{4 - 0}{-1 - 2} = -\dfrac{4}{3}$

$$y - 0 = -\frac{4}{3}(x - 2)$$

$$3y = -4x + 8$$

$$4x + 3y - 8 = 0$$

121. Point: $(0, -5)$ Slope: $m = \dfrac{3}{2}$

$$y - (-5) = \frac{3}{2}(x - 0)$$

$$2y + 10 = 3x$$

$$3x - 2y - 10 = 0$$

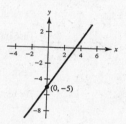

123. Point: $(-2, 6)$

Slope: $m = 0$

$$y - 6 = 0(x - (-2))$$

$$y - 6 = 0$$

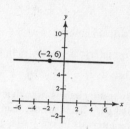

125. Point: $(-8, 5)$

Slope: Undefined

$$x = -8$$

$$x + 8 = 0$$

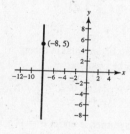

127. $x - 5 = 0 \implies x = 5$ and m is undefined.

(a) $(2, -1)$, m is undefined.

$$x = 2 \quad \text{or} \quad x - 2 = 0$$

(b) $(2, -1)$, $m = 0$

$$y = -1 \quad \text{or} \quad y + 1 = 0$$

129. $x + 4 = 0 \implies x = -4$ and m is undefined.

(a) $(3, 2)$, m is undefined.

$$x = 3 \quad \text{or} \quad x - 3 = 0$$

(b) $(3, 2)$, $m = 0$

$$y = 2 \quad \text{or} \quad y - 2 = 0$$

131. Point: $(3, -2)$, Line: $5x - 4y = 8$

(a) Parallel line: $m = \dfrac{5}{4}$

$$y - (-2) = \frac{5}{4}(x - 3)$$

$$4y + 8 = 5x - 15$$

$$5x - 4y - 23 = 0$$

(b) Perpendicular line: $m = -\dfrac{4}{5}$

$$y - (-2) = -\frac{4}{5}(x - 3)$$

$$5y + 10 = -4x + 12$$

$$4x + 5y - 2 = 0$$

133. 2005 *Value*: $12,500

Slope: $850 increase per year

$$V = 850(t - 5) + 12,500$$

$$V = 850t + 8250$$

Problem Solving for Chapter P

1. (a) $3(x + 4)^2 + (x + 4) - 2 = 0$

Let $u = x + 4$.

$$3u^2 + u - 2 = 0$$

$$(3u - 2)(u + 1) = 0$$

$$u = \tfrac{2}{3}, -1$$

$$x = u - 4 = -\tfrac{10}{3}, -5$$

(b) $3(x^2 + 8x + 16) + x + 4 - 2 = 0$

$$3x^2 + 24x + 48 + x + 4 - 2 = 0$$

$$3x^2 + 25x + 50 = 0$$

$$(3x + 10)(x + 5) = 0$$

$$x = -\tfrac{10}{3}, -5$$

(c) Answers will vary. Method (a) is slightly easier since there are less algebraic steps.

3. (a) $x^2 + bx + 4 = 0$

To have at least one real solution,

$$b^2 - 4(1)(4) \geq 0$$

$$b^2 - 16 \geq 0.$$

Critical numbers: $b = \pm 4$

Test intervals: $(-\infty, -4), (-4, 4), (4, \infty)$

Test: Is $b^2 - 16 \geq 0$?

By testing values in each test interval, we see that $b^2 - 16$ is greater than or equal to zero on the intervals $(-\infty, -4] \cup [4, \infty)$.

(b) $x^2 + bx - 4 = 0$

To have at least one real solution,

$$b^2 - 4(1)(-4) \geq 0$$

$$b^2 + 16 \geq 0.$$

This is true for all values of b:

$$-\infty < b < \infty$$

(c) $3x^2 + bx + 10 = 0$

To have at least one real solution,

$$b^2 - 4(3)(10) \geq 0$$

$$b^2 - 120 \geq 0.$$

Critical numbers: $b = \pm\sqrt{120} = \pm 2\sqrt{30}$

Test intervals:
$(-\infty, -2\sqrt{30}), (-2\sqrt{30}, 2\sqrt{30}), (2\sqrt{30}, \infty)$

Test: Is $b^2 - 120 \geq 0$?

By testing values in each test interval, we see that $b^2 - 120$ is greater than or equal to zero on the intervals $(-\infty, -2\sqrt{30}] \cup [2\sqrt{30}, \infty)$.

(d) $2x^2 + bx + 5 = 0$

To have at least one real solution,

$$b^2 - 4(2)(5) \geq 0$$

$$b^2 - 40 \geq 0.$$

This is true for $b \leq -2\sqrt{10}$ or $b \geq 2\sqrt{10}$,

$$(-\infty, -2\sqrt{10}] \cup [2\sqrt{10}, \infty).$$

(e) If $a > 0$ and $c \leq 0$, then b can be any real number since $b^2 - 4ac$ would always be positive.

If $a > 0$ and $c > 0$, then $b \leq -2\sqrt{ac}$ or $b \geq 2\sqrt{ac}$, as in parts (a), (e), and (d).

(f) Since the intervals for b are symmetric about $b = 0$, the center of the interval is $b = 0$.

5. $y = ax^2 + bx^3$

(a) If the graph is symmetric with respect to the y-axis, then you can replace x with $-x$ and the result is an equivalent equation. This happens when $b = 0$ and a is any real number.

(b) If the graph is symmetric with respect to the origin, then you can replace x with $-x$ and y with $-y$ and the result is an equivalent equation. This happens when $a = 0$ and b is any real number.

7. No. The slope cannot be determined without knowing the scale on the y-axis. The slopes could be the same.

9.

$$d_1 d_2 = 1$$

$$[(x+1)^2 + y^2][(x-1)^2 + y^2] = 1$$

$$(x+1)^2(x-1)^2 + y^2[(x+1)^2 + (x-1)^2] + y^4 = 1$$

$$(x^2-1)^2 + y^2[2x^2 + 2] + y^4 = 1$$

$$x^4 - 2x^2 + 1 + 2x^2y^2 + 2y^2 + y^4 = 1$$

$$(x^4 + 2x^2y^2 + y^4) - 2x^2 + 2y^2 = 0$$

$$(x^2 + y^2)^2 = 2(x^2 - y^2)$$

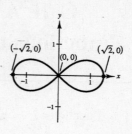

Let $y = 0$. Then $x^4 = 2x^2 \Rightarrow x = 0$ or $x^2 = 2$. Thus, $(0, 0)$, $\left(\sqrt{2}, 0\right)$ and $\left(-\sqrt{2}, 0\right)$ are on the curve.

11. (a) $\dfrac{7000 - 5500}{10} = 150$ students per year

(b) 1993: $5500 + 3(150) = 5950$ students

1997: $5500 + 7(150) = 6550$ students

1999: $5500 + 9(150) = 6850$ students

(c) Equation: $y = 150x + 5500$ where $x = 0$ represents 1990.

Slope: $m = 150$

This means that enrollments increase by approximately 150 students per year.

13. (a) and (b)

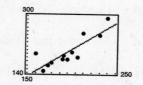

(c) $200 = 1.266x - 35.766$

$235.766 = 1.266x$

$x \approx 186.23$ pounds

(d) Answers will vary.

(c) An ambitious salesperson who can sell more than $20,000 per month would be wise to select choice 1. A more conservative choice for a salesperson who is unsure of the market for this product would be choice 2.

15. (a) Choice 1: $W = 3000 + 0.07s$

Choice 2: $W = 3400 + 0.05s$

(b) $3000 + 0.07s = 3400 + 0.05s$

$0.02s = 400$

$s = \$20,000$

The salaries are the same ($\$4400$ per month) when sales equal $\$20,000$.

C H A P T E R 1
Functions and Their Graphs

CHAPTER 1
Functions and Their Graphs

Section 1.1   Functions

1. Yes, the relationship is a function. Each domain value is matched with only one range value.

3. No, the relationship is not a function. The domain values are each matched with two range values.

5. No, the relationship is not a function. Each input value is matched with three output values.

7. Yes, it does represent a function. Each input value is matched with only one output value.

9. No, it does not represent a function. The input values of 10 and 7 are each matched with two output values.

11. (a) Each element of A is matched with exactly one element of B, so it does represent a function.

    (b) The element 1 in A is matched with two elements, -2 and 1 of B, so it does not represent a function.

    (c) Each element of A is matched with exactly one element of B, so it does represent a function.

    (d) The element 2 in A is not matched with an element of B, so it does not represent a function.

13. Each is a function. For each year there corresponds one and only one circulation.

15. $x^2 + y^2 = 4 \implies y = \pm\sqrt{4 - x^2}$

    No, y is *not* a function of x.

17. $x^2 + y = 4 \implies y = 4 - x^2$

    Yes, y is a function of x.

19. $2x + 3y = 4 \implies y = \dfrac{1}{3}(4 - 2x)$

    Yes, y is a function of x.

21. $y^2 = x^2 - 1 \implies y = \pm\sqrt{x^2 - 1}$

    No, y is *not* a function of x.

23. $y = |4 - x|$

    Yes, y is a function of x.

25. $|x| - y = 1 \implies y = |x| - 1$

    Yes, y is a function of x.

27. $x = y^2 \implies y = \pm\sqrt{x}$

    No, y is *not* a function of x.

29. $f(s) = \dfrac{1}{s + 1}$

    (a) $f(4) = \dfrac{1}{(4) + 1}$ (b) $f(0) = \dfrac{1}{(0) + 1}$ (c) $f(4x) = \dfrac{1}{(4x) + 1}$ (d) $f(x + c) = \dfrac{1}{(x + c) + 1}$

31. $f(x) = 2x - 3$

    (a) $f(1) = 2(1) - 3 = -1$

    (b) $f(-3) = 2(-3) - 3 = -9$

    (c) $f(x - 1) = 2(x - 1) - 3 = 2x - 5$

33. $V(r) = \frac{4}{3}\pi r^3$

    (a) $V(3) = \frac{4}{3}\pi(3)^3 = \frac{4}{3}\pi(27) = 36\pi$

    (b) $V\left(\frac{3}{2}\right) = \frac{4}{3}\pi\left(\frac{3}{2}\right)^3 = \frac{4}{3}\pi\left(\frac{27}{8}\right) = \frac{9}{2}\pi$

    (c) $V(2r) = \frac{4}{3}\pi(2r)^3 = \frac{4}{3}\pi(8r^3) = \frac{32}{3}\pi r^3$

35. $f(y) = 3 - \sqrt{y}$

    (a) $f(4) = 3 - \sqrt{4} = 1$

    (b) $f(0.25) = 3 - \sqrt{0.25} = 2.5$

    (c) $f(4x^2) = 3 - \sqrt{4x^2} = 3 - 2|x|$

37. $q(x) = \dfrac{1}{x^2 - 9}$

    (a) $q(0) = \dfrac{1}{0^2 - 9} = -\dfrac{1}{9}$

    (b) $q(3) = \dfrac{1}{3^2 - 9}$ is undefined.

    (c) $q(y + 3) = \dfrac{1}{(y + 3)^2 - 9} = \dfrac{1}{y^2 + 6y}$

39. $f(x) = \dfrac{|x|}{x}$

    (a) $f(2) = \dfrac{|2|}{2} = 1$

    (b) $f(-2) = \dfrac{|-2|}{-2} = -1$

    (c) $f(x - 1) = \dfrac{|x - 1|}{x - 1}$

41. $f(x) = \begin{cases} 2x - 1, & x < 0 \\ 2x + 2, & x \geq 0 \end{cases}$

    (a) $f(-1) = 2(-1) - 1 = -3$

    (b) $f(0) = 2(0) + 2 = 2$

    (c) $f(2) = 2(2) + 2 = 6$

43. $f(x) = x^2 - 3$

x	-2	-1	0	1	2
$f(x)$	1	-2	-3	-2	1

45. $h(t) = \frac{1}{2}|t + 3|$

t	-5	-4	-3	-2	-1
$h(t)$	1	$\frac{1}{2}$	0	$\frac{1}{2}$	1

47. $f(x) = \begin{cases} -\frac{1}{2}x + 4, & x \leq 0 \\ (x - 2)^2, & x > 0 \end{cases}$

x	-2	-1	0	1	2
$f(x)$	5	$\frac{9}{2}$	4	1	0

49. $f(x) = 15 - 3x$

$$15 - 3x = 0$$
$$3x = 15$$
$$x = 5$$

51. $\quad f(x) = \dfrac{x^2 - 16}{x - 2}$

$$\dfrac{x^2 - 16}{x - 2} = 0$$
$$x^2 - 16 = 0$$
$$x^2 = 16$$
$$x = \pm 4$$

53. $\quad f(x) = x^2 - 5x + 4$

$$x^2 - 5x + 4 = 0$$
$$(x - 1)(x - 4) = 0$$
$$x - 1 = 0 \implies x = 1$$
$$x - 4 = 0 \implies x = 4$$

55. $\quad f(x) = x^3 - 4x^2 - x + 4$

$$x^3 - 4x^2 - x + 4 = 0$$
$$x^2(x - 4) - 1(x - 4) = 0$$
$$(x^2 - 1)(x - 4) = 0$$
$$(x + 1)(x - 1)(x - 4) = 0$$
$$x + 1 = 0 \implies x = -1$$
$$x - 1 = 0 \implies x = 1$$
$$x - 4 = 0 \implies x = 4$$

57.
$$f(x) = g(x)$$
$$x^2 = x + 2$$
$$x^2 - x - 2 = 0$$
$$(x + 1)(x - 2) = 0$$
$$x + 1 = 0 \implies x = -1$$
$$x - 2 = 0 \implies x = 2$$

59.
$$f(x) = g(x)$$
$$x^3 - 3x + 1 = x + 1$$
$$x^3 - 4x = 0$$
$$x(x^2 - 4) = 0$$
$$x \, (x + 2)(x - 2) = 0$$
$$x = 0$$
$$x + 2 = 0 \implies x = -2$$
$$x - 2 = 0 \implies x = 2$$

61. $f(x) = 5x^2 + 2x - 1$

Since $f(x)$ is a polynomial, the domain is all real numbers x.

63. $h(t) = \dfrac{4t - 3}{t}$

Domain: All real numbers $t \neq 0$.

65. $g(y) = \sqrt{y - 10}$

Domain: $y - 10 \geq 0$
$$y \geq 10$$

67. $f(x) = \sqrt[4]{1 - x^2}$

Domain: $1 - x^2 \geq 0$
$$-x^2 \geq -1$$
$$x^2 \leq 1$$
$$x^2 - 1 \leq 0$$

Critical Numbers: $x = \pm 1$

Test Intervals: $(-\infty, -1), (-1, 1), (1, \infty)$

Test: Is $x^2 - 1 \leq 0$?

Solution: $[-1, 1]$ or $-1 \leq x \leq 1$

69. $g(x) = \dfrac{1}{x} - \dfrac{3}{x + 2}$

Domain: All real numbers except $x = 0$, $x = -2$

71. $f(s) = \dfrac{\sqrt{s - 1}}{s - 4}$

Domain: $s - 1 \geq 0 \implies s \geq 1$ and $s \neq 4$

The domain consists of all real numbers s, such that $s \geq 1$ and $s \neq 4$.

73. $f(x) = \dfrac{\sqrt[3]{x - 4}}{x}$

The domain is all real numbers except $x = 0$.

75. $f(x) = x^2$

$\{(-2, 4), (-1, 1), (0, 0), (1, 1), (2, 4)\}$

77. $f(x) = \sqrt{x + 2}$

$\{(-2, 0), (-1, 1), (0, \sqrt{2}), (1, \sqrt{3}), (2, 2)\}$

79. By plotting the points, we have a parabola, so $g(x) = cx^2$. Since $(-4, -32)$ is on the graph, we have $-32 = c(-4)^2 \implies c = -2$. Thus, $g(x) = -2x^2$.

81. Since the function is undefined at 0, we have $r(x) = c/x$. Since $(-4, -8)$ is on the graph, we have $-8 = c/-4 \implies c = 32$. Thus, $r(x) = 32/x$.

83.
$$f(x) = x^2 - x + 1$$
$$f(2 + \Delta x) = (2 + \Delta x)^2 - (2 + \Delta x) + 1$$
$$= 4 + 4\Delta x + (\Delta x)^2 - 2 - \Delta x + 1$$
$$= (\Delta x)^2 + 3\Delta x + 3$$
$$f(2) = (2)^2 - 2 + 1 = 3$$
$$f(2 + \Delta x) - f(2) = (\Delta x)^2 + 3\Delta x$$
$$\frac{f(2 + \Delta x) - f(2)}{\Delta x} = \frac{(\Delta x)^2 + 3\Delta x}{\Delta x} = \Delta x + 3, \ \Delta x \neq 0$$

85.
$$f(x) = x^3 + 2x - 1$$
$$\frac{f(x + c) - f(x)}{c} = \frac{[(x + c)^3 + 2(x + c) - 1] - (x^3 + 2x - 1)}{c}$$
$$= \frac{x^3 + 3x^2c + 3xc^2 + c^3 + 2x + 2c - 1 - x^3 - 2x + 1}{c}$$
$$= \frac{3x^2c + 3xc^2 + c^3 + 2c}{c} = \frac{c(3x^2 + 3xc + c^2 + 2)}{c}$$
$$= 3x^2 + 3xc + c^2 + 2, \ c \neq 0$$

87. $g(x) = 3x - 1$
$$\frac{g(x) - g(3)}{x - 3} = \frac{(3x - 1) - 8}{x - 3} = \frac{3x - 9}{x - 3} = \frac{3(x - 3)}{x - 3} = 3, \ x \neq 3$$

89. $f(x) = \sqrt{5x}$
$$\frac{f(x) - f(5)}{x - 5} = \frac{\sqrt{5x} - 5}{x - 5}$$

91. No. The element 3 in the domain corresponds to two elements in the range.

93. An advantage to function notation is that it gives a name to the relationship so it can be easily referenced. When evaluating a function, you see both the input and output values.

95. $A = \pi r^2, \quad C = 2\pi r$
$$r = \frac{C}{2\pi}$$
$$A = \pi\left(\frac{C}{2\pi}\right)^2 = \frac{C^2}{4\pi}$$

97. $A = \frac{1}{2}bh$, and in an equilateral triangle $b = s$
$$s^2 = h^2 + \left(\frac{s}{2}\right)^2$$
$$h = \sqrt{s^2 - \left(\frac{s}{2}\right)^2}$$
$$h = \sqrt{\frac{4s^2}{4} - \frac{s^2}{4}} = \frac{\sqrt{3}s}{2}$$
$$A = \frac{1}{2}s \cdot \frac{\sqrt{3}s}{2} = \frac{\sqrt{3}s^2}{4}$$

99. (a)

Units x	Price p	Profit P
110	90 − 10(0.15)	110[90 − 10(0.15)] − 110(60) = 3135
120	90 − 20(0.15)	120[90 − 20(0.15)] − 120(60) = 3240
130	90 − 30(0.15)	130[90 − 30(0.15)] − 130(60) = 3315
140	90 − 40(0.15)	140[90 − 40(0.15)] − 140(60) = 3360
150	90 − 50(0.15)	150[90 − 50(0.15)] − 150(60) = 3375
160	90 − 60(0.15)	160[90 − 60(0.15)] − 160(60) = 3360

The maximum profit is $3375.

(b)

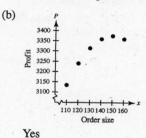

Yes

(c) Profit = Revenue − Cost

$$= \text{(price per unit)(number of units)} - \text{(cost)(number of units)}$$
$$= [90 - (x - 100)(0.15)]x - 60x, \ x > 100$$
$$= (90 - 0.15x + 15)x - 60x$$
$$= (105 - 0.15x)x - 60x$$
$$= 105x - 0.15x^2 - 60x$$
$$= 45x - 0.15x^2, \ x > 100$$

101. (a) $V = l \cdot w \cdot h = x \cdot y \cdot x = x^2y$ where $4x + y = 108$.

Thus, $y = 108 - 4x$ and $V = x^2(108 - 4x) = 108x^2 - 4x^3$

(b) Domain: $0 < x < 27$

(c)

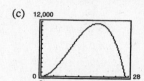

(d) The dimensions that will maximize the volume of the package are $18 \times 18 \times 36$. From the graph, the maximum volume occurs when $x = 18$. To find the dimension for y, use the equation $y = 108 - 4x$.

$$y = 108 - 4x = 108 - 4(18) = 108 - 72 = 36.$$

103. $C = 0.95x + 6000$

$$\overline{C} = \frac{C}{x} = \frac{0.95x + 6000}{x} = 0.95 + \frac{6000}{x}$$

105. $F(y) = 149.76\sqrt{10}y^{5/2}$

(a)

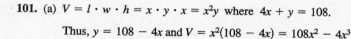

y	5	10	20	30	40
$F(y)$	26,474.08	149,760.00	847,170.49	2,334,527.36	4,792,320

The force, in tons, of the water against the dam increases with the depth of the water.

(b) It appears that approximately 21 feet of water would produce 1,000,000 tons of force.
You can find a better estimate by creating a new table with y at smaller intervals between 20 and 25.

107. Answers may vary slightly.

$$\frac{f(2001) - f(1993)}{2001 - 1993} \approx \frac{14,200 - 11,000}{8} = 400$$

The number of kidney transplants increased by approximately 400 transplants per year.

109. False. The range is $[-1, \infty)$.

111. True.
As long as **all** elements in the domain are matched with elements in the range, even if it is the same element, then the relation is a function.

Section 1.2   Analyzing Graphs of Functions

1. $f(x) = \frac{2}{3}x - 4$

Domain: All real numbers
$(-\infty, \infty)$

Range: All real numbers
$(-\infty, \infty)$

3. $f(x) = \sqrt{x^2 - 1}$

Domain: $(-\infty, -1] \cup [1, \infty)$

Range: $[0, \infty)$

5. $f(x) = 1 - x^2$

Domain: All real numbers

Range: $(-\infty, 1]$

7. $h(x) = \sqrt{16 - x^2}$

Domain: $[-4, 4]$

Range: $[0, 4]$

9. $f(x) = \dfrac{1}{x^2 + 1}$

Domain: All real numbers

Range: $(0, 1]$

11. $y = \frac{1}{2}x^2$

A vertical line intersects the graph just once, so y is a function of x.

13. $x - y^2 = 1 \implies y = \pm\sqrt{x - 1}$

y is not a function of x.
Some vertical lines cross the graph twice.

15. $x^2 = 2xy - 1$

A vertical line intersects the graph just once, so y is a function of x.

17. $2x^2 - 7x - 30 = 0$

$(2x + 5)(x - 6) = 0$

$2x + 5 = 0$ or $x - 6 = 0$

$x = -\dfrac{5}{2}$ or $x = 6$

19. $f(x) = \dfrac{9x^2 - 4}{x}$

$\dfrac{9x^2 - 4}{x} = 0$

$9x^2 - 4 = 0$

$x^2 = \dfrac{4}{9}$

$x = \pm\dfrac{2}{3}$

21. $f(x) = \frac{1}{2}x^3 - 2x$

$\frac{1}{2}x^3 - 2x = 0$

$x^3 - 4x = 0$

$x(x^2 - 4) = 0$

$x(x - 2)(x + 2) = 0$

$x = 0$ or $x = \pm 2$

23. $f(x) = x^3 - 4x^2 - 9x + 36$

$0 = x^3 - 4x^2 - 9x + 36$

$0 = x^2(x - 4) - 9(x - 4)$

$0 = (x - 4)(x^2 - 9)$

$x - 4 = 0 \implies x = 4$

$x^2 - 9 = 0 \implies x = \pm 3$

25. $3 + \dfrac{5}{x} = 0$

$3x + 5 = 0$

$x = -\dfrac{5}{3}$

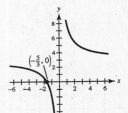

27. $\sqrt{2x + 11} = 0$

$2x + 11 = 0$

$x = -\dfrac{11}{2}$

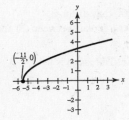

29. $\dfrac{3x^2 - 1}{x - 6} = 0$

$3x^2 - 1 = 0$

$x^2 = \dfrac{1}{3}$

$x = \pm\dfrac{1}{\sqrt{3}}$

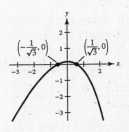

31. $f(x) = \frac{3}{2}x$

    (a) f is increasing on $(-\infty, \infty)$.

    (b) Since $f(-x) = -f(x)$, f is odd.

33. $f(x) = x^3 - 3x^2 + 2$

    (a) f is increasing on $(-\infty, 0)$ and $(2, \infty)$.

        f is decreasing on $(0, 2)$.

    (b) $f(-x) = (-x)^3 - 3(-x)^2 + 2$

            $= -x^3 - 3x^2 + 2$

        $f(-x) \neq -f(x)$

        $f(-x) \neq f(x)$

    f is neither odd nor even.

35. $f(x) = 3$

    (a)

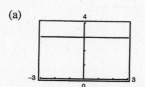

    Constant on $(-\infty, \infty)$

    (b)

x	-2	-1	0	1	2
$f(x)$	3	3	3	3	3

37. $f(x) = 5 - 3x$

    (a)

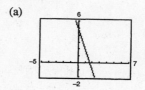

    Decreasing on $(-\infty, \infty)$

    (b)

x	-2	-1	0	1	2
$f(x)$	11	8	5	2	-1

39. $g(s) = \dfrac{s^2}{4} - 1$

    (a)

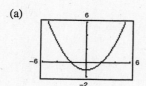

    Decreasing on $(-\infty, 0)$

    Increasing on $(0, \infty)$

    (b)

s	-4	-2	0	2	4
$g(s)$	3	0	-1	0	3

41. $f(t) = -t^4 + 2t^2$

    (a)

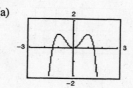

    Increasing on $(-\infty, -1)$ and $(0, 1)$

    Decreasing on $(-1, 0)$ and $(1, \infty)$

    (b)

t	-2	-1	0	1	2
$f(t)$	-8	1	0	1	-8

43. $f(x) = x\sqrt{1-x}$

    (a)

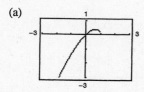

    Increasing on $\left(-\infty, \frac{2}{3}\right)$

    Decreasing on $\left(\frac{2}{3}, 1\right)$

    (b)

x	-3	-2	-1	0	1
$f(x)$	-6	$-2\sqrt{3}$	$-\sqrt{2}$	0	0

45. $f(x) = x^{3/2} - 1$

    (a)

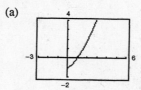

    Increasing on $(0, \infty)$

    (b)

x	0	1	2	3	4
$f(x)$	-1	0	1.83	4.2	7

47. $g(t) = \sqrt[3]{t-1}$

(a)

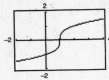

Increasing on $(-\infty, \infty)$

(b)

t	-2	-1	0	1	2
$g(t)$	-1.44	-1.26	-1	0	1

49. $f(x) = |x + 2| - |x - 2|$

(a)

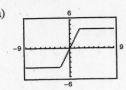

Increasing on $(-2, 2)$

Constant on $(-\infty, -2)$ and $(2, \infty)$

(b)

x	-2	-1	0	1	4
$f(x)$	-4	-2	0	2	4

51. $f(x) = \begin{cases} x + 3, & x \le 0 \\ 3, & 0 < x \le 2 \\ 2x - 1, & x > 2 \end{cases}$

(a)

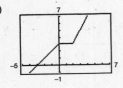

Increasing on $(-\infty, 0)$ and $(2, \infty)$

Constant on $(0, 2)$

(b)

x	-2	-1	0	1	2	3	4
$f(x)$	1	2	3	3	3	5	7

53. $f(x) = (x - 4)(x + 2)$

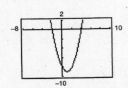

Relative Minimum at $(1, -9)$

55. $f(x) = x(x - 2)(x + 3)$

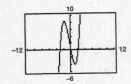

Relative Minimum at $(1.12, -4.06)$

Relative Maximum at $(-1.79, 8.21)$

57. $f(x) = 2x^3 - 5x^2 - 4x - 1$

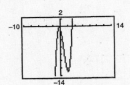

Relative Minimum at $(2, -13)$

Relative Maximum at $(-0.33, -0.30)$

59. $f(x) = 1 - 2x$

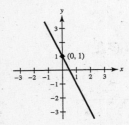

61. $f(x) = -x - \frac{3}{4}$

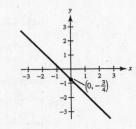

63. $f(x) = -\frac{1}{6}x - \frac{5}{2}$

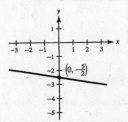

65. $f(x) = -1.8 + 2.5x$

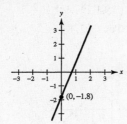

(0, -1.8)

67. $f(1) = 4, f(0) = 6$

(1, 4) and (0, 6)

$m = \dfrac{6 - 4}{0 - 1} = -2$

$y - 6 = -2(x - 0)$

$y = -2x + 6$

$f(x) = -2x + 6$

69. $f(5) = -4, f(-2) = 17$

(5, -4) and (-2, 17)

$m = \dfrac{17 - (-4)}{-2 - 5} = \dfrac{21}{-7} = -3$

$y - (-4) = -3(x - 5)$

$y + 4 = -3x + 15$

$y = -3x + 11$

$f(x) = -3x + 11$

71. $f(-5) = -5, f(5) = -1$

(-5, -5) and (5, -1)

$m = \dfrac{-1 - (-5)}{5 - (-5)} = \dfrac{4}{10} = \dfrac{2}{5}$

$y - (-5) = \dfrac{2}{5}(x - (-5))$

$y + 5 = \dfrac{2}{5}x + 2$

$y = \dfrac{2}{5}x - 3$

$f(x) = \dfrac{2}{5}x - 3$

73. $f\left(\dfrac{1}{2}\right) = -6, f(4) = -3$

$\left(\dfrac{1}{2}, -6\right)$ and (4, -3)

$m = \dfrac{-3 - (-6)}{4 - \dfrac{1}{2}} = \dfrac{3}{7/2} = \dfrac{6}{7}$

$y - (-3) = \dfrac{6}{7}(x - 4)$

$y + 3 = \dfrac{6}{7}x - \dfrac{24}{7}$

$y = \dfrac{6}{7}x - \dfrac{45}{7}$

$f(x) = \dfrac{6}{7}x - \dfrac{45}{7}$

75. Vertical shift 2 units downward.

$f(x) = [\![x]\!] - 2$

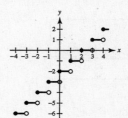

77. Horizontal shift 1 unit right.

$f(x) = [\![x - 1]\!]$

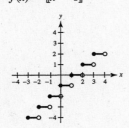

79. $f(x) = \begin{cases} 2x + 3, & x < 0 \\ 3 - x, & x \ge 0 \end{cases}$

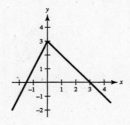

81. $f(x) = \begin{cases} x^2 + 5, & x \le 1 \\ -x^2 + 4x + 3, & x > 1 \end{cases}$

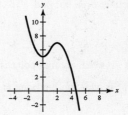

83. $f(x) = 4 - x$

$f(x) \ge 0$ on $(-\infty, 4]$.

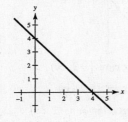

85. $f(x) = x^2 - 9$

$f(x) \ge 0$ on $(-\infty, -3]$ and $[3, \infty)$.

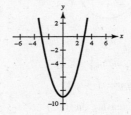

87. $f(x) = 1 - x^4$

$f(x) \ge 0$ on $[-1, 1]$.

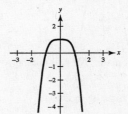

89. $f(x) = x^2 + 1$

$f(x) \geq 0$ on $(-\infty, \infty)$.

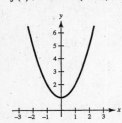

91. $f(x) = -5$, $f(x) < 0$ for all x.

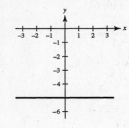

93. $f(x) = \begin{cases} 1 - 2x^2, & x \leq -2 \\ -x + 8, & x > -2 \end{cases}$

$f(x) \geq 0$ on $(-2, 8]$

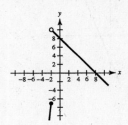

95. $s(x) = 2\left(\frac{1}{4}x - \left[\!\left[\frac{1}{4}x\right]\!\right]\right)$

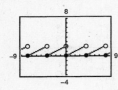

Domain: $(-\infty, \infty)$

Range: $[0, 2)$

Sawtooth pattern

97. $f(x) = x^6 - 2x^2 + 3$

$f(-x) = (-x)^6 - 2(-x)^2 + 3$

$= x^6 - 2x^2 + 3$

$= f(x)$

f is even.

99. $g(x) = x^3 + 5x - 2$

$g(-x) = (-x)^3 + 5(-x) - 2$

$= -x^3 - 5x - 2$

$= -(x^3 + 5x + 2)$

$\neq g(x), \neq -g(x)$

g is neither even nor odd.

101. $f(t) = 2t^{3/2} = 2\sqrt{t^3}$ or $2\left(\sqrt{t}\right)^3$

Domain: $t \geq 0$

$f(-t) = 2\sqrt{(-t)^3} = 2\sqrt{-t^3}, t \leq 0$

Note: $2\sqrt{-t^3} \neq -2\sqrt{t^3}$ thus $f(-t) \neq -f(t)$, $\neq f(t)$

f is neither even nor odd

103. $\left(-\frac{3}{2}, 4\right)$

(a) If f is even, another point is $\left(\frac{3}{2}, 4\right)$.

(b) If f is odd, another point is $\left(\frac{3}{2}, -4\right)$.

105. $(4, 9)$

(a) If f is even, another point is $(-4, 9)$.

(b) If f is odd, another point is $(-4, -9)$.

107.

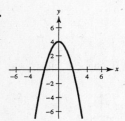

(a) Domain: All real numbers or $(-\infty, \infty)$

(b) Range: $(-\infty, 4]$

(c) Increasing on $(-\infty, 0)$
Decreasing on $(0, \infty)$

109.

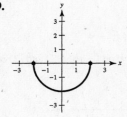

(a) Domain: $[-2, 2]$

(b) Range: $[-2, 0]$

(c) Increasing on $(0, 2)$
Decreasing on $(-2, 0)$

111.

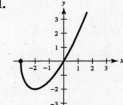

(a) Domain: $[-3, \infty)$

(b) Range: $[-2, \infty)$

(c) Increasing on $(-2, \infty)$
Decreasing on $(-3, -2)$

113.

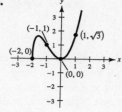

(a) $f(-1) = 1$

(b) $f(1) = \sqrt{3}$

(c) f is increasing on $(-2, -1.6)$ and $(0, \infty)$
f is decreasing on $(-1.6, 0)$.

115. (a) $C_2(t) = 1.05 - 0.38[\![-(t-1)]\!]$ is the appropriate
model since the cost does not increase until after
the next minute of conversation has started.

(b)

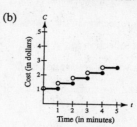

$C = 1.05 - 0.38[\![-17.75]\!] = \$7.89$

117. $h = \text{top} - \text{bottom}$
$= (-x^2 + 4x - 1) - 2$
$= -x^2 + 4x - 3$

119. $h = \text{top} - \text{bottom}$
$= (4x - x^2) - 2x$
$= 2x - x^2$

121. $L = \text{right} - \text{left}$
$= \frac{1}{2}y^2 - 0$
$= \frac{1}{2}y^2$

123. $L = \text{right} - \text{left}$
$= 4 - y^2$

125.

Year	Doctors, y
1992	33.5
1993	33.4
1994	35.0
1995	35.7
1996	37.3
1997	38.9
1998	40.8
1999	43.5
2000	44.9
2001	47.0

(a) Let $t = 2$ corresponds to 1992.
$y \approx -0.01336t^3 + 0.3621t^2 - 1.234t + 34.55$

(b) Domain: All real numbers

(c)

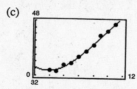

(d) The model is most accurate for the year 1996 and
least accurate for the year 1999.

127. (a) For the average salaries of college professors, a
scale of $10,000 would be appropriate.

(b) For the population of the United States, use a scale
of 50,000,000.

(c) For the percent of the civilian workforce that is
unemployed, use a scale of 1%.

129. $N = -0.01842t^4 + 0.6098t^3 - 7.062t^2$
$+ 26.59t + 46.9$, $t = 0$ corresponds to 1990

(a) The numbers of new cases was increasing from
1990-1993 and decreasing from 1993-2001.

(b) There was a maximum number of approximately
78,000 new cases in 1993.

131. False. The function $f(x) = \sqrt{x^2 + 1}$ has a domain of all real numbers.

133. True. A horizontal line, $y = k$, k a constant, is even.

135. (a) Even. The graph is a reflection in the x-axis.

   (b) Even. The graph is a reflection in the y-axis.

   (c) Even. The graph is a vertical translation of f.

   (d) Neither. The graph is a horizontal translation of f.

137. Yes, each y-value corresponds to only one x-value, so the graph does represent y as a function of x. Some values of y near 0 "appear" to correspond to more than one value of x, but that is a result of the vertical scaling.

Section 1.3   Shifting, Reflecting, and Stretching Graphs

1. (a) $f(x) = x^3 + c$

   $c = -2 : f(x) = x^3 - 2$ Vertical shift 2 units downward

   $c = 0 : f(x) = x^3$ Basic cubic function

   $c = 2 : f(x) = x^3 + 2$ Vertical shift 2 units upward

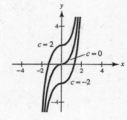

   (b) $f(x) = (x - c)^3$

   $c = -2 : f(x) = (x + 2)^3$ Horizontal shift 2 units to the left

   $c = 0 : f(x) = x^3$ Basic cubic function

   $c = 2 : f(x) = (x - 2)^3$ Horizontal shift 2 units to the right

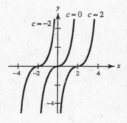

   (c) $f(x) = (x + 1)^3 + c$ Horizontal shift 1 unit to the left and a vertical shift

   $c = -2: f(x) = (x + 1)^3 - 2$ 2 units down

   $c = 0: f(x) = (x + 1)^3$

   $c = 2: f(x) = (x + 1)^3 + 2$ 2 units up

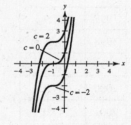

3. (a) $f(x) = \sqrt{x} + c$ Vertical shifts

   $c = -3 : f(x) = \sqrt{x} - 3$ 3 units downward

   $c = -1 : f(x) = \sqrt{x} - 1$ 1 unit downward

   $c = 1 : f(x) = \sqrt{x} + 1$ 1 unit upward

   $c = 3 : f(x) = \sqrt{x} + 3$ 3 units upward

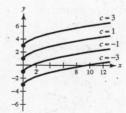

—CONTINUED—

3. —CONTINUED—

(b) $f(x) = \sqrt{x - c}$ Horizontal shifts

$c = -3 : f(x) = \sqrt{x + 3}$ 3 units to the left

$c = -1 : f(x) = \sqrt{x + 1}$ 1 unit to the left

$c = 1 : f(x) = \sqrt{x - 1}$ 1 unit to the right

$c = 3 : f(x) = \sqrt{x - 3}$ 3 units to the right

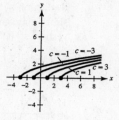

(c) $f(x) = \sqrt{x - 3} + c$ Horizontal shift 3 units to the right and a vertical shift

$c = -3 : f(x) = \sqrt{x - 3} - 3$ 3 units downward

$c = -1 : f(x) = \sqrt{x - 3} - 1$ 1 unit downward

$c = 1 : f(x) = \sqrt{x - 3} + 1$ 1 unit upward

$c = 3 : f(x) = \sqrt{x - 3} + 3$ 3 units upward

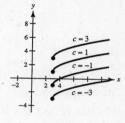

5. (a) Vertical shift 1 unit downward.

$f(x) = x^2 - 1$

(b) Reflection about the x-axis, horizontal shift 1 unit to the left, and a vertical shift 1 unit upward.

$f(x) = -(x + 1)^2 + 1$

(c) Reflection about the x-axis, horizontal shift 2 units to the right, and a vertical shift 6 units upward.

$f(x) = -(x - 2)^2 + 6$

(d) Horizontal shift 5 units to the right and a vertical shift 3 units downward.

$f(x) = (x - 5)^2 - 3$

7. (a) Vertical shift 5 units upward.

$f(x) = |x| + 5$

(b) Reflection in the x-axis and a horizontal shift 3 units to the left.

$f(x) = -|x + 3|$

(c) Horizontal shift 2 units to the right and a vertical shift 4 units downward.

$f(x) = |x - 2| - 4$

(d) Reflection in the x-axis, horizontal shift 6 units to the right, and a vertical shift 1 unit downward.

$f(x) = -|x - 6| - 1$

9. Common function: $f(x) = x^3$

Horizontal shift 2 units to the right: $y = (x - 2)^3$

11. Common function: $f(x) = x^2$

Reflection in the x-axis: $y = -x^2$

13. Common function: $f(x) = \sqrt{x}$

Reflection in the x-axis and a vertical shift 1 unit upward: $y = -\sqrt{x} + 1$

15. $f(x) = 12 - x^2$

Common function: $g(x) = x^2$

Reflection in the x-axis and a vertical shift 12 units upward.

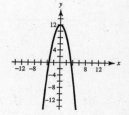

17. $f(x) = x^3 + 7$

Common function: $g(x) = x^3$

Vertical shift 7 units upward.

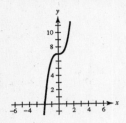

19. $f(x) = 2 - (x + 5)^2$

Common function: $g(x) = x^2$

Reflection in the *x*-axis, horizontal shift 5 units to the left, and a vertical shift 2 units upward.

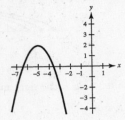

21. $f(x) = (x - 1)^3 + 2$

Common function: $g(x) = x^3$

Horizontal shift 1 unit to the right and a vertical shift 2 units upward.

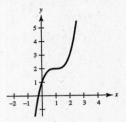

23. $f(x) = -|x| - 2$

Common function: $g(x) = |x|$

Reflection in the *x*-axis and a vertical shift 2 units downward.

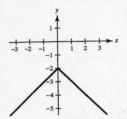

25. $f(x) = -|x + 4| + 8$

Common function: $g(x) = |x|$

Reflection in the *x*-axis, horizontal shift 4 units to the left, and a vertical shift 8 units upward.

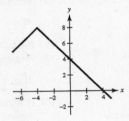

27. $f(x) = \sqrt{x - 9}$

Common function: $g(x) = \sqrt{x}$

Horizontal shift 9 units to the right.

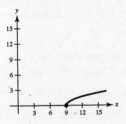

29. $f(x) = \sqrt{7 - x} - 2$ or

$f(x) = \sqrt{-(x - 7)} - 2$

Common function: $g(x) = \sqrt{x}$

Reflection in the *y*-axis, horizontal shift 7 units to the right, and a vertical shift 2 units downward.

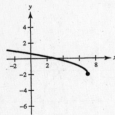

31. $f(x) = x^2$ moved 2 units to the right and 8 units down.

$g(x) = (x - 2)^2 - 8$

33. $f(x) = x^3$ moved 13 units to the right.

$g(x) = (x - 13)^3$

35. $f(x) = |x|$ moved 10 units up and reflected about the *x*-axis.

$g(x) = -(|x| + 10) = -|x| - 10$

37. $f(x) = \sqrt{x}$ moved 6 units to the left and reflected in both the *x* and *y* axes.

$g(x) = -\sqrt{-x + 6}$

39. $f(x) = x^2$

    (a) Reflection in the x-axis and a vertical stretch by a factor of 3.

        $g(x) = -3x^2$

    (b) A vertical stretch by a factor of 4 and a vertical shift 3 units upward.

        $g(x) = 4x^2 + 3$

41. $f(x) = |x|$

    (a) Reflection in the x-axis and a vertical shrink by a factor of $\frac{1}{2}$.

        $g(x) = -\frac{1}{2}|x|$

    (b) Vertical stretch by a factor of 3 and a vertical shift 3 units downward.

        $g(x) = 3|x| - 3$

43. Common function: $f(x) = x^3$

Vertical stretch by a factor of 2:
$g(x) = 2x^3$

45. Common function: $f(x) = x^2$

Reflection in the x-axis and a vertical shrink by a factor of $\frac{1}{2}$:
$g(x) = -\frac{1}{2}x^2$

47. Common function: $f(x) = \sqrt{x}$

Reflection in the y-axis and a vertical shrink by a factor of $\frac{1}{2}$:
$g(x) = \frac{1}{2}\sqrt{-x}$

49. Common function: $f(x) = x^3$

Reflection in the x-axis, horizontal shift 2 units to the right and a vertical shift 2 units upward:
$g(x) = -(x - 2)^3 + 2$

51. Common function: $f(x) = \sqrt{x}$

Reflection in the x-axis and a vertical shift 3 units downward: $g(x) = -\sqrt{x} - 3$

53. (a) $y = f(x) + 2$

Vertical shift 2 units upward.

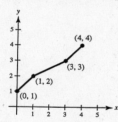

(b) $y = f(x - 2)$

Horizontal shift 2 units to the right.

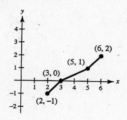

(c) $y = 2f(x)$

Vertical stretch by a factor of 2.

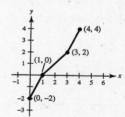

(d) $y = -f(x)$

Reflection in the x-axis.

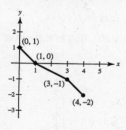

(e) $y = f(x + 3)$

Horizontal shift 3 units to the left.

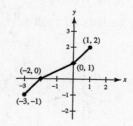

(f) $y = f(-x)$

Reflection in the y-axis.

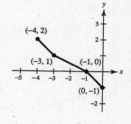

(g) $y = f(2x)$

Horizontal shrink. Each x-value is divided by 2.

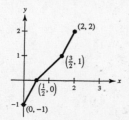

55. (a) $y = f(x) - 1$

Vertical shift 1 unit downward.

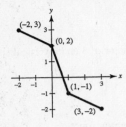

(b) $y = f(x - 1)$

Horizontal shift 1 unit
to the right.

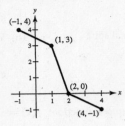

(c) $y = f(-x)$

Reflection about the y-axis.

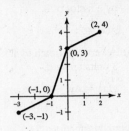

(d) $y = f(x + 1)$

Horizontal shift 1 unit
to the left.

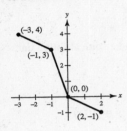

(e) $y = -f(x - 2)$

Reflection about the x-axis
and a horizontal shift 2 units
to the right.

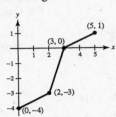

(f) $y = \frac{1}{2} f(x)$

Vertical shrink by a factor of $\frac{1}{2}$.

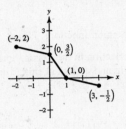

(g) $y = f\left(\frac{1}{2}x\right)$

Horizontal stretch. Each x-value is multiplied by 2.

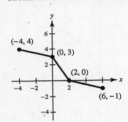

57. (a) $g(x) = f(x) + 2$

Vertical shift 2 units upward.

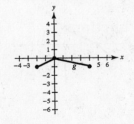

(b) $g(x) = f(x) - 1$

Vertical shift 1 unit downward.

(c) $g(x) = f(-x)$

Reflection in the y-axis.

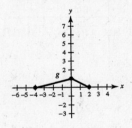

—CONTINUED—

57. —CONTINUED—

(d) $g(x) = -2f(x)$

Reflection in the x-axis and a
vertical stretch by a factor of 2.

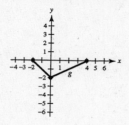

(e) $g(x) = f(2x)$

Horizontal shrink.
Each x-value is divided by 2.

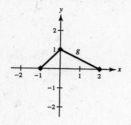

(f) $g(x) = f\left(\frac{1}{2}x\right)$

Horizontal stretch.
Each x-value is multiplied by 2.

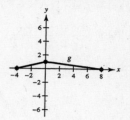

59. $T = f(t) = 29.1\sqrt{t} + 152.4$, $t = 4$ corresponds to 1994

(a) Vertical stretch of 29.1 and a vertical shift of
152.4 units upward

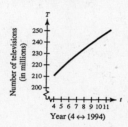

(b) $g(t) = 29.1\sqrt{t + 10} + 152.4$; By shifting the graph
10 units to the left, $t = 0$ would represent 2000.

61. True, since $|x| = |-x|$, the graphs of $f(x) = |x| + 6$ and
$f(x) = |-x| + 6$ are identical.

63. True

Let $g(x) = -f(x)$.

Then $g(-x) = -f(-x)$

$= -[-f(x)]$ since f is odd.

$= f(x)$

$= -g(x)$

Thus, $g(x) = -f(x)$ is also odd.

65. (a) and (b); Answers will vary.

Section 1.4   Combinations of Functions

1. $f(x) = x + 2, g(x) = x - 2$

(a) $(f + g)(x) = f(x) + g(x) = (x + 2) + (x - 2) = 2x$

(b) $(f - g)(x) = f(x) - g(x) = (x + 2) - (x - 2) = 4$

(c) $(fg)(x) = f(x) \cdot g(x) = (x + 2)(x - 2) = x^2 - 4$

(d) $\left(\dfrac{f}{g}\right)(x) = \dfrac{f(x)}{g(x)} = \dfrac{x + 2}{x - 2}$

Domain: all real numbers except $x = 2$

3. $f(x) = x^2, g(x) = 2 - x$

$(f + g)(x) = f(x) + g(x) = x^2 + (2 - x) = x^2 - x + 2$

$(f - g)(x) = f(x) - g(x) = x^2 - (2 - x) = x^2 + x - 2$

$(fg)(x) = f(x) \cdot g(x) = x^2(2 - x) = 2x^2 - x^3$

$\left(\dfrac{f}{g}\right)(x) = \dfrac{f(x)}{g(x)} = \dfrac{x^2}{2 - x}$,

Domain: all real numbers except $x = 2$

5. $f(x) = x^2 + 6, g(x) = \sqrt{1 - x}$

$(f + g)(x) = f(x) + g(x) = (x^2 + 6) + \sqrt{1 - x}$

$(f - g)(x) = f(x) - g(x) = (x^2 + 6) - \sqrt{1 - x}$

$(fg)(x) = f(x) \cdot g(x) = (x^2 + 6)\sqrt{1 - x}$

$\left(\dfrac{f}{g}\right)(x) = \dfrac{f(x)}{g(x)} = \dfrac{x^2 + 6}{\sqrt{1 - x}},$

Domain: $x < 1$

7. $f(x) = \dfrac{1}{x}, g(x) = \dfrac{1}{x^2}$

$(f + g)(x) = f(x) + g(x) = \dfrac{1}{x} + \dfrac{1}{x^2} = \dfrac{x + 1}{x^2}$

$(f - g)(x) = f(x) - g(x) = \dfrac{1}{x} - \dfrac{1}{x^2} = \dfrac{x - 1}{x^2}$

$(fg)(x) = f(x) \cdot g(x) = \dfrac{1}{x}\left(\dfrac{1}{x^2}\right) = \dfrac{1}{x^3}$

$\left(\dfrac{f}{g}\right)(x) = \dfrac{f(x)}{g(x)} = \dfrac{1/x}{1/x^2} = \dfrac{x^2}{x} = x, \; x \neq 0$

For Exercises 9–18, $f(x) = x^2 + 1$ and $g(x) = x - 4$

9. $(f + g)(2) = f(2) + g(2) = (2^2 + 1) + (2 - 4) = 3$

11. $(f - g)(3t) = f(3t) - g(3t) = [(3t)^2 + 1] - (3t - 4)$
$$= 9t^2 - 3t + 5$$

13. $(fg)(6) = f(6)g(6) = (6^2 + 1)(6 - 4) = 74$

15. $\left(\dfrac{f}{g}\right)(5) = \dfrac{f(5)}{g(5)} = \dfrac{5^2 + 1}{5 - 4} = 26$

17. $\left(\dfrac{f}{g}\right)(-1) - 2g(3) = \dfrac{f(-1)}{g(-1)} - 2g(3)$

$\qquad = \dfrac{2}{-5} - 2(-1)$

$\qquad = \dfrac{8}{5}$

19. $f(x) = \dfrac{1}{2}x, g(x) = x - 1, (f + g)(x) = \dfrac{3}{2}x - 1$

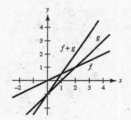

21. $f(x) = x^2, g(x) = -2x, (f + g)(x) = x^2 - 2x$

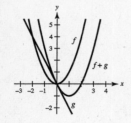

23.

x	0	1	2	3
f	2	3	1	2
g	-1	0	$\frac{1}{2}$	0
$f + g$	1	3	$\frac{3}{2}$	2

25.

x	-2	0	1	2	4
f	2	0	1	2	4
g	4	2	1	0	2
$f+g$	6	2	2	2	6

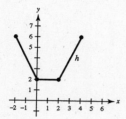

29. (a) $(f \circ g)(2) = f(g(2)) = f(2) = 0$

   (b) $(g \circ f)(2) = g(f(2)) = g(0) = 4$

33. $T(x) = R(x) + B(x) = \frac{3}{4}x + \frac{1}{15}x^2$

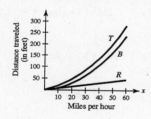

37. (a) T is a function of t since for each time t there corresponds one and only one temperature T.

   (b) $T(4) = 60°$

   $T(15) = 72°$

   (c) $H(t) = T(t - 1)$; All the temperature changes would be one hour later.

   (d) $H(t) = T(t) - 1$; The temperature would be decreased by one degree.

27. (a) $(f + g)(3) = f(3) + g(3) = 2 + 1 = 3$

   (b) $\left(\dfrac{f}{g}\right)(2) = \dfrac{f(2)}{g(2)} = \dfrac{0}{2} = 0$

31. $f(x) = 3x, g(x) = -\dfrac{x^3}{10}, (f + g)(x) = 3x - \dfrac{x^3}{10}$

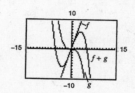

For $0 \le x \le 2$, $f(x)$ contributes most to the magnitude. For $x > 6$, $g(x)$ contributes most to the magnitude.

35.

Year	y_1	y_2
1995	1510	395
1996	1273	432
1997	1366	571
1998	1378	646
1999	1715	763
2000	2019	841
2001	2099	978

$y_1 \approx -14.056t^3 + 376.50t^2 - 3098.1t + 9351$

$y_2 \approx -0.500t^3 + 15.32t^2 - 47.1t + 300$

39. $f(x) = x^2, g(x) = x - 1$

   (a) $(f \circ g)(x) = f(g(x)) = f(x - 1) = (x - 1)^2$

   (b) $(g \circ f)(x) = g(f(x)) = g(x^2) = x^2 - 1$

   (c) $(f \circ f)(x) = f(f(x)) = f(x^2) = (x^2)^2 = x^4$

41. $f(x) = 3x + 5, g(x) = 5 - x$

(a) $(f \circ g)(x) = f(g(x)) = f(5 - x) = 3(5 - x) + 5 = 20 - 3x$

(b) $(g \circ f)(x) = g(f(x)) = g(3x + 5) = 5 - (3x + 5) = -3x$

(c) $(f \circ f)(x) = f(f(x)) = f(3x + 5) = 3(3x + 5) + 5 = 9x + 20$

43. $f(x) = \sqrt{x + 4}$ Domain: $x \geq -4$

$g(x) = x^2$ Domain: all real numbers

(a) $(f \circ g)(x) = f(g(x)) = f(x^2) = \sqrt{x^2 + 4}$ Domain: all real numbers

(b) $(g \circ f)(x) = g(f(x)) = g(\sqrt{x + 4}) = (\sqrt{x + 4})^2 = x + 4$ Domain: $x \geq -4$

45. $f(x) = \frac{1}{3}x - 3$ Domain: all real numbers

$g(x) = 3x + 1$ Domain: all real numbers

(a) $(f \circ g)(x) = f(g(x)) = f(3x + 1) = \frac{1}{3}(3x + 1) - 3 = x - \frac{8}{3}$ Domain: all real numbers

(b) $(g \circ f)(x) = g(f(x)) = g(\frac{1}{3}x - 3) = 3(\frac{1}{3}x - 3) + 1 = x - 8$ Domain: all real numbers

47. $f(x) = x^4$ Domain: all real numbers

$g(x) = x^4$ Domain: all real numbers

(a) and (b) $(f \circ g)(x) = (g \circ f)(x) = (x^4)^4 = x^{16}$

    Domain: all real numbers

49. $f(x) = |x|$ Domain: all real numbers

$g(x) = x + 6$ Domain: all real numbers

(a) $(f \circ g)(x) = f(g(x)) = f(x + 6) = |x + 6|$

    Domain: all real numbers

(b) $(g \circ f)(x) = g(f(x)) = g(|x|) = |x| + 6$

    Domain: all real numbers

51. $f(x) = \dfrac{1}{x}$ Domain: all real numbers except $x = 0$

$g(x) = x + 3$ Domain: all real numbers

(a) $(f \circ g)(x) = f(g(x)) = f(x + 3) = \dfrac{1}{x + 3}$

    Domain: all real numbers except $x = -3$

(b) $(g \circ f)(x) = g(f(x)) = g\left(\dfrac{1}{x}\right) = \dfrac{1}{x} + 3$

    Domain: all real numbers except $x = 0$

53. Let $f(x) = x^2$ and $g(x) = 2x + 1$, then $(f \circ g)(x) = h(x)$.

This is not a unique solution.

For example, if $f(x) = (x + 1)^2$ and $g(x) = 2x$, then $(f \circ g)(x) = h(x)$ as well.

55. Let $f(x) = \sqrt[3]{x}$ and $g(x) = x^2 - 4$, then $(f \circ g)(x) = h(x)$.

This answer is not unique. Other possibilities may be:

$f(x) = \sqrt[3]{x - 4}$ and $g(x) = x^2$

or $f(x) = \sqrt[3]{-x}$ and $g(x) = 4 - x^2$

or $f(x) = \sqrt[9]{x}$ and $g(x) = (x^2 - 4)^3$

57. Let $f(x) = 1/x$ and $g(x) = x + 2$, then $(f \circ g)(x) = h(x)$.

This is not a unique solution.

Other possibilities may be:

$$f(x) = \dfrac{1}{x + 2} \text{ and } g(x) = x$$

$$\text{or } f(x) = \dfrac{1}{x + 1} \text{ and } g(x) = x + 1$$

59. Let $f(x) = \dfrac{x+3}{4+x}$ and $g(x) = -x^2$, then $(f \circ g)(x) = h(x)$.

This answer is not unique.

Another possibility may be:

$$f(x) = \frac{x+1}{x+2} \text{ and } g(x) = -x^2 + 2$$

61. (a) $r(x) = \dfrac{x}{2}$

(b) $A(r) = \pi r^2$

(c) $(A \circ r)(x) = A(r(x)) = A\left(\dfrac{x}{2}\right) = \pi\left(\dfrac{x}{2}\right)^2$

$(A \circ r)(x)$ represents the area of the circular base of the tank on the square foundation with side length x.

63. $(C \circ x)(t) = C(x(t))$

$\qquad = 60(50t) + 750$

$\qquad = 3000t + 750$

$(C \circ x)(t)$ represents the cost after t production hours.

65. False. $(f \circ g)(x) = 6x + 1$ and $(g \circ f)(x) = 6x + 6$.

67. True

Let $\quad h(x) = f(x) + g(x)$.

Then $h(-x) = f(-x) + g(-x)$

$\qquad\qquad = f(x) + g(x) \qquad$ since f and g are even

$\qquad\qquad = h(x)$.

Thus, $h(x) = f(x) + g(x)$ is also even.

69. (a) $f(g(x)) = f(0.03x) = 0.03x - 500{,}000$

(b) $g(f(x)) = g(x - 500{,}000) = 0.03(x - 500{,}000)$

$g(f(x))$ represents 3% of an amount over \$500,000.

Section 1.5   Inverse Functions

1. $f^{-1}(x) = \dfrac{x}{6} = \dfrac{1}{6}x$

$$f(f^{-1}(x)) = f\left(\frac{x}{6}\right) = 6\left(\frac{x}{6}\right) = x$$

$$f^{-1}(f(x)) = f^{-1}(6x) = \frac{6x}{6} = x$$

3. $f^{-1}(x) = x - 9$

$f(f^{-1}(x)) = f(x - 9) = (x - 9) + 9 = x$

$f^{-1}(f(x)) = f^{-1}(x + 9) = (x + 9) - 9 = x$

5. $f^{-1}(x) = \dfrac{x-1}{3}$

$f(f^{-1}(x)) = f\left(\dfrac{x-1}{3}\right) = 3\left(\dfrac{x-1}{3}\right) + 1 = x$

$f^{-1}(f(x)) = f^{-1}(3x + 1) = \dfrac{(3x+1) - 1}{3} = x$

7. $\quad f(x) = \dfrac{x-1}{5}$

$f^{-1}(x) = 5x + 1$

$f(f^{-1}(x)) = f(5x + 1) = \dfrac{5x + 1 - 1}{5} = \dfrac{5x}{5} = x$

$f^{-1}(f(x)) = f^{-1}\left(\dfrac{x-1}{5}\right) = 5\left(\dfrac{x-1}{5}\right) + 1 = x - 1 + 1 = x$

9. $f^{-1}(x) = x^3$

$f(f^{-1}(x)) = f(x^3) = \sqrt[3]{x^3} = x$

$f^{-1}(f(x)) = f^{-1}(\sqrt[3]{x}) = (\sqrt[3]{x})^3 = x$

11. (a) $f(g(x)) = f\left(\dfrac{x}{2}\right) = 2\left(\dfrac{x}{2}\right) = x$

$g(f(x)) = g(2x) = \dfrac{2x}{2} = x$

(b)

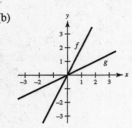

13. (a) $f(g(x)) = f\left(\dfrac{x-1}{5}\right) = 5\left(\dfrac{x-1}{5}\right) + 1 = x$

$g(f(x)) = g(5x+1) = \dfrac{(5x+1)-1}{5} = x$

(b)

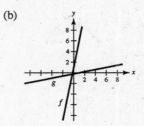

15. (a) $f(g(x)) = f\left(\sqrt[3]{x}\right) = \left(\sqrt[3]{x}\right)^3 = x$

$g(f(x)) = g(x^3) = \sqrt[3]{x^3} = x$

(b)

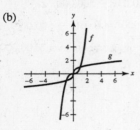

17. (a) $f(g(x)) = f(x^2 + 4),\ x \ge 0$

$\qquad = \sqrt{(x^2+4) - 4} = x$

$g(f(x)) = g\left(\sqrt{x-4}\right)$

$\qquad = \left(\sqrt{x-4}\right)^2 + 4 = x$

(b)

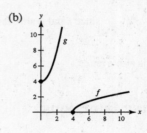

19. (a) $f(g(x)) = f\left(\sqrt{9-x}\right),\ x \le 9$

$\qquad = 9 - \left(\sqrt{9-x}\right)^2 = x$

$g(f(x)) = g(9 - x^2),\ x \ge 0$

$\qquad = \sqrt{9 - (9 - x^2)} = x$

(b)

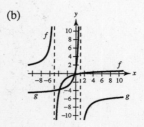

21. (a) $f(g(x)) = f\left(-\dfrac{5x+1}{x-1}\right) = \dfrac{\left(-\dfrac{5x+1}{x-1} - 1\right)}{\left(-\dfrac{5x+1}{x-1} + 5\right)} \cdot \dfrac{x-1}{x-1} = \dfrac{-(5x+1) - (x-1)}{-(5x+1) + 5(x-1)} = \dfrac{-6x}{-6} = x$

$g(f(x)) = g\left(\dfrac{x-1}{x+5}\right) = -\dfrac{\left[5\left(\dfrac{x-1}{x+5}\right) + 1\right]}{\left[\dfrac{x-1}{x+5} - 1\right]} \cdot \dfrac{x+5}{x+5} = -\dfrac{5(x-1) + (x+5)}{(x-1) - (x+5)} = -\dfrac{6x}{-6} = x$

(b)

23. No, $\{(-2, -1), (1, 0), (2, 1), (1, 2), (-2, 3), (-6, 4)\}$ does not represent a function. -2 and 1 are paired with two different values.

25. Since no horizontal line crosses the graph of f at more than one point, f **has** an inverse.

27. Since some horizontal lines cross the graph of f twice, f does **not** have an inverse.

29. $g(x) = \dfrac{4 - x}{6}$

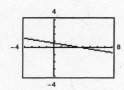

g passes the horizontal line test, so g **has** an inverse.

31. $h(x) = |x + 4| - |x - 4|$

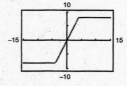

h does not pass the horizontal line test, so h does **not** have an inverse.

33. $f(x) = -2x\sqrt{16 - x^2}$

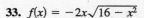
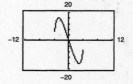

f does not pass the horizontal line test, so f does **not** have an inverse.

35. $f(x) = 2x - 3$

$y = 2x - 3$

$x = 2y - 3$

$y = \dfrac{x + 3}{2}$

$f^{-1}(x) = \dfrac{x + 3}{2}$

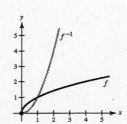

37. $f(x) = x^5 - 2$

$y = x^5 - 2$

$x = y^5 - 2$

$y = \sqrt[5]{x + 2}$

$f^{-1}(x) = \sqrt[5]{x + 2}$

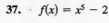
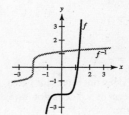

39. $f(x) = \sqrt{x}$

$y = \sqrt{x}$

$x = \sqrt{y}$

$y = x^2$

$f^{-1}(x) = x^2,\ x \geq 0$

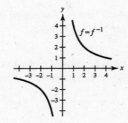

41. $f(x) = \sqrt{4 - x^2},\ 0 \leq x \leq 2$

$y = \sqrt{4 - x^2}$

$x = \sqrt{4 - y^2}$

$f^{-1}(x) = \sqrt{4 - x^2},\ 0 \leq x \leq 2$

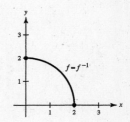

43. $f(x) = \dfrac{4}{x}$

$y = \dfrac{4}{x}$

$x = \dfrac{4}{y}$

$xy = 4$

$y = \dfrac{4}{x}$

$f^{-1}(x) = \dfrac{4}{x}$

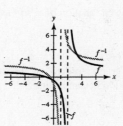

45. $f(x) = \dfrac{x + 1}{x - 2}$

$y = \dfrac{x + 1}{x - 2}$

$x = \dfrac{y + 1}{y - 2}$

$x(y - 2) = y + 1$

$xy - 2x = y + 1$

$xy - y = 2x + 1$

$y(x - 1) = 2x + 1$

$y = \dfrac{2x + 1}{x - 1}$

$f^{-1}(x) = \dfrac{2x + 1}{x - 1}$

47. $f(x) = \sqrt[3]{x} - 1$
   $y = \sqrt[3]{x} - 1$
   $x = \sqrt[3]{y} - 1$
   $x^3 = y - 1$
   $y = x^3 + 1$
   $f^{-1}(x) = x^3 + 1$

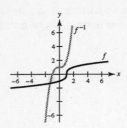

49. $f(x) = \dfrac{6x + 4}{4x + 5}$

   $y = \dfrac{6x + 4}{4x + 5}$

   $x = \dfrac{6y + 4}{4y + 5}$

   $x(4y + 5) = 6y + 4$

   $4xy + 5x = 6y + 4$

   $4xy - 6y = -5x + 4$

   $y(4x - 6) = -5x + 4$

   $y = \dfrac{-5x + 4}{4x - 6}$

   $f^{-1}(x) = \dfrac{-5x + 4}{4x - 6} = \dfrac{5x - 4}{6 - 4x}$

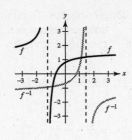

51. $f(x) = x^4$
   $y = x^4$
   $x = y^4$
   $y = \pm\sqrt[4]{x}$

This does not represent y as a function of x. f does not have an inverse.

53. $g(x) = \dfrac{x}{8}$

   $y = \dfrac{x}{8}$

   $x = \dfrac{y}{8}$

   $y = 8x$

This is a function of x, so g has an inverse.

   $g^{-1}(x) = 8x$

55. $p(x) = -4$
   $y = -4$

Since $y = -4$ for all x, the graph is a horizontal line and fails the horizontal line test. p does not have an inverse.

57. $f(x) = (x + 3)^2,\ x \ge -3 \implies y \ge 0$
   $y = (x + 3)^2,\ x \ge -3,\ y \ge 0$
   $x = (y + 3)^2,\ y \ge -3,\ x \ge 0$
   $\sqrt{x} = y + 3,\ y \ge -3,\ x \ge 0$
   $y = \sqrt{x} - 3,\ x \ge 0,\ y \ge -3$

This is a function of x, so f has an inverse.
   $f^{-1}(x) = \sqrt{x} - 3,\ x \ge 0$

59. $f(x) = \begin{cases} x + 3, & x < 0 \\ 6 - x, & x \ge 0 \end{cases}$

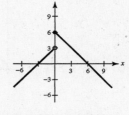

The graph fails the horizontal line test, so $f(x)$ does not have an inverse.

61.   The graph represents a one-to-one function, so the function has an inverse.

   $h(x) = \dfrac{1}{x}$

   $y = \dfrac{1}{x}$

   $xy = 1$

   $y = \dfrac{1}{x}$

   $h^{-1}(x) = \dfrac{1}{x}$

63.   The graph represents a one-to-one function, so the function has an inverse.

   $f(x) = \sqrt{2x + 3} \implies x \ge -\dfrac{3}{2},\ y \ge 0$

   $y = \sqrt{2x + 3},\ x \ge -\dfrac{3}{2},\ y \ge 0$

   $x = \sqrt{2y + 3},\ y \ge -\dfrac{3}{2},\ x \ge 0$

   $x^2 = 2y + 3,\ x \ge 0,\ y \ge -\dfrac{3}{2}$

   $y = \dfrac{x^2 - 3}{2},\ x \ge 0,\ y \ge -\dfrac{3}{2}$

   $f^{-1}(x) = \dfrac{x^2 - 3}{2},\ x \ge 0$

In Exercises 65–70, $f(x) = \frac{1}{8}x - 3$, $f^{-1}(x) = 8(x + 3)$, $g(x) = x^3$, $g^{-1}(x) = \sqrt[3]{x}$.

65. $(f^{-1} \circ g^{-1})(1) = f^{-1}(g^{-1}(1)) = f^{-1}(\sqrt[3]{1}) = 8(\sqrt[3]{1} + 3) = 32$

67. $(f^{-1} \circ f^{-1})(6) = f^{-1}(f^{-1}(6)) = f^{-1}(8[6 + 3]) = 8[8(6 + 3) + 3] = 600$

69. $\quad (f \circ g)(x) = f(g(x)) = f(x^3) = \frac{1}{8}x^3 - 3$

$$y = \frac{1}{8}x^3 - 3$$

$$x = \frac{1}{8}y^3 - 3$$

$$x + 3 = \frac{1}{8}y^3$$

$$8(x + 3) = y^3$$

$$\sqrt[3]{8(x + 3)} = y$$

$$(f \circ g)^{-1}(x) = 2\sqrt[3]{x + 3}$$

In Exercises 71–74, $f(x) = x + 4$, $f^{-1}(x) = x - 4$, $g(x) = 2x - 5$, $g^{-1}(x) = \dfrac{x + 5}{2}$.

71. $(g^{-1} \circ f^{-1})(x) = g^{-1}(f^{-1}(x)) = g^{-1}(x - 4) = \dfrac{(x - 4) + 5}{2} = \dfrac{x + 1}{2}$

73. $(f \circ g)(x) = f(g(x)) = f(2x - 5) = (2x - 5) + 4 = 2x - 1$

$$(f \circ g)^{-1}(x) = \frac{x + 1}{2}$$

Note: Comparing Exercises 71 and 73, we see that $(f \circ g)^{-1}(x) = (g^{-1} \circ f^{-1})(x)$.

75. The inverse is a line through $(-1, 0)$.

Matches graph (c).

77. The inverse is half a parabola starting at $(1, 0)$.

Matches graph (a).

79.

x	-2	0	2	4	6	8
$f^{-1}(x)$	-2	-1	0	1	2	3

81. (a) $\quad y = 8 + 0.75x$

$$x = 8 + 0.75y$$

$$x - 8 = 0.75y$$

$$\frac{x - 8}{0.75} = y$$

$$f^{-1}(x) = \frac{x - 8}{0.75}$$

(b) $x =$ hourly wage

$y =$ number of units produced

(c) $y = \dfrac{22.25 - 8}{0.75} = 19$ units

83. (a)
$$y = 0.03x^2 + 245.50, \ 0 < x < 100$$
$$x = 0.03y^2 + 245.50$$
$$x - 245.50 = 0.03y^2$$
$$\frac{x - 245.50}{0.03} = y^2$$
$$\sqrt{\frac{x - 245.50}{0.03}} = y, \ 245.50 < x < 545.50$$
$$f^{-1}(x) = \sqrt{\frac{x - 245.50}{0.03}}$$
$$x = \text{temperature in degrees Fahrenheit}$$
$$y = \text{percent load for a diesel engine}$$

(b)

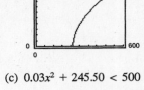

(c)
$$0.03x^2 + 245.50 < 500$$
$$0.03x^2 < 254.50$$
$$x^2 < 8483.333$$
$$x < 92.11$$

Thus, $0 < x < 92.11$.

85. (a) $f^{-1}(59.196) = 10$

(b) f^{-1} yields the year for a given amount spent on wireless communications services in the United States.

(c) $f(t) = 12.016t - 57.366$

(d) $f^{-1}(t) = \dfrac{t + 57.366}{12.016}$

(e) $f^{-1}(170.938) = 19$ which represents 2009.

87. False. $f(x) = x^2$ is even and does not have an inverse.

89. True. If $f(x) = x^n$, then $f^{-1}(x) = \sqrt[n]{x}$.

Section 1.6   Mathematical Modeling

1.

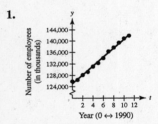

The model is a good fit for the actual data.

3. The graph appears to represent $y = 4/x$, so y varies inversely as x.

5. $k = 1$

x	2	4	6	8	10
$y = kx^2$	4	16	36	64	100

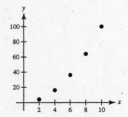

7. $k = \frac{1}{2}$

x	2	4	6	8	10
$y = kx^2$	2	8	18	32	50

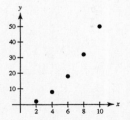

9. $k = 2$

x	2	4	6	8	10
$y = \dfrac{k}{x^2}$	$\dfrac{1}{2}$	$\dfrac{1}{8}$	$\dfrac{1}{18}$	$\dfrac{1}{32}$	$\dfrac{1}{50}$

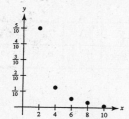

11. $k = 10$

x	2	4	6	8	10
$y = \dfrac{k}{x^2}$	$\dfrac{5}{2}$	$\dfrac{5}{8}$	$\dfrac{5}{18}$	$\dfrac{5}{32}$	$\dfrac{1}{10}$

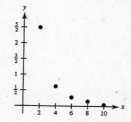

13. The table represents the equation $y = \dfrac{5}{x}$.

15.
$y = kx$
$-7 = k(10)$
$-\dfrac{7}{10} = k$
$y = -\dfrac{7}{10}x$

This equation checks with the other points given in the table.

17.
$y = kx$
$12 = k(5)$
$\dfrac{12}{5} = k$
$y = \dfrac{12}{5}x$

19.
$y = kx$
$2050 = k(10)$
$205 = k$
$y = 205x$

21. $A = kr^2$

23. $y = \dfrac{k}{x^2}$

25. $F = \dfrac{kg}{r^2}$

27. $A = \dfrac{1}{2}bh$

The area of a triangle is jointly proportional to its base and height.

29. $V = \dfrac{4}{3}\pi r^3$

The volume of a sphere varies directly as the cube of its radius.

31. $r = \dfrac{d}{t}$

Average speed is directly proportional to the distance and inversely proportional to the time.

33. The points do not follow a linear pattern. A linear model would be a poor approximation. A quadratic model would be better.

35. The data shown could be represented by a linear model which would be a good approximation.

37.

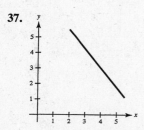

The line appears to pass through $(2, 5.5)$ and $(6, 0.5)$, so its equation is $y = -\dfrac{5}{4}x + 8$.

39.

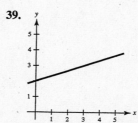

The line appears to pass through $(0, 2)$ and $(3, 3)$ so its equation is $y = \dfrac{1}{3}x + 2$.

41. $I = kP$

$187.50 = k(5000)$

$0.0375 = k$

$I = 0.0375P$

43. $y = kx$

$53 = k(14)$

$\frac{53}{14} = k$

$y = \frac{53}{14}x$

Gallons	5	10	15	20	25	30
Liters	18.9	37.9	56.8	75.7	94.6	113.6

45. $d = kF$

$0.15 = k(265)$

$\frac{3}{5300} = k$

$d = \frac{3}{5300}F$

(a) $d = \frac{3}{5300}(90) \approx 0.05$ meter

(b) $0.1 = \frac{3}{5300}F$

$\frac{530}{3} = F$

$F = 176\frac{2}{3}$ newtons

47. $d = kF$

$1.9 = k(25) \implies k = 0.076$

$d = 0.076F$

When the distance compressed
is 3 inches, we have

$3 = 0.076F$

$F \approx 39.47.$

No child over 39.47 pounds
should use the toy.

49. $R = k(T - T_e)$

51. $F = \dfrac{km_1m_2}{r^2}$

53. $y = kx$

$10.22 = k(145.99)$

$0.07 \approx k$

$y = 0.07x$

$y = 0.07(540.50)$

$y \approx 37.84$

The sales tax is $37.84.

55. $A = kr^2$

$9\pi = k(3)^2$

$\pi = k$

$A = \pi r^2$

57. $y = \dfrac{k}{x}$

$7 = \dfrac{k}{4}$

$28 = k$

$y = \dfrac{28}{x}$

59. $F = krs^3$

$4158 = k(11)(3)^3$

$k = 14$

$F = 14rs^3$

61. $z = \dfrac{kx^2}{y}$

$6 = \dfrac{k(6)^2}{4}$

$\dfrac{24}{36} = k$

$\dfrac{2}{3} = k$

$z = \dfrac{\frac{2}{3}x^2}{y} = \dfrac{2x^2}{3y}$

63. $d = kv^2$

$$0.02 = k\left(\frac{1}{4}\right)^2$$

$$k = 0.32$$

$$d = 0.32v^2$$

$$0.12 = 0.32v^2$$

$$v^2 = \frac{0.12}{0.32} = \frac{3}{8}$$

$$v = \frac{\sqrt{3}}{2\sqrt{2}} = \frac{\sqrt{6}}{4} \approx 0.61 \text{ mi/hr}$$

65. $r = \dfrac{kl}{A}, \ A = \pi r^2 = \dfrac{\pi d^2}{4}$

$$r = \frac{4kl}{\pi d^2}$$

$$64.9 = \frac{4(1000)k}{\pi\left(\dfrac{0.0126}{12}\right)^2}$$

$$k \approx 5.62 \times 10^{-8}$$

$$r = \frac{4(5.62 \times 10^{-8})l}{\pi\left(\dfrac{0.0126}{12}\right)^2}$$

$$33.5 = \frac{4(5.62 \times 10^{-8})l}{\pi\left(\dfrac{0.0126}{12}\right)^2}$$

$$\frac{33.5\pi\left(\dfrac{0.0126}{12}\right)^2}{4(5.62 \times 10^{-8})} = l$$

$$l \approx 516 \text{ feet}$$

67. $s = kt^2$

$$144 = k(3)^2$$

$$16 = k$$

$$s = 16t^2$$

$$s = 16(5)^2 = 400 \text{ feet}$$

69. $P = kA = k(\pi r^2) = k\pi\left(\dfrac{d}{2}\right)^2$

$$8.78 = k\pi\left(\frac{9}{2}\right)^2$$

$$\frac{4(8.78)}{81\pi} = k$$

$$k \approx 0.138$$

However, we do not obtain $11.78 when $d = 12$ inches.

$$P = 0.138\pi\left(\frac{12}{2}\right)^2 \approx \$15.61$$

Instead, $k = \dfrac{11.78}{36\pi} \approx 0.104$.

For the 15-inch pizza, we have $k = \dfrac{4(14.18)}{225\pi} \approx 0.080$.

The price is not directly proportional to the surface area. The best buy is the 15-inch pizza.

71. $v = \dfrac{k}{A}$

(a) $v = \dfrac{k}{0.75A} = \dfrac{4}{3}\left(\dfrac{k}{A}\right)$

The velocity is increased by one-third.

(b) $v = \dfrac{k}{\left(1 + \frac{1}{3}\right)A} = \dfrac{k}{\frac{4}{3}A} = \dfrac{3}{4}\left(\dfrac{k}{A}\right)$

The velocity is decreased by one-fourth.

73. (a)

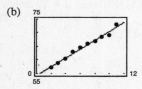

(b) $C = \dfrac{k}{d} \Rightarrow k = dC$

(1000, 4.2) : $k = 4200$

(2000, 1.9) : $k = 3800$

(3000, 1.4) : $k = 4200$

(4000, 1.2) : $k = 4800$

(5000, 0.9) : $k = 4500$

(c) $\dfrac{4200 + 3800 + 4200 + 4800 + 4500}{5} = 4300$

$C = \dfrac{4300}{d}$

(d)

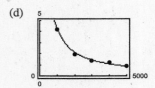

(e) $3 = \dfrac{4300}{d}$

$d = \dfrac{4300}{3} \approx 1433$ meters

75. $y = \dfrac{k}{d^2}$

When d is doubled,

$$y = \frac{k}{(2d)^2} = \frac{k}{4d^2} = \frac{1}{4}\left(\frac{k}{d^2}\right)$$

Thus the illumination is $\frac{1}{4}$ of what it was before.

77. (a) $y = 2.1560t - 4.8904$

(b)

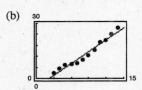

(c) $y(15) \approx 27.45$ billion dollars

(d) Answers will vary.

79. (a) $y \approx 1.63t + 54.2$

(b)

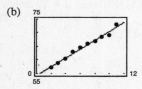

(c) 2005 : $y(15) = 78.65$ million households

2010 : $y(20) = 86.80$ million households

(d) The number of households with cable televisions is increasing at a rate of 1.63 million per year.

81. (a) $y \approx 0.45x - 4.9$

(b)

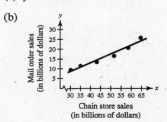

(c) $y \approx 0.45(75) - 4.9 \approx 28.85$ billion dollars

83. False. y will increase if k is positive and y will decrease if k is negative.

85. False.

"y varies directly as x" means $y = kx$ for some nonzero constant k.

"y is inversely proportional to x" means $y = \dfrac{k}{x}$ for some nonzero constant k.

Review Exercises for Chapter 1

1. $16x - y^4 = 0$

No. This determines x as a function of y.

3. $y = \sqrt{1 - x}$

Yes. This determines y as a function of x.

5. $f(x) = x^2 + 1$

    (a) $f(2) = 2^2 + 1 = 5$

    (b) $f(-4) = (-4)^2 + 1 = 17$

    (c) $f(t^2) = (t^2)^2 + 1 = t^4 + 1$

    (d) $-f(x) = -(x^2 + 1)$

            $= -x^2 - 1$

7. $f(x) = \sqrt{25 - x^2}$

    Domain: $25 - x^2 \geq 0$

            $-5 \leq x \leq 5$

9. $h(x) = \dfrac{x}{x^2 - x - 6}$

    $x^2 - x - 6 = 0$

    $(x - 3)(x + 2) = 0$

    Domain: All real numbers

            $x \neq -2, 3$

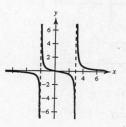

11. $v(t) = -32t + 48$

    (a) $v(1) = -32(1) + 48 = 16$
        feet per second

    (b) $v(t) = 0$

        $-32t + 48 = 0$

            $t = \frac{48}{32} = 1.5$ seconds

    (c) $v(2) = -32t + 48 = -32(2) + 48$

            $= -16$ feet per second

13. (a) Let $x =$ the width of the rectangle and $y =$ the length of the rectangle.

        $y = 4x + 2$

        $P(x) = 2x + 2y$

              $= 2x + 2(4x + 2)$

              $= 10x + 4$

    (b) $A(x) = xy$

             $= x(4x + 2)$

             $= 4x^2 + 2x$

    (c) Since x is the width of a rectangle, x must be positive.

        Domain of $P(x)$: $x > 0$

        Domain of $A(x)$: $x > 0$

    (d)

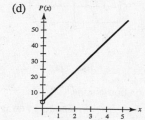 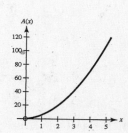

15. $y = (x - 3)^2$

All vertical lines intersect graph exactly once $\implies$ y is a function of x.

17. $x - 4 = y^2$

Some vertical lines intersect graph more than once $\implies$ y is **not** a function of x.

19. $f(x) = 5x^2 + 4x - 1$

    $5x^2 + 4x - 1 = 0$

    $(5x - 1)(x + 1) = 0$

    Zeros: $x = \frac{1}{5}, -1$

21. $f(x) = \dfrac{8x + 3}{11 - x}$

$\dfrac{8x + 3}{11 - x} = 0$

$8x + 3 = 0$

Zero: $x = -\dfrac{3}{8}$

23. $g(x) = |x + 2| - |x - 2|$

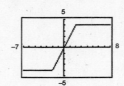

Increasing: $(-2, 2)$

Constant: $(-\infty, -2), (2, \infty)$

25. $h(x) = 4x^3 - x^4$

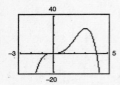

Increasing: $(-\infty, 3)$

Decreasing: $(3, \infty)$

27. $f(2) = -6,\ f(-1) = 3$

Two points: $(2, -6), (-1, 3)$

Slope: $m = \dfrac{-6 - 3}{2 - (-1)} = \dfrac{-9}{3} = -3$

$y - 3 = -3(x + 1)$

$y = -3x$

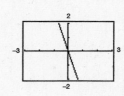

29. $f\!\left(-\dfrac{4}{5}\right) = 2,\ f\!\left(\dfrac{11}{5}\right) = 7$

Two points: $\left(-\dfrac{4}{5}, 2\right), \left(\dfrac{11}{5}, 7\right)$

Slope: $m = \dfrac{2 - 7}{-\frac{4}{5} - \frac{11}{5}} = \dfrac{-5}{-\frac{15}{5}} = \dfrac{5}{3}$

$y - 7 = \dfrac{5}{3}\left(x - \dfrac{11}{5}\right)$

$y = \dfrac{5}{3}x + \dfrac{10}{3}$

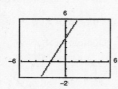

31. $f(x) = \begin{cases} 5x - 3, & x \ge -1 \\ -4x + 5, & x < -1 \end{cases}$

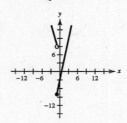

33. $f(x) = x^5 + 4x - 7$

$f(-x) = -x^5 - 4x - 7$

$f(-x) \ne f(x)$

$f(-x) \ne -f(x)$

Neither.

35. $f(x) = 2x\sqrt{x^2 + 3}$

$f(-x) = -2x\sqrt{x^2 + 3}$

The function is odd.

37. Common Function: $f(x) = x^3$

Transformation: Horizontal shift 4 units to the left, and a vertical shift 4 units upward.

$g(x) = (x + 4)^3 + 4$

39. Common Function: $f(x) = |x|$

Transformation: Reflection in the x-axis, a horizontal shift 2 units to the left, and a vertical shift 1 unit upward.

$g(x) = -|x + 2| + 1$

41. $f(x) = x^2$

$h(x) = x^2 - 9$

Transformation: Vertical shift 9 units downward.

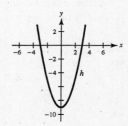

43. $f(x) = |x|$

$h(x) = |x + 3| - 5$

Transformation: Horizontal shift 3 units to the left and a vertical shift 5 units downward.

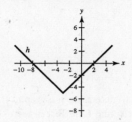

45. $f(x) = \sqrt{x}$

$h(x) = -\sqrt{x + 1} + 9$

Transformation: Reflection in the x-axis, horizontal shift 1 unit to the left, and a vertical shift 9 units upward.

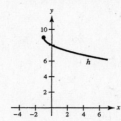

For Exercises 47–50, let $f(x) = 3 - 2x$, $g(x) = \sqrt{x}$, and $h(x) = 3x^2 + 2$.

47. $(f - g)(4) = f(4) - g(4)$

$\qquad = 3 - 2(4) - \sqrt{4}$

$\qquad = -7$

49. $(h \circ g)(7) = 3(g(7))^2 + 2$

$\qquad = 3(\sqrt{7})^2 + 2$

$\qquad = 23$

51. Let $t = 5$ represent 1995.

$y_1 \approx 1.5t^2 - 8.674t + 70.057$

$y_2 \approx 2.775t^2 - 25.468t + 205.371$

53. $f(x) = 6x$

$f^{-1}(x) = \dfrac{x}{6}$

$f(f^{-1}(x)) = 6\left(\dfrac{x}{6}\right) = x$

$f^{-1}(f(x)) = \dfrac{6x}{6} = x$

55. $f(x) = x - 7$

$f^{-1}(x) = x + 7$

$f(f^{-1}(x)) = (x + 7) - 7 = x$

$f^{-1}(f(x)) = (x - 7) + 7 = x$

57. $f(x) = 3x^3 - 5$

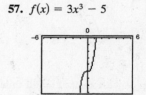

$f(x)$ passes Horizontal Line Test $\implies$ it has an inverse.

59. $f(x) = -\sqrt{4 - x}$

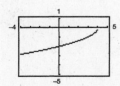

$f(x)$ passes Horizontal Line Test $\implies$ it has an inverse.

61. $f(x) = \frac{1}{2}x - 3$

(a) $\quad y = \frac{1}{2}x - 3$

$\qquad x = \frac{1}{2}y - 3$

$\quad 2x + 6 = y$

$\qquad f^{-1}(x) = 2x + 6$

(b)

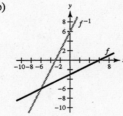

(c) $f^{-1}(f(x)) = 2\left[\frac{1}{2}x - 3\right] + 6 = x$

$\quad f(f^{-1}(x)) = \frac{1}{2}[2x + 6] - 3 = x$

63. $f(x) = \sqrt{x+1}$

(a) $\quad y = \sqrt{x+1},\ x \geq -1,\ y \geq 0$

$\qquad x = \sqrt{y+1},\ y \geq -1,\ x \geq 0$

$\quad x^2 - 1 = y$

$\quad f^{-1}(x) = x^2 - 1,\ x \geq 0$

(b)

(c) $f^{-1}(f(x)) = \left[\sqrt{x+1}\right]^2 - 1 = x$

$\qquad f(f^{-1}(x)) = \sqrt{[x^2 - 1] + 1} = x$

65. $f(x) = 2(x-4)^2$

$f(x)$ is increasing on the interval $(4, \infty)$.

$\qquad y = 2(x-4)^2$

$\qquad x = 2(y-4)^2$

$\pm\sqrt{\dfrac{x}{2}} = y - 4$

Choose positive root, corresponding to $(4, \infty)$ as the domain of $f(x)$.

$4 + \sqrt{\dfrac{x}{2}} = y$

$\qquad f^{-1}(x) = \sqrt{\dfrac{x}{2}} + 4$

67. The Vertical Line Test is used to determine if a graph represents a function. The Horizontal Line Test is used to determine if a function has an inverse.

69. 2.5 miles = 4 kilometers $\Longrightarrow$ 1 mile $= \dfrac{4}{2.5} = 1.6$ kilometers

$y = 1.6x$

Miles, x	2	5	10	12
Kilometers, $1.6x$	3.2	8	16	19.2

71. $F = ks^2$

If s is doubled; $F = k(2s)^2 = 4ks^2$.

The force is changed by a factor of 4.

73. y is inversely proportional to x: $y = \dfrac{k}{x}$

$y = 9$ when $x = 5.5$

$9 = \dfrac{k}{5.5} \Longrightarrow k = (9)(5.5) = 49.5$

So, $y = \dfrac{49.5}{x}$

75. If $y = kx$, then the y-intercept is $(0, 0)$.

Problem Solving for Chapter 1

1. $A = \frac{1}{2}bh = \frac{1}{2}xy$

Since $(0, y)$, $(2, 1)$, and $(x, 0)$ all lie on the same line, the slopes between any pair are equal.

$$\frac{1 - y}{2 - 0} = \frac{0 - 1}{x - 2}$$

$$\frac{1 - y}{2} = \frac{-1}{x - 2}$$

$$y = \frac{2}{x - 2} + 1$$

$$y = \frac{x}{x - 2}$$

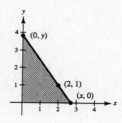

Therefore, $A = \frac{1}{2}x\left(\frac{x}{x - 2}\right) = \frac{x^2}{2(x - 2)}$.

The domain of A includes x-values such that $x^2/[2(x - 2)] > 0$. The domain is $x > 2$.

3. $f(x) = a_{2n+1}x^{2n+1} + a_{2n-1}x^{2n-1} + \cdots + a_3x^3 + a_1x$

$f(-x) = a_{2n+1}(-x)^{2n+1} + a_{2n-1}(-x)^{2n-1} + \cdots + a_3(-x)^3 + a_1(-x)$

$\quad = -a_{2n+1}x^{2n+1} - a_{2n-1}x^{2n-1} - \cdots - a_3x^3 - a_1x = -f(x)$

Therefore, $f(x)$ is odd.

5. (a) $y = x$ (b) $y = x^2$ (c) $y = x^3$

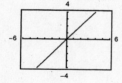 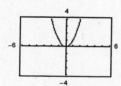 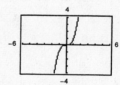

(d) $y = x^4$ (e) $y = x^5$ (f) $y = x^6$

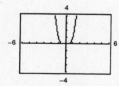 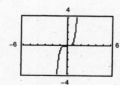 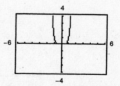

All the graphs pass through the origin. The graphs of the odd powers of x are symmetric with respect to the origin and the graphs of the even powers are symmetric with respect to the y-axis. As the powers increase, the graphs become flatter in the interval $-1 < x < 1$.

(g) The graph of $y = x^7$ will pass through the origin and will be symmetric with the origin.
   The graph of $y = x^8$ will pass through the origin and will be symmetric with respect to the y-axis.

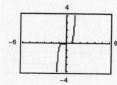 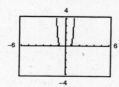

7. $y = f(x + 2) - 1$

Horizontal shift 2 units to the left and a vertical shift 1 unit downward.

$(0, 1) \rightarrow (0 - 2, 1 - 1) = (-2, 0)$

$(1, 2) \rightarrow (1 - 2, 2 - 1) = (-1, 1)$

$(2, 3) \rightarrow (2 - 2, 3 - 1) = (0, 2)$

9. Let $f(x)$ be an odd function, $g(x)$ be an even function and define $h(x) = f(x)g(x)$. Then

$$h(-x) = f(-x)g(-x)$$
$$= [-f(x)]g(x) \qquad \text{Since } f \text{ is odd and } g \text{ is even.}$$
$$= -f(x)g(x)$$
$$= -h(x)$$

Thus, h is odd.

11.

x	1	3	4	6
f	1	2	6	7

x	1	2	6	7
$f^{-1}(x)$	1	3	4	6

13.

x	-2	-1	3	4
f	6	0	-2	-3

x	-3	-2	0	6
$f^{-1}(x)$	4	3	-1	-2

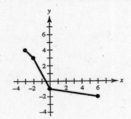

15. If $f(x) = k(2 - x - x^3)$ has an inverse and $f^{-1}(3) = -2$, then $f(-2) = 3$. Thus,

$$f(-2) = k(2 - (-2) - (-2)^3) = 3$$
$$k(2 + 2 + 8) = 3$$
$$12k = 3$$
$$k = \frac{3}{12} = \frac{1}{4}$$

Thus, $k = \frac{1}{4}$.

17. $H(x) = \begin{cases} 1, & x \geq 0 \\ 0, & x < 0 \end{cases}$

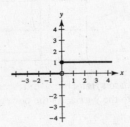

(a) $H(x) - 2$

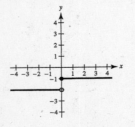

(b) $H(x - 2)$

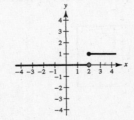

(c) $-H(x)$

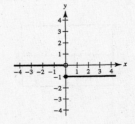

—CONTINUED—

17. **—CONTINUED—**

(d) $H(-x)$

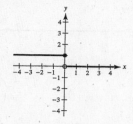

(e) $\frac{1}{2}H(x)$

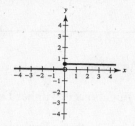

(f) $-H(x-2)+2$

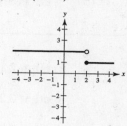

19. $(f \circ (g \circ h))(x) = f((g \circ h)(x))$

$\qquad\qquad\quad = f(g(h(x)))$

$\qquad\qquad\quad = (f \circ g \circ h)(x)$

$((f \circ g) \circ h)(x) = (f \circ g)(h(x))$

$\qquad\qquad\quad = f(g(h(x)))$

$\qquad\qquad\quad = (f \circ g \circ h)(x)$

21.

x	$f(x)$	$(f+g)(x)$
-2	-2	0
-1	0	-2
0	1	0
1	0	1
2	1	1

This chart was derived from the graphs of f and $f+g$.

(a)

x	$g(x) = (f+g)(x) - f(x)$
-2	2
-1	-2
0	-1
1	1
2	0

(b)

x	$(g-f)(x) = g(x) - f(x)$
-2	4
-1	-2
0	-2
1	1
2	-1

(c)

x	$g(f(x))$
-2	$g(f(-2)) = g(-2) = 2$
-1	$g(f(-1)) = g(0) = -1$
0	$g(f(0)) = g(1) = 1$
1	$g(f(1)) = g(0) = -1$
2	$g(f(2)) = g(1) = 1$

(d)

x	$f(g(-x))$
-2	$f(g(2)) = f(0) = 1$
-1	$f(g(1)) = f(1) = 0$
0	$f(g(0)) = f(-1) = 0$
1	$f(g(-1)) = f(-2) = -2$
2	$f(g(-2)) = f(2) = 1$

C H A P T E R 2
Polynomial and Rational Functions

CHAPTER 2
Polynomial and Rational Functions

Section 2.1   Quadratic Functions

1. $f(x) = (x - 2)^2$ opens upward and has vertex $(2, 0)$. Matches graph (g).

3. $f(x) = x^2 - 2$ opens upward and has vertex $(0, -2)$. Matches graph (b).

5. $f(x) = 4 - (x - 2)^2 = -(x - 2)^2 + 4$ opens downward and has vertex $(2, 4)$. Matches graph (f).

7. $f(x) = -(x - 3)^2 - 2$ opens downward and has vertex $(3, -2)$. Matches graph (e).

9. $y = x^2$

(a) $f(x) = \frac{1}{2}x^2$

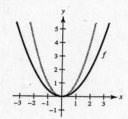

Vertical shrink by a factor of $\frac{1}{2}$.

(b) $g(x) = x^2 - 1$

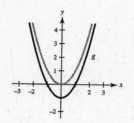

Vertical shift one unit down.

(c) $h(x) = (x - 1)^2$

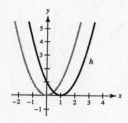

Horizontal shift one unit to the right.

(d) $k(x) = -\frac{1}{2}(x - 2)^2 - 1$

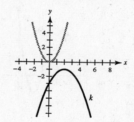

Vertical shrink by a factor of $\frac{1}{2}$, a reflection in the x-axis, a horizontal shift 2 units to the right, and a vertical shift one unit down.

11. $y = x^2$

(a) $f(x) = -x^2$

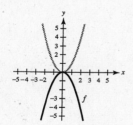

A reflection in the x-axis.

(b) $g(x) = -x^2 - 2$

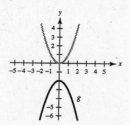

A reflection in the x-axis and a vertical shift 2 units down.

—CONTINUED—

85

11. **—CONTINUED—**

(c) $h(x) = -(x - 3)^2$

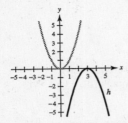

A reflection in the x-axis and a horizontal shift 3 units to the right.

(d) $k(x) = -\frac{1}{2}(x - 3)^2 - 2$

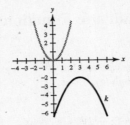

A reflection in the x-axis, a vertical shrink by a factor of $\frac{1}{2}$, a horizontal shift 3 units to the right and a vertical shift 2 units down.

13. $f(x) = x^2 - 16$

Vertex: $(0, -16)$

Find x-intercepts: $x^2 - 16 = 0$

$$x^2 = 16$$
$$x = \pm 4$$

x-intercepts: $(\pm 4, 0)$

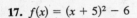

15. $f(x) = 16 - \frac{1}{4}x^2 = -\frac{1}{4}x^2 + 16$

Vertex: $(0, 16)$

Find x-intercepts: $16 - \frac{1}{4}x^2 = 0$

$$x^2 = 64$$
$$x = \pm 8$$

x-intercepts: $(\pm 8, 0)$

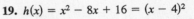

17. $f(x) = (x + 5)^2 - 6$

Vertex: $(-5, -6)$

Find x-intercepts:

$$(x + 5)^2 - 6 = 0$$
$$(x + 5)^2 = 6$$
$$x + 5 = \pm\sqrt{6}$$
$$x = -5 \pm \sqrt{6}$$

x-intercepts: $\left(-5 - \sqrt{6}, 0\right), \left(-5 + \sqrt{6}, 0\right)$

19. $h(x) = x^2 - 8x + 16 = (x - 4)^2$

Vertex: $(4, 0)$

x-intercept: $(4, 0)$

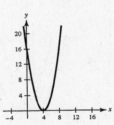

21. $f(x) = x^2 - x + \dfrac{5}{4}$

$$= \left(x^2 - x + \frac{1}{4}\right) - \frac{1}{4} + \frac{5}{4}$$

$$= \left(x - \frac{1}{2}\right)^2 + 1$$

Vertex: $\left(\dfrac{1}{2}, 1\right)$

Find x-intercepts:

$$x^2 - x + \frac{5}{4} = 0$$

$$x = \frac{1 \pm \sqrt{1 - 5}}{2}$$

Not a real number $\Rightarrow$ No x-intercepts

23. $f(x) = -x^2 + 2x + 5$

$\quad = -(x^2 - 2x + 1) - (-1) + 5$

$\quad = -(x - 1)^2 + 6$

Vertex: $(1, 6)$

Find x-intercepts:

$\quad -x^2 + 2x + 5 = 0$

$\quad x^2 - 2x - 5 = 0$

$\quad x = \dfrac{2 \pm \sqrt{4 + 20}}{2}$

$\quad = 1 \pm \sqrt{6}$

x-intercepts: $\left(1 - \sqrt{6}, 0\right), \left(1 + \sqrt{6}, 0\right)$

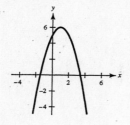

25. $h(x) = 4x^2 - 4x + 21$

$\quad = 4\left(x^2 - x + \dfrac{1}{4}\right) - 4\left(\dfrac{1}{4}\right) + 21$

$\quad = 4\left(x - \dfrac{1}{2}\right)^2 + 20$

Vertex: $\left(\dfrac{1}{2}, 20\right)$

Find x-intercepts:

$\quad 4x^2 - 4x + 21 = 0$

$\quad x = \dfrac{4 \pm \sqrt{16 - 336}}{2(4)}$

Not a real number $\Rightarrow$ No x-intercepts

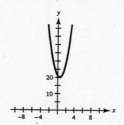

27. $f(x) = \frac{1}{4}x^2 - 2x - 12$

$\quad = \frac{1}{4}(x^2 - 8x + 16) - \frac{1}{4}(16) - 12$

$\quad = \frac{1}{4}(x - 4)^2 - 16$

Vertex: $(4, -16)$

Find x-intercepts:

$\quad \frac{1}{4}x^2 - 2x - 12 = 0$

$\quad x^2 - 8x - 48 = 0$

$\quad (x + 4)(x - 12) = 0$

$\quad x = -4 \quad \text{or} \quad x = 12$

x-intercepts: $(-4, 0), (12, 0)$

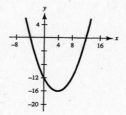

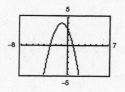

29. $f(x) = -(x^2 + 2x - 3) = -(x + 1)^2 + 4$

Vertex: $(-1, 4)$

x-intercepts: $(-3, 0), (1, 0)$

31. $g(x) = x^2 + 8x + 11 = (x + 4)^2 - 5$

Vertex: $(-4, -5)$

x-intercepts: $\left(-4 \pm \sqrt{5}, 0\right)$

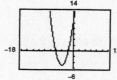

33. $f(x) = 2x^2 - 16x + 31$

$= 2(x - 4)^2 - 1$

Vertex: $(4, -1)$

x-intercepts: $\left(4 \pm \frac{1}{2}\sqrt{2}, 0\right)$

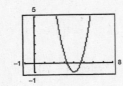

35. $g(x) = \frac{1}{2}(x^2 + 4x - 2) = \frac{1}{2}(x + 2)^2 - 3$

Vertex: $(-2, -3)$

x-intercepts: $(-2 \pm \sqrt{6}, 0)$

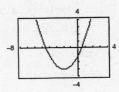

37. $(1, 0)$ is the vertex.

$y = a(x - 1)^2 + 0 = a(x - 1)^2$

Since the graph passes through the point $(0, 1)$, we have:

$1 = a(0 - 1)^2$

$1 = a$

$y = 1(x - 1)^2 = (x - 1)^2$

39. $(-1, 4)$ is the vertex.

$y = a(x + 1)^2 + 4$

Since the graph passes through the point $(1, 0)$, we have:

$0 = a(1 + 1)^2 + 4$

$-4 = 4a$

$-1 = a$

$y = -1(x + 1)^2 + 4 = -(x + 1)^2 + 4$

41. $(-2, 2)$ is the vertex.

$y = a(x + 2)^2 + 2$

Since the graph passes through the point $(-1, 0)$, we have:

$0 = a(-1 + 2)^2 + 2$

$-2 = a$

$y = -2(x + 2)^2 + 2$

43. $(-2, 5)$ is the vertex.

$f(x) = a(x + 2)^2 + 5$

Since the graph passes through the point $(0, 9)$, we have:

$9 = a(0 + 2)^2 + 5$

$4 = 4a$

$1 = a$

$f(x) = 1(x + 2)^2 + 5 = (x + 2)^2 + 5$

45. $(3, 4)$ is the vertex.

$f(x) = a(x - 3)^2 + 4$

Since the graph passes through the point $(1, 2)$, we have:

$2 = a(1 - 3)^2 + 4$

$-2 = 4a$

$-\frac{1}{2} = a$

$f(x) = -\frac{1}{2}(x - 3)^2 + 4$

47. $(5, 12)$ is the vertex.

$f(x) = a(x - 5)^2 + 12$

Since the graph passes through the point $(7, 15)$, we have:

$15 = a(7 - 5)^2 + 12$

$3 = 4a \implies a = \frac{3}{4}$

$f(x) = \frac{3}{4}(x - 5)^2 + 12$

49. $\left(-\frac{1}{4}, \frac{3}{2}\right)$ is the vertex.

$f(x) = a\left(x + \frac{1}{4}\right)^2 + \frac{3}{2}$

Since the graph passes through the point $(-2, 0)$, we have:

$0 = a\left(-2 + \frac{1}{4}\right)^2 + \frac{3}{2}$

$-\frac{3}{2} = \frac{49}{16}a \implies a = -\frac{24}{49}$

$f(x) = -\frac{24}{49}\left(x + \frac{1}{4}\right)^2 + \frac{3}{2}$

51. $\left(-\frac{5}{2}, 0\right)$ is the vertex.

$f(x) = a\left(x + \frac{5}{2}\right)^2$

Since the graph passes through the point $\left(-\frac{7}{2}, -\frac{16}{3}\right)$, we have:

$-\frac{16}{3} = a\left(-\frac{7}{2} + \frac{5}{2}\right)^2$

$-\frac{16}{3} = a$

$f(x) = -\frac{16}{3}\left(x + \frac{5}{2}\right)^2$

53. $y = x^2 - 16$

    (a) x-intercepts: $(\pm 4, 0)$

    (b) The x-intercepts and the solutions of the equation are the same.

    (c) $0 = x^2 - 16$

        $x^2 = 16$

        $x = \pm 4 \implies$ The x-intercepts are $(\pm 4, 0)$.

55. $y = -x^2 - 2x - 1$

    (a) From the graph it appears that the x- intercept is $(-1, 0)$.

    (b) The x-intercept and the solution to $-x^2 - 2x - 1 = 0$ are the same.

    (c) $-x^2 - 2x - 1 = 0$

        $x^2 + 2x + 1 = 0$

        $(x + 1)^2 = 0$

        $x + 1 = 0$

        $x = -1 \implies$ The x-intercept is at $(-1, 0)$.

57. $y = x^2 - 4x - 5$

    (a) x-intercepts: $(5, 0), (-1, 0)$

    (b) The x-intercepts and the solutions of the equation are the same.

    (c) $0 = x^2 - 4x - 5$

        $0 = (x - 5)(x + 1)$

        $x = 5$ or $x = -1$

    The x-intercepts are $(5, 0)$ and $(-1, 0)$.

59. $y = -2x^2 + 3x + 2$

    (a) From the graph it appears that the x-intercepts are $\left(-\frac{1}{2}, 0\right)$ and $(2, 0)$.

    (b) The x-intercepts and the solutions of the equation are the same.

    (c) $0 = -2x^2 + 3x + 2 \implies 2x^2 - 3x - 2 = 0$

        $(2x + 1)(x - 2) = 0$

        $x = -\frac{1}{2}$ or $x = 2 \implies$ The x-intercepts are $\left(-\frac{1}{2}, 0\right)$ and $(2, 0)$.

61. $y = x^2 + 2x + 2$

    (a) From the graph we see that there are no x-intercepts.

    (b) $0 = x^2 + 2x + 2$ has no real solutions.

    (c) Completing the square yields $(x + 1)^2 = -1$ which has no real solutions.

63. $f(x) = ax^2 + bx + c$

$$= a\left(x^2 + \frac{b}{a}x\right) + c$$

$$= a\left(x^2 + \frac{b}{a}x + \frac{b^2}{4a^2} - \frac{b^2}{4a^2}\right) + c$$

$$= a\left(x^2 + \frac{b}{a}x + \frac{b^2}{4a^2}\right) - a\left(\frac{b^2}{4a^2}\right) + c$$

$$= a\left(x + \frac{b}{2a}\right)^2 - \frac{b^2}{4a} + \frac{4ac}{4a}$$

$$= a\left(x + \frac{b}{2a}\right)^2 + \frac{4ac - b^2}{4a}$$

The vertex is $\left(-\dfrac{b}{2a}, \dfrac{4ac - b^2}{4a}\right)$.

$$f\left(-\frac{b}{2a}\right) = a\left(-\frac{b}{2a}\right)^2 + b\left(-\frac{b}{2a}\right) + c$$

$$= a\left(\frac{b^2}{4a^2}\right) - \frac{b^2}{2a} + c$$

$$= \frac{b^2}{4a} - \frac{2b^2}{4a} + \frac{4ac}{4a}$$

$$= \frac{-b^2 + 4ac}{4a}$$

$$= \frac{4ac - b^2}{4a}$$

Thus, the vertex occurs at $\left(-\dfrac{b}{2a}, f\left(-\dfrac{b}{2a}\right)\right)$.

65. $f(x) = x^2 - 4x$

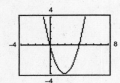

x-intercepts: $(0, 0), (4, 0)$

$0 = x^2 - 4x$

$0 = x(x - 4)$

$x = 0$ or $x = 4$

67. $f(x) = x^2 - 9x + 18$

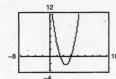

x-intercepts: $(3, 0), (6, 0)$

$0 = x^2 - 9x + 18$

$0 = (x - 3)(x - 6)$

$x = 3$ or $x = 6$

69. $f(x) = 2x^2 - 7x - 30$

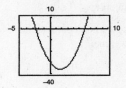

x-intercepts: $\left(-\frac{5}{2}, 0\right), (6, 0)$

$0 = 2x^2 - 7x - 30$

$0 = (2x + 5)(x - 6)$

$x = -\frac{5}{2}$ or $x = 6$

71. $f(x) = -\frac{1}{2}(x^2 - 6x - 7)$

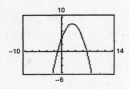

x-intercepts: $(-1, 0), (7, 0)$

$0 = -\frac{1}{2}(x^2 - 6x - 7)$

$0 = x^2 - 6x - 7$

$0 = (x + 1)(x - 7)$

$x = -1$ or $x = 7$

73. $f(x) = [x - (-1)](x - 3)$ opens upward

      $= (x + 1)(x - 3)$

      $= x^2 - 2x - 3$

      $g(x) = -[x - (-1)](x - 3)$ opens downward

      $= -(x + 1)(x - 3)$

      $= -(x^2 - 2x - 3)$

      $= -x^2 + 2x + 3$

Note: $f(x) = a(x + 1)(x - 3)$ has x-intercepts $(-1, 0)$ and $(3, 0)$ for all real numbers $a \neq 0$.

75. $f(x) = (x - 0)(x - 10)$ opens upward

      $= x^2 - 10x$

      $g(x) = -(x - 0)(x - 10)$ opens downward

      $= -x^2 + 10x$

Note: $f(x) = a(x - 0)(x - 10) = ax(x - 10)$ has x-intercepts $(0, 0)$ and $(10, 0)$ for all real numbers $a \neq 0$.

77. $f(x) = [x - (-3)]\left[x - \left(-\frac{1}{2}\right)\right](2)$ opens upward

      $= (x + 3)\left(x + \frac{1}{2}\right)(2)$

      $= (x + 3)(2x + 1)$

      $= 2x^2 + 7x + 3$

      $g(x) = -(2x^2 + 7x + 3)$ opens downward

      $= -2x^2 - 7x - 3$

Note: $f(x) = a(x + 3)(2x + 1)$ has x-intercepts $(-3, 0)$ and $\left(-\frac{1}{2}, 0\right)$ for all real numbers $a \neq 0$.

79. Let x = the first number and y = the second number. Then the sum is $x + y = 156 \implies y = 156 - x$.

The product is $P(x) = xy = x(156 - x)$

$$= -x^2 + 156x$$
$$= -(x^2 - 156x)$$
$$= -(x^2 - 156x + 6084 - 6084)$$
$$= -(x - 78)^2 + 6084.$$

The maximum value of the product occurs at the vertex and is 6084. This happens when $x = 78$ and $y = 156 - 78 = 78$. Thus, the numbers are 78 and 78.

81. Let x = the first number and y = the second number. Then the sum is

$$x + 2y = 24 \implies y = \frac{24 - x}{2}.$$

The product is $P(x) = xy = x\left(\frac{24 - x}{2}\right)$.

$$P(x) = \frac{1}{2}(-x^2 + 24x)$$
$$= -\frac{1}{2}(x^2 - 24x + 144 - 144)$$
$$= -\frac{1}{2}[(x - 12)^2 - 144] = -\frac{1}{2}(x - 12)^2 + 72$$

The maximum value of the product occurs at the vertex of $P(x)$ and is 72. This happens when $x = 12$ and $y = (24 - 12)/2 = 6$. Thus, the numbers are 12 and 6.

83.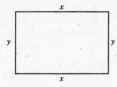

$2x + 2y = 100$

$y = 50 - x$

(a) $A(x) = xy = x(50 - x)$

Domain: $0 < x < 50$

(b)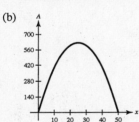

(c) The area is maximum (625 square feet) when $x = y = 25$. The rectangle has dimensions 25 ft $\times$ 25 ft.

85. (a) $4x + 3y = 200 \implies y = \frac{1}{3}(200 - 4x)$

x	y	Area
2	$\frac{1}{3}[200 - 4(2)]$	$2xy = (2)(2)(\frac{1}{3})[200 - 4(2)] = 256$
4	$\frac{1}{3}[200 - 4(4)]$	$2xy = (2)(4)(\frac{1}{3})[200 - 4(4)] \approx 491$
6	$\frac{1}{3}[200 - 4(6)]$	$2xy = (2)(6)(\frac{1}{3})[200 - 4(6)] = 704$
8	$\frac{1}{3}[200 - 4(8)]$	$2xy = (2)(8)(\frac{1}{3})[200 - 4(8)] = 896$
10	$\frac{1}{3}[200 - 4(10)]$	$2xy = (2)(10)(\frac{1}{3})[200 - 4(10)] \approx 1067$
12	$\frac{1}{3}[200 - 4(12)]$	$2xy = (2)(12)(\frac{1}{3})[200 - 4(12)] = 1216$

—CONTINUED—

85. —CONTINUED—

(b)

x	y	Area
20	$\frac{1}{3}[200 - 4(20)]$	$2xy = (2)(20)(\frac{1}{3})[200 - 4(20)] = 1600$
22	$\frac{1}{3}[200 - 4(22)]$	$2xy = (2)(22)(\frac{1}{3})[200 - 4(22)] \approx 1643$
24	$\frac{1}{3}[200 - 4(24)]$	$2xy = (2)(24)(\frac{1}{3})[200 - 4(24)] = 1664$
26	$\frac{1}{3}[200 - 4(26)]$	$2xy = (2)(26)(\frac{1}{3})[200 - 4(26)] = 1664$
28	$\frac{1}{3}[200 - 4(28)]$	$2xy = (2)(28)(\frac{1}{3})[200 - 4(28)] \approx 1643$
30	$\frac{1}{3}[200 - 4(30)]$	$2xy = (2)(30)(\frac{1}{3})[200 - 4(30)] = 1600$

The table indicates that the area is maximum when x is between 24 and 26 feet and y is between $34\frac{2}{3}$ and 32 feet.

(c) $A = 2xy = 2x\left(\dfrac{200 - 4x}{3}\right) = \dfrac{2x(4)(50 - x)}{3}$

$\quad = \dfrac{8x(50 - x)}{3}$

(d)

This area is maximum when $x = 25$ feet and $y = \dfrac{100}{3} = 33\dfrac{1}{3}$ feet.

(e) $A = \dfrac{8}{3}x(50 - x)$

$\quad = -\dfrac{8}{3}(x^2 - 50x)$

$\quad = -\dfrac{8}{3}(x^2 - 50x + 625 - 625)$

$\quad = -\dfrac{8}{3}[(x - 25)^2 - 625]$

$\quad = -\dfrac{8}{3}(x - 25)^2 + \dfrac{5000}{3}$

The maximum area occurs at the vertex and is 5000/3 square feet. This happens when $x = 25$ feet and $y = (200 - 4(25))/3 = 100/3$ feet. The dimensions are $2x = 50$ feet by $33\frac{1}{3}$ feet.

87. $R = 900x - 0.1x^2 = -0.1x^2 + 900x$

The vertex occurs at $x = -\dfrac{b}{2a} = -\dfrac{900}{2(-0.1)} = 4500.$

The revenue is maximum when $x = 4500$ units.

89. $C = 800 - 10x + 0.25x^2 = 0.25x^2 - 10x + 800$

The vertex occurs at $x = -\dfrac{b}{2a} = -\dfrac{-10}{2(0.25)} = 20.$

The cost is minimum when $x = 20$ fixtures.

91. $P = -0.0002x^2 + 140x - 250,000$

The vertex occurs at $x = -\dfrac{b}{2a} = -\dfrac{140}{2(-0.0002)} = 350,000.$

The profit is maximum when $x = 350,000$ units.

93. $y = -\dfrac{1}{12}x^2 + 2x + 4$

(a) When $x = 0$, $y = 4$ feet.

(b) The vertex occurs at $x = -\dfrac{b}{2a} = -\dfrac{2}{2(-1/12)} = 12$. The maximum height is

$$y = -\frac{1}{12}(12)^2 + 2(12) + 4 = 16 \text{ feet.}$$

(c) When the ball strikes the ground, $y = 0$.

$$0 = -\frac{1}{12}x^2 + 2x + 4$$

$$0 = x^2 - 24x - 48 \qquad \text{Multiply both sides by } -12.$$

$$x = \frac{-(-24) \pm \sqrt{(-24)^2 - 4(1)(-48)}}{2(1)}$$

$$= \frac{24 \pm \sqrt{768}}{2} = \frac{24 \pm 16\sqrt{3}}{2} = 12 \pm 8\sqrt{3}$$

Using the positive value for x, we have $x = 12 + 8\sqrt{3} \approx 25.86$ feet.

95. $V = 0.77x^2 - 1.32x - 9.31$, $5 \le x \le 40$

(a)

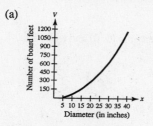

(b) $V(16) = 166.69$ board feet

(c) $500 = 0.77x^2 - 1.32x - 9.31$

$$0 = 0.77x^2 - 1.32x - 509.31$$

Using the Quadratic Formula and selecting the positive value for x, we have $x \approx 26.6$ inches in diameter.

97. Let $t = 0$ correspond to 1960.

$$C = -1.43t^2 + 0.5t + 4287, \quad 0 \le t \le 42$$

(a)

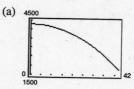

(b) The vertex occurs at $-\dfrac{b}{2a} = \dfrac{-0.5}{2(-1.43)} \approx 0.1748$.

$C(0.1748) \approx 4287$, which is the maximum average annual consumption. The warnings may not have had an immediate effect, and consumption was decreasing by 1966 without them, but over time they and other findings about the health risks of cigarettes have had an effect.

(c) $C(40) = 2019$

$$\frac{209{,}128{,}094(2019)}{48{,}308{,}590} \approx 8740 \text{ cigarettes per smoker per year.}$$

$$\frac{8740}{365} \approx 24 \text{ cigarettes per smoker per day.}$$

99. (a) and (c)

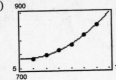

The model is a good fit for the data.

(b) $y \approx 3.55t^2 - 36.9t + 831$ where $t = 6$ corresponds to 1996.

(d) The number of hairdressers and cosmetologists is minimum when $t \approx 5$, which corresponds to 1995.

(e) $y(20) = 1513$ which corresponds to 1,513,000 hairdressers and cosmetologists in 2010.

101. True. The vertex of $f(x)$ is $\left(-\frac{5}{4}, \frac{53}{4}\right)$ and the vertex of $g(x)$ is $\left(-\frac{5}{4}, -\frac{71}{4}\right)$.

103. True. All quadratic functions of the form $y = a(x-1)(x-2)$, $a \neq 0$, have x-intercepts at $(1, 0)$ and $(2, 0)$.

Section 2.2   Polynomial Functions of Higher Degree

1. $f(x) = -2x + 3$ is a line with y-intercept $(0, 3)$.

Matches graph (c).

3. $f(x) = -2x^2 - 5x$ is a parabola with x-intercepts $(0, 0)$ and $\left(-\frac{5}{2}, 0\right)$ and opens downward.

Matches graph (h).

5. $f(x) = -\frac{1}{4}x^4 + 3x^2$ has intercepts $(0, 0)$ and $\left(\pm 2\sqrt{3}, 0\right)$.

Matches graph (a).

7. $f(x) = x^4 + 2x^3$ has intercepts $(0, 0)$ and $(-2, 0)$.

Matches graph (d).

9. $y = x^3$

(a) $f(x) = (x - 2)^3$

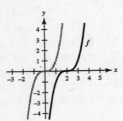

Horizontal shift two units to the right

(b) $f(x) = x^3 - 2$

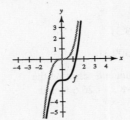

Vertical shift two units downward

(c) $f(x) = -\frac{1}{2}x^3$

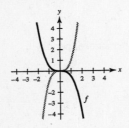

Reflection in the x-axis and a vertical shrink

(d) $f(x) = (x - 2)^3 - 2$

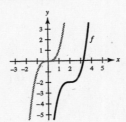

Horizontal shift two units to the right and a vertical shift two units downward

11. $y = x^4$

    (a) $f(x) = (x + 3)^4$

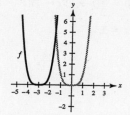

Horizontal shift three units to the left

    (b) $f(x) = x^4 - 3$

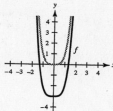

Vertical shift three units downward

    (c) $f(x) = 4 - x^4$

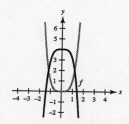

Reflection in the x-axis and then a
vertical shift four units upward

    (d) $f(x) = \frac{1}{2}(x - 1)^4$

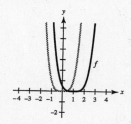

Horizontal shift one unit to the right
and a vertical shrink

13. $f(x) = \frac{1}{3}x^3 + 5x$

Degree: 3

Leading coefficient: $\frac{1}{3}$

The degree is odd and the leading coefficient is positive.
The graph falls to the left and rises to the right.

15. $g(x) = 5 - \frac{7}{2}x - 3x^2$

Degree: 2

Leading coefficient: -3

The degree is even and the leading coefficient is negative.
The graph falls to the left and falls to the right.

17. $f(x) = -2.1x^5 + 4x^3 - 2$

Degree: 5

Leading coefficient: -2.1

The degree is odd and the
leading coefficient is negative.
The graph rises to the left and
falls to the right.

19. $f(x) = 6 - 2x + 4x^2 - 5x^3$

Degree: 3

Leading coefficient: -5

The degree is odd and the
leading coefficient is negative.
The graph rises to the left and
falls to the right.

21. $h(t) = -\frac{2}{3}(t^2 - 5t + 3)$

Degree: 2

Leading coefficient: $-\frac{2}{3}$

The degree is even and the
leading coefficient is negative.
The graph falls to the left and
falls to the right.

23. $f(x) = 3x^3 - 9x + 1;\ g(x) = 3x^3$

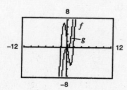

25. $f(x) = -(x^4 - 4x^3 + 16x);\ g(x) = -x^4$

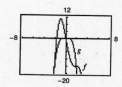

27. $f(x) = x^2 - 25$

    $0 = (x + 5)(x - 5)$

    $x = \pm 5$

29. $h(t) = t^2 - 6t + 9$

    $0 = (t - 3)^2$

    $t = 3$

31. $f(x) = \frac{1}{3}x^2 + \frac{1}{3}x - \frac{2}{3}$

    $0 = \frac{1}{3}(x + 2)(x - 1)$

    $x = -2, 1$

33. $f(x) = 3x^2 - 12x + 3$

$0 = 3(x^2 - 4x + 1)$

$x = \dfrac{4 \pm \sqrt{16 - 4}}{2} = 2 \pm \sqrt{3}$

35. $f(t) = t^3 - 4t^2 + 4t$

$0 = t(t - 2)^2$

$t = 0, 2$

37. $g(t) = \dfrac{1}{2}t^4 - \dfrac{1}{2}$

$0 = \dfrac{1}{2}(t + 1)(t - 1)(t^2 + 1)$

$t = \pm 1$

39. $g(t) = t^5 - 6t^3 + 9t$

$0 = t(t^2 - 3)^2$

$0 = t(t + \sqrt{3})^2(t - \sqrt{3})^2$

$t = 0, \pm\sqrt{3}$

41. $f(x) = 5x^4 + 15x^2 + 10$

$0 = 5(x^4 + 3x^2 + 2)$

$0 = 5(x^2 + 2)(x^2 + 1)$

No real zeros

43. $y = 4x^3 - 20x^2 + 25x$

(a)

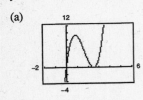

(b) x-intercepts: $(0, 0)$, $\left(\dfrac{5}{2}, 0\right)$

(c) $0 = 4x^3 - 20x^2 + 25x$

$0 = x(2x - 5)^2$

$x = 0$ or $x = \dfrac{5}{2}$

The solutions are the same as the x-coordinates of the x-intercepts.

45. $y = x^5 - 5x^3 + 4x$

(a)

(b) x-intercepts: $(0, 0)$, $(\pm 1, 0)$, $(\pm 2, 0)$

(c) $0 = x^5 - 5x^3 + 4x$

$0 = x(x^2 - 1)(x^2 - 4)$

$0 = x(x + 1)(x - 1)(x + 2)(x - 2)$

$x = 0, \pm 1, \pm 2$

The solutions are the same as the x-coordinates of the x-intercepts.

47. $f(x) = (x - 0)(x - 10)$

$f(x) = x^2 - 10x$

Note: $f(x) = a(x - 0)(x - 10) = ax(x - 10)$ has zeros 0 and 10 for all real numbers $a \neq 0$.

49. $f(x) = (x - 2)(x - (-6))$

$= (x - 2)(x + 6)$

$= x^2 + 4x - 12$

Note: $f(x) = a(x - 2)(x + 6)$ has zeros 2 and -6 for all real numbers $a \neq 0$.

51. $f(x) = (x - 0)(x - (-2))(x - (-3))$

$= x(x + 2)(x + 3)$

$= x^3 + 5x^2 + 6x$

Note: $f(x) = ax(x + 2)(x + 3)$ has zeros $0, -2, -3$ for all real numbers $a \neq 0$.

53. $f(x) = (x - 4)(x + 3)(x - 3)(x - 0)$

$= (x - 4)(x^2 - 9)x$

$= x^4 - 4x^3 - 9x^2 + 36x$

Note: $f(x) = a(x^4 - 4x^3 - 9x^2 + 36x)$ has these zeros for all real numbers $a \neq 0$.

55. $f(x) = \left[x - \left(1 + \sqrt{3}\right)\right]\left[x - \left(1 - \sqrt{3}\right)\right]$

$= \left[(x - 1) - \sqrt{3}\right]\left[(x - 1) + \sqrt{3}\right]$

$= (x - 1)^2 - \left(\sqrt{3}\right)^2$

$= x^2 - 2x + 1 - 3$

$= x^2 - 2x - 2$

Note: $f(x) = a(x^2 - 2x - 2)$ has these zeros for all real numbers $a \neq 0$.

57. $f(x) = (x - (-2))(x - (-2)) = (x + 2)^2 = x^2 + 4x + 4$

Note: If $a \neq 0$, then $f(x) = a(x^2 + 4x + 4)$ has degree 2 and zero $x = -2$, as does $f(x) = a(x + 2)(x - b)$ for any b value.

59. $f(x) = (x - (-3))(x - 0)(x - 1) = x(x + 3)(x - 1) = x^3 + 2x^2 - 3x$

Note: $f(x) = a(x^3 + 2x^2 - 3x)$, $a \neq 0$, has degree 3 and zeros $x = -3, 0, 1$.

61. $f(x) = (x - 0)\left(x - \sqrt{3}\right)\left(x - \left(-\sqrt{3}\right)\right)$

$\quad = x\left(x - \sqrt{3}\right)\left(x + \sqrt{3}\right)$

$\quad = x^3 - 3x$

Note: $f(x) = a(x^3 - 3x)$, $a \neq 0$, has degree 3 and zeros $x = 0, \sqrt{3}, -\sqrt{3}$.

63. $f(x) = (x - 1)[x - (-2)]\left[x - \left(1 + \sqrt{3}\right)\right]\left[x - \left(1 - \sqrt{3}\right)\right]$

$\quad = (x - 1)(x + 2)\left[(x - 1) - \sqrt{3}\right]\left[(x - 1) + \sqrt{3}\right]$

$\quad = (x^2 + x - 2)[(x - 1)^2 - 3]$

$\quad = (x^2 + x - 2)(x^2 - 2x - 2)$

$\quad = x^4 - x^3 - 6x^2 + 2x + 4$

Note: $f(x) = a(x^4 - x^3 - 6x^2 + 2x + 4)$ has these zeros for all real numbers $a \neq 0$.

65. For $n = 5$ there are many possibilities.

$f(x) = (x - 0)^3(x + 4)(x - 3) = x^5 + x^4 - 12x^3$

$f(x) = x^2(x + 4)^2(x - 3) = x^5 + 5x^4 - 8x^3 - 48x^2$

$f(x) = x^2(x + 4)(x - 3)^2 = x^5 - 2x^4 - 15x^3 + 36x^2$

$f(x) = x(x + 4)^3(x - 3) = x^5 + 9x^4 + 12x^3 - 80x^2 - 192x$

$f(x) = x(x + 4)^2(x - 3)^2 = x^5 + 2x^4 - 23x^3 - 24x^2 + 144x$

$f(x) = x(x + 4)(x - 3)^3 = x^5 - 5x^4 - 9x^3 + 81x^2 - 108x$

Also, any non-zero multiple of the right side of these functions would have these zeros and a degree of 5.

67. $f(x) = x^3 - 9x = x(x^2 - 9) = x(x + 3)(x - 3)$

(a) Falls to the left

Rises to the right

(b) Zeros: $0, -3, 3$

(c)

x	-3	-2	-1	0	1	2	3
$f(x)$	0	10	8	0	-8	-10	0

(d)

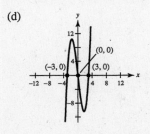

69. $f(t) = \frac{1}{4}(t^2 - 2t + 15) = \frac{1}{4}(t - 1)^2 + \frac{7}{2}$

(a) Rises to the left

Rises to the right

(b) No real zero (no x-intercepts)

(c)

t	-1	0	1	2	3
$f(t)$	4.5	3.75	3.5	3.75	4.5

(d) The graph is a parabola with vertex $\left(1, \frac{7}{2}\right)$.

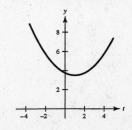

71. $f(x) = x^3 - 3x^2 = x^2(x - 3)$

(a) Falls to the left

Rises to the right

(b) Zeros: 0 and 3

(c)

x	-1	0	1	2	3
$f(x)$	-4	0	-2	-4	0

(d)

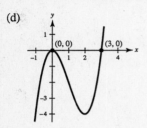

73. $f(x) = 3x^3 - 15x^2 + 18x = 3x(x - 2)(x - 3)$

(a) Falls to the left

Rises to the right

(b) Zeros: 0, 2, 3

(c)

x	0	1	2	2.5	3	3.5
$f(x)$	0	6	0	-1.875	0	7.875

(d)

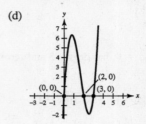

75. $f(x) = -5x^2 - x^3 = -x^2(5 + x)$

(a) Rises to the left

Falls to the right

(c)

x	-5	-4	-3	-2	-1	0	1
$f(x)$	0	-16	-18	-12	-4	0	-6

(b) Zeros: 0, -5

(d)

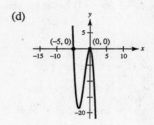

77. $f(x) = x^2(x - 4)$

(a) Falls to the left

Rises to the right

(b) Zeros: 0, 4

(c)

x	-1	0	1	2	3	4	5
$f(x)$	-5	0	-3	-8	-9	0	25

(d)

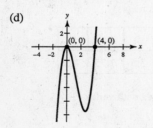

79. $g(t) = -\frac{1}{4}(t - 2)^2(t + 2)^2$

(a) Falls to the left

Falls to the right

(b) Zeros: 2 and -2

(c)

t	-3	-2	-1	0	1	2	3
$g(t)$	$-\frac{25}{4}$	0	$-\frac{9}{4}$	-4	$-\frac{9}{4}$	0	$-\frac{25}{4}$

(d)

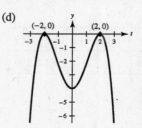

81. $f(x) = x^3 - 4x = x(x + 2)(x - 2)$

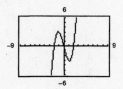

Zeros: $0, -2, 2$ all of multiplicity 1

83. $g(x) = \frac{1}{5}(x + 1)^2(x - 3)(2x - 9)$

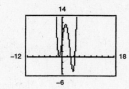

Zeros: -1 of multiplicity 2, 3 of multiplicity 1, and $\frac{9}{2}$ of multiplicity 1

85. (a) Degree: 3

   Leading coefficient: Positive

(b) Degree: 2

   Leading coefficient: Positive

(c) Degree: 4

   Leading coefficient: Positive

(d) Degree: 5

   Leading coefficient: Positive

87. (a) and (b)

Box Height	Box Width	Box Volume, V
1	$36 - 2(1)$	$1[36 - 2(1)]^2 = 1156$
2	$36 - 2(2)$	$2[36 - 2(2)]^2 = 2048$
3	$36 - 2(3)$	$3[36 - 2(3)]^2 = 2700$
4	$36 - 2(4)$	$4[36 - 2(4)]^2 = 3136$
5	$36 - 2(5)$	$5[36 - 2(5)]^2 = 3380$
6	$36 - 2(6)$	$6[36 - 2(6)]^2 = 3456$
7	$36 - 2(7)$	$7[36 - 2(7)]^2 = 3388$

Volume is maximum at 3456 cubic inches when the height is 6 inches and the length and width are each 24 inches. So the dimensions are $6 \times 24 \times 24$ inches.

(c) Volume = length × width × height

   height = x

   length = width = $36 - 2x$

   Thus, $V(x) = (36 - 2x)(36 - 2x)(x) = x(36 - 2x)^2$

   Domain: $0 < x < 18$

(d)

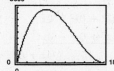

The maximum point on the graph occurs at $x = 6$. This agrees with the maximum found in part (b).

89. $R = \dfrac{1}{100,000}(-x^3 + 600x^2)$

The point of diminishing returns (where the graph changes from curving upward to curving downward) occurs when $x = 200$. The point is $(200, 160)$ which corresponds to spending \$2,000,000 on advertising to obtain a revenue of \$160 million.

91. False. A fifth degree polynomial can have at most four turning points.

93. False.

If the leading coefficient of a third-degree polynomial function is greater than zero, then the graph falls to the left and rises to the right. If the leading coefficient is less than zero, then the graph rises to the left and falls to the right.

Section 2.3   Polynomial and Synthetic Division

1. $y_1 = \dfrac{4x}{x-1}$ and $y_2 = 4 + \dfrac{4}{x-1}$

$$
\begin{array}{r}
4 \\
x-1 \overline{) 4x + 0} \\
\underline{4x - 4} \\
4
\end{array}
$$

Thus, $\dfrac{4x}{x-1} = 4 + \dfrac{4}{x-1}$ and $y_1 = y_2$.

3. $y_1 = \dfrac{x^2}{x+2}$ and $y_2 = x - 2 + \dfrac{4}{x+2}$

$$
\begin{array}{r}
x - 2 \\
x+2 \overline{) x^2 + 0x + 0} \\
\underline{x^2 + 2x} \\
-2x + 0 \\
\underline{-2x - 4} \\
4
\end{array}
$$

Thus, $\dfrac{x^2}{x+2} = x - 2 + \dfrac{4}{x+2}$ and $y_1 = y_2$.

5. $y_1 = \dfrac{x^5 - 3x^3}{x^2 + 1}$ and $y_2 = x^3 - 4x + \dfrac{4x}{x^2 + 1}$

$$
\begin{array}{r}
x^3 - 4x \\
x^2 + 0x + 1 \overline{) x^5 + 0x^4 - 3x^3 + 0x^2 + 0x + 0} \\
\underline{x^5 + 0x^4 + x^3} \\
-4x^3 + 0x^2 + 0x \\
\underline{-4x^3 + 0x^2 - 4x} \\
4x + 0
\end{array}
$$

Thus, $\dfrac{x^5 - 3x^3}{x^2 + 1} = x^3 - 4x + \dfrac{4x}{x^2 + 1}$ and $y_1 = y_2$.

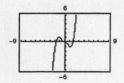

7.
$$
\begin{array}{r}
2x + 4 \\
x+3 \overline{) 2x^2 + 10x + 12} \\
\underline{2x^2 + 6x} \\
4x + 12 \\
\underline{4x + 12} \\
0
\end{array}
$$

$$\dfrac{2x^2 + 10x + 12}{x + 3} = 2x + 4$$

9.
$$
\begin{array}{r}
x^2 - 3x + 1 \\
4x+5 \overline{) 4x^3 - 7x^2 - 11x + 5} \\
\underline{4x^3 + 5x^2} \\
-12x^2 - 11x \\
\underline{-12x^2 - 15x} \\
4x + 5 \\
\underline{4x + 5} \\
0
\end{array}
$$

$$\dfrac{4x^3 - 7x^2 - 11x + 5}{4x + 5} = x^2 - 3x + 1$$

11.
$$
\begin{array}{r}
x^3 + 3x^2 \quad\ - 1 \\
x+2 \overline{) x^4 + 5x^3 + 6x^2 - x - 2} \\
\underline{x^4 + 2x^3} \\
3x^3 + 6x^2 \\
\underline{3x^3 + 6x^2} \\
- x - 2 \\
\underline{- x - 2} \\
0
\end{array}
$$

$$\dfrac{x^4 + 5x^3 + 6x^2 - x - 2}{x + 2} = x^3 + 3x^2 - 1$$

13.
$$
\begin{array}{r}
7 \\
x+2 \overline{) 7x + 3} \\
\underline{7x + 14} \\
-11
\end{array}
$$

$$\dfrac{7x + 3}{x + 2} = 7 - \dfrac{11}{x + 2}$$

15.

$$
\begin{array}{r}
3x + 5 \\
2x^2 + 0x + 1 \overline{\smash{)}\ 6x^3 + 10x^2 +\ x + 8} \\
\underline{6x^3 +\ 0x^2 + 3x} \\
10x^2 - 2x + 8 \\
\underline{10x^2 + 0x + 5} \\
-2x + 3
\end{array}
$$

$$
\frac{6x^3 + 10x^2 + x + 8}{2x^2 + 1} = 3x + 5 + \frac{-2x + 3}{2x^2 + 1}
$$

$$
= 3x + 5 - \frac{2x - 3}{2x^2 + 1}
$$

17.

$$
\begin{array}{r}
x^2 + 2x +\ 4 \\
x^2 - 2x + 3 \overline{\smash{)}\ x^4 + 0x^3 + 3x^2 + 0x +\ 1} \\
\underline{x^4 - 2x^3 + 3x^2} \\
2x^3 + 0x^2 + 0x \\
\underline{2x^3 - 4x^2 + 6x} \\
4x^2 - 6x +\ 1 \\
\underline{4x^2 - 8x + 12} \\
2x - 11
\end{array}
$$

$\Longrightarrow$

$$
\frac{x^4 + 3x^2 + 1}{x^2 - 2x + 3} = x^2 + 2x + 4 + \frac{2x - 11}{x^2 - 2x + 3}
$$

19.

$$
\begin{array}{r}
x + 3 \\
x^3 - 3x^2 + 3x - 1 \overline{\smash{)}\ x^4 + 0x^3 + 0x^2 + 0x + 0} \\
\underline{x^4 - 3x^3 + 3x^2 -\ x} \\
3x^3 - 3x^2 +\ x + 0 \\
\underline{3x^3 - 9x^2 + 9x - 3} \\
6x^2 - 8x + 3
\end{array}
$$

$$
\frac{x^4}{(x - 1)^3} = x + 3 + \frac{6x^2 - 8x + 3}{(x - 1)^3}
$$

21.

$$
\begin{array}{r|rrrr}
5 & 3 & -17 & 15 & -25 \\
& & 15 & -10 & 25 \\
\hline
& 3 & -2 & 5 & 0
\end{array}
$$

$$
\frac{3x^3 - 17x^2 + 15x - 25}{x - 5} = 3x^2 - 2x + 5
$$

23.

$$
\begin{array}{r|rrrr}
-2 & 4 & 8 & -9 & -18 \\
& & -8 & 0 & 18 \\
\hline
& 4 & 0 & -9 & 0
\end{array}
$$

$$
\frac{4x^3 + 8x^2 - 9x - 18}{x + 2} = 4x^2 - 9
$$

25.

$$
\begin{array}{r|rrrr}
-10 & -1 & 0 & 75 & -250 \\
& & 10 & -100 & 250 \\
\hline
& -1 & 10 & -25 & 0
\end{array}
$$

$$
\frac{-x^3 + 75x - 250}{x + 10} = -x^2 + 10x - 25
$$

27.

$$
\begin{array}{r|rrrr}
4 & 5 & -6 & 0 & 8 \\
& & 20 & 56 & 224 \\
\hline
& 5 & 14 & 56 & 232
\end{array}
$$

$$
\frac{5x^3 - 6x^2 + 8}{x - 4} = 5x^2 + 14x + 56 + \frac{232}{x - 4}
$$

29.

$$
\begin{array}{r|rrrrr}
6 & 10 & -50 & 0 & 0 & -800 \\
& & 60 & 60 & 360 & 2160 \\
\hline
& 10 & 10 & 60 & 360 & 1360
\end{array}
$$

$$
\frac{10x^4 - 50x^3 - 800}{x - 6} = 10x^3 + 10x^2 + 60x + 360 + \frac{1360}{x - 6}
$$

31.

$$
\begin{array}{r|rrrr}
-8 & 1 & 0 & 0 & 512 \\
& & -8 & 64 & -512 \\
\hline
& 1 & -8 & 64 & 0
\end{array}
$$

$$
\frac{x^3 + 512}{x + 8} = x^2 - 8x + 64
$$

33.

$$
\begin{array}{r|rrrrr}
2 & -3 & 0 & 0 & 0 & 0 \\
& & -6 & -12 & -24 & -48 \\
\hline
& -3 & -6 & -12 & -24 & -48
\end{array}
$$

$$
\frac{-3x^4}{x - 2} = -3x^3 - 6x^2 - 12x - 24 - \frac{48}{x - 2}
$$

35.

$$6 \begin{array}{|rrrrr} -1 & 0 & 0 & 180 & 0 \\ & -6 & -36 & -216 & -216 \\ \hline -1 & -6 & -36 & -36 & -216 \end{array}$$

$$\frac{180x - x^4}{x - 6} = -x^3 - 6x^2 - 36x - 36 - \frac{216}{x - 6}$$

37.

$$-\tfrac{1}{2} \begin{array}{|rrrr} 4 & 16 & -23 & -15 \\ & -2 & -7 & 15 \\ \hline 4 & 14 & -30 & 0 \end{array}$$

$$\frac{4x^3 + 16x^2 - 23x - 15}{x + \frac{1}{2}} = 4x^2 + 14x - 30$$

39. $f(x) = x^3 - x^2 - 14x + 11, \; k = 4$

$$4 \begin{array}{|rrrr} 1 & -1 & -14 & 11 \\ & 4 & 12 & -8 \\ \hline 1 & 3 & -2 & 3 \end{array}$$

$$f(x) = (x - 4)(x^2 + 3x - 2) + 3$$
$$f(4) = 4^3 - 4^2 - 14(4) + 11 = 3$$

41. $f(x) = 15x^4 + 10x^3 - 6x^2 + 14, \; k = -\tfrac{2}{3}$

$$-\tfrac{2}{3} \begin{array}{|rrrrr} 15 & 10 & -6 & 0 & 14 \\ & -10 & 0 & 4 & -\frac{8}{3} \\ \hline 15 & 0 & -6 & 4 & \frac{34}{3} \end{array}$$

$$f(x) = \left(x + \tfrac{2}{3}\right)(15x^3 - 6x + 4) + \tfrac{34}{3}$$
$$f\left(-\tfrac{2}{3}\right) = 15\left(-\tfrac{2}{3}\right)^4 + 10\left(-\tfrac{2}{3}\right)^3 - 6\left(-\tfrac{2}{3}\right)^2 + 14 = \tfrac{34}{3}$$

43. $f(x) = x^3 + 3x^2 - 2x - 14, \; k = \sqrt{2}$

$$\sqrt{2} \begin{array}{|rrrr} 1 & 3 & -2 & -14 \\ & \sqrt{2} & 2 + 3\sqrt{2} & 6 \\ \hline 1 & 3 + \sqrt{2} & 3\sqrt{2} & -8 \end{array}$$

$$f(x) = \left(x - \sqrt{2}\right)\!\left[x^2 + \left(3 + \sqrt{2}\right)x + 3\sqrt{2}\right] - 8$$
$$f\left(\sqrt{2}\right) = \left(\sqrt{2}\right)^3 + 3\left(\sqrt{2}\right)^2 - 2\sqrt{2} - 14 = -8$$

45. $f(x) = -4x^3 + 6x^2 + 12x + 4, \; k = 1 - \sqrt{3}$

$$1 - \sqrt{3} \begin{array}{|rrrr} -4 & 6 & 12 & 4 \\ & -4 + 4\sqrt{3} & -10 + 2\sqrt{3} & -4 \\ \hline -4 & 2 + 4\sqrt{3} & 2 + 2\sqrt{3} & 0 \end{array}$$

$$f(x) = \left[x - \left(1 - \sqrt{3}\right)\right]\!\left[-4x^2 + \left(2 + 4\sqrt{3}\right)x + \left(2 + 2\sqrt{3}\right)\right]$$
$$f\left(1 - \sqrt{3}\right) = -4\left(1 - \sqrt{3}\right)^3 + 6\left(1 - \sqrt{3}\right)^2 + 12\left(1 - \sqrt{3}\right) + 4 = 0$$

47. $f(x) = 4x^3 - 13x + 10$

(a) $$1 \begin{array}{|rrrr} 4 & 0 & -13 & 10 \\ & 4 & 4 & -9 \\ \hline 4 & 4 & -9 & \underline{1} = f(1) \end{array}$$

(b) $$-2 \begin{array}{|rrrr} 4 & 0 & -13 & 10 \\ & -8 & 16 & -6 \\ \hline 4 & -8 & 3 & \underline{4} = f(-2) \end{array}$$

(c) $$\tfrac{1}{2} \begin{array}{|rrrr} 4 & 0 & -13 & 10 \\ & 2 & 1 & -6 \\ \hline 4 & 2 & -12 & \underline{4} = f\left(\tfrac{1}{2}\right) \end{array}$$

(d) $$8 \begin{array}{|rrrr} 4 & 0 & -13 & 10 \\ & 32 & 256 & 1944 \\ \hline 4 & 32 & 243 & \underline{1954} = f(8) \end{array}$$

49. $h(x) = 3x^3 + 5x^2 - 10x + 1$

(a) $$3 \begin{array}{|rrrr} 3 & 5 & -10 & 1 \\ & 9 & 42 & 96 \\ \hline 3 & 14 & 32 & \underline{97} = h(3) \end{array}$$

(b) $$\tfrac{1}{3} \begin{array}{|rrrr} 3 & 5 & -10 & 1 \\ & 1 & 2 & -\frac{8}{3} \\ \hline 3 & 6 & -8 & \underline{-\frac{5}{3}} = h\left(\tfrac{1}{3}\right) \end{array}$$

(c) $$-2 \begin{array}{|rrrr} 3 & 5 & -10 & 1 \\ & -6 & 2 & 16 \\ \hline 3 & -1 & -8 & \underline{17} = h(-2) \end{array}$$

(d) $$-5 \begin{array}{|rrrr} 3 & 5 & -10 & 1 \\ & -15 & 50 & -200 \\ \hline 3 & -10 & 40 & \underline{-199} = h(-5) \end{array}$$

51.
$$2 \,\big|\; \begin{array}{rrrr} 1 & 0 & -7 & 6 \\ & 2 & 4 & -6 \\ \hline 1 & 2 & -3 & 0 \end{array}$$

$$x^3 - 7x + 6 = (x-2)(x^2 + 2x - 3)$$
$$= (x-2)(x+3)(x-1)$$

Zeros: $2, -3, 1$

53.
$$\tfrac{1}{2} \,\big|\; \begin{array}{rrrr} 2 & -15 & 27 & -10 \\ & 1 & -7 & 10 \\ \hline 2 & -14 & 20 & 0 \end{array}$$

$$2x^3 - 15x^2 + 27x - 10$$
$$= \left(x - \tfrac{1}{2}\right)(2x^2 - 14x + 20)$$
$$= (2x-1)(x-2)(x-5)$$

Zeros: $\tfrac{1}{2}, 2, 5$

55.
$$\sqrt{3} \,\big|\; \begin{array}{rrrr} 1 & 2 & -3 & -6 \\ & \sqrt{3} & 3 + 2\sqrt{3} & 6 \\ \hline 1 & 2 + \sqrt{3} & 2\sqrt{3} & 0 \end{array}$$
$$-\sqrt{3} \,\big|\; \begin{array}{rrr} 1 & 2 + \sqrt{3} & 2\sqrt{3} \\ & -\sqrt{3} & -2\sqrt{3} \\ \hline 1 & 2 & 0 \end{array}$$

$$x^3 + 2x^2 - 3x - 6 = (x - \sqrt{3})(x + \sqrt{3})(x + 2)$$

Zeros: $\pm\sqrt{3}, -2$

57.
$$1 + \sqrt{3} \,\big|\; \begin{array}{rrrr} 1 & -3 & 0 & 2 \\ & 1 + \sqrt{3} & 1 - \sqrt{3} & -2 \\ \hline 1 & -2 + \sqrt{3} & 1 - \sqrt{3} & 0 \end{array}$$
$$1 - \sqrt{3} \,\big|\; \begin{array}{rrr} 1 & -2 + \sqrt{3} & 1 - \sqrt{3} \\ & 1 - \sqrt{3} & -1 + \sqrt{3} \\ \hline 1 & -1 & 0 \end{array}$$

$$x^3 - 3x^2 + 2 = \left[x - \left(1 + \sqrt{3}\right)\right]\left[x - \left(1 - \sqrt{3}\right)\right](x - 1)$$
$$= (x - 1)(x - 1 - \sqrt{3})(x - 1 + \sqrt{3})$$

Zeros: $1, 1 \pm \sqrt{3}$

59. $f(x) = 2x^3 + x^2 - 5x + 2;$

Factors: $(x + 2), (x - 1)$

(a)
$$-2 \,\big|\; \begin{array}{rrrr} 2 & 1 & -5 & 2 \\ & -4 & 6 & -2 \\ \hline 2 & -3 & 1 & 0 \end{array}$$
$$1 \,\big|\; \begin{array}{rrr} 2 & -3 & 1 \\ & 2 & -1 \\ \hline 2 & -1 & 0 \end{array}$$

Both are factors of $f(x)$ since the remainders are zero.

(b) The remaining factor of $f(x)$ is $(2x - 1)$.

(c) $f(x) = (2x - 1)(x + 2)(x - 1)$

(d) Zeros: $\tfrac{1}{2}, -2, 1$

(e)

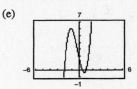

61. $f(x) = x^4 - 4x^3 - 15x^2 + 58x - 40;$

Factors: $(x - 5), (x + 4)$

(a)
$$5 \,\big|\; \begin{array}{rrrrr} 1 & -4 & -15 & 58 & -40 \\ & 5 & 5 & -50 & 40 \\ \hline 1 & 1 & -10 & 8 & 0 \end{array}$$
$$-4 \,\big|\; \begin{array}{rrrr} 1 & 1 & -10 & 8 \\ & -4 & 12 & -8 \\ \hline 1 & -3 & 2 & 0 \end{array}$$

Both are factors of $f(x)$ since the remainders are zero.

(b) $x^2 - 3x + 2 = (x - 1)(x - 2)$

The remaining factors are $(x - 1)$ and $(x - 2)$.

(c) $f(x) = (x - 1)(x - 2)(x - 5)(x + 4)$

(d) Zeros: $1, 2, 5, -4$

(e)

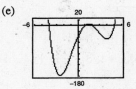

63. $f(x) = 6x^3 + 41x^2 - 9x - 14$; Factors: $(2x + 1), (3x - 2)$

(a)

$$-\tfrac{1}{2} \, \begin{array}{|rrrr} 6 & 41 & -9 & -14 \\ & -3 & -19 & 14 \\ \hline 6 & 38 & -28 & 0 \end{array}$$

$$\tfrac{2}{3} \, \begin{array}{|rrr} 6 & 38 & -28 \\ & 4 & 28 \\ \hline 6 & 42 & 0 \end{array}$$

Both are factors since the remainders are zero.

(b) $6x + 42 = 6(x + 7)$

This shows that $\dfrac{f(x)}{\left(x + \frac{1}{2}\right)\left(x - \frac{2}{3}\right)} = 6(x + 7)$, so $\dfrac{f(x)}{(2x + 1)(3x - 2)} = x + 7$.

The remaining factor is $(x + 7)$.

(c) $f(x) = (x + 7)(2x + 1)(3x - 2)$

(d) Zeros: $-7, -\tfrac{1}{2}, \tfrac{2}{3}$

(e)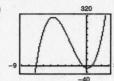

65. $f(x) = 2x^3 - x^2 - 10x + 5$;

Factors: $(2x - 1), \left(x + \sqrt{5}\right)$

(a)

$$\tfrac{1}{2} \, \begin{array}{|rrrr} 2 & -1 & -10 & 5 \\ & 1 & 0 & -5 \\ \hline 2 & 0 & -10 & 0 \end{array}$$

$$-\sqrt{5} \, \begin{array}{|rrr} 2 & 0 & -10 \\ & -2\sqrt{5} & 10 \\ \hline 2 & -2\sqrt{5} & 0 \end{array}$$

Both are factors since the remainders are zero.

(b) $2x - 2\sqrt{5} = 2\left(x - \sqrt{5}\right)$

This shows that $\dfrac{f(x)}{\left(x - \frac{1}{2}\right)\left(x + \sqrt{5}\right)} = 2\left(x - \sqrt{5}\right)$,

so $\dfrac{f(x)}{\left(2x - 1\right)\left(x + \sqrt{5}\right)} = x - \sqrt{5}$.

The remaining factor is $\left(x - \sqrt{5}\right)$.

(c) $f(x) = \left(x + \sqrt{5}\right)\left(x - \sqrt{5}\right)(2x - 1)$

(d) Zeros: $-\sqrt{5}, \sqrt{5}, \tfrac{1}{2}$

(e)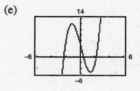

67. $f(x) = x^3 - 2x^2 - 5x + 10$

(a) The zeros of f are 2 and $\approx \pm 2.236$.

(b)

$$2 \, \begin{array}{|rrrr} 1 & -2 & -5 & 10 \\ & 2 & 0 & -10 \\ \hline 1 & 0 & -5 & 0 \end{array}$$

(c) $f(x) = (x - 2)(x^2 - 5)$

$\qquad = (x - 2)\left(x - \sqrt{5}\right)\left(x + \sqrt{5}\right)$

69. $h(t) = t^3 - 2t^2 - 7t + 2$

(a) The zeros of h are -2, ≈ 3.732, ≈ 0.268.

(b)

$$-2 \, \begin{array}{|rrrr} 1 & -2 & -7 & 2 \\ & -2 & 8 & -2 \\ \hline 1 & -4 & 1 & 0 \end{array}$$

$h(t) = (t + 2)(t^2 - 4t + 1)$

By the Quadratic Formula, the zeros of

$\qquad t^2 - 4t + 1$ are $2 \pm \sqrt{3}$.

(c) Thus, $h(t) = (t + 2)\left[t - \left(2 + \sqrt{3}\right)\right]\left[t - \left(2 - \sqrt{3}\right)\right]$.

71.
$$\frac{3}{2} \, \big| \begin{array}{rrrr} 4 & -8 & 1 & 3 \\ & 6 & -3 & -3 \\ \hline 4 & -2 & -2 & 0 \end{array}$$

$$\frac{4x^3 - 8x^2 + x + 3}{x - \frac{3}{2}} = 4x^2 - 2x - 2$$

Thus, $\dfrac{4x^3 - 8x^2 + x + 3}{2x - 3} = 2x^2 - x - 1,\ x \neq \dfrac{3}{2}.$

73.
$$-1 \, \big| \begin{array}{rrrr} 1 & 3 & -1 & -3 \\ & -1 & -2 & 3 \\ \hline 1 & 2 & -3 & 0 \end{array}$$

$$\frac{x^3 + 3x^2 - x - 3}{x + 1} = x^2 + 2x - 3,\ x \neq -1$$

75. Note that $x^2 + 3x + 2 = (x + 1)(x + 2)$.

$$-1 \, \big| \begin{array}{rrrrr} 1 & 6 & 11 & 6 & 0 \\ & -1 & -5 & -6 & 0 \\ \hline 1 & 5 & 6 & 0 & 0 \end{array}$$

$$-2 \, \big| \begin{array}{rrrr} 1 & 5 & 6 & 0 \\ & -2 & -6 & 0 \\ \hline 1 & 3 & 0 & 0 \end{array}$$

$$\frac{x^4 + 6x^3 + 11x^2 + 6x}{(x + 1)(x + 2)} = x^2 + 3x,\ x \neq -2, -1$$

77.
$$\begin{array}{r} x^{2n} + 6x^n + 9 \\ x^n + 3 \, \overline{)\, x^{3n} + 9x^{2n} + 27x^n + 27} \\ \underline{x^{3n} + 3x^{2n}} \\ 6x^{2n} + 27x^n \\ \underline{6x^{2n} + 18x^n} \\ 9x^n + 27 \\ \underline{9x^n + 27} \\ 0 \end{array}$$

$$\frac{x^{3n} + 9x^{2n} + 27x^n + 27}{x^n + 3} = x^{2n} + 6x^n + 9$$

79. A divisor divides evenly into a dividend if the remainder is zero.

81.
$$5 \, \big| \begin{array}{rrrr} 1 & 4 & -3 & c \\ & 5 & 45 & 210 \\ \hline 1 & 9 & 42 & c + 210 \end{array}$$

For $c + 210$ to equal zero, c must equal -210.

83. $f(x) = (x + 3)^2(x - 3)(x + 1)^3$

The remainder when $k = -3$ is zero since $(x + 3)$ is a factor of $f(x)$.

85. (a) and (b)

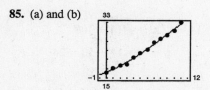

$y \approx -0.00320t^3 + 0.0856t^2 + 0.865t + 16.87$, where $t = 0$ corresponds to 1990.

(c)

t	0	1	2	3	4	5
R	16.87	17.82	18.92	20.15	21.49	22.94

t	6	7	8	9	10	11
R	24.45	26.02	27.63	29.26	30.88	32.48

The model is a good fit to the actual data.

(d)
$$15 \, \big| \begin{array}{rrrr} -0.00320 & 0.0856 & 0.865 & 16.87 \\ & -0.048 & 0.564 & 21.435 \\ \hline -0.00320 & 0.0376 & 1.429 & 38.305 \end{array}$$

Thus, $y(15) \approx \$38.31.$

No, because the model will eventually climb too fast.

87. False. If $(7x + 4)$ is a factor of f, then $-\frac{4}{7}$ is a zero of f.

89. True. If $x = k$ is a zero of $f(x)$, then $(x - k)$ is a factor of $f(x)$, and $f(k) = 0$.

Section 2.4   Complex Numbers

1. $a + bi = -10 + 6i$
 $a = -10$
 $b = 6$

3. $(a - 1) + (b + 3)i = 5 + 8i$
 $a - 1 = 5 \implies a = 6$
 $b + 3 = 8 \implies b = 5$

5. $4 + \sqrt{-9} = 4 + 3i$

7. $2 - \sqrt{-27} = 2 - \sqrt{27}i = 2 - 3\sqrt{3}i$

9. $\sqrt{-75} = \sqrt{75}i = 5\sqrt{3}i$

11. $8 = 8 + 0i = 8$

13. $-6i + i^2 = -6i - 1 = -1 - 6i$

15. $\sqrt{-0.09} = \sqrt{0.09}i = 0.3i$

17. $(5 + i) + (6 - 2i) = 11 - i$

19. $(8 - i) - (4 - i) = 8 - i - 4 + i = 4$

21. $\left(-2 + \sqrt{-8}\right) + \left(5 - \sqrt{-50}\right) = -2 + 2\sqrt{2}i + 5 - 5\sqrt{2}i = 3 - 3\sqrt{2}i$

23. $13i - (14 - 7i) = 13i - 14 + 7i = -14 + 20i$

25. $-\left(\frac{3}{2} + \frac{5}{2}i\right) + \left(\frac{5}{3} + \frac{11}{3}i\right) = -\frac{3}{2} - \frac{5}{2}i + \frac{5}{3} + \frac{11}{3}i$
$= -\frac{9}{6} - \frac{15}{6}i + \frac{10}{6} + \frac{22}{6}i$
$= \frac{1}{6} + \frac{7}{6}i$

27. $\sqrt{-6} \cdot \sqrt{-2} = \left(\sqrt{6}i\right)\left(\sqrt{2}i\right) = \sqrt{12}i^2 = \left(2\sqrt{3}\right)(-1)$
$= -2\sqrt{3}$

29. $\left(\sqrt{-10}\right)^2 = \left(\sqrt{10}i\right)^2 = 10i^2 = -10$

31. $(1 + i)(3 - 2i) = 3 - 2i + 3i - 2i^2 = 3 + i + 2 = 5 + i$

33. $6i(5 - 2i) = 30i - 12i^2 = 30i + 12 = 12 + 30i$

35. $\left(\sqrt{14} + \sqrt{10}i\right)\left(\sqrt{14} - \sqrt{10}i\right) = 14 - 10i^2$
$= 14 + 10 = 24$

37. $(4 + 5i)^2 = 16 + 40i + 25i^2$
$= 16 + 40i - 25$
$= -9 + 40i$

39. $(2 + 3i)^2 + (2 - 3i)^2 = 4 + 12i + 9i^2 + 4 - 12i + 9i^2$
$= 4 + 12i - 9 + 4 - 12i - 9$
$= -10$

41. The complex conjugate of $6 + 3i$ is $6 - 3i$.
 $(6 + 3i)(6 - 3i) = 36 - (3i)^2 = 36 + 9 = 45$

43. The complex conjugate of $-1 - \sqrt{5}i$ is $-1 + \sqrt{5}i$.
 $\left(-1 - \sqrt{5}i\right)\left(-1 + \sqrt{5}i\right) = (-1)^2 - \left(\sqrt{5}i\right)^2$
$= 1 + 5 = 6$

45. The complex conjugate of $\sqrt{-20} = 2\sqrt{5}i$ is $-2\sqrt{5}i$.
 $\left(2\sqrt{5}i\right)\left(-2\sqrt{5}i\right) = -20i^2 = 20$

47. The complex conjugate of $\sqrt{8}$ is $\sqrt{8}$.
 $\left(\sqrt{8}\right)\left(\sqrt{8}\right) = 8$

49. $\dfrac{5}{i} = \dfrac{5}{i} \cdot \dfrac{-i}{-i} = \dfrac{-5i}{1} = -5i$

51. $\dfrac{2}{4-5i} = \dfrac{2}{4-5i} \cdot \dfrac{4+5i}{4+5i} = \dfrac{2(4+5i)}{16+25} = \dfrac{8+10i}{41} = \dfrac{8}{41} + \dfrac{10}{41}i$

53. $\dfrac{3+i}{3-i} = \dfrac{3+i}{3-i} \cdot \dfrac{3+i}{3+i}$

$\quad\quad = \dfrac{9+6i+i^2}{9+1}$

$\quad\quad = \dfrac{8+6i}{10}$

$\quad\quad = \dfrac{4}{5} + \dfrac{3}{5}i$

55. $\dfrac{6-5i}{i} = \dfrac{6-5i}{i} \cdot \dfrac{-i}{-i}$

$\quad\quad = \dfrac{-6i+5i^2}{1}$

$\quad\quad = -5 - 6i$

57. $\dfrac{3i}{(4-5i)^2} = \dfrac{3i}{16-40i+25i^2} = \dfrac{3i}{-9-40i} \cdot \dfrac{-9+40i}{-9+40i}$

$\quad\quad = \dfrac{-27i+120i^2}{81+1600} = \dfrac{-120-27i}{1681}$

$\quad\quad = -\dfrac{120}{1681} - \dfrac{27}{1681}i$

59. $\dfrac{2}{1+i} - \dfrac{3}{1-i} = \dfrac{2(1-i)-3(1+i)}{(1+i)(1-i)}$

$\quad\quad = \dfrac{2-2i-3-3i}{1+1}$

$\quad\quad = \dfrac{-1-5i}{2}$

$\quad\quad = -\dfrac{1}{2} - \dfrac{5}{2}i$

61. $\dfrac{i}{3-2i} + \dfrac{2i}{3+8i} = \dfrac{i(3+8i)+2i(3-2i)}{(3-2i)(3+8i)}$

$\quad\quad = \dfrac{3i+8i^2+6i-4i^2}{9+24i-6i-16i^2}$

$\quad\quad = \dfrac{4i^2+9i}{9+18i+16}$

$\quad\quad = \dfrac{-4+9i}{25+18i} \cdot \dfrac{25-18i}{25-18i}$

$\quad\quad = \dfrac{-100+72i+225i-162i^2}{625+324}$

$\quad\quad = \dfrac{-100+297i+162}{949}$

$\quad\quad = \dfrac{62+297i}{949}$

$\quad\quad = \dfrac{62}{949} + \dfrac{297}{949}i$

63. $x^2 - 2x + 2 = 0;\ \ a = 1,\ \ b = -2,\ \ c = 2$

$\quad x = \dfrac{-(-2) \pm \sqrt{(-2)^2 - 4(1)(2)}}{2(1)}$

$\quad\quad = \dfrac{2 \pm \sqrt{-4}}{2}$

$\quad\quad = \dfrac{2 \pm 2i}{2}$

$\quad\quad = 1 \pm i$

65. $4x^2 + 16x + 17 = 0;\ \ a = 4,\ \ b = 16,\ \ c = 17$

$\quad x = \dfrac{-16 \pm \sqrt{(16)^2 - 4(4)(17)}}{2(4)} = \dfrac{-16 \pm \sqrt{-16}}{8} = \dfrac{-16 \pm 4i}{8} = -2 \pm \dfrac{1}{2}i$

67. $4x^2 + 16x + 15 = 0$; $a = 4$, $b = 16$, $c = 15$

$$x = \frac{-16 \pm \sqrt{(16)^2 - 4(4)(15)}}{2(4)}$$

$$= \frac{-16 \pm \sqrt{16}}{8}$$

$$= \frac{-16 \pm 4}{8}$$

$$x = -\frac{12}{8} = -\frac{3}{2} \quad \text{or} \quad x = -\frac{20}{8} = -\frac{5}{2}$$

69. $\frac{3}{2}x^2 - 6x + 9 = 0$ Multiply both sides by 2.

$$3x^2 - 12x + 18 = 0$$

$$x = \frac{-(-12) \pm \sqrt{(-12)^2 - 4(3)(18)}}{2(3)}$$

$$= \frac{12 \pm \sqrt{-72}}{6}$$

$$= \frac{12 \pm 6\sqrt{2}i}{6}$$

$$= 2 \pm \sqrt{2}i$$

71. $1.4x^2 - 2x - 10 = 0$ Multiply both sides by 5

$$7x^2 - 10x - 50 = 0$$

$$x = \frac{-(-10) \pm \sqrt{(-10)^2 - 4(7)(-50)}}{2(7)}$$

$$= \frac{10 \pm \sqrt{1500}}{14}$$

$$= \frac{10 \pm 10\sqrt{15}}{14}$$

$$= \frac{5 \pm 5\sqrt{15}}{7}$$

$$= \frac{5}{7} \pm \frac{5\sqrt{15}}{7}$$

73. $i^{40} = i^4 \cdot i^4 \cdot i^4 \cdot i^4 \cdot i^4 \cdot i^4 \cdot i^4 \cdot i^4 \cdot i^4 \cdot i^4$

$$= 1 \cdot 1 \cdot 1 \cdot 1 \cdot 1 \cdot 1 \cdot 1 \cdot 1 \cdot 1 \cdot 1$$

$$= 1$$

75. $i^{50} = i^{25} \cdot i^{25} = i \cdot i = i^2 = -1$

77. $4i^2 - 2i^3 = -4 + 2i$

79. $(-i)^3 = (-1)(i^3) = (-1)(-i) = i$

81. $\left(\sqrt{-2}\right)^6 = \left(\sqrt{2}i\right)^6 = 8i^6 = 8i^4i^2 = -8$

83. $\dfrac{1}{(2i)^3} = \dfrac{1}{8i^3} = \dfrac{1}{-8i} \cdot \dfrac{8i}{8i} = \dfrac{8i}{-64i^2} = \dfrac{1}{8}i$

85. False, if $b = 0$ then $a + bi = a - bi = a$.
That is, if the complex number is real, the number equals its conjugate.

87. False.

$$i^{44} + i^{150} - i^{74} - i^{109} + i^{61} = (i^4)^{11} + (i^4)^{37}(i^2) - (i^4)^{18}(i^2) - (i^4)^{27}(i) + (i^4)^{15}(i)$$

$$= (1)^{11} + (1)^{37}(-1) - (1)^{18}(-1) - (1)^{27}(i) + (1)^{15}(i)$$

$$= 1 + (-1) + 1 - i + i = 1$$

89. $(a + bi)(a - bi) = a^2 + abi - abi - b^2i^2$

$$= a^2 - b^2(-1)$$

$$= a^2 + b^2$$

which is a real number since a and b are real numbers.
Thus, the product of a complex number and its conjugate is a real number.

91. Answers will vary.

93. (a) $z_1 = 5 + 2i, z_2 = 3 - 4i$

$$\frac{1}{z} = \frac{1}{z_1} + \frac{1}{z_2}$$

$$= \frac{1}{5 + 2i} + \frac{1}{3 - 4i}$$

$$= \frac{3 - 4i + 5 + 2i}{(5 + 2i)(3 - 4i)}$$

$$= \frac{8 - 2i}{23 - 14i}$$

$$z = \left(\frac{23 - 14i}{8 - 2i}\right)\left(\frac{8 + 2i}{8 + 2i}\right)$$

$$= \frac{212 - 66i}{68} = \frac{53}{17} - \frac{33}{34}i$$

(b) $z_1 = 9 + 16i, z_2 = 20 - 10i$

$$\frac{1}{z} = \frac{1}{z_1} + \frac{1}{z_2}$$

$$= \frac{1}{9 + 16i} + \frac{1}{20 - 10i}$$

$$= \frac{20 - 10i + 9 + 16i}{(9 + 16i)(20 - 10i)}$$

$$= \frac{29 + 6i}{340 + 230i}$$

$$z = \left(\frac{340 + 230i}{29 + 6i}\right)\left(\frac{29 - 6i}{29 - 6i}\right)$$

$$= \frac{11,240 + 4630i}{877} = \frac{11,240}{877} + \frac{4630}{877}i$$

Section 2.5   The Fundamental Theorem of Algebra

1. $f(x) = x(x - 6)^2 = x(x - 6)(x - 6)$

The three zeros are: $x = 0, x = 6, x = 6$.

3. $g(x) = (x - 2)(x + 4)^3 = (x - 2)(x + 4)(x + 4)(x + 4)$

The four zeros are: $x = 2, x = -4, x = -4, x = -4$.

5. $f(x) = (x + 6)(x + i)(x - i)$

The three zeros are: $x = -6, x = -i, x = i$.

7. $f(x) = x^3 + 3x^2 - x - 3$

Possible rational zeros: $\pm 1, \pm 3$

Zeros shown on graph: $-3, -1, 1$

9. $f(x) = 2x^4 - 17x^3 + 35x^2 + 9x - 45$

Possible rational zeros: $\pm 1, \pm 3, \pm 5, \pm 9, \pm 15, \pm 45, \pm \frac{1}{2}, \pm \frac{3}{2}, \pm \frac{5}{2}, \pm \frac{9}{2}, \pm \frac{15}{2}, \pm \frac{45}{2}$

Zeros shown on graph: $-1, \frac{3}{2}, 3, 5$

11. $f(x) = x^3 - 6x^2 + 11x - 6$

Possible rational zeros: $\pm 1, \pm 2, \pm 3, \pm 6$

$$
\begin{array}{r|rrrr}
1 & 1 & -6 & 11 & -6 \\
  &   & 1 & -5 & 6 \\
\hline
  & 1 & -5 & 6 & 0
\end{array}
$$

$x^3 - 6x^2 + 11x - 6 = (x - 1)(x^2 - 5x + 6) = (x - 1)(x - 2)(x - 3)$

Thus, the real zeros are 1, 2, and 3.

13. $f(x) = x^3 - 7x - 6$

Possible rational zeros: $\pm 1, \pm 2, \pm 3, \pm 6$

$$
\begin{array}{r|rrrr}
3 & 1 & 0 & -7 & -6 \\
  &   & 3 & 9 & 6 \\
\hline
  & 1 & 3 & 2 & 0
\end{array}
$$

$f(x) = (x - 3)(x^2 + 3x + 2) = (x - 3)(x + 2)(x + 1)$

Thus, the real zeros are $-2, -1, 3$.

15. $h(t) = t^3 + 12t^2 + 21t + 10$

Possible rational zeros: $\pm 1, \pm 2, \pm 5, \pm 10$

$$
\begin{array}{r|rrrr}
-1 & 1 & 12 & 21 & 10 \\
   &   & -1 & -11 & -10 \\
\hline
   & 1 & 11 & 10 & 0
\end{array}
$$

$t^3 + 12t^2 + 21t + 10 = (t + 1)(t^2 + 11t + 10)$

$$= (t + 1)(t + 1)(t + 10)$$

$$= (t + 1)^2(t + 10)$$

Thus, the zeros are -1 and -10.

17. $h(x) = x^3 - 9x^2 + 20x - 12$

Possible rational zeros: $\pm 1, \pm 2, \pm 3, \pm 4, \pm 6, \pm 12$

$$
\begin{array}{r|rrrr}
1 & 1 & -9 & 20 & -12 \\
  &   & 1 & -8 & 12 \\
\hline
  & 1 & -8 & 12 & 0
\end{array}
$$

$h(x) = (x - 1)(x^2 - 8x + 12)$

$\qquad = (x - 1)(x - 2)(x - 6)$

Thus, the real zeros are 1, 2, 6.

19. $f(x) = 9x^4 - 9x^3 - 58x^2 + 4x + 24$

Possible rational zeros: $\pm 1, \pm 2, \pm 3, \pm 4, \pm 6, \pm 8, \pm 12, \pm 24, \pm \frac{1}{3}, \pm \frac{2}{3},$
$\pm \frac{4}{3}, \pm \frac{8}{3}, \pm \frac{1}{9}, \pm \frac{2}{9}, \pm \frac{4}{9}, \pm \frac{8}{9}$

$$
\begin{array}{r|rrrrr}
-2 & 9 & -9 & -58 & 4 & 24 \\
   &   & -18 & 54 & 8 & -24 \\
\hline
   & 9 & -27 & -4 & 12 & 0
\end{array}
$$

$$
\begin{array}{r|rrrr}
3 & 9 & -27 & -4 & 12 \\
  &   & 27 & 0 & -12 \\
\hline
  & 9 & 0 & -4 & 0
\end{array}
$$

$9x^4 - 9x^3 - 58x^2 + 4x - 24 = (x + 2)(x - 3)(9x^2 - 4)$

$\qquad\qquad\qquad\qquad\qquad = (x + 2)(x - 3)(3x - 2)(3x + 2)$

Thus, the zeros are -2, 3, and $\pm \frac{2}{3}$.

21. $z^4 - z^3 - 2z - 4 = 0$

Possible rational zeros: $\pm 1, \pm 2, \pm 4$

$$
\begin{array}{r|rrrrr}
-1 & 1 & -1 & 0 & -2 & -4 \\
   &   & -1 & 2 & -2 & 4 \\
\hline
   & 1 & -2 & 2 & -4 & 0
\end{array}
$$

$$
\begin{array}{r|rrrr}
2 & 1 & -2 & 2 & -4 \\
  &   & 2 & 0 & 4 \\
\hline
  & 1 & 0 & 2 & 0
\end{array}
$$

$z^4 - z^3 - 2z - 4 = (z + 1)(z - 2)(z^2 + 2)$

The only real zeros are -1 and 2.

23. $2y^4 + 7y^3 - 26y^2 + 23y - 6 = 0$

Possible rational zeros: $\pm 1, \pm 2, \pm 3, \pm 6, \pm \frac{1}{2}, \pm \frac{3}{2}$

$$
\begin{array}{r|rrrrr}
1 & 2 & 7 & -26 & 23 & -6 \\
  &   & 2 & 9 & -17 & 6 \\
\hline
  & 2 & 9 & -17 & 6 & 0
\end{array}
$$

$$
\begin{array}{r|rrrr}
-6 & 2 & 9 & -17 & 6 \\
   &   & -12 & 18 & -6 \\
\hline
   & 2 & -3 & 1 & 0
\end{array}
$$

$2y^4 + 7y^3 - 26y^2 + 23y - 6 = (y - 1)(y + 6)(2y^2 - 3y + 1)$

$\qquad\qquad\qquad\qquad\qquad = (y - 1)(y + 6)(2y - 1)(y - 1)$

$\qquad\qquad\qquad\qquad\qquad = (y - 1)^2(y + 6)(2y - 1)$

The only real zeros are 1, -6, and $\frac{1}{2}$.

25. $f(x) = x^3 + x^2 - 4x - 4$

(a) Possible rational zeros: $\pm 1, \pm 2, \pm 4$

(b)

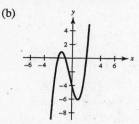

(c) The zeros are: $-2, -1, 2.$

27. $f(x) = -4x^3 + 15x^2 - 8x - 3$

(a) Possible rational zeros: $\pm 1, \pm 3, \pm\frac{1}{2}, \pm\frac{3}{2}, \pm\frac{1}{4}, \pm\frac{3}{4}$

(b)

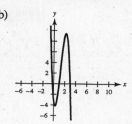

(c) The zeros are: $-\frac{1}{4},\ 1, 3.$

29. $f(x) = -2x^4 + 13x^3 - 21x^2 + 2x + 8$

(a) Possible rational zeros: $\pm 1, \pm 2, \pm 4, \pm 8, \pm\frac{1}{2}$

(b)

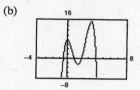

(c) The zeros are: $-\frac{1}{2},\ 1,\ 2,\ 4$

31. $f(x) = 32x^3 - 52x^2 + 17x + 3$

(a) Possible rational zeros: $\pm 1,\ \pm 3,\ \pm\frac{1}{2},\ \pm\frac{3}{2},\ \pm\frac{1}{4},\ \pm\frac{3}{4},$
$\pm\frac{1}{8},\ \pm\frac{3}{8},\ \pm\frac{1}{16},\ \pm\frac{3}{16},\ \pm\frac{1}{32},\ \pm\frac{3}{32}$

(b)

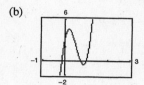

(c) The zeros are: $-\dfrac{1}{8},\ \dfrac{3}{4},\ 1$

33. $f(x) = x^4 - 3x^2 + 2$

(a) From the calculator we have
$x = \pm 1$ and $x \approx \pm 1.414.$

(b)

$$
\begin{array}{r|rrrrr}
1 & 1 & 0 & -3 & 0 & 2 \\
  &   & 1 & 1 & -2 & -2 \\
\hline
  & 1 & 1 & -2 & -2 & 0
\end{array}
$$

$$
\begin{array}{r|rrrr}
-1 & 1 & 1 & -2 & -2 \\
   &   & -1 & 0 & 2 \\
\hline
   & 1 & 0 & -2 & 0
\end{array}
$$

$f(x) = (x - 1)(x + 1)(x^2 - 2)$
$\quad = (x - 1)(x + 1)(x - \sqrt{2})(x + \sqrt{2})$

The exact roots are $x = \pm 1,\ \pm\sqrt{2}.$

35. $h(x) = x^5 - 7x^4 + 10x^3 + 14x^2 - 24x$

(a) $h(x) = x(x^4 - 7x^3 + 10x^2 + 14x - 24)$

From the calculator we have
$x = 0,\ 3,\ 4$ and $x \approx \pm 1.414.$

(b)
$$
\begin{array}{r|rrrrr}
3 & 1 & -7 & 10 & 14 & -24 \\
  &   & 3 & -12 & -6 & 24 \\
\hline
  & 1 & -4 & -2 & 8 & 0
\end{array}
$$

$$
\begin{array}{r|rrrr}
4 & 1 & -4 & -2 & 8 \\
  &   & 4 & 0 & -8 \\
\hline
  & 1 & 0 & -2 & 0
\end{array}
$$

$f(x) = x(x - 3)(x - 4)(x^2 - 2)$
$\quad = x(x - 3)(x - 4)(x - \sqrt{2})(x + \sqrt{2})$

The exact roots are $x = 0,\ 3,\ 4,\ \pm\sqrt{2}.$

37. $f(x) = (x - 1)(x - 5i)(x + 5i)$

$\quad = (x - 1)(x^2 + 25)$

$\quad = x^3 - x^2 + 25x - 25$

Note: $f(x) = a(x^3 - x^2 + 25x - 25)$.
where a is any nonzero real number, has
the zeros 1 and $\pm 5i$.

39. $f(x) = (x - 6)[x - (-5 + 2i)][x - (-5 - 2i)]$

$\quad = (x - 6)[(x + 5) - 2i][(x + 5) + 2i]$

$\quad = (x - 6)[(x + 5)^2 - (2i)^2]$

$\quad = (x - 6)(x^2 + 10x + 25 + 4)$

$\quad = (x - 6)(x^2 + 10x + 29)$

$\quad = x^3 + 4x^2 - 31x - 174$

Note: $f(x) = a(x^3 + 4x^2 - 31x - 174)$, where
a is any nonzero real number, has the zeros 6, and
$-5 \pm 2i$.

41. If $3 + \sqrt{2}i$ is a zero, so is its conjugate, $3 - \sqrt{2}i$.

$f(x) = (3x - 2)(x + 1)\left[x - \left(3 + \sqrt{2}i\right)\right]\left[x - \left(3 - \sqrt{2}i\right)\right]$

$\quad = (3x - 2)(x + 1)\left[(x - 3) - \sqrt{2}i\right]\left[(x - 3) + \sqrt{2}i\right]$

$\quad = (3x^2 + x - 2)\left[(x - 3)^2 - \left(\sqrt{2}i\right)^2\right]$

$\quad = (3x^2 + x - 2)(x^2 - 6x + 9 + 2)$

$\quad = (3x^2 + x - 2)(x^2 - 6x + 11)$

$\quad = 3x^4 - 17x^3 + 25x^2 + 23x - 22$

Note: $f(x) = a(3x^4 - 17x^3 + 25x^2 + 23x - 22)$,
where a is any nonzero real number, has the zeros $\frac{2}{3}$, -1,
and $3 \pm \sqrt{2}i$.

43. $f(x) = x^4 + 6x^2 - 27$

(a) $f(x) = (x^2 + 9)(x^2 - 3)$

(b) $f(x) = (x^2 + 9)(x + \sqrt{3})(x - \sqrt{3})$

(c) $f(x) = (x + 3i)(x - 3i)(x + \sqrt{3})(x - \sqrt{3})$

45.

$$\begin{array}{r} x^2 - 2x + 3 \\ x^2 - 2x - 2 \overline{\smash{)}x^4 - 4x^3 + 5x^2 - 2x - 6} \\ \underline{x^4 - 2x^3 - 2x^2} \\ -2x^3 + 7x^2 - 2x \\ \underline{-2x^3 + 4x^2 + 4x} \\ 3x^2 - 6x - 6 \\ \underline{3x^2 - 6x - 6} \\ 0 \end{array}$$

$f(x) = (x^2 - 2x - 2)(x^2 - 2x + 3)$

(a) $f(x) = (x^2 - 2x - 2)(x^2 - 2x + 3)$

(b) $f(x) = \left(x - 1 + \sqrt{3}\right)\left(x - 1 - \sqrt{3}\right)(x^2 - 2x + 3)$

(c) $f(x) = \left(x - 1 + \sqrt{3}\right)\left(x - 1 - \sqrt{3}\right)\left(x - 1 + \sqrt{2}\,i\right)\left(x - 1 - \sqrt{2}\,i\right)$

Note: Use the Quadratic Formula for (b) and (c).

47. $f(x) = 2x^3 + 3x^2 + 50x + 75$

Since $5i$ is a zero, so is $-5i$.

$$
\begin{array}{r|rrrr}
5i & 2 & 3 & 50 & 75 \\
& & 10i & -50 + 15i & -75 \\
\hline
& 2 & 3 + 10i & 15i & 0
\end{array}
$$

$$
\begin{array}{r|rrr}
-5i & 2 & 3 + 10i & 15i \\
& & -10i & -15i \\
\hline
& 2 & 3 & 0
\end{array}
$$

The zero of $2x + 3$ is $x = -\frac{3}{2}$.

The zeros of $f(x)$ are $x = -\frac{3}{2}$ and $x = \pm 5i$.

Alternate Solution

Since $x = \pm 5i$ are zeros of $f(x)$, $(x + 5i)(x - 5i) = x^2 + 25$ is a factor of $f(x)$.
By long division we have:

$$
\begin{array}{r}
2x + 3 \\
x^2 + 0x + 25 \overline{)\, 2x^3 + 3x^2 + 50x + 75} \\
\underline{2x^3 + 0x^2 + 50x} \\
3x^2 + 0x + 75 \\
\underline{3x^2 + 0x + 75} \\
0
\end{array}
$$

Thus, $f(x) = (x^2 + 25)(2x + 3)$ and the zeros of f are $x = \pm 5i$ and $x = -\frac{3}{2}$.

49. $f(x) = 2x^4 - x^3 + 7x^2 - 4x - 4$

Since $2i$ is a zero, so is $-2i$.

$$
\begin{array}{r|rrrrr}
2i & 2 & -1 & 7 & -4 & -4 \\
& & 4i & -8 - 2i & 4 - 2i & 4 \\
\hline
& 2 & -1 + 4i & -1 - 2i & -2i & 0
\end{array}
$$

$$
\begin{array}{r|rrrr}
-2i & 2 & -1 + 4i & -1 - 2i & -2i \\
& & -4i & 2i & 2i \\
\hline
& 2 & -1 & -1 & 0
\end{array}
$$

The zeros of $2x^2 - x - 1 = (2x + 1)(x - 1)$ are $x = -\frac{1}{2}$ and $x = 1$.

The zeros of $f(x)$ are $x = \pm 2i$, $x = -\frac{1}{2}$, and $x = 1$.

Alternate Solution

Since $x = \pm 2i$ are zeros of $f(x)$, $(x + 2i)(x - 2i) = x^2 + 4$ is a factor of $f(x)$.
By long division we have:

$$
\begin{array}{r}
2x^2 - x - 1 \\
x^2 + 0x + 4 \overline{)\, 2x^4 - x^3 + 7x^2 - 4x - 4} \\
\underline{2x^4 + 0x^3 + 8x^2} \\
-x^3 - x^2 - 4x \\
\underline{-x^3 + 0x^2 - 4x} \\
-x^2 + 0x - 4 \\
\underline{-x^2 + 0x - 4} \\
0
\end{array}
$$

Thus, $f(x) = (x^2 + 4)(2x^2 - x - 1)$

$$= (x + 2i)(x - 2i)(2x + 1)(x - 1)$$

and the zeros of $f(x)$ are $x = \pm 2i$, $x = -\frac{1}{2}$, and $x = 1$.

51. $g(x) = 4x^3 + 23x^2 + 34x - 10$

Since $-3 + i$ is a zero, so is $-3 - i$.

$$
\begin{array}{r|rrrr}
-3 + i & 4 & 23 & 34 & -10 \\
 & & -12 + 4i & -37 - i & 10 \\
\hline
 & 4 & 11 + 4i & -3 - i & 0
\end{array}
$$

$$
\begin{array}{r|rrr}
-3 - i & 4 & 11 + 4i & -3 - i \\
 & & -12 - 4i & 3 + i \\
\hline
 & 4 & -1 & 0
\end{array}
$$

The zero of $4x - 1$ is $x = \frac{1}{4}$. The zeros of $g(x)$ are

$x = -3 \pm i$ and $x = \frac{1}{4}$.

Alternate Solution

Since $-3 \pm i$ are zeros of $g(x)$,
$[x - (-3 + i)][x - (-3 - i)] = [(x + 3) - i][(x + 3) + i]$
$\qquad\qquad\qquad\qquad = (x + 3)^2 - i^2$
$\qquad\qquad\qquad\qquad = x^2 + 6x + 10$

is a factor of $g(x)$. By long division we have:

$$
\begin{array}{r}
4x - 1 \\
x^2 + 6x + 10 \overline{)\, 4x^3 + 23x^2 + 34x - 10} \\
\underline{4x^3 + 24x^2 + 40x} \\
-x^2 - 6x - 10 \\
\underline{-x^2 - 6x - 10} \\
0
\end{array}
$$

Thus, $g(x) = (x^2 + 6x + 10)(4x - 1)$ and the zeros of $g(x)$ are $x = -3 \pm i$ and $x = \frac{1}{4}$.

55. $f(x) = x^2 + 25$

$\qquad = (x + 5i)(x - 5i)$

The zeros of $f(x)$ are $x = \pm 5i$.

57. $h(x) = x^2 - 4x + 1$

h has no rational zeros.

By the Quadratic Formula, the zeros are $x = \dfrac{4 \pm \sqrt{16 - 4}}{2} = 2 \pm \sqrt{3}$.

$h(x) = \left[x - \left(2 + \sqrt{3}\right)\right]\left[x - \left(2 - \sqrt{3}\right)\right] = \left(x - 2 - \sqrt{3}\right)\left(x - 2 + \sqrt{3}\right)$

59. $f(x) = x^4 - 81$

$\qquad = (x^2 - 9)(x^2 + 9)$

$\qquad = (x + 3)(x - 3)(x + 3i)(x - 3i)$

The zeros of $f(x)$ are $x = \pm 3$ and $x = \pm 3i$.

53. Since $-3 + \sqrt{2}\, i$ is a zero, so is $-3 - \sqrt{2}\, i$, and

$\left[x - \left(-3 + \sqrt{2}\, i\right)\right]\left[x - \left(-3 - \sqrt{2}\, i\right)\right]$

$\qquad = \left[(x + 3) - \sqrt{2}\, i\right]\left[(x + 3) + \sqrt{2}\, i\right]$

$\qquad = (x + 3)^2 - \left(\sqrt{2}\, i\right)^2$

$\qquad = x^2 + 6x + 11$

is a factor of $f(x)$. By long division, we have:

$$
\begin{array}{r}
x^2 - 3x + 2 \\
x^2 + 6x + 11 \overline{)\, x^4 + 3x^3 - 5x^2 - 21x + 22} \\
\underline{x^4 + 6x^3 + 11x^2} \\
-3x^3 - 16x^2 - 21x \\
\underline{-3x^3 - 18x^2 - 33x} \\
2x^2 + 12x + 22 \\
\underline{2x^2 + 12x + 22} \\
0
\end{array}
$$

Thus, $f(x) = (x^2 + 6x + 11)(x^2 - 3x + 2)$

$\qquad = (x^2 + 6x + 11)(x - 1)(x - 2)$

and the zeros of f are $x = -3 \pm \sqrt{2}\, i, x = 1$, and $x = 2$.

61. $f(z) = z^2 - 2z + 2$

f has no rational zeros.

By the Quadratic Formula,

the zeros are $z = \dfrac{2 \pm \sqrt{4 - 8}}{2} = 1 \pm i$.

$f(z) = [z - (1 + i)][z - (1 - i)] = (z - 1 - i)(z - 1 + i)$

63. $g(x) = x^3 - 6x^2 + 13x - 10$

Possible rational zeros: $\pm 1, \pm 2, \pm 5, \pm 10$

$$\begin{array}{r|rrrr}
2 & 1 & -6 & 13 & -10 \\
  &   & 2 & -8 & 10 \\
\hline
  & 1 & -4 & 5 & 0
\end{array}$$

By the Quadratic Formula, the zeros of

$x^2 - 4x + 5$ are $x = \dfrac{4 \pm \sqrt{16 - 20}}{2} = 2 \pm i.$

The zeros of $g(x)$ are $x = 2$ and $x = 2 \pm i.$

$g(x) = (x - 2)[x - (2 + i)][x - (2 - i)]$

$\quad = (x - 2)(x - 2 - i)(x - 2 + i)$

67. $f(x) = 5x^3 - 9x^2 + 28x + 6$

Possible rational zeros: $\pm 1, \pm 2, \pm 3, \pm 6, \pm \frac{1}{5}, \pm \frac{2}{5}, \pm \frac{3}{5}, \pm \frac{6}{5}$

$$\begin{array}{r|rrrr}
-\frac{1}{5} & 5 & -9 & 28 & 6 \\
  &   & -1 & 2 & -6 \\
\hline
  & 5 & -10 & 30 & 0
\end{array}$$

By the Quadratic Formula, the zeros of
$5x^2 - 10x + 30 = 5(x^2 - 2x + 6)$ are

$x = \dfrac{2 \pm \sqrt{4 - 24}}{2} = 1 \pm \sqrt{5}\,i.$

The zeros of $f(x)$ are $x = -\frac{1}{5}$ and $x = 1 \pm \sqrt{5}\,i.$

$f(x) = [x - (-\frac{1}{5})](5)[x - (1 + \sqrt{5}\,i)][x - (1 - \sqrt{5}\,i)]$

$\quad = (5x + 1)(x - 1 - \sqrt{5}\,i)(x - 1 + \sqrt{5}\,i)$

71. $f(x) = x^4 + 10x^2 + 9$

$\quad = (x^2 + 1)(x^2 + 9)$

$\quad = (x + i)(x - i)(x + 3i)(x - 3i)$

The zeros of $f(x)$ are $x = \pm i$ and $x = \pm 3i.$

65. $h(x) = x^3 - x + 6$

Possible rational zeros: $\pm 1, \pm 2, \pm 3, \pm 6$

$$\begin{array}{r|rrrr}
-2 & 1 & 0 & -1 & 6 \\
   &   & -2 & 4 & -6 \\
\hline
   & 1 & -2 & 3 & 0
\end{array}$$

By the Quadratic Formula, the zeros of $x^2 - 2x + 3$ are

$x = \dfrac{2 \pm \sqrt{4 - 12}}{2} = 1 \pm \sqrt{2}\,i.$

The zeros of $h(x)$ are $x = -2$ and $x = 1 \pm \sqrt{2}\,i.$

$h(x) = [x - (-2)][x - (1 + \sqrt{2}\,i)][x - (1 - \sqrt{2}\,i)]$

$\quad = (x + 2)(x - 1 - \sqrt{2}\,i)(x - 1 + \sqrt{2}\,i)$

69. $g(x) = x^4 - 4x^3 + 8x^2 - 16x + 16$

Possible rational zeros: $\pm 1, \pm 2, \pm 4, \pm 8, \pm 16$

$$\begin{array}{r|rrrrr}
2 & 1 & -4 & 8 & -16 & 16 \\
  &   & 2 & -4 & 8 & -16 \\
\hline
  & 1 & -2 & 4 & -8 & 0
\end{array}$$

$$\begin{array}{r|rrrr}
2 & 1 & -2 & 4 & -8 \\
  &   & 2 & 0 & 8 \\
\hline
  & 1 & 0 & 4 & 0
\end{array}$$

$g(x) = (x - 2)(x - 2)(x^2 + 4) = (x - 2)^2(x + 2i)(x - 2i)$

The zeros of $g(x)$ are 2 and $\pm 2i.$

73. $f(x) = x^3 + 24x^2 + 214x + 740$

Possible rational zeros: $\pm 1, \pm 2, \pm 4, \pm 5, \pm 10, \pm 20, \pm 37,$
$\pm 74, \pm 148, \pm 185, \pm 370, \pm 740$

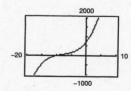

Based on the graph, try $x = -10.$

$$\begin{array}{r|rrrr}
-10 & 1 & 24 & 214 & 740 \\
    &   & -10 & -140 & -740 \\
\hline
    & 1 & 14 & 74 & 0
\end{array}$$

By the Quadratic Formula, the zeros of $x^2 + 14x + 74$ are

$x = \dfrac{-14 \pm \sqrt{196 - 296}}{2} = -7 \pm 5i.$

The zeros of $f(x)$ are $x = -10$ and $x = -7 \pm 5i.$

75. $f(x) = 16x^3 - 20x^2 - 4x + 15$

Possible rational zeros: $\pm 1, \pm 3, \pm 5, \pm 15, \pm\frac{1}{2}, \pm\frac{3}{2}, \pm\frac{5}{2}, \pm\frac{15}{2},$

$$\pm\frac{1}{4}, \pm\frac{3}{4}, \pm\frac{5}{4}, \pm\frac{15}{4}, \pm\frac{1}{8}, \pm\frac{3}{8}, \pm\frac{5}{8}, \pm\frac{15}{8},$$

$$\pm\frac{1}{16}, \pm\frac{3}{16}, \pm\frac{5}{16}, \pm\frac{15}{16}$$

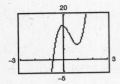

Based on the graph, try $x = -\frac{3}{4}$.

$$
\begin{array}{r|rrrr}
-\frac{3}{4} & 16 & -20 & -4 & 15 \\
 &   & -12 & 24 & -15 \\
\hline
 & 16 & -32 & 20 & 0
\end{array}
$$

By the Quadratic Formula, the zeros of
$16x^2 - 32x + 20 = 4(4x^2 - 8x + 5)$ are

$$x = \frac{8 \pm \sqrt{64 - 80}}{8} = 1 \pm \frac{1}{2}i.$$

The zeros of $f(x)$ are $x = -\frac{3}{4}$ and $x = 1 \pm \frac{1}{2}i$.

77. $f(x) = 2x^4 + 5x^3 + 4x^2 + 5x + 2$

Possible rational zeros: $\pm 1, \pm 2, \pm\frac{1}{2}$

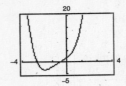

Based on the graph, try $x = -2$ and $x = -\frac{1}{2}$.

$$
\begin{array}{r|rrrrr}
-2 & 2 & 5 & 4 & 5 & 2 \\
 &   & -4 & -2 & -4 & -2 \\
\hline
 & 2 & 1 & 2 & 1 & 0
\end{array}
$$

$$
\begin{array}{r|rrrr}
-\frac{1}{2} & 2 & 1 & 2 & 1 \\
 &   & -1 & 0 & -1 \\
\hline
 & 2 & 0 & 2 & 0
\end{array}
$$

The zeros of $2x^2 + 2 = 2(x^2 + 1)$ are $x = \pm i$.

The zeros of $f(x)$ are $x = -2$, $x = -\frac{1}{2}$, and $x = \pm i$.

79. $f(x) = 4x^3 - 3x - 1$

Possible rational zeros: $\pm 1, \pm\frac{1}{2}, \pm\frac{1}{4}$

$$
\begin{array}{r|rrrr}
1 & 4 & 0 & -3 & -1 \\
 &   & 4 & 4 & 1 \\
\hline
 & 4 & 4 & 1 & 0
\end{array}
$$

$4x^3 - 3x - 1 = (x - 1)(4x^2 + 4x + 1) = (x - 1)(2x + 1)^2$

Thus, the zeros are 1 and $-\frac{1}{2}$.

81. $f(y) = 4y^3 + 3y^2 + 8y + 6$

Possible rational zeros: $\pm 1, \pm 2, \pm 3, \pm 6, \pm\frac{1}{2}, \pm\frac{3}{2}, \pm\frac{1}{4}, \pm\frac{3}{4}$

$$
\begin{array}{r|rrrr}
-\frac{3}{4} & 4 & 3 & 8 & 6 \\
 &   & -3 & 0 & -6 \\
\hline
 & 4 & 0 & 8 & 0
\end{array}
$$

$4y^3 + 3y^2 + 8y + 6 = \left(y + \frac{3}{4}\right)(4y^2 + 8) = \left(y + \frac{3}{4}\right)4(y^2 + 2) = (4y + 3)(y^2 + 2)$

Thus, the only real zero is $-\frac{3}{4}$.

83. $P(x) = x^4 - \frac{25}{4}x^2 + 9$

$= \frac{1}{4}(4x^4 - 25x^2 + 36)$

$= \frac{1}{4}(4x^2 - 9)(x^2 - 4)$

$= \frac{1}{4}(2x + 3)(2x - 3)(x + 2)(x - 2)$

The zeros are $\pm\frac{3}{2}$ and ± 2.

85. $f(x) = x^3 - \frac{1}{4}x^2 - x + \frac{1}{4}$

$= \frac{1}{4}(4x^3 - x^2 - 4x + 1)$

$= \frac{1}{4}[x^2(4x - 1) - 1(4x - 1)]$

$= \frac{1}{4}(4x - 1)(x^2 - 1)$

$= \frac{1}{4}(4x - 1)(x + 1)(x - 1)$

The zeros are $\frac{1}{4}$ and ± 1.

87. $f(x) = x^3 - 1 = (x - 1)(x^2 + x + 1)$

Rational zeros: 1 $(x = 1)$

Irrational zeros: 0

Matches (d).

89. $f(x) = x^3 - x = x(x + 1)(x - 1)$

Rational zeros: 3 $(x = 0, \pm 1)$

Irrational zeros: 0

Matches (b).

91. Zeros: $-2, \frac{1}{2}, 3$

$f(x) = -(x + 2)(2x - 1)(x - 3)$

$\quad = -2x^3 + 3x^2 + 11x - 6$

Any nonzero scalar multiple of f would have the same three zeros.

Let $g(x) = af(x)$, $a > 0$.

There are infinitely many possible functions for f.

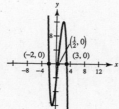

93.

Interval: $(-\infty, -2)$, $(-2, 1)$, $(1, 4)$, $(4, \infty)$

Value of $f(x)$:   Positive   Negative Negative Positive

(a) Zeros of $f(x)$: $x = -2, x = 1, x = 4$.

(b) The graph touches the x-axis at $x = 1$.

(c) The least possible degree of the function is 4 because there are at least four real zeros (1 is repeated) and a function can have at most the number of real zeros equal to the degree of the function. The degree cannot be odd by the behavior at $\pm\infty$.

(d) The leading coefficient of f is positive. From the information in the table, you can conclude that the graph will eventually rise to the left and to the right.

(e) $f(x) = (x + 2)(x - 1)^2(x - 4)$

$\quad = x^4 - 4x^3 - 3x^2 + 14x - 8$

(Any nonzero multiple of $f(x)$ is also a solution.)

(f)

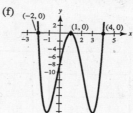

95. (a) Combined length and width:

$4x + y = 120 \Longrightarrow y = 120 - 4x$

Volume $= l \cdot w \cdot h = x^2 y$

$\quad = x^2(120 - 4x)$

$\quad = 4x^2(30 - x)$

(b)

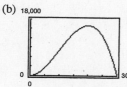

Dimensions with maximum volume:

20 in. $\times$ 20 in. $\times$ 40 in.

(c) $13,500 = 4x^2(30 - x)$

$4x^3 - 120x^2 + 13,500 = 0$

$x^3 - 30x^2 + 3375 = 0$

$$
\begin{array}{r|rrrr}
15 & 1 & -30 & 0 & 3375 \\
   &   & 15 & -225 & -3375 \\
\hline
   & 1 & -15 & -225 & 0
\end{array}
$$

$(x - 15)(x^2 - 15x - 225) = 0$

Using the Quadratic Formula,

$x = 15, \dfrac{15 \pm 15\sqrt{5}}{2}$.

The value of $\dfrac{15 - 15\sqrt{5}}{2}$ is not possible because it is negative.

97. $C = 100\left(\dfrac{200}{x^2} + \dfrac{x}{x + 30}\right), 1 \le x$

C is minimum when $3x^3 - 40x^2 - 2400x - 36000 = 0$. The only real zero is $x \approx 40$.

99. $P = -45x^3 + 2500x^2 - 275,000$

$800,000 = -45x^3 + 2500x^2 - 275,000$

$0 = 45x^3 - 2500x^2 + 1,075,000$

$0 = 9x^3 - 500x^2 + 215,000$

The zeros of this equation are $x \approx -18.0$, $x \approx 31.5$, and $x \approx 42.0$. Because $0 \le x \le 50$, disregard $x \approx -18.0$. The smaller remaining solution is $x \approx 31.5$, or \$315,000.

101. False. The most nonreal complex zeros it can have is two and the Linear Factorization Theorem guarantees that there are 3 linear factors, so one zero must be real.

103. (a) $f(x) = \left(x - \sqrt{b}\,i\right)\left(x + \sqrt{b}\,i\right) = x^2 + b$

(b) $f(x) = [x - (a + bi)][x - (a - bi)]$

$= [(x - a) - bi][(x - a) + bi]$

$= (x - a)^2 - (bi)^2$

$= x^2 - 2ax + a^2 + b^2$

Section 2.6   Rational Functions

1. $f(x) = \dfrac{1}{x - 1}$

(a)

x	$f(x)$	x	$f(x)$	x	$f(x)$	x	$f(x)$
0.5	-2	1.5	2	5	0.25	-5	$-0.1\overline{6}$
0.9	-10	1.1	10	10	$0.\overline{1}$	-10	$-0.\overline{09}$
0.99	-100	1.01	100	100	$0.\overline{01}$	-100	$-0.\overline{0099}$
0.999	-1000	1.001	1000	1000	$0.\overline{001}$	-1000	$-0.\overline{000999}$

(b) The zero of the denominator is $x = 1$, so $x = 1$ is a vertical asymptote. The degree of the numerator is less than the degree of the denominator so the x-axis, or $y = 0$, is a horizontal asymptote.

(c) The domain is all real numbers except $x = 1$.

3. $f(x) = \dfrac{4x}{|x - 1|}$

(a)

x	$f(x)$	x	$f(x)$	x	$f(x)$	x	$f(x)$
0.5	4	1.5	12	5	5	-5	$-3.\overline{3}$
0.9	36	1.1	44	10	$4.\overline{4}$	-10	$-3.\overline{63}$
0.99	396	1.01	404	100	$4.\overline{04}$	-100	$-3.\overline{9603}$
0.999	3996	1.001	4004	1000	$4.\overline{004}$	-1000	$-3.\overline{996003}$

(b) The zero of the denominator is $x = 1$, so $x = 1$ is a vertical asymptote. Since $f(x) \to 4$ as $x \to \infty$ and $f(x) \to -4$ as $x \to -\infty$, both $y = 4$ and $y = -4$ are horizontal asymptotes.

(c) The domain is all real numbers except $x = 1$.

5. $f(x) = \dfrac{3x^2}{x^2 - 1}$

(a)

x	$f(x)$
0.5	-1
0.9	-12.79
0.99	-147.8
0.999	-1498

x	$f(x)$
1.5	5.4
1.1	17.29
1.01	152.3
1.001	1502

x	$f(x)$
5	3.125
10	$3.\overline{03}$
100	$3.\overline{0003}$
1000	3

x	$f(x)$
-5	3.125
-10	$3.\overline{03}$
-100	$3.\overline{0003}$
-1000	$3.\overline{000003}$

(b) The zeros of the denominator are $x = \pm 1$ so both $x = 1$ and $x = -1$ are vertical asymptotes. Since the degree of the numerator equals the degree of the denominator, $y = \frac{3}{1} = 3$ is a horizontal asymptote.

(c) The domain is all real numbers except $x = \pm 1$.

7. $f(x) = \dfrac{1}{x^2}$

Domain: all real numbers except $x = 0$

Vertical asymptote: $x = 0$

Horizontal asymptote: $y = 0$

[Degree of $N(x) <$ degree of $D(x)$]

9. $f(x) = \dfrac{2 + x}{2 - x} = \dfrac{x + 2}{-x + 2}$

Domain: all real numbers except $x = 2$

Vertical asymptote: $x = 2$

Horizontal asymptote: $y = -1$

[Degree of $N(x) =$ degree of $D(x)$]

11. $f(x) = \dfrac{x^3}{x^2 - 1}$

Domain: all real numbers except $x = \pm 1$

Vertical asymptotes: $x = \pm 1$

Horizontal asymptote: None

[Degree of $N(x) >$ degree of $D(x)$]

13. $f(x) = \dfrac{3x^2 + 1}{x^2 + x + 9}$

Domain: All real numbers. The denominator has no real zeros. [Try the Quadratic Formula on the denominator.]

Vertical asymptote: None

Horizontal asymptote: $y = 3$

[Degree of $N(x) =$ degree of $D(x)$]

15. $f(x) = \dfrac{2}{x + 3}$

Vertical asymptote: $x = -3$

Matches graph (d).

17. $f(x) = \dfrac{3x + 1}{x}$

Horizontal asymptote: $y = 3$

Matches graph (f).

19. $f(x) = \dfrac{x - 1}{x - 4}$

Vertical asymptote: $x = 4$

Horizontal asymptote: $y = 1$

Matches graph (e).

21. $g(x) = 1 - \dfrac{1}{x - 1}$

$1 - \dfrac{1}{x - 1} = 0$

$1 = \dfrac{1}{x - 1}$

$x - 1 = 1$

$x = 2$ is a zero of $g(x)$.

23. $f(x) = 1 - \dfrac{3}{x - 3}$

$1 - \dfrac{3}{x - 3} = 0$

$1 = \dfrac{3}{x - 3}$

$x - 3 = 3$

$x = 6$ is a zero of $f(x)$.

25. $f(x) = \dfrac{1}{x+2}$

(a) y-intercept: $\left(0, \frac{1}{2}\right)$

(b) Vertical asymptote: $x = -2$
    Horizontal asymptote: $y = 0$

(c) No origin or axis symmetry

(d)

x	-4	-3	-1	0	1
y	$-\frac{1}{2}$	-1	1	$\frac{1}{2}$	$\frac{1}{3}$

(e)

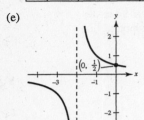

27. $h(x) = \dfrac{-1}{x+2}$

(a) y-intercept: $\left(0, -\frac{1}{2}\right)$

(b) Vertical asymptote: $x = -2$
    Horizontal asymptote: $y = 0$

(c) No origin or axis symmetry

(d)

x	-4	-3	-1	0
y	$\frac{1}{2}$	1	-1	$-\frac{1}{2}$

(e)

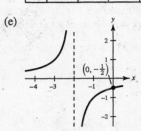

Note: This is the graph of $f(x) = \dfrac{1}{x+2}$
(Exercise 25) reflected about the x-axis.

29. $C(x) = \dfrac{5 + 2x}{1 + x} = \dfrac{2x + 5}{x + 1}$

(a) x-intercept: $\left(-\frac{5}{2}, 0\right)$

   y-intercept: $(0, 5)$

(b) Vertical asymptote: $x = -1$

   Horizontal asymptote: $y = 2$

(c) No origin or axis symmetry

(d)

x	-4	-3	-2	0	1	2
$C(x)$	1	$\frac{1}{2}$	-1	5	$\frac{7}{2}$	3

(e)

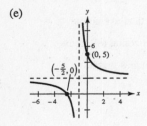

31. $g(x) = \dfrac{1}{x+2} + 2 = \dfrac{2x+5}{x+2}$

(a) x-intercept: $\left(-\frac{5}{2}, 0\right)$

   y-intercept: $\left(0, \frac{5}{2}\right)$

(b) Vertical asymptote: $x = -2$

   Horizontal asymptote: $y = 2$

(c) No origin or axis symmetry

(d)

x	-4	-3	-1	0	1
y	$\frac{3}{2}$	1	3	$\frac{5}{2}$	$\frac{7}{3}$

(e)

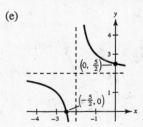

Note: This is the graph of $f(x) = \dfrac{1}{x+2}$
(Exercise 25) shifted upward two units.

33. $f(x) = \dfrac{x^2}{x^2 + 9}$

  (a) Intercept: $(0, 0)$

  (b) Horizontal asymptote: $y = 1$

  (c) y-axis symmetry

  (d)

x	± 1	± 2	± 3
y	$\frac{1}{10}$	$\frac{4}{13}$	$\frac{1}{2}$

  (e)

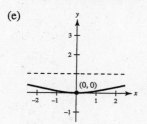

35. $h(x) = \dfrac{x^2}{x^2 - 9}$

  (a) Intercept: $(0, 0)$

  (b) Vertical asymptotes: $x = \pm 3$

     Horizontal asymptote: $y = 1$

  (c) y-axis symmetry

  (d)

x	± 5	± 4	± 2	± 1	0
y	$\frac{25}{16}$	$\frac{16}{7}$	$-\frac{4}{5}$	$-\frac{1}{8}$	0

  (e)

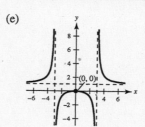

37. $g(s) = \dfrac{s}{s^2 + 1}$

  (a) Intercept: $(0, 0)$

  (b) Horizontal asymptote: $y = 0$

  (c) Origin symmetry

  (d)

s	-2	-1	0	1	2
$g(s)$	$-\frac{2}{5}$	$-\frac{1}{2}$	0	$\frac{1}{2}$	$\frac{2}{5}$

  (e)

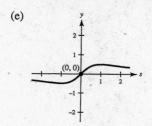

39. $g(x) = \dfrac{4(x + 1)}{x(x - 4)}$

  (a) Intercept: $(-1, 0)$

  (b) Vertical asymptotes: $x = 0$ and $x = 4$

     Horizontal asymptote: $y = 0$

  (c) No origin or axis symmetry

  (d)

x	-2	-1	1	2	3	5	6
y	$-\frac{1}{3}$	0	$-\frac{8}{3}$	-3	$-\frac{16}{3}$	$\frac{24}{5}$	$\frac{7}{3}$

  (e)

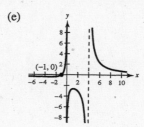

41. $f(x) = \dfrac{3x}{x^2 - x - 2} = \dfrac{3x}{(x + 1)(x - 2)}$

  (a) Intercept: $(0, 0)$

  (b) Vertical asymptotes: $x = -1$ and $x = 2$

     Horizontal asymptote: $y = 0$

  (c) No origin or axis symmetry

  (d)

x	-3	0	1	3	4
y	$-\frac{9}{10}$	0	$-\frac{3}{2}$	$\frac{9}{4}$	$\frac{6}{5}$

  (e)

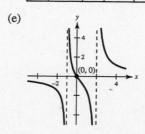

43. $f(x) = \dfrac{6x}{x^2 - 5x - 14} = \dfrac{6x}{(x + 2)(x - 7)}$

(a) Intercept: $(0, 0)$

(b) Vertical asymptotes: $x = -2$, and $x = 7$

   Horizontal asymptotes: $y = 0$

(c) No origin or axis symmetry

(d)

x	-6	-4	0	2	4	6	8	10
$f(x)$	$-\frac{9}{13}$	$-\frac{12}{11}$	0	$-\frac{3}{5}$	$-\frac{4}{3}$	$-\frac{9}{2}$	$\frac{24}{5}$	$\frac{5}{3}$

(e)

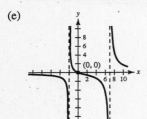

45. $h(t) = \dfrac{4}{t^2 + 1}$

Domain: all real numbers

Horizontal asymptote: $y = 0$

y-axis symmetry

t	± 2	± 1	0
$h(t)$	$\frac{4}{5}$	2	4

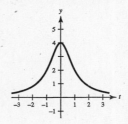

47. $f(t) = \dfrac{2t^2}{t^2 - 4}$

Domain: all real numbers except ± 2,

Vertical asymptotes: $t = \pm 2$

Horizontal asymptote: $y = 2$

y-axis symmetry

t	± 4	± 3	± 1	0
$f(t)$	$\frac{8}{3}$	$\frac{18}{5}$	$-\frac{2}{3}$	0

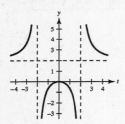

49. $f(x) = \dfrac{20x}{x^2 + 1} - \dfrac{1}{x} = \dfrac{19x^2 - 1}{x(x^2 + 1)}$

Domain: all real numbers except 0,

Vertical asymptote: $x = 0$

Horizontal asymptote: $y = 0$

Origin symmetry

x	-2	-1	1	2
y	$-\frac{15}{2}$	-9	9	$\frac{15}{2}$

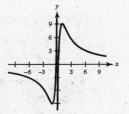

51. $f(x) = \dfrac{2x^2 + 1}{x} = 2x + \dfrac{1}{x}$

(a) No intercepts

(b) Vertical asymptote: $x = 0$
   Slant asymptote: $y = 2x$

(c) Origin symmetry

(d)

x	-4	-2	2	4	6
$f(x)$	$-\frac{33}{4}$	$-\frac{9}{2}$	$\frac{9}{2}$	$\frac{33}{4}$	$\frac{73}{6}$

(e)

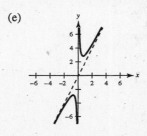

53. $g(x) = \dfrac{x^2 + 1}{x} = x + \dfrac{1}{x}$

(a) No intercepts

(b) Vertical asymptote: $x = 0$
Slant asymptote: $y = x$

(c) Origin symmetry

(d)

x	-4	-2	2	4	6
$g(x)$	$-\dfrac{17}{4}$	$-\dfrac{5}{2}$	$\dfrac{5}{2}$	$\dfrac{17}{4}$	$\dfrac{37}{6}$

(e)

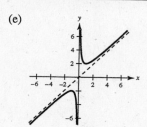

55. $f(x) = \dfrac{x^3}{x^2 - 1} = x + \dfrac{x}{x^2 - 1}$

(a) Intercept: $(0, 0)$

(b) Vertical asymptotes: $x = \pm 1$
Slant asymptote: $y = x$

(c) Origin symmetry

(d)

x	-4	-2	0	2	4
$f(x)$	$-\dfrac{64}{15}$	$-\dfrac{8}{3}$	0	$\dfrac{8}{3}$	$\dfrac{64}{15}$

(e)

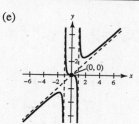

57. $f(x) = \dfrac{x^2 - x + 1}{x - 1} = x + \dfrac{1}{x - 1}$

(a) y-intercept: $(0, -1)$

(b) Vertical asymptote: $x = 1$
Slant asymptote: $y = x$

(c) No axis or origin symmetry

(d)

x	-4	-2	0	2	4
$f(x)$	$-\dfrac{21}{5}$	$-\dfrac{7}{3}$	-1	3	$\dfrac{13}{3}$

(e)

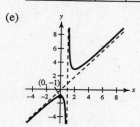

59. Vertical asymptotes: $x = -2, x = 1$

The denominator is zero when $x = -2$ and $x = 1$, so $(x + 2)$ and $(x - 1)$ are factors of the denominator.

One possibility: $f(x) = \dfrac{1}{(x + 2)(x - 1)} = \dfrac{1}{x^2 + x - 2}$
(Answers may vary).

61. Vertical asymptotes: None

Horizontal asymptote: $y = 2$

Since there are no vertical asymptotes, the denominator is never zero (if it is reduced), and since there is a horizontal asymptote, the degree of the numerator equals the degree of the denominator. Also, $\dfrac{a_n}{b_n} = 2$.

One possibility: $f(x) = \dfrac{2x^2}{x^2 + 1}$ (Answers may vary).

63. Domain: All real numbers

One possibility: $f(x) = \dfrac{1}{x^2 + 2}$

Domain: All real numbers except $x = 2$.

One possibility: $f(x) = \dfrac{1}{x - 2}$

(Answers are not unique).

65. $f(x) = \dfrac{x^2 + 5x + 8}{x + 3} = x + 2 + \dfrac{2}{x + 3}$

Domain: all real numbers except -3

y-intercept: $\left(0, \frac{8}{3}\right)$

Vertical asymptote: $x = -3$

Slant asymptote: $y = x + 2$

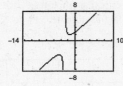

67. $g(x) = \dfrac{1 + 3x^2 - x^3}{x^2} = \dfrac{1}{x^2} + 3 - x = -x + 3 + \dfrac{1}{x^2}$

Domain: all real numbers except 0

Vertical asymptote: $x = 0$

Slant asymptote: $y = -x + 3$

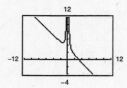

69. (a) x-intercept: $(-1, 0)$

(b) $0 = \dfrac{x + 1}{x - 3}$

$0 = x + 1$

$-1 = x$

71. (a) x-intercepts: $(\pm 1, 0)$

(b) $0 = \dfrac{1}{x} - x$

$x = \dfrac{1}{x}$

$x^2 = 1$

$x = \pm 1$

73. $C = \dfrac{255p}{100 - p}, \ 0 \le p < 100$

(a) $C(10) = \dfrac{255(10)}{100 - 10} \approx 28.33$ million dollars

(b) $C(40) = \dfrac{255(40)}{100 - 40} = 170$ million dollars

(c) $C(75) = \dfrac{255(75)}{100 - 75} = 765$ million dollars

(d) $C \to \infty$ as $x \to 100$. No, it would not be possible to remove 100% of the pollutants.

75. $N = \dfrac{20(5 + 3t)}{1 + 0.04t}, \ 0 \le t$

(a) $N(5) \approx 333$ deer

$N(10) = 500$ deer

$N(25) = 800$ deer

(b) The herd is limited by the horizontal asymptote:

$N = \dfrac{60}{0.04} = 1500$ deer

77. Let w be the width of the page and h be the height. The total of the horizontal margins is 3 inches and the total of the vertical margins is 2 inches. Thus,

64 square inches $= (w - 3)(h - 2)$.

Solving $64 = (w - 3)(h - 2)$ for h gives

$h = 2 + \dfrac{64}{w - 3}$.

The total area of the paper is

$A = w \cdot h = w\left(2 + \dfrac{64}{w - 3}\right) = \dfrac{2w^2 + 58w}{w - 3}$.

Plotting A versus w using a graphing utility, the graph has a minimum, for $w > 3$, at $w \approx 12.80$ inches.

$h \approx 2 + \dfrac{64}{12.80 - 3} \approx 8.53$ inches

Width: 12.80 inches

Height: 8.53 inches

79. $C = \dfrac{3t^2 + t}{t^3 + 50}, \ t > 0$

(a) Since the degree of $N(t) <$ degree of $D(t)$, the horizontal asymptote is $C = 0$. The chemical will eventually dissipate.

(b)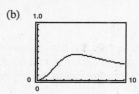

C is maximum when $t \approx 4.6$ hours

81. $f(x) = \dfrac{x^2 - 1}{x + 1}$, $g(x) = x - 1$

    (a) Domain of f: all real numbers except -1

        Domain of g: all real numbers

    (b) Because $(x + 1)$ is a factor of both the numerator and the denominator of f, $x = -1$ is not a vertical asymptote. f has no vertical asymptotes.

    (c)

x	-3	-2	-1.5	-1	-0.5	0	1
$f(x)$	-4	-3	-2.5	Undef.	-1.5	-1	0
$g(x)$	-4	-3	-2.5	-2	-1.5	-1	0

    (d)

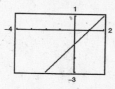

    (e) Because there are only a finite number of pixels, the utility may not attempt to evaluate the function where it does not exist.

85. False. Polynomials do not have vertical asymptotes.

87. True.

    If $f(x) = \dfrac{N(x)}{D(x)}$, where $N(x)$ and $D(x)$ are polynomials, and $D(x) \neq 0$ for any real number, then $f(x)$ has no vertical asymptotes.

91. $P = \dfrac{1419.96 + 11.802t^2}{1.0 + 0.0027t^2}$, where $t = 7$ corresponds to 1997.

    (a)

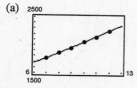

    (b) $P(18) \approx 2797$ thousand

    (c) Answers will vary. It will be somewhat useful since realistically, business partnerships will not continue to infinity. Likewise, the model levels off. However, the model is the best approximation only for the years in the data range, and as you choose a year farther from the data, the approximation is less accurate.

83. $f(x) = \dfrac{x - 2}{x^2 - 2x}$, $g(x) = \dfrac{1}{x}$

    (a) Domain of f: all real numbers except 0 and 2

        Domain of g: all real numbers except 0

    (b) Because $(x - 2)$ is a factor of both the numerator and the denominator of f, $x = 2$ is not a vertical asymptote. The only vertical asymptote of f is $x = 0$.

    (c)

x	-0.5	0	0.5	1	1.5	2	3
$f(x)$	-2	Undef.	2	1	$\frac{2}{3}$	Undef.	$\frac{1}{3}$
$g(x)$	-2	Undef.	2	1	$\frac{2}{3}$	$\frac{1}{2}$	$\frac{1}{3}$

    (d)

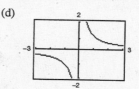

    (e) Because there are only a finite number of pixels, the utility may not attempt to evaluate the function where it does not exist.

89. $h(x) = \dfrac{6 - 2x}{3 - x}$

    Although $x = 3$ is not in the domain of $h(x)$,

    $h(x) = \dfrac{2(3 - x)}{3 - x} = 2$ for all real numbers except 3.

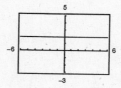

    (d) $y \approx 95.2t + 1090$

    (e) Answers will vary. The linear approximation fits the data better (it has an average error of 4.6 partnerships/year as opposed to the model given, which has an average error of 7 partnerships /year). However, the linear model does not interpolate past the data as well as the model given and the linear model continues to infinity as t increases.

Review Exercises for Chapter 2

1. (a) $y = 2x^2$

Vertical stretch

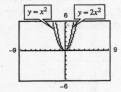

(b) $y = -2x^2$

Vertical stretch and a reflection in the x-axis

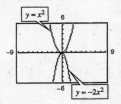

(c) $y = x^2 + 2$

Vertical shift two units upward

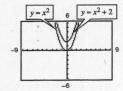

(d) $y = (x + 2)^2$

Horizontal shift two units to the left

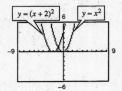

3. Vertex: $(4, 1)$ $\Rightarrow$ $f(x) = a(x - 4)^2 + 1$

Point: $(2, -1)$ $\Rightarrow$ $-1 = a(2 - 4)^2 + 1$

$$-2 = 4a$$
$$-\tfrac{1}{2} = a$$

Thus, $f(x) = -\tfrac{1}{2}(x - 4)^2 + 1$.

5. Vertex: $(1, -4)$ $\Rightarrow$ $f(x) = a(x - 1)^2 - 4$

Point: $(2, -3)$ $\Rightarrow$ $-3 = a(2 - 1)^2 - 4$

$$1 = a$$

Thus, $f(x) = (x - 1)^2 - 4$.

7. $g(x) = x^2 - 2x$

$\quad = x^2 - 2x + 1 - 1$

$\quad = (x - 1)^2 - 1$

Vertex: $(1, -1)$

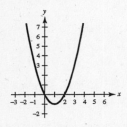

9. $f(x) = x^2 + 8x + 10$

$\quad = x^2 + 8x + 16 - 16 + 10$

$\quad = (x + 4)^2 - 6$

Vertex: $(-4, -6)$

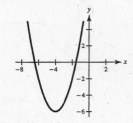

11. $f(t) = -2t^2 + 4t + 1$

$\quad = -2(t^2 - 2t + 1 - 1) + 1$

$\quad = -2[(t - 1)^2 - 1] + 1$

$\quad = -2(t - 1)^2 + 3$

Vertex: $(1, 3)$

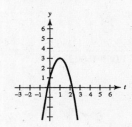

13. $h(x) = 4x^2 + 4x + 13$

$\quad = 4(x^2 + x) + 13$

$\quad = 4\left(x^2 + x + \frac{1}{4} - \frac{1}{4}\right) + 13$

$\quad = 4\left(x^2 + x + \frac{1}{4}\right) - 1 + 13$

$\quad = 4\left(x + \frac{1}{2}\right)^2 + 12$

Vertex: $\left(-\frac{1}{2}, 12\right)$

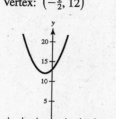

15. $f(x) = \frac{1}{3}(x^2 + 5x - 4)$

$\quad = \frac{1}{3}\left(x^2 + 5x + \frac{25}{4} - \frac{25}{4} - 4\right)$

$\quad = \frac{1}{3}\left[\left(x + \frac{5}{2}\right)^2 - \frac{41}{4}\right]$

$\quad = \frac{1}{3}\left(x + \frac{5}{2}\right)^2 - \frac{41}{12}$

Vertex: $\left(-\frac{5}{2}, -\frac{41}{12}\right)$

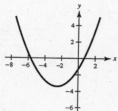

17. (a)

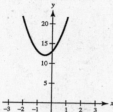

(b) $2x + 2y = 200$

$\qquad y = \frac{200 - 2x}{2}$

$\qquad y = 100 - x$

$\quad A = xy = x(100 - x)$

$\qquad = 100x - x^2$

(c) $A = -x^2 + 100x$

$\quad = -(x^2 - 100x)$

$\quad = -(x^2 - 100x + 2500 - 2500)$

$\quad = -[(x - 50)^2 - 2500]$

$\quad = -(x - 50)^2 + 2500 \implies$ Vertex is $(50, 2500)$

Area is maximum when $x = 50$ meters and $y = 50$ meters.

19. $C = 0.055x^2 - 120x + 20{,}000$

$\quad = 0.055\left(x^2 - \frac{24{,}000}{11}x\right) + 20{,}000$

$\quad = 0.055\left(x^2 - \frac{24{,}000}{11}x + \frac{144{,}000{,}000}{121}\right) - 0.055\left(\frac{144{,}000{,}000}{121}\right) + 20{,}000 \approx 0.055\left(x - \frac{12{,}000}{11}\right)^2 - 45{,}454.55$

C is minimum when $x = \dfrac{12{,}000}{11} \approx 1091$ units

21. $y = x^3$, $f(x) = -(x - 4)^3$

$f(x)$ is a reflection in the x-axis and a horizontal shift 4 units to the right of the graph of $y = x^3$.

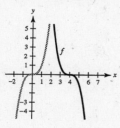

23. $y = x^4$, $f(x) = 2 - x^4$

$f(x)$ is a reflection in the x-axis and a vertical shift 2 units upward of the graph of $y = x^4$.

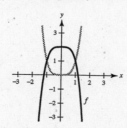

25. $y = x^5$, $f(x) = (x - 3)^5$

$f(x)$ is a horizontal shift 3 units to the right of the graph of $y = x^5$.

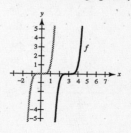

27. $f(x) = -x^2 + 6x + 9$

The degree is even and the leading coefficient is negative. The graph falls to the left and falls to the right.

29. $g(x) = \frac{3}{4}(x^4 + 3x^2 + 2)$

The degree is even and the leading coefficient is positive. The graph rises to the left and rises to the right.

31. $f(x) = 2x^2 + 11x - 21$

$\qquad = 2\left(x + \frac{11}{4}\right)^2 - \frac{289}{8}$

Vertex: $(-2.75, -36.125)$

Zeros: $2x^2 + 11x - 21 = 0$

$\qquad (2x - 3)(x + 7) = 0$

$\qquad 2x - 3 = 0 \implies x = \frac{3}{2}$

$\qquad x + 7 = 0 \implies x = -7$

The graph rises to the left and rises to the right.

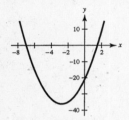

33. $f(t) = t^3 - 3t = t(t^2 - 3)$

Zeros: $t = 0, \pm\sqrt{3}$

The graph rises to the right and falls to the left.

x	-2	-1	0	1	2
y	-2	2	0	-2	2

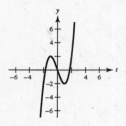

35. $f(x) = -12x^3 + 20x^2 = -4x^2(3x - 5)$

Zeros: $x = 0, \frac{5}{3}$

The graph rises to the left and falls to the right.

x	-2	-1	0	1	2
y	176	32	0	8	-16

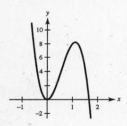

37.

$$
\begin{array}{r}
8x + 5 \\
3x - 2 \overline{\smash{)}\ 24x^2 - x - 8} \\
\underline{24x^2 - 16x} \\
15x - 8 \\
\underline{15x - 10} \\
2
\end{array}
$$

Thus, $\dfrac{24x^2 - x - 8}{3x - 2} = 8x + 5 + \dfrac{2}{3x - 2}.$

39.

$$
\begin{array}{r}
5x + 2 \\
x^2 - 3x + 1 \overline{\smash{)}\ 5x^3 - 13x^2 - x + 2} \\
\underline{5x^3 - 15x^2 + 5x} \\
2x^2 - 6x + 2 \\
\underline{2x^2 - 6x + 2} \\
0
\end{array}
$$

Thus, $\dfrac{5x^3 - 13x^2 - x + 2}{x^2 - 3x + 1} = 5x + 2.$

41.

$$
\begin{array}{r}
x^2 - 3x + 2 \\
x^2 + 0x + 2 \overline{\smash{)}\ x^4 - 3x^3 + 4x^2 - 6x + 3} \\
\underline{x^4 + 0x^3 + 2x^2} \\
-3x^3 + 2x^2 - 6x \\
\underline{-3x^3 + 0x^2 - 6x} \\
2x^2 + 0x + 3 \\
\underline{2x^2 + 0x + 4} \\
-1
\end{array}
$$

Thus,

$\dfrac{x^4 - 3x^3 + 4x^2 - 6x + 3}{x^2 + 2} = x^2 - 3x + 2 - \dfrac{1}{x^2 + 2}.$

43. $\dfrac{2}{3}$ $\begin{array}{r|rrrrr} & 6 & -4 & -27 & 18 & 0 \\ & & 4 & 0 & -18 & 0 \\ \hline & 6 & 0 & -27 & 0 & 0 \end{array}$

Thus, $\dfrac{6x^4 - 4x^3 - 27x^2 + 18x}{x - \frac{2}{3}} = 6x^3 - 27x.$

45. 4 $\begin{array}{r|rrrr} & 2 & -19 & 38 & 24 \\ & & 8 & -44 & -24 \\ \hline & 2 & -11 & -6 & 0 \end{array}$

Thus, $\dfrac{2x^3 - 19x^2 + 38x + 24}{x - 4} = 2x^2 - 11x - 6.$

47. $f(x) = 20x^4 + 9x^3 - 14x^2 - 3x$

(a) -1 $\begin{array}{r|rrrrr} & 20 & 9 & -14 & -3 & 0 \\ & & -20 & 11 & 3 & 0 \\ \hline & 20 & -11 & -3 & 0 & 0 \end{array}$

Yes, $x = -1$ is a zero of f.

(c) 0 $\begin{array}{r|rrrrr} & 20 & 9 & -14 & -3 & 0 \\ & & 0 & 0 & 0 & 0 \\ \hline & 20 & 9 & -14 & -3 & 0 \end{array}$

Yes, $x = 0$ is a zero of f.

(b) $\dfrac{3}{4}$ $\begin{array}{r|rrrrr} & 20 & 9 & -14 & -3 & 0 \\ & & 15 & 18 & 3 & 0 \\ \hline & 20 & 24 & 4 & 0 & 0 \end{array}$

Yes, $x = \frac{3}{4}$ is a zero of f.

(d) 1 $\begin{array}{r|rrrrr} & 20 & 9 & -14 & -3 & 0 \\ & & 20 & 29 & 15 & 12 \\ \hline & 20 & 29 & 15 & 12 & 12 \end{array}$

No, $x = 1$ is not a zero of f.

49. $f(x) = x^4 + 10x^3 - 24x^2 + 20x + 44$

(a) -3 $\begin{array}{r|rrrrr} & 1 & 10 & -24 & 20 & 44 \\ & & -3 & -21 & 135 & -465 \\ \hline & 1 & 7 & -45 & 155 & -421 \end{array}$

Thus, $f(-3) = -421.$

(b) -1 $\begin{array}{r|rrrrr} & 1 & 10 & -24 & 20 & 44 \\ & & -1 & -9 & 33 & -53 \\ \hline & 1 & 9 & -33 & 53 & -9 \end{array}$

Thus, $f(-1) = -9.$

51. $f(x) = x^3 + 4x^2 - 25x - 28$

(a)
$$
\begin{array}{r}
x^2 + 8x + 7 \\
x - 4 \overline{)\, x^3 + 4x^2 - 25x - 28} \\
\underline{x^3 - 4x^2} \\
8x^2 - 25x \\
\underline{8x^2 - 32x} \\
7x - 28 \\
\underline{7x - 28} \\
0
\end{array}
$$

Since the remainder is 0, $(x - 4)$ is a factor of $f(x)$.

(b) $f(x) = (x - 4)(x^2 + 8x + 7)$

$ = (x - 4)(x + 7)(x + 1)$

The remaining factors of $f(x)$ are $(x + 7)$ and $(x + 1)$.

(c) $f(x) = (x + 7)(x + 1)(x - 4)$

(d) The zeros of $f(x)$ are: $x = -7, -1, 4.$

(e)

53. $f(x) = x^4 - 4x^3 - 7x^2 + 22x + 24$

(a) -2 $\begin{array}{r|rrrrr} & 1 & -4 & -7 & 22 & 24 \\ & & -2 & 12 & -10 & -24 \\ \hline & 1 & -6 & 5 & 12 & 0 \end{array}$

3 $\begin{array}{r|rrrr} & 1 & -6 & 5 & 12 \\ & & 3 & -9 & -12 \\ \hline & 1 & -3 & -4 & 0 \end{array}$

Since the remainders are zero, both $(x + 2)$ and $(x - 3)$ are factors of $f(x)$.

(b) $f(x) = (x + 2)(x - 3)(x^2 - 3x - 4)$

$ = (x + 2)(x - 3)(x + 1)(x - 4)$

(c) $f(x) = (x + 1)(x - 4)(x + 2)(x - 3)$

(d) The zeros of $f(x)$ are: $x = -1, 4, -2, 3$

(e)

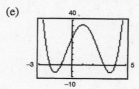

55. $A \approx -0.0022t^3 + 0.044t^2 + 0.17t + 2.3$

57.

t	A, actual	A, cubic model
0	2.3	2.3
1	2.4	2.5
2	2.9	2.8
3	3.2	3.1
4	3.6	3.5
5	4.0	4.0
6	4.2	4.4
7	4.9	4.9
8	5.4	5.3
9	5.8	5.8
10	6.4	6.2
11	6.5	6.6
12	6.9	6.9

59. $3 - \sqrt{-25} = 3 - 5i$

61. $-5i + i^2 = -5i - 1 = -1 - 5i$

63. $\left(\dfrac{\sqrt{2}}{2} - \dfrac{\sqrt{2}}{2}i\right) - \left(\dfrac{\sqrt{2}}{2} + \dfrac{\sqrt{2}}{2}i\right) = \dfrac{\sqrt{2}}{2} - \dfrac{\sqrt{2}}{2}i - \dfrac{\sqrt{2}}{2} - \dfrac{\sqrt{2}}{2}i = -2\dfrac{\sqrt{2}}{2}i = -\sqrt{2}i$

65. $(1 + 6i)(5 - 2i) = 5 - 2i + 30i - 12i^2 = 17 + 28i$

67. $i(6 + i)(3 - 2i) = i(18 - 12i + 3i - 2i^2) = i(20 - 9i)$
$$= 20i - 9i^2 = 9 + 20i$$

69. $\dfrac{3 + 2i}{5 + i} = \dfrac{3 + 2i}{5 + i} \cdot \dfrac{5 - i}{5 - i} = \dfrac{15 - 3i + 10i - 2i^2}{25 - i^2}$
$$= \dfrac{17 + 7i}{26} = \dfrac{17}{26} + \dfrac{7}{26}i$$

71. $\dfrac{1}{2 + i} - \dfrac{5}{1 + 4i} = \dfrac{1 + 4i - 5(2 + i)}{(2 + i)(1 + 4i)} = \dfrac{-9 - i}{-2 + 9i}$
$$= \dfrac{9 + i}{2 - 9i} \cdot \dfrac{2 + 9i}{2 + 9i} = \dfrac{18 + 81i + 2i + 9i^2}{4 - 81i^2}$$
$$= \dfrac{9 + 83i}{85} = \dfrac{9}{85} + \dfrac{83}{85}i$$

73. $2 + 8x^2 = 0$
$$8x^2 = -2$$
$$x^2 = -\dfrac{1}{4}$$
$$x = \pm\sqrt{-\dfrac{1}{4}} = \pm\dfrac{1}{2}i$$

75. $6x^2 + 3x + 27 = 0$
$$3(2x^2 + x + 9) = 0$$
By the Quadratic Formula: $x = \dfrac{-1 \pm \sqrt{-71}}{4}$
$$= -\dfrac{1}{4} \pm \dfrac{\sqrt{71}}{4}i$$

77. $f(x) = (x - 4)(x + 9)^2$

There are three zeros (one repeating zero at $x = -9$).

The zeros are: $x = 4, -9$

79. $f(x) = x^3 + 6x$

There are three zeros.

$f(x) = x(x^2 + 6) = x(x + \sqrt{6}i)(x - \sqrt{6}i)$

The zeros are: $x = 0, \pm\sqrt{6}i$

81. $f(x) = (x - 8)(x - 5)^2(x - 3 + i)(x - 3 - i)$

There are five zeros (one repeating zero).

The zeros are: $x = 8, 5, 3 \pm i$

83. $f(x) = 3x^4 + 4x^3 - 5x^2 - 8$

Possible rational zeros: $\pm 1, \pm 2, \pm 4, \pm 8, \pm\frac{1}{3}, \pm\frac{2}{3}, \pm\frac{4}{3}, \pm\frac{8}{3}$

85. $f(x) = 3x^3 - 20x^2 + 7x + 30$

Possible rational zeros: $\pm 1, \pm 2, \pm 3, \pm 5, \pm 6, \pm 10, \pm 15,$
$\pm 30, \pm\frac{1}{3}, \pm\frac{2}{3}, \pm\frac{5}{3}, \pm\frac{10}{3}$

$x = -1$ is a zero.

$$
\begin{array}{r|rrrr}
-1 & 3 & -20 & 7 & 30 \\
   &   & -3 & 23 & -30 \\
\hline
   & 3 & -23 & 30 & 0
\end{array}
$$

$f(x) = (x + 1)(3x^2 - 23x + 30)$

$\quad\quad = (x + 1)(3x - 5)(x - 6)$

The zeros of $f(x)$ are: $x = -1, \frac{5}{3}, 6$.

87. $f(x) = 25x^4 + 25x^3 - 154x^2 - 4x + 24$

Possible rational zeros:
$\pm 1, \pm 2, \pm 3, \pm 4, \pm 6, \pm 8, \pm 12, \pm 24, \pm\frac{1}{5}, \pm\frac{2}{5}, \pm\frac{3}{5}, \pm\frac{4}{5},$
$\pm\frac{6}{5}, \pm\frac{8}{5}, \pm\frac{12}{5}, \pm\frac{24}{5}, \pm\frac{1}{25}, \pm\frac{2}{25}, \pm\frac{3}{25}, \pm\frac{4}{25}, \pm\frac{6}{25}, \pm\frac{8}{25}, \pm\frac{12}{25}, \pm\frac{24}{25}$

$x = 2$ and $x = -3$ are zeros.

$$
\begin{array}{r|rrrrr}
2 & 25 & 25 & -154 & -4 & 24 \\
  &   & 50 & 150 & -8 & -24 \\
\hline
  & 25 & 75 & -4   & -12 & 0
\end{array}
$$

$$
\begin{array}{r|rrrr}
-3 & 25 & 75 & -4 & -12 \\
   &   & -75 & 0 & 12 \\
\hline
   & 25 & 0 & -4 & 0
\end{array}
$$

$f(x) = (x - 2)(x + 3)(25x^2 - 4)$

$\quad\quad = (x - 2)(x + 3)(5x - 2)(5x + 2)$

The zeros of $f(x)$ are: $x = 2, -3, \pm\frac{2}{5}$.

89. Zeros: $2, -3, 1 - 2i$ (Since $1 - 2i$ is a zero, so is $1 + 2i$.)

$(x - 2)(x + 3)(x - (1 - 2i)[x - (1 + 2i)]$

$= (x^2 + x - 6)(x^2 - 2x + 5)$

$= x^4 - x^3 - 3x^2 + 17x - 30$

Any non-zero multiple of this polynomial would also have these zeros.

91. $f(x) = x^4 - 2x^3 - 2x^2 - 2x - 3$

$$
\require{enclose}
\begin{array}{r}
x^2 - 2x - 3 \\
x^2 + 0x + 1 \enclose{longdiv}{x^4 - 2x^3 - 2x^2 - 2x - 3} \\
\underline{x^4 + 0x^3 + x^2} \\
-2x^3 - 3x^2 - 2x \\
\underline{-2x^3 + 0x^2 - 2x} \\
-3x^2 + 0x - 3 \\
\underline{-3x^2 + 0x - 3} \\
0
\end{array}
$$

(a) Rationals: $f(x) = (x^2 + 1)(x^2 - 2x - 3)$

$\quad\quad\quad\quad\quad = (x^2 + 1)(x + 1)(x - 3)$

(b) Linear and Quadratic Factors:

$\quad f(x) = (x^2 + 1)(x + 1)(x - 3)$

(c) Completely factored:

$\quad f(x) = (x + i)(x - i)(x + 1)(x - 3)$

93. Answers will vary.

$ax^2 + bx + c = 0$

(a) $b^2 - 4ac > 0$

$x^2 - 3x + 2 = 0$

$x = 1, 2$

(b) $b^2 - 4ac < 0$

$x^2 + 2x + 4 = 0$

$x = -1 \pm \sqrt{3}\,i$

(c) $b^2 - 4ac < 0$

$x^2 + 1 = 0$

$x = \pm i$

95. $f(x) = \dfrac{3x^2}{1 + 3x}$

$1 + 3x \neq 0$

$3x = -1$

$x = -\dfrac{1}{3}$

Domain: all real numbers except $x = -\frac{1}{3}$.

97. $f(x) = \dfrac{x^2 + x - 2}{x^2 + 4}$

Domain: all real numbers

99. $f(x) = \dfrac{2x^2 + 5x - 3}{x^2 + 2}$

Vertical asymptote: none

Horizontal asymptote: $y = 2$

101. $g(x) = \dfrac{1}{(x - 3)^2}$

Vertical asymptote: $x = 3$

Horizontal asymptote: $y = 0$

103. $f(x) = \dfrac{4}{x}$

No intercepts

Origin symmetry

Vertical asymptote: $x = 0$

Horizontal asymptote: $y = 0$

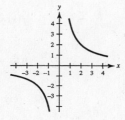

x	-3	-2	-1	1	2	3
y	$-\frac{4}{3}$	-2	-4	4	2	$\frac{4}{3}$

105. $h(x) = \dfrac{x - 3}{x - 2}$

x-intercept: $(3, 0)$

y-intercept: $\left(0, \dfrac{3}{2}\right)$

No axis or origin symmetry

Vertical asymptote: $x = 2$

Horizontal asymptote: $y = 1$

x	-1	0	1	3	4	5
y	$\frac{4}{3}$	$\frac{3}{2}$	2	0	$\frac{1}{2}$	$\frac{2}{3}$

107. $p(x) = \dfrac{x^2}{x^2 + 1}$

Intercept: $(0, 0)$

Symmetry: y-axis

Vertical asymptote: none

Horizontal asymptote: $y = 1$

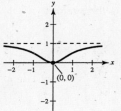

109. $y = \dfrac{2x^2}{x^2 - 4}$

Intercept: $(0, 0)$

y-axis symmetry

Vertical asymptotes: $x = 2, x = -2$

Horizontal asymptote: $y = 2$

x	± 5	± 4	± 3	± 1	0
y	$\frac{50}{21}$	$\frac{8}{3}$	$\frac{18}{5}$	$-\frac{2}{3}$	0

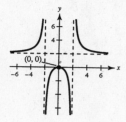

111. $g(x) = \dfrac{-2}{(x + 3)^2}$

y-intercept: $\left(0, -\dfrac{2}{9}\right)$

No axis or origin symmetry

Vertical asymptote: $x = -3$

Horizontal asymptote: $y = 0$

x	-6	-5	-4	-2	-1	0
y	$-\frac{2}{9}$	$-\frac{1}{2}$	-2	-2	$-\frac{1}{2}$	$-\frac{2}{9}$

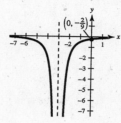

113. $f(x) = \dfrac{x^2 + 1}{x + 1}$

Using long division, $f(x) = \dfrac{x^2 + 1}{x + 1} = x - 1 + \dfrac{2}{x + 1}$

Slant asymptote: $y = x - 1$

Vertical asymptote: $x = -1$

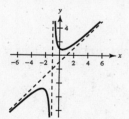

115. $f(x) = \dfrac{x^3}{x^2 - 4}$

Using long division, $f(x) = \dfrac{x^3}{x^2 - 4} = x + \dfrac{4x}{x^2 - 4}$

Slant asymptote: $y = x$

Vertical asymptotes: $x = \pm 2$

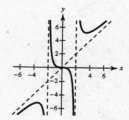

117. $\overline{C} = \dfrac{C}{x} = \dfrac{0.5x + 500}{x}, x > 0$

Horizontal asymptote: $y = 0.5$

Average cost per unit as the number of units increases without bound is 0.5.

119. (a)

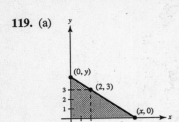

(b) $A = \frac{1}{2}bh$ where $b = x$ and $h = y$.

Slope: $\dfrac{y - 0}{0 - x} = \dfrac{0 - 3}{x - 2}$

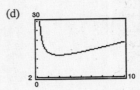

$$y = -x\left(\dfrac{-3}{x - 2}\right) = \dfrac{3x}{x - 2}$$

Area: $A = \dfrac{1}{2}xy = \dfrac{1}{2}x\left(\dfrac{3x}{x - 2}\right) = \dfrac{3x^2}{2(x - 2)}, \; x > 2$

(c)

x	2.5	3	3.5	4	4.5
A	18.75	13.50	12.25	12	12.15

The area is minimum when

$$x = 4 \text{ and } y = \dfrac{3(4)}{4 - 2} = 6.$$

(d)

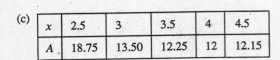

The area is minimum (12) when $x = 4$ and $y = 6$.

(e) $A = \dfrac{3x^2}{2x - 4} = \dfrac{3}{2}x + 3 + \dfrac{12}{2x - 4} = \dfrac{3}{2}(x + 2) + \dfrac{6}{x - 2}$

The slant asymptote is $y = \frac{3}{2}(x + 2)$. The area increases without bound as x increases.

Problem Solving for Chapter 2

1. Conditions (a) and (d) are preferable because profits would be increasing in the long run.

3. If $h = 0$ and $k = 0$, then $a < -1$ produces a stretch that is reflected in the x-axis, and $-1 < a < 0$ produces a shrink that is reflected in the x-axis.

5. $f(x) = (x - k)q(x) + r$

    (a) $k = 2$, $r = 5$, $q(x) =$ any quadratic $ax^2 + bx + c$ where $a > 0$.

    One example: $f(x) = (x - 2)x^2 + 5 = x^3 - 2x^2 + 5$

    (b) $k = -3$, $r = 1$, $q(x) =$ any quadratic $ax^2 + bx + c$ where $a < 0$.

    One example:
$$f(x) = (x + 3)(-x^2) + 1 = -x^3 - 3x^2 + 1$$

7. $f(x) = \dfrac{2x^2 + x - 1}{x + 1}$

    (a)

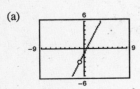

The graph has a "hole" when $x = -1$. There are no vertical asymptotes.

(b) $\dfrac{2x^2 + x - 1}{x + 1} = \dfrac{(2x - 1)(x + 1)}{x + 1} = 2x - 1, \, x \neq -1$

(c) As $x \to -1, \dfrac{2x^2 + x - 1}{x + 1} \to -3$

9. $y = ax^2 + bx + c$

(a) $(0, -4)$: $-4 = c$

$\quad (1, 0)$: $\quad 0 = a + b + c \implies a + b = 4$

$\quad (2, 2)$: $\quad 2 = 4a + 2b + c \implies 4a + 2b = 6$

Solve the system of equations:

$$4a + 2b = 6 \implies 2a + b = 3$$

$$a + b = 4 \implies \underline{-a - b = -4}$$

$$a = -1$$

$$-1 + b = 4$$

$$b = 5$$

Thus, $y = -x^2 + 5x - 4$.

Check: $(4, 0)$: $\quad 0 = -(4)^2 + 5(4) - 4$ ✓

$\quad\quad\quad (6, -10)$: $-10 = -(6)^2 + 5(6) - 4$ ✓

(b)

L_1	L_2
0	-4
1	0
2	2
4	0
6	-10

Use the "Quad Reg" feature of your graphing utility to obtain

$$y = -x^2 + 5x - 4$$

11. (a) Slope $= \dfrac{9 - 4}{3 - 2} = 5$. Slope of tangent line is less than 5.

(b) Slope $= \dfrac{4 - 1}{2 - 1} = 3$. Slope of tangent line is greater than 3.

(c) Slope $= \dfrac{4.41 - 4}{2.1 - 2} = 4.1$. Slope of tangent line is less than 4.1.

(d) Slope $= \dfrac{f(2 + h) - f(2)}{(2 + h) - 2}$

$\quad\quad = \dfrac{(2 + h)^2 - 4}{h}$

$\quad\quad = \dfrac{4h + h^2}{h}$

$\quad\quad = 4 + h, h \neq 0$

(e) $\quad$ Slope $= 4 + h, \quad h \neq 0$

$\quad\quad 4 + (-1) = 3$

$\quad\quad 4 + 1 = 5$

$\quad\quad 4 + 0.1 = 4.1$

The results are the same as in (a)–(c).

(f) Letting h get closer and closer to 0, the slope approaches 4. Hence, the slope at $(2, 4)$ is 4.

13. $f(x) = \dfrac{ax + b}{cx + d}$

$f(x)$ has a vertical asymptote at $x = -\dfrac{d}{c}$ and a horizontal asymptote at $y = \dfrac{a}{c}$.

(i) $a > 0$

$\quad b < 0$

$\quad c > 0$

$\quad d < 0$

$x = -\dfrac{d}{c}$ is positive.

$y = \dfrac{a}{c}$ is positive.

Both asymptotes are positive on graph (d).

(ii) $a > 0$

$\quad b > 0$

$\quad c < 0$

$\quad d < 0$

$x = -\dfrac{d}{c}$ is negative.

$y = \dfrac{a}{c}$ is negative.

Both asymptotes are negative on graph (b).

(iii) $a < 0$

$\quad b > 0$

$\quad c > 0$

$\quad d < 0$

$x = -\dfrac{d}{c}$ is positive.

$y = \dfrac{a}{c}$ is negative.

The vertical asymptote is positive and the horizontal asymptote is negative on graph (a).

(iv) $a > 0$

$\quad b < 0$

$\quad c > 0$

$\quad d > 0$

$x = -\dfrac{d}{c}$ is negative.

$y = \dfrac{a}{c}$ is positive.

The vertical asymptote is negative and the horizontal asymptote is positive on graph (c).

C H A P T E R   3
Limits and Their Properties

C H A P T E R 3
Limits and Their Properties

Section 3.1　A Preview of Calculus

1. Precalculus: $(20 \text{ ft/sec})(15 \text{ seconds}) = 300 \text{ feet}$

3. Calculus required: slope of tangent line at $x = 2$ is rate of change, and equals about 0.16.

5. Precalculus: Area $= \frac{1}{2}bh = \frac{1}{2}(5)(3) = \frac{15}{2}$ sq. units

7. $f(x) = 4x - x^2$

(a)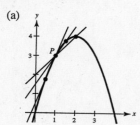

(b) slope $= m = \dfrac{(4x - x^2) - 3}{x - 1}$

$= \dfrac{(x - 1)(3 - x)}{x - 1} = 3 - x, \quad x \neq 1$

$x = 2$: $m = 3 - 2 = 1$

$x = 1.5$: $m = 3 - 1.5 = 1.5$

$x = 0.5$: $m = 3 - 0.5 = 2.5$

(c) At $P(1, 3)$ the slope is 2.

You can improve your approximation of the slope at $x = 1$ by considering x-values very close to 1.

9. Area $\approx 5 + \dfrac{5}{2} + \dfrac{5}{3} + \dfrac{5}{4} \approx 10.417$

Area $\approx \dfrac{1}{2}\left(5 + \dfrac{5}{1.5} + \dfrac{5}{2} + \dfrac{5}{2.5} + \dfrac{5}{3} + \dfrac{5}{3.5} + \dfrac{5}{4} + \dfrac{5}{4.5}\right) \approx 9.145$

11. (a) $D_1 = \sqrt{(5 - 1)^2 + (1 - 5)^2} = \sqrt{16 + 16} \approx 5.66$

(b) $D_2 = \sqrt{1 + \left(\frac{5}{2}\right)^2} + \sqrt{1 + \left(\frac{5}{2} - \frac{5}{3}\right)^2} + \sqrt{1 + \left(\frac{5}{3} - \frac{5}{4}\right)^2} + \sqrt{1 + \left(\frac{5}{4} - 1\right)^2}$

$\approx 2.693 + 1.302 + 1.083 + 1.031 \approx 6.11$

(c) Increase the number of line segments.

Section 3.2   Finding Limits Graphically and Numerically

1.

x	1.9	1.99	1.999	2.001	2.01	2.1
$f(x)$	0.3448	0.3344	0.3334	0.3332	0.3322	0.3226

$$\lim_{x \to 2} \frac{x-2}{x^2-x-2} \approx 0.3333 \quad \left(\text{Actual limit is } \tfrac{1}{3}.\right)$$

3.

x	-0.1	-0.01	-0.001	0.001	0.01	0.1
$f(x)$	0.2911	0.2889	0.2887	0.2887	0.2884	0.2863

$$\lim_{x \to 0} \frac{\sqrt{x+3} - \sqrt{3}}{x} \approx 0.2887 \quad \left(\text{Actual limit is } 1/(2\sqrt{3}).\right)$$

5.

x	2.9	2.99	2.999	3.001	3.01	3.1
$f(x)$	-0.0641	-0.0627	-0.0625	-0.0625	-0.0623	-0.0610

$$\lim_{x \to 3} \frac{[1/(x+1)] - (1/4)}{x-3} \approx -0.0625 \quad \left(\text{Actual limit is } -\tfrac{1}{16}.\right)$$

7. $\displaystyle \lim_{x \to 3} (4 - x) = 1$

9. $\displaystyle \lim_{x \to 2} f(x) = \lim_{x \to 2} (4 - x) = 2$

11. $\displaystyle \lim_{x \to 5} \frac{|x-5|}{x-5}$ does not exist. For values of x to the left of 5,

$|x-5|/(x-5)$ equals -1, whereas for values of x to the right of 5, $|x-5|/(x-5)$ equals 1.

13. $C(t) = 0.75 - 0.50 [\![-(t-1)]\!]$

(a)

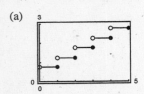

(b)

t	3	3.3	3.4	3.5	3.6	3.7	4
C	1.75	2.25	2.25	2.25	2.25	2.25	2.25

$$\lim_{t \to 3.5} C(t) = 2.25$$

(c)

t	2	2.5	2.9	3	3.1	3.5	4
C	1.25	1.75	1.75	1.75	2.25	2.25	2.25

$\displaystyle \lim_{t \to 3} C(t)$ does not exist. The values of C jump from 1.75 to 2.25 at $t = 3$.

15. You need to find δ such that $0 < |x - 1| < \delta$ implies

$$|f(x) - 1| = \left|\frac{1}{x} - 1\right| < 0.1. \text{ That is,}$$

$$-0.1 < \frac{1}{x} - 1 < 0.1$$

$$1 - 0.1 < \frac{1}{x} < 1 + 0.1$$

$$\frac{9}{10} < \frac{1}{x} < \frac{11}{10}$$

$$\frac{10}{9} > x > \frac{10}{11}$$

$$\frac{10}{9} - 1 > x - 1 > \frac{10}{11} - 1$$

$$\frac{1}{9} > x - 1 > -\frac{1}{11}.$$

So take $\delta = \frac{1}{11}$. Then $0 < |x - 1| < \delta$ implies

$$-\frac{1}{11} < x - 1 < \frac{1}{11}$$

$$-\frac{1}{11} < x - 1 < \frac{1}{9}.$$

Using the first series of equivalent inequalities, you obtain

$$|f(x) - 1| = \left|\frac{1}{x} - 1\right| < \varepsilon < 0.1.$$

17. $\lim\limits_{x \to 2} (3x + 2) = 8 = L$

$$|(3x + 2) - 8| < 0.01$$

$$|3x - 6| < 0.01$$

$$3|x - 2| < 0.01$$

$$0 < |x - 2| < \frac{0.01}{3} \approx 0.0033 = \delta$$

Hence, if $0 < |x - 2| < \delta = \dfrac{0.01}{3}$, you have

$$3|x - 2| < 0.01$$

$$|3x - 6| < 0.01$$

$$|(3x + 2) - 8| < 0.01$$

$$|f(x) - L| < 0.01.$$

19. $\lim\limits_{x \to 2} (x^2 - 3) = 1 = L$

$$|(x^2 - 3) - 1| < 0.01$$

$$|x^2 - 4| < 0.01$$

$$|(x + 2)(x - 2)| < 0.01$$

$$|x + 2|\,|x - 2| < 0.01$$

$$|x - 2| < \frac{0.01}{|x + 2|}$$

If we assume $1 < x < 3$, then $\delta = 0.01/5 = 0.002$.

Hence, if $0 < |x - 2| < \delta = 0.002$, you have

$$|x - 2| < 0.002 = \frac{1}{5}(0.01) < \frac{1}{|x + 2|}(0.01)$$

$$|x + 2|\,|x - 2| < 0.01$$

$$|x^2 - 4| < 0.01$$

$$|(x^2 - 3) - 1| < 0.01$$

$$|f(x) - L| < 0.01.$$

21. $\lim\limits_{x \to 2} (x + 3) = 5$

Given $\varepsilon > 0$:

$$|(x + 3) - 5| < \varepsilon$$

$$|x - 2| < \varepsilon = \delta$$

Hence, let $\delta = \varepsilon$.

Hence, if $0 < |x - 2| < \delta = \varepsilon$, you have

$$|x - 2| < \varepsilon$$

$$|(x + 3) - 5| < \varepsilon$$

$$|f(x) - L| < \varepsilon.$$

23. $\lim_{x \to -4} \left(\frac{1}{2}x - 1\right) = \frac{1}{2}(-4) - 1 = -3$

Given $\varepsilon > 0$:

$$\left|\left(\frac{1}{2}x - 1\right) - (-3)\right| < \varepsilon$$

$$\left|\frac{1}{2}x + 2\right| < \varepsilon$$

$$\frac{1}{2}|x - (-4)| < \varepsilon$$

$$|x - (-4)| < 2\varepsilon$$

Hence, let $\delta = 2\varepsilon$.

Hence, if $0 < |x - (-4)| < \delta = 2\varepsilon$, you have

$$|x - (-4)| < 2\varepsilon$$

$$\left|\frac{1}{2}x + 2\right| < \varepsilon$$

$$\left|\left(\frac{1}{2}x - 1\right) + 3\right| < \varepsilon$$

$$|f(x) - L| < \varepsilon.$$

25. $\lim_{x \to 6} 3 = 3$

Given $\varepsilon > 0$:

$$|3 - 3| < \varepsilon$$

$$0 < \varepsilon$$

Hence, any $\delta > 0$ will work.

Hence, for any $\delta > 0$, you have

$$|3 - 3| < \varepsilon$$

$$|f(x) - L| < \varepsilon.$$

27. $\lim_{x \to 0} \sqrt[3]{x} = 0$

Given $\varepsilon > 0$: $\left|\sqrt[3]{x} - 0\right| < \varepsilon$

$$\left|\sqrt[3]{x}\right| < \varepsilon$$

$$|x| < \varepsilon^3 = \delta$$

Hence, let $\delta = \varepsilon^3$.

Hence for $0 < |x - 0| < \delta = \varepsilon^3$, you have

$$|x| < \varepsilon^3$$

$$\left|\sqrt[3]{x}\right| < \varepsilon$$

$$\left|\sqrt[3]{x} - 0\right| < \varepsilon$$

$$|f(x) - L| < \varepsilon.$$

29. $\lim_{x \to -2} |x - 2| = |(-2) - 2| = 4$

Given $\varepsilon > 0$:

$$||x - 2| - 4| < \varepsilon$$

$$|-(x - 2) - 4| < \varepsilon$$

$$(x - 2 < 0)$$

$$|-x - 2| = |x + 2| = |x - (-2)| < \varepsilon$$

Hence, $\delta = \varepsilon$.

Hence for $0 < |x - (-2)| < \delta = \varepsilon$, you have

$$|x + 2| < \varepsilon$$

$$|-(x + 2)| < \varepsilon$$

$$|-(x - 2) - 4| < \varepsilon$$

$$||x - 2| - 4| < \varepsilon \quad \text{(because } x - 2 < 0\text{)}$$

$$|f(x) - L| < \varepsilon.$$

31. $\lim_{x \to 1} (x^2 + 1) = 2$

Given $\varepsilon > 0$:

$$|(x^2 + 1) - 2| < \varepsilon$$

$$|x^2 - 1| < \varepsilon$$

$$|(x + 1)(x - 1)| < \varepsilon$$

$$|x - 1| < \frac{\varepsilon}{|x + 1|}$$

If we assume $0 < x < 2$, then $\delta = \varepsilon/3$.

Hence for $0 < |x - 1| < \delta = \frac{\varepsilon}{3}$, you have

$$|x - 1| < \frac{1}{3}\varepsilon < \frac{1}{|x + 1|}\varepsilon$$

$$|x^2 - 1| < \varepsilon$$

$$|(x^2 + 1) - 2| < \varepsilon$$

$$|f(x) - 2| < \varepsilon.$$

33. $f(x) = \dfrac{\sqrt{x + 5} - 3}{x - 4}$

$\lim_{x \to 4} f(x) = \dfrac{1}{6}$

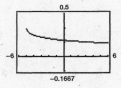

The domain is $[-5, 4) \cup (4, \infty)$.

The graphing utility does not show the hole at $\left(4, \frac{1}{6}\right)$.

35. $f(x) = \dfrac{x-9}{\sqrt{x}-3}$

$\lim\limits_{x \to 9} f(x) = 6$

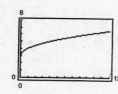

The domain is all $x \geq 0$ except $x = 9$. The graphing utility does not show the hole at $(9, 6)$.

37. $\lim\limits_{x \to 8} f(x) = 25$ means that the values of f approach 25 as x gets closer and closer to 8.

39. (i) The values of f approach different numbers as x approaches c from different sides of c:

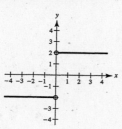

(ii) The values of f increase without bound as x approaches c:

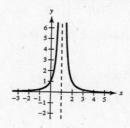

(iii) The values of f oscillate between two fixed numbers as x approaches c:

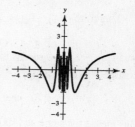

41. False. For example, let $f(x) = \dfrac{x-2}{x^2-4}$. Then $f(2)$ does not exist, but $\lim\limits_{x \to 2} f(x) = \dfrac{1}{4}$ (See Exercise 2, Section 3.2.)

43. False. Let $f(x) = \begin{cases} 4-x, & x \neq 2 \\ 0, & x = 2 \end{cases}$

$\lim\limits_{x \to 2} f(x) = 2$, but $f(2) = 0$ (See Exercise 9.)

45. (a) $C = 2\pi r$

$r = \dfrac{C}{2\pi} = \dfrac{6}{2\pi} = \dfrac{3}{\pi} \approx 0.9549$ cm

(b) If $C = 5.5$, $r = \dfrac{5.5}{2\pi} \approx 0.87535$ cm.

If $C = 6.5$, $r = \dfrac{6.5}{2\pi} \approx 1.03451$ cm.

Thus, $0.87535 < r < 1.03451$.

(c) $\lim\limits_{r \to 3/\pi} (2\pi r) = 6$; $\varepsilon = 0.5$; $\delta \approx 0.0796$

47. $f(x) = (1+x)^{1/x}$

$\lim\limits_{x \to 0} (1+x)^{1/x} = e \approx 2.71828$

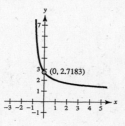

x	$f(x)$	x	$f(x)$
-0.1	2.867972	0.1	2.593742
-0.01	2.731999	0.01	2.704814
-0.001	2.719642	0.001	2.716924
-0.0001	2.718418	0.0001	2.718146
-0.00001	2.718295	0.00001	2.718268
-0.000001	2.718283	0.000001	2.718280

49. $f(x) = \sqrt{x}$

(a) $\lim_{x \to 0.25} \sqrt{x} = 0.5$ is true.

As x approaches $0.25 = \frac{1}{4}$, $f(x) = \sqrt{x}$ approaches $\frac{1}{2} = 0.5$

(b) $\lim_{x \to 0} \sqrt{x} = 0$ is false.

$f(x) = \sqrt{x}$ is not defined on an open interval containing 0 because the domain of f is $x \geq 0$.

51. If $\lim_{x \to c} f(x) = L_1$ and $\lim_{x \to c} f(x) = L_2$, then for every $\varepsilon > 0$, there exists $\delta_1 > 0$ and $\delta_2 > 0$ such that

$|x - c| < \delta_1 \implies |f(x) - L_1| < \varepsilon$ and $|x - c| < \delta_2 \implies |f(x) - L_2| < \varepsilon$. Let δ equal the smaller of δ_1 and δ_2. Then for

$|x - c| < \delta$, we have

$$|L_1 - L_2| = |L_1 - f(x) + f(x) - L_2| \leq |L_1 - f(x)| + |f(x) - L_2| < \varepsilon + \varepsilon.$$

Therefore, $|L_1 - L_2| < 2\varepsilon$. Since $\varepsilon > 0$ is arbitrary, it follows that $L_1 = L_2$.

53. $\lim_{x \to c} [f(x) - L] = 0$ means that for every $\varepsilon > 0$ there exists $\delta > 0$ such that if

$$0 < |x - c| < \delta, \quad \text{then} \quad |(f(x) - L) - 0| < \varepsilon.$$

This means the same as $|f(x) - L| < \varepsilon$ when

$$0 < |x - c| < \delta.$$

Thus, $\lim_{x \to c} f(x) = L$.

Section 3.3   Evaluating Limits Analytically

1.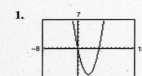

(a) $\lim_{x \to 5} h(x) = 0$

(b) $\lim_{x \to -1} h(x) = 6$

$h(x) = x^2 - 5x$

3.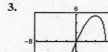

(a) $\lim_{x \to 0} f(x) = 0$

(b) $\lim_{x \to 2} f(x) = 4$

5. $\lim_{x \to 2} x^4 = 2^4 = 16$

7. $\lim_{x \to 0} (2x - 1) = 2(0) - 1 = -1$

9. $\lim_{x \to -3} (x^2 + 3x) = (-3)^2 + 3(-3) = 9 - 9 = 0$

11. $\lim_{x \to -3} (2x^2 + 4x + 1) = 2(-3)^2 + 4(-3) + 1 = 18 - 12 + 1 = 7$

13. $\lim_{x \to 2} \dfrac{1}{x} = \dfrac{1}{2}$

15. $\lim_{x \to 1} \dfrac{x - 3}{x^2 + 4} = \dfrac{1 - 3}{1^2 + 4} = \dfrac{-2}{5} = -\dfrac{2}{5}$

17. $\lim_{x \to 7} \dfrac{5x}{\sqrt{x + 2}} = \dfrac{5(7)}{\sqrt{7 + 2}} = \dfrac{35}{\sqrt{9}} = \dfrac{35}{3}$

19. $\lim_{x \to 3} \sqrt{x + 1} = \sqrt{3 + 1} = 2$

21. $\lim_{x \to -4} (x + 3)^2 = (-4 + 3)^2 = 1$

23. (a) $\lim_{x \to 1} f(x) = 5 - 1 = 4$

(b) $\lim_{x \to 4} g(x) = 4^3 = 64$

(c) $\lim_{x \to 1} g(f(x)) = g(f(1)) = g(4) = 64$

25. (a) $\lim\limits_{x \to 1} f(x) = 4 - 1 = 3$

(b) $\lim\limits_{x \to 3} g(x) = \sqrt{3 + 1} = 2$

(c) $\lim\limits_{x \to 1} g(f(x)) = g(3) = 2$

27. (a) $\lim\limits_{x \to c} [5g(x)] = 5 \lim\limits_{x \to c} g(x) = 5(3) = 15$

(b) $\lim\limits_{x \to c} [f(x) + g(x)] = \lim\limits_{x \to c} f(x) + \lim\limits_{x \to c} g(x) = 2 + 3 = 5$

(c) $\lim\limits_{x \to c} [f(x)g(x)] = \left[\lim\limits_{x \to c} f(x)\right]\left[\lim\limits_{x \to c} g(x)\right] = (2)(3) = 6$

(d) $\lim\limits_{x \to c} \dfrac{f(x)}{g(x)} = \dfrac{\lim\limits_{x \to c} f(x)}{\lim\limits_{x \to c} g(x)} = \dfrac{2}{3}$

29. (a) $\lim\limits_{x \to c} [f(x)]^3 = \left[\lim\limits_{x \to c} f(x)\right]^3 = (4)^3 = 64$

(b) $\lim\limits_{x \to c} \sqrt{f(x)} = \sqrt{\lim\limits_{x \to c} f(x)} = \sqrt{4} = 2$

(c) $\lim\limits_{x \to c} [3 f(x)] = 3 \lim\limits_{x \to c} f(x) = 3(4) = 12$

(d) $\lim\limits_{x \to c} [f(x)]^{3/2} = \left[\lim\limits_{x \to c} f(x)\right]^{3/2} = (4)^{3/2} = 8$

31. $f(x) = -2x + 1$ and $g(x) = \dfrac{-2x^2 + x}{x}$ agree except at $x = 0$.

(a) $\lim\limits_{x \to 0} g(x) = \lim\limits_{x \to 0} f(x) = 1$

(b) $\lim\limits_{x \to -1} g(x) = \lim\limits_{x \to -1} f(x) = 3$

33. $f(x) = x(x + 1)$ and $g(x) = \dfrac{x^3 - x}{x - 1}$ agree except at $x = 1$.

(a) $\lim\limits_{x \to 1} g(x) = \lim\limits_{x \to 1} f(x) = 2$

(b) $\lim\limits_{x \to -1} g(x) = \lim\limits_{x \to -1} f(x) = 0$

35. $f(x) = \dfrac{x^2 - 1}{x + 1}$ and $g(x) = x - 1$ agree except at $x = -1$.

$$\lim\limits_{x \to -1} f(x) = \lim\limits_{x \to -1} g(x) = -2$$

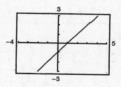

37. $f(x) = \dfrac{x^3 - 8}{x - 2}$ and $g(x) = x^2 + 2x + 4$ agree except at $x = 2$.

$$\lim\limits_{x \to 2} f(x) = \lim\limits_{x \to 2} g(x) = 12$$

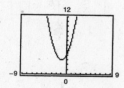

39. $\lim\limits_{x \to 5} \dfrac{x - 5}{x^2 - 25} = \lim\limits_{x \to 5} \dfrac{x - 5}{(x + 5)(x - 5)}$

$$= \lim\limits_{x \to 5} \dfrac{1}{x + 5} = \dfrac{1}{10}$$

41. $\lim\limits_{x \to -3} \dfrac{x^2 + x - 6}{x^2 - 9} = \lim\limits_{x \to -3} \dfrac{(x + 3)(x - 2)}{(x + 3)(x - 3)}$

$$= \lim\limits_{x \to -3} \dfrac{x - 2}{x - 3} = \dfrac{-5}{-6} = \dfrac{5}{6}$$

43. $\lim\limits_{x \to 0} \dfrac{\sqrt{x + 5} - \sqrt{5}}{x} = \lim\limits_{x \to 0} \dfrac{\sqrt{x + 5} - \sqrt{5}}{x} \cdot \dfrac{\sqrt{x + 5} + \sqrt{5}}{\sqrt{x + 5} + \sqrt{5}}$

$$= \lim\limits_{x \to 0} \dfrac{(x + 5) - 5}{x\left(\sqrt{x + 5} + \sqrt{5}\right)} = \lim\limits_{x \to 0} \dfrac{1}{\sqrt{x + 5} + \sqrt{5}} = \dfrac{1}{2\sqrt{5}} = \dfrac{\sqrt{5}}{10}$$

45. $\lim\limits_{x\to 4} \dfrac{\sqrt{x+5}-3}{x-4} = \lim\limits_{x\to 4} \dfrac{\sqrt{x+5}-3}{x-4}\cdot\dfrac{\sqrt{x+5}+3}{\sqrt{x+5}+3}$

$\qquad = \lim\limits_{x\to 4} \dfrac{(x+5)-9}{(x-4)(\sqrt{x+5}+3)} = \lim\limits_{x\to 4} \dfrac{1}{\sqrt{x+5}+3} = \dfrac{1}{\sqrt{9}+3} = \dfrac{1}{6}$

47. $\lim\limits_{x\to 0} \dfrac{\dfrac{1}{3+x}-\dfrac{1}{3}}{x} = \lim\limits_{x\to 0} \dfrac{3-(3+x)}{3(3+x)x} = \lim\limits_{x\to 0} \dfrac{-1}{3(3+x)} = \dfrac{-1}{9}$

49. $\lim\limits_{\Delta x\to 0} \dfrac{2(x+\Delta x)-2x}{\Delta x} = \lim\limits_{\Delta x\to 0} \dfrac{2x+2\Delta x-2x}{\Delta x} = \lim\limits_{\Delta x\to 0} 2 = 2$

51. $\lim\limits_{\Delta x\to 0} \dfrac{(x+\Delta x)^2-2(x+\Delta x)+1-(x^2-2x+1)}{\Delta x} = \lim\limits_{\Delta x\to 0} \dfrac{x^2+2x\Delta x+(\Delta x)^2-2x-2\Delta x+1-x^2+2x-1}{\Delta x}$

$\qquad = \lim\limits_{\Delta x\to 0} (2x+\Delta x-2) = 2x-2$

53. $\lim\limits_{x\to 0} \dfrac{\sqrt{x+2}-\sqrt{2}}{x} \approx 0.354$

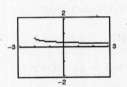

x	-0.1	-0.01	-0.001	0	0.001	0.01	0.1
$f(x)$	0.358	0.354	0.354	?	0.354	0.353	0.349

Analytically, $\lim\limits_{x\to 0} \dfrac{\sqrt{x+2}-\sqrt{2}}{x} = \lim\limits_{x\to 0} \dfrac{\sqrt{x+2}-\sqrt{2}}{x}\cdot\dfrac{\sqrt{x+2}+\sqrt{2}}{\sqrt{x+2}+\sqrt{2}}$

$\qquad = \lim\limits_{x\to 0} \dfrac{x+2-2}{x(\sqrt{x+2}+\sqrt{2})} = \lim\limits_{x\to 0} \dfrac{1}{\sqrt{x+2}+\sqrt{2}} = \dfrac{1}{2\sqrt{2}} = \dfrac{\sqrt{2}}{4} \approx 0.354.$

55. $\lim\limits_{x\to 0} \dfrac{\dfrac{1}{2+x}-\dfrac{1}{2}}{x} = -\dfrac{1}{4}$

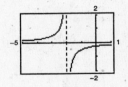

x	-0.1	-0.01	-0.001	0	0.001	0.01	0.1
$f(x)$	-0.263	-0.251	-0.250	?	-0.250	-0.249	-0.238

Analytically, $\lim\limits_{x\to 0} \dfrac{\dfrac{1}{2+x}-\dfrac{1}{2}}{x} = \lim\limits_{x\to 0} \dfrac{2-(2+x)}{2(2+x)}\cdot\dfrac{1}{x} = \lim\limits_{x\to 0} \dfrac{-x}{2(2+x)}\cdot\dfrac{1}{x} = \lim\limits_{x\to 0} \dfrac{-1}{2(2+x)} = -\dfrac{1}{4}.$

57. $f(x) = 2x+3$

$\lim\limits_{\Delta x\to 0} \dfrac{f(x+\Delta x)-f(x)}{\Delta x} = \lim\limits_{\Delta x\to 0} \dfrac{2(x+\Delta x)+3-(2x+3)}{\Delta x} = \lim\limits_{\Delta x\to 0} \dfrac{2x+2\Delta x-2x}{\Delta x} = \lim\limits_{\Delta x\to 0} \dfrac{2\Delta x}{\Delta x} = \lim\limits_{\Delta x\to 0} 2 = 2$

59. $f(x) = \dfrac{4}{x}$

$\lim\limits_{\Delta x\to 0} \dfrac{f(x+\Delta x)-f(x)}{\Delta x} = \lim\limits_{\Delta x\to 0} \dfrac{\dfrac{4}{x+\Delta x}-\dfrac{4}{x}}{\Delta x} = \lim\limits_{\Delta x\to 0} \dfrac{4x-4(x+\Delta x)}{(x+\Delta x)x\,\Delta x}$

$\qquad = \lim\limits_{\Delta x\to 0} \dfrac{-4\Delta x}{(x+\Delta x)x\,\Delta x} = \lim\limits_{\Delta x\to 0} \dfrac{-4}{(x+\Delta x)x} = \dfrac{-4}{x^2}$

61. $\lim_{x \to 0}(4 - x^2) = 4$

$\lim_{x \to 0}(4 + x^2) = 4$

Since $4 - x^2 \le f(x) \le 4 + x^2$ the Squeeze Theorem says $\lim_{x \to 0} f(x) = 4$.

63. We say that two functions f and g agree at all but one point (on an open interval) if $f(x) = g(x)$ for all x in the interval except for $x = c$, where c is in the interval.

65. An indeterminant form is obtained when evaluating a limit using direct substitution produces a meaningless fractional expression such as $0/0$. That is,

$$\lim_{x \to c} \frac{f(x)}{g(x)}$$

for which $\lim_{x \to c} f(x) = \lim_{x \to c} g(x) = 0$.

67. $s(t) = -16t^2 + 1000$

$\lim_{t \to 5} \dfrac{s(5) - s(t)}{5 - t} = \lim_{t \to 5} \dfrac{600 - (-16t^2 + 1000)}{5 - t} = \lim_{t \to 5} \dfrac{16(t + 5)(t - 5)}{-(t - 5)} = \lim_{t \to 5} -16(t + 5) = -160 \text{ ft/sec.}$

Speed $= 160$ ft/sec

69. $s(t) = -4.9t^2 + 150$

$\lim_{t \to 3} \dfrac{s(3) - s(t)}{3 - t} = \lim_{t \to 3} \dfrac{-4.9(3^2) + 150 - (-4.9t^2 + 150)}{3 - t}$

$= \lim_{t \to 3} \dfrac{-4.9(9 - t^2)}{3 - t}$

$= \lim_{t \to 3} \dfrac{-4.9(3 - t)(3 + t)}{3 - t}$

$= \lim_{t \to 3} -4.9(3 + t) = -29.4 \text{ m/sec}$

71. False. As x approaches 0 from the left, $\dfrac{|x|}{x} = -1$.

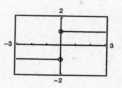

73. True

75. False. The limit does not exist.

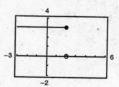

77. Let $f(x) = 1/x$ and $g(x) = -1/x$. $\lim_{x \to 0} f(x)$ and $\lim_{x \to 0} g(x)$ do not exist.

$\lim_{x \to 0} [f(x) \div g(x)] = \lim_{x \to 0} \left[\dfrac{1}{x} \div \left(-\dfrac{1}{x} \right) \right]$

$= \lim_{x \to 0} (-1) = -1$

79. Given $f(x) = b$, show that for every $\varepsilon > 0$ there exists a $\delta > 0$ such that $|f(x) - b| < \varepsilon$ whenever $|x - c| < \delta$. Since $|f(x) - b| = |b - b| = 0 < \epsilon$ for any $\varepsilon > 0$, then any value of $\delta > 0$ will work.

81. If $b = 0$, then the property is true because both sides are equal to 0. If $b \neq 0$, let $\varepsilon > 0$ be given. Since $\lim_{x \to c} f(x) = L$,

there exists $\delta > 0$ such that $|f(x) - L| < \varepsilon/|b|$ whenever $0 < |x - c| < \delta$. Hence, wherever $0 < |x - c| < \delta$,

we have

$$|b||f(x) - L| < \varepsilon \quad \text{or} \quad |bf(x) - bL| < \varepsilon$$

which implies that $\lim_{x \to c} [bf(x)] = bL$.

Section 3.4   Continuity and One-Sided Limits

1. (a) $\lim_{x \to 3^+} f(x) = 1$

    (b) $\lim_{x \to 3^-} f(x) = 1$

    (c) $\lim_{x \to 3} f(x) = 1$

The function is continuous at $x = 3$.

3. (a) $\lim_{x \to 3^+} f(x) = 0$

    (b) $\lim_{x \to 3^-} f(x) = 0$

    (c) $\lim_{x \to 3} f(x) = 0$

The function is NOT continuous at $x = 3$.

5. (a) $\lim_{x \to 4^+} f(x) = 2$

    (b) $\lim_{x \to 4^-} f(x) = -2$

    (c) $\lim_{x \to 4} f(x)$ does not exist

The function is NOT continuous at $x = 4$.

7. $\lim_{x \to 5^+} \dfrac{x - 5}{x^2 - 25} = \lim_{x \to 5^+} \dfrac{1}{x + 5} = \dfrac{1}{10}$

9. $\lim_{x \to -3^-} \dfrac{x}{\sqrt{x^2 - 9}}$ does not exist because $\dfrac{x}{\sqrt{x^2 - 9}}$ grows without bound as $x \to -3^-$.

11. $\lim_{x \to 0^-} \dfrac{|x|}{x} = \lim_{x \to 0^-} \dfrac{-x}{x} = -1$

13. $\lim_{\Delta x \to 0^-} \dfrac{\dfrac{1}{x + \Delta x} - \dfrac{1}{x}}{\Delta x} = \lim_{\Delta x \to 0^-} \dfrac{x - (x + \Delta x)}{x(x + \Delta x)} \cdot \dfrac{1}{\Delta x} = \lim_{\Delta x \to 0^-} \dfrac{-\Delta x}{x(x + \Delta x)} \cdot \dfrac{1}{\Delta x}$

$$= \lim_{\Delta x \to 0^-} \dfrac{-1}{x(x + \Delta x)}$$

$$= \dfrac{-1}{x(x + 0)} = -\dfrac{1}{x^2}$$

15. $\lim_{x \to 3^-} f(x) = \lim_{x \to 3^-} \dfrac{x + 2}{2} = \dfrac{5}{2}$

17. $\lim_{x \to 1^+} f(x) = \lim_{x \to 1^+} (x + 1) = 2$

    $\lim_{x \to 1^-} f(x) = \lim_{x \to 1^-} (x^3 + 1) = 2$

    $\lim_{x \to 1} f(x) = 2$

19. $\lim_{x \to 4^-} (3[\![x]\!] - 5) = 3(3) - 5 = 4$

    $([\![x]\!] = 3$ for $3 < x < 4)$

21. $\lim\limits_{x \to 3} (2 - [\![-x]\!])$ does not exist

because

$\lim\limits_{x \to 3^-} (2 - [\![-x]\!]) = 2 - (-3) = 5$

and

$\lim\limits_{x \to 3^+} (2 - [\![-x]\!]) = 2 - (-4) = 6.$

23. $f(x) = \dfrac{1}{x^2 - 4}$

has discontinuities at $x = -2$ and $x = 2$ since $f(-2)$ and $f(2)$ are not defined.

25. $f(x) = \dfrac{[\![x]\!]}{2} + x$

has discontinuities at each integer k since
$\lim\limits_{x \to k^-} f(x) \neq \lim\limits_{x \to k^+} f(x).$

27. $g(x) = \sqrt{25 - x^2}$ is continuous on $[-5, 5]$.

29. $\lim\limits_{x \to 0^-} f(x) = 3 = \lim\limits_{x \to 0^+} f(x).$
f is continuous on $[-1, 4]$.

31. $f(x) = x^2 - 2x + 1$ is continuous for all real x.

33. $f(x) = \dfrac{x}{x^2 - x}$ is not continuous at $x = 0, 1$.

Since $\dfrac{x}{x^2 - x} = \dfrac{1}{x - 1}$ for $x \neq 0$, $x = 0$ is a removable discontinuity, whereas $x = 1$ is a nonremovable discontinuity.

35. $f(x) = \dfrac{x}{x^2 + 1}$ is continuous for all real x.

37. $f(x) = \dfrac{x + 2}{(x + 2)(x - 5)}$

has a nonremovable discontinuity at $x = 5$ since $\lim\limits_{x \to 5} f(x)$ does not exist, and has a removable discontinuity at $x = -2$ since

$$\lim\limits_{x \to -2} f(x) = \lim\limits_{x \to -2} \dfrac{1}{x - 5} = -\dfrac{1}{7}.$$

39. $f(x) = \dfrac{|x + 2|}{x + 2}$ has a nonremovable discontinuity at $x = -2$ since $\lim\limits_{x \to -2} f(x)$ does not exist.

41. $f(x) = \begin{cases} x, & x \leq 1 \\ x^2, & x > 1 \end{cases}$

has a **possible** discontinuity at $x = 1$.

1. $f(1) = 1$

2. $\left.\begin{array}{l} \lim\limits_{x \to 1^-} f(x) = \lim\limits_{x \to 1^-} x = 1 \\ \lim\limits_{x \to 1^+} f(x) = \lim\limits_{x \to 1^+} x^2 = 1 \end{array}\right\} \lim\limits_{x \to 1} f(x) = 1$

3. $f(1) = \lim\limits_{x \to 1} f(x)$

f is continuous at $x = 1$, therefore, f is continuous for all real x.

43. $f(x) = \begin{cases} \dfrac{x}{2} + 1, & x \leq 2 \\ 3 - x, & x > 2 \end{cases}$ has a **possible** discontinuity at $x = 2$.

1. $f(2) = \dfrac{2}{2} + 1 = 2$

2. $\left.\begin{array}{l} \lim\limits_{x \to 2^-} f(x) = \lim\limits_{x \to 2^-} \left(\dfrac{x}{2} + 1\right) = 2 \\ \lim\limits_{x \to 2^+} f(x) = \lim\limits_{x \to 2^+} (3 - x) = 1 \end{array}\right\} \lim\limits_{x \to 2} f(x)$ does not exist.

Therefore, f has a nonremovable discontinuity at $x = 2$.

45. $f(x) = [\![x - 1]\!]$ has nonremovable discontinuities at each integer k.

47. $\lim_{x \to 0^+} f(x) = 0$

$\lim_{x \to 0^-} f(x) = 0$

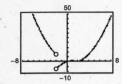

f is not continuous at $x = -2$.

49. Find a such that

$$\lim_{x \to 2^+} ax^2 = 4a$$

$$= \lim_{x \to 2^-} x^3 = 8.$$

$$4a = 8$$

$$a = 2$$

51. $f(g(x)) = (x - 1)^2$

Continuous for all real x.

53. $h(x) = f(g(x)) = f\left(\dfrac{1}{x}\right) = \dfrac{1}{\sqrt{1/x}} = \sqrt{x}, \; x \neq 0$

Continuous on $(0, \infty)$.

55. $f(g(x)) = \dfrac{1}{(x^2 + 5) - 6} = \dfrac{1}{x^2 - 1}$

Nonremovable discontinuities at $x = \pm 1$.

57. $f(x) = [\![x]\!] - x$

Nonremovable discontinuity at each integer.

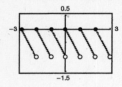

59. $g(x) = \begin{cases} 2x - 4, & x \leq 3 \\ x^2 - 2x, & x > 3 \end{cases}$

Nonremovable discontinuity at $x = 3$.

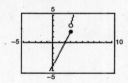

61. $f(x) = \dfrac{x}{x^2 + 1}$

Continuous on $(-\infty, \infty)$.

63. $f(x) = \dfrac{x^2}{x^2 - 36}$

Continuous on: $(-\infty, -6), (-6, 6), (6, \infty)$.

65. $f(x) = \dfrac{x^2 - x - 2}{x + 1}$

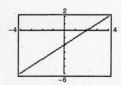

67. $f(x) = \frac{1}{16}x^4 - x^3 + 3$ is continuous on $[1, 2]$. $f(1) = \frac{33}{16}$ and $f(2) = -4$. By the Intermediate Value Theorem, $f(c) = 0$ for at least one value of c between 1 and 2.

The graph **appears** to be continuous on the interval $[-4, 4]$. Since $f(-1)$ is not defined, we know that f has a discontinuity at $x = -1$. This discontinuity is removable so it does not show up on the graph.

69. $f(x) = x^3 + x - 1$

$f(x)$ is continuous on $[0, 1]$.

$f(0) = -1$ and $f(1) = 1$.

By the Intermediate Value Theorem, $f(x) = 0$ for at least one value of c between 0 and 1.

Using a graphing utility, we find that $x \approx 0.6823$.

71. $g(t) = 3\sqrt{t^2 + 1} - 4$. g is continuous on $[0, 1]$.

$g(0) = 3 - 4 = -1$ and $g(1) = 3\sqrt{2} - 4 \approx 0.2426$.

By the Intermediate Value Theorem, $g(t) = 0$ for at least one value of c between 0 and 1.

Using the zoom feature, we find that $c \approx 0.88$. Using a graphing utility, we find that $c \approx 0.8819$.

73. $f(x) = x^2 + x - 1$

f is continuous on $[0, 5]$.

$f(0) = -1$ and $f(5) = 29$.

$$-1 < 11 < 29$$

The Intermediate Value Theorem applies.

$$x^2 + x - 1 = 11$$
$$x^2 + x - 12 = 0$$
$$(x + 4)(x - 3) = 0$$
$$x = -4 \text{ or } x = 3$$
$$c = 3 \ (x = -4 \text{ is not in the interval.})$$

Thus, $f(3) = 11$.

75. $f(x) = x^3 - x^2 + x - 2$

f is continuous on $[0, 3]$.

$f(0) = -2$ and $f(3) = 19$.

$$-2 < 4 < 19$$

The Intermediate Value Theorem applies.

$$x^3 - x^2 + x - 2 = 4$$
$$x^3 - x^2 + x - 6 = 0$$
$$(x - 2)(x^2 + x + 3) = 0$$
$$x = 2$$
$$(x^2 + x + 3 = 0 \text{ has no real solution.})$$
$$c = 2$$

Thus, $f(2) = 4$.

77. (a) The limit does not exist at $x = c$.

(b) The function is not defined at $x = c$.

(c) The limit exists at $x = c$, but it is not equal to the value of the function at $x = c$.

(d) The limit does not exist at $x = c$.

79.

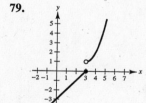

The function is not continuous at $x = 3$ because

$$\lim_{x \to 3^+} f(x) = 1 \neq 0 = \lim_{x \to 3^-} f(x).$$

81. The functions agree for integer values of x:

$$\left. \begin{array}{l} g(x) = 3 - [\![-x]\!] = 3 - (-x) = 3 + x \\ f(x) = 3 + [\![x]\!] = 3 + x \end{array} \right\} \text{ for } x \text{ an integer}$$

However, for non-integer values of x, the functions differ by 1.

$$f(x) = 3 + [\![x]\!] = g(x) - 1 = 2 - [\![-x]\!].$$

For example, $f\left(\frac{1}{2}\right) = 3 + 0 = 3, g\left(\frac{1}{2}\right) = 3 - (-1) = 4$.

83. $N(t) = 25\left(2\left[\!\!\left[\dfrac{t+2}{2}\right]\!\!\right] - t\right)$

t	0	1	1.8	2	3	3.8
$N(t)$	50	25	5	50	25	5

Discontinuous at every positive even integer. The company replenishes its inventory every two months.

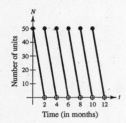

85. Let $V = \dfrac{4}{3}\pi r^3$ be the volume of a sphere of radius r.

V is continuous on $[1, 5]$.

$V(1) = \dfrac{4}{3}\pi \approx 4.19$

$V(5) = \dfrac{4}{3}\pi(5^3) \approx 523.6$

Since $4.19 < 275 < 523.6$ the Intermediate Value Theorem implies that there is at least one value r between 1 and 5 such that $V(r) = 275$. (In fact, $r \approx 4.0341$.)

87. Let c be any real number. Then $\lim\limits_{x \to c} f(x)$ does not exist since there are both rational and irrational numbers arbitrarily close to c. Therefore, f is not continuous at c.

89. True; if $f(x) = g(x)$, $x \neq c$, then $\lim\limits_{x \to c} f(x) = \lim\limits_{x \to c} g(x)$ (if they exist) and at least one of these limits then does not equal the corresponding function value at $x = c$.

91. False; $f(1)$ is not defined and $\lim\limits_{x \to 1} f(x)$ does not exist.

93. (a) $f(x) = \begin{cases} 0 & 0 \le x < b \\ b & b < x \le 2b \end{cases}$

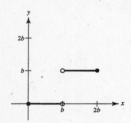

NOT continuous at $x = b$.

(b) $g(x) = \begin{cases} \dfrac{x}{2} & 0 \le x \le b \\ b - \dfrac{x}{2} & b < x \le 2b \end{cases}$

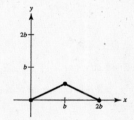

Continuous on $[0, 2b]$.

95. $f(x) = \dfrac{\sqrt{x + c^2} - c}{x}$, $c > 0$

Domain: $x + c^2 \ge 0 \implies x \ge -c^2$ and $x \neq 0$, $[-c^2, 0) \cup (0, \infty)$

$\lim\limits_{x \to 0} \dfrac{\sqrt{x + c^2} - c}{x} = \lim\limits_{x \to 0} \dfrac{\sqrt{x + c^2} - c}{x} \cdot \dfrac{\sqrt{x + c^2} + c}{\sqrt{x + c^2} + c}$

$\qquad = \lim\limits_{x \to 0} \dfrac{(x + c^2) - c^2}{x\left[\sqrt{x + c^2} + c\right]} = \lim\limits_{x \to 0} \dfrac{1}{\sqrt{x + c^2} + c} = \dfrac{1}{2c}$

Define $f(0) = 1/(2c)$ to make f continuous at $x = 0$.

Section 3.5   Infinite Limits

1. $\displaystyle\lim_{x \to -2^+} 2\left|\frac{x}{x^2 - 4}\right| = \infty$

$\displaystyle\lim_{x \to -2^-} 2\left|\frac{x}{x^2 - 4}\right| = \infty$

3. $f(x) = \dfrac{1}{x^2 - 9}$

x	-3.5	-3.1	-3.01	-3.001	-2.999	-2.99	-2.9	-2.5
$f(x)$	0.308	1.639	16.64	166.6	-166.7	-16.69	-1.695	-0.364

$\displaystyle\lim_{x \to -3^-} f(x) = \infty$

$\displaystyle\lim_{x \to -3^+} f(x) = -\infty$

5. $f(x) = \dfrac{x^2}{x^2 - 9}$

x	-3.5	-3.1	-3.01	-3.001	-2.999	-2.99	-2.9	-2.5
$f(x)$	3.769	15.75	150.8	1501	-1499	-149.3	-14.25	-2.273

$\displaystyle\lim_{x \to -3^-} f(x) = \infty$

$\displaystyle\lim_{x \to -3^+} f(x) = -\infty$

7. $\displaystyle\lim_{x \to 2^+} \frac{x^2 - 2}{(x - 2)(x + 1)} = \infty$

$\displaystyle\lim_{x \to 2^-} \frac{x^2 - 2}{(x - 2)(x + 1)} = -\infty$

Therefore, $x = 2$ is a vertical asymptote.

$\displaystyle\lim_{x \to -1^+} \frac{x^2 - 2}{(x - 2)(x + 1)} = \infty$

$\displaystyle\lim_{x \to -1^-} \frac{x^2 - 2}{(x - 2)(x + 1)} = -\infty$

Therefore, $x = -1$ is a vertical asymptote.

9. $\displaystyle\lim_{x \to 0^+} \frac{1}{x^2} = \infty = \lim_{x \to 0^-} \frac{1}{x^2}$

Therefore, $x = 0$ is a vertical asymptote.

11. $f(x) = \dfrac{x^2}{x^2 + x - 6} = \dfrac{x^2}{(x + 3)(x - 2)}$

Vertical asymptotes: $x = -3, 2$

13. $\displaystyle\lim_{x \to -2^-} \frac{x^2}{x^2 - 4} = \infty$ and $\displaystyle\lim_{x \to -2^+} \frac{x^2}{x^2 - 4} = -\infty$.

Therefore, $x = -2$ is a vertical asymptote.

$\displaystyle\lim_{x \to 2^-} \frac{x^2}{x^2 - 4} = -\infty$ and $\displaystyle\lim_{x \to 2^+} \frac{x^2}{x^2 - 4} = \infty$.

Therefore, $x = 2$ is a vertical asymptote.

15. $g(t) = \dfrac{t-1}{t^2+1}$

$t^2 + 1 \neq 0$. No vertical asymptotes.

19. $\displaystyle\lim_{x \to -2^+} \dfrac{x}{(x+2)(x-1)} = \infty$

$\displaystyle\lim_{x \to -2^-} \dfrac{x}{(x+2)(x-1)} = -\infty$

Therefore, $x = -2$ is a vertical asymptote.

$\displaystyle\lim_{x \to 1^+} \dfrac{x}{(x+2)(x-1)} = \infty$

$\displaystyle\lim_{x \to 1^-} \dfrac{x}{(x+2)(x-1)} = -\infty$

Therefore, $x = 1$ is a vertical asymptote.

23. $f(x) = \dfrac{(x-5)(x+3)}{(x-5)(x^2+1)} = \dfrac{x+3}{x^2+1}, \ x \neq 5$

No vertical asymptotes. The graph has a hole at $x = 5$.

17. $\displaystyle\lim_{t \to 0^+} \left(1 - \dfrac{4}{t^2}\right) = -\infty = \lim_{t \to 0^-} \left(1 - \dfrac{4}{t^2}\right)$

Therefore, $t = 0$ is a vertical asymptote.

21. $g(x) = \dfrac{x^3+1}{x+1} = \dfrac{(x+1)(x^2-x+1)}{x+1}$

has no vertical asymptote since

$\displaystyle\lim_{x \to -1} f(x) = \lim_{x \to -1} (x^2 - x + 1) = 3.$

25. $\displaystyle\lim_{x \to -1} \dfrac{x^2-1}{x+1} = \lim_{x \to -1} (x-1) = -2$

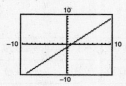

Removable discontinuity at $x = -1$.

27. $\displaystyle\lim_{x \to -1^+} \dfrac{x^2+1}{x+1} = \infty$

$\displaystyle\lim_{x \to -1^-} \dfrac{x^2+1}{x+1} = -\infty$

Vertical asymptote at $x = -1$.

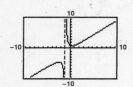

29. $\displaystyle\lim_{x \to 2^+} \dfrac{x-3}{x-2} = -\infty$

31. $\displaystyle\lim_{x \to 3^+} \dfrac{x^2}{(x-3)(x+3)} = \infty$

33. $\displaystyle\lim_{x \to -3^-} \dfrac{x^2+2x-3}{x^2+x-6} = \lim_{x \to -3^-} \dfrac{x-1}{x-2} = \dfrac{4}{5}$

35. $\displaystyle\lim_{x \to 1} \dfrac{x^2-x}{(x^2+1)(x-1)} = \lim_{x \to 1} \dfrac{x}{x^2+1} = \dfrac{1}{2}$

37. $\displaystyle\lim_{x \to 0^-} \left(1 + \dfrac{1}{x}\right) = -\infty$

39. $\displaystyle\lim_{x \to 4} f(x) = \lim_{x \to 4} \dfrac{1}{(x-4)^2} = \infty$

41. $\displaystyle\lim_{x \to 4} [f(x) + g(x)] = \lim_{x \to 4} \left[\dfrac{1}{(x-4)^2} + x^2 - 5x\right] = \infty$

43. $\displaystyle\lim_{x \to 4} \left[\dfrac{f(x)}{g(x)}\right] = \lim_{x \to 4} \dfrac{1}{(x-4)^2(x^2-5x)} = -\infty$

45. $f(x) = \dfrac{x^2 + x + 1}{x^3 - 1}$

$\lim\limits_{x \to 1^+} f(x) = \lim\limits_{x \to 1^+} \dfrac{1}{x - 1} = \infty$

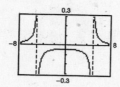

47. $f(x) = \dfrac{1}{x^2 - 25}$

$\lim\limits_{x \to 5^-} f(x) = -\infty$

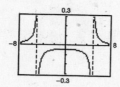

49. A limit in which $f(x)$ increases or decreases without bound as x approaches c is called an infinite limit. ∞ is not a number. Rather, the symbol

$$\lim_{x \to c} f(x) = \infty$$

says how the limit fails to exist.

51. One answer is $f(x) = \dfrac{x - 3}{(x - 6)(x + 2)} = \dfrac{x - 3}{x^2 - 4x - 12}$.

53.

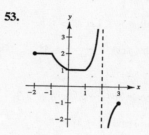

55. $C = \dfrac{80,000p}{100 - p}, \ 0 \le p < 100$

   (a) $C(15) \approx \$14,118$

   (b) $C(50) = \$80,000$

   (c) $C(90) = \$720,000$

   (d) $\lim\limits_{p \to 100^-} C = \infty$

   The cost increases without bound.

57. $m = \dfrac{m_0}{\sqrt{1 - (v^2/c^2)}}$

$\lim\limits_{v \to c^-} m = \lim\limits_{v \to c^-} \dfrac{m_0}{\sqrt{1 - (v^2/c^2)}} = \infty$

59. (a) Average speed $= \dfrac{\text{Total distance}}{\text{Total time}}$

$$50 = \dfrac{2d}{(d/x) + (d/y)}$$

$$50 = \dfrac{2xy}{y + x}$$

$$50y + 50x = 2xy$$

$$50x = 2xy - 50y$$

$$50x = 2y(x - 25)$$

$$\dfrac{25x}{x - 25} = y$$

Domain: $x > 25$

(b)

x	30	40	50	60
y	150	66.667	50	42.857

(c) $\lim\limits_{x \to 25^+} \dfrac{25x}{x - 25} = \infty$

As x gets close to 25 mph, y becomes larger and larger.

61. False; for instance, let

$$f(x) = \dfrac{x^2 - 1}{x - 1}.$$

The graph of f has a hole at $(1, 2)$, not a vertical asymptote.

63. True

65. Let $f(x) = \dfrac{1}{x^2}$ and $g(x) = \dfrac{1}{x^4}$, and $c = 0$.

$$\lim_{x\to 0} \frac{1}{x^2} = \infty \text{ and } \lim_{x\to 0} \frac{1}{x^4} = \infty, \text{ but } \lim_{x\to 0}\left(\frac{1}{x^2} - \frac{1}{x^4}\right) = \lim_{x\to 0}\left(\frac{x^2 - 1}{x^4}\right) = -\infty \neq 0.$$

67. Given $\displaystyle\lim_{x\to c} f(x) = \infty$, let $g(x) = 1$. Then $\displaystyle\lim_{x\to c} \frac{g(x)}{f(x)} = 0$ by Theorem 3.13.

69. $f(x) = \dfrac{1}{x-3}$ is defined for all $x > 3$. Let $M > 0$ be

given. We need $\delta > 0$ such that $f(x) = \dfrac{1}{x-3} > M$

whenever $3 < x < 3 + \delta$.

Equivalently, $x - 3 < \dfrac{1}{M}$ whenever $|x - 3| < \delta,\ x > 3$.

So take $\delta = \dfrac{1}{M}$. Then for $x > 3$ and $|x - 3| < \delta$,

$\dfrac{1}{x-3} > \dfrac{1}{\delta} = M$ and hence $f(x) > M$.

Review Exercises for Chapter 3

1. Calculus required. Using a graphing utility, you can estimate the length to be 8.3.
Or, the length is slightly longer than the distance between the two points, approximately 8.25.

3.

x	-0.1	-0.01	-0.001	0.001	0.01	0.1
$f(x)$	-1.053	-1.005	-1.001	-0.9995	-0.995	-0.9524

$\displaystyle\lim_{x\to 0} f(x) \approx -1$

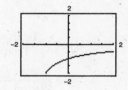

5. $h(x) = \dfrac{x^2 - 2x}{x}$

  (a) $\displaystyle\lim_{x\to 0} h(x) = -2$

  (b) $\displaystyle\lim_{x\to -1} h(x) = -3$

7. $\displaystyle\lim_{x\to 1}(3 - x) = 3 - 1 = 2$

Let $\varepsilon > 0$ be given. Choose $\delta = \varepsilon$. Then for

$0 < |x - 1| < \delta = \varepsilon$, you have

$$|x - 1| < \varepsilon$$
$$|1 - x| < \varepsilon$$
$$|(3 - x) - 2| < \varepsilon$$
$$|f(x) - L| < \varepsilon.$$

9. $\lim_{x \to 2} (x^2 - 3) = 1$

Let $\varepsilon > 0$ be given. We need $|x^2 - 3 - 1| < \varepsilon \Rightarrow |x^2 - 4| = |(x - 2)(x + 2)| < \varepsilon \Rightarrow |x - 2| < \dfrac{1}{|x + 2|}\varepsilon$.

Assuming, $1 < x < 3$, you can choose $\delta = \varepsilon/5$. Hence, for $0 < |x - 2| < \delta = \varepsilon/5$ you have

$$|x - 2| < \frac{\varepsilon}{5} < \frac{1}{|x + 2|}\varepsilon$$
$$|x - 2||x + 2| < \varepsilon$$
$$|x^2 - 4| < \varepsilon$$
$$|(x^2 - 3) - 1| < \varepsilon$$
$$|f(x) - L| < \varepsilon.$$

11. $\lim_{t \to 4} \sqrt{t + 2} = \sqrt{4 + 2} = \sqrt{6} \approx 2.45$

13. $\lim_{t \to -2} \dfrac{t + 2}{t^2 - 4} = \lim_{t \to -2} \dfrac{1}{t - 2} = -\dfrac{1}{4}$

15. $\lim_{x \to 4} \dfrac{\sqrt{x} - 2}{x - 4} = \lim_{x \to 4} \dfrac{\sqrt{x} - 2}{(\sqrt{x} - 2)(\sqrt{x} + 2)}$

$\qquad = \lim_{x \to 4} \dfrac{1}{\sqrt{x} + 2} = \dfrac{1}{\sqrt{4} + 2} = \dfrac{1}{4}$

17. $\lim_{x \to 0} \dfrac{[1/(x + 1)] - 1}{x} = \lim_{x \to 0} \dfrac{1 - (x + 1)}{x(x + 1)}$

$\qquad = \lim_{x \to 0} \dfrac{-1}{x + 1} = -1$

19. $\lim_{x \to -5} \dfrac{x^3 + 125}{x + 5} = \lim_{x \to -5} \dfrac{(x + 5)(x^2 - 5x + 25)}{x + 5}$

$\qquad = \lim_{x \to -5} (x^2 - 5x + 25)$

$\qquad = 75$

21. $\lim_{x \to c} [f(x) \cdot g(x)] = \left(-\dfrac{3}{4}\right)\left(\dfrac{2}{3}\right) = -\dfrac{1}{2}$

23. $\lim_{t \to a} \dfrac{s(a) - s(t)}{a - t} = \lim_{t \to 4} \dfrac{(-4.9(4)^2 + 200) - (-4.9t^2 + 200)}{4 - t}$

$\qquad = \lim_{t \to 4} \dfrac{4.9(t - 4)(t + 4)}{4 - t}$

$\qquad = \lim_{t \to 4} -4.9(t + 4) = -39.2 \text{ m/sec}$

25. $\lim_{x \to 3^-} \dfrac{|x - 3|}{x - 3} = \lim_{x \to 3^-} \dfrac{-(x - 3)}{x - 3} = -1$

27. $\lim_{x \to 2} f(x) = 0$

29. $\lim_{t \to 1} h(t)$ does not exist because $\lim_{t \to 1^-} h(t) = 1 + 1 = 2$ and $\lim_{t \to 1^+} h(t) = \frac{1}{2}(1 + 1) = 1$.

31. $f(x) = [\![x + 3]\!]$

$\lim_{x \to k^+} [\![x + 3]\!] = k + 3$ where k is an integer.

$\lim_{x \to k^-} [\![x + 3]\!] = k + 2$ where k is an integer.

Nonremovable discontinuity at each integer k.

Continuous on $(k, k + 1)$ for all integers k.

33. $f(x) = \dfrac{3x^2 - x - 2}{x - 1} = \dfrac{(3x + 2)(x - 1)}{x - 1}$

$\lim_{x \to 1} f(x) = \lim_{x \to 1} (3x + 2) = 5$

Removable discontinuity at $x = 1$.

Continuous on $(-\infty, 1) \cup (1, \infty)$.

35. $f(x) = \dfrac{1}{(x-2)^2}$

$\lim\limits_{x \to 2} \dfrac{1}{(x-2)^2} = \infty$

Nonremovable discontinuity at $x = 2$.

Continuous on $(-\infty, 2) \cup (2, \infty)$.

37. $f(x) = \dfrac{3}{x+1}$

$\lim\limits_{x \to -1^-} f(x) = -\infty$

$\lim\limits_{x \to -1^+} f(x) = \infty$

Nonremovable discontinuity at $x = -1$.

Continuous on $(-\infty, -1) \cup (-1, \infty)$.

39. $f(x) = \dfrac{\sqrt{2x+1} - \sqrt{3}}{x-1}$

(a)

x	1.1	1.01	1.001	1.0001
$f(x)$	0.5680	0.5764	0.5773	0.5773

(b)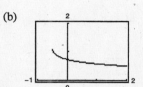

$\lim\limits_{x \to 1^+} \dfrac{\sqrt{2x+1} - \sqrt{3}}{x-1} \approx 0.577$ $\left(\text{Actual limit is } \sqrt{3}/3.\right)$

(c) $\lim\limits_{x \to 1^+} \dfrac{\sqrt{2x+1} - \sqrt{3}}{x-1} = \lim\limits_{x \to 1^+} \dfrac{\sqrt{2x+1} - \sqrt{3}}{x-1} \cdot \dfrac{\sqrt{2x+1} + \sqrt{3}}{\sqrt{2x+1} + \sqrt{3}}$

$= \lim\limits_{x \to 1^+} \dfrac{(2x+1) - 3}{(x-1)\left(\sqrt{2x+1} + \sqrt{3}\right)}$

$= \lim\limits_{x \to 1^+} \dfrac{2}{\sqrt{2x+1} + \sqrt{3}}$

$= \dfrac{2}{2\sqrt{3}} = \dfrac{1}{\sqrt{3}} = \dfrac{\sqrt{3}}{3}$

41. $f(2) = 5$

Find c so that $\lim\limits_{x \to 2^+} (cx + 6) = 5$.

$c(2) + 6 = 5$

$2c = -1$

$c = -\frac{1}{2}$

43. f is continuous on $[1, 2]$. $f(1) = -1 < 0$ and $f(2) = 13 > 0$. Therefore by the Intermediate Value Theorem, there is at least one value c in $(1, 2)$ such that $2c^3 - 3 = 0$.

45. $f(x) = \dfrac{x^2 - 4}{|x-2|} = (x+2)\left[\dfrac{x-2}{|x-2|}\right]$

(a) $\lim\limits_{x \to 2^-} f(x) = -4$

(b) $\lim\limits_{x \to 2^+} f(x) = 4$

(c) $\lim\limits_{x \to 2} f(x)$ does not exist.

47. $g(x) = 1 + \dfrac{2}{x}$

Vertical asymptote at $x = 0$.

49. $f(x) = \dfrac{8}{(x-10)^2}$

Vertical asymptote at $x = 10$.

51. $\lim\limits_{x \to -2^-} \dfrac{2x^2 + x + 1}{x+2} = -\infty$

53. $\lim\limits_{x \to -1^+} \dfrac{x+1}{x^3+1} = \lim\limits_{x \to -1^+} \dfrac{1}{x^2 - x + 1} = \dfrac{1}{3}$

55. $\lim\limits_{x \to 1^-} \dfrac{x^2 + 2x + 1}{x-1} = -\infty$

57. $\lim\limits_{x \to 0^+} \left(x - \dfrac{1}{x^3}\right) = -\infty$

59. $r = \dfrac{2L}{\sqrt{L^2 - 144}}$

    (a) When $L = 25$, $r = \dfrac{2(25)}{\sqrt{25^2 - 144}} = \dfrac{50\sqrt{481}}{481} \approx 2.28$ ft/sec

    (b) When $L = 13$, $r = \dfrac{2(13)}{\sqrt{13^2 - 144}} = \dfrac{26}{5} = 5.2$ ft/sec

    (c) $\lim\limits_{L \to 12^+} r = \infty$

Problem Solving for Chapter 3

1. (a) Perimeter $\triangle PAO = \sqrt{x^2 + (y-1)^2} + \sqrt{x^2 + y^2} + 1$

                         $= \sqrt{x^2 + (x^2-1)^2} + \sqrt{x^2 + x^4} + 1$

       Perimeter $\triangle PBO = \sqrt{(x-1)^2 + y^2} + \sqrt{x^2 + y^2} + 1$

                         $= \sqrt{(x-1)^2 + x^4} + \sqrt{x^2 + x^4} + 1$

  (b) $r(x) = \dfrac{\sqrt{x^2 + (x^2-1)^2} + \sqrt{x^2 + x^4} + 1}{\sqrt{(x-1)^2 + x^4} + \sqrt{x^2 + x^4} + 1}$

x	4	2	1	0.1	0.01
Perimeter $\triangle PAO$	33.02	9.08	3.41	2.10	2.01
Perimeter $\triangle PBO$	33.77	9.60	3.41	2.00	2.00
$r(x)$		0.98	0.95	1	1.05	1.005

  (c) $\lim\limits_{x \to 0^+} r(x) = \dfrac{1 + 0 + 1}{1 + 0 + 1} = \dfrac{2}{2} = 1$

3. (a) Slope $= \dfrac{4 - 0}{3 - 0} = \dfrac{4}{3}$

  (b) Slope $= -\dfrac{3}{4}$ Tangent line: $y - 4 = -\dfrac{3}{4}(x - 3)$

                                   $y = -\dfrac{3}{4}x + \dfrac{25}{4}$

  (c) Let $Q = (x, y) = \left(x, \sqrt{25 - x^2}\right)$

       $m_x = \dfrac{\sqrt{25 - x^2} - 4}{x - 3}$

  (d) $\lim\limits_{x \to 3} m_x = \lim\limits_{x \to 3} \dfrac{\sqrt{25 - x^2} - 4}{x - 3} \cdot \dfrac{\sqrt{25 - x^2} + 4}{\sqrt{25 - x^2} + 4}$

                $= \lim\limits_{x \to 3} \dfrac{25 - x^2 - 16}{(x - 3)\left(\sqrt{25 - x^2} + 4\right)}$

                $= \lim\limits_{x \to 3} \dfrac{(3 - x)(3 + x)}{(x - 3)\left(\sqrt{25 - x^2} + 4\right)}$

                $= \lim\limits_{x \to 3} \dfrac{-(3 + x)}{\sqrt{25 - x^2} + 4} = \dfrac{-6}{4 + 4} = -\dfrac{3}{4}$

  This is the slope of the tangent line at P.

5. $\dfrac{\sqrt{a + bx} - \sqrt{3}}{x} = \dfrac{\sqrt{a + bx} - \sqrt{3}}{x} \cdot \dfrac{\sqrt{a + bx} + \sqrt{3}}{\sqrt{a + bx} + \sqrt{3}}$

                $= \dfrac{(a + bx) - 3}{x\left(\sqrt{a + bx} + \sqrt{3}\right)}$

Letting $a = 3$ simplifies the numerator.

Thus,

$$\lim_{x \to 0} \dfrac{\sqrt{3 + bx} - \sqrt{3}}{x} = \lim_{x \to 0} \dfrac{bx}{x\left(\sqrt{3 + bx} + \sqrt{3}\right)}$$

$$= \lim_{x \to 0} \dfrac{b}{\sqrt{3 + bx} + \sqrt{3}}.$$

Setting $\dfrac{b}{\sqrt{3} + \sqrt{3}} = \sqrt{3}$, you obtain $b = 6$.

Thus, $a = 3$ and $b = 6$.

7. (a) $\lim\limits_{x \to 2} f(x) = 3$: g_1, g_4

  (b) f continuous at 2: g_1

  (c) $\lim\limits_{x \to 2^-} f(x) = 3$: g_1, g_3, g_4

9.

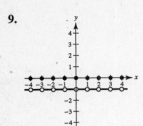

(a) $f(1) = [\![1]\!] + [\![-1]\!] = 1 + (-1) = 0$

     $f(0) = 0$

     $f\left(\frac{1}{2}\right) = 0 + (-1) = -1$

   $f(-2.7) = -3 + 2 = -1$

(b) $\lim\limits_{x \to 1^-} f(x) = -1$

     $\lim\limits_{x \to 1^+} f(x) = -1$

     $\lim\limits_{x \to 1/2} f(x) = -1$

(c) f is continuous for all real numbers except

   $x = 0, \pm 1, \pm 2, \pm 3, \ldots$

11. (a) $\quad v^2 = \dfrac{192{,}000}{r} + v_0{}^2 - 48$

     $\dfrac{192{,}000}{r} = v^2 - v_0{}^2 + 48$

        $r = \dfrac{192{,}000}{v^2 - v_0{}^2 + 48}$

     $\lim\limits_{v \to 0} r = \dfrac{192{,}000}{48 - v_0{}^2}$

Let $v_0 = \sqrt{48} = 4\sqrt{3}$ miles/sec.

(b) $\quad v^2 = \dfrac{1920}{r} + v_0{}^2 - 2.17$

     $\dfrac{1920}{r} = v^2 - v_0{}^2 + 2.17$

        $r = \dfrac{1920}{v^2 - v_0{}^2 + 2.17}$

     $\lim\limits_{v \to 0} r = \dfrac{1920}{2.17 - v_0{}^2}$

Let $v_0 = \sqrt{2.17}$ mi/sec $(\approx 1.47$ mi/sec$)$.

(c) $\quad r = \dfrac{10{,}600}{v^2 - v_0{}^2 + 6.99}$

     $\lim\limits_{v \to 0} r = \dfrac{10{,}600}{6.99 - v_0{}^2}$

Let $v_0 = \sqrt{6.99} \approx 2.64$ mi/sec.

Since this is smaller than the escape velocity for earth, the mass is less.

C H A P T E R   4
Differentiation

CHAPTER 4
Differentiation

Section 4.1   The Derivative and the Tangent Line Problem

1. (a) At (x_1, y_1), slope $= 0$.

   At (x_2, y_2), slope $\approx \frac{5}{2}$.

   (b) At (x_1, y_1), slope $\approx -\frac{5}{2}$.

   At (x_2, y_2), slope ≈ 2.

3. (a), (b)

$$y = \frac{f(4) - f(1)}{4 - 1}(x - 1) + f(1) = x + 1$$

(c) $y = \dfrac{f(4) - f(1)}{4 - 1}(x - 1) + f(1)$

$\qquad = \dfrac{3}{3}(x - 1) + 2$

$\qquad = 1(x - 1) + 2$

$\qquad = x + 1$

5. $f(x) = 3 - 2x$ is a line. Slope $= -2$

7. Slope at $(1, -3) = \lim\limits_{\Delta x \to 0} \dfrac{g(1 + \Delta x) - g(1)}{\Delta x}$

$\qquad = \lim\limits_{\Delta x \to 0} \dfrac{(1 + \Delta x)^2 - 4 - (-3)}{\Delta x}$

$\qquad = \lim\limits_{\Delta x \to 0} \dfrac{1 + 2(\Delta x) + (\Delta x)^2 - 1}{\Delta x}$

$\qquad = \lim\limits_{\Delta x \to 0} [2 + 2(\Delta x)] = 2$

9. Slope at $(0, 0) = \lim\limits_{\Delta t \to 0} \dfrac{f(0 + \Delta t) - f(0)}{\Delta t}$

$\qquad = \lim\limits_{\Delta t \to 0} \dfrac{3(\Delta t) - (\Delta t)^2 - 0}{\Delta t}$

$\qquad = \lim\limits_{\Delta t \to 0} (3 - \Delta t) = 3$

11. $f(x) = 3$

$\qquad f'(x) = \lim\limits_{\Delta x \to 0} \dfrac{f(x + \Delta x) - f(x)}{\Delta x}$

$\qquad\qquad = \lim\limits_{\Delta x \to 0} \dfrac{3 - 3}{\Delta x}$

$\qquad\qquad = \lim\limits_{\Delta x \to 0} 0 = 0$

13. $f(x) = -5x$

$\qquad f'(x) = \lim\limits_{\Delta x \to 0} \dfrac{f(x + \Delta x) - f(x)}{\Delta x}$

$\qquad\qquad = \lim\limits_{\Delta x \to 0} \dfrac{-5(x + \Delta x) - (-5x)}{\Delta x}$

$\qquad\qquad = \lim\limits_{\Delta x \to 0} -5 = -5$

15. $h(s) = 3 + \dfrac{2}{3}s$

$\qquad h'(s) = \lim\limits_{\Delta s \to 0} \dfrac{h(s + \Delta s) - h(s)}{\Delta s}$

$\qquad\qquad = \lim\limits_{\Delta s \to 0} \dfrac{3 + \dfrac{2}{3}(s + \Delta s) - \left(3 + \dfrac{2}{3}s\right)}{\Delta s}$

$\qquad\qquad = \lim\limits_{\Delta s \to 0} \dfrac{\dfrac{2}{3}\Delta s}{\Delta s} = \dfrac{2}{3}$

17. $f(x) = 2x^2 + x - 1$

$$f'(x) = \lim_{\Delta x \to 0} \frac{f(x + \Delta x) - f(x)}{\Delta x}$$

$$= \lim_{\Delta x \to 0} \frac{[2(x + \Delta x)^2 + (x + \Delta x) - 1] - [2x^2 + x - 1]}{\Delta x}$$

$$= \lim_{\Delta x 0} \frac{(2x^2 + 4x\,\Delta x + 2(\Delta x)^2 + x + \Delta x - 1) - (2x^2 + x - 1)}{\Delta x}$$

$$= \lim_{\Delta x \to 0} \frac{4x\,\Delta x + 2(\Delta x)^2 + \Delta x}{\Delta x} = \lim_{\Delta x \to 0} (4x + 2\,\Delta x + 1) = 4x + 1$$

19. $f(x) = x^3 - 12x$

$$f'(x) = \lim_{\Delta x \to 0} \frac{f(x + \Delta x) - f(x)}{\Delta x}$$

$$= \lim_{\Delta x \to 0} \frac{[(x + \Delta x)^3 - 12(x + \Delta x)] - [x^3 - 12x]}{\Delta x}$$

$$= \lim_{\Delta x \to 0} \frac{x^3 + 3x^2\Delta x + 3x(\Delta x)^2 + (\Delta x)^3 - 12x - 12\,\Delta x - x^3 + 12x}{\Delta x}$$

$$= \lim_{\Delta x \to 0} \frac{3x^2\,\Delta x + 3x(\Delta x)^2 + (\Delta x)^3 - 12\,\Delta x}{\Delta x}$$

$$= \lim_{\Delta x \to 0} (3x^2 + 3x\,\Delta x + (\Delta x)^2 - 12) = 3x^2 - 12$$

21. $f(x) = \dfrac{1}{x - 1}$

$$f'(x) = \lim_{\Delta x \to 0} \frac{f(x + \Delta x) - f(x)}{\Delta x}$$

$$= \lim_{\Delta x \to 0} \frac{\dfrac{1}{x + \Delta x - 1} - \dfrac{1}{x - 1}}{\Delta x}$$

$$= \lim_{\Delta x \to 0} \frac{(x - 1) - (x + \Delta x - 1)}{\Delta x(x + \Delta x - 1)(x - 1)}$$

$$= \lim_{\Delta x \to 0} \frac{-\Delta x}{\Delta x(x + \Delta x - 1)(x - 1)}$$

$$= \lim_{\Delta x \to 0} \frac{-1}{(x + \Delta x - 1)(x - 1)}$$

$$= -\frac{1}{(x - 1)^2}$$

23. $f(x) = \sqrt{x + 1}$

$$f'(x) = \lim_{\Delta x \to 0} \frac{f(x + \Delta x) - f(x)}{\Delta x}$$

$$= \lim_{\Delta x \to 0} \frac{\sqrt{x + \Delta x + 1} - \sqrt{x + 1}}{\Delta x} \cdot \left(\frac{\sqrt{x + \Delta x + 1} + \sqrt{x + 1}}{\sqrt{x + \Delta x + 1} + \sqrt{x + 1}} \right)$$

$$= \lim_{\Delta x \to 0} \frac{(x + \Delta x + 1) - (x + 1)}{\Delta x \left[\sqrt{x + \Delta x + 1} + \sqrt{x + 1} \right]}$$

$$= \lim_{\Delta x \to 0} \frac{1}{\sqrt{x + \Delta x + 1} + \sqrt{x + 1}}$$

$$= \frac{1}{\sqrt{x + 1} + \sqrt{x + 1}} = \frac{1}{2\sqrt{x + 1}}$$

25. (a) $f(x) = x^2 + 1$

$$f'(x) = \lim_{\Delta x \to 0} \frac{f(x + \Delta x) - f(x)}{\Delta x}$$

$$= \lim_{\Delta x \to 0} \frac{[(x + \Delta x)^2 + 1] - [x^2 + 1]}{\Delta x}$$

$$= \lim_{\Delta x \to 0} \frac{2x \, \Delta x + (\Delta x)^2}{\Delta x}$$

$$= \lim_{\Delta x \to 0} (2x + \Delta x) = 2x$$

At $(2, 5)$, the slope of the tangent line is $m = 2(2) = 4$. The equation of the tangent line is

$$y - 5 = 4(x - 2)$$

$$y - 5 = 4x - 8$$

$$y = 4x - 3.$$

(b)

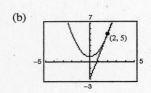

27. (a) $f(x) = x^3$

$$f'(x) = \lim_{\Delta x \to 0} \frac{f(x + \Delta x) - f(x)}{\Delta x}$$

$$= \lim_{\Delta x \to 0} \frac{(x + \Delta x)^3 - x^3}{\Delta x}$$

$$= \lim_{\Delta x \to 0} \frac{3x^2 \, \Delta x + 3x(\Delta x)^2 + (\Delta x)^3}{\Delta x}$$

$$= \lim_{\Delta x \to 0} (3x^2 + 3x\Delta x + (\Delta x)^2) = 3x^2$$

At $(2, 8)$, the slope of the tangent is $m = 3(2)^2 = 12$. The equation of the tangent line is

$$y - 8 = 12(x - 2)$$

$$y = 12x - 16.$$

(b)

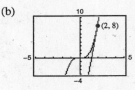

29. (a) $f(x) = x + \dfrac{4}{x}$

(b)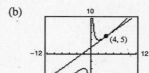

$$f'(x) = \lim_{\Delta x \to 0} \frac{f(x + \Delta x) - f(x)}{\Delta x}$$

$$= \lim_{\Delta x \to 0} \frac{(x + \Delta x) + \dfrac{4}{x + \Delta x} - \left(x + \dfrac{4}{x}\right)}{\Delta x}$$

$$= \lim_{\Delta x \to 0} \frac{x(x + \Delta x)(x + \Delta x) + 4x - x^2(x + \Delta x) - 4(x + \Delta x)}{x(\Delta x)(x + \Delta x)}$$

$$= \lim_{\Delta x \to 0} \frac{x^3 + 2x^2(\Delta x) + x(\Delta x)^2 - x^3 - x^2(\Delta x) - 4(\Delta x)}{x(\Delta x)(x + \Delta x)}$$

$$= \lim_{\Delta x \to 0} \frac{x^2(\Delta x) + x(\Delta x)^2 - 4(\Delta x)}{x(\Delta x)(x + \Delta x)}$$

$$= \lim_{\Delta x \to 0} \frac{x^2 + x(\Delta x) - 4}{x(x + \Delta x)}$$

$$= \frac{x^2 - 4}{x^2} = 1 - \frac{4}{x^2}$$

At $(4, 5)$, the slope of the tangent line is $m = 1 - \dfrac{4}{16} = \dfrac{3}{4}$.

The equation of the tangent line is $y - 5 = \dfrac{3}{4}(x - 4)$

$$y = \frac{3}{4}x + 2.$$

31. From Exercise 27 we know that $f'(x) = 3x^2$. Since the slope of the given line is 3, we have

$$3x^2 = 3$$

$$x = \pm 1.$$

Therefore, at the points $(1, 1)$ and $(-1, -1)$ the tangent lines are parallel to $3x - y + 1 = 0$. These lines have equations

$$y - 1 = 3(x - 1) \qquad \text{and} \qquad y + 1 = 3(x + 1)$$
$$y = 3x - 2 \qquad\qquad\qquad y = 3x + 2.$$

33. $g(5) = 2$ because the tangent line passes through $(5, 2)$.

$$g'(5) = \frac{2 - 0}{5 - 9} = \frac{2}{-4} = -\frac{1}{2}$$

35. $f(x) = x \implies f'(x) = 1$ (b)

37. $f(x) = \sqrt{x} \implies f'(x)$ matches (a)

(decreasing slope as $x \to \infty$)

39.

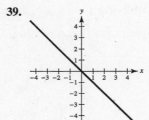

Answers will vary.

Sample answer: $y = -x$

41. (a) If $f'(c) = 3$ and f is odd, then $f'(-c) = f'(c) = 3$.

(b) If $f'(c) = 3$ and f is even, then $f'(-c) = -f'(c) = -3$.

43. The slope of the graph of f is $1 \implies f'(x) = 1$.

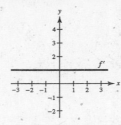

45. The slope of the graph of f is negative for $x < 4$, positive for $x > 4$, and 0 at $x = 4$.

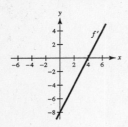

47. Let (x_0, y_0) be a point of tangency on the graph of f. By the limit definition for the derivative, $f'(x) = 4 - 2x$. The slope of the line through $(2, 5)$ and (x_0, y_0) equals the derivative of f at x_0:

$$\frac{5 - y_0}{2 - x_0} = 4 - 2x_0$$

$$5 - y_0 = (2 - x_0)(4 - 2x_0)$$

$$5 - (4x_0 - x_0^2) = 8 - 8x_0 + 2x_0^2$$

$$0 = x_0^2 - 4x_0 + 3$$

$$0 = (x_0 - 1)(x_0 - 3) \implies x_0 = 1, 3$$

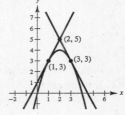

Therefore, the points of tangency are $(1, 3)$ and $(3, 3)$, and the corresponding slopes are 2 and -2. The equations of the tangent lines are

$$y - 5 = 2(x - 2) \qquad y - 5 = -2(x - 2)$$
$$y = 2x + 1 \qquad\qquad y = -2x + 9.$$

49. (a) $g'(0) = -3$

(b) $g'(3) = 0$

(c) Because $g'(1) = -\frac{8}{3}$, g is decreasing (falling) at $x = 1$.

(d) Because $g'(-4) = \frac{7}{3}$, g is increasing (rising) at $x = -4$.

(e) Because $g'(x) > 0$ for all x between 4 and 6, $g(6)$ is greater than $g(4)$, and $g(6) - g(4) > 0$.

(f) No, it is not possible. All you can say is that g is decreasing (falling) at $x = 2$.

51. $f(x) = \frac{1}{4}x^3$

By the limit definition of the derivative we have $f'(x) = \frac{3}{4}x^2$.

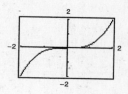

x	-2	-1.5	-1	-0.5	0	0.5	1	1.5	2
$f(x)$	-2	$-\frac{27}{32}$	$-\frac{1}{4}$	$-\frac{1}{32}$	0	$\frac{1}{32}$	$\frac{1}{4}$	$\frac{27}{32}$	2
$f'(x)$	3	$\frac{27}{16}$	$\frac{3}{4}$	$\frac{3}{16}$	0	$\frac{3}{16}$	$\frac{3}{4}$	$\frac{27}{16}$	3

53. $g(x) = \dfrac{f(x + 0.01) - f(x)}{0.01}$

$= [2(x + 0.01) - (x + 0.01)^2 - 2x + x^2]100$

$= 1.99 - 2x$

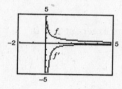

The graph of $g(x)$ is approximately the graph of $f'(x)$.

55. $f(2) = 2(4 - 2) = 4$, $f(2.1) = 2.1(4 - 2.1) = 3.99$

$f'(2) \approx \dfrac{3.99 - 4}{2.1 - 2} = -0.1$ [Exact: $f'(2) = 0$]

57. $f(x) = \dfrac{1}{\sqrt{x}}$ and $f'(x) = \dfrac{-1}{2x^{3/2}}$.

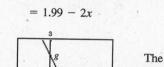

As $x \to \infty$, f is nearly horizontal and thus $f' \approx 0$.

59. $f(x) = 4 - (x - 3)^2$

$S_{\Delta x}(x) = \dfrac{f(2 + \Delta x) - f(2)}{\Delta x}(x - 2) + f(2)$

$= \dfrac{4 - (2 + \Delta x - 3)^2 - 3}{\Delta x}(x - 2) + 3 = \dfrac{1 - (\Delta x - 1)^2}{\Delta x}(x - 2) + 3 = (-\Delta x + 2)(x - 2) + 3$

(a) $\Delta x = 1$: $S_{\Delta x} = (x - 2) + 3 = x + 1$

$\Delta x = 0.5$: $S_{\Delta x} = \left(\frac{3}{2}\right)(x - 2) + 3 = \frac{3}{2}x$

$\Delta x = 0.1$: $S_{\Delta x} = \left(\frac{19}{10}\right)(x - 2) + 3 = \frac{19}{10}x - \frac{4}{5}$

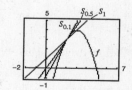

(b) As $\Delta x \to 0$, the line approaches the tangent line to f at $(2, 3)$.

61. $f(x) = x^2 - 1$, $c = 2$

$f'(2) = \lim_{x \to 2} \dfrac{f(x) - f(2)}{x - 2} = \lim_{x \to 2} \dfrac{(x^2 - 1) - 3}{x - 2} = \lim_{x \to 2} \dfrac{(x - 2)(x + 2)}{x - 2} = \lim_{x \to 2} (x + 2) = 4$

63. $f(x) = x^3 + 2x^2 + 1$, $c = -2$

$f'(-2) = \lim_{x \to -2} \dfrac{f(x) - f(-2)}{x + 2} = \lim_{x \to -2} \dfrac{(x^3 + 2x^2 + 1) - 1}{x + 2} = \lim_{x \to -2} \dfrac{x^2(x + 2)}{x + 2} = \lim_{x \to -2} x^2 = 4$

65. $g(x) = \sqrt{|x|}, c = 0$

$g'(0) = \lim\limits_{x \to 0} \dfrac{g(x) - g(0)}{x - 0} = \lim\limits_{x \to 0} \dfrac{\sqrt{|x|}}{x}$. Does not exist.

As $x \to 0^-$, $\dfrac{\sqrt{|x|}}{x} = \dfrac{-1}{\sqrt{|x|}} \to -\infty$

As $x \to 0^+$, $\dfrac{\sqrt{|x|}}{x} = \dfrac{1}{\sqrt{x}} \to \infty$

67. $f(x) = (x - 6)^{2/3}, c = 6$

$f'(6) = \lim\limits_{x \to 6} \dfrac{f(x) - f(6)}{x - 6}$

$\qquad = \lim\limits_{x \to 6} \dfrac{(x - 6)^{2/3} - 0}{x - 6}$

$\qquad = \lim\limits_{x \to 6} \dfrac{1}{(x - 6)^{1/3}}$

Does not exist.

69. $h(x) = |x + 5|, c = -5$

$h'(-5) = \lim\limits_{x \to -5} \dfrac{h(x) - h(-5)}{x - (-5)}$

$\qquad = \lim\limits_{x \to -5} \dfrac{|x + 5| - 0}{x + 5}$

$\qquad = \lim\limits_{x \to -5} \dfrac{|x + 5|}{x + 5}$

Does not exist.

71. $f(x)$ is differentiable everywhere except at $x = -3$. (Sharp turn in the graph.)

73. $f(x)$ is differentiable everywhere except at $x = -1$. (Discontinuity)

75. $f(x)$ is differentiable everywhere except at $x = 3$. (Sharp turn in the graph.)

77. $f(x)$ is differentiable on the interval $(1, \infty)$. (At $x = 1$ the tangent line is vertical.)

79. $f(x)$ is differentiable everywhere except at $x = 0$. (Discontinuity)

81. $f(x) = |x - 1|$

The derivative from the left is $\lim\limits_{x \to 1^-} \dfrac{f(x) - f(1)}{x - 1} = \lim\limits_{x \to 1^-} \dfrac{|x - 1| - 0}{x - 1} = -1$.

The derivative from the right is $\lim\limits_{x \to 1^+} \dfrac{f(x) - f(1)}{x - 1} = \lim\limits_{x \to 1^+} \dfrac{|x - 1| - 0}{x - 1} = 1$.

The one-sided limits are not equal. Therefore, f is not differentiable at $x = 1$.

83. The function is continuous at $x = 1$.

$$f(x) = \begin{cases} (x - 1)^3, & x \le 1 \\ (x - 1)^2, & x > 1 \end{cases}$$

The derivative from the left is $\lim\limits_{x \to 1^-} \dfrac{f(x) - f(1)}{x - 1} = \lim\limits_{x \to 1^-} \dfrac{(x - 1)^3 - 0}{x - 1} = \lim\limits_{x \to 1^-} (x - 1)^2 = 0$.

The derivative from the right is $\lim\limits_{x \to 1^+} \dfrac{f(x) - f(1)}{x - 1} = \lim\limits_{x \to 1^+} \dfrac{(x - 1)^2 - 0}{x - 1} = \lim\limits_{x \to 1^+} (x - 1) = 0$.

These one-sided limits are equal. Therefore, f is differentiable at $x = 1$. $\left(f'(1) = 0\right)$

85. Note that f is continuous at $x = 2$. $f(x) = \begin{cases} x^2 + 1, & x \le 2 \\ 4x - 3, & x > 2 \end{cases}$

The derivative from the left is $\lim\limits_{x \to 2^-} \dfrac{f(x) - f(2)}{x - 2} = \lim\limits_{x \to 2^-} \dfrac{(x^2 + 1) - 5}{x - 2} = \lim\limits_{x \to 2^-} (x + 2) = 4$.

The derivative from the right is $\lim\limits_{x \to 2^+} \dfrac{f(x) - f(2)}{x - 2} = \lim\limits_{x \to 2^+} \dfrac{(4x - 3) - 5}{x - 2} = \lim\limits_{x \to 2^+} 4 = 4$.

The one-sided limits are equal. Therefore, f is differentiable at $x = 2$. $(f'(2) = 4)$

87. False. the slope is $\lim\limits_{\Delta x \to 0} \dfrac{f(2 + \Delta x) - f(2)}{\Delta x}$.

89. False. If the derivative from the left of a point does not equal the derivative from the right of a point, then the derivative does not exist at that point. For example, if $f(x) = |x|$, then the derivative from the left at $x = 0$ is -1 and the derivative from the right at $x = 0$ is 1. At $x = 0$, the derivative does not exist.

91.

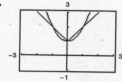

As you zoom in, the graph of $y_1 = x^2 + 1$ appears to be locally the graph of a horizontal line, whereas the graph of $y_2 = |x| + 1$ always has a sharp corner at $(0, 1)$. y_2 is not differentiable at $(0, 1)$.

Section 4.2   Basic Differentiation Rules and Rates of Change

1. (a) $\quad y = x^{1/2}$

$y' = \frac{1}{2}x^{-1/2}$

$y'(1) = \frac{1}{2}$

(b) $\quad y = x^3$

$y' = 3x^2$

$y'(1) = 3$

3. $y = 8$

$y' = 0$

5. $y = x^6$

$y' = 6x^5$

7. $y = \dfrac{1}{x^7} = x^{-7}$

$y' = -7x^{-8} = \dfrac{-7}{x^8}$

9. $f(x) = \sqrt[5]{x} = x^{1/5}$

$f'(x) = \dfrac{1}{5}x^{-4/5} = \dfrac{1}{5x^{4/5}}$

11. $f(x) = x + 1$

$f'(x) = 1$

13. $f(t) = -2t^2 + 3t - 6$

$f'(t) = -4t + 3$

15. $y = 16 - 3x - \frac{1}{2}x^2$

$y' = -3 - x$

17. $g(x) = x^2 + 4x^3$

$g'(x) = 2x + 12x^2$

19. $s(t) = t^3 - 2t + 4$

$s'(t) = 3t^2 - 2$

Function	Rewrite	Derivative	Simplify
21. $y = \dfrac{5}{2x^2}$	$y = \dfrac{5}{2}x^{-2}$	$y' = -5x^{-3}$	$y' = \dfrac{-5}{x^3}$
23. $y = \dfrac{3}{(2x)^3}$	$y = \dfrac{3}{8}x^{-3}$	$y' = \dfrac{-9}{8}x^{-4}$	$y' = \dfrac{-9}{8x^4}$
25. $y = \dfrac{\sqrt{x}}{x}$	$y = x^{-1/2}$	$y' = -\dfrac{1}{2}x^{-3/2}$	$y' = -\dfrac{1}{2x^{3/2}}$

27. $f(x) = \dfrac{3}{x^2} = 3x^{-2}, \ (1, 3)$

$f'(x) = -6x^{-3} = \dfrac{-6}{x^3}$

$f'(1) = -6$

29. $f(x) = -\dfrac{1}{2} + \dfrac{7}{5}x^3, \ \left(0, -\dfrac{1}{2}\right)$

$f'(x) = \dfrac{21}{5}x^2$

$f'(0) = 0$

31. $y = (2x + 1)^2, \ (0, 1)$

$\quad = 4x^2 + 4x + 1$

$\quad y' = 8x + 4$

$\quad y'(0) = 4$

33. $f(x) = x^2 + 5 - 3x^{-2}$

$f'(x) = 2x + 6x^{-3} = 2x + \dfrac{6}{x^3}$

35. $g(t) = t^2 - \dfrac{4}{t^3} = t^2 - 4t^{-3}$

$g'(t) = 2t + 12t^{-4} = 2t + \dfrac{12}{t^4}$

37. $f(x) = \dfrac{x^3 - 3x^2 + 4}{x^2} = x - 3 + 4x^{-2}$

$f'(x) = 1 - \dfrac{8}{x^3} = \dfrac{x^3 - 8}{x^3}$

39. $y = x(x^2 + 1) = x^3 + x$

$y' = 3x^2 + 1$

41. $f(x) = \sqrt{x} - 6\sqrt[3]{x} = x^{1/2} - 6x^{1/3}$

$f'(x) = \dfrac{1}{2}x^{-1/2} - 2x^{-2/3} = \dfrac{1}{2\sqrt{x}} - \dfrac{2}{x^{2/3}}$

43. $h(s) = s^{4/5} - s^{2/3}$

$h'(s) = \dfrac{4}{5}s^{-1/5} - \dfrac{2}{3}s^{-1/3} = \dfrac{4}{5s^{1/5}} - \dfrac{2}{3s^{1/3}}$

45. (a) $y = x^4 - 3x^2 + 2$

$y' = 4x^3 - 6x$

At $(1, 0)$: $y' = 4(1)^3 - 6(1) = -2$

Tangent line: $y - 0 = -2(x - 1)$

$2x + y - 2 = 0$

(b)

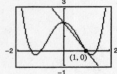

47. (a) $f(x) = \dfrac{2}{\sqrt[4]{x^3}} = 2x^{-3/4}$

$f'(x) = \dfrac{-3}{2}x^{-7/4} = \dfrac{-3}{2x^{7/4}}$

At $(1, 2)$: $f'(1) = \dfrac{-3}{2}$

Tangent line: $y - 2 = -\dfrac{3}{2}(x - 1)$

$y = -\dfrac{3}{2}x + \dfrac{7}{2}$

$3x + 2y - 7 = 0$

(b)

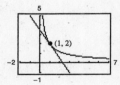

49. $y = x^4 - 8x^2 + 2$

$y' = 4x^3 - 16x$

$\quad = 4x(x^2 - 4)$

$\quad = 4x(x - 2)(x + 2)$

$y' = 0 \implies x = 0, \pm 2$

Horizontal tangents: $(0, 2), (2, -14), (-2, -14)$

51. $y = \dfrac{1}{x^2} = x^{-2}$

$y' = -2x^{-3} = \dfrac{-2}{x^3}$ cannot equal zero.

Therefore, there are no horizontal tangents.

53. $x^2 - kx = 4x - 9$ Equate functions

$2x - k = 4$ Equate derivatives

Hence, $k = 2x - 4$ and

$x^2 - (2x - 4)x = 4x - 9 \implies -x^2 = -9 \implies x = \pm 3$.

For $x = 3, k = 2$ and for $x = -3, k = -10$.

55. $\dfrac{k}{x} = -\dfrac{3}{4}x + 3$ Equate functions

$-\dfrac{k}{x^2} = -\dfrac{3}{4}$ Equate derivatives

Hence, $k = \dfrac{3}{4}x^2$ and $\dfrac{\frac{3}{4}x^2}{x} = \dfrac{-3}{4}x + 3 \implies \dfrac{3}{4}x = -\dfrac{3}{4}x + 3 \implies \dfrac{3}{2}x = 3 \implies x = 2 \implies k = 3$.

57. Let (x_1, y_1) and (x_2, y_2) be the points of tangency on $y = x^2$ and $y = -x^2 + 6x - 5$, respectively. The derivatives of these functions are

$y' = 2x \implies m = 2x_1$ and $y' = -2x + 6 \implies m = -2x_2 + 6$.

$m = 2x_1 = -2x_2 + 6$

$x_1 = -x_2 + 3$

Since $y_1 = x_1^2$ and $y_2 = -x_2^2 + 6x_2 - 5$,

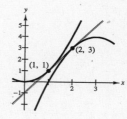

$m = \dfrac{y_2 - y_1}{x_2 - x_1} = \dfrac{(-x_2^2 + 6x_2 - 5) - (x_1^2)}{x_2 - x_1} = -2x_2 + 6$.

$\dfrac{(-x_2^2 + 6x_2 - 5) - (-x_2 + 3)^2}{x_2 - (-x_2 + 3)} = -2x_2 + 6$

$(-x_2^2 + 6x_2 - 5) - (x_2^2 - 6x_2 + 9) = (-2x_2 + 6)(2x_2 - 3)$

$-2x_2^2 + 12x_2 - 14 = -4x_2^2 + 18x_2 - 18$

$2x_2^2 - 6x_2 + 4 = 0$

$2(x_2 - 2)(x_2 - 1) = 0$

$x_2 = 1$ or 2

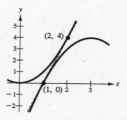

$x_2 = 1 \implies y_2 = 0, x_1 = 2$ and $y_1 = 4$

Thus, the tangent line through $(1, 0)$ and $(2, 4)$ is

$y - 0 = \left(\dfrac{4 - 0}{2 - 1}\right)(x - 1) \implies y = 4x - 4$.

$x_2 = 2 \implies y_2 = 3, x_1 = 1$ and $y_1 = 1$

Thus, the tangent line through $(2, 3)$ and $(1, 1)$ is

$y - 1 = \left(\dfrac{3 - 1}{2 - 1}\right)(x - 1) \implies y = 2x - 1$.

59. (a) The slope appears to be steepest between A and B.

(b) The average rate of change between A and B is **greater** than the instantaneous rate of change at B.

(c)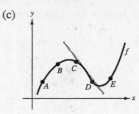

61. $g(x) = f(x) + 6 \implies g'(x) = f'(x)$

63.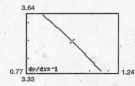

If f is linear then its derivative is a constant function.

$$f(x) = ax + b$$
$$f'(x) = a$$

65. $f(x) = \sqrt{x}, \ (-4, 0)$

$$f'(x) = \frac{1}{2}x^{-1/2} = \frac{1}{2\sqrt{x}}$$

$$\frac{1}{2\sqrt{x}} = \frac{0 - y}{-4 - x}$$

$$4 + x = 2\sqrt{x}y$$

$$4 + x = 2\sqrt{x}\sqrt{x}$$

$$4 + x = 2x$$

$$x = 4, y = 2$$

The point $(4, 2)$ is on the graph of f.

Tangent line: $y - 2 = \dfrac{0 - 2}{-4 - 4}(x - 4)$

$$4y - 8 = x - 4$$

$$0 = x - 4y + 4$$

67. $f'(1) = -1$

69. False. Let $f(x) = x^2$ and $g(x) = x^2 + 4$.
Then $f'(x) = g'(x) = 2x$, but $f(x) \neq g(x)$.

71. False. If $y = \pi^2$, then $dy/dx = 0$. (π^2 is a constant.)

73. True. If $g(x) = 3f(x)$, then $g'(x) = 3f'(x)$.

75. (a) One possible secant is between (3.9, 7.7019) and (4, 8):

$$y - 8 = \frac{8 - 7.7019}{4 - 3.9}(x - 4)$$

$$y - 8 = 2.981(x - 4)$$

$$y = S(x) = 2.981x - 3.924$$

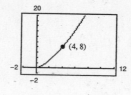

(b) $f'(x) = \frac{3}{2}x^{1/2} \implies f'(4) = \frac{3}{2}(2) = 3$

$$T(x) = 3(x - 4) + 8 = 3x - 4$$

$S(x)$ is an approximation of the tangent line $T(x)$.

(c) As you move further away from (4, 8), the accuracy of the approximation T gets worse.

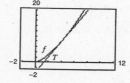

(d)

Δx	-3	-2	-1	-0.5	-0.1	0	0.1	0.5	1	2	3
$f(4 + \Delta x)$	1	2.828	5.196	6.548	7.702	8	8.302	9.546	11.180	14.697	18.520
$T(4 + \Delta x)$	-1	2	5	6.5	7.7	8	8.3	9.5	11	14	17

77. $f(t) = 2t + 7, [1, 2]$

$f'(t) = 2$

Instantaneous rate of change is the constant 2.
Average rate of change:

$$\frac{f(2) - f(1)}{2 - 1} = \frac{[2(2) + 7] - [2(1) + 7]}{1} = 2$$

(These are the same because f is a line of slope 2.)

79. $f(x) = -\frac{1}{x}, [1, 2]$

$$f'(x) = \frac{1}{x^2}$$

Instantaneous rate of change:

$$(1, -1) \implies f'(1) = 1$$

$$\left(2, -\frac{1}{2}\right) \implies f'(2) = \frac{1}{4}$$

Average rate of change:

$$\frac{f(2) - f(1)}{2 - 1} = \frac{(-1/2) - (-1)}{2 - 1} = \frac{1}{2}$$

81. (a) $s(t) = -16t^2 + 1362$

$v(t) = -32t$

(b) $\frac{s(2) - s(1)}{2 - 1} = 1298 - 1346 = -48$ ft/sec

(c) $v(t) = s'(t) = -32t$

When $t = 1$: $v(1) = -32$ ft/sec.

When $t = 2$: $v(2) = -64$ ft/sec.

(d) $-16t^2 + 1362 = 0$

$$t^2 = \frac{1362}{16} \implies t = \frac{\sqrt{1362}}{4} \approx 9.226 \text{ sec}$$

$$v\left(\frac{\sqrt{1362}}{4}\right) = -32\left(\frac{\sqrt{1362}}{4}\right)$$

$$= -8\sqrt{1362} \approx -295.242 \text{ ft/sec}$$

83. $s(t) = -4.9t^2 + v_0 t + s_0$

$$= -4.9t^2 + 120t$$

$$v(t) = -9.8t + 120$$

$$v(5) = -9.8(5) + 120 = 71 \text{ m/sec}$$

$$v(10) = -9.8(10) + 120 = 22 \text{ m/sec}$$

85.

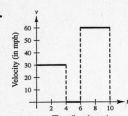

(The velocity has been converted to miles per hour.)

87. $v = 40 \text{ mph} = \frac{2}{3} \text{ mi/min}$

$\left(\frac{2}{3} \text{ mi/min}\right)(6 \text{ min}) = 4 \text{ mi}$

$v = 0 \text{ mph} = 0 \text{ mi/min}$

$(0 \text{ mi/min})(2 \text{ min}) = 0 \text{ mi}$

$v = 60 \text{ mph} = 1 \text{mi/min}$

$(1 \text{ mi/min})(2 \text{ min}) = 2 \text{ mi}$

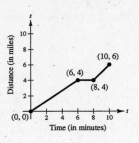

89. (a) $R(v) = 0.417v - 0.02$

(b) $B(v) = 0.00557v^2 + 0.0014v + 0.04$

(c) $T(v) = R(v) + B(v) = 0.00557v^2 + 0.418v + 0.02$

(d)

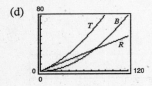

(e) $T'(v) = 0.01114v + 0.418$ (Answers will vary.)

$T'(40) = 0.01114(40) + 0.418 \approx 0.864$

$T'(80) = 0.01114(80) + 0.418 \approx 1.309$

$T'(100) = 0.01114(100) + 0.418 \approx 1.532$

(f) Stopping distances increase at an increasing rate.

91. $A = s^2, \dfrac{dA}{ds} = 2s$

When $s = 4 \text{ m}$,

$\dfrac{dA}{ds} = 8$ square meters per meter change in s.

93.

$C = \dfrac{1{,}008{,}000}{Q} + 6.3Q$

$\dfrac{dC}{dQ} = -\dfrac{1{,}008{,}000}{Q^2} + 6.3$

$C(351) - C(350) \approx 5083.095 - 5085 \approx -\$1.91$

When $Q = 350, \dfrac{dC}{dQ} \approx -\$1.93$.

95. (a) $f'(1.479)$ is the rate of change of the amount of gasoline sold when the price is \$1.479 per gallon.

(b) $f'(1.479)$ is usually negative. As prices go up, sales go down.

97. $f(x) = \begin{cases} ax^3, & x \le 2 \\ x^2 + b, & x > 2 \end{cases}$

f must be continuous at $x = 2$ to be differentiable at $x = 2$.

$\left. \begin{array}{l} \displaystyle\lim_{x \to 2^-} f(x) = \lim_{x \to 2^-} ax^3 = 8a \\[2mm] \displaystyle\lim_{x \to 2^+} f(x) = \lim_{x \to 2^+} (x^2 + b) = 4 + b \end{array} \right\} \quad \begin{array}{l} 8a = 4 + b \\[2mm] 8a - 4 = b \end{array}$

$f'(x) = \begin{cases} 3ax^2, & x < 2 \\ 2x, & x > 2 \end{cases}$

For f to be differentiable at $x = 2$, the left derivative must equal the right derivative.

$3a(2)^2 = 2(2)$

$12a = 4$

$a = \frac{1}{3}$

$b = 8a - 4 = -\frac{4}{3}$

99. $y = x^3 - 9x$

$y' = 3x^2 - 9$

Tangent lines through $(1, -9)$:

$$y + 9 = (3x^2 - 9)(x - 1)$$

$$(x^3 - 9x) + 9 = 3x^3 - 3x^2 - 9x + 9$$

$$0 = 2x^3 - 3x^2 = x^2(2x - 3)$$

$$x = 0 \text{ or } x = \tfrac{3}{2}$$

The points of tangency are $(0, 0)$ and $\left(\tfrac{3}{2}, -\tfrac{81}{8}\right)$. At $(0, 0)$ the slope is $y'(0) = -9$. At $\left(\tfrac{3}{2}, -\tfrac{81}{8}\right)$ the slope is $y'\left(\tfrac{3}{2}\right) = -\tfrac{9}{4}$.

Tangent lines:

$$y - 0 = -9(x - 0) \quad \text{and} \qquad y + \tfrac{81}{8} = -\tfrac{9}{4}\left(x - \tfrac{3}{2}\right)$$

$$y = -9x \qquad\qquad\qquad y = -\tfrac{9}{4}x - \tfrac{27}{4}$$

$$9x + y = 0 \qquad\qquad 9x + 4y + 27 = 0$$

Section 4.3   The Product and Quotient Rules and Higher-Order Derivatives

1. $g(x) = (x^2 + 1)(x^2 - 2x)$

$g'(x) = (x^2 + 1)(2x - 2) + (x^2 - 2x)(2x)$

$\quad = 2x^3 - 2x^2 + 2x - 2 + 2x^3 - 4x^2$

$\quad = 4x^3 - 6x^2 + 2x - 2$

3. $h(t) = \sqrt[3]{t}(t^2 + 4) = t^{1/3}(t^2 + 4)$

$h'(t) = t^{1/3}(2t) + (t^2 + 4)\tfrac{1}{3}t^{-2/3}$

$\quad = 2t^{4/3} + \dfrac{t^2 + 4}{3t^{2/3}}$

$\quad = \dfrac{7t^2 + 4}{3t^{2/3}}$

5. $g(t) = (2t^2 - 3)(4 - t^2 - t^4)$

$g'(t) = (2t^2 - 3)(-2t - 4t^3) + (4 - t^2 - t^4)(4t)$

$\quad = -4t^3 + 6t - 8t^5 + 12t^3 + 16t - 4t^3 - 4t^5$

$\quad = -2t(6t^4 - 2t^2 - 11)$

7. $f(x) = \dfrac{x}{x^2 + 1}$

$f'(x) = \dfrac{(x^2 + 1)(1) - x(2x)}{(x^2 + 1)^2} = \dfrac{1 - x^2}{(x^2 + 1)^2}$

9. $h(x) = \dfrac{\sqrt[3]{x}}{x^3 + 1} = \dfrac{x^{1/3}}{x^3 + 1}$

$h'(x) = \dfrac{(x^3 + 1)\tfrac{1}{3}x^{-2/3} - x^{1/3}(3x^2)}{(x^3 + 1)^2}$

$\quad = \dfrac{(x^3 + 1) - x(9x^2)}{3x^{2/3}(x^3 + 1)^2}$

$\quad = \dfrac{1 - 8x^3}{3x^{2/3}(x^3 + 1)^2}$

11. $f(x) = \dfrac{x^3 + 3x + 2}{x^2 - 1}$

$f'(x) = \dfrac{(x^2 - 1)(3x^2 + 3) - (x^3 + 3x + 2)(2x)}{(x^2 - 1)^2}$

$\quad = \dfrac{x^4 - 6x^2 - 4x - 3}{(x^2 - 1)^2}$

13. $f(x) = \dfrac{5}{x^2}(x + 3) = \dfrac{5}{x} + \dfrac{15}{x^2} = 5x^{-1} + 15x^{-2}$

$f'(x) = -\dfrac{5}{x^2} - \dfrac{30}{x^3} = -\dfrac{5(x + 6)}{x^3}$

$f'(1) = -5 - 30 = -35$

15. $f(x) = \dfrac{x^2 - 4}{x - 3}$

$$f'(x) = \frac{(x-3)(2x) - (x^2-4)(1)}{(x-3)^2} = \frac{2x^2 - 6x - x^2 + 4}{(x-3)^2}$$

$$= \frac{x^2 - 6x + 4}{(x-3)^2}$$

$$f'(1) = \frac{1 - 6 + 4}{(1-3)^2} = -\frac{1}{4}$$

17. $f(x) = (x^3 - 3x)(2x^2 + 3x + 5)$

$$f'(x) = (x^3 - 3x)(4x + 3) + (2x^2 + 3x + 5)(3x^2 - 3)$$

$$= 10x^4 + 12x^3 - 3x^2 - 18x - 15$$

$$f'(0) = -15$$

19. $f(x) = (x^5 - 3x)\left(\dfrac{1}{x^2}\right) = x^3 - \dfrac{3}{x}$

$$f'(x) = 3x^2 + \frac{3}{x^2} = \frac{3(x^4 + 1)}{x^2}$$

$$f'(-1) = 3 + 3 = 6$$

Function	*Rewrite*	*Derivative*	*Simplify*
21. $y = \dfrac{x^2 + 2x}{3}$	$y = \dfrac{1}{3}x^2 + \dfrac{2}{3}x$	$y' = \dfrac{2}{3}x + \dfrac{2}{3}$	$y' = \dfrac{2x + 2}{3}$
23. $y = \dfrac{4x^{3/2}}{x}$	$y = 4\sqrt{x},\, x > 0$	$y' = 2x^{-1/2}$	$y' = \dfrac{2}{\sqrt{x}}$

25. $f(x) = \dfrac{3 - 2x - x^2}{x^2 - 1}$

$$f'(x) = \frac{(x^2 - 1)(-2 - 2x) - (3 - 2x - x^2)(2x)}{(x^2 - 1)^2}$$

$$= \frac{2x^2 - 4x + 2}{(x^2 - 1)^2} = \frac{2(x - 1)^2}{(x^2 - 1)^2}$$

$$= \frac{2}{(x + 1)^2},\, x \neq 1$$

27. $f(x) = x\left(1 - \dfrac{4}{x + 3}\right) = x - \dfrac{4x}{x + 3}$

$$f'(x) = 1 - \frac{(x + 3)4 - 4x(1)}{(x + 3)^2} = \frac{(x^2 + 6x + 9) - 12}{(x + 3)^2}$$

$$= \frac{x^2 + 6x - 3}{(x + 3)^2}$$

29. $f(x) = \dfrac{2x + 5}{\sqrt{x}} = 2x^{1/2} + 5x^{-1/2}$

$$f'(x) = x^{-1/2} - \frac{5}{2}x^{-3/2} = x^{-3/2}\left[x - \frac{5}{2}\right]$$

$$= \frac{2x - 5}{2x\sqrt{x}} = \frac{2x - 5}{2x^{3/2}}$$

31. $h(s) = (s^3 - 2)^2 = s^6 - 4s^3 + 4$

$$h'(s) = 6s^5 - 12s^2 = 6s^2(s^3 - 2)$$

33. $f(x) = \dfrac{2 - \dfrac{1}{x}}{x - 3} = \dfrac{2x - 1}{x(x - 3)} = \dfrac{2x - 1}{x^2 - 3x}$

$$f'(x) = \frac{(x^2 - 3x)2 - (2x - 1)(2x - 3)}{(x^2 - 3x)^2} = \frac{2x^2 - 6x - 4x^2 + 8x - 3}{(x^2 - 3x)^2}$$

$$= \frac{-2x^2 + 2x - 3}{(x^2 - 3x)^2} = -\frac{2x^2 - 2x + 3}{x^2(x - 3)^2}$$

35. $f(x) = (3x^3 + 4x)(x - 5)(x + 1)$

$f'(x) = (9x^2 + 4)(x - 5)(x + 1) + (3x^3 + 4x)(1)(x + 1) + (3x^3 + 4x)(x - 5)(1)$

$\quad = (9x^2 + 4)(x^2 - 4x - 5) + 3x^4 + 3x^3 + 4x^2 + 4x + 3x^4 - 15x^3 + 4x^2 - 20x$

$\quad = 9x^4 - 36x^3 - 41x^2 - 16x - 20 + 6x^4 - 12x^3 + 8x^2 - 16x$

$\quad = 15x^4 - 48x^3 - 33x^2 - 32x - 20$

37. $f(x) = \dfrac{x^2 + c^2}{x^2 - c^2}$

$f'(x) = \dfrac{(x^2 - c^2)(2x) - (x^2 + c^2)(2x)}{(x^2 - c^2)^2}$

$\quad = \dfrac{-4xc^2}{(x^2 - c^2)^2}$

39. $g(x) = \left(\dfrac{x + 1}{x + 2}\right)(2x - 5)$

$g'(x) = \dfrac{2x^2 + 8x - 1}{(x + 2)^2}$ (form of answer may vary)

41. (a) $f(x) = (x^3 - 3x + 1)(x + 2), \quad (1, -3)$

$f'(x) = (x^3 - 3x + 1)(1) + (x + 2)(3x^2 - 3)$

$\quad = 4x^3 + 6x^2 - 6x - 5$

$f'(1) = -1 =$ slope at $(1, -3)$.

Tangent line: $y + 3 = -1(x - 1) \implies y = -x - 2$

(b)

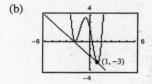

43. (a) $f(x) = \dfrac{x}{x - 1}, \ (2, 2)$

$f'(x) = \dfrac{(x - 1)(1) - x(1)}{(x - 1)^2} = \dfrac{-1}{(x - 1)^2}$

$f'(2) = \dfrac{-1}{(2 - 1)^2} = -1 =$ slope at $(2, 2)$.

Tangent line: $y - 2 = -1(x - 2) \implies y = -x + 4$

(b)

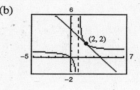

45. $f(x) = \dfrac{x^2}{x - 1}$

$f'(x) = \dfrac{(x - 1)(2x) - x^2(1)}{(x - 1)^2}$

$\quad = \dfrac{x^2 - 2x}{(x - 1)^2} = \dfrac{x(x - 2)}{(x - 1)^2}$

$f'(x) = 0$ when $x = 0$ or $x = 2$.

Horizontal tangents are at $(0, 0)$ and $(2, 4)$.

47. $f(x) = \dfrac{8}{x^2 + 4}; \quad (2, 1)$

$f'(x) = \dfrac{(x^2 + 4)(0) - 8(2x)}{(x^2 + 4)^2} = \dfrac{-16x}{(x^2 + 4)^2}$

$f'(2) = \dfrac{-16(2)}{(4 + 4)^2} = -\dfrac{1}{2}$

$y - 1 = -\dfrac{1}{2}(x - 2)$

$y = -\dfrac{1}{2}x + 2$

$2y + x - 4 = 0$

49. $f(x) = \dfrac{16x}{x^2 + 16}, \ \left(-2, -\dfrac{8}{5}\right)$

$f'(x) = \dfrac{(x^2 + 16)(16) - 16x(2x)}{(x^2 + 16)^2} = \dfrac{256 - 16x^2}{(x^2 + 16)^2}$

$f'(-2) = \dfrac{256 - 16(4)}{20^2} = \dfrac{12}{25}$

$y + \dfrac{8}{5} = \dfrac{12}{25}(x + 2)$

$y = \dfrac{12}{25}x - \dfrac{16}{25}$

$25y - 12x + 16 = 0$

51. $f(x) = \dfrac{x + 1}{x - 1}$

$f'(x) = \dfrac{(x - 1) - (x + 1)}{(x - 1)^2} = \dfrac{-2}{(x - 1)^2}$

$2y + x = 6 \implies y = -\dfrac{1}{2}x + 3; \ \text{Slope: } -\dfrac{1}{2}$

$\dfrac{-2}{(x - 1)^2} = -\dfrac{1}{2}$

$(x - 1)^2 = 4$

$x - 1 = \pm 2$

$x = -1, 3; \ f(-1) = 0, \ f(3) = 2$

$y - 0 = -\dfrac{1}{2}(x + 1) \implies y = -\dfrac{1}{2}x - \dfrac{1}{2}$

$y - 2 = -\dfrac{1}{2}(x - 3) \implies y = -\dfrac{1}{2}x + \dfrac{7}{2}$

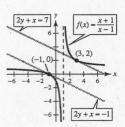

53. (a) $p'(x) = f'(x)g(x) + f(x)g'(x)$

$p'(1) = f'(1)g(1) + f(1)g'(1) = 1(4) + 6\left(-\dfrac{1}{2}\right) = 1$

(b) $q'(x) = \dfrac{g(x)f'(x) - f(x)g'(x)}{g(x)^2}$

$q'(4) = \dfrac{3(-1) - 7(0)}{3^2} = -\dfrac{1}{3}$

55. $f'(x) = \dfrac{(x + 2)3 - 3x(1)}{(x + 2)^2} = \dfrac{6}{(x + 2)^2}$

$g'(x) = \dfrac{(x + 2)5 - (5x + 4)(1)}{(x + 2)^2} = \dfrac{6}{(x + 2)^2}$

$g(x) = \dfrac{5x + 4}{(x + 2)} = \dfrac{3x}{(x + 2)} + \dfrac{2x + 4}{(x + 2)} = f(x) + 2$

f and g differ by a constant.

57. Area $= A(t) = (2t + 1)\sqrt{t} = 2t^{3/2} + t^{1/2}$

$A'(t) = 2\left(\dfrac{3}{2}t^{1/2}\right) + \dfrac{1}{2}t^{-1/2}$

$= 3t^{1/2} + \dfrac{1}{2}t^{-1/2}$

$= \dfrac{6t + 1}{2\sqrt{t}} \ \text{cm}^2/\text{sec}$

59. $C = 100\left(\dfrac{200}{x^2} + \dfrac{x}{x + 30}\right), \ 1 \le x$

$\dfrac{dC}{dx} = 100\left(-\dfrac{400}{x^3} + \dfrac{30}{(x + 30)^2}\right)$

(a) When $x = 10$: $\dfrac{dC}{dx} = -\$38.13.$

(b) When $x = 15$: $\dfrac{dC}{dx} = -\$10.37.$

(c) When $x = 20$: $\dfrac{dC}{dx} = -\$3.80.$

As the order size increases, the cost per item decreases.

61. (a) $n(t) = -3.5806t^3 + 82.577t^2 - 603.60t + 1667.5$

$v(t) = -0.1361t^3 + 3.165t^2 - 23.02t + 59.8$

(b)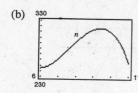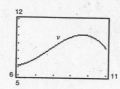

(c) $A(t) = \dfrac{v(t)}{n(t)} = \dfrac{-0.1361t^3 + 3.165t^2 - 23.02t + 59.8}{-3.5806t^3 + 82.577t^2 - 603.60t + 1667.5}$

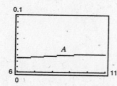

$A(t)$ represents the average retail value (in billions of dollars) per 1000 motor homes.

(d) $A'(t)$ is the rate of change of the average retail value (in billions of dollars) per 1000 motor homes, per year.

63. $f(x) = 4x^{3/2}$

$f'(x) = 6x^{1/2}$

$f''(x) = 3x^{-1/2} = \dfrac{3}{\sqrt{x}}$

65. $f(x) = \dfrac{x}{x-1}$

$f'(x) = \dfrac{(x-1)(1) - x(1)}{(x-1)^2} = \dfrac{-1}{(x-1)^2}$

$f''(x) = \dfrac{2}{(x-1)^3}$

67. $f'(x) = x^2$

$f''(x) = 2x$

69. $f''(x) = 2 - 2x^{-1}$

$f'''(x) = 2x^{-2} = \dfrac{2}{x^2}$

71. $f(x) = 2x^2 - 2$

$f'(x) = 4x$

$f''(x) = 4$

73.

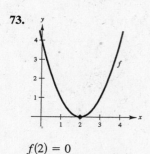

$f(2) = 0$

One such function is $f(x) = (x-2)^2$.

75. $f(x) = 2g(x) + h(x)$

$f'(x) = 2g'(x) + h'(x)$

$f'(2) = 2g'(2) + h'(2)$

$\qquad = 2(-2) + 4$

$\qquad = 0$

77. $f(x) = \dfrac{g(x)}{h(x)}$

$f'(x) = \dfrac{h(x)g'(x) - g(x)h'(x)}{[h(x)]^2}$

$f'(2) = \dfrac{h(2)g'(2) - g(2)h'(2)}{[h(2)]^2}$

$\quad = \dfrac{(-1)(-2) - (3)(4)}{(-1)^2}$

$\quad = -10$

79.

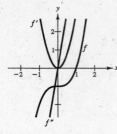

It appears that f is cubic; so f' would be quadratic and f'' would be linear.

81.

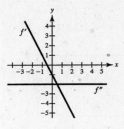

83.

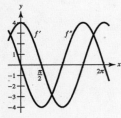

85. $y = \dfrac{1}{x}, \ y' = -\dfrac{1}{x^2}, \ y'' = \dfrac{2}{x^3}$

$x^3 y'' + 2x^2 y' = x^3 \left[\dfrac{2}{x^3}\right] + 2x^2 \left[-\dfrac{1}{x^2}\right] = 2 - 2 = 0$

87. $v(t) = 36 - t^2, \ 0 \le t \le 6$

$a(t) = -2t$

$v(3) = 27 \ \text{m/sec}$

$a(3) = -6 \ \text{m/sec}^2$

The speed of the object is decreasing.

89. $v(t) = \dfrac{100t}{2t + 15}$

$a(t) = \dfrac{(2t + 15)(100) - (100t)(2)}{(2t + 15)^2}$

$\quad = \dfrac{1500}{(2t + 15)^2}$

(a) $a(5) = \dfrac{1500}{[2(5) + 15]^2} = 2.4 \ \text{ft/sec}^2$

(b) $a(10) = \dfrac{1500}{[2(10) + 15]^2} \approx 1.2 \ \text{ft/sec}^2$

(c) $a(20) = \dfrac{1500}{[2(20) + 15]^2} \approx 0.5 \ \text{ft/sec}^2$

91. $f(x) = x^n$

$f^{(n)}(x) = n(n - 1)(n - 2) \cdots (2)(1) = n!$

93. $f(x) = g(x)h(x)$

(a) $f'(x) = g(x)h'(x) + h(x)g'(x)$

$f''(x) = g(x)h''(x) + g'(x)h'(x) + h(x)g''(x) + h'(x)g'(x)$

$\qquad = g(x)h''(x) + 2g'(x)h'(x) + h(x)g''(x)$

$f'''(x) = g(x)h'''(x) + g'(x)h''(x) + 2g'(x)h''(x) + 2g''(x)h'(x) + h(x)g'''(x) + h'(x)g''(x)$

$\qquad = g(x)h'''(x) + 3g'(x)h''(x) + 3g''(x)h'(x) + g'''(x)h(x)$

$f^{(4)}(x) = g(x)h^{(4)}(x) + g'(x)h'''(x) + 3g'(x)h'''(x) + 3g''(x)h''(x) + 3g''(x)h''(x) + 3g'''(x)h'(x)$

$\qquad\quad + g'''(x)h'(x) + g^{(4)}(x)h(x)$

$\qquad = g(x)h^{(4)}(x) + 4g'(x)h'''(x) + 6g''(x)h''(x) + 4g'''(x)h'(x) + g^{(4)}(x)h(x)$

(b) $f^{(n)}(x) = g(x)h^{(n)}(x) + \dfrac{n(n-1)(n-2)\cdots(2)(1)}{1[(n-1)(n-2)\cdots(2)(1)]}g'(x)h^{(n-1)}(x) + \dfrac{n(n-1)(n-2)\cdots(2)(1)}{(2)(1)[(n-2)(n-3)\cdots(2)(1)]}g''(x)h^{(n-2)}(x)$

$\qquad\quad + \dfrac{n(n-1)(n-2)\cdots(2)(1)}{(3)(2)(1)[(n-3)(n-4)\cdots(2)(1)]}g'''(x)h^{(n-3)}(x) + \cdots$

$\qquad\quad + \dfrac{n(n-1)(n-2)\cdots(2)(1)}{[(n-1)(n-2)\cdots(2)(1)](1)}g^{(n-1)}(x)h'(x) + g^{(n)}(x)h(x)$

$\qquad = g(x)h^{(n)}(x) + \dfrac{n!}{1!(n-1)!}g'(x)h^{(n-1)}(x) + \dfrac{n!}{2!(n-2)!}g''(x)h^{(n-2)}(x) + \cdots$

$\qquad\quad + \dfrac{n!}{(n-1)!1!}g^{(n-1)}(x)h'(x) + g^{(n)}(x)h(x)$

Note: $n! = n(n-1)\ldots 3\cdot 2\cdot 1$ (read "*n* factorial.")

95. (a) $\quad f(x) = \sqrt{x} = x^{1/2} \qquad\qquad\qquad f(1) = 1$

$\qquad f'(x) = \dfrac{1}{2}x^{-1/2} = \dfrac{1}{2\sqrt{x}} \qquad\qquad f'(1) = \dfrac{1}{2}$

$\qquad f''(x) = -\dfrac{1}{4}x^{-3/2} = -\dfrac{1}{4x^{3/2}} \qquad\quad f''(1) = -\dfrac{1}{4}$

$\qquad P_1(x) = f'(1)(x-1) + f(1) = \dfrac{1}{2}(x-1) + 1 = \dfrac{1}{2}x + \dfrac{1}{2}$

$\qquad P_2(x) = \dfrac{1}{2}f''(1)(x-1)^2 + f'(1)(x-1) + f(1)$

$\qquad\qquad = \dfrac{1}{2}\left(-\dfrac{1}{4}\right)(x-1)^2 + \dfrac{1}{2}(x-1) + 1$

$\qquad\qquad = -\dfrac{1}{8}(x-1)^2 + \dfrac{1}{2}(x-1) + 1$

(b)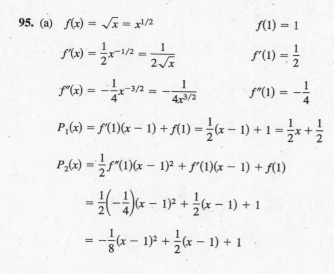

$\qquad\qquad f(1) = 1$

$\qquad\qquad f'(1) = \dfrac{1}{2}$

$\qquad\qquad f''(1) = -\dfrac{1}{4}$

(c) P_2 is a better approximation.

(d) P_1 and P_2 become less accurate as you move farther away from $x = a = 1$.

97. False. If $y = f(x)g(x)$, then

$\dfrac{dy}{dx} = f(x)g'(x) + g(x)f'(x).$

99. True

$h'(c) = f(c)g'(c) + g(c)f'(c)$

$\qquad = f(c)(0) + g(c)(0)$

$\qquad = 0$

101. True

103. $f(x) = x|x| = \begin{cases} x^2, & \text{if } x \geq 0 \\ -x^2, & \text{if } x < 0 \end{cases}$

$f'(x) = \begin{cases} 2x, & \text{if } x \geq 0 \\ -2x, & \text{if } x < 0 \end{cases} = 2|x|$

$f''(x) = \begin{cases} 2, & \text{if } x > 0 \\ -2, & \text{if } x < 0 \end{cases}$

$f''(0)$ does not exist since the left and right derivatives are not equal.

Section 4.4　The Chain Rule

$y = f(g(x))$	$u = g(x)$	$y = f(u)$
1. $y = (6x - 5)^4$	$u = 6x - 5$	$y = u^4$
3. $y = (x^2 - 3x + 4)^6$	$u = x^2 - 3x + 4$	$y = u^6$
5. $y = \sqrt{x^2 - 1}$	$u = x^2 - 1$	$y = \sqrt{u}$

7. $y = (2x - 7)^3$

$y' = 3(2x - 7)^2(2) = 6(2x - 7)^2$

9. $g(x) = 3(4 - 9x)^4$

$g'(x) = 12(4 - 9x)^3(-9) = -108(4 - 9x)^3$

11. $f(t) = (1 - t)^{1/2}$

$f'(t) = \frac{1}{2}(1 - t)^{-1/2}(-1) = -\frac{1}{2\sqrt{1 - t}}$

13. $y = (9x^2 + 4)^{1/3}$

$y' = \frac{1}{3}(9x^2 + 4)^{-2/3}(18x) = \frac{6x}{(9x^2 + 4)^{2/3}}$

15. $y = 2(4 - x^2)^{1/4}$

$y' = 2\left(\frac{1}{4}\right)(4 - x^2)^{-3/4}(-2x)$

$= \frac{-x}{\sqrt[4]{(4 - x^2)^3}}$

17. $y = (x - 2)^{-1}$

$y' = -1(x - 2)^{-2}(1) = \frac{-1}{(x - 2)^2}$

19. $f(t) = (t - 3)^{-2}$

$f'(t) = -2(t - 3)^{-3} = \frac{-2}{(t - 3)^3}$

21. $y = (x + 2)^{-1/2}$

$\frac{dy}{dx} = -\frac{1}{2}(x + 2)^{-3/2} = -\frac{1}{2(x + 2)^{3/2}}$

23. $f(x) = x^2(x - 2)^4$

$f'(x) = x^2[4(x - 2)^3(1)] + (x - 2)^4(2x)$

$= 2x(x - 2)^3[2x + (x - 2)]$

$= 2x(x - 2)^3(3x - 2)$

25. $y = x\sqrt{1 - x^2} = x(1 - x^2)^{1/2}$

$y' = x\left[\frac{1}{2}(1 - x^2)^{-1/2}(-2x)\right] + (1 - x^2)^{1/2}(1)$

$= -x^2(1 - x^2)^{-1/2} + (1 - x^2)^{1/2}$

$= (1 - x^2)^{-1/2}[-x^2 + (1 - x^2)]$

$= \frac{1 - 2x^2}{\sqrt{1 - x^2}}$

27. $y = \dfrac{x}{\sqrt{x^2 + 1}} = x(x^2 + 1)^{-1/2}$

$y' = x\left[-\dfrac{1}{2}(x^2 + 1)^{-3/2}(2x)\right] + (x^2 + 1)^{-1/2}(1)$

$\quad = -x^2(x^2 + 1)^{-3/2} + (x^2 + 1)^{-1/2}$

$\quad = (x^2 + 1)^{-3/2}[-x^2 + (x^2 + 1)]$

$\quad = \dfrac{1}{(x^2 + 1)^{3/2}}$

29. $g(x) = \left(\dfrac{x + 5}{x^2 + 2}\right)^2$

$g'(x) = 2\left(\dfrac{x + 5}{x^2 + 2}\right)\left(\dfrac{(x^2 + 2) - (x + 5)(2x)}{(x^2 + 2)^2}\right)$

$\quad = \dfrac{2(x + 5)(2 - 10x - x^2)}{(x^2 + 2)^3}$

31. $f(v) = \left(\dfrac{1 - 2v}{1 + v}\right)^3$

$f'(v) = 3\left(\dfrac{1 - 2v}{1 + v}\right)^2\left(\dfrac{(1 + v)(-2) - (1 - 2v)}{(1 + v)^2}\right)$

$\quad = \dfrac{-9(1 - 2v)^2}{(1 + v)^4}$

33. $y = \dfrac{\sqrt{x} + 1}{x^2 + 1}$

$y' = \dfrac{1 - 3x^2 - 4x^{3/2}}{2\sqrt{x}(x^2 + 1)^2}$

The zero of y' corresponds to the point on the graph of y where the tangent line is horizontal.

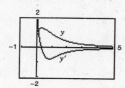

35. $g(t) = \dfrac{3t^2}{\sqrt{t^2 + 2t - 1}}$

$g'(t) = \dfrac{3t(t^2 + 3t - 2)}{(t^2 + 2t - 1)^{3/2}}$

The zeros of g' correspond to the points on the graph of g where the tangent lines are horizontal.

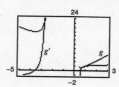

37. $y = \sqrt{\dfrac{x + 1}{x}}$

$y' = -\dfrac{\sqrt{(x + 1)/x}}{2x(x + 1)}$

y' has no zeros.

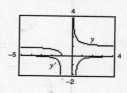

39. $s(t) = \dfrac{-2(2 - t)\sqrt{1 + t}}{3}$

$s'(t) = \dfrac{t}{\sqrt{1 + t}}$

The zero of $s'(t)$ corresponds to the point on the graph of $s(t)$ where the tangent line is horizontal.

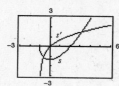

41. $s(t) = (t^2 + 2t + 8)^{1/2}, \quad (2, 4)$

$s'(t) = \dfrac{1}{2}(t^2 + 2t + 8)^{-1/2}(2t + 2)$

$\quad = \dfrac{t + 1}{\sqrt{t^2 + 2t + 8}}$

$s'(2) = \dfrac{3}{4}$

43. $f(x) = \dfrac{3}{x^3 - 4} = 3(x^3 - 4)^{-1}, \quad \left(-1, -\dfrac{3}{5}\right)$

$f'(x) = -3(x^3 - 4)^{-2}(3x^2) = -\dfrac{9x^2}{(x^3 - 4)^2}$

$f'(-1) = -\dfrac{9}{25}$

45. $f(t) = \dfrac{3t + 2}{t - 1}, \quad (0, -2)$

$f'(t) = \dfrac{(t - 1)(3) - (3t + 2)(1)}{(t - 1)^2} = \dfrac{-5}{(t - 1)^2}$

$f'(0) = -5$

47. (a) $f(x) = \sqrt{3x^2 - 2}, \quad (3, 5)$

$f'(x) = \dfrac{1}{2}(3x^2 - 2)^{-1/2}(6x)$

$\quad = \dfrac{3x}{\sqrt{3x^2 - 2}}$

$f'(3) = \dfrac{9}{5}$

Tangent line:

$y - 5 = \dfrac{9}{5}(x - 3) \implies 9x - 5y - 2 = 0$

(b)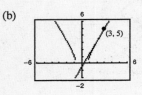

49. (a) $y = (2x^3 + 1)^2, \quad (-1, 1)$

$y' = 2(2x^3 + 1)(6x^2) = 12x^2(2x^3 + 1)$

$y'(-1) = 12(-1) = -12$

Tangent line:

$y - 1 = -12(x + 1)$

$y = -12x - 11$

(b)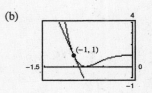

51. $f(x) = 2(x^2 - 1)^3$

$f'(x) = 6(x^2 - 1)^2(2x)$

$\quad = 12x(x^4 - 2x^2 + 1)$

$\quad = 12x^5 - 24x^3 + 12x$

$f''(x) = 60x^4 - 72x^2 + 12$

$\quad = 12(5x^2 - 1)(x^2 - 1)$

53. $f(t) = \dfrac{\sqrt{t^2 + 1}}{t} = \dfrac{(t^2 + 1)^{1/2}}{t}$

$f'(t) = \dfrac{t^2(t^2 + 1)^{-1/2} - (t^2 + 1)^{1/2}}{t^2} = (t^2 + 1)^{-1/2} - \dfrac{(t^2 + 1)^{1/2}}{t^2}$

$f''(t) = -(t^2 + 1)^{-3/2}(t) - \left[\dfrac{t^3(t^2 + 1)^{-1/2} - (t^2 + 1)^{1/2}(2t)}{t^4}\right]$

$\quad = \dfrac{-t^4 - t^2(t^2 + 1) + 2(t^2 + 1)^2}{t^3(t^2 + 1)^{3/2}}$

$\quad = \dfrac{3t^2 + 2}{t^3(t^2 + 1)^{3/2}}$

55. $h(x) = \frac{1}{9}(3x + 1)^3$, $\left(1, \frac{64}{9}\right)$

  $h'(x) = \frac{1}{9}3(3x + 1)^2(3) = (3x + 1)^2$

  $h''(x) = 2(3x + 1)(3) = 6(3x + 1)$

  $h''(1) = 24$

57. The zeros of f' correspond to the points where the graph of f has horizontal tangents.

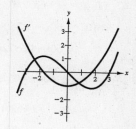

59. $g(x) = f(3x)$

  $g'(x) = f'(3x)(3) \implies g'(x) = 3f'(3x)$

61. (a) $f(x) = g(x)h(x)$

    $f'(x) = g(x)h'(x) + g'(x)h(x)$

    $f'(5) = (-3)(-2) + (6)(3) = 24$

  (b) $f(x) = g(h(x))$

    $f'(x) = g'(h(x))h'(x)$

    $f'(5) = g'(3)(-2) = -2g'(3)$

    Need $g'(3)$ to find $f'(5)$.

  (c) $f(x) = \frac{g(x)}{h(x)}$

    $f'(x) = \frac{h(x)g'(x) - g(x)h'(x)}{[h(x)]^2}$

    $f'(5) = \frac{(3)(6) - (-3)(-2)}{(3)^2} = \frac{12}{9} = \frac{4}{3}$

  (d) $f(x) = [g(x)]^3$

    $f'(x) = 3[g(x)]^2g'(x)$

    $f'(5) = 3(-3)^2(6) = 162$

63. (a) $h(x) = f(g(x))$, $g(1) = 4$, $g'(1) = -\frac{1}{2}$, $f'(4) = -1$, $h'(x) = f'(g(x))g'(x)$

    $h'(1) = f'(g(1))g'(1)$

     $= f'(4)g'(1)$

     $= (-1)\left(-\frac{1}{2}\right) = \frac{1}{2}$

  (b) $s(x) = g(f(x))$, $f(5) = 6$, $f'(5) = -1$, $g'(6)$ does not exist.

    $s'(x) = g'(f(x))f'(x)$

    $s'(5) = g'(f(5))f'(5) = g'(6)(-1)$

    Since $g'(6)$ does not exist, $s'(5)$ is not defined.

65. (a) $F = 132,400(331 - v)^{-1}$

    $F' = (-1)(132,400)(331 - v)^{-2}(-1)$

     $= \frac{132,400}{(331 - v)^2}$

    When $v = 30$, $F' \approx 1.461$.

  (b) $F = 132,400(331 + v)^{-1}$

    $F' = (-1)(132,400)(331 + v)^{-2}(1)$

     $= \frac{-132,400}{(331 + v)^2}$

    When $v = 30$, $F' \approx -1.016$.

67. (a) $x = -1.6372t^3 + 19.3120t^2 - 0.5082t - 0.6162$

(b) $C = 60x + 1350$

$$= 60(-1.6372t^3 + 19.3120t^2 - 0.5082t - 0.6162) + 1350$$

$$\frac{dC}{dt} = 60(-4.9116t^2 + 38.624t - 0.5082)$$

$$= -294.696t^2 + 2317.44t - 30.492$$

(c) The function $\dfrac{dC}{dt}$ is quadratic, not linear. The cost function levels off at the end of the day, perhaps due to fatigue.

69. (a) $r'(x) = f'(g(x))g'(x)$

$r'(1) = f'(g(1))g'(1)$

Note that $g(1) = 4$ and $f'(4) = \dfrac{5-0}{6-2} = \dfrac{5}{4}$.

Also, $g'(1) = 0$. Thus, $r'(1) = 0$.

(b) $s'(x) = g'(f(x))f'(x)$

$s'(4) = g'(f(4))f'(4)$

Note that $f(4) = \dfrac{5}{2}$, $g'\left(\dfrac{5}{2}\right) = \dfrac{6-4}{6-2} = \dfrac{1}{2}$ and

$f'(4) = \dfrac{5}{4}$.

Thus, $s'(4) = \dfrac{1}{2}\left(\dfrac{5}{4}\right) = \dfrac{5}{8}$.

71. $g(x) = |2x - 3|$

$$g'(x) = 2\left(\frac{2x-3}{|2x-3|}\right), \quad x \neq \frac{3}{2}$$

73. (a) $f(x) = \dfrac{1}{\sqrt{x^2 - 3}} = (x^2 - 3)^{-1/2}, \qquad f(2) = 1$

$$f'(x) = -\frac{1}{2}(x^2 - 3)^{-3/2}(2x) = -\frac{x}{(x^2 - 3)^{3/2}}, \ f'(2) = -2$$

$$f''(x) = \frac{(x^2 - 3)^{3/2}(-1) + x\left(\dfrac{3}{2}\right)(x^2 - 3)^{1/2}(2x)}{(x^2 - 3)^3}$$

$$= \frac{-(x^2 - 3) + 3x^2}{(x^2 - 3)^{5/2}} = \frac{2x^2 + 3}{(x^2 - 3)^{5/2}}, \qquad f''(2) = 11$$

$$P_1(x) = f'(2)(x - 2) + f(2) = -2(x - 2) + 1 = -2x + 5$$

$$P_2(x) = \frac{1}{2}f''(2)(x - 2)^2 + f'(2)(x - 2) + f(2)$$

$$= \frac{1}{2}(11)(x - 2)^2 - 2(x - 2) + 1$$

$$= \frac{11}{2}(x - 2)^2 - 2(x - 2) + 1$$

(b)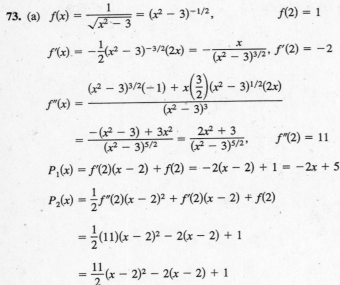

(c) P_2 is the better approximation.

(d) The accuracy worsens as you move away from $x = a = 2$.

75. False. If $y = (1 - x)^{1/2}$, then $y' = \frac{1}{2}(1 - x)^{-1/2}(-1)$.

Section 4.5   Implicit Differentiation

1. $x^2 + y^2 = 36$

$2x + 2yy' = 0$

$$y' = \frac{-x}{y}$$

3. $x^{1/2} + y^{1/2} = 9$

$\dfrac{1}{2}x^{-1/2} + \dfrac{1}{2}y^{-1/2}y' = 0$

$$y' = -\frac{x^{-1/2}}{y^{-1/2}} = -\sqrt{\frac{y}{x}}$$

5. $x^3 - xy + y^2 = 4$

$3x^2 - xy' - y + 2yy' = 0$

$(2y - x)y' = y - 3x^2$

$$y' = \frac{y - 3x^2}{2y - x}$$

7. $x^3y^3 - y - x = 0$

$3x^3y^2y' + 3x^2y^3 - y' - 1 = 0$

$(3x^3y^2 - 1)y' = 1 - 3x^2y^3$

$$y' = \frac{1 - 3x^2y^3}{3x^3y^2 - 1}$$

9. $x^3 - 3x^2y + 2xy^2 = 12$

$3x^2 - 3x^2y' - 6xy + 4xyy' + 2y^2 = 0$

$(4xy - 3x^2)y' = 6xy - 3x^2 - 2y^2$

$$y' = \frac{6xy - 3x^2 - 2y^2}{4xy - 3x^2}$$

11. (a) $x^2 + y^2 = 16$

$\qquad y^2 = 16 - x^2$

$\qquad y = \pm\sqrt{16 - x^2}$

(b)

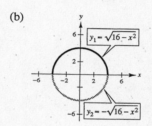

(c) Explicitly:

$$\frac{dy}{dx} = \pm\frac{1}{2}(16 - x^2)^{-1/2}(-2x)$$

$$= \frac{\mp x}{\sqrt{16 - x^2}} = \frac{-x}{\pm\sqrt{16 - x^2}} = \frac{-x}{y}$$

(d) Implicitly:

$2x + 2yy' = 0$

$$y' = -\frac{x}{y}$$

13. (a) $16y^2 = 144 - 9x^2$

$\qquad y^2 = \dfrac{1}{16}(144 - 9x^2) = \dfrac{9}{16}(16 - x^2)$

$\qquad y = \pm\dfrac{3}{4}\sqrt{16 - x^2}$

(b)

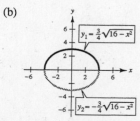

(c) Explicitly:

$$\frac{dy}{dx} = \pm\frac{3}{8}(16 - x^2)^{-1/2}(-2x)$$

$$= \mp\frac{3x}{4\sqrt{16 - x^2}} = \frac{-3x}{4(4/3)y} = \frac{-9x}{16y}$$

(d) Implicitly:

$18x + 32yy' = 0$

$$y' = \frac{-9x}{16y}$$

15.
$$xy = 4$$
$$xy' + y(1) = 0$$
$$xy' = -y$$
$$y' = \frac{-y}{x}$$

At $(-4, -1)$: $y' = -\frac{1}{4}$.

17.
$$y^2 = \frac{x^2 - 4}{x^2 + 4}$$
$$2yy' = \frac{(x^2 + 4)(2x) - (x^2 - 4)(2x)}{(x^2 + 4)^2}$$
$$2yy' = \frac{16x}{(x^2 + 4)^2}$$
$$y' = \frac{8x}{y(x^2 + 4)^2}$$

At $(2, 0)$, y' is undefined.

19.
$$x^{2/3} + y^{2/3} = 5$$
$$\frac{2}{3}x^{-1/3} + \frac{2}{3}y^{-1/3}y' = 0$$
$$y' = \frac{-x^{-1/3}}{y^{-1/3}} = -\sqrt[3]{\frac{y}{x}}$$

At $(8, 1)$: $y' = -\frac{1}{2}$.

21.
$$(x^2 + 4)y = 8$$
$$(x^2 + 4)y' + y(2x) = 0$$
$$y' = \frac{-2xy}{x^2 + 4}$$

At $(2, 1)$: $y' = -\frac{1}{2}$

$\left(\text{Or, you could just solve for } y: \; y = \frac{8}{x^2 + 4}\right)$

23.
$$(x^2 + y^2)^2 = 4x^2y$$
$$2(x^2 + y^2)(2x + 2yy') = 4x^2y' + y(8x)$$
$$4x^3 + 4x^2yy' + 4xy^2 + 4y^3y' = 4x^2y' + 8xy$$
$$4x^2yy' + 4y^3y' - 4x^2y' = 8xy - 4x^3 - 4xy^2$$
$$4y'(x^2y + y^3 - x^2) = 4(2xy - x^3 - xy^2)$$
$$y' = \frac{2xy - x^3 - xy^2}{x^2y + y^3 - x^2}$$

At $(1, 1)$: $y' = 0$.

25. $(y - 2)^2 = 4(x - 3), \quad (4, 0)$
$$2(y - 2)y' = 4$$
$$y' = \frac{2}{y - 2}$$

At $(4, 0)$: $y' = -1$.

Tangent line: $y - 0 = -1(x - 4)$
$$y = -x + 4$$

27. $xy = 1, \quad (1, 1)$
$$xy' + y = 0$$
$$y' = \frac{-y}{x}$$

At $(1, 1)$: $y' = -1$.

Tangent line: $y - 1 = -1(x - 1)$
$$y = -x + 2$$

29. $x^2y^2 - 9x^2 - 4y^2 = 0, \quad \left(-4, 2\sqrt{3}\right)$
$$x^2\,2yy' + 2xy^2 - 18x - 8yy' = 0$$
$$y' = \frac{18x - 2xy^2}{2x^2y - 8y}$$

At $\left(-4, 2\sqrt{3}\right)$: $y' = \dfrac{18(-4) - 2(-4)(12)}{2(16)(2\sqrt{3}) - 16\sqrt{3}}$

$$= \frac{24}{48\sqrt{3}} = \frac{1}{2\sqrt{3}} = \frac{\sqrt{3}}{6}.$$

Tangent line: $y - 2\sqrt{3} = \dfrac{\sqrt{3}}{6}(x + 4)$

$$y = \frac{\sqrt{3}}{6}x + \frac{8}{3}\sqrt{3}$$

31. $$3(x^2 + y^2)^2 = 100(x^2 - y^2), \quad (4, 2)$$
$$6(x^2 + y^2)(2x + 2yy') = 100(2x - 2yy')$$
At (4, 2):
$$6(16 + 4)(8 + 4y') = 100(8 - 4y')$$
$$960 + 480y' = 800 - 400y'$$
$$880y' = -160$$
$$y' = -\tfrac{2}{11}.$$
Tangent line: $\quad y - 2 = -\tfrac{2}{11}(x - 4)$
$$11y + 2x - 30 = 0$$
$$y = -\tfrac{2}{11}x + \tfrac{30}{11}$$

33. (a) $\dfrac{x^2}{2} + \dfrac{y^2}{8} = 1, \quad (1, 2)$
$$x + \frac{yy'}{4} = 0$$
$$y' = -\frac{4x}{y}$$
At (1, 2): $y' = -2.$
Tangent line: $y - 2 = -2(x - 1)$
$$y = -2x + 4$$
Note: From part (a), $\dfrac{1(x)}{2} + \dfrac{2(y)}{8} = 1 \Rightarrow \dfrac{1}{4}y = -\dfrac{1}{2}x + 1 \Rightarrow y = -2x + 4, \quad$ Tangent line.

(b) $\dfrac{x^2}{a^2} + \dfrac{y^2}{b^2} = 1 \Rightarrow \dfrac{2x}{a^2} + \dfrac{2yy'}{b^2} = 0 \Rightarrow y' = \dfrac{-b^2x}{a^2y}$
$$y - y_0 = \frac{-b^2x_0}{a^2y_0}(x - x_0), \quad \text{Tangent line at } (x_0, y_0)$$
$$\frac{y_0y}{b^2} - \frac{y_0^2}{b^2} = \frac{-x_0x}{a^2} + \frac{x_0^2}{a^2}$$
Since $\dfrac{x_0^2}{a^2} + \dfrac{y_0^2}{b^2} = 1$, you have $\dfrac{y_0y}{b^2} + \dfrac{x_0x}{a^2} = 1.$

35. $x^2 + y^2 = 36$
$$2x + 2yy' = 0$$
$$y' = \frac{-x}{y}$$
$$y'' = \frac{y(-1) + xy'}{y^2} = \frac{-y + x\left(-\frac{x}{y}\right)}{y^2} = \frac{-y^2 - x^2}{y^3} = \frac{-36}{y^3}$$

37. $x^2 - y^2 = 16$
$$2x - 2yy' = 0$$
$$y' = \frac{x}{y}$$
$$x - yy' = 0$$
$$1 - yy'' - (y')^2 = 0$$
$$1 - yy'' - \left(\frac{x}{y}\right)^2 = 0$$
$$y^2 - y^3y'' = x^2$$
$$y'' = \frac{y^2 - x^2}{y^3} = \frac{-16}{y^3}$$

39. $y^2 = x^3$

$2yy' = 3x^2$

$y' = \dfrac{3x^2}{2y} = \dfrac{3x^2}{2y} \cdot \dfrac{xy}{xy} = \dfrac{3y}{2x} \cdot \dfrac{x^3}{y^2} = \dfrac{3y}{2x}$

$y'' = \dfrac{2x(3y') - 3y(2)}{4x^2}$

$\quad = \dfrac{2x[3 \cdot (3y/2x)] - 6y}{4x^2}$

$\quad = \dfrac{3y}{4x^2} = \dfrac{3x}{4y}$

41. $\sqrt{x} + \sqrt{y} = 4$

$\dfrac{1}{2}x^{-1/2} + \dfrac{1}{2}y^{-1/2}y' = 0$

$y' = \dfrac{-\sqrt{y}}{\sqrt{x}}$

At $(9, 1)$: $y' = -\dfrac{1}{3}$.

Tangent line: $y - 1 = -\dfrac{1}{3}(x - 9)$

$y = -\dfrac{1}{3}x + 4$

$x + 3y - 12 = 0$

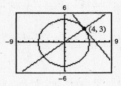

43. $x^2 + y^2 = 25$

$2x + 2yy' = 0$

$y' = \dfrac{-x}{y}$

At $(4, 3)$:

Tangent line: $y - 3 = \dfrac{-4}{3}(x - 4) \Longrightarrow 4x + 3y - 25 = 0$

Normal line: $y - 3 = \dfrac{3}{4}(x - 4) \Longrightarrow 3x - 4y = 0$.

At $(-3, 4)$:

Tangent line: $y - 4 = \dfrac{3}{4}(x + 3) \Longrightarrow 3x - 4y + 25 = 0$

Normal line: $y - 4 = \dfrac{-4}{3}(x + 3) \Longrightarrow 4x + 3y = 0$.

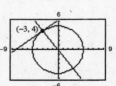

45. $x^2 + y^2 = r^2$

$2x + 2yy' = 0$

$y' = \dfrac{-x}{y} =$ slope of tangent line

$\dfrac{y}{x} =$ slope of normal line

Let (x_0, y_0) be a point on the circle. If $x_0 = 0$, then the tangent line is horizontal, the normal line is vertical and, hence, passes through the origin. If $x_0 \neq 0$, then the equation of the normal line is

$y - y_0 = \dfrac{y_0}{x_0}(x - x_0)$

$y = \dfrac{y_0}{x_0}x$

which passes through the origin.

47. $25x^2 + 16y^2 + 200x - 160y + 400 = 0$

$$50x + 32yy' + 200 - 160y' = 0$$

$$y' = \frac{200 + 50x}{160 - 32y}$$

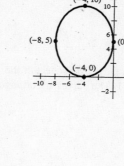

Horizontal tangents occur when $x = -4$:

$$25(16) + 16y^2 + 200(-4) - 160y + 400 = 0$$

$$y(y - 10) = 0 \implies y = 0, 10$$

Horizontal tangents: $(-4, 0), (-4, 10)$.

Vertical tangents occur when $y = 5$:

$$25x^2 + 400 + 200x - 800 + 400 = 0$$

$$25x(x + 8) = 0 \implies x = 0, -8$$

Vertical tangents: $(0, 5), (-8, 5)$

49. In the explicit form of a function, the variable is explicitly written as a function of x. In an implicit equation, the function is only implied by the equation. An example of an implicit function is $x^2 + xy = 5$. In explicit form, it would be

$$y = \frac{5 - x^2}{x}.$$

51. Find the points of intersection by letting $y^2 = 4x$ in the equation $2x^2 + y^2 = 6$.

$$2x^2 + 4x = 6 \quad \text{and} \quad (x + 3)(x - 1) = 0$$

The curves intersect at $(1, \pm 2)$.

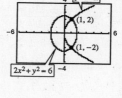

Ellipse:	*Parabola:*
$4x + 2yy' = 0$	$2yy' = 4$
$y' = -\dfrac{2x}{y}$	$y' = \dfrac{2}{y}$

At $(1, 2)$, the slopes are:

$y' = -1$	$y' = 1.$

At $(1, -2)$, the slopes are:

$y' = 1$	$y' = -1.$

Tangents are perpendicular.

53. $x + y = 0$

$x^2 + y^2 = 4$

The line and circle intersect at $\left(\sqrt{2}, -\sqrt{2}\right)$ and $\left(-\sqrt{2}, \sqrt{2}\right)$.

$$x + y = 0 \implies y' = -1$$

$$x^2 + y^2 = 4 \implies 2x + 2y\frac{dy}{dx} = 0 \implies \frac{dy}{dx} = -\frac{x}{y}$$

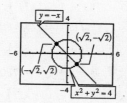

At $\left(-\sqrt{2}, \sqrt{2}\right)$, slope of line is -1, slope of circle is 1.

At $\left(\sqrt{2}, -\sqrt{2}\right)$, slope of line is -1, slope of circle is 1.

55.

$$xy = C \qquad\qquad x^2 - y^2 = K$$

$$xy' + y = 0 \qquad\qquad 2x - 2yy' = 0$$

$$y' = -\frac{y}{x} \qquad\qquad y' = \frac{x}{y}$$

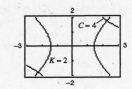

At any point of intersection (x, y) the product of the slopes is $(-y/x)(x/y) = -1$. The curves are orthogonal.

57. $2y^2 - 3x^4 = 0$

(a) $4yy' - 12x^3 = 0$

$$4yy' = 12x^3$$

$$y' = \frac{12x^3}{4y} = \frac{3x^3}{y}$$

(b) $4y\dfrac{dy}{dt} - 12x^3\dfrac{dx}{dt} = 0$

$$y\frac{dy}{dt} = 3x^3\frac{dx}{dt}$$

59. (a) $x^4 = 4(4x^2 - y^2)$

$$4y^2 = 16x^2 - x^4$$

$$y^2 = 4x^2 - \frac{1}{4}x^4$$

$$y = \pm\sqrt{4x^2 - \frac{1}{4}x^4}$$

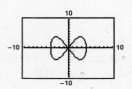

(b) $\quad y = 3 \implies 9 = 4x^2 - \dfrac{1}{4}x^4$

$$36 = 16x^2 - x^4$$

$$x^4 - 16x^2 + 36 = 0$$

$$x^2 = \frac{16 \pm \sqrt{256 - 144}}{2} = 8 \pm \sqrt{28}$$

Note that $x^2 = 8 \pm \sqrt{28} = 8 \pm 2\sqrt{7} = \left(1 \pm \sqrt{7}\right)^2$.

Hence, there are four values of x:

$$-1-\sqrt{7},\, 1-\sqrt{7},\, -1+\sqrt{7},\, 1 + \sqrt{7}$$

To find the slope, $2yy' = 8x - x^3 \implies y' = \dfrac{x(8 - x^2)}{2(3)}.$

For $x = -1 - \sqrt{7}$, $y' = \frac{1}{3}\left(\sqrt{7} + 7\right)$, and the line is

$$y_1 = \tfrac{1}{3}\left(\sqrt{7} + 7\right)\left(x + 1 + \sqrt{7}\right) + 3 = \tfrac{1}{3}\left[\left(\sqrt{7} + 7\right)x + 8\sqrt{7} + 23\right].$$

For $x = 1 - \sqrt{7}$, $y' = \frac{1}{3}\left(\sqrt{7} - 7\right)$, and the line is

$$y_2 = \tfrac{1}{3}\left(\sqrt{7} - 7\right)\left(x - 1 + \sqrt{7}\right) + 3 = \tfrac{1}{3}\left[\left(\sqrt{7} - 7\right)x + 23 - 8\sqrt{7}\right].$$

For $x = -1 + \sqrt{7}$, $y' = -\frac{1}{3}\left(\sqrt{7} - 7\right)$, and the line is

$$y_3 = -\tfrac{1}{3}\left(\sqrt{7} - 7\right)\left(x + 1 - \sqrt{7}\right) + 3 = -\tfrac{1}{3}\left[\left(\sqrt{7} - 7\right)x - \left(23 - 8\sqrt{7}\right)\right].$$

For $x = 1 + \sqrt{7}$, $y' = -\frac{1}{3}\left(\sqrt{7} + 7\right)$, and the line is

$$y_4 = -\tfrac{1}{3}\left(\sqrt{7} + 7\right)\left(x - 1 - \sqrt{7}\right) + 3 = -\tfrac{1}{3}\left[\left(\sqrt{7} + 7\right)x - \left(8\sqrt{7} + 23\right)\right].$$

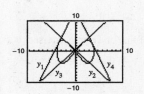

—CONTINUED—

59. —CONTINUED—

(c) Equating y_3 and y_4,

$$-\frac{1}{3}(\sqrt{7}-7)(x+1-\sqrt{7})+3 = -\frac{1}{3}(\sqrt{7}+7)(x-1-\sqrt{7})+3$$

$$(\sqrt{7}-7)(x+1-\sqrt{7}) = (\sqrt{7}+7)(x-1-\sqrt{7})$$

$$\sqrt{7}x+\sqrt{7}-7-7x-7+7\sqrt{7} = \sqrt{7}x-\sqrt{7}-7+7x-7-7\sqrt{7}$$

$$16\sqrt{7} = 14x$$

$$x = \frac{8\sqrt{7}}{7}$$

If $x = \frac{8\sqrt{7}}{7}$, then $y = 5$ and the lines intersect at $\left(\frac{8\sqrt{7}}{7}, 5\right)$.

61. Suppose $n = \frac{p}{q}$ with p, q integers $q > 0$. If $q = 1$, then

$n = p$ and $\frac{d}{dx}[x^n] = nx^{n-1}$ by the integer case.

So suppose $q > 1$,

$y = x^{p/q}$ implies $y^q = x^p$ so

$$qy^{q-1}y' = px^{p-1}$$

$$y' = \frac{p}{q}\frac{x^{p-1}}{y^{q-1}}$$

$$= \frac{p}{q}\frac{x^{p-1}}{(x^{p/q})^{q-1}}$$

$$= \frac{p}{q}x^{p/q-1}$$

$$= nx^{n-1}.$$

Section 4.6   Related Rates

1. $y = \sqrt{x}$

$$\frac{dy}{dt} = \left(\frac{1}{2\sqrt{x}}\right)\frac{dx}{dt}$$

$$\frac{dx}{dt} = 2\sqrt{x}\frac{dy}{dt}$$

(a) When $x = 4$ and $dx/dt = 3$,

$$\frac{dy}{dt} = \frac{1}{2\sqrt{4}}(3) = \frac{3}{4}.$$

(b) When $x = 25$ and $dy/dt = 2$,

$$\frac{dx}{dt} = 2\sqrt{25}(2) = 20.$$

3. $xy = 4$

$$x\frac{dy}{dt}+y\frac{dx}{dt} = 0$$

$$\frac{dy}{dt} = \left(-\frac{y}{x}\right)\frac{dx}{dt}$$

$$\frac{dx}{dt} = \left(-\frac{x}{y}\right)\frac{dy}{dt}$$

(a) When $x = 8$, $y = 1/2$, and $dx/dt = 10$,

$$\frac{dy}{dt} = -\frac{1/2}{8}(10) = -\frac{5}{8}.$$

(b) When $x = 1$, $y = 4$, and $dy/dt = -6$,

$$\frac{dx}{dt} = -\frac{1}{4}(-6) = \frac{3}{2}.$$

5. $y = x^2 + 1$

$\dfrac{dx}{dt} = 2$

$\dfrac{dy}{dt} = 2x\dfrac{dx}{dt}$

(a) When $x = -1$,

$\dfrac{dy}{dt} = 2(-1)(2) = -4$ cm/sec.

(b) When $x = 0$,

$\dfrac{dy}{dt} = 2(0)(2) = 0$ cm/sec.

(c) When $x = 1$,

$\dfrac{dy}{dt} = 2(1)(2) = 4$ cm/sec.

9. (a) $\dfrac{dx}{dt}$ negative $\implies \dfrac{dy}{dt}$ positive

(b) $\dfrac{dy}{dt}$ positive $\implies \dfrac{dx}{dt}$ negative

13. $D = \sqrt{x^2 + y^2} = \sqrt{x^2 + (x^2+1)^2} = \sqrt{x^4 + 3x^2 + 1}$

$\dfrac{dx}{dt} = 2$

$\dfrac{dD}{dt} = \dfrac{1}{2}(x^4 + 3x^2 + 1)^{-1/2}(4x^3 + 6x)\dfrac{dx}{dt} = \dfrac{2x^3 + 3x}{\sqrt{x^4 + 3x^2 + 1}}\dfrac{dx}{dt} = \dfrac{4x^3 + 6x}{\sqrt{x^4 + 3x^2 + 1}}$

15. $A = \pi r^2$

$\dfrac{dr}{dt} = 3$

$\dfrac{dA}{dt} = 2\pi r\dfrac{dr}{dt}$

(a) When $r = 6$,

$\dfrac{dA}{dt} = 2\pi(6)(3) = 36\pi$ cm²/min.

(b) When $r = 24$,

$\dfrac{dA}{dt} = 2\pi(24)(3) = 144\pi$ cm²/min.

7. $\sqrt{x} + \sqrt{y} = 4$

$\dfrac{1}{2\sqrt{x}}\dfrac{dx}{dt} + \dfrac{1}{2\sqrt{y}}\dfrac{dy}{dt} = 0$

$\dfrac{1}{2\sqrt{y}}\dfrac{dy}{dt} = \dfrac{-1}{2\sqrt{x}}\dfrac{dx}{dt}$

$\dfrac{dy}{dt} = -\dfrac{\sqrt{y}}{\sqrt{x}}\dfrac{dx}{dt} = -\dfrac{2\sqrt{y}}{\sqrt{x}}$

(a) $x = 1, y = 9, \dfrac{dy}{dt} = -\dfrac{2(3)}{1} = -6$ cm/sec

(b) $x = 4, y = 4, \dfrac{dy}{dt} = -2\left(\dfrac{2}{2}\right) = -2$ cm/sec

(c) $x = 9, y = 1, \dfrac{dy}{dt} = -\dfrac{2(1)}{3} = -\dfrac{2}{3}$ cm/sec

11. Yes, y changes at a constant rate: $\dfrac{dy}{dt} = a \cdot \dfrac{dx}{dt}$.

No, the rate $\dfrac{dy}{dt}$ is a multiple of $\dfrac{dx}{dt}$.

17. (a) $h^2 + \left(\dfrac{b}{2}\right)^2 = 30^2$

$h = \sqrt{900 - b^2/4}$

$A = \dfrac{1}{2}bh = \dfrac{1}{2}b\sqrt{900 - b^2/4}$

$= \dfrac{b\sqrt{3600 - b^2}}{4}$

(b) $\dfrac{dA}{dt} = \dfrac{1}{4}\left[-b^2(3600 - b^2)^{-1/2} + (3600 - b^2)^{1/2}\right]\dfrac{db}{dt}$

$= \dfrac{-b^2 + (3600 - b^2)}{4\sqrt{3600 - b^2}}(3)$

$= \dfrac{5400 - 3b^2}{2\sqrt{3600 - b^2}}$

If $b = 20, \dfrac{dA}{dt} = \dfrac{105\sqrt{2}}{4} \approx 37.1$ cm/sec.

If $b = 56, \dfrac{dA}{dt} = -\dfrac{501\sqrt{29}}{29} \approx -93.0$ cm/sec.

(c) If $\dfrac{db}{dt}$ is constant, $\dfrac{dA}{dt}$ is a nonconstant function of b.

19. $V = \frac{4}{3}\pi r^3, \frac{dV}{dt} = 800$

$$\frac{dV}{dt} = 4\pi r^2 \frac{dr}{dt}$$

$$\frac{dr}{dt} = \frac{1}{4\pi r^2}\left(\frac{dV}{dt}\right) = \frac{1}{4\pi r^2}(800)$$

(a) When $r = 30, \frac{dr}{dt} = \frac{1}{4\pi(30)^2}(800) = \frac{2}{9\pi}$ cm/min.

(b) When $r = 60, \frac{dr}{dt} = \frac{1}{4\pi(60)^2}(800) = \frac{1}{18\pi}$ cm/min.

21. $s = 6x^2$

$$\frac{dx}{dt} = 3$$

$$\frac{ds}{dt} = 12x \frac{dx}{dt}$$

(a) When $x = 1, \frac{ds}{dt} = 12(1)(3) = 36$ cm²/sec.

(b) When $x = 10, \frac{ds}{dt} = 12(10)(3) = 360$ cm²/sec.

23. $V = \frac{1}{3}\pi r^2 h = \frac{1}{3}\pi\left(\frac{9}{4}h^2\right)h$ [since $2r = 3h$]

$$= \frac{3\pi}{4}h^3$$

$$\frac{dV}{dt} = 10$$

$$\frac{dV}{dt} = \frac{9\pi}{4}h^2 \frac{dh}{dt} \implies \frac{dh}{dt} = \frac{4(dV/dt)}{9\pi h^2}$$

When $h = 15, \frac{dh}{dt} = \frac{4(10)}{9\pi(15)^2} = \frac{8}{405\pi}$ ft/min.

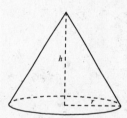

25.

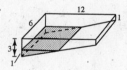

(a) Total volume of pool $= \frac{1}{2}(2)(12)(6) + (1)(6)(12) = 144$ m³

Volume of 1m. of water $= \frac{1}{2}(1)(6)(6) = 18$ m³

(see similar triangle diagram)

% pool filled $= \frac{18}{144}(100\%) = 12.5\%$

(b) Since for $0 \le h \le 2, b = 6h$, you have

$$V = \frac{1}{2}bh(6) = 3bh = 3(6h)h = 18h^2$$

$$\frac{dV}{dt} = 36h \frac{dh}{dt} = \frac{1}{4} \implies \frac{dh}{dt} = \frac{1}{144h} = \frac{1}{144(1)} = \frac{1}{144}$$ m/min.

27. $x^2 + y^2 = 25^2$

$2x\dfrac{dx}{dt} + 2y\dfrac{dy}{dt} = 0$

$\dfrac{dy}{dt} = \dfrac{-x}{y}\cdot\dfrac{dx}{dt} = \dfrac{-2x}{y}$ since $\dfrac{dx}{dt} = 2$.

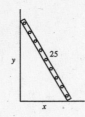

(a) When $x = 7, y = \sqrt{576} = 24, \dfrac{dy}{dt} = \dfrac{-2(7)}{24} = \dfrac{-7}{12}$ ft/sec.

When $x = 15, y = \sqrt{400} = 20, \dfrac{dy}{dt} = \dfrac{-2(15)}{20} = \dfrac{-3}{2}$ ft/sec.

When $x = 24, y = 7, \dfrac{dy}{dt} = \dfrac{-2(24)}{7} = \dfrac{-48}{7}$ ft/sec.

(b) $A = \dfrac{1}{2}xy$

$\dfrac{dA}{dt} = \dfrac{1}{2}\left(x\dfrac{dy}{dt} + y\dfrac{dx}{dt}\right)$

From part (a) we have $x = 7, y = 24, \dfrac{dx}{dt} = 2$, and $\dfrac{dy}{dt} = -\dfrac{7}{12}$.

Thus, $\dfrac{dA}{dt} = \dfrac{1}{2}\left[7\left(-\dfrac{7}{12}\right) + 24(2)\right] = \dfrac{527}{24} \approx 21.96$ ft²/sec.

29. When $y = 6, x = \sqrt{12^2 - 6^2} = 6\sqrt{3}$, and

$s = \sqrt{x^2 + (12 - y)^2}$

$= \sqrt{108 + 36} = 12.$

$x^2 + (12 - y)^2 = s^2$

$2x\dfrac{dx}{dt} + 2(12 - y)(-1)\dfrac{dy}{dt} = 2s\dfrac{ds}{dt}$

$x\dfrac{dx}{dt} + (y - 12)\dfrac{dy}{dt} = s\dfrac{ds}{dt}$

Also, $x^2 + y^2 = 12^2$

$2x\dfrac{dx}{dt} + 2y\dfrac{dy}{dt} = 0 \implies \dfrac{dy}{dt} = \dfrac{-x}{y}\dfrac{dx}{dt}.$

Thus, $x\dfrac{dx}{dt} + (y - 12)\left(\dfrac{-x}{y}\dfrac{dx}{dt}\right) = s\dfrac{ds}{dt}$

$\dfrac{dx}{dt}\left[x - x + \dfrac{12x}{y}\right] = s\dfrac{ds}{dt} \implies \dfrac{dx}{dt} = \dfrac{sy}{12x}\cdot\dfrac{ds}{dt} = \dfrac{(12)(6)}{(12)(6\sqrt3)}(-0.2) = \dfrac{-1}{5\sqrt3} = \dfrac{-\sqrt3}{15}$ m/sec (horizontal)

$\dfrac{dy}{dt} = \dfrac{-x}{y}\dfrac{dx}{dt} = \dfrac{-6\sqrt3}{6}\cdot\dfrac{(-\sqrt3)}{15} = \dfrac{1}{5}$ m/sec (vertical).

31. (a) $s^2 = x^2 + y^2$

$$\frac{dx}{dt} = -450$$

$$\frac{dy}{dt} = -600$$

$$2s\,\frac{ds}{dt} = 2x\,\frac{dx}{dt} + 2y\,\frac{dy}{dt}$$

$$\frac{ds}{dt} = \frac{x(dx/dt) + y(dy/dt)}{s}$$

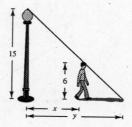

When $x = 150$ and $y = 200$, $s = 250$ and

$$\frac{ds}{dt} = \frac{150(-450) + 200(-600)}{250} = -750 \text{ mph.}$$

(b) $t = \dfrac{250}{750} = \dfrac{1}{3}$ hr $= 20$ min

33. $s^2 = 90^2 + x^2$

$$x = 30$$

$$\frac{dx}{dt} = -28$$

$$2s\,\frac{ds}{dt} = 2x\,\frac{dx}{dt} \Longrightarrow \frac{ds}{dt} = \frac{x}{s}\cdot\frac{dx}{dt}$$

When $x = 30$,

$$s = \sqrt{90^2 + 30^2} = 30\sqrt{10}$$

$$\frac{ds}{dt} = \frac{30}{30\sqrt{10}}(-28) = \frac{-28}{\sqrt{10}} \approx -8.85 \text{ ft/sec.}$$

35. (a) $\dfrac{15}{6} = \dfrac{y}{y - x} \Longrightarrow 15y - 15x = 6y$

$$y = \frac{5}{3}x$$

$$\frac{dx}{dt} = 5$$

$$\frac{dy}{dt} = \frac{5}{3}\cdot\frac{dx}{dt} = \frac{5}{3}(5) = \frac{25}{3} \text{ ft/sec}$$

(b) $\dfrac{d(y - x)}{dt} = \dfrac{dy}{dt} - \dfrac{dx}{dt} = \dfrac{25}{3} - 5 = \dfrac{10}{3}$ ft/sec

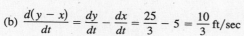

37. Since the evaporation rate is proportional to the surface area, $dV/dt = k(4\pi r^2)$. However, since $V = (4/3)\pi r^3$, we have

$$\frac{dV}{dt} = 4\pi r^2\,\frac{dr}{dt}.$$

Therefore,

$$k(4\pi r^2) = 4\pi r^2\,\frac{dr}{dt} \Longrightarrow k = \frac{dr}{dt}.$$

39. $x^2 + y^2 = 25$; acceleration of the top of the ladder $= \dfrac{d^2y}{dt^2}$

First derivative: $2x\dfrac{dx}{dt} + 2y\dfrac{dy}{dt} = 0$

$$x\dfrac{dx}{dt} + y\dfrac{dy}{dt} = 0$$

Second derivative: $x\dfrac{d^2x}{dt^2} + \dfrac{dx}{dt}\cdot\dfrac{dx}{dt} + y\dfrac{d^2y}{dt^2} + \dfrac{dy}{dt}\cdot\dfrac{dy}{dt} = 0$

$$\dfrac{d^2y}{dt^2} = \left(\dfrac{1}{y}\right)\left[-x\dfrac{d^2x}{dt^2} - \left(\dfrac{dx}{dt}\right)^2 - \left(\dfrac{dy}{dt}\right)^2\right]$$

When $x = 7, y = 24, \dfrac{dy}{dt} = -\dfrac{7}{12}$, and $\dfrac{dx}{dt} = 2$ (see Exercise 27). Since $\dfrac{dx}{dt}$ is constant, $\dfrac{d^2x}{dt^2} = 0$.

$$\dfrac{d^2y}{dt^2} = \dfrac{1}{24}\left[-7(0) - (2)^2 - \left(-\dfrac{7}{12}\right)^2\right] = \dfrac{1}{24}\left[-4 - \dfrac{49}{144}\right] = \dfrac{1}{24}\left[-\dfrac{625}{144}\right] \approx -0.1808 \text{ ft/sec}^2$$

41. $y(t) = -4.9t^2 + 20$

$\dfrac{dy}{dt} = -9.8t$

$y(1) = -4.9 + 20 = 15.1$

$y'(1) = -9.8$

By similar triangles, $\dfrac{20}{x} = \dfrac{y}{x - 12}$

$20x - 240 = xy.$

When $y = 15.1, 20x - 240 = x(15.1)$

$(20 - 15.1)x = 240$

$x = \dfrac{240}{4.9}.$

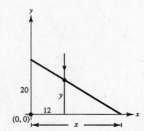

$20x - 240 = xy$

$20\dfrac{dx}{dt} = x\dfrac{dy}{dt} + y\dfrac{dx}{dt}$

$\dfrac{dx}{dt} = \dfrac{x}{20 - y}\dfrac{dy}{dt}$

At $t = 1, \dfrac{dx}{dt} = \dfrac{240/4.9}{20 - 15.1}(-9.8) \approx -97.96 \text{ m/sec.}$

Review Exercises for Chapter 4

1. $f(x) = x^2 - 2x + 3$

$f'(x) = \displaystyle\lim_{\Delta x \to 0} \dfrac{f(x + \Delta x) - f(x)}{\Delta x}$

$= \displaystyle\lim_{\Delta x \to 0} \dfrac{[(x + \Delta x)^2 - 2(x + \Delta x) + 3] - [x^2 - 2x + 3]}{\Delta x}$

$= \displaystyle\lim_{\Delta x \to 0} \dfrac{(x^2 + 2x(\Delta x) + (\Delta x)^2 - 2x - 2(\Delta x) + 3) - (x^2 - 2x + 3)}{\Delta x}$

$= \displaystyle\lim_{\Delta x \to 0} \dfrac{2x(\Delta x) + (\Delta x)^2 - 2(\Delta x)}{\Delta x} = \displaystyle\lim_{\Delta x \to 0} (2x + \Delta x - 2) = 2x - 2$

3. $f(x) = \sqrt{x} + 1$

$$f'(x) = \lim_{\Delta x \to 0} \frac{f(x + \Delta x) - f(x)}{\Delta x}$$

$$= \lim_{\Delta x \to 0} \frac{\left(\sqrt{x + \Delta x} + 1\right) - \left(\sqrt{x} + 1\right)}{\Delta x}$$

$$= \lim_{\Delta x \to 0} \frac{\sqrt{x + \Delta x} - \sqrt{x}}{\Delta x} \cdot \frac{\sqrt{x + \Delta x} + \sqrt{x}}{\sqrt{x + \Delta x} + \sqrt{x}}$$

$$= \lim_{\Delta x \to 0} \frac{(x + \Delta x) - x}{\Delta x\left(\sqrt{x + \Delta x} + \sqrt{x}\right)}$$

$$= \lim_{\Delta x \to 0} \frac{1}{\sqrt{x + \Delta x} + \sqrt{x}} = \frac{1}{2\sqrt{x}}$$

5. f is differentiable for all $x \neq -1$.

7. $f(x) = 4 - |x - 2|$

    (a) Continuous at $x = 2$.

    (b) Not differentiable at $x = 2$ because of the sharp turn in the graph.

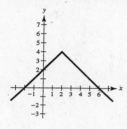

9. Using the limit definition, you obtain $g'(x) = \frac{4}{3}x - \frac{1}{6}$.

At $x = -1$, $g'(-1) = -\frac{4}{3} - \frac{1}{6} = \frac{-3}{2}$.

11. (a) Using the limit definition, $f'(x) = 3x^2$.

At $x = -1$, $f'(-1) = 3$. The tangent line is

$$y - (-2) = 3(x - (-1))$$

$$y = 3x + 1.$$

(b)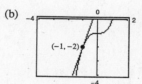

13. $g'(2) = \lim_{x \to 2} \dfrac{g(x) - g(2)}{x - 2}$

$$= \lim_{x \to 2} \frac{x^2(x - 1) - 4}{x - 2}$$

$$= \lim_{x \to 2} \frac{x^3 - x^2 - 4}{x - 2}$$

$$= \lim_{x \to 2} \frac{(x - 2)(x^2 + x + 2)}{x - 2}$$

$$= \lim_{x \to 2} (x^2 + x + 2) = 8$$

15. $y = 25$

    $y' = 0$

17. $f(x) = x^8$

    $f'(x) = 8x^7$

19. $h(t) = 3t^4$

    $h'(t) = 12t^3$

21. $f(x) = x^3 - 3x^2$

    $f'(x) = 3x^2 - 6x = 3x(x - 2)$

23. $h(x) = 6\sqrt{x} + 3\sqrt[3]{x} = 6x^{1/2} + 3x^{1/3}$

$h'(x) = 3x^{-1/2} + x^{-2/3} = \dfrac{3}{\sqrt{x}} + \dfrac{1}{\sqrt[3]{x^2}}$

25. $g(t) = \dfrac{2}{3}t^{-2}$

$g'(t) = \dfrac{-4}{3}t^{-3} = \dfrac{-4}{3t^3}$

27. f is always increasing so $f' > 0$.

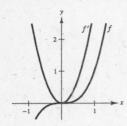

29. $F = 200\sqrt{T}$

$F'(T) = \dfrac{100}{\sqrt{T}}$

(a) When $T = 4$, $F'(4) = 50$ vibrations/sec/lb.

(b) When $T = 9$, $F'(9) = 33\frac{1}{3}$ vibrations/sec/lb.

31. $s(t) = -16t^2 + s_0$

$s(9.2) = -16(9.2)^2 + s_0 = 0$

$s_0 = 1354.24$

The building is approximately 1354 feet high (or 415 m).

33. (a)

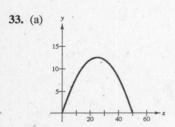

(b) $0 = x - 0.02x^2$

$0 = x\left(1 - \dfrac{x}{50}\right)$ implies $x = 50$.

Total horizontal distance: 50

(c) Ball reaches maximum height when $x = 25$.

(d) $\quad y = x - 0.02x^2$

$\quad y' = 1 - 0.04x$

$y'(0) = 1$

$y'(10) = 0.6$

$y'(25) = 0$

$y'(30) = -0.2$

$y'(50) = -1$

(e) $y'(25) = 0$

35. $x(t) = t^2 - 3t + 2 = (t - 2)(t - 1)$

(a) $v(t) = x'(t) = 2t - 3$

(b) $v(t) < 0$ for $t < \frac{3}{2}$.

(c) $v(t) = 0$ for $t = \frac{3}{2}$.

$x = \left(\frac{3}{2} - 2\right)\left(\frac{3}{2} - 1\right) = \left(-\frac{1}{2}\right)\left(\frac{1}{2}\right) = -\frac{1}{4}$

(d) $x(t) = 0$ for $t = 1, 2$.

$|v(1)| = |2(1) - 3| = 1$

$|v(2)| = |2(2) - 3| = 1$

The speed is 1 when the position is 0.

37. $f(x) = (3x^2 + 7)(x^2 - 2x + 3)$

$\quad f'(x) = (3x^2 + 7)(2x - 2) + (x^2 - 2x + 3)(6x)$

$\qquad\quad = 2(6x^3 - 9x^2 + 16x - 7)$

39. $h(t) = \sqrt{t}(t^3 + 4t - 1) = t^{7/2} + 4t^{3/2} - t^{1/2}$

$\quad h'(t) = \dfrac{7}{2}t^{5/2} + 6t^{1/2} - \dfrac{1}{2t^{1/2}}$

$\qquad\quad = \dfrac{7t^3 + 12t - 1}{2\sqrt{t}}$

41. $f(x) = \dfrac{x^2 + x - 1}{x^2 - 1}$

$\quad f'(x) = \dfrac{(x^2 - 1)(2x + 1) - (x^2 + x - 1)(2x)}{(x^2 - 1)^2}$

$\qquad\quad = \dfrac{-(x^2 + 1)}{(x^2 - 1)^2}$

43. $f(x) = (4 - 3x^2)^{-1}$

$\quad f'(x) = -(4 - 3x^2)^{-2}(-6x) = \dfrac{6x}{(4 - 3x^2)^2}$

45. $f(x) = \dfrac{2x^3 - 1}{x^2} = 2x - x^{-2}, \; (1, 1)$

$\quad f'(x) = 2 + 2x^{-3} = 2 + \dfrac{2}{x^3}$

$\quad f'(1) = 4$

Tangent line: $y - 1 = 4(x - 1)$

$\qquad\qquad\qquad y = 4x - 3$

47. $g(t) = t^3 - 3t + 2$

$\quad g'(t) = 3t^2 - 3$

$\quad g''(t) = 6t$

49. $f(x) = (3x^2 + 7)(x^2 - 2x + 3)$

$\quad f'(x) = (3x^2 + 7)(2x - 2) + (x^2 - 2x + 3)(6x)$

$\qquad\quad = 6x^3 - 6x^2 + 14x - 14 + 6x^3 - 12x^2 + 18x$

$\qquad\quad = 12x^3 - 18x^2 + 32x - 14$

$\quad f''(x) = 36x^2 - 36x + 32$

$\qquad\quad = 4(9x^2 - 9x + 8)$

51. $h(x) = (x^2 - 4x)^3$

$\quad h'(x) = 3(x^2 - 4x)^2(2x - 4)$

$\qquad\quad = 6x^2(x - 4)^2(x - 2)$

53. $f(x) = -2(1 - 4x^2)^2$

$\quad f'(x) = -4(1 - 4x^2)(-8x)$

$\qquad\quad = 32x(1 - 4x^2)$

$\qquad\quad = 32x - 128x^3$

55. $h(x) = \left(\dfrac{x - 3}{x^2 + 1}\right)^2$

$\quad h'(x) = 2\left(\dfrac{x - 3}{x^2 + 1}\right)\left(\dfrac{(x^2 + 1)(1) - (x - 3)(2x)}{(x^2 + 1)^2}\right)$

$\qquad\quad = \dfrac{2(x - 3)(-x^2 + 6x + 1)}{(x^2 + 1)^3}$

57. $f(s) = (s^2 - 1)^{5/2}(s^3 + 5)$

$\quad f'(s) = (s^2 - 1)^{5/2}(3s^2) + (s^3 + 5)\left(\dfrac{5}{2}\right)(s^2 - 1)^{3/2}(2s)$

$\qquad\quad = s(s^2 - 1)^{3/2}[3s(s^2 - 1) + 5(s^3 + 5)]$

$\qquad\quad = s(s^2 - 1)^{3/2}(8s^3 - 3s + 25)$

59. $f(x) = \sqrt{1 - x^3} = (1 - x^3)^{1/2}$

$$f'(x) = \frac{1}{2}(1 - x^3)^{-1/2}(-3x^2) = \frac{-3x^2}{2\sqrt{1 - x^3}}$$

$$f'(-2) = -\frac{-3(-2)^2}{2\sqrt{9}} = -2$$

61. $f(t) = t^2(t - 1)^5$

$\quad f'(t) = t(t - 1)^4(7t - 2)$

The zeros of f' correspond to the points on the graph of f where the tangent line is horizontal.

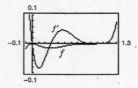

63. $g(x) = 2x(x + 1)^{-1/2}$

$$g'(x) = \frac{x + 2}{(x + 1)^{3/2}}$$

g' does not equal zero for any value of x in the domain. The graph of g has no horizontal tangent lines.

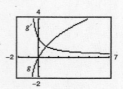

65. $f(t) = \sqrt{t + 1}\sqrt[3]{t + 1}$

$\quad f(t) = (t + 1)^{1/2}(t + 1)^{1/3} = (t + 1)^{5/6}$

$$f'(t) = \frac{5}{6(t + 1)^{1/6}}$$

f' does not equal zero for any t in the domain. The graph of f has no horizontal tangent lines.

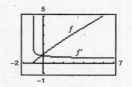

67. $f(x) = \sqrt{x^2 + 9} = (x^2 + 9)^{1/2}$

$$f'(x) = \frac{1}{2}(x^2 + 9)^{-1/2}(2x) = \frac{x}{(x^2 + 9)^{1/2}}$$

$$f''(x) = \frac{(x^2 + 9)^{1/2} - x\frac{1}{2}(x^2 + 9)^{-1/2}(2x)}{(x^2 + 9)}$$

$$= \frac{(x^2 + 9) - x^2}{(x^2 + 9)^{3/2}} = \frac{9}{(x^2 + 9)^{3/2}}$$

69. $y = \dfrac{4}{(x - 2)^2} = 4(x - 2)^{-2}$

$\quad y' = -8(x - 2)^{-3}$

$\quad y'' = 24(x - 2)^{-4} = \dfrac{24}{(x - 2)^4}$

71. $f(t) = \dfrac{t}{(1 - t)^2}$

$\quad f'(t) = \dfrac{t + 1}{(1 - t)^3}$

$\quad f''(t) = \dfrac{2(t + 2)}{(1 - t)^4}$

73. $g(x) = \dfrac{x}{\sqrt{x + 3}}$

$\quad g'(x) = \dfrac{x + 6}{2(x + 3)^{3/2}}$

$\quad g''(x) = -\dfrac{x + 12}{4(x + 3)^{5/2}}$

75. (a) $y = (x + 3)^3$, $(-2, 1)$

$y' = 3(x + 3)^2$, $y'(-2) = 3$

$y - 1 = 3(x + 2)$

$3x - y + 7 = 0$ or $y = 3x + 7$

(b)

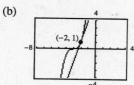

77. (a) $y = \sqrt[3]{(x - 2)^2} = (x - 2)^{2/3}$, $(3, 1)$

$y' = \frac{2}{3}(x - 2)^{-1/3}$, $y'(3) = \frac{2}{3}$

$y - 1 = \frac{2}{3}(x - 3)$

$3y - 3 = 2x - 6$

$2x - 3y - 3 = 0$ or $y = \frac{2}{3}x - 1$

(b)

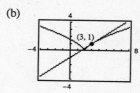

79. $T = \dfrac{700}{t^2 + 4t + 10} = 700(t^2 + 4t + 10)^{-1}$

$T' = \dfrac{-1400(t + 2)}{(t^2 + 4t + 10)^2}$

(a) When $t = 1$,

$T' = \dfrac{-1400(1 + 2)}{(1 + 4 + 10)^2} \approx -18.667$ deg/hr.

(b) When $t = 3$,

$T' = \dfrac{-1400(3 + 2)}{(9 + 12 + 10)^2} \approx -7.284$ deg/hr.

(c) When $t = 5$,

$T' = \dfrac{-1400(5 + 2)}{(25 + 20 + 10)^2} \approx -3.240$ deg/hr.

(d) When $t = 10$,

$T' = \dfrac{-1400(10 + 2)}{(100 + 40 + 10)^2} \approx -0.747$ deg/hr.

81. $x^2 + 3xy + y^3 = 10$

$2x + 3xy' + 3y + 3y^2y' = 0$

$3(x + y^2)y' = -(2x + 3y)$

$y' = \dfrac{-(2x + 3y)}{3(x + y^2)}$

83. $y\sqrt{x} - x\sqrt{y} = 16$

$y\left(\dfrac{1}{2}x^{-1/2}\right) + x^{1/2}y' - x\left(\dfrac{1}{2}y^{-1/2}y'\right) - y^{1/2} = 0$

$\left(\sqrt{x} - \dfrac{x}{2\sqrt{y}}\right)y' = \sqrt{y} - \dfrac{y}{2\sqrt{x}}$

$\dfrac{2\sqrt{xy} - x}{2\sqrt{y}}y' = \dfrac{2\sqrt{xy} - y}{2\sqrt{x}}$

$y' = \dfrac{2\sqrt{xy} - y}{2\sqrt{x}} \cdot \dfrac{2\sqrt{y}}{2\sqrt{xy} - x} = \dfrac{2y\sqrt{x} - y\sqrt{y}}{2x\sqrt{y} - x\sqrt{x}}$

85. $x^2 + y^2 = 20$

$2x + 2yy' = 0$

$y' = -\dfrac{x}{y}$

At $(2, 4)$: $y' = -\dfrac{1}{2}$.

Tangent line: $y - 4 = -\dfrac{1}{2}(x - 2)$

$x + 2y - 10 = 0$

Normal line: $y - 4 = 2(x - 2)$

$2x - y = 0$

87. $y = \sqrt{x}$

$\dfrac{dy}{dt} = 2$ units/sec

$\dfrac{dy}{dt} = \dfrac{1}{2\sqrt{x}}\dfrac{dx}{dt} \Rightarrow \dfrac{dx}{dt} = 2\sqrt{x}\dfrac{dy}{dt} = 4\sqrt{x}$

(a) When $x = \dfrac{1}{2}$, $\dfrac{dx}{dt} = 2\sqrt{2}$ units/sec.

(b) When $x = 1$, $\dfrac{dx}{dt} = 4$ units/sec.

(c) When $x = 4$, $\dfrac{dx}{dt} = 8$ units/sec.

89. Surface area $= A = 6x^2$, x length of edge.

$\dfrac{dx}{dt} = 5$

$\dfrac{da}{dt} = 12x\dfrac{dx}{dt} = 12(4.5)(5) = 270 \text{ cm}^2/\text{sec}$

91. $D = \sqrt{x^2 + y^2}$

$\dfrac{dD}{dt} = \dfrac{1}{2}(x^2 + y^2)^{-1/2}\left(2x\dfrac{dx}{dt} + 2y\dfrac{dy}{dt}\right)$

$= \dfrac{x\dfrac{dx}{dt} + y\dfrac{dy}{dt}}{\sqrt{x^2 + y^2}}$

When $t = \dfrac{1}{2}$, $x = 32.5$ and $y = 25$.

$\dfrac{dD}{dt} = \dfrac{(32.5)(65) + (25)(50)}{\sqrt{32.5^2 + 25^2}}$

$= \dfrac{3362.5}{\sqrt{1681.25}} \approx 82 \text{ mph}$

93. For $0 \le h \le 5$,

$\dfrac{b}{40} = \dfrac{h}{5}$

$b = 8h.$

$V = \dfrac{1}{2}bh(20) = 10bh = 10(8h)h = 80h^2$

$\dfrac{dV}{dt} = 160h\dfrac{dh}{dt} = 10 \Rightarrow \dfrac{dh}{dt} = \dfrac{1}{16h}$

When $h = 4$, $\dfrac{dh}{dt} = \dfrac{1}{16(4)} = \dfrac{1}{64}$ ft/min

Problem Solving for Chapter 4

1. (a) $x^2 + (y - r)^2 = r^2$ Circle

$\qquad\qquad x^2 = y$ Parabola

Substituting,

$$(y - r)^2 = r^2 - y$$

$$y^2 - 2ry + r^2 = r^2 - y$$

$$y^2 - 2ry + y = 0$$

$$y(y - 2r + 1) = 0$$

Since you want only one solution, let $1 - 2r = 0 \implies r = \frac{1}{2}$.

Graph $y = x^2$ and $x^2 + \left(y - \frac{1}{2}\right)^2 = \frac{1}{4}$.

(b) Let (x, y) be a point of tangency: $x^2 + (y - b)^2 = 1 \implies 2x + 2(y - b)y' = 0 \implies y' = \dfrac{x}{b - y}$ (circle).

$y = x^2 \implies y' = 2x$ (parabola). Equating,

$$2x = \frac{x}{b - y}$$

$$2(b - y) = 1$$

$$b - y = \frac{1}{2} \implies b = y + \frac{1}{2}.$$

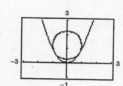

Also, $x^2 + (y - b)^2 = 1$ and $y = x^2$ imply

$$y + (y - b)^2 = 1 \implies y + \left[y - \left(y + \frac{1}{2}\right)\right]^2 = 1 \implies y + \frac{1}{4} = 1 \implies y = \frac{3}{4} \text{ and } b = \frac{5}{4}.$$

Center: $\left(0, \dfrac{5}{4}\right)$

Graph $y = x^2$ and $x^2 + \left(y - \dfrac{5}{4}\right)^2 = 1$.

3. (a) $f(x) = \sqrt{x + 1}, f'(x) = \dfrac{1}{2\sqrt{x + 1}}, f(0) = 1, f'(0) = \dfrac{1}{2}$

$$y - 1 = \frac{1}{2}(x - 0) \implies y = \frac{1}{2}x + 1$$

$$P_1(x) = \frac{1}{2}x + 1$$

(b) $f''(x) = -\dfrac{1}{4\sqrt{(x + 1)^3}}, \quad f''(0) = -\dfrac{1}{4}$

$$P_2(x) = a_0 + a_1 x + a_2 x^2, \quad P_2{}'(x) = a_1 + 2a_2 x, \quad P_2{}''(x) = 2a_2 = -\frac{1}{4}$$

Hence, $a_2 = -\dfrac{1}{8}$ and $P_2(x) = 1 + \dfrac{1}{2}x - \dfrac{1}{8}x^2$.

(c)

x	-1.0	-0.1	-0.001	0	0.001	0.1	1.0
$\sqrt{x + 1}$	0	0.9487	0.9995	1	1.0005	1.0488	1.4142
$P_2(x)$	0.3750	0.9488	0.9995	1	1.0005	1.0488	1.3750

The accuracy is better near $x = 0$.

—CONTINUED—

3. —CONTINUED—

(d)

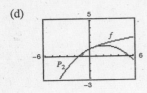

The graphs are close together near $x = 0$.

5. Let $p(x) = Ax^3 + Bx^2 + Cx + D$.

$\quad p'(x) = 3Ax^2 + 2Bx + C$

At $(1, 1)$: $A + B + C + D = 1$ Equation 1

$\qquad\qquad 3A + 2B + C \quad = 14$ Equation 2

At $(-1, -3)$: $-A + B - C + D = -3$ Equation 3

$\qquad\qquad 3A - 2B + C \quad = -2$ Equation 4

Adding Equations 1 and 3: $2B + 2D = -2$

Subtracting Equations 1 and 3: $2A + 2C = 4$

Adding Equations 2 and 4: $6A + 2C = 12$

Subtracting Equations 2 and 4: $4B = 16$

Hence, $B = 4$ and $D = \frac{1}{2}(-2 - 2B) = -5$.

Subtracting $2A + 2C = 4$ and $6A + 2C = 12$, you obtain $4A = 8 \implies A = 2$. Finally, $C = \frac{1}{2}(4 - 2A) = 0$.

Thus, $p(x) = 2x^3 + 4x^2 - 5$.

7. (a) $x^4 = a^2x^2 - a^2y^2$

$\qquad a^2y^2 = a^2x^2 - x^4$

$\qquad\quad y = \dfrac{\pm\sqrt{a^2x^2 - x^4}}{a}$

Graph: $y_1 = \dfrac{\sqrt{a^2x^2 - x^4}}{a}$ and $y_2 = -\dfrac{\sqrt{a^2x^2 - x^4}}{a}$.

(b)

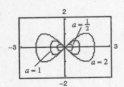

$(\pm a, 0)$ are the x-intercepts, along with $(0, 0)$.

—CONTINUED—

7. —CONTINUED—

(c) Differentiating implicitly,

$$4x^3 = 2a^2x - 2a^2yy'$$

$$y' = \frac{2a^2x - 4x^3}{2a^2y} = \frac{x(a^2 - 2x^2)}{a^2y} = 0 \implies 2x^2 = a^2 \implies x = \frac{\pm a}{\sqrt{2}}.$$

$$\left(\frac{a^2}{2}\right)^2 = a^2\left(\frac{a^2}{2}\right) - a^2y^2$$

$$\frac{a^4}{4} = \frac{a^4}{2} - a^2y^2$$

$$a^2y^2 = \frac{a^4}{4}$$

$$y^2 = \frac{a^2}{4}$$

$$y = \pm\frac{a}{2}$$

Four points: $\left(\frac{a}{\sqrt{2}}, \frac{a}{2}\right), \left(\frac{a}{\sqrt{2}}, -\frac{a}{2}\right), \left(-\frac{a}{\sqrt{2}}, \frac{a}{2}\right), \left(\frac{-a}{\sqrt{2}}, -\frac{a}{2}\right)$

9. (a)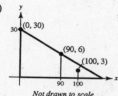

Not drawn to scale

Line determined by $(0, 30)$ and $(90, 6)$:

$$y - 30 = \frac{30 - 6}{0 - 90}(x - 0) = -\frac{24}{90}x = -\frac{4}{15}x \implies y = -\frac{4}{15}x + 30$$

When $x = 100$, $y = \frac{-4}{15}(100) + 30 = \frac{10}{3} > 3 \implies$ Shadow determined by man.

(b)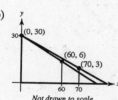

Not drawn to scale

Line determined by $(0, 30)$ and $(60, 6)$:

$$y - 30 = \frac{30 - 6}{0 - 60}(x - 0) = -\frac{2}{5}x \implies y = -\frac{2}{5}x + 30$$

When $x = 70$, $y = \frac{-2}{5}(70) + 30 = 2 < 3 \implies$ Shadow determined by child.

(c) Need $(0, 30)$, $(d, 6)$, $(d + 10, 3)$ collinear.

$$\frac{30 - 6}{0 - d} = \frac{6 - 3}{d - (d + 10)} \implies \frac{24}{d} = \frac{3}{10} \implies d = 80 \text{ feet}$$

(d) Let y be the distance from the base of the street light to the tip of the shadow. We know that $\frac{dx}{dt} = -5$.

For $x > 80$, the shadow is determined by the man.

$$\frac{y}{30} = \frac{y - x}{6} \implies y = \frac{5}{4}x \text{ and } \frac{dy}{dt} = \frac{5}{4}\frac{dx}{dt} = \frac{-25}{4}.$$

For $x < 80$, the shadow is determined by the child.

$$\frac{y}{30} = \frac{y - x - 10}{3} \implies y = \frac{10}{9}x + \frac{100}{9} \text{ and } \frac{dy}{dt} = \frac{10}{9}\frac{dx}{dt} = \frac{-50}{9}.$$

Therefore,

$$\frac{dy}{dt} = \begin{cases} \frac{-25}{4} & x > 80 \\ \frac{-50}{9} & 0 < x < 80 \end{cases}.$$

$\frac{dy}{dt}$ is not continuous at $x = 80$.

11. (a) $y = \dfrac{1}{x}, y' = -\dfrac{1}{x^2}$. At $(1, 1)$, slope $= -1$.

Tangent line: $y - 1 = -1(x - 1)$ or $y = -x + 2$

Point A: $y = 0, x = 2$: $A = (2, 0)$

Point B $= (0, 2)$

Distance AP $=$ Distance BP $= \sqrt{2}$

(c) At $\left(2, \dfrac{1}{2}\right)$, slope $= -\dfrac{1}{4}$.

Tangent line $y - \dfrac{1}{2} = -\dfrac{1}{4}(x - 2)$ or $y = -\dfrac{1}{4}x + 1$

Point A: $y = 0, x = 4$: $A = (4, 0)$

Point B $= (0, 1)$

Distance AP $= \sqrt{2^2 + (1/2)^2} = \sqrt{17}/2$

Distance BP $= \sqrt{2^2 + (1/2)^2} = \sqrt{17}/2$

Area $= \dfrac{1}{2}(4)(1) = 2$ square units

(b) Area $= \dfrac{1}{2}(\text{base})(\text{height}) = \dfrac{1}{2}(2)(2) = 2$ square units

(d) $y = a/x, y' = -a/x^2. P = (c, a/c)$ on graph.

Tangent line $y - \dfrac{a}{c} = \dfrac{-a}{c^2}(x - c)$ or $y = \dfrac{-a}{c^2}x + \dfrac{2a}{c}$

Point A: $y = 0, x = 2c. A = (2c, 0)$

Point B $= (0, 2a/c)$

Distance AP $= \sqrt{c^2 + (a/c)^2} =$ Distance BP

Area $= \dfrac{1}{2}(\text{base})(\text{height}) = \dfrac{1}{2}(2c)\left(\dfrac{2a}{c}\right) = 2a,$

independent of point.

13. (a) $v(t) = -\dfrac{27}{5}t + 27$ ft/sec

$a(t) = -\dfrac{27}{5}$ ft/sec^2

(b) $v(t) = -\dfrac{27}{5}t + 27 = 0 \implies \dfrac{27}{5}t$

$= 27 \implies t = 5$ seconds

$s(5) = -\dfrac{27}{10}(5)^2 + 27(5) + 6 = 73.5$ feet

(c) The acceleration due to gravity on Earth is greater in magnitude than that on the moon.

CHAPTER 5
Applications of Differentiation

CHAPTER 5
Applications of Differentiation

Section 5.1   Extrema on an Interval

1. A: neither
   B: absolute maximum (and relative maximum)
   C: neither
   D: neither
   E: relative maximum
   F: relative minimum
   G: neither

3. $f(x) = \dfrac{x^2}{x^2 + 4}$

$f'(x) = \dfrac{(x^2 + 4)(2x) - (x^2)(2x)}{(x^2 + 4)^2} = \dfrac{8x}{(x^2 + 4)^2}$

$f'(0) = 0$

5. $f(x) = x + \dfrac{27}{2x^2} = x + \dfrac{27}{2}x^{-2}$

$f'(x) = 1 - 27x^{-3} = 1 - \dfrac{27}{x^3}$

$f'(3) = 1 - \dfrac{27}{3^3} = 1 - 1 = 0$

7. $f(x) = (x + 2)^{2/3}$

$f'(x) = \dfrac{2}{3}(x + 2)^{-1/3}$

$f'(-2)$ is undefined.

9. Critical numbers: $x = 2$

   $x = 2$: absolute maximum

11. Critical numbers: $x = 1, 2, 3$

   $x = 1, 3$: absolute maximum

   $x = 2$: absolute minimum

13. $f(x) = x^2(x - 3) = x^3 - 3x^2$

   $f'(x) = 3x^2 - 6x = 3x(x - 2)$

   Critical numbers: $x = 0, x = 2$

15. $g(t) = t\sqrt{4 - t},\ t < 3$

$g'(t) = t\left[\dfrac{1}{2}(4 - t)^{-1/2}(-1)\right] + (4 - t)^{1/2}$

$= \dfrac{1}{2}(4 - t)^{-1/2}[-t + 2(4 - t)]$

$= \dfrac{8 - 3t}{2\sqrt{4 - t}}$

Critical number is $t = \dfrac{8}{3}$.

17. $f(x) = 2(3 - x), [-1, 2]$

   $f'(x) = -2 \implies$ No critical numbers

   Left endpoint: $(-1, 8)$ Maximum

   Right endpoint: $(2, 2)$ Minimum

19. $f(x) = -x^2 + 3x, [0, 3]$

   $f'(x) = -2x + 3$

   Left endpoint: $(0, 0)$ Minimum

   Critical number: $\left(\dfrac{3}{2}, \dfrac{9}{4}\right)$ Maximum

   Right endpoint: $(3, 0)$ Minimum

21. $f(x) = x^3 - \dfrac{3}{2}x^2,\ [-1, 2]$

   $f'(x) = 3x^2 - 3x = 3x(x - 1)$

   Left endpoint: $\left(-1, -\dfrac{5}{2}\right)$ Minimum

   Right endpoint: $(2, 2)$ Maximum

   Critical number: $(0, 0)$

   Critical number: $\left(1, -\dfrac{1}{2}\right)$

23. $f(x) = 3x^{2/3} - 2x, [-1, 1]$

$f'(x) = 2x^{-1/3} - 2 = \dfrac{2(1 - \sqrt[3]{x})}{\sqrt[3]{x}}$

Left endpoint: $(-1, 5)$ Maximum

Critical number: $(0, 0)$ Minimum

Right endpoint: $(1, 1)$

25. $y = 3 - |t - 3|, [-1, 5]$

From the graph, you see that $t = 3$ is a critical number.

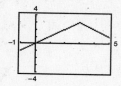

Left endpoint: $(-1, -1)$ Minimum

Right endpoint: $(5, 1)$

Critical number: $(3, 3)$ Maximum

27. $h(s) = \dfrac{1}{s - 2}, [0, 1]$

$h'(s) = \dfrac{-1}{(s - 2)^2}$

Left endpoint: $\left(0, -\dfrac{1}{2}\right)$ Maximum

Right endpoint: $(1, -1)$ Minimum

29. (a) Minimum: $(0, -3)$

   Maximum: $(2, 1)$

   (b) Minimum: $(0, -3)$

   (c) Maximum: $(2, 1)$

   (d) No extrema

31. $f(x) = x^2 - 2x$

   (a) Minimum: $(1, -1)$

   Maximum: $(-1, 3)$

   (b) Maximum: $(3, 3)$

   (c) Minimum: $(1, -1)$

   (d) Minimum: $(1, -1)$

33. $f(x) = \begin{cases} 2x + 2, & 0 \le x \le 1 \\ 4x^2, & 1 < x \le 3 \end{cases}$

Left endpoint: $(0, 2)$ Minimum

Right endpoint: $(3, 36)$ Maximum

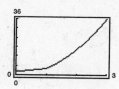

35. $f(x) = \dfrac{3}{x - 1}, (1, 4]$

Right endpoint: $(4, 1)$ Minimum

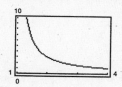

37. (a)

Maximum: $(1, 4.7)$ (endpoint)

Minimum: $(0.4398, -1.0613)$

(b)
$$f(x) = 3.2x^5 + 5x^3 - 3.5x, [0, 1]$$
$$f'(x) = 16x^4 + 15x^2 - 3.5$$
$$16x^4 + 15x^2 - 3.5 = 0$$
$$x^2 = \frac{-15 \pm \sqrt{(15)^2 - 4(16)(-3.5)}}{2(16)}$$
$$= \frac{-15 \pm \sqrt{449}}{32}$$
$$x = \sqrt{\frac{-15 + \sqrt{449}}{32}} \approx 0.4398$$
$$f(0) = 0$$
$$f(1) = 4.7 \text{ Maximum (endpoint)}$$
$$f\left(\sqrt{\frac{-15 + \sqrt{449}}{32}}\right) \approx -1.0613$$

Minimum: $(0.4398, -1.0613)$

39. $f(x) = (1 + x^3)^{1/2}, [0, 2]$

$f'(x) = \dfrac{3}{2}x^2(1 + x^3)^{-1/2}$

$f''(x) = \dfrac{3}{4}(x^4 + 4x)(1 + x^3)^{-3/2}$

$f'''(x) = -\dfrac{3}{8}(x^6 + 20x^3 - 8)(1 + x^3)^{-5/2}$

Setting $f''' = 0$, we have $x^6 + 20x^3 - 8 = 0$.

$x^3 = \dfrac{-20 \pm \sqrt{400 - 4(1)(-8)}}{2}$

$x = \sqrt[3]{-10 + \sqrt{108}} = \sqrt{3} - 1$

In the interval $[0, 2]$, choose

$x = \sqrt[3]{-10 + \sqrt{108}} = \sqrt{3} - 1 \approx 0.732.$

$\left| f''\left(\sqrt[3]{-10 + \sqrt{108}} \right) \right| \approx 1.47$ is the maximum value.

41. The function f is continuous on $[3, 5]$, so it has a maximum by Theorem 5.1. The function is not continuous on $[1, 3]$.

43. $f(x) = (x + 1)^{2/3}, [0, 2]$

$f'(x) = \dfrac{2}{3}(x + 1)^{-1/3}$

$f''(x) = -\dfrac{2}{9}(x + 1)^{-4/3}$

$f'''(x) = \dfrac{8}{27}(x + 1)^{-7/3}$

$f^{(4)}(x) = -\dfrac{56}{81}(x + 1)^{-10/3}$

$f^{(5)}(x) = \dfrac{560}{243}(x + 1)^{-13/3}$

$\left| f^{(4)}(0) \right| = \dfrac{56}{81}$ is the maximum value.

45.

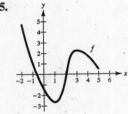

47. (a) Yes

(b) No

49. (a) No

(b) Yes

51. $P = VI - RI^2 = 12I - 0.5I^2, 0 \le I \le 15$

$P = 0$ when $I = 0$.

$P = 67.5$ when $I = 15$.

$P' = 12 - I = 0$

Critical number: $I = 12$ amps

When $I = 12$ amps, $P = 72$, the maximum output.

No, a 20-amp fuse would not increase the power output. P is decreasing for $I > 12$.

53. (a) $y = ax^2 + bx + c$

$y' = 2ax + b$

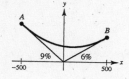

The coordinates of B are $(500, 30)$, and those of A are $(-500, 45)$. From the slopes at A and B,

$$-1000a + b = -0.09$$

$$1000a + b = 0.06.$$

Solving these two equations, you obtain $a = 3/40000$ and $b = -3/200$. From the points $(500, 30)$ and $(-500, 45)$, you obtain

$$30 = \frac{3}{40000}500^2 + 500\left(\frac{-3}{200}\right) + c$$

$$45 = \frac{3}{40000}500^2 - 500\left(\frac{-3}{200}\right) + c.$$

In both cases, $c = 18.75 = \frac{75}{4}$. Thus,

$$y = \frac{3}{40000}x^2 - \frac{3}{200}x + \frac{75}{4}.$$

(b)

x	-500	-400	-300	-200	-100	0	100	200	300	400	500
d	0	$.75$	3	6.75	12	18.75	12	6.75	3	$.75$	0

For $-500 \le x \le 0$, $d = (ax^2 + bx + c) - (-0.09x)$.

For $0 \le x \le 500$, $d = (ax^2 + bx + c) - (0.06x)$.

(c) The lowest point on the highway is $(100, 18)$, which is not directly over the point where the two hillsides come together.

55. True.

57. True.

Section 5.2   Rolle's Theorem and the Mean Value Theorem

1. Rolle's Theorem does not apply to $f(x) = 1 - |x - 1|$ over $[0, 2]$ since f is not differentiable at $x = 1$.

3. $f(x) = \left|\frac{1}{x}\right|$

$f(-1) = f(1) = 1$. But, f is not continuous on $[-1, 1]$.

5. $f(x) = x^2 - x - 2 = (x - 2)(x + 1)$

x-intercepts: $(-1, 0), (2, 0)$

$f'(x) = 2x - 1 = 0$ at $x = \frac{1}{2}$.

7. $f(x) = x\sqrt{x + 4}$

x-intercepts: $(-4, 0), (0, 0)$

$f'(x) = x\frac{1}{2}(x + 4)^{-1/2} + (x + 4)^{1/2}$

$\qquad = (x + 4)^{-1/2}\left(\frac{x}{2} + (x + 4)\right)$

$f'(x) = \left(\frac{3}{2}x + 4\right)(x + 4)^{-1/2} = 0$ at $x = -\frac{8}{3}$

9. $f(x) = x^2 - 2x$, $[0, 2]$

$f(0) = f(2) = 0$

f is continuous on $[0, 2]$. f is differentiable on $(0, 2)$. Rolle's Theorem applies.

$$f'(x) = 2x - 2$$

$$2x - 2 = 0 \implies x = 1$$

c value: 1

11. $f(x) = (x - 1)(x - 2)(x - 3)$, $[1, 3]$

$f(1) = f(3) = 0$

f is continuous on $[1, 3]$. f is differentiable on $(1, 3)$. Rolle's Theorem applies.

$$f(x) = x^3 - 6x^2 + 11x - 6$$

$$f'(x) = 3x^2 - 12x + 11$$

$$3x^2 - 12x + 11 = 0 \implies x = \frac{6 \pm \sqrt{3}}{3}$$

$$c = \frac{6 - \sqrt{3}}{3}, c = \frac{6 + \sqrt{3}}{3}$$

13. $f(x) = x^{2/3} - 1$, $[-8, 8]$

$f(-8) = f(8) = 3$

f is continuous on $[-8, 8]$. f is not differentiable on $(-8, 8)$ since $f'(0)$ does not exist. Rolle's Theorem does not apply.

15. $f(x) = \dfrac{x^2 - 2x - 3}{x + 2}$, $[-1, 3]$

$f(-1) = f(3) = 0$

f is continuous on $[-1, 3]$. (**Note:** The discontinuity, $x = -2$, is not in the interval.) f is differentiable on $(-1, 3)$. Rolle's Theorem applies.

$$f'(x) = \frac{(x + 2)(2x - 2) - (x^2 - 2x - 3)(1)}{(x + 2)^2} = 0$$

$$\frac{x^2 + 4x - 1}{(x + 2)^2} = 0$$

$$x = \frac{-4 \pm 2\sqrt{5}}{2} = -2 \pm \sqrt{5}$$

c value: $-2 + \sqrt{5}$

17. $f(x) = |x| - 1$, $[-1, 1]$

$f(-1) = f(1) = 0$

f is continuous on $[-1, 1]$. f is not differentiable on $(-1, 1)$ since $f'(0)$ does not exist. Rolle's Theorem does not apply.

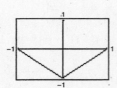

19. $C(x) = 10\left(\dfrac{1}{x} + \dfrac{x}{x + 3}\right)$

(a) $C(3) = C(6) = \dfrac{25}{3}$

(b) $\qquad C'(x) = 10\left(-\dfrac{1}{x^2} + \dfrac{3}{(x + 3)^2}\right) = 0$

$$\frac{3}{x^2 + 6x + 9} = \frac{1}{x^2}$$

$$2x^2 - 6x - 9 = 0$$

$$x = \frac{6 \pm \sqrt{108}}{4}$$

$$= \frac{6 \pm 6\sqrt{3}}{4} = \frac{3 \pm 3\sqrt{3}}{2}$$

In the interval $(3, 6)$: $c = \dfrac{3 + 3\sqrt{3}}{2} \approx 4.098$.

21. $f(x) = x^2 + 1$

(a) slope $= \dfrac{5 - 2}{2 + 1} = 1$

secant line: $y - 2 = 1(x + 1)$

$$y = x + 3$$

(b) $f'(x) = 2x = 1 \Rightarrow c = \dfrac{1}{2}$

$$f'\left(\dfrac{1}{2}\right) = 1, \quad f\left(\dfrac{1}{2}\right) = \dfrac{5}{4}$$

(c) Tangent line: $y - \dfrac{5}{4} = 1\left(x - \dfrac{1}{2}\right)$

$$y = x + \dfrac{3}{4}$$

(d)

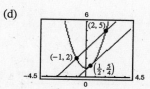

23.

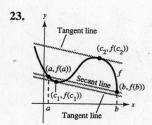

25. $f(x) = \dfrac{1}{x - 3},\ [0, 6]$

f has a discontinuity at $x = 3$.

27. $f(x) = x^2$ is continuous on $[-2, 1]$ and differentiable on $(-2, 1)$.

$$\dfrac{f(1) - f(-2)}{1 - (-2)} = \dfrac{1 - 4}{3} = -1$$

$f'(x) = 2x = -1$ when $x = -\dfrac{1}{2}$. Therefore,

$$c = -\dfrac{1}{2}.$$

29. $f(x) = x^{2/3}$ is continuous on $[0, 1]$ and differentiable on $(0, 1)$.

$$\dfrac{f(1) - f(0)}{1 - 0} = 1$$

$$f'(x) = \dfrac{2}{3}x^{-1/3} = 1$$

$$x = \left(\dfrac{2}{3}\right)^3 = \dfrac{8}{27}$$

$$c = \dfrac{8}{27}$$

31. $f(x) = \sqrt{2 - x}$ is continuous on $[-7, 2]$ and differentiable on $(-7, 2)$.

$$\dfrac{f(2) - f(-7)}{2 - (-7)} = \dfrac{0 - 3}{9} = -\dfrac{1}{3}$$

$$f'(x) = \dfrac{-1}{2\sqrt{2 - x}} = -\dfrac{1}{3}$$

$$2\sqrt{2 - x} = 3$$

$$\sqrt{2 - x} = \dfrac{3}{2}$$

$$2 - x = \dfrac{9}{4}$$

$$x = -\dfrac{1}{4}$$

$$c = -\dfrac{1}{4}$$

33. $f(x) = \dfrac{x}{x+1}$ on $\left[-\dfrac{1}{2}, 2\right]$.

(a)

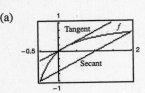

(b) Secant line:

$$\text{slope} = \frac{f(2) - f(-1/2)}{2 - (-1/2)} = \frac{2/3 - (-1)}{5/2} = \frac{2}{3}$$

$$y - \frac{2}{3} = \frac{2}{3}(x - 2)$$

$$3y - 2 = 2x - 4$$

$$3y - 2x + 2 = 0$$

(c) $f'(x) = \dfrac{1}{(x+1)^2} = \dfrac{2}{3}$

$$(x + 1)^2 = \frac{3}{2}$$

$$x = -1 \pm \sqrt{\frac{3}{2}} = -1 \pm \frac{\sqrt{6}}{2}$$

In the interval $[-1/2, 2]$, $c = -1 + (\sqrt{6}/2)$.

$$f(c) = \frac{-1 + (\sqrt{6}/2)}{[-1 + (\sqrt{6}/2)] + 1} = \frac{-2 + \sqrt{6}}{\sqrt{6}} = \frac{-2}{\sqrt{6}} + 1$$

Tangent line: $y - 1 + \dfrac{2}{\sqrt{6}} = \dfrac{2}{3}\left(x - \dfrac{\sqrt{6}}{2} + 1\right)$

$$y - 1 + \frac{\sqrt{6}}{3} = \frac{2}{3}x - \frac{\sqrt{6}}{3} + \frac{2}{3}$$

$$3y - 2x - 5 + 2\sqrt{6} = 0$$

35. $f(x) = \sqrt{x}$, $[1, 9]$

$(1, 1), (9, 3)$

$m = \dfrac{3 - 1}{9 - 1} = \dfrac{1}{4}$

(a)

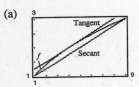

(b) Secant line: $y - 1 = \dfrac{1}{4}(x - 1)$

$$y = \frac{1}{4}x + \frac{3}{4}$$

$$0 = x - 4y + 3$$

(c) $f'(x) = \dfrac{1}{2\sqrt{x}}$

$$\frac{f(9) - f(1)}{9 - 1} = \frac{1}{4}$$

$$\frac{1}{2\sqrt{c}} = \frac{1}{4}$$

$$\sqrt{c} = 2$$

$$c = 4$$

$$(c, f(c)) = (4, 2)$$

$$m = f'(4) = \frac{1}{4}$$

Tangent line: $y - 2 = \dfrac{1}{4}(x - 4)$

$$y = \frac{1}{4}x + 1$$

$$0 = x - 4y + 4$$

37. $s(t) = -4.9t^2 + 500$

(a) $v_{\text{avg}} = \dfrac{s(3) - s(0)}{3 - 0} = \dfrac{455.9 - 500}{3} = -14.7 \text{ m/sec}$

(b) $s(t)$ is continuous on $[0, 3]$ and differentiable on $(0, 3)$. Therefore, the Mean Value Theorem applies.

$$v(t) = s'(t) = -9.8t = -14.7 \text{ m/sec}$$

$$t = \frac{-14.7}{-9.8} = 1.5 \text{ seconds}$$

39. No. Let $f(x) = x^2$ on $[-1, 2]$.

$$f'(x) = 2x$$

$f'(0) = 0$ and zero is in the interval $(-1, 2)$ but $f(-1) \neq f(2)$.

41. Let $S(t)$ be the position function of the plane. If $t = 0$ corresponds to 2 P.M., $S(0) = 0$, $S(5.5) = 2500$ and the Mean Value Theorem says that there exists a time t_0, $0 < t_0 < 5.5$, such that

$$S'(t_0) = v(t_0) = \frac{2500 - 0}{5.5 - 0} \approx 454.55.$$

Applying the Intermediate Value Theorem to the velocity function on the intervals $[0, t_0]$ and $[t_0, 5.5]$, you see that there are at least two times during the flight when the speed was 400 miles per hour. $(0 < 400 < 454.54)$

43. (a) $f(x) = |9 - x^2|$

$$f(x) = \begin{cases} 9 - x^2, & -3 \leq x \leq 3 \\ x^2 - 9, & x > 3 \text{ or } x < -3 \end{cases}$$

$$f'(x) = \begin{cases} -2x, & -3 < x < 3 \\ 2x, & x > 3 \text{ or } x < -3 \end{cases}$$

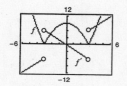

(b) f is continuous.

    f' is not continuous.

(c) Since $f(-1) = f(1) = 8$ and f is differentiable on $(-1, 1)$, Rolle's Theorem applies on $[-1, 1]$. Since $f(2) = 5$ and $f(4) = 7$, and f is not differentiable at $x = 3$, Rolle's Theorem does not apply on $[2, 4]$.

(d) $\lim\limits_{x \to 3^-} f'(x) = -6$ and $\lim\limits_{x \to 3^+} f'(x) = 6$.

45. $f'(x) = 0$

    $f(x) = c$

    $f(2) = 5$

    Hence, $f(x) = 5$.

47. $f'(x) = 2x$

    $f(x) = x^2 + c$

    $f(1) = 0 \implies 0 = 1 + c \implies c = -1$

    Hence, $f(x) = x^2 - 1$.

49. f is continuous on $[-5, 5]$ and does not satisfy the conditions of the Mean Value Theorem.
$\implies f$ is not differentiable on $(-5, 5)$.
Example: $f(x) = |x|$

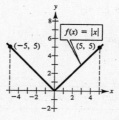

51. False. $f(x) = 1/x$ has a discontinuity at $x = 0$.

53. True. A polynomial is continuous and differentiable everywhere.

55. Suppose $f(x)$ is not constant on (a, b). Then there exists x_1 and x_2 in (a, b) such that $f(x_1) \neq f(x_2)$. Then by the Mean Value Theorem, there exists c in (a, b) such that

$$f'(c) = \frac{f(x_2) - f(x_1)}{x_2 - x_1} \neq 0.$$

This contradicts the fact that $f'(x) = 0$ for all x in (a, b).

Section 5.3   Increasing and Decreasing Functions and the First Derivative Test

1. $f(x) = x^2 - 6x + 8$

Increasing on: $(3, \infty)$

Decreasing on: $(-\infty, 3)$

3. $y = \dfrac{x^3}{4} - 3x$

Increasing on: $(-\infty, -2), (2, \infty)$

Decreasing on: $(-2, 2)$

5. $f(x) = \dfrac{1}{x^2} = x^{-2}$

$f'(x) = \dfrac{-2}{x^3}$

Discontinuity: $x = 0$

Test intervals:	$-\infty < x < 0$	$0 < x < \infty$
Sign of $f'(x)$:	$f' > 0$	$f' < 0$
Conclusion:	Increasing	Decreasing

Increasing on: $(-\infty, 0)$

Decreasing on: $(0, \infty)$

7. $g(x) = x^2 - 2x - 8$

$g'(x) = 2x - 2$

Critical number: $x = 1$

Test intervals:	$-\infty < x < 1$	$1 < x < \infty$
Sign of $g'(x)$:	$g' < 0$	$g' > 0$
Conclusion:	Decreasing	Increasing

Increasing on: $(1, \infty)$

Decreasing on: $(-\infty, 1)$

9. $y = x\sqrt{16 - x^2}$ Domain: $[-4, 4]$

$y' = \dfrac{-2(x^2 - 8)}{\sqrt{16 - x^2}} = \dfrac{-2}{\sqrt{16 - x^2}}\left(x - 2\sqrt{2}\right)\left(x + 2\sqrt{2}\right)$

Critical numbers: $x = \pm 2\sqrt{2}$

Test intervals:	$-4 < x < -2\sqrt{2}$	$-2\sqrt{2} < x < 2\sqrt{2}$	$2\sqrt{2} < x < 4$
Sign of y':	$y' < 0$	$y' > 0$	$y' < 0$
Conclusion:	Decreasing	Increasing	Decreasing

Increasing on: $\left(-2\sqrt{2}, 2\sqrt{2}\right)$

Decreasing on: $\left(-4, -2\sqrt{2}\right), \left(2\sqrt{2}, 4\right)$

11. $f(x) = x^2 - 6x$

$f'(x) = 2x - 6 = 0$

Critical number: $x = 3$

Test intervals:	$-\infty < x < 3$	$3 < x < \infty$
Sign of $f'(x)$:	$f' < 0$	$f' > 0$
Conclusion:	Decreasing	Increasing

Increasing on: $(3, \infty)$

Decreasing on: $(-\infty, 3)$

Relative minimum: $(3, -9)$

13. $f(x) = -2x^2 + 4x + 3$

$f'(x) = -4x + 4 = 0$

Critical number: $x = 1$

Test intervals:	$-\infty < x < 1$	$1 < x < \infty$
Sign of $f'(x)$:	$f' > 0$	$f' < 0$
Conclusion:	Increasing	Decreasing

Increasing on: $(-\infty, 1)$

Decreasing on: $(1, \infty)$

Relative maximum: $(1, 5)$

15. $f(x) = x^3 - 6x^2 + 15$

$f'(x) = 3x^2 - 12x = 3x(x - 4)$

Critical numbers: $x = 0, 4$

Test intervals:	$-\infty < x < 0$	$0 < x < 4$	$4 < x < \infty$
Sign of $f'(x)$:	$f' > 0$	$f' < 0$	$f' > 0$
Conclusion:	Increasing	Decreasing	Increasing

Increasing on: $(-\infty, 0), (4, \infty)$

Decreasing on: $(0, 4)$

Relative maximum: $(0, 15)$

Relative minimum: $(4, -17)$

17. $f(x) = x^2(3 - x) = 3x^2 - x^3$

$f'(x) = 6x - 3x^2 = 3x(2 - x)$

Critical numbers: $x = 0, 2$

Test intervals:	$-\infty < x < 0$	$0 < x < 2$	$2 < x < \infty$
Sign of $f'(x)$:	$f' < 0$	$f' > 0$	$f' < 0$
Conclusion:	Decreasing	Increasing	Decreasing

Increasing on: $(0, 2)$

Decreasing on: $(-\infty, 0), (2, \infty)$

Relative maximum: $(2, 4)$

Relative minimum: $(0, 0)$

19. $f(x) = 2x^3 + 3x^2 - 12x$

$f'(x) = 6x^2 + 6x - 12 = 6(x + 2)(x - 1) = 0$

Critical numbers: $x = -2, 1$

Test intervals:	$-\infty < x < -2$	$-2 < x < 1$	$1 < x < \infty$
Sign of $f'(x)$:	$f' > 0$	$f' < 0$	$f' > 0$
Conclusion:	Increasing	Decreasing	Increasing

Increasing on: $(-\infty, -2), (1, \infty)$

Decreasing on: $(-2, 1)$

Relative maximum: $(-2, 20)$

Relative minimum: $(1, -7)$

21. $f(x) = \dfrac{x^5 - 5x}{5}$

$f'(x) = x^4 - 1$

Critical numbers: $x = -1, 1$

Test intervals:	$-\infty < x < -1$	$-1 < x < 1$	$1 < x < \infty$
Sign of $f'(x)$:	$f' > 0$	$f' < 0$	$f' > 0$
Conclusion:	Increasing	Decreasing	Increasing

Increasing on: $(-\infty, -1), (1, \infty)$

Decreasing on: $(-1, 1)$

Relative maximum: $\left(-1, \frac{4}{5}\right)$

Relative minimum: $\left(1, -\frac{4}{5}\right)$

23. $f(x) = x^{1/3} + 1$

$f'(x) = \dfrac{1}{3}x^{-2/3} = \dfrac{1}{3x^{2/3}}$

Critical number: $x = 0$

Test intervals:	$-\infty < x < 0$	$0 < x < \infty$
Sign of $f'(x)$:	$f' > 0$	$f' > 0$
Conclusion:	Increasing	Increasing

Increasing on: $(-\infty, \infty)$

No relative extrema.

25. $f(x) = (x - 1)^{2/3}$

$f'(x) = \dfrac{2}{3(x - 1)^{1/3}}$

Critical number: $x = 1$

Test intervals:	$-\infty < x < 1$	$1 < x < \infty$
Sign of $f'(x)$:	$f' < 0$	$f' > 0$
Conclusion:	Decreasing	Increasing

Increasing on: $(1, \infty)$

Decreasing on: $(-\infty, 1)$

Relative minimum: $(1, 0)$

27. $f(x) = 5 - |x - 5|$

$f'(x) = -\dfrac{x-5}{|x-5|} = \begin{cases} 1, & x < 5 \\ -1, & x > 5 \end{cases}$

Critical number: $x = 5$

Test intervals:	$-\infty < x < 5$	$5 < x < \infty$
Sign of $f'(x)$:	$f' > 0$	$f' < 0$
Conclusion:	Increasing	Decreasing

Increasing on: $(-\infty, 5)$

Decreasing on: $(5, \infty)$

Relative maximum: $(5, 5)$

29. $f(x) = x + \dfrac{1}{x}$

$f'(x) = 1 - \dfrac{1}{x^2} = \dfrac{x^2 - 1}{x^2}$

Critical numbers: $x = -1, 1$

Discontinuity: $x = 0$

Test intervals:	$-\infty < x < -1$	$-1 < x < 0$	$0 < x < 1$	$1 < x < \infty$
Sign of $f'(x)$:	$f' > 0$	$f' < 0$	$f' < 0$	$f' > 0$
Conclusion:	Increasing	Decreasing	Decreasing	Increasing

Increasing on: $(-\infty, -1), (1, \infty)$

Decreasing on: $(-1, 0), (0, 1)$

Relative maximum: $(-1, -2)$

Relative minimum: $(1, 2)$

31. $f(x) = \dfrac{x^2}{x^2 - 9}$

$f'(x) = \dfrac{(x^2 - 9)(2x) - (x^2)(2x)}{(x^2 - 9)^2} = \dfrac{-18x}{(x^2 - 9)^2}$

Critical number: $x = 0$

Discontinuities: $x = -3, 3$

Test intervals:	$-\infty < x < -3$	$-3 < x < 0$	$0 < x < 3$	$3 < x < \infty$
Sign of $f'(x)$:	$f' > 0$	$f' > 0$	$f' < 0$	$f' < 0$
Conclusion:	Increasing	Increasing	Decreasing	Decreasing

Increasing on: $(-\infty, -3), (-3, 0)$

Decreasing on: $(0, 3), (3, \infty)$

Relative maximum: $(0, 0)$

33. $f(x) = \dfrac{x^2 - 2x + 1}{x + 1}$

$f'(x) = \dfrac{(x + 1)(2x - 2) - (x^2 - 2x + 1)(1)}{(x + 1)^2} = \dfrac{x^2 + 2x - 3}{(x + 1)^2} = \dfrac{(x + 3)(x - 1)}{(x + 1)^2}$

Critical numbers: $x = -3, 1$

Discontinuity: $x = -1$

Test intervals:	$-\infty < x < -3$	$-3 < x < -1$	$-1 < x < 1$	$1 < x < \infty$
Sign of $f'(x)$:	$f' > 0$	$f' < 0$	$f' < 0$	$f' > 0$
Conclusion:	Increasing	Decreasing	Decreasing	Increasing

Increasing on: $(-\infty, -3), (1, \infty)$

Decreasing on: $(-3, -1), (-1, 1)$

Relative maximum: $(-3, -8)$

Relative minimum: $(1, 0)$

35. $f(x) = 2x\sqrt{9 - x^2}, [-3, 3]$

(a) $f'(x) = \dfrac{2(9 - 2x^2)}{\sqrt{9 - x^2}}$

(b)

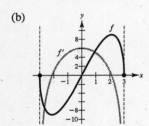

(c) $\dfrac{2(9 - 2x^2)}{\sqrt{9 - x^2}} = 0$

Critical numbers: $x = \pm\dfrac{3}{\sqrt{2}} = \pm\dfrac{3\sqrt{2}}{2}$

(d) Intervals:

$$\left(-3, -\dfrac{3\sqrt{2}}{2}\right) \quad \left(-\dfrac{3\sqrt{2}}{2}, \dfrac{3\sqrt{2}}{2}\right) \quad \left(\dfrac{3\sqrt{2}}{2}, 3\right)$$

$\qquad f'(x) < 0 \qquad\qquad f'(x) > 0 \qquad\qquad f'(x) < 0$

$\quad$ Decreasing $\qquad\qquad$ Increasing $\qquad\qquad$ Decreasing

f is increasing when f' is positive and decreasing when f' is negative.

37. $f(x) = \dfrac{x^5 - 4x^3 + 3x}{x^2 - 1} = \dfrac{(x^2 - 1)(x^3 - 3x)}{x^2 - 1} = x^3 - 3x, \ x \neq \pm 1$

$f(x) = g(x) = x^3 - 3x$ for all $x \neq \pm 1$.

$f'(x) = 3x^2 - 3 = 3(x^2 - 1), \ x \neq \pm 1 \quad f'(x) \neq 0$

f symmetric about origin

zeros of f: $(0, 0), \left(\pm\sqrt{3}, 0\right)$

No relative extrema.

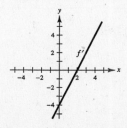

Holes at $(-1, 2)$ and $(1, -2)$.

39. $f(x) = c$ is constant $\Rightarrow f'(x) = 0$.

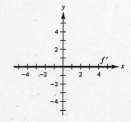

41. f is quadratic $\Rightarrow f'$ is a line.

43. f has positive, but decreasing slope.

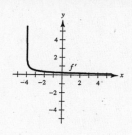

47. (a) f increasing on $(-\infty, 0)$ and $(1, \infty)$ because
   $f' > 0$ there

   f decreasing on $(0, 1)$ because $f' < 0$ there

   (b) f has a relative maximum at $x = 0$, and a relative
   minimum at $x = 1$.

45. (a) f increasing on $(2, \infty)$ because $f' > 0$ on $(2, \infty)$.

   f decreasing on $(-\infty, 2)$ because $f' < 0$ on $(-\infty, 2)$.

   (b) f has a relative minimum at $x = 2$.

In Exercises 49–54, $f'(x) > 0$ on $(-\infty, -4)$, $f'(x) < 0$ on $(-4, 6)$ and $f'(x) > 0$ on $(6, \infty)$.

49. $g(x) = f(x) + 5$

$g'(x) = f'(x)$

$g'(0) = f'(0) < 0$

51. $g(x) = -f(x)$

$g'(x) = -f'(x)$

$g'(-6) = -f'(-6) < 0$

53. $g(x) = f(x - 10)$

$g'(x) = f'(x - 10)$

$g'(0) = f'(-10) > 0$

55. $f'(x) = \begin{cases} > 0, & x < 4 \Rightarrow f \text{ is increasing on } (-\infty, 4). \\ \text{undefined}, & x = 4 \\ < 0, & x > 4 \Rightarrow f \text{ is decreasing on } (4, \infty). \end{cases}$

Two possibilities for $f(x)$ are given below.

(a)

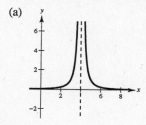

(b)

57. The critical numbers are in intervals $(-0.50, -0.25)$ and
$(0.25, 0.50)$ since the sign of f' changes in these intervals.
f is decreasing on approximately $(-1, -0.40)$, $(0.48, 1)$,
and increasing on $(-0.40, 0.48)$.

Relative minimum when $x \approx -0.40$.

Relative maximum when $x \approx 0.48$.

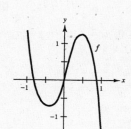

One possible answer. Others
could be vertical shifts of this
curve.

59. $v = k(R - r)r^2 = k(Rr^2 - r^3)$

$v' = k(2Rr - 3r^2)$

$\quad = kr(2R - 3r) = 0$

$r = 0 \text{ or } \frac{2}{3}R$

Maximum when $r = \frac{2}{3}R$.

61. $P = \dfrac{vR_1R_2}{(R_1 + R_2)^2}$, v and R_1 are constant.

$\dfrac{dP}{dR_2} = \dfrac{(R_1 + R_2)^2(vR_1) - vR_1R_2[2(R_1 + R_2)(1)]}{(R_1 + R_2)^4}$

$\quad = \dfrac{vR_1(R_1 - R_2)}{(R_1 + R_2)^3} = 0 \Rightarrow R_2 = R_1$

Maximum when $R_1 = R_2$.

63. (a) $B = -0.14649t^4 + 8.5979t^3 - 182.185t^2 + 1710.31t - 5250.7$

(b)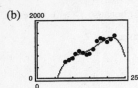

(c) $B' = 0$ when $t \approx 21.1$

Maximum is $(21.1, 1458)$.

65. Suppose $f'(x)$ changes from positive to negative at c. Then there exists a and b in I such that $f'(x) > 0$ for all x in (a, c) and $f'(x) < 0$ for all x in (c, b). By Theorem 5.5, f is increasing on (a, c) and decreasing on (c, b). Therefore, $f(c)$ is a maximum of f on (a, b) and thus, a relative maximum of f.

67. (a) Use a cubic polynomial
$f(x) = a_3x^3 + a_2x^2 + a_1x + a_0$.

(b) $f'(x) = 3a_3x^2 + 2a_2x + a_1$

$(0, 0)$: $0 = a_0$ $(f(0) = 0)$

$\qquad\quad 0 = a_1$ $(f'(0) = 0)$

$(2, 2)$: $2 = 8a_3 + 4a_2$ $(f(2) = 2)$

$\qquad\quad 0 = 12a_3 + 4a_2$ $(f'(2) = 0)$

(c) The solution is $a_0 = a_1 = 0, a_2 = \dfrac{3}{2}, a_3 = -\dfrac{1}{2}$:

$$f(x) = -\frac{1}{2}x^3 + \frac{3}{2}x^2.$$

(d)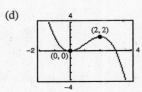

69. (a) Use a fourth degree polynomial $f(x) = a_4x^4 + a_3x^3 + a_2x^2 + a_1x + a_0$.

(b) $f'(x) = 4a_4x^3 + 3a_3x^2 + 2a_2x + a_1$

$(0, 0)$: $0 = a_0$ $(f(0) = 0)$

$\qquad\quad 0 = a_1$ $(f'(0) = 0)$

$(4, 0)$: $0 = 256a_4 + 64a_3 + 16a_2$ $(f(4) = 0)$

$\qquad\quad 0 = 256a_4 + 48a_3 + 8a_2$ $(f'(4) = 0)$

$(2, 4)$: $4 = 16a_4 + 8a_3 + 4a_2$ $(f(2) = 4)$

$\qquad\quad 0 = 32a_4 + 12a_3 + 4a_2$ $(f'(2) = 0)$

(c) The solution is $a_0 = a_1 = 0, a_2 = 4, a_3 = -2, a_4 = \dfrac{1}{4}$.

$$f(x) = \frac{1}{4}x^4 - 2x^3 + 4x^2$$

(d)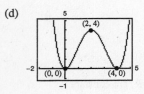

71. True.

Let $h(x) = f(x) + g(x)$ where f and g are increasing. Then $h'(x) = f'(x) + g'(x) > 0$ since $f'(x) > 0$ and $g'(x) > 0$.

73. False.

Let $f(x) = x^3$, then $f'(x) = 3x^2$ and f only has one critical number. Or, let $f(x) = x^3 + 3x + 1$, then $f'(x) = 3(x^2 + 1)$ has no critical numbers.

75. False. For example, $f(x) = x^3$ does not have a relative extrema at the critical number $x = 0$.

Section 5.4   Concavity and the Second Derivative Test

1. $y = x^2 - x - 2$, $y'' = 2$

Concave upward: $(-\infty, \infty)$

3. $f(x) = \dfrac{24}{x^2 + 12}$, $y'' = \dfrac{-144(4 - x^2)}{(x^2 + 12)^3}$

Concave upward: $(-\infty, -2), (2, \infty)$

Concave downward: $(-2, 2)$

5. $f(x) = \dfrac{x^2 + 1}{x^2 - 1}$, $y'' = \dfrac{4(3x^2 + 1)}{(x^2 - 1)^3}$

Concave upward: $(-\infty, -1), (1, \infty)$

Concave downward: $(-1, 1)$

7. $g(x) = 3x^2 - x^3$

$g'(x) = 6x - 3x^2$

$g''(x) = 6 - 6x$

Concave upward: $(-\infty, 1)$

Concave downward: $(1, \infty)$

9. $y = 2x$

$y' = 2$

$y'' = 0$

No concavity

11. $f(x) = x^3 - 6x^2 + 12x$

$f'(x) = 3x^2 - 12x + 12$

$f''(x) = 6(x - 2) = 0$ when $x = 2$.

The concavity changes at $x = 2$. $(2, 8)$ is a point of inflection.

Concave upward: $(2, \infty)$

Concave downward: $(-\infty, 2)$

13. $f(x) = \dfrac{1}{4}x^4 - 2x^2$

$f'(x) = x^3 - 4x$

$f''(x) = 3x^2 - 4$

$f''(x) = 3x^2 - 4 = 0$ when $x = \pm\dfrac{2}{\sqrt{3}}$.

Test interval:	$-\infty < x < -\dfrac{2}{\sqrt{3}}$	$-\dfrac{2}{\sqrt{3}} < x < \dfrac{2}{\sqrt{3}}$	$\dfrac{2}{\sqrt{3}} < x < \infty$
Sign of $f''(x)$:	$f''(x) > 0$	$f''(x) < 0$	$f''(x) > 0$
Conclusion:	Concave upward	Concave downward	Concave upward

Points of inflection: $\left(\pm\dfrac{2}{\sqrt{3}}, -\dfrac{20}{9}\right)$

15. $f(x) = x(x - 4)^3$

$f'(x) = x[3(x - 4)^2] + (x - 4)^3$

$\quad = (x - 4)^2(4x - 4)$

$f''(x) = 4(x - 1)[2(x - 4)] + 4(x - 4)^2$

$\quad = 4(x - 4)[2(x - 1) + (x - 4)]$

$\quad = 4(x - 4)(3x - 6) = 12(x - 4)(x - 2)$

$f''(x) = 12(x - 4)(x - 2) = 0$ when $x = 2, 4$.

Test interval:	$-\infty < x < 2$	$2 < x < 4$	$4 < x < \infty$
Sign of $f''(x)$:	$f''(x) > 0$	$f''(x) < 0$	$f''(x) > 0$
Conclusion:	Concave upward	Concave downward	Concave upward

Points of inflection: $(2, -16), (4, 0)$

17. $f(x) = x\sqrt{x + 3}$, Domain: $[-3, \infty)$

$f'(x) = x\left(\dfrac{1}{2}\right)(x + 3)^{-1/2} + \sqrt{x + 3} = \dfrac{3(x + 2)}{2\sqrt{x + 3}}$

$f''(x) = \dfrac{6\sqrt{x + 3} - 3(x + 2)(x + 3)^{-1/2}}{4(x + 3)} = \dfrac{3(x + 4)}{4(x + 3)^{3/2}}$

$f''(x) > 0$ on the entire domain of f (except for $x = -3$, for which $f''(x)$ is undefined). There are no points of inflection. Concave upward on $(-3, \infty)$.

19. $f(x) = \dfrac{x}{x^2 + 1}$

$f'(x) = \dfrac{1 - x^2}{(x^2 + 1)^2}$

$f''(x) = \dfrac{2x(x^2 - 3)}{(x^2 + 1)^3} = 0$ when $x = 0, \pm\sqrt{3}$.

Test intervals:	$-\infty < x < -\sqrt{3}$	$-\sqrt{3} < x < 0$	$0 < x < \sqrt{3}$	$\sqrt{3} < x < \infty$
Sign of $f'(x)$:	$f'' < 0$	$f'' > 0$	$f'' < 0$	$f'' > 0$
Conclusion:	Concave downward	Concave upward	Concave downward	Concave upward

Points of inflection: $\left(-\sqrt{3}, -\dfrac{\sqrt{3}}{4}\right), (0, 0), \left(\sqrt{3}, \dfrac{\sqrt{3}}{4}\right)$

21. $f(x) = 6x - x^2$

$f'(x) = 6 - 2x$ Critical number: $x = 3$

$f''(x) = -2$

$f''(3) = -2 < 0$

f has a relative maximum at $(3, 9)$.

23. $f(x) = (x - 5)^2$

$f'(x) = 2(x - 5)$

$f''(x) = 2$

Critical number: $x = 5$

$f''(5) > 0$

Therefore, $(5, 0)$ is a relative minimum.

25. $g(x) = x^2(6 - x) = 6x^2 - x^3$

$g'(x) = 12x - 3x^2 = 3x(4 - x)$ Critical numbers: $x = 0, 4$

$g''(x) = 12 - 6x$

$g''(0) = 12 > 0$ Relative minimum at $(0, 0)$.

$g''(4) = -12 < 0$ Relative maximum at $(4, 32)$.

27. $f(x) = x^3 - 3x^2 + 3$

$f'(x) = 3x^2 - 6x = 3x(x - 2)$

$f''(x) = 6x - 6 = 6(x - 1)$

Critical numbers: $x = 0, x = 2$

$f''(0) = -6 < 0$

Therefore, $(0, 3)$ is a relative maximum.

$f''(2) = 6 > 0$

Therefore, $(2, -1)$ is a relative minimum.

29. $f(x) = x^4 - 4x^3 + 2$

$f'(x) = 4x^3 - 12x^2 = 4x^2(x - 3)$

$f''(x) = 12x^2 - 24x = 12x(x - 2)$

Critical numbers: $x = 0, x = 3$

However, $f''(0) = 0$, so we must use the First Derivative Test. $f'(x) < 0$ on the intervals $(-\infty, 0)$ and $(0, 3)$; hence, $(0, 2)$ is not an extremum. $f''(3) > 0$ so $(3, -25)$ is a relative minimum.

31. $f(x) = x^{2/3} - 3$

$f'(x) = \dfrac{2}{3x^{1/3}}$

$f''(x) = \dfrac{-2}{9x^{4/3}}$

Critical number: $x = 0$

However, $f''(0)$ is undefined, so we must use the First Derivative Test. Since $f'(x) < 0$ on $(-\infty, 0)$ and $f'(x) > 0$ on $(0, \infty)$, $(0, -3)$ is a relative minimum.

33. $f(x) = x + \dfrac{4}{x}$

$f'(x) = 1 - \dfrac{4}{x^2} = \dfrac{x^2 - 4}{x^2}$

$f''(x) = \dfrac{8}{x^3}$

Critical numbers: $x = \pm 2$

$f''(-2) < 0$

Therefore, $(-2, -4)$ is a relative maximum.

$f''(2) > 0$

Therefore, $(2, 4)$ is a relative minimum.

35. $f(x) = 0.2x^2(x - 3)^3, [-1, 4]$

(a) $f'(x) = 0.2x(5x - 6)(x - 3)^2$

$f''(x) = (x - 3)(4x^2 - 9.6x + 3.6)$

$\quad\quad = 0.4(x - 3)(10x^2 - 24x + 9)$

(b) $f''(0) < 0 \implies (0, 0)$ is a relative maximum.

$f''\left(\dfrac{6}{5}\right) > 0 \implies (1.2, -1.6796)$ is a relative minimum.

Points of inflection:

$\quad (3, 0), (0.4652, -0.7048), (1.9348, -0.9048)$

(c)

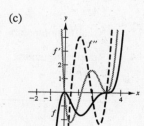

f is increasing when $f' > 0$ and decreasing when $f' < 0$. f is concave upward when $f'' > 0$ and concave downward when $f'' < 0$.

37. (a)

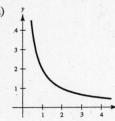

$f' < 0$ means f decreasing

f' increasing means concave upward

39. Let $f(x) = x^4$.

$$f''(x) = 12x^2$$

$f''(0) = 0$, but $(0, 0)$ is not a point of inflection.

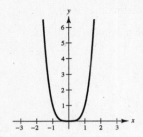

(b)

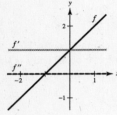

$f' > 0$ means f increasing

f' increasing means concave upward

41.

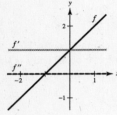

43.

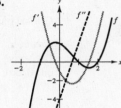

45.

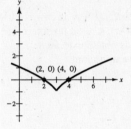

47.

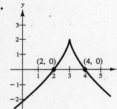

49. f'' is linear.

f' is quadratic.

f is cubic.

f concave upwards on $(-\infty, 3)$, downward on $(3, \infty)$.

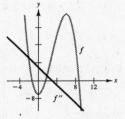

51. (a) $n = 1$:

$$f(x) = x - 2$$
$$f'(x) = 1$$
$$f''(x) = 0$$

No inflection points

$n = 2$:

$$f(x) = (x - 2)^2$$
$$f'(x) = 2(x - 2)$$
$$f''(x) = 2$$

No inflection points

Relative minimum: $(2, 0)$

$n = 3$:

$$f(x) = (x - 2)^3$$
$$f'(x) = 3(x - 2)^2$$
$$f''(x) = 6(x - 2)$$

Inflection point: $(2, 0)$

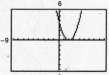

Point of inflection

$n = 4$:

$$f(x) = (x - 2)^4$$
$$f'(x) = 4(x - 2)^3$$
$$f''(x) = 12(x - 2)^2$$

No inflection points:

Relative minimum: $(2, 0)$

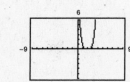

—CONTINUED—

51. —CONTINUED—

> **Conclusion:** If $n \geq 3$ and n is odd, then $(2, 0)$ is an inflection point. If $n \geq 2$ and n is even, then $(2, 0)$ is a relative minimum.
>
> (b) Let $f(x) = (x - 2)^n$, $f'(x) = n(x - 2)^{n-1}$, $f''(x) = n(n - 1)(x - 2)^{n-2}$.
>
> For $n \geq 3$ and odd, $n - 2$ is also odd and the concavity changes at $x = 2$.
>
> For $n \geq 4$ and even, $n - 2$ is also even and the concavity does not change at $x = 2$.
>
> Thus, $x = 2$ is an inflection point if and only if $n \geq 3$ is odd.

53. $f(x) = ax^3 + bx^2 + cx + d$

Relative maximum: $(3, 3)$

Relative minimum: $(5, 1)$

Point of inflection: $(4, 2)$

$f'(x) = 3ax^2 + 2bx + c, f''(x) = 6ax + 2b$

$\left.\begin{array}{l} f(3) = 27a + 9b + 3c + d = 3 \\ f(5) = 125a + 25b + 5c + d = 1 \end{array}\right\}$ $98a + 16b + 2c = -2 \Rightarrow 49a + 8b + c = -1$

$f'(3) = 27a + 6b + c = 0$, $f''(4) = 24a + 2b = 0$

$\begin{array}{ll} 49a + 8b + c = -1 & 24a + 2b = 0 \\ 27a + 6b + c = 0 & 22a + 2b = -1 \\ 22a + 2b \quad = -1 & 2a \quad = 1 \end{array}$

$a = \frac{1}{2}, b = -6, c = \frac{45}{2}, d = -24$

$f(x) = \frac{1}{2}x^3 - 6x^2 + \frac{45}{2}x - 24$

55. $f(x) = ax^3 + bx^2 + cx + d$

Maximum: $(-4, 1)$

Minimum: $(0, 0)$

(a) $f'(x) = 3ax^2 + 2bx + c$, $f''(x) = 6ax + 2b$

$f(0) = 0 \Rightarrow d = 0$

$f(-4) = 1 \Rightarrow -64a + 16b - 4c = 1$

$f'(-4) = 0 \Rightarrow 48a - 8b + c = 0$

$f'(0) = 0 \Rightarrow c = 0$

Solving this system yields $a = \frac{1}{32}$ and $b = 6a = \frac{3}{16}$.

$f(x) = \frac{1}{32}x^3 + \frac{3}{16}x^2$

(b) The plane would be descending at the greatest rate at the point of inflection.

$f''(x) = 6ax + 2b = \frac{3}{16}x + \frac{3}{8} = 0 \Rightarrow x = -2$

Two miles from touchdown.

57. (a) line *OA*: $y = -0.06x$ slope: -0.06

line *CB*: $y = 0.04x + 50$ slope: 0.04

$f(x) = ax^3 + bx^2 + cx + d$

$f'(x) = 3ax^2 + 2bx + c$

$(-1000, 60)$: $60 = (-1000)^3 a + (1000)^2 b - 1000c + d$

$-0.06 = (1000)^2 3a - 2000b + c$

$(1000, 90)$: $90 = (1000)^3 a + (1000)^2 b + 1000c + d$

$0.04 = (1000)^2 3a + 2000b + c$

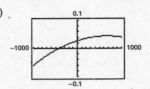

The solution to this system of 4 equations is $a = -1.25 \times 10^{-8}$, $b = 0.000025$, $c = 0.0275$, and $d = 50$.

(b) $y = -1.25 \times 10^{-8}x^3 + 0.000025x^2 + 0.0275x + 50$

(c)

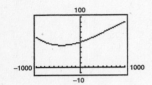

(d) The steepest part of the road is 6% at the point *A*.

59. $C = 0.5x^2 + 15x + 5000$

$\overline{C} = \dfrac{C}{x} = 0.5x + 15 + \dfrac{5000}{x}$

$\overline{C}$ = average cost per unit

$\dfrac{d\overline{C}}{dx} = 0.5 - \dfrac{5000}{x^2} = 0$ when $x = 100$.

By the First Derivative Test, $\overline{C}$ is minimized when $x = 100$ units.

61. $f(x) = \sqrt{1 - x}$, $f(0) = 1$

$f'(x) = -\dfrac{1}{2\sqrt{1 - x}}$, $f'(0) = -\dfrac{1}{2}$

$f''(x) = -\dfrac{1}{4(1 - x)^{3/2}}$, $f''(0) = -\dfrac{1}{4}$

$P_1(x) = 1 + \left(-\dfrac{1}{2}\right)(x - 0) = 1 - \dfrac{x}{2}$

$P_1'(x) = -\dfrac{1}{2}$

$P_2(x) = 1 + \left(-\dfrac{1}{2}\right)(x - 0) + \dfrac{1}{2}\left(-\dfrac{1}{4}\right)(x - 0)^2 = 1 - \dfrac{x}{2} - \dfrac{x^2}{8}$

$P_2'(x) = -\dfrac{1}{2} - \dfrac{x}{4}$

$P_2''(x) = -\dfrac{1}{4}$

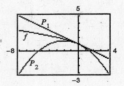

The values of f, P_1, P_2, and their first derivatives are equal at $x = 0$. The values of the second derivatives of f and P_2 are equal at $x = 0$. The approximations worsen as you move away from $x = 0$.

63. $S = \dfrac{100t^2}{65 + t^2}$, $t > 0$

(a)

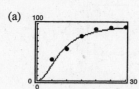

(b) $S'(t) = \dfrac{13,000t}{(65 + t^2)^2}$

$S''(t) = \dfrac{13,000(65 - 3t^2)}{(65 + t^2)^3} = 0 \implies t = 4.65$

S is concave upward on $(0, 4.65)$, concave downward on $(4.65, 30)$.

(c) $S'(t) > 0$ for $t > 0$.

As t increases, the speed increases, but at a slower rate.

67. False. Concavity is determined by f''.

65. True. Let $y = ax^3 + bx^2 + cx + d$, $a \neq 0$.
Then $y'' = 6ax + 2b = 0$ when $x = -(b/3a)$,
and the concavity changes at this point.

Section 5.5   Limits at Infinity

1. $f(x) = \dfrac{3x^2}{x^2 + 2}$

No vertical asymptotes

Horizontal asymptote: $y = 3$

Matches (f)

3. $f(x) = \dfrac{x}{x^2 + 2}$

No vertical asymptotes

Horizontal asymptote: $y = 0$

Matches (d)

5. Horizontal asymptote: $y = 0$

Intercept: $(0, 4)$

Matches (b)

7. $f(x) = \dfrac{4x + 3}{2x - 1}$

x	10^0	10^1	10^2	10^3	10^4	10^5	10^6
$f(x)$	7	2.26	2.025	2.0025	2.0003	2	2

$\lim\limits_{x \to \infty} f(x) = 2$

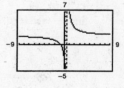

9. $f(x) = \dfrac{-6x}{\sqrt{4x^2 + 5}}$

x	10^0	10^1	10^2	10^3	10^4	10^5	10^6
$f(x)$	-2	-2.98	-2.9998	-3	-3	-3	-3

$\lim\limits_{x \to \infty} f(x) = -3$

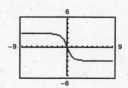

11. $f(x) = 5 - \dfrac{1}{x^2 + 1}$

x	10^0	10^1	10^2	10^3	10^4	10^5	10^6
$f(x)$	4.5	4.99	4.9999	4.999999	5	5	5

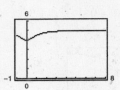

$\lim\limits_{x \to \infty} f(x) = 5$

13. (a) $h(x) = \dfrac{f(x)}{x^2} = \dfrac{5x^3 - 3x^2 + 10}{x^2} = 5x - 3 + \dfrac{10}{x^2}$

$\lim\limits_{x \to \infty} h(x) = \infty$ (Limit does not exist)

(b) $h(x) = \dfrac{f(x)}{x^3} = \dfrac{5x^3 - 3x^2 + 10}{x^3} = 5 - \dfrac{3}{x} + \dfrac{10}{x^3}$

$\lim\limits_{x \to \infty} h(x) = 5$

15. (a) $\lim\limits_{x \to \infty} \dfrac{x^2 + 2}{x^3 - 1} = 0$

(b) $\lim\limits_{x \to \infty} \dfrac{x^2 + 2}{x^2 - 1} = 1$

(c) $\lim\limits_{x \to \infty} \dfrac{x^2 + 2}{x - 1} = \infty$ (Limit does not exist.)

17. (a) $\lim\limits_{x \to \infty} \dfrac{5 - 2x^{3/2}}{3x^2 - 4} = 0$

(b) $\lim\limits_{x \to \infty} \dfrac{5 - 2x^{3/2}}{3x^{3/2} - 4} = -\dfrac{2}{3}$

(c) $\lim\limits_{x \to \infty} \dfrac{5 - 2x^{3/2}}{3x - 4} = -\infty$ (Limit does not exist.)

19. $\lim\limits_{x \to \infty} \dfrac{2x - 1}{3x + 2} = \lim\limits_{x \to \infty} \dfrac{2 - (1/x)}{3 + (2/x)} = \dfrac{2 - 0}{3 + 0} = \dfrac{2}{3}$

21. $\lim\limits_{x \to \infty} \dfrac{x}{x^2 - 1} = \lim\limits_{x \to \infty} \dfrac{1/x}{1 - (1/x^2)} = \dfrac{0}{1} = 0$

23. $\lim\limits_{x \to \infty} \left(10 - \dfrac{2}{x^2}\right) = 10 - 0 = 10$

25. $\lim\limits_{x \to -\infty} \dfrac{5x^2}{x + 3} = \lim\limits_{x \to -\infty} \dfrac{5x}{1 + (3/x)} = -\infty$

Limit does not exist.

27. $\lim\limits_{x \to -\infty} \dfrac{x}{\sqrt{x^2 - x}} = \lim\limits_{x \to -\infty} \dfrac{1}{\dfrac{\sqrt{x^2 - x}}{-\sqrt{x^2}}}, \left(\text{for } x < 0 \text{ we have } x = -\sqrt{x^2}\right)$

$= \lim\limits_{x \to -\infty} \dfrac{-1}{\sqrt{1 - (1/x)}} = -1$

29. $\lim\limits_{x \to \infty} \dfrac{2x + 1}{\sqrt{x^2 - x}} = \lim\limits_{x \to \infty} \dfrac{2 + (1/x)}{\sqrt{1 - (1/x)}} = \dfrac{2}{1} = 2$

31. (a) $f(x) = \dfrac{|x|}{x + 1}$

$\lim\limits_{x \to \infty} \dfrac{|x|}{x + 1} = 1$

$\lim\limits_{x \to -\infty} \dfrac{|x|}{x + 1} = -1$

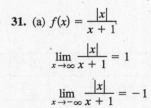
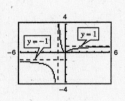

Therefore, $y = 1$ and $y = -1$ are both horizontal asymptotes.

33. $\displaystyle\lim_{x\to-\infty}\left(x+\sqrt{x^2+3}\right)=\lim_{x\to-\infty}\left[\left(x+\sqrt{x^2+3}\right)\cdot\frac{x-\sqrt{x^2+3}}{x-\sqrt{x^2+3}}\right]=\lim_{x\to-\infty}\frac{-3}{x-\sqrt{x^2+3}}=0$

35. $\displaystyle\lim_{x\to\infty}\left(x-\sqrt{x^2+x}\right)=\lim_{x\to\infty}\left[\left(x-\sqrt{x^2+x}\right)\cdot\frac{x+\sqrt{x^2+x}}{x+\sqrt{x^2+x}}\right]$

$$=\lim_{x\to\infty}\frac{-x}{x+\sqrt{x^2+x}}=\lim_{x\to\infty}\frac{-1}{1+\sqrt{1+(1/x)}}=-\frac{1}{2}$$

37.

x	10^0	10^1	10^2	10^3	10^4	10^5	10^6
$f(x)$	1	0.513	0.501	0.500	0.500	0.500	0.500

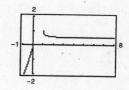

$\displaystyle\lim_{x\to\infty}\left(x-\sqrt{x(x-1)}\right)=\lim_{x\to\infty}\frac{x-\sqrt{x^2-x}}{1}\cdot\frac{x+\sqrt{x^2-x}}{x+\sqrt{x^2-x}}$

$\displaystyle=\lim_{x\to\infty}\frac{x}{x+\sqrt{x^2-x}}$

$\displaystyle=\lim_{x\to\infty}\frac{1}{1+\sqrt{1-(1/x)}}$

$\displaystyle=\frac{1}{2}$

39.

x	10^0	10^1	10^2	10^3	10^4	10^5	10^6
$f(x)$	-0.236	-0.025	-0.002	-2.5×10^{-4}	-2.5×10^{-5}	-2.5×10^{-6}	0

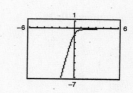

$2x-\sqrt{4x^2+1}=2x-\sqrt{4x^2+1}\left(\dfrac{2x+\sqrt{4x^2+1}}{2x+\sqrt{4x^2+1}}\right)$

$=\dfrac{4x^2-(4x^2+1)}{2x+\sqrt{4x^2+1}}$

$=\dfrac{-1}{2x+\sqrt{4x^2+1}}$

Hence, $\displaystyle\lim_{x\to\infty}f(x)=\lim_{x\to\infty}\frac{-1}{2x+\sqrt{4x^2+1}}=0$

41. (a)

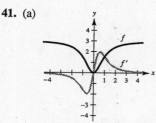

43. Yes. For example, let $f(x)=\dfrac{6|x-2|}{\sqrt{(x-2)^2+1}}$.

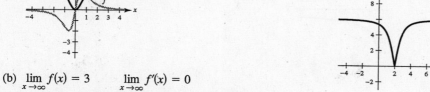

(b) $\displaystyle\lim_{x\to\infty}f(x)=3\qquad\lim_{x\to\infty}f'(x)=0$

(c) Since $\displaystyle\lim_{x\to\infty}f(x)=3$, the graph approaches that of a

horizontal line, $\displaystyle\lim_{x\to\infty}f'(x)=0$.

45. $y = \dfrac{2 + x}{1 - x}$

Intercepts: $(-2, 0), (0, 2)$

Symmetry: none

Horizontal asymptote: $y = -1$ since

$$\lim_{x \to -\infty} \frac{2 + x}{1 - x} = -1 = \lim_{x \to \infty} \frac{2 + x}{1 - x}.$$

Discontinuity: $x = 1$ (Vertical asymptote)

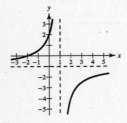

47. $y = \dfrac{x}{x^2 - 4}$

Intercept: $(0, 0)$

Symmetry: origin

Horizontal asymptote: $y = 0$

Vertical asymptote: $x = \pm 2$

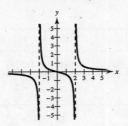

49. $y = \dfrac{x^2}{x^2 + 9}$

Intercept: $(0, 0)$

Symmetry: y-axis

Horizontal asymptote: $y = 1$ since

$$\lim_{x \to -\infty} \frac{x^2}{x^2 + 9} = 1 = \lim_{x \to \infty} \frac{x^2}{x^2 + 9}.$$

Relative minimum: $(0, 0)$

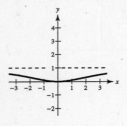

51. $y = \dfrac{2x^2}{x^2 - 4}$

Intercept: $(0, 0)$

Symmetry: y-axis

Horizontal asymptote: $y = 2$

Vertical asymptotes: $x = \pm 2$

Relative maximum: $(0, 0)$

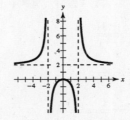

53. $xy^2 = 4$

Domain: $x > 0$

Intercepts: none

Symmetry: x-axis

Horizontal asymptote: $y = 0$ since

$$\lim_{x \to \infty} \frac{2}{\sqrt{x}} = 0 = \lim_{x \to \infty} -\frac{2}{\sqrt{x}}.$$

Discontinuity: $x = 0$ (Vertical asymptote)

55. $y = \dfrac{2x}{1-x}$

Intercept: $(0, 0)$

Symmetry: none

Horizontal asymptote: $y = -2$ since

$$\lim_{x \to -\infty} \frac{2x}{1-x} = -2 = \lim_{x \to \infty} \frac{2x}{1-x}.$$

Discontinuity: $x = 1$ (Vertical asymptote)

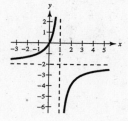

59. $y = \dfrac{x^3}{\sqrt{x^2-4}}$

Domain: $(-\infty, -2), (2, \infty)$

Intercepts: none

Symmetry: origin

Horizontal asymptote: none

Vertical asymptotes: $x = \pm 2$ (discontinuities)

Relative minimum: $\left(\sqrt{6}, 6\sqrt{3}\right)$

Relative maximum: $\left(-\sqrt{6}, -6\sqrt{3}\right)$

61. $f(x) = 5 - \dfrac{1}{x^2} = \dfrac{5x^2 - 1}{x^2}$

Domain: $(-\infty, 0), (0, \infty)$

$f'(x) = \dfrac{2}{x^3} \implies$ No relative extrema

$f''(x) = -\dfrac{6}{x^4} \implies$ No points of inflection

Vertical asymptote: $x = 0$

Horizontal asymptote: $y = 5$

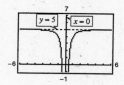

57. $y = 2 - \dfrac{3}{x^2}$

Intercepts: $\left(\pm\sqrt{3/2}, 0\right)$

Symmetry: y-axis

Horizontal asymptote: $y = 2$ since

$$\lim_{x \to -\infty} \left(2 - \frac{3}{x^2}\right) = 2 = \lim_{x \to \infty} \left(2 - \frac{3}{x^2}\right).$$

Discontinuity: $x = 0$ (Vertical asymptote)

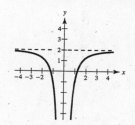

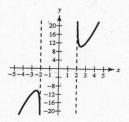

63. $f(x) = \dfrac{x}{x^2 - 4}$

$f'(x) = \dfrac{(x^2 - 4) - x(2x)}{(x^2 - 4)^2}$

$= \dfrac{-(x^2 + 4)}{(x^2 - 4)^2} \neq 0$ for any x in the domain of f.

$f''(x) = \dfrac{(x^2 - 4)^2(-2x) + (x^2 + 4)(2)(x^2 - 4)(2x)}{(x^2 - 4)^4}$

$= \dfrac{2x(x^2 + 12)}{(x^2 - 4)^3} = 0$ when $x = 0$.

Since $f''(x) > 0$ on $(-2, 0)$ and $f''(x) < 0$ on $(0, 2)$, then $(0, 0)$ is a point of inflection.

Vertical asymptotes: $x = \pm 2$

Horizontal asymptote: $y = 0$

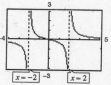

65. $f(x) = \dfrac{x - 2}{x^2 - 4x + 3} = \dfrac{x - 2}{(x - 1)(x - 3)}$

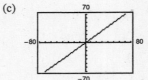

$f'(x) = \dfrac{(x^2 - 4x + 3) - (x - 2)(2x - 4)}{(x^2 - 4x + 3)^2} = \dfrac{-x^2 + 4x - 5}{(x^2 - 4x + 3)^2} \neq 0$

$f''(x) = \dfrac{(x^2 - 4x + 3)^2(-2x + 4) - (-x^2 + 4x - 5)(2)(x^2 - 4x + 3)(2x - 4)}{(x^2 - 4x + 3)^4}$

$= \dfrac{2(x^3 - 6x^2 + 15x - 14)}{(x^2 - 4x + 3)^3} = 0$ when $x = 2$.

Since $f''(x) > 0$ on $(1, 2)$ and $f''(x) < 0$ on $(2, 3)$, then $(2, 0)$ is a point of inflection.

Vertical asymptote: $x = 1, x = 3$

Horizontal asymptote: $y = 0$

67. $f(x) = \dfrac{x^3 - 3x^2 + 2}{x(x - 3)}, g(x) = x + \dfrac{2}{x(x - 3)}$

(a)

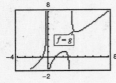

(b) $f(x) = \dfrac{x^3 - 3x^2 + 2}{x(x - 3)}$

$= \dfrac{x^2(x - 3)}{x(x - 3)} + \dfrac{2}{x(x - 3)}$

$= x + \dfrac{2}{x(x - 3)} = g(x)$

(c)

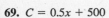

The graph appears as the slant asymptote $y = x$.

69. $C = 0.5x + 500$

$\overline{C} = \dfrac{C}{x}$

$\overline{C} = 0.5 + \dfrac{500}{x}$

$\lim\limits_{x \to \infty} \left(0.5 + \dfrac{500}{x}\right) = 0.5$

71. (a) $T_1(t) = -0.003t^2 + 0.677t + 26.564$

(b)

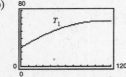

(c)

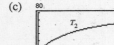

$T_2 = \dfrac{1451 + 86t}{58 + t}$

(d) $T_1(0) \approx 26.6$

$T_2(0) \approx 25.0$

(e) $\lim\limits_{t \to \infty} T_2 = \dfrac{86}{1} = 86$

(f) The limiting temperature is 86. T_1 has no horizontal asymptote.

73. $y = \dfrac{3.351t^2 + 42.461t - 543.730}{t^2}$

(a)

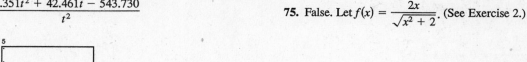

(b) Yes. $\displaystyle\lim_{t \to \infty} y = 3.351$

75. False. Let $f(x) = \dfrac{2x}{\sqrt{x^2 + 2}}$. (See Exercise 2.)

77. $\displaystyle\lim_{x \to \infty} \dfrac{p(x)}{q(x)} = \lim_{x \to \infty} \dfrac{a_n x^n + \cdots + a_1 x + a_0}{b_m x^m + \cdots + b_1 x + b_0}$

Divide $p(x)$ and $q(x)$ by x^m.

Case 1: If $n < m$: $\displaystyle\lim_{x \to \infty} \dfrac{p(x)}{q(x)} = \lim_{x \to \infty} \dfrac{\dfrac{a_n}{x^{m-n}} + \cdots + \dfrac{a_1}{x^{m-1}} + \dfrac{a_0}{x^m}}{b_m + \cdots + \dfrac{b_1}{x^{m-1}} + \dfrac{b_0}{x^m}} = \dfrac{0 + \cdots + 0 + 0}{b_m + \cdots + 0 + 0} = \dfrac{0}{b_m} = 0.$

Case 2: If $m = n$: $\displaystyle\lim_{x \to \infty} \dfrac{p(x)}{q(x)} = \lim_{x \to \infty} \dfrac{a_n + \cdots + \dfrac{a_1}{x^{m-1}} + \dfrac{a_0}{x^m}}{b_m + \cdots + \dfrac{b_1}{x^{m-1}} + \dfrac{b_0}{x^m}} = \dfrac{a_n + \cdots + 0 + 0}{b_m + \cdots + 0 + 0} = \dfrac{a_n}{b_m}.$

Case 3: If $n > m$: $\displaystyle\lim_{x \to \infty} \dfrac{p(x)}{q(x)} = \lim_{x \to \infty} \dfrac{a_n x^{n-m} + \cdots + \dfrac{a_1}{x^{m-1}} + \dfrac{a_0}{x^m}}{b_m + \cdots + \dfrac{b_1}{x^{m-1}} + \dfrac{b_0}{x^m}} = \dfrac{\pm\infty + \cdots + 0}{b_m + \cdots + 0} = \pm\infty.$

Section 5.6   A Summary of Curve Sketching

1. f has constant negative slope. Matches (d)

3. $f'(x) = 0$ at $x \approx \pm\dfrac{1}{2}$

Matches graph (a)

5. (a) $f'(x) = 0$ for $x = -2$ and $x = 2$.

f' is negative for $-2 < x < 2$ (decreasing function).

f' is positive for $x > 2$ and $x < -2$ (increasing function).

(b) $f''(x) = 0$ at $x = 0$ (Inflection point).

f'' is positive for $x > 0$ (Concave upward).

f'' is negative for $x < 0$ (Concave downward).

(c) f' is increasing on $(0, \infty)$.   ($f'' > 0$)

(d) $f'(x)$ is minimum at $x = 0$. The rate of change of f at $x = 0$ is less than the rate of change of f for all other values of x.

7. $y = \dfrac{x^2}{x^2 + 3}$

$y' = \dfrac{6x}{(x^2 + 3)^2} = 0$ when $x = 0$.

$y'' = \dfrac{18(1 - x^2)}{(x^2 + 3)^3} = 0$ when $x = \pm 1$.

Horizontal asymptote: $y = 1$

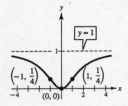

	y	y'	y''	Conclusion
$-\infty < x < -1$		$-$	$-$	Decreasing, concave down
$x = -1$	$\frac{1}{4}$	$-$	0	Point of inflection
$-1 < x < 0$		$-$	$+$	Decreasing, concave up
$x = 0$	0	0	$+$	Relative minimum
$0 < x < 1$		$+$	$+$	Increasing, concave up
$x = 1$	$\frac{1}{4}$	$+$	0	Point of inflection
$1 < x < \infty$		$+$	$-$	Increasing, concave down

9. $y = \dfrac{1}{x - 2} - 3$

$y' = -\dfrac{1}{(x - 2)^2} < 0$ when $x \neq 2$.

$y'' = \dfrac{2}{(x - 2)^3}$

No relative extrema, no points of inflection

Intercepts: $\left(\dfrac{7}{3}, 0\right), \left(0, -\dfrac{7}{2}\right)$

Vertical asymptote: $x = 2$

Horizontal asymptote: $y = -3$

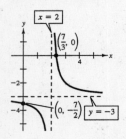

11. $y = \dfrac{2x}{x^2 - 1}$

$y' = \dfrac{-2(x^2 + 1)}{(x^2 - 1)^2} < 0$ if $x \neq \pm 1$.

$y'' = \dfrac{4x(x^2 + 3)}{(x^2 - 1)^3} = 0$ if $x = 0$.

Inflection point: $(0, 0)$

Intercept: $(0, 0)$

Vertical asymptote: $x = \pm 1$

Horizontal asymptote: $y = 0$

Symmetry with respect to the origin

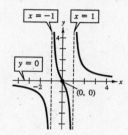

13. $g(x) = x + \dfrac{4}{x^2 + 1}$

$g'(x) = 1 - \dfrac{8x}{(x^2 + 1)^2} = \dfrac{x^4 + 2x^2 - 8x + 1}{(x^2 + 1)^2} = 0$ when $x \approx 0.1292,\ 1.6085$.

$g''(x) = \dfrac{8(3x^2 - 1)}{(x^2 + 1)^3} = 0$ when $x = \pm\dfrac{\sqrt{3}}{3}$.

$g''(0.1292) < 0$, therefore, $(0.1292, 4.064)$ is relative maximum.

$g''(1.6085) > 0$, therefore, $(1.6085, 2.724)$ is a relative minimum.

Points of inflection: $\left(-\dfrac{\sqrt{3}}{3}, 2.423\right), \left(\dfrac{\sqrt{3}}{3}, 3.577\right)$

Intercepts: $(0, 4), (-1.3788, 0)$

Slant asymptote: $y = x$

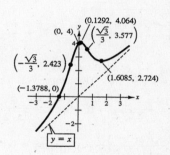

15. $f(x) = \dfrac{x^2 + 1}{x} = x + \dfrac{1}{x}$

$f'(x) = 1 - \dfrac{1}{x^2} = 0$ when $x = \pm 1$.

$f''(x) = \dfrac{2}{x^3} \neq 0$

Relative maximum: $(-1, -2)$

Relative minimum: $(1, 2)$

Vertical asymptote: $x = 0$

Slant asymptote: $y = x$

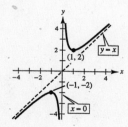

17. $y = \dfrac{x^2 - 6x + 12}{x - 4} = x - 2 + \dfrac{4}{x - 4}$

$y' = 1 - \dfrac{4}{(x - 4)^2}$

$\qquad = \dfrac{(x - 2)(x - 6)}{(x - 4)^2} = 0$ when $x = 2, 6$.

$y'' = \dfrac{8}{(x - 4)^3}$

$y'' < 0$ when $x = 2$.

Therefore, $(2, -2)$ is a relative maximum.

$y'' > 0$ when $x = 6$.

Therefore, $(6, 6)$ is a relative minimum.

Vertical asymptote: $x = 4$

Slant asymptote: $y = x - 2$

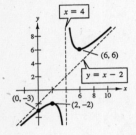

19. $y = x\sqrt{4 - x}$

$y' = \dfrac{8 - 3x}{2\sqrt{4 - x}}$

$y'' = \dfrac{3x - 16}{4(4 - x)^{3/2}}$

Domain: $x \le 4$

Intercepts: $(0, 0)$, $(4, 0)$

Relative maximum: $\left(\dfrac{8}{3}, \dfrac{16}{3\sqrt{3}}\right) \approx \left(\dfrac{8}{3}, 3.079\right)$

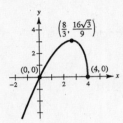

21. $h(x) = x\sqrt{9 - x^2}$ Domain: $-3 \le x \le 3$

$h'(x) = \dfrac{9 - 2x^2}{\sqrt{9 - x^2}} = 0$ when $x = \pm\dfrac{3}{\sqrt{2}} = \pm\dfrac{3\sqrt{2}}{2}$.

$h''(x) = \dfrac{x(2x^2 - 27)}{(9 - x^2)^{3/2}} = 0$ when $x = 0$.

Relative maximum: $\left(\dfrac{3\sqrt{2}}{2}, \dfrac{9}{2}\right)$

Relative minimum: $\left(-\dfrac{3\sqrt{2}}{2}, -\dfrac{9}{2}\right)$

Intercepts: $(0, 0)$, $(\pm 3, 0)$

Symmetric with respect to the origin

Point of inflection: $(0, 0)$

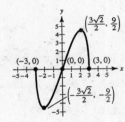

23. $y = 3x^{2/3} - 2x$

$y' = 2x^{-1/3} - 2 = \dfrac{2(1 - x^{1/3})}{x^{1/3}}$

    $= 0$ when $x = 1$ and undefined when $x = 0$.

$y'' = \dfrac{-2}{3x^{4/3}} < 0$ when $x \neq 0$.

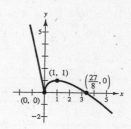

	y	y'	y''	Conclusion
$-\infty < x < 0$		$-$	$-$	Decreasing, concave down
$x = 0$	0	Undefined	Undefined	Relative minimum
$0 < x < 1$		$+$	$-$	Increasing, concave down
$x = 1$	1	0	$-$	Relative maximum
$1 < x < \infty$		$-$	$-$	Decreasing, concave down

25. $y = x^3 - 3x^2 + 3$

$y' = 3x^2 - 6x = 3x(x - 2) = 0$ when $x = 0$, $x = 2$.

$y'' = 6x - 6 = 6(x - 1) = 0$ when $x = 1$.

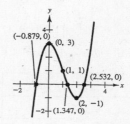

	y	y'	y''	Conclusion
$-\infty < x < 0$		$+$	$-$	Increasing, concave down
$x = 0$	3	0	$-$	Relative maximum
$0 < x < 1$		$-$	$-$	Decreasing, concave down
$x = 1$	1	$-$	0	Point of inflection
$1 < x < 2$		$-$	$+$	Decreasing, concave up
$x = 2$	-1	0	$+$	Relative minimum
$2 < x < \infty$		$+$	$+$	Increasing, concave up

27. $y = 2 - x - x^3$

$y' = -1 - 3x^2$

No critical numbers

$y'' = -6x = 0$ when $x = 0$.

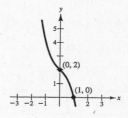

	y	y'	y''	Conclusion
$-\infty < x < 0$		$-$	$+$	Decreasing, concave up
$x = 0$	2	$-$	0	Point of inflection
$0 < x < \infty$		$-$	$-$	Decreasing, concave down

29. $f(x) = 3x^3 - 9x + 1$

$f'(x) = 9x^2 - 9 = 9(x^2 - 1) = 0$ when $x = \pm 1$.

$f''(x) = 18x = 0$ when $x = 0$.

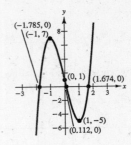

	$f(x)$	$f'(x)$	$f''(x)$	Conclusion
$-\infty < x < -1$		+	−	Increasing, concave down
$x = -1$	7	0	−	Relative maximum
$-1 < x < 0$		−	−	Decreasing, concave down
$x = 0$	1	−	0	Point of inflection
$0 < x < 1$		−	+	Decreasing, concave up
$x = 1$	−5	0	+	Relative minimum
$1 < x < \infty$		+	+	Increasing, concave up

31. $y = 3x^4 + 4x^3$

$y' = 12x^3 + 12x^2 = 12x^2(x + 1) = 0$ when $x = 0, x = -1$.

$y'' = 36x^2 + 24x = 12x(3x + 2) = 0$ when $x = 0, x = -\frac{2}{3}$.

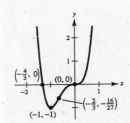

	y	y'	y''	Conclusion
$-\infty < x < -1$		−	+	Decreasing, concave up
$x = -1$	−1	0	+	Relative minimum
$-1 < x < -\frac{2}{3}$		+	+	Increasing, concave up
$x = -\frac{2}{3}$	$-\frac{16}{27}$	+	0	Point of inflection
$-\frac{2}{3} < x < 0$		+	−	Increasing, concave down
$x = 0$	0	0	0	Point of inflection
$0 < x < \infty$		+	+	Increasing, concave up

33. $f(x) = x^4 - 4x^3 + 16x$

$f'(x) = 4x^3 - 12x^2 + 16 = 4(x + 1)(x - 2)^2 = 0$ when $x = -1, x = 2$.

$f''(x) = 12x^2 - 24x = 12x(x - 2) = 0$ when $x = 0, x = 2$.

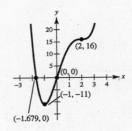

	$f(x)$	$f'(x)$	$f''(x)$	Conclusion
$-\infty < x < -1$		−	+	Decreasing, concave up
$x = -1$	−11	0	+	Relative minimum
$-1 < x < 0$		+	+	Increasing, concave up
$x = 0$	0	+	0	Point of inflection
$0 < x < 2$		+	−	Increasing, concave down
$x = 2$	16	0	0	Point of inflection
$2 < x < \infty$		+	+	Increasing, concave up

35. $y = x^5 - 5x$

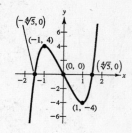

$y' = 5x^4 - 5 = 5(x^4 - 1) = 0$ when $x = \pm 1$.

$y'' = 20x^3 = 0$ when $x = 0$.

	y	y'	y''	Conclusion
$-\infty < x < -1$		$+$	$-$	Increasing, concave down
$x = -1$	4	0	$-$	Relative maximum
$-1 < x < 0$		$-$	$-$	Decreasing, concave down
$x = 0$	0	$-$	0	Point of inflection
$0 < x < 1$		$-$	$+$	Decreasing, concave up
$x = 1$	-4	0	$+$	Relative minimum
$1 < x < \infty$		$+$	$+$	Increasing, concave up

37. $y = |2x - 3|$

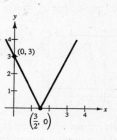

$y' = \dfrac{2(2x - 3)}{|2x - 3|}$ undefined at $x = \dfrac{3}{2}$.

$y'' = 0$

	y	y'	Conclusion
$-\infty < x < \frac{3}{2}$		$-$	Decreasing
$x = \frac{3}{2}$	0	Undefined	Relative minimum
$\frac{3}{2} < x < \infty$		$+$	Increasing

39. $f(x) = \dfrac{20x}{x^2 + 1} - \dfrac{1}{x} = \dfrac{19x^2 - 1}{x(x^2 + 1)}$

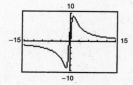

$x = 0$ vertical asymptote

$y = 0$ horizontal asymptote

Minimum: $(-1.10, -9.05)$

Maximum: $(1.10, 9.05)$

Points of inflection: $(-1.84, -7.86), (1.84, 7.86)$

41. $y = \dfrac{x}{\sqrt{x^2 + 7}}$

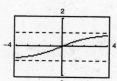

$(0, 0)$ point of inflection

$y = \pm 1$ horizontal asymptotes

43. f is cubic.

f' is quadratic.

f'' is linear.

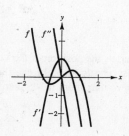

45.

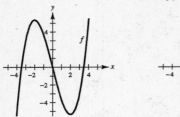 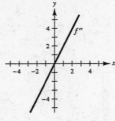

(any vertical translate of f will do)

47.

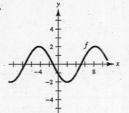 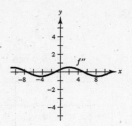

(any vertical translate of f will do)

49. Since the slope is negative, the function is decreasing on $(2, 8)$, and hence $f(3) > f(5)$.

51. $f(x) = \dfrac{4(x-1)^2}{x^2 - 4x + 5}$

Vertical asymptote: none

Horizontal asymptote: $y = 4$

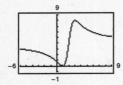

The graph crosses the horizontal asymptote $y = 4$. If a function has a vertical asymptote at $x = c$, the graph would not cross it since $f(c)$ is undefined.

53. $h(x) = \dfrac{6 - 2x}{3 - x}$

$= \dfrac{2(3-x)}{3-x} = \begin{cases} 2, & \text{if } x \neq 3 \\ \text{Undefined}, & \text{if } x = 3 \end{cases}$

The rational function is not reduced to lowest terms.

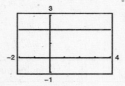

hole at $(3, 2)$

55. $f(x) = -\dfrac{x^2 - 3x - 1}{x - 2} = -x + 1 + \dfrac{3}{x - 2}$

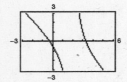

The graph appears to approach the slant asymptote $y = -x + 1$.

57. $f(x) = \dfrac{x^3}{x^2 + 1} = x - \dfrac{x}{x^2 + 1}$

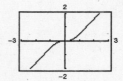

The graph appears to approach the slant asymptote $y = x$.

59. Vertical asymptote: $x = 5$

Horizontal asymptote: $y = 0$

$y = \dfrac{1}{x - 5}$

61. Vertical asymptote: $x = 5$

Slant asymptote: $y = 3x + 2$

$y = 3x + 2 + \dfrac{1}{x - 5} = \dfrac{3x^2 - 13x - 9}{x - 5}$

63. $f(x) = \dfrac{ax}{(x - b)^2}$

(a) The graph has a vertical asymptote at $x = b$. If $a > 0$, the graph approaches ∞ as $x \to b$. If $a < 0$, the graph approaches $-\infty$ as $x \to b$. The graph approaches its vertical asymptote faster as $|a| \to 0$.

(b) As b varies, the position of the vertical asymptote changes: $x = b$. Also, the coordinates of the minimum $(a > 0)$ or maximum $(a < 0)$ are changed.

65. $f(x) = \dfrac{3x^n}{x^4 + 1}$

(a) For n even, f is symmetric about the y-axis. For n odd, f is symmetric about the origin.

(b) The x-axis will be the horizontal asymptote if the degree of the numerator is less than 4. That is, $n = 0, 1, 2, 3$.

(c) $n = 4$ gives $y = 3$ as the horizontal asymptote.

(d) There is a slant asymptote $y = 3x$ if $n = 5$:

$$\frac{3x^5}{x^4 + 1} = 3x - \frac{3x}{x^4 + 1}.$$

(e)

n	0	1	2	3	4	5
M	1	2	3	2	1	0
N	2	3	4	5	2	3

67. (a)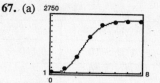

(b) When $t = 10$, $N(10) \approx 2434$ bacteria.

(c) N is a maximum when $t \approx 7.2$ (seventh day).

(d) $N''(t) = 0$ for $t \approx 3.2$

(e) $\displaystyle\lim_{t\to\infty} N(t) = \frac{13{,}250}{7} \approx 1892.86$

69. $y = \sqrt{x^2 + 6x} = \sqrt{(x + 3)^2 - 9}$

$y \to x + 3$ as $x \to \infty$, and $y \to -x - 3$ as $x \to -\infty$.

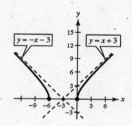

Section 5.7   Optimization Problems

1. (a)

First Number, x	Second Number	Product, P
10	$110 - 10$	$10(110 - 10) = 1000$
20	$110 - 20$	$20(110 - 20) = 1800$
30	$110 - 30$	$30(110 - 30) = 2400$
40	$110 - 40$	$40(110 - 40) = 2800$
50	$110 - 50$	$50(110 - 50) = 3000$
60	$110 - 60$	$60(110 - 60) = 3000$

—CONTINUED—

1. —CONTINUED—

(b)

First Number, x	Second Number	Product, P
10	$110 - 10$	$10(110 - 10) = 1000$
20	$110 - 20$	$20(110 - 20) = 1800$
30	$110 - 30$	$30(110 - 30) = 2400$
40	$110 - 40$	$40(110 - 40) = 2800$
50	$110 - 50$	$50(110 - 50) = 3000$
60	$110 - 60$	$60(110 - 60) = 3000$
70	$110 - 70$	$70(110 - 70) = 2800$
80	$110 - 80$	$80(110 - 80) = 2400$
90	$110 - 90$	$90(110 - 90) = 1800$
100	$110 - 100$	$100(110 - 100) = 1000$

The maximum is attained near $x = 50$ and 60.

(c) $P = x(110 - x) = 110x - x^2$

(d)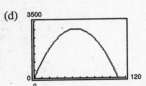

The solution appears to be $x = 55$.

(e) $\dfrac{dP}{dx} = 110 - 2x = 0$ when $x = 55$.

$\dfrac{d^2P}{dx^2} = -2 < 0$

P is a maximum when $x = 110 - x = 55$.
The two numbers are 55 and 55.

3. Let x and y be two positive numbers such that $x + y = S$.

$$P = xy = x(S - x) = Sx - x^2$$

$$\frac{dP}{dx} = S - 2x = 0 \text{ when } x = \frac{S}{2}.$$

$$\frac{d^2P}{dx^2} = -2 < 0 \text{ when } x = \frac{S}{2}.$$

P is a maximum when $x = y = S/2$.

5. Let x and y be two positive numbers such that $xy = 192$.

$$S = x + 3y = \frac{192}{y} + 3y$$

$$\frac{dS}{dy} = 3 - \frac{192}{y^2} = 0 \text{ when } y = 8.$$

$$\frac{d^2S}{dy^2} = \frac{384}{y^3} > 0 \text{ when } y = 8.$$

S is minimum when $y = 8$ and $x = 24$.

7. Let x and y be two positive numbers such that $x + 2y = 100$.

$$P = xy = y(100 - 2y) = 100y - 2y^2$$

$$\frac{dP}{dy} = 100 - 4y = 0 \text{ when } y = 25.$$

$$\frac{d^2P}{dy^2} = -4 < 0 \text{ when } y = 25.$$

P is a maximum when $x = 50$ and $y = 25$.

9. Let x be the length and y the width of the rectangle.

$$2x + 2y = 100$$

$$y = 50 - x$$

$$A = xy = x(50 - x)$$

$$\frac{dA}{dx} = 50 - 2x = 0 \text{ when } x = 25.$$

$$\frac{d^2A}{dx^2} = -2 < 0 \text{ when } x = 25.$$

A is maximum when $x = y = 25$ meters.

11. Let x be the length and y the width of the rectangle.

$$xy = 64$$

$$y = \frac{64}{x}$$

$$P = 2x + 2y = 2x + 2\left(\frac{64}{x}\right) = 2x + \frac{128}{x}$$

$$\frac{dP}{dx} = 2 - \frac{128}{x^2} = 0 \text{ when } x = 8.$$

$$\frac{d^2P}{dx^2} = \frac{256}{x^3} > 0 \text{ when } x = 8.$$

P is minimum when $x = y = 8$ feet.

13. $d = \sqrt{(x-4)^2 + (\sqrt{x} - 0)^2}$

$\quad = \sqrt{x^2 - 7x + 16}$

Since d is smallest when the expression inside the radical is smallest, you need only find the critical numbers of

$$f(x) = x^2 - 7x + 16.$$

$$f'(x) = 2x - 7 = 0$$

$$x = \frac{7}{2}$$

By the First Derivative Test, the point nearest to $(4, 0)$ is $\left(7/2, \sqrt{7/2}\right)$.

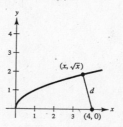

15. $d = \sqrt{(x-2)^2 + [x^2 - (1/2)]^2}$

$\quad = \sqrt{x^4 - 4x + (17/4)}$

Since d is smallest when the expression inside the radical is smallest, you need only find the critical numbers of

$$f(x) = x^4 - 4x + \frac{17}{4}.$$

$$f'(x) = 4x^3 - 4 = 0$$

$$x = 1$$

By the First Derivative Test, the point nearest to $\left(2, \frac{1}{2}\right)$ is $(1, 1)$.

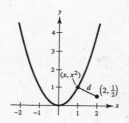

17. $\frac{dQ}{dx} = kx(Q_0 - x) = kQ_0 x - kx^2$

$$\frac{d^2Q}{dx^2} = kQ_0 - 2kx$$

$$= k(Q_0 - 2x) = 0 \text{ when } x = \frac{Q_0}{2}.$$

$$\frac{d^3Q}{dx^3} = -2k < 0 \text{ when } x = \frac{Q_0}{2}.$$

dQ/dx is maximum when $x = Q_0/2$.

19. $xy = 180,000$ (see figure)

$$S = x + 2y = \left(x + \frac{360,000}{x}\right) \text{ where } S \text{ is the length}$$
of fence needed.

$$\frac{dS}{dx} = 1 - \frac{360,000}{x^2} = 0 \text{ when } x = 600.$$

$$\frac{d^2S}{dx^2} = \frac{720,000}{x^3} > 0 \text{ when } x = 600.$$

S is a minimum when $x = 600$ meters and $y = 300$ meters.

21. (a) $A = 4(\text{area of side}) + 2(\text{area of Top})$

    (a) $A = 4(3)(11) + 2(3)(3) = 150$ square inches

    (b) $A = 4(5)(5) + 2(5)(5) = 150$ square inches

    (c) $A = 4(3.25)(6) + 2(6)(6) = 150$ square inches

  (b) $V = (\text{length})(\text{width})(\text{height})$

    (a) $V = (3)(3)(11) = 99$ cubic inches

    (b) $V = (5)(5)(5) = 125$ cubic inches

    (c) $V = (6)(6)(3.25) = 117$ cubic inches

(c) $S = 4xy + 2x^2 = 150 \implies y = \dfrac{150 - 2x^2}{4x}$

$$V = x^2 y = x^2\left(\frac{150 - 2x^2}{4x}\right) = \frac{75}{2}x - \frac{1}{2}x^3$$

$$V' = \frac{75}{2} - \frac{3}{2}x^2 = 0 \implies x = \pm 5$$

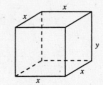

By the First Derivative Test, $x = 5$ yields the maximum volume.
Dimensions: $5 \times 5 \times 5$. (A cube!)

23. $16 = 2y + x + \pi\left(\dfrac{x}{2}\right)$

$$32 = 4y + 2x + \pi x$$

$$y = \frac{32 - 2x - \pi x}{4}$$

$$A = xy + \frac{\pi}{2}\left(\frac{x}{2}\right)^2 = \left(\frac{32 - 2x - \pi x}{4}\right)x + \frac{\pi x^2}{8}$$

$$= 8x - \frac{1}{2}x^2 - \frac{\pi}{4}x^2 + \frac{\pi}{8}x^2$$

$$\frac{dA}{dx} = 8 - x - \frac{\pi}{2}x + \frac{\pi}{4}x = 8 - x\left(1 + \frac{\pi}{4}\right)$$

$$= 0 \text{ when } x = \frac{8}{1 + (\pi/4)} = \frac{32}{4 + \pi}.$$

$$\frac{d^2 A}{dx^2} = -\left(1 + \frac{\pi}{4}\right) < 0 \text{ when } x = \frac{32}{4 + \pi}$$

$$y = \frac{32 - 2[32/(4 + \pi)] - \pi[32/(4 + \pi)]}{4} = \frac{16}{4 + \pi}$$

The area is maximum when $y = \dfrac{16}{4 + \pi}$ feet and $x = \dfrac{32}{4 + \pi}$ feet.

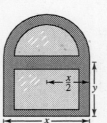

25. (a) $\dfrac{y - 2}{0 - 1} = \dfrac{0 - 2}{x - 1}$

$$y = 2 + \frac{2}{x - 1}$$

$$L = \sqrt{x^2 + y^2} = \sqrt{x^2 + \left(2 + \frac{2}{x - 1}\right)^2}$$

$$= \sqrt{x^2 + 4 + \frac{8}{x - 1} + \frac{4}{(x - 1)^2}}, \quad x > 1$$

  (b)

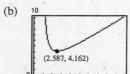

L is minimum when $x \approx 2.587$ and $L \approx 4.162$.

(c) Area $= A(x) = \dfrac{1}{2}xy = \dfrac{1}{2}x\left(2 + \dfrac{2}{x - 1}\right) = x + \dfrac{x}{x - 1}$

$$A'(x) = 1 + \frac{(x - 1) - x}{(x - 1)^2} = 1 - \frac{1}{(x - 1)^2} = 0$$

$$(x - 1)^2 = 1$$

$$x - 1 = \pm 1$$

$$x = 0, 2 \text{ (select } x = 2)$$

Then $y = 4$ and $A = 4$.

Vertices: $(0, 0), (2, 0), (0, 4)$

27. $A = 2xy = 2x\sqrt{25 - x^2}$ (see figure)

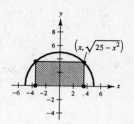

$$\frac{dA}{dx} = 2x\left(\frac{1}{2}\right)\left(\frac{-2x}{\sqrt{25 - x^2}}\right) + 2\sqrt{25 - x^2}$$

$$= 2\left(\frac{25 - 2x^2}{\sqrt{25 - x^2}}\right) = 0 \text{ when } x = y = \frac{5\sqrt{2}}{2} \approx 3.54.$$

By the First Derivative Test, the inscribed rectangle of maximum area has vertices

$$\left(\pm\frac{5\sqrt{2}}{2}, 0\right), \left(\pm\frac{5\sqrt{2}}{2}, \frac{5\sqrt{2}}{2}\right).$$

Width: $\dfrac{5\sqrt{2}}{2}$; Length: $5\sqrt{2}$

29. $xy = 30 \implies y = \dfrac{30}{x}$

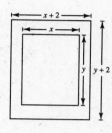

$$A = (x + 2)\left(\frac{30}{x} + 2\right) \quad \text{(see figure)}$$

$$\frac{dA}{dx} = (x + 2)\left(\frac{-30}{x^2}\right) + \left(\frac{30}{x} + 2\right) = \frac{2(x^2 - 30)}{x^2} = 0 \text{ when } x = \sqrt{30}.$$

$$y = \frac{30}{\sqrt{30}} = \sqrt{30}$$

By the First Derivative Test, the dimensions $(x + 2)$ by $(y + 2)$ are $\left(2 + \sqrt{30}\right)$ by $\left(2 + \sqrt{30}\right)$ (approximately 7.477 by 7.477). These dimensions yield a minimum area.

31. (a) $P = 2x + 2\pi r$

$$= 2x + 2\pi\left(\frac{y}{2}\right)$$

$$= 2x + \pi y = 200$$

$$\implies y = \frac{200 - 2x}{\pi} = \frac{2}{\pi}(100 - x)$$

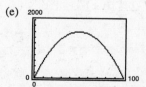

(b)

Length, x	Width, y	Area, xy
10	$\frac{2}{\pi}(100 - 10)$	$(10)\frac{2}{\pi}(100 - 10) \approx 573$
20	$\frac{2}{\pi}(100 - 20)$	$(20)\frac{2}{\pi}(100 - 20) \approx 1019$
30	$\frac{2}{\pi}(100 - 30)$	$(30)\frac{2}{\pi}(100 - 30) \approx 1337$
40	$\frac{2}{\pi}(100 - 40)$	$(40)\frac{2}{\pi}(100 - 40) \approx 1528$
50	$\frac{2}{\pi}(100 - 50)$	$(50)\frac{2}{\pi}(100 - 50) \approx 1592$
60	$\frac{2}{\pi}(100 - 60)$	$(60)\frac{2}{\pi}(100 - 60) = 1528$

The maximum area of the rectangle is approximately 1592 m².

(c) $A = xy = x\dfrac{2}{\pi}(100 - x) = \dfrac{2}{\pi}(100x - x^2)$

(d) $A' = \dfrac{2}{\pi}(100 - 2x)$

$A' = 0$ when $x = 50$.

Maximum area $= A(50) = \dfrac{5000}{\pi}$

(e)

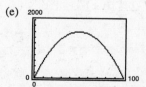

Maximum area is approximately

1591.55 m² ($x = 50$ m).

33. $V = \pi r^2 h = V_0$ cubic units or $h = \dfrac{V_0}{\pi r^2}$

$$S = 2\pi r^2 + 2\pi rh = 2\left(\pi r^2 + \dfrac{V_0}{r}\right)$$

$$\dfrac{dS}{dr} = 2\left(2\pi r - \dfrac{V_0}{r^2}\right) = 0 \text{ when } r = \sqrt[3]{\dfrac{V_0}{2\pi}} \text{ units.}$$

$$h = \dfrac{V_0}{\pi\left(\sqrt[3]{V_0/2\pi}\right)^2} = \dfrac{V_0(2\pi)^{2/3}}{\pi V_0^{2/3}} = \dfrac{2V_0^{1/3}}{(2\pi)^{1/3}} = 2r$$

By the First Derivative Test, this will yield the minimum surface area.

35. $V = \pi r^2 x$

$x + 2\pi r = 108 \implies x = 108 - 2\pi r$ (see figure)

$$V = \pi r^2(108 - 2\pi r) = \pi(108r^2 - 2\pi r^3)$$

$$\dfrac{dV}{dr} = \pi(216r - 6\pi r^2) = 6\pi r(36 - \pi r)$$

$$= 0 \text{ when } r = \dfrac{36}{\pi} \text{ and } x = 36.$$

$$\dfrac{d^2V}{dr^2} = \pi(216 - 12\pi r) < 0 \text{ when } r = \dfrac{36}{\pi}.$$

Volume is maximum when $x = 36$ inches and $r = 36/\pi \approx 11.459$ inches.

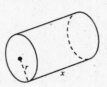

37. $V = \pi x^2 h = \pi x^2\left(2\sqrt{r^2 - x^2}\right) = 2\pi x^2\sqrt{r^2 - x^2}$ (see figure)

$$\dfrac{dV}{dx} = 2\pi\left[x^2\left(\dfrac{1}{2}\right)(r^2 - x^2)^{-1/2}(-2x) + 2x\sqrt{r^2 - x^2}\right] = \dfrac{2\pi x}{\sqrt{r^2 - x^2}}(2r^2 - 3x^2)$$

$$= 0 \text{ when } x = 0 \text{ and } x^2 = \dfrac{2r^2}{3} \implies x = \dfrac{\sqrt{6}r}{3}.$$

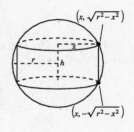

By the First Derivative Test, the volume is a maximum when

$$x = \dfrac{\sqrt{6}r}{3} \text{ and } h = \dfrac{2r}{\sqrt{3}}.$$

Thus, the maximum volume is

$$V = \pi\left(\dfrac{2}{3}r^2\right)\left(\dfrac{2r}{\sqrt{3}}\right) = \dfrac{4\pi r^3}{3\sqrt{3}}.$$

39. No, there is no minimum area. If the sides are x and y, then $2x + 2y = 20 \implies y = 10 - x$.
The area is $A(x) = x(10 - x) = 10x - x^2$. This can be made arbitrarily small by selecting $x \approx 0$.

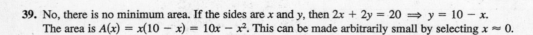

41. $V = 12 = \dfrac{4}{3}\pi r^3 + \pi r^2 h$

$$h = \dfrac{12 - (4/3)\pi r^3}{\pi r^2} = \dfrac{12}{\pi r^2} - \dfrac{4}{3}r$$

$$S = 4\pi r^2 + 2\pi rh = 4\pi r^2 + 2\pi r\left(\dfrac{12}{\pi r^2} - \dfrac{4}{3}r\right)$$

$$= 4\pi r^2 + \dfrac{24}{r} - \dfrac{8}{3}\pi r^2 = \dfrac{4}{3}\pi r^2 + \dfrac{24}{r}$$

$$\dfrac{dS}{dr} = \dfrac{8}{3}\pi r - \dfrac{24}{r^2} = 0 \text{ when } r = \sqrt[3]{9/\pi} \approx 1.42 \text{ cm.}$$

$$\dfrac{d^2S}{dr^2} = \dfrac{8}{3}\pi + \dfrac{48}{r^3} > 0 \text{ when } r = \sqrt[3]{9/\pi} \text{ cm.}$$

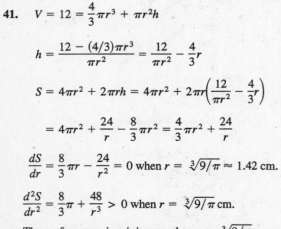

The surface area is minimum when $r = \sqrt[3]{9/\pi}$ cm and $h = 0$. The resulting solid is a sphere of radius $r \approx 1.42$ cm.

43. Let x be the length of a side of the square and y the length of a side of the triangle.

$$4x + 3y = 10$$

$$A = x^2 + \frac{1}{2}y\left(\frac{\sqrt{3}}{2}y\right)$$

$$= \frac{(10 - 3y)^2}{16} + \frac{\sqrt{3}}{4}y^2$$

$$\frac{dA}{dy} = \frac{1}{8}(10 - 3y)(-3) + \frac{\sqrt{3}}{2}y = 0$$

$$-30 + 9y + 4\sqrt{3}y = 0$$

$$y = \frac{30}{9 + 4\sqrt{3}}$$

$$\frac{d^2A}{dy^2} = \frac{9 + 4\sqrt{3}}{8} > 0$$

A is minimum when

$$y = \frac{30}{9 + 4\sqrt{3}} \text{ and } x = \frac{10\sqrt{3}}{9 + 4\sqrt{3}}.$$

45. Let S be the strength and k the constant of proportionality. Given $h^2 + w^2 = 24^2$, $h^2 = 24^2 - w^2$,

$$S = kwh^2$$

$$S = kw(576 - w^2) = k(576w - w^3)$$

$$\frac{dS}{dw} = k(576 - 3w^2) = 0 \text{ when } w = 8\sqrt{3}, h = 8\sqrt{6}.$$

$$\frac{d^2S}{dw^2} = -6kw < 0 \text{ when } w = 8\sqrt{3}.$$

These values yield a maximum.

47. $f(x) = \frac{1}{2}x^2$ $\qquad g(x) = \frac{1}{16}x^4 - \frac{1}{2}x^2$ on $[0, 4]$

(a)

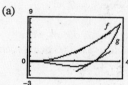

(b) $d(x) = f(x) - g(x) = \frac{1}{2}x^2 - \left(\frac{1}{16}x^4 - \frac{1}{2}x^2\right) = x^2 - \frac{1}{16}x^4$

$d'(x) = 2x - \frac{1}{4}x^3 = 0 \implies 8x = x^3$

$\implies x = 0, 2\sqrt{2}$ (in $[0, 4]$)

The maximum distance is $d = 4$ when $x = 2\sqrt{2}$.

(c) $f'(x) = x$, Tangent line at $\left(2\sqrt{2}, 4\right)$ is

$$y - 4 = 2\sqrt{2}\left(x - 2\sqrt{2}\right)$$

$$y = 2\sqrt{2}x - 4.$$

$g'(x) = \frac{1}{4}x^3 - x$, Tangent line at $\left(2\sqrt{2}, 0\right)$ is

$$y - 0 = \left(\frac{1}{4}\left(2\sqrt{2}\right)^3 - 2\sqrt{2}\right)\left(x - 2\sqrt{2}\right)$$

$$y = 2\sqrt{2}x - 8.$$

The tangent lines are parallel and 4 vertical units apart.

(d) The tangent lines will be parallel. If $d(x) = f(x) - g(x)$, then $d'(x) = 0 = f'(x) - g'(x)$ implies that $f'(x) = g'(x)$ at the point x where the distance is maximum.

49. Let F be the illumination at point P which is x units from source 1.

$$F = \frac{kI_1}{x^2} + \frac{kI_2}{(d-x)^2}$$

$$\frac{dF}{dx} = \frac{-2kI_1}{x^3} + \frac{2kI_2}{(d-x)^3} = 0 \text{ when } \frac{2kI_1}{x^3} = \frac{2kI_2}{(d-x)^3}.$$

$$\frac{\sqrt[3]{I_1}}{\sqrt[3]{I_2}} = \frac{x}{d-x}$$

$$(d-x)\sqrt[3]{I_1} = x\sqrt[3]{I_2}$$

$$d\sqrt[3]{I_1} = x\left(\sqrt[3]{I_1} + \sqrt[3]{I_2}\right)$$

$$x = \frac{d\sqrt[3]{I_1}}{\sqrt[3]{I_1} + \sqrt[3]{I_2}}$$

$$\frac{d^2F}{dx^2} = \frac{6kI_1}{x^4} + \frac{6kI_2}{(d-x)^4} > 0 \text{ when } x = \frac{d\sqrt[3]{I_1}}{\sqrt[3]{I_1} + \sqrt[3]{I_2}}.$$

This is the minimum point.

51. Let d be the amount deposited in the bank, i be the interest rate paid by the bank, and P be the profit.

$$P = (0.12)d - id$$

$$d = ki^2 \text{ (since d is proportional to i^2)}$$

$$P = (0.12)(ki^2) - i(ki^2) = k(0.12i^2 - i^3)$$

$$\frac{dP}{di} = k(0.24i - 3i^2) = 0 \text{ when } i = \frac{0.24}{3} = 0.08.$$

$$\frac{d^2P}{di^2} = k(0.24 - 6i) < 0 \text{ when } i = 0.08 \text{ (\textbf{Note:} $k > 0$)}.$$

The profit is a maximum when $i = 8\%$.

53. $\quad P = -\frac{1}{10}s^3 + 6s^2 + 400$

(a) $\quad \dfrac{dP}{ds} = -\dfrac{3}{10}s^2 + 12s = -\dfrac{3}{10}s(s-40) = 0$

when $x = 0, s = 40.$

$$\frac{d^2P}{ds^2} = -\frac{3}{5}s + 12$$

$$\frac{d^2P}{ds^2}(0) > 0 \implies s = 0 \text{ yields a minimum.}$$

$$\frac{d^2P}{ds^2}(40) < 0 \implies s = 40 \text{ yields a maximum.}$$

The maximum profit occurs when $s = 40$, which corresponds to $\$40,000$ ($P = \$3,600,000$).

(b) $\dfrac{d^2P}{ds^2} = -\dfrac{3}{5}s + 12 = 0$ when $s = 20.$

The point of diminishing returns occurs when $s = 20$, which corresponds to $\$20,000$ being spent on advertising.

55. $\quad S_1 = (4m-1)^2 + (5m-6)^2 + (10m-3)^2$

$$\frac{dS_1}{dm} = 2(4m-1)(4) + 2(5m-6)(5) + 2(10m-3)(10) = 282m - 128 = 0 \text{ when } m = \frac{64}{141}.$$

Line: $y = \dfrac{64}{141}x$

$$S = \left|4\left(\frac{64}{141}\right) - 1\right| + \left|5\left(\frac{64}{141}\right) - 6\right| + \left|10\left(\frac{64}{141}\right) - 3\right|$$

$$= \left|\frac{256}{141} - 1\right| + \left|\frac{320}{141} - 6\right| + \left|\frac{640}{141} - 3\right| = \frac{858}{141} \approx 6.1 \text{ mi}$$

57. $S_3 = \dfrac{|4m - 1|}{\sqrt{m^2 + 1}} + \dfrac{|5m - 6|}{\sqrt{m^2 + 1}} + \dfrac{|10m - 3|}{\sqrt{m^2 + 1}}$

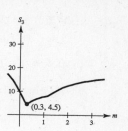

Using a graphing utility, you can see that the minimum occurs when $m \approx 0.3$.

Line: $y \approx 0.3x$

$S_3 = \dfrac{|4(0.3) - 1| + |5(0.3) - 6| + |10(0.3) - 3|}{\sqrt{(0.3)^2 + 1}} \approx 4.5$ mi.

Section 5.8   Differentials

1. $f(x) = x^2$

$f'(x) = 2x$

Tangent line at $(2, 4)$: $y - f(2) = f'(2)(x - 2)$

$\qquad\qquad\qquad\qquad y - 4 = 4(x - 2)$

$\qquad\qquad\qquad\qquad\qquad y = 4x - 4$

x	1.9	1.99	2	2.01	2.1
$f(x) = x^2$	3.6100	3.9601	4	4.0401	4.4100
$T(x) = 4x - 4$	3.6000	3.9600	4	4.0400	4.4000

3. $f(x) = x^5$

$f'(x) = 5x^4$

Tangent line at $(2, 32)$: $y - f(2) = f'(2)(x - 2)$

$\qquad\qquad\qquad\qquad y - 32 = 80(x - 2)$

$\qquad\qquad\qquad\qquad\qquad y = 80x - 128$

x	1.9	1.99	2	2.01	2.1
$f(x) = x^5$	24.7610	31.2080	32	32.8080	40.8410
$T(x) = 80x - 128$	24.0000	31.2000	32	32.8000	40.0000

5. $y = f(x) = \frac{1}{2}x^3, f'(x) = \frac{3}{2}x^2, x = 2, \Delta x = dx = 0.1$

$\Delta y = f(x + \Delta x) - f(x)$

$\quad = f(2.1) - f(2)$

$\quad = 0.6305$

$dy = f'(x)dx$

$\quad = f'(2)(0.1)$

$\quad = 6(0.1) = 0.6$

7. $y = f(x) = x^4 + 1, f'(x) = 4x^3, x = -1, \Delta x = dx = 0.01$

$\Delta y = f(x + \Delta x) - f(x)$

$\quad = f(-0.99) - f(-1)$

$\quad = [(-0.99)^4 + 1] - [(-1)^4 + 1] \approx -0.0394$

$dy = f'(x)\, dx$

$\quad = f'(-1)(0.01)$

$\quad = (-4)(0.01) = -0.04$

9. $y = 3x^2 - 4$

$dy = 6x\, dx$

11. $y = 4x^3$

$\dfrac{dy}{dx} = 12x^2$

$dy = 12x^2\, dx$

13. $y = \dfrac{x + 1}{2x - 1}$

$dy = \dfrac{-3}{(2x - 1)^2}\, dx$

15. $y = \sqrt{x} = x^{1/2}$

$\dfrac{dy}{dx} = \dfrac{1}{2\sqrt{x}}$

$dy = \dfrac{1}{2\sqrt{x}}\, dx$

17. $y = x\sqrt{1 - x^2}$

$dy = \left(x\dfrac{-x}{\sqrt{1 - x^2}} + \sqrt{1 - x^2} \right) dx = \dfrac{1 - 2x^2}{\sqrt{1 - x^2}}\, dx$

19. (a) $f(1.9) = f(2 - 0.1) \approx f(2) + f'(2)(-0.1)$

$\qquad\qquad\qquad\qquad \approx 1 + (1)(-0.1) = 0.9$

(b) $f(2.04) = f(2 + 0.04) \approx f(2) + f'(2)(0.04)$

$\qquad\qquad\qquad\qquad \approx 1 + (1)(0.04) = 1.04$

21. (a) $f(1.9) = f(2 - 0.1) \approx f(2) + f'(2)(-0.1)$

$\qquad \approx 1 + \left(-\frac{1}{2}\right)(-0.1) = 1.05$

(b) $f(2.04) = f(2 + 0.04) \approx f(2) + f'(2)(0.04)$

$\qquad \approx 1 + \left(-\frac{1}{2}\right)(0.04) = 0.98$

23. (a) $g(2.93) = g(3 - 0.07) \approx g(3) + g'(3)(-0.07)$

$\qquad \approx 8 + \left(-\frac{1}{2}\right)(-0.07) = 8.035$

(b) $g(3.1) = g(3 + 0.1) \approx g(3) + g'(3)(0.1)$

$\qquad \approx 8 + \left(-\frac{1}{2}\right)(0.1) = 7.95$

25. $A = x^2$

$x = 12$

$\Delta x = dx = \pm\frac{1}{64}$

$dA = 2x\,dx$

$\Delta A \approx dA = 2(12)\left(\pm\frac{1}{64}\right)$

$= \pm\frac{3}{8}$ square inches

27. $A = \pi r^2$

$r = 14$

$\Delta r = dr = \pm\frac{1}{4}$

$\Delta A \approx dA = 2\pi r\,dr = \pi(28)\left(\pm\frac{1}{4}\right)$

$= \pm 7\pi$ square inches

29. $r = 6$ inches

$\Delta r = dr = \pm 0.02$ inches

(a) $V = \frac{4}{3}\pi r^3$

$dV = 4\pi r^2\,dr = 4\pi(6)^2(\pm 0.02) = \pm 2.88\pi$ cubic inches

(b) $S = 4\pi r^2$

$dS = 8\pi r\,dr = 8\pi(6)(\pm 0.02) = \pm 0.96\pi$ square inches

(c) Relative error: $\dfrac{dV}{V} = \dfrac{4\pi r^2\,dr}{(4/3)\pi r^3} = \dfrac{3dr}{r}$

$= \dfrac{3}{6}(0.02) = 0.01 = 1\%$

Relative error: $\dfrac{dS}{S} = \dfrac{8\pi r\,dr}{4\pi r^2} = \dfrac{2dr}{r}$

$= \dfrac{2(0.02)}{6} = 0.00666\ldots = \dfrac{2}{3}\%$

31. (a) $C = 56$ centimeters

$\Delta C = dC = \pm 1.2$ centimeters

$C = 2\pi r \Rightarrow r = \dfrac{C}{2\pi}$

$A = \pi r^2 = \pi\left(\dfrac{C}{2\pi}\right)^2 = \dfrac{1}{4\pi}C^2$

$dA = \dfrac{1}{2\pi}C\,dC = \dfrac{1}{2\pi}(56)(\pm 1.2) = \dfrac{33.6}{\pi}$

$\dfrac{dA}{A} = \dfrac{33.6/\pi}{[1/(4\pi)](56)^2} \approx 0.042857 = 4.2857\%$

(b) $\dfrac{dA}{A} = \dfrac{(1/2\pi)C\,dC}{(1/4\pi)C^2} = \dfrac{2dC}{C} \le 0.03$

$\dfrac{dC}{C} \le \dfrac{0.03}{2} = 0.015 = 1.5\%$

33. $V = \pi r^2 h = 40\pi r^2,\ r = 5$ cm, $h = 40$ cm, $dr = 0.2$ cm

$\Delta V \approx dV = 80\pi r\,dr = 80\pi(5)(0.2) = 80\pi$ cm^3

35. $E = IR$

$R = \dfrac{E}{I}$

$dR = -\dfrac{E}{I^2}dI$

$\dfrac{dR}{R} = \dfrac{-(E/I^2)dI}{E/I} = -\dfrac{dI}{I}$

$\left|\dfrac{dR}{R}\right| = \left|-\dfrac{dI}{I}\right| = \left|\dfrac{dI}{I}\right|$

37. Let $f(x) = \sqrt{x}$, $x = 100$, $dx = -0.6$.

$$f(x + \Delta x) \approx f(x) + f'(x)\,dx$$

$$= \sqrt{x} + \frac{1}{2\sqrt{x}}\,dx$$

$$f(x + \Delta x) = \sqrt{99.4}$$

$$\approx \sqrt{100} + \frac{1}{2\sqrt{100}}(-0.6) = 9.97$$

Using a calculator: $\sqrt{99.4} \approx 9.96995$

41. Let $f(x) = \sqrt{x}$, $x = 4$, $dx = 0.02$, $f'(x) = 1/(2\sqrt{x})$.

Then

$$f(4.02) \approx f(4) + f'(4)\,dx$$

$$\sqrt{4.02} \approx \sqrt{4} + \frac{1}{2\sqrt{4}}(0.02) = 2 + \frac{1}{4}(0.02).$$

39. Let $f(x) = \sqrt[4]{x}$, $x = 625$, $dx = -1$.

$$f(x + \Delta x) \approx f(x) + f'(x)\,dx = \sqrt[4]{x} + \frac{1}{4\sqrt[4]{x^3}}\,dx$$

$$f(x + \Delta x) = \sqrt[4]{624} \approx \sqrt[4]{625} + \frac{1}{4(\sqrt[4]{625})^3}(-1)$$

$$= 5 - \frac{1}{500} = 4.998$$

Using a calculator, $\sqrt[4]{624} \approx 4.9980$.

43. $f(x) = \sqrt{x + 4}$

$$f'(x) = \frac{1}{2\sqrt{x + 4}}$$

At $(0, 2)$, $f(0) = 2$, $f'(0) = \frac{1}{4}$.

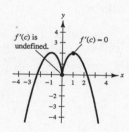

Tangent line: $y - 2 = \frac{1}{4}(x - 0)$

$$y = \frac{1}{4}x + 2$$

45. In general, when $\Delta x \to 0$, dy approaches Δy.

47. True

49. True

Review Exercises for Chapter 5

1. A number c in the domain of f is a critical number if $f'(c) = 0$ or f' is undefined at c.

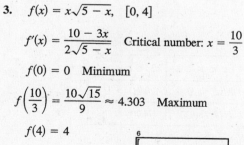

3. $f(x) = x\sqrt{5 - x}$, $[0, 4]$

$$f'(x) = \frac{10 - 3x}{2\sqrt{5 - x}}$$ Critical number: $x = \frac{10}{3}$

$f(0) = 0$ Minimum

$f\left(\frac{10}{3}\right) = \frac{10\sqrt{15}}{9} \approx 4.303$ Maximum

$f(4) = 4$

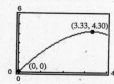

5. Yes. $f(-3) = f(2) = 0$. f is continuous on $[-3, 2]$, differentiable on $(-3, 2)$.

$$f'(x) = (x + 3)(3x - 1) = 0 \text{ for } x = \frac{1}{3}.$$

$c = \frac{1}{3}$ satisfies $f'(c) = 0$.

7. $f(x) = 3 - |x - 4|$

(a)

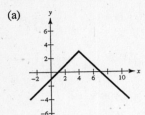

$f(1) = f(7) = 0$

(b) f is not differentiable at $x = 4$.

9. $f(x) = x^{2/3}, 1 \le x \le 8$

$$f'(x) = \frac{2}{3}x^{-1/3}$$

$$\frac{f(b) - f(a)}{b - a} = \frac{4 - 1}{8 - 1} = \frac{3}{7}$$

$$f'(c) = \frac{2}{3}c^{-1/3} = \frac{3}{7}$$

$$c = \left(\frac{14}{9}\right)^3 = \frac{2744}{729} \approx 3.764$$

11. On $[0, 2], f(x) = 9 - x^2, f'(x) = -2x$.

$$f'(x) = \frac{f(2) - f(0)}{2 - 0} = \frac{5 - 9}{2} = -2$$

$$-2x = -2$$

$$x = 1$$

$$f'(1) = -2$$

13. $f(x) = Ax^2 + Bx + C$

$$f'(x) = 2Ax + B$$

$$\frac{f(x_2) - f(x_1)}{x_2 - x_1} = \frac{A(x_2^2 - x_1^2) + B(x_2 - x_1)}{x_2 - x_1}$$

$$= A(x_1 + x_2) + B$$

$$f'(c) = 2Ac + B = A(x_1 + x_2) + B$$

$$2Ac = A(x_1 + x_2)$$

$$c = \frac{x_1 + x_2}{2} = \text{Midpoint of } [x_1, x_2]$$

15. $f(x) = (x - 1)^2(x - 3)$

$$f'(x) = (x - 1)^2(1) + (x - 3)(2)(x - 1)$$

$$= (x - 1)(3x - 7)$$

Critical numbers: $x = 1$ and $x = \frac{7}{3}$

Interval:	$-\infty < x < 1$	$1 < x < \frac{7}{3}$	$\frac{7}{3} < x < \infty$
Sign of $f'(x)$:	$f'(x) > 0$	$f'(x) < 0$	$f'(x) > 0$
Conclusion:	Increasing	Decreasing	Increasing

17. $h(x) = \sqrt{x}(x - 3) = x^{3/2} - 3x^{1/2}$

Domain: $(0, \infty)$

$$h'(x) = \frac{3}{2}x^{1/2} - \frac{3}{2}x^{-1/2}$$

$$= \frac{3}{2}x^{-1/2}(x - 1) = \frac{3(x - 1)}{2\sqrt{x}}$$

Critical number: $x = 1$

Interval:	$0 < x < 1$	$1 < x < \infty$
Sign of $h'(x)$:	$h'(x) < 0$	$h'(x) > 0$
Conclusion:	Decreasing	Increasing

19. $h(t) = \frac{1}{4}t^4 - 8t$

$h'(t) = t^3 - 8 = 0$ when $t = 2$.

Relative minimum: $(2, -12)$

Test Interval:	$-\infty < t < 2$	$2 < t < \infty$
Sign of $h'(t)$:	$h'(t) < 0$	$h'(t) > 0$
Conclusion:	Decreasing	Increasing

21. $f(x) = -x^3 + 6x^2 - 9x + 1$

$$f'(x) = -3x^2 + 12x - 9$$

$$f''(x) = -6x + 12 = 0 \implies x = 2.$$

Because f'' changes sign at $x = 2$, $(2, -1)$ is a point of inflection.

23. $g(x) = 2x^2(1 - x^2)$

$g'(x) = -4x(2x^2 - 1)$ Critical numbers: $x = 0, \pm\dfrac{1}{\sqrt{2}}$

$g''(x) = 4 - 24x^2$

$g''(0) = 4 > 0$ Relative minimum at $(0, 0)$

$g''\left(\pm\dfrac{1}{\sqrt{2}}\right) = -8 < 0$ Relative maximums at $\left(\pm\dfrac{1}{\sqrt{2}}, \dfrac{1}{2}\right)$

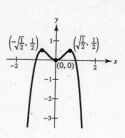

25.

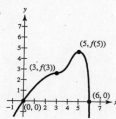

27. The first derivative is positive and the second derivative is negative. The graph is increasing and is concave down.

29. (a) $D = 0.00375t^4 - 0.2517t^3 + 5.179t^2 - 21.89t + 94.6$

(b)

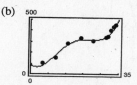

(c) Maximum in 2004

Minimum in 1972 ($t \approx 2.6$)

(d) The increase is greatest in 2004.

31. $\displaystyle\lim_{x \to \infty} \frac{2x^2}{3x^2 + 5} = \lim_{x \to \infty} \frac{2}{3 + 5/x^2} = \frac{2}{3}$

33. $\displaystyle\lim_{x \to -\infty} \left(7 + \frac{1}{x}\right) = 7 + 0 = 7$

35. $h(x) = \dfrac{2x + 3}{x - 4}$

Discontinuity: $x = 4$

$\displaystyle\lim_{x \to \infty} \frac{2x + 3}{x - 4} = \lim_{x \to \infty} \frac{2 + (3/x)}{1 - (4/x)} = 2$

Vertical asymptote: $x = 4$

Horizontal asymptote: $y = 2$

37. $f(x) = \dfrac{3}{x} - 2$

Discontinuity: $x = 0$

$\displaystyle\lim_{x \to \infty} \left(\frac{3}{x} - 2\right) = -2$

Vertical asymptote: $x = 0$

Horizontal asymptote: $y = -2$

39. $f(x) = x^3 + \dfrac{243}{x}$

Relative minimum: $(3, 108)$

Relative maximum: $(-3, -108)$

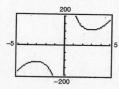

Vertical asymptote: $x = 0$

41. $f(x) = \dfrac{x - 1}{1 + 3x^2}$

Relative minimum: $(-0.155, -1.077)$

Relative maximum: $(2.155, 0.077)$

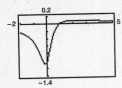

Horizontal asymptote: $y = 0$

43. $f(x) = 4x - x^2 = x(4 - x)$

Domain: $(-\infty, \infty)$; Range: $(-\infty, 4]$

$f'(x) = 4 - 2x = 0$ when $x = 2$.

$f''(x) = -2$

Therefore, $(2, 4)$ is a relative maximum.

Intercepts: $(0, 0)$, $(4, 0)$

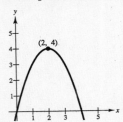

45. $f(x) = x\sqrt{16 - x^2}$, Domain: $[-4, 4]$, Range: $[-8, 8]$

Domain: $[-4, 4]$; Range: $[-8, 8]$

$f'(x) = \dfrac{16 - 2x^2}{\sqrt{16 - x^2}} = 0$ when $x = \pm 2\sqrt{2}$ and undefined when $x = \pm 4$.

$f''(x) = \dfrac{2x(x^2 - 24)}{(16 - x^2)^{3/2}}$

$f''(-2\sqrt{2}) > 0$

Therefore, $\left(-2\sqrt{2}, -8\right)$ is a relative minimum.

$f''(2\sqrt{2}) < 0$

Therefore, $\left(2\sqrt{2}, 8\right)$ is a relative maximum.

Point of inflection: $(0, 0)$

Intercepts: $(-4, 0)$, $(0, 0)$, $(4, 0)$

Symmetry with respect to origin

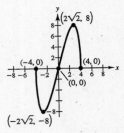

47. $f(x) = (x - 1)^3(x - 3)^2$

Domain: $(-\infty, \infty)$; Range: $(-\infty, \infty)$

$f'(x) = (x - 1)^2(x - 3)(5x - 11) = 0$ when $x = 1, \dfrac{11}{5}, 3$.

$f''(x) = 4(x - 1)(5x^2 - 22x + 23) = 0$ when $x = 1, \dfrac{11 \pm \sqrt{6}}{5}$.

$f''(3) > 0$

Therefore, $(3, 0)$ is a relative minimum.

$f''\left(\dfrac{11}{5}\right) < 0$

Therefore, $\left(\dfrac{11}{5}, \dfrac{3456}{3125}\right)$ is a relative maximum.

Points of inflection: $(1, 0)$, $\left(\dfrac{11 - \sqrt{6}}{5}, 0.60\right)$, $\left(\dfrac{11 + \sqrt{6}}{5}, 0.46\right)$

Intercepts: $(0, -9)$, $(1, 0)$, $(3, 0)$

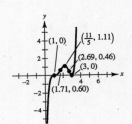

49. $f(x) = x^{1/3}(x + 3)^{2/3}$

Domain: $(-\infty, \infty)$; Range: $(-\infty, \infty)$

$f'(x) = \dfrac{x + 1}{(x + 3)^{1/3}x^{2/3}} = 0$ when $x = -1$ and undefined when $x = -3, 0$.

$f''(x) = \dfrac{-2}{x^{5/3}(x + 3)^{4/3}}$ is undefined when $x = 0, -3$.

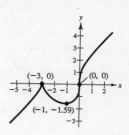

By the First Derivative Test $(-3, 0)$ is a relative maximum and $\left(-1, -\sqrt[3]{4}\right)$ is a relative minimum. $(0, 0)$ is a point of inflection.

Intercepts: $(-3, 0), (0, 0)$

51. $f(x) = \dfrac{x + 1}{x - 1}$

Domain: $(-\infty, 1), (1, \infty)$; Range: $(-\infty, 1), (1, \infty)$

$f'(x) = \dfrac{-2}{(x - 1)^2} < 0$ if $x \neq 1$.

$f''(x) = \dfrac{4}{(x - 1)^3}$

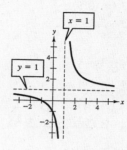

Horizontal asymptote: $y = 1$

Vertical asymptote: $x = 1$

Intercepts: $(-1, 0), (0, -1)$

53. $f(x) = \dfrac{4}{1 + x^2}$

Domain: $(-\infty, \infty)$; Range: $(0, 4]$

$f'(x) = \dfrac{-8x}{(1 + x^2)^2} = 0$ when $x = 0$.

$f''(x) = \dfrac{-8(1 - 3x^2)}{(1 + x^2)^3} = 0$ when $x = \pm\dfrac{\sqrt{3}}{3}$.

$f''(0) < 0$

Therefore, $(0, 4)$ is a relative maximum.

Points of inflection: $\left(\pm\sqrt{3}/3, 3\right)$

Intercept: $(0, 4)$

Symmetric to the y-axis

Horizontal asymptote: $y = 0$

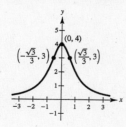

55. $f(x) = x^3 + x + \dfrac{4}{x}$

Domain: $(-\infty, 0), (0, \infty)$; Range: $(-\infty, -6], [6, \infty)$

$f'(x) = 3x^2 + 1 - \dfrac{4}{x^2} = \dfrac{3x^4 + x^2 - 4}{x^2} = 0$ when $x = \pm 1$.

$f''(x) = 6x + \dfrac{8}{x^3} = \dfrac{6x^4 + 8}{x^3} \neq 0$

$f''(-1) < 0$

Therefore, $(-1, -6)$ is a relative maximum.

$f''(1) > 0$

Therefore, $(1, 6)$ is a relative minimum.

Vertical asymptote: $x = 0$

Symmetric with respect to origin

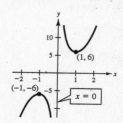

57. $f(x) = |x^2 - 9|$

Domain: $(-\infty, \infty)$; Range: $[0, \infty)$

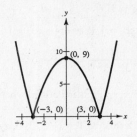

$f'(x) = \dfrac{2x(x^2 - 9)}{|x^2 - 9|} = 0$ when $x = 0$ and is undefined when $x = \pm 3$.

$f''(x) = \dfrac{2(x^2 - 9)}{|x^2 - 9|}$ is undefined at $x = \pm 3$.

$\quad f''(0) < 0$

Therefore, $(0, 9)$ is a relative maximum.

Relative minima: $(\pm 3, 0)$

Intercepts: $(\pm 3, 0)$, $(0, 9)$

Symmetric to the y-axis

59. $x^2 + 4y^2 - 2x - 16y + 13 = 0$

(a) $(x^2 - 2x + 1) + 4(y^2 - 4y + 4) = -13 + 1 + 16$

$\qquad (x - 1)^2 + 4(y - 2)^2 = 4$

$\qquad \dfrac{(x - 1)^2}{4} + \dfrac{(y - 2)^2}{1} = 1$

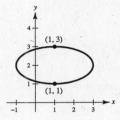

The graph is an ellipse:

Maximum: $(1, 3)$

Minimum: $(1, 1)$

(b) $x^2 + 4y^2 - 2x - 16y + 13 = 0$

$\quad 2x + 8y\dfrac{dy}{dx} - 2 - 16\dfrac{dy}{dx} = 0$

$\qquad \dfrac{dy}{dx}(8y - 16) = 2 - 2x$

$\qquad\quad \dfrac{dy}{dx} = \dfrac{2 - 2x}{8y - 16} = \dfrac{1 - x}{4y - 8}$

The critical numbers are $x = 1$ and $y = 2$. These correspond to the points $(1, 1)$, $(1, 3)$, $(2, -1)$, and $(2, 3)$. Hence, the maximum is $(1, 3)$ and the minimum is $(1, 1)$.

61. Let $t = 0$ at noon.

$\qquad L = d^2 = (100 - 12t)^2 + (-10t)^2 = 10,000 - 2400t + 244t^2$

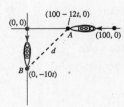

$\qquad \dfrac{dL}{dt} = -2400 + 488t = 0$ when $t = \dfrac{300}{61} \approx 4.92$ hr.

Ship A at $(40.98, 0)$; Ship B at $(0, -49.18)$

$\qquad d^2 = 10,000 - 2400t + 244t^2$

$\qquad\quad \approx 4098.36$ when $t \approx 4.92 \approx 4{:}55$ P.M..

$\qquad d \approx 64$ km

63. We have points $(0, y)$, $(x, 0)$, and $(1, 8)$. Thus,

$$m = \frac{y-8}{0-1} = \frac{0-8}{x-1} \text{ or } y = \frac{8x}{x-1}.$$

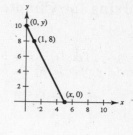

Let $f(x) = L^2 = x^2 + \left(\frac{8x}{x-1}\right)^2.$

$$f'(x) = 2x + 128\left(\frac{x}{x-1}\right)\left[\frac{(x-1)-x}{(x-1)^2}\right] = 0$$

$$x - \frac{64x}{(x-1)^3} = 0$$

$x[(x-1)^3 - 64] = 0$ when $x = 0, 5$ (minimum).

Vertices of triangle: $(0, 0)$, $(5, 0)$, $(0, 10)$

65. $A = $ (Average of bases)(Height)

$$= \left(\frac{x+s}{2}\right)\frac{\sqrt{3s^2 + 2sx - x^2}}{2} \text{ (see figure)}$$

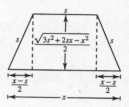

$$\frac{dA}{dx} = \frac{1}{4}\left[\frac{(s-x)(s+x)}{\sqrt{3s^2 + 2sx - x^2}} + \sqrt{3s^2 + 2sx - x^2}\right]$$

$$= \frac{2(2s-x)(s+x)}{4\sqrt{3s^2 + 2sx - x^2}} = 0 \text{ when } x = 2s.$$

A is a maximum when $x = 2s$.

67. Total cost = (Cost per hour)(Number of hours)

$$T = \left(\frac{v^2}{600} + 5\right)\left(\frac{110}{v}\right) = \frac{11v}{60} + \frac{550}{v}$$

$$\frac{dT}{dv} = \frac{11}{60} - \frac{550}{v^2} = \frac{11v^2 - 33,000}{60v^2}$$

$$= 0 \text{ when } v = \sqrt{3000} = 10\sqrt{30} \approx 54.8 \text{ mph.}$$

$$\frac{d^2T}{dv^2} = \frac{1100}{v^3} > 0 \text{ when } v = 10\sqrt{30} \text{ so this value yields a minimum.}$$

69. $y = (3x^2 - 2)^3$

$$\frac{dy}{dx} = 3(3x^2 - 2)^2(6x)$$

$$dy = 18x(3x^2 - 2)^2\, dx$$

71.
$$S = 4\pi r^2.\ dr = \Delta r = \pm 0.025$$

$$dS = 8\pi r\, dr = 8\pi(9)(\pm 0.025)$$

$$= \pm 1.8\pi \text{ square cm}$$

$$\frac{dS}{S}(100) = \frac{8\pi r\, dr}{4\pi r^2}(100) = \frac{2\, dr}{r}(100)$$

$$= \frac{2(\pm 0.025)}{9}(100) \approx \pm 0.56\%$$

$$V = \frac{4}{3}\pi r^3$$

$$dV = 4\pi r^2\, dr = 4\pi(9)^2(\pm 0.025)$$

$$= \pm 8.1\pi \text{ cubic cm}$$

$$\frac{dV}{V}(100) = \frac{4\pi r^2\, dr}{(4/3)\pi r^3}(100) = \frac{3\, dr}{r}(100)$$

$$= \frac{3(\pm 0.025)}{9}(100) \approx \pm 0.83\%$$

Problem Solving for Chapter 5

1. Assume $y_1 < d < y_2$. Let $g(x) = f(x) - d(x - a)$. g is continuous on $[a, b]$ and therefore has a minimum $(c, g(c))$ on $[a, b]$. The point c cannot be an endpoint of $[a, b]$ because

   $$g'(a) = f'(a) - d = y_1 - d < 0$$
   $$g'(b) = f'(b) - d = y_2 - d > 0$$

   Hence, $a < c < b$ and $g'(c) = 0 \implies f'(c) = d$.

3. (a) For $a = -3, -2, -1, 0, p$ has a relative maximum at $(0, 0)$.

   For $a = 1, 2, 3, p$ has a relative maximum at $(0, 0)$ and 2 relative minima.

   (b) $p'(x) = 4ax^3 - 12x = 4x(ax^2 - 3) = 0 \implies x = 0, \pm\sqrt{\dfrac{3}{a}}$

   $$p''(x) = 12ax^2 - 12 = 12(ax^2 - 1)$$

   For $x = 0, p''(0) = -12 < 0 \implies p$ has a relative maximum at $(0, 0)$.

   (c) If $a > 0, x = \pm\sqrt{\dfrac{3}{a}}$ are the remaining critical numbers.

   $$p''\left(\pm\sqrt{\dfrac{3}{a}}\right) = 12a\left(\dfrac{3}{a}\right) - 12 = 24 > 0 \implies p \text{ has relative minima for } a > 0.$$

   (d) $(0, 0)$ lies on $y = -3x^2$.

   Let $x = \pm\sqrt{\dfrac{3}{a}}$. Then

   $$p(x) = a\left(\dfrac{3}{a}\right)^2 - 6\left(\dfrac{3}{a}\right) = \dfrac{9}{a} - \dfrac{18}{a} = -\dfrac{9}{a}.$$

   Thus, $y = -\dfrac{9}{a} = -3\left(\pm\sqrt{\dfrac{3}{a}}\right)^2 = -3x^2$ is satisfied by all the relative extrema of p.

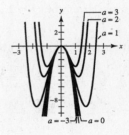

5. $p(x) = x^4 + ax^2 + 1$

   (a) $p'(x) = 4x^3 + 2ax = 2x(2x^2 + a)$

   $$p''(x) = 12x^2 + 2a$$

   For $a \geq 0$, there is one relative minimum at $(0, 1)$.

   (b) For $a < 0$, there is a relative maximum at $(0, 1)$.

   (c) For $a < 0$, there are two relative minima at $x = \pm\sqrt{-\dfrac{a}{2}}$.

   (d) There are either 1 or 3 critical points. The above analysis shows that there cannot be exactly two relative extrema.

7. $f(x) = \dfrac{c}{x} + x^2$

$$f'(x) = -\frac{c}{x^2} + 2x = 0 \implies \frac{c}{x^2} = 2x \implies x^3 = \frac{c}{2} \implies x = \sqrt[3]{\frac{c}{2}}$$

$$f''(x) = \frac{2c}{x^3} + 2$$

If $c = 0, f(x) = x^2$ has a relative minimum, but no relative maximum.

If $c > 0, x = \sqrt[3]{\dfrac{c}{2}}$ is a relative minimum, because $f''\left(\sqrt[3]{\dfrac{c}{2}}\right) > 0$.

If $c < 0, x = \sqrt[3]{\dfrac{c}{2}}$ is a relative minimum too.

Answer: all c.

9. (a) Let $M > 0$ be given. Take $N = \sqrt{M}$. Then whenever $x > N = \sqrt{M}$, you have

$$f(x) = x^2 > M.$$

(b) Let $\varepsilon > 0$ be given. Let $M = \sqrt{\dfrac{1}{\varepsilon}}$. Then whenever $x > M = \sqrt{\dfrac{1}{\varepsilon}}$, you have

$$x^2 > \frac{1}{\varepsilon} \implies \frac{1}{x^2} < \varepsilon \implies \left|\frac{1}{x^2} - 0\right| < \varepsilon.$$

(c) Let $\varepsilon > 0$ be given. There exists $N > 0$ such that $|f(x) - L| < \varepsilon$ whenever $x > N$.
Let $\delta = \dfrac{1}{N}$. Let $x = \dfrac{1}{y}$.

If $0 < y < \delta = \dfrac{1}{N}$, then $\dfrac{1}{x} < \dfrac{1}{N} \implies x > N$ and

$$|f(x) - L| = \left|f\left(\frac{1}{y}\right) - L\right| < \varepsilon.$$

11. Let m be the slope of the line, $m < 0$.

$y - q = m(x - p)$

$y = mx + q - mp$

Intercepts: $y = 0 \implies x = \dfrac{mp - q}{m} = p - \dfrac{q}{m} \quad A = \left(p - \dfrac{q}{m}, 0\right)$

$\qquad\qquad x = 0 \implies y = q - mp \quad B = (0, q - mp)$

(a) $f(m) = OA + OB = p - \dfrac{q}{m} + q - mp$

$\qquad f'(m) = \dfrac{q}{m^2} - p = 0 \implies m^2 = \dfrac{q}{p} \implies m = -\sqrt{(q/p)}.$

$\qquad$ Thus $f\left(-\sqrt{(q/p)}\right) = p + q\sqrt{(p/q)} + q + p\sqrt{(q/p)}$

$\qquad\qquad\qquad\qquad\quad = p + q + 2\sqrt{pq}.$

—CONTINUED—

11. **—CONTINUED—**

(b) $g(m) = OA \cdot OB = \left(p - \dfrac{q}{m}\right)(q - mp)$

$$= pq - mp^2 - \dfrac{q^2}{m} + pq$$

$g'(m) = -p^2 + \dfrac{q^2}{m^2} = 0 \implies m^2 = \dfrac{q^2}{p^2} \implies m = -\dfrac{q}{p}.$

$g\left(-\dfrac{q}{p}\right) = pq + \left(\dfrac{q}{p}\right)p^2 + q^2\left(\dfrac{p}{q}\right) + pq$

$\qquad = 4pq$

(c) Let $h(m) = (AB)^2 = \left(p - \dfrac{q}{m}\right)^2 + (q - mp)^2$

$$= p^2 - \dfrac{2pq}{m} + \dfrac{q^2}{m^2} + q^2 - 2pqm + m^2p^2$$

$h'(m) = \dfrac{2pq}{m^2} - \dfrac{2q^2}{m^3} - 2pq + 2mp^2 = 0$

$2pqm - 2q^2 - 2pqm^3 + 2m^4p^2 = 0$

$(pm - q)(pm^3 + q) = 0$

Thus, $m = \left(\dfrac{-q}{p}\right)^{1/3}$ and

$AB^2 = \left(p - \dfrac{q}{m}\right)^2 + (q - mp)^2$

$\qquad = (p + p^{1/3}q^{2/3})^2 + (q + p^{2/3}q^{1/3})^2$

$\qquad = p^2 + 2p^{4/3}q^{2/3} + p^{2/3}q^{4/3} + q^2 + 2p^{2/3}q^{4/3} + p^{4/3}q^{2/3}$

$\qquad = p^2 + q^2 + 3p^{4/3}q^{2/3} + 3p^{2/3}q^{4/3}$

$\qquad = (p^{2/3} + q^{2/3})^3$

$\implies AB = (p^{2/3} + q^{2/3})^{3/2}.$

13. Place the triangle as indicated in the figure. Let (x_0, y_0) be between A and B.

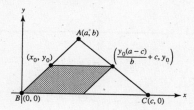

$$\text{Area} = f(y_0) = (\text{base})(\text{height})$$

$$= \left(\frac{y_0(a-c)}{b} + c - x_0\right)(y_0)$$

$$= \left(\frac{y_0(a-c)}{b} + c - \frac{ay_0}{b}\right)y_0$$

$$= \left(c - \frac{cy_0}{b}\right)y_0$$

$$= cy_0 - \frac{cy_0^2}{b}$$

$$f'(y_0) = c - \frac{2cy_0}{b} = 0 \implies y_0 = \frac{b}{2}, \quad x_0 = \frac{a}{2}$$

and area $= \frac{1}{4}bc = \frac{1}{2}$ area $\triangle ABC$

Without calculus, observe from the figure that the maximum area is obtained by using the midpoints of each side.

17. (a)

x	0	0.5	1	2
$\sqrt{1+x}$	1	1.2247	1.4142	1.7321
$\frac{1}{2}x + 1$	1	1.25	1.5	2

15. The line has equation $\frac{x}{3} + \frac{y}{4} = 1$ or $y = -\frac{4}{3}x + 4$.

Rectangle:

$$\text{Area} = A = xy = x\left(-\frac{4}{3}x + 4\right) = -\frac{4}{3}x^2 + 4x.$$

$$A'(x) = -\frac{8}{3}x + 4 = 0 \implies \frac{8}{3}x = 4 \implies x = \frac{3}{2}$$

Dimensions: $\frac{3}{2} \times 2$ Calculus was helpful.

Circle: The distance from the center (r, r) to the line $\frac{x}{3} + \frac{y}{4} - 1 = 0$ must be r:

$$r = \frac{\left|\frac{r}{3} + \frac{r}{4} - 1\right|}{\sqrt{\frac{1}{9} + \frac{1}{16}}} = \frac{12}{5}\left|\frac{7r - 12}{12}\right| = \frac{|7r - 12|}{5}$$

$$5r = |7r - 12| \implies r = 1 \text{ or } r = 6.$$

Clearly, $r = 1$.

Semicircle: The center lies on the line $\frac{x}{3} + \frac{y}{4} = 1$ and satisfies $x = y = r$.

Thus $\frac{r}{3} + \frac{r}{4} = 1 \implies \frac{7}{12}r = 1 \implies r = \frac{12}{7}$.

No calculus necessary.

(b) Let $f(x) = \sqrt{1 + x}$. Using the Mean Value Theorem on the interval $[0, x]$, there exists c, $0 < c < x$, satisfying

$$f'(c) = \frac{1}{2\sqrt{1 + c}} = \frac{f(x) - f(0)}{x - 0} = \frac{\sqrt{1 + x} - 1}{x}.$$

Thus $\sqrt{1 + x} = \frac{x}{2\sqrt{1 + c}} + 1 < \frac{x}{2} + 1$

(because $\sqrt{1 + c} > 1$).

CHAPTER 6
Integration

CHAPTER 6
Integration

Section 6.1 Antiderivatives and Indefinite Integration

1. $\dfrac{d}{dx}\left(\dfrac{3}{x^3}+C\right)=\dfrac{d}{dx}(3x^{-3}+C)=-9x^{-4}=\dfrac{-9}{x^4}$

3. $\dfrac{d}{dx}\left(\dfrac{1}{3}x^3-4x+C\right)=x^2-4=(x-2)(x+2)$

5. $\dfrac{dy}{dt}=3t^2$

$y=t^3+C$

Check: $\dfrac{d}{dt}[t^3+C]=3t^2$

7. $\dfrac{dy}{dx}=x^{3/2}$

$y=\dfrac{2}{5}x^{5/2}+C$

Check: $\dfrac{d}{dx}\left[\dfrac{2}{5}x^{5/2}+C\right]=x^{3/2}$

	Given	*Rewrite*	*Integrate*	*Simplify*
9.	$\displaystyle\int\sqrt[3]{x}\,dx$	$\displaystyle\int x^{1/3}\,dx$	$\dfrac{x^{4/3}}{4/3}+C$	$\dfrac{3}{4}x^{4/3}+C$
11.	$\displaystyle\int\dfrac{1}{x\sqrt{x}}\,dx$	$\displaystyle\int x^{-3/2}\,dx$	$\dfrac{x^{-1/2}}{-1/2}+C$	$-\dfrac{2}{\sqrt{x}}+C$
13.	$\displaystyle\int\dfrac{1}{2x^3}\,dx$	$\dfrac{1}{2}\displaystyle\int x^{-3}\,dx$	$\dfrac{1}{2}\left(\dfrac{x^{-2}}{-2}\right)+C$	$-\dfrac{1}{4x^2}+C$

15. $\displaystyle\int(x+3)dx=\dfrac{x^2}{2}+3x+C$

Check: $\dfrac{d}{dx}\left[\dfrac{x^2}{2}+3x+C\right]=x+3$

17. $\displaystyle\int(2x-3x^2)dx=x^2-x^3+C$

Check: $\dfrac{d}{dx}[x^2-x^3+C]=2x-3x^2$

19. $\displaystyle\int(x^3+2)\,dx=\dfrac{1}{4}x^4+2x+C$

Check: $\dfrac{d}{dx}\left(\dfrac{1}{4}x^4+2x+C\right)=x^3+2$

21. $\displaystyle\int(x^{3/2}+2x+1)\,dx=\dfrac{2}{5}x^{5/2}+x^2+x+C$

Check: $\dfrac{d}{dx}\left(\dfrac{2}{5}x^{5/2}+x^2+x+C\right)=x^{3/2}+2x+1$

23. $\displaystyle\int\sqrt[3]{x^2}\,dx=\int x^{2/3}\,dx=\dfrac{x^{5/3}}{5/3}+C=\dfrac{3}{5}x^{5/3}+C$

Check: $\dfrac{d}{dx}\left(\dfrac{3}{5}x^{5/3}+C\right)=x^{2/3}=\sqrt[3]{x^2}$

25. $\displaystyle\int\dfrac{1}{x^3}\,dx=\int x^{-3}\,dx=\dfrac{x^{-2}}{-2}+C=-\dfrac{1}{2x^2}+C$

Check: $\dfrac{d}{dx}\left(-\dfrac{1}{2x^2}+C\right)=\dfrac{1}{x^3}$

27. $\displaystyle\int\dfrac{x^2+x+1}{\sqrt{x}}\,dx=\int(x^{3/2}+x^{1/2}+x^{-1/2})\,dx=\dfrac{2}{5}x^{5/2}+\dfrac{2}{3}x^{3/2}+2x^{1/2}+C=\dfrac{2}{15}x^{1/2}(3x^2+5x+15)+C$

Check: $\dfrac{d}{dx}\left(\dfrac{2}{5}x^{5/2}+\dfrac{2}{3}x^{3/2}+2x^{1/2}+C\right)=x^{3/2}+x^{1/2}+x^{-1/2}=\dfrac{x^2+x+1}{\sqrt{x}}$

29. $\displaystyle\int (x + 1)(3x - 2)\, dx = \int (3x^2 + x - 2)\, dx$

$$= x^3 + \frac{1}{2}x^2 - 2x + C$$

Check: $\dfrac{d}{dx}\left(x^3 + \dfrac{1}{2}x^2 - 2x + C\right) = 3x^2 + x - 2$

$$= (x + 1)(3x - 2)$$

31. $\displaystyle\int y^2 \sqrt{y}\, dy = \int y^{5/2}\, dy = \frac{2}{7}y^{7/2} + C$

Check: $\dfrac{d}{dy}\left(\dfrac{2}{7}y^{7/2} + C\right) = y^{5/2} = y^2\sqrt{y}$

33. $\displaystyle\int dx = \int 1\, dx = x + C$

Check: $\dfrac{d}{dx}(x + C) = 1$

35.

(graph showing curves labeled $C = 3$, $C = 0$, $C = -2$)

37. $f'(x) = 2$

$\quad f(x) = 2x + C$

(graph showing lines $f(x) = 2x + 2$, $f(x) = 2x$, and f')

Answers will vary.

39. $f'(x) = 1 - x^2$

$\quad f(x) = x - \dfrac{x^3}{3} + C$

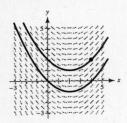

(graph labeled $f(x) = -\dfrac{x^3}{3} + x$, $f(x) = -\dfrac{x^3}{3} + x + 3$, and f')

Answers will vary.

41. $\dfrac{dy}{dx} = 2x - 1,\ (1, 1)$

$\quad y = \displaystyle\int (2x - 1)\, dx = x^2 - x + C$

$\quad 1 = (1)^2 - (1) + C \implies C = 1$

$\quad y = x^2 - x + 1$

43. $\dfrac{dy}{dx} = 3x^2 - 1$

$\quad y = \displaystyle\int (3x^2 - 1)\, dx = x^3 - x + C$

$\quad 2 = 0^3 - 0 + C \implies C = 2$

$\quad y = x^3 - x + 2$

45. (a) Answers will vary.

(slope field graph)

(b) $\dfrac{dy}{dx} = \dfrac{1}{2}x - 1,\ (4, 2)$

$\quad y = \dfrac{x^2}{4} - x + C$

$\quad 2 = \dfrac{4^2}{4} - 4 + C$

$\quad 2 = C$

$\quad y = \dfrac{x^2}{4} - x + 2$

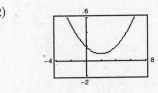

47. $f'(x) = 4x, f(0) = 6$

$$f(x) = \int 4x \, dx = 2x^2 + C$$

$$f(0) = 6 = 2(0)^2 + C \implies C = 6$$

$$f(x) = 2x^2 + 6$$

49. $h'(t) = 8t^3 + 5, h(1) = -4$

$$h(t) = \int (8t^3 + 5) dt = 2t^4 + 5t + C$$

$$h(1) = -4 = 2 + 5 + C \implies C = -11$$

$$h(t) = 2t^4 + 5t - 11$$

51. $f''(x) = 2$

$$f'(2) = 5$$

$$f(2) = 10$$

$$f'(x) = \int 2 \, dx = 2x + C_1$$

$$f'(2) = 4 + C_1 = 5 \implies C_1 = 1$$

$$f'(x) = 2x + 1$$

$$f(x) = \int (2x + 1) \, dx = x^2 + x + C_2$$

$$f(2) = 6 + C_2 = 10 \implies C_2 = 4$$

$$f(x) = x^2 + x + 4$$

53. (a) $h(t) = \int (1.5t + 5) \, dt = 0.75t^2 + 5t + C$

$$h(0) = 0 + 0 + C = 12 \implies C = 12$$

$$h(t) = 0.75t^2 + 5t + 12$$

(b) $h(6) = 0.75(6)^2 + 5(6) + 12 = 69$ cm

55. $f(0) = -4$. Graph of f' is given.

(a) $f'(4) \approx -1.0$

(b) No. The slopes of the tangent lines are greater than 2 on $[0, 2]$. Therefore, f must increase more than 4 units on $[0, 2]$.

(c) No, $f(5) < f(4)$ because f is decreasing on $[4, 5]$.

(d) f has a relative minimum at $x \approx -0.75$ and $x \approx 6.25$. $f' = 0$ at both these values of x, and f' is negative to the left and positive to the right of each.

(e) f is concave upward when f' is increasing on $(-\infty, 1)$ and $(5, \infty)$. f is concave downward on $(1, 5)$. Points of inflection at $x = 1, 5$.

(f) f'' is a minimum at $x = 3$.

(g)

57. $a(t) = -32$ ft/sec^2

$$v(t) = \int -32 \, dt = -32t + C_1$$

$$v(0) = 60 = C_1$$

$$s(t) = \int (-32t + 60) dt = -16t^2 + 60t + C_2$$

$$s(0) = 6 = C_2$$

$$s(t) = -16t^2 + 60t + 6 \text{ Position function}$$

The ball reaches its maximim height when

$$v(t) = -32t + 60 = 0$$

$$32t = 60$$

$$t = \frac{15}{8} \text{ seconds.}$$

$$s\left(\frac{15}{8}\right) = -16\left(\frac{15}{8}\right)^2 + 60\left(\frac{15}{8}\right) + 6 = 62.25 \text{ feet}$$

59. From Exercise 58, we have:

$$s(t) = -16t^2 + v_0 t$$

$$s'(t) = -32t + v_0 = 0 \text{ when } t = \frac{v_0}{32} = \text{ time to reach maximum height.}$$

$$s\left(\frac{v_0}{32}\right) = -16\left(\frac{v_0}{32}\right)^2 + v_0\left(\frac{v_0}{32}\right) = 550$$

$$-\frac{v_0^2}{64} + \frac{v_0^2}{32} = 550$$

$$v_0^2 = 35{,}200$$

$$v_0 \approx 187.617 \text{ ft/sec}$$

61. $a(t) = -9.8$

$$v(t) = \int -9.8 \, dt = -9.8t + C_1$$

$$v(0) = v_0 = C_1 \implies v(t) = -9.8t + v_0$$

$$f(t) = \int (-9.8t + v_0) \, dt = -4.9t^2 + v_0 t + C_2$$

$$f(0) = s_0 = C_2 \implies f(t) = -4.9t^2 + v_0 t + s_0$$

65. $a = -1.6$

$$v(t) = \int -1.6 \, dt = -1.6t + v_0 = -1.6t,$$

since the stone was dropped, $v_0 = 0$.

$$s(t) = \int (-1.6t) \, dt = -0.8t^2 + s_0$$

$$s(20) = 0 \implies -0.8(20)^2 + s_0 = 0$$

$$s_0 = 320$$

Thus, the height of the cliff is 320 meters.

$$v(t) = -1.6t$$

$$v(20) = -32 \text{ m/sec}$$

69. $v(t) = \dfrac{1}{\sqrt{t}} = t^{-1/2} \quad t > 0$

$$x(t) = \int v(t) \, dt = 2t^{1/2} + C$$

$$x(1) = 4 = 2(1) + C \implies C = 2$$

$$x(t) = 2t^{1/2} + 2 \text{ position function}$$

$$a(t) = v'(t) = -\frac{1}{2}t^{-3/2} = \frac{-1}{2t^{3/2}} \text{ acceleration}$$

71. $v(0) = 45 \text{ mph} = 66 \text{ ft/sec}$

$\quad\quad 30 \text{ mph} = 44 \text{ ft/sec}$

$\quad\quad 15 \text{ mph} = 22 \text{ ft/sec}$

$$a(t) = -a$$

$$v(t) = -at + 66$$

$$s(t) = -\frac{a}{2}t^2 + 66t \text{ (Let } s(0) = 0.)$$

$v(t) = 0$ after car moves 132 ft.

$-at + 66 = 0$ when $t = \dfrac{66}{a}$.

$$s\left(\frac{66}{a}\right) = -\frac{a}{2}\left(\frac{66}{a}\right)^2 + 66\left(\frac{66}{a}\right)$$

$$= 132 \text{ when } a = \frac{33}{2} = 16.5.$$

$$a(t) = -16.5$$

$$v(t) = -16.5t + 66$$

$$s(t) = -8.25t^2 + 66t$$

63. From Exercise 61, $f(t) = -4.9t^2 + 10t + 2$.

$$v(t) = -9.8t + 10 = 0 \text{ (Maximum height when } v = 0.)$$

$$9.8t = 10$$

$$t = \frac{10}{9.8}$$

$$f\left(\frac{10}{9.8}\right) \approx 7.1 \text{ m}$$

67. $x(t) = t^3 - 6t^2 + 9t - 2 \quad 0 \le t \le 5$

(a) $v(t) = x'(t) = 3t^2 - 12t + 9$

$\quad\quad = 3(t^2 - 4t + 3) = 3(t-1)(t-3)$

$\quad\quad a(t) = v'(t) = 6t - 12 = 6(t-2)$

(b) $v(t) > 0$ when $0 < t < 1$ or $3 < t < 5$.

(c) $a(t) = 6(t-2) = 0$ when $t = 2$.

$\quad\quad v(2) = 3(1)(-1) = -3$

(a) $-16.5t + 66 = 44$

$$t = \frac{22}{16.5} \approx 1.333$$

$$s\left(\frac{22}{16.5}\right) \approx 73.33 \text{ ft}$$

(b) $-16.5t + 66 = 22$

$$t = \frac{44}{16.5} \approx 2.667$$

$$s\left(\frac{44}{16.5}\right) \approx 117.33 \text{ ft}$$

(c)

It takes 1.333 seconds to reduce the speed from 45 mph to 30 mph, 1.333 seconds to reduce the speed from 30 mph to 15 mph, and 1.333 seconds to reduce the speed from 15 mph to 0 mph. Each time, less distance is needed to reach the next speed reduction.

73. (a) $v = 0.6139t^3 - 5.525t^2 + 0.0492t + 65.9881$

(b) $s(t) = \int v(t)dt = \frac{0.6139t^4}{4} - \frac{5.525t^3}{3} + \frac{0.0492t^2}{2} + 65.9881t$

(**Note:** Assume $s(0) = 0$ is initial position)

$s(6) \approx 197.9$ feet

75. True

77. True

79. False. For example, $\int x \cdot x \, dx \neq \int x \, dx \cdot \int x \, dx$ because $\frac{x^3}{3} + C \neq \left(\frac{x^2}{2} + C_1\right)\left(\frac{x^2}{2} + C_2\right)$.

81. $f''(x) = 2x$

$f'(x) = x^2 + C$

$f'(2) = 0 \Rightarrow 4 + C = 0 \Rightarrow C = -4$

$f(x) = \frac{x^3}{3} - 4x + C_1$

$f(2) = 0 \Rightarrow \frac{8}{3} - 8 + C_1 = 0 \Rightarrow C_1 = \frac{16}{3}$

Answer: $f(x) = \frac{x^3}{3} - 4x + \frac{16}{3}$

Section 6.2   Area

1. $\displaystyle\sum_{i=1}^{5}(2i + 1) = 2\sum_{i=1}^{5}i + \sum_{i=1}^{5}1 = 2(1 + 2 + 3 + 4 + 5) + 5 = 35$

3. $\displaystyle\sum_{k=0}^{4}\frac{1}{k^2 + 1} = 1 + \frac{1}{2} + \frac{1}{5} + \frac{1}{10} + \frac{1}{17} = \frac{158}{85}$

5. $\displaystyle\sum_{k=1}^{4}c = c + c + c + c = 4c$

7. $\displaystyle\sum_{i=1}^{9}\frac{1}{3i}$

9. $\displaystyle\sum_{j=1}^{8}\left[5\left(\frac{j}{8}\right) + 3\right]$

11. $\displaystyle\frac{2}{n}\sum_{i=1}^{n}\left[\left(\frac{2i}{n}\right)^3 - \left(\frac{2i}{n}\right)\right]$

13. $\displaystyle\frac{3}{n}\sum_{i=1}^{n}\left[2\left(1 + \frac{3i}{n}\right)^2\right]$

15. $\displaystyle\sum_{i=1}^{20}2i = 2\sum_{i=1}^{20}i = 2\left[\frac{20(21)}{2}\right] = 420$

17. $\displaystyle\sum_{i=1}^{20}(i - 1)^2 = \sum_{i=1}^{19}i^2$

$= \left[\frac{19(20)(39)}{6}\right] = 2470$

19. $\displaystyle\sum_{i=1}^{15}i(i - 1)^2 = \sum_{i=1}^{15}i^3 - 2\sum_{i=1}^{15}i^2 + \sum_{i=1}^{15}i$

$= \frac{15^2(16)^2}{4} - 2\frac{15(16)(31)}{6} + \frac{15(16)}{2}$

$= 14{,}400 - 2{,}480 + 120$

$= 12{,}040$

21. sum seq$(x \boxed{\wedge} 2 + 3, x, 1, 20, 1) = 2930$ (*TI-82*)

$\displaystyle\sum_{i=1}^{20}(i^2 + 3) = \frac{20(20 + 1)(2(20) + 1)}{6} + 3(20)$

$= \frac{(20)(21)(41)}{6} + 60 = 2930$

23. $S = \left[3 + 4 + \dfrac{9}{2} + 5\right](1) = \dfrac{33}{2} = 16.5$

$\quad\; s = \left[1 + 3 + 4 + \dfrac{9}{2}\right](1) = \dfrac{25}{2} = 12.5$

25. $S = [3 + 3 + 5](1) = 11$

$\quad\; s = [2 + 2 + 3](1) = 7$

27. $S(4) = \sqrt{\dfrac{1}{4}}\left(\dfrac{1}{4}\right) + \sqrt{\dfrac{1}{2}}\left(\dfrac{1}{4}\right) + \sqrt{\dfrac{3}{4}}\left(\dfrac{1}{4}\right) + \sqrt{1}\left(\dfrac{1}{4}\right) = \dfrac{1 + \sqrt{2} + \sqrt{3} + 2}{8} \approx 0.768$

$\quad\; s(4) = 0\left(\dfrac{1}{4}\right) + \sqrt{\dfrac{1}{4}}\left(\dfrac{1}{4}\right) + \sqrt{\dfrac{1}{2}}\left(\dfrac{1}{4}\right) + \sqrt{\dfrac{3}{4}}\left(\dfrac{1}{4}\right) = \dfrac{1 + \sqrt{2} + \sqrt{3}}{8} \approx 0.518$

29. $S(5) = 1\left(\dfrac{1}{5}\right) + \dfrac{1}{6/5}\left(\dfrac{1}{5}\right) + \dfrac{1}{7/5}\left(\dfrac{1}{5}\right) + \dfrac{1}{8/5}\left(\dfrac{1}{5}\right) + \dfrac{1}{9/5}\left(\dfrac{1}{5}\right) = \dfrac{1}{5} + \dfrac{1}{6} + \dfrac{1}{7} + \dfrac{1}{8} + \dfrac{1}{9} \approx 0.746$

$\quad\; s(5) = \dfrac{1}{6/5}\left(\dfrac{1}{5}\right) + \dfrac{1}{7/5}\left(\dfrac{1}{5}\right) + \dfrac{1}{8/5}\left(\dfrac{1}{5}\right) + \dfrac{1}{9/5}\left(\dfrac{1}{5}\right) + \dfrac{1}{2}\left(\dfrac{1}{5}\right) = \dfrac{1}{6} + \dfrac{1}{7} + \dfrac{1}{8} + \dfrac{1}{9} + \dfrac{1}{10} \approx 0.646$

31. $\displaystyle\lim_{n\to\infty}\left[\left(\dfrac{4}{3n^3}\right)(2n^3 + 3n^2 + n)\right] = \dfrac{4}{3}\lim_{n\to\infty}\left[\dfrac{2n^3 + 3n^2 + n}{n^3}\right]$

$\qquad\qquad\qquad\qquad\qquad\qquad = \dfrac{4}{3}(2) = \dfrac{8}{3}$

33. $\displaystyle\lim_{n\to\infty}\left[\left(\dfrac{81}{n^4}\right)\dfrac{n^2(n+1)^2}{4}\right] = \dfrac{81}{4}\lim_{n\to\infty}\left[\dfrac{n^4 + 2n^3 + n^2}{n^4}\right]$

$\qquad\qquad\qquad\qquad\qquad\qquad = \dfrac{81}{4}(1) = \dfrac{81}{4}$

35. $\displaystyle\lim_{n\to\infty}\left[\left(\dfrac{18}{n^2}\right)\dfrac{n(n+1)}{2}\right] = \dfrac{18}{2}\lim_{n\to\infty}\left[\dfrac{n^2 + n}{n^2}\right] = \dfrac{18}{2}(1) = 9$

37. $\displaystyle\sum_{i=1}^{n}\dfrac{2i+1}{n^2} = \dfrac{1}{n^2}\sum_{i=1}^{n}(2i+1) = \dfrac{1}{n^2}\left[2\dfrac{n(n+1)}{2} + n\right] = \dfrac{n+2}{n} = S(n)$

$\qquad S(10) = \dfrac{12}{10} = 1.2$

$\qquad S(100) = 1.02$

$\qquad S(1000) = 1.002$

$\qquad S(10,000) = 1.0002$

39. $\displaystyle\sum_{k=1}^{n}\dfrac{6k(k-1)}{n^3} = \dfrac{6}{n^3}\sum_{k=1}^{n}(k^2 - k) = \dfrac{6}{n^3}\left[\dfrac{n(n+1)(2n+1)}{6} - \dfrac{n(n+1)}{2}\right]$

$\qquad\qquad\qquad\quad = \dfrac{6}{n^2}\left[\dfrac{2n^2 + 3n + 1 - 3n - 3}{6}\right] = \dfrac{1}{n^2}[2n^2 - 2] = S(n)$

$\qquad S(10) = 1.98$

$\qquad S(100) = 1.9998$

$\qquad S(1000) = 1.999998$

$\qquad S(10,000) = 1.99999998$

41. $\displaystyle\lim_{n\to\infty}\sum_{i=1}^{n}\left(\dfrac{16i}{n^2}\right) = \lim_{n\to\infty}\dfrac{16}{n^2}\sum_{i=1}^{n}i = \lim_{n\to\infty}\dfrac{16}{n^2}\left(\dfrac{n(n+1)}{2}\right) = \lim_{n\to\infty}\left[8\left(\dfrac{n^2 + n}{n^2}\right)\right] = 8\lim_{n\to\infty}\left(1 + \dfrac{1}{n}\right) = 8$

43. $\displaystyle\lim_{n\to\infty}\sum_{i=1}^{n}\dfrac{1}{n^3}(i-1)^2 = \lim_{n\to\infty}\dfrac{1}{n^3}\sum_{i=1}^{n-1}i^2 = \lim_{n\to\infty}\dfrac{1}{n^3}\left[\dfrac{(n-1)(n)(2n-1)}{6}\right]$

$\qquad\qquad\qquad\qquad = \lim_{n\to\infty}\dfrac{1}{6}\left[\dfrac{2n^3 - 3n^2 + n}{n^3}\right] = \lim_{n\to\infty}\left[\dfrac{1}{6}\left(\dfrac{2 - (3/n) + (1/n^2)}{1}\right)\right] = \dfrac{1}{3}$

45. $\lim\limits_{n\to\infty}\sum\limits_{i=1}^{n}\left(1+\dfrac{i}{n}\right)\left(\dfrac{2}{n}\right)=2\lim\limits_{n\to\infty}\dfrac{1}{n}\left[\sum\limits_{i=1}^{n}1+\dfrac{1}{n}\sum\limits_{i=1}^{n}i\right]=2\lim\limits_{n\to\infty}\dfrac{1}{n}\left[n+\dfrac{1}{n}\left(\dfrac{n(n+1)}{2}\right)\right]=2\lim\limits_{n\to\infty}\left[1+\dfrac{n^2+n}{2n^2}\right]=2\left(1+\dfrac{1}{2}\right)=3$

47. (a)

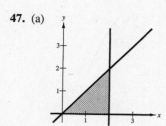

(e)

x	5	10	50	100
$s(n)$	1.6	1.8	1.96	1.98
$S(n)$	2.4	2.2	2.04	2.02

(b) $\Delta x=\dfrac{2-0}{n}=\dfrac{2}{n}$

Endpoints:

$$0<1\left(\dfrac{2}{n}\right)<2\left(\dfrac{2}{n}\right)<\cdots<(n-1)\left(\dfrac{2}{n}\right)<n\left(\dfrac{2}{n}\right)=2$$

(c) Since $y=x$ is increasing, $f(m_i)=f(x_{i-1})$ on $[x_{i-1},x_i]$.

$$s(n)=\sum\limits_{i=1}^{n}f(x_{i-1})\,\Delta x$$

$$=\sum\limits_{i=1}^{n}f\left(\dfrac{2i-2}{n}\right)\left(\dfrac{2}{n}\right)=\sum\limits_{i=1}^{n}\left[(i-1)\left(\dfrac{2}{n}\right)\right]\left(\dfrac{2}{n}\right)$$

(d) $f(M_i)=f(x_i)$ on $[x_{i-1},x_i]$

$$S(n)=\sum\limits_{i=1}^{n}f(x_i)\,\Delta x=\sum\limits_{i=1}^{n}f\left(\dfrac{2i}{n}\right)\dfrac{2}{n}=\sum\limits_{i=1}^{n}\left[i\left(\dfrac{2}{n}\right)\right]\left(\dfrac{2}{n}\right)$$

(f) $\lim\limits_{n\to\infty}\sum\limits_{i=1}^{n}\left[(i-1)\left(\dfrac{2}{n}\right)\right]\left(\dfrac{2}{n}\right)=\lim\limits_{n\to\infty}\dfrac{4}{n^2}\sum\limits_{i=1}^{n}(i-1)$

$$=\lim\limits_{n\to\infty}\dfrac{4}{n^2}\left[\dfrac{n(n+1)}{2}-n\right]$$

$$=\lim\limits_{n\to\infty}\left[\dfrac{2(n+1)}{n}-\dfrac{4}{n}\right]=2$$

$\lim\limits_{n\to\infty}\sum\limits_{i=1}^{n}\left[i\left(\dfrac{2}{n}\right)\right]\left(\dfrac{2}{n}\right)=\lim\limits_{n\to\infty}\dfrac{4}{n^2}\sum\limits_{i=1}^{n}i$

$$=\lim\limits_{n\to\infty}\left(\dfrac{4}{n^2}\right)\dfrac{n(n+1)}{2}$$

$$=\lim\limits_{n\to\infty}\dfrac{2(n+1)}{n}=2$$

49. $y=-2x+3$ on $[0,1]$. $\left(\textbf{Note: }\Delta x=\dfrac{1-0}{n}=\dfrac{1}{n}\right)$

$$s(n)=\sum\limits_{i=1}^{n}f\left(\dfrac{i}{n}\right)\left(\dfrac{1}{n}\right)=\sum\limits_{i=1}^{n}\left[-2\left(\dfrac{i}{n}\right)+3\right]\left(\dfrac{1}{n}\right)$$

$$=3-\dfrac{2}{n^2}\sum\limits_{i=1}^{n}i=3-\dfrac{2(n+1)n}{2n^2}=2-\dfrac{1}{n}$$

Area $=\lim\limits_{n\to\infty}s(n)=2$

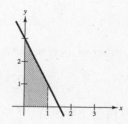

51. $y=x^2+2$ on $[0,1]$. $\left(\textbf{Note: }\Delta x=\dfrac{1}{n}\right)$

$$S(n)=\sum\limits_{i=1}^{n}f\left(\dfrac{i}{n}\right)\left(\dfrac{1}{n}\right)=\sum\limits_{i=1}^{n}\left[\left(\dfrac{i}{n}\right)^2+2\right]\left(\dfrac{1}{n}\right)$$

$$=\left[\dfrac{1}{n^3}\sum\limits_{i=1}^{n}i^2\right]+2=\dfrac{n(n+1)(2n+1)}{6n^3}+2=\dfrac{1}{6}\left(2+\dfrac{3}{n}+\dfrac{1}{n^2}\right)+2$$

Area $=\lim\limits_{n\to\infty}S(n)=\dfrac{7}{3}$

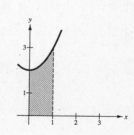

53. $y = 16 - x^2$ on $[1, 3]$. $\left(\text{Note: } \Delta x = \dfrac{2}{n}\right)$

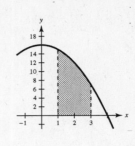

$$s(n) = \sum_{i=1}^{n} f\left(1 + \frac{2i}{n}\right)\left(\frac{2}{n}\right) = \sum_{i=1}^{n} \left[16 - \left(1 + \frac{2i}{n}\right)^2\right]\left(\frac{2}{n}\right)$$

$$= \frac{2}{n}\sum_{i=1}^{n}\left[15 - \frac{4i^2}{n^2} - \frac{4i}{n}\right]$$

$$= \frac{2}{n}\left[15n - \frac{4}{n^2}\frac{n(n+1)(2n+1)}{6} - \frac{4}{n}\frac{n(n+1)}{2}\right]$$

$$= 30 - \frac{8}{6n^2}(n+1)(2n+1) - \frac{4}{n}(n+1)$$

$$\text{Area} = \lim_{n \to \infty} s(n) = 30 - \frac{8}{3} - 4 = \frac{70}{3} = 23\frac{1}{3}$$

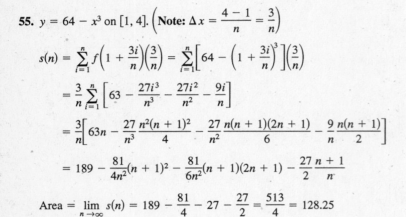

55. $y = 64 - x^3$ on $[1, 4]$. $\left(\text{Note: } \Delta x = \dfrac{4-1}{n} = \dfrac{3}{n}\right)$

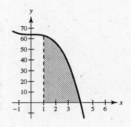

$$s(n) = \sum_{i=1}^{n} f\left(1 + \frac{3i}{n}\right)\left(\frac{3}{n}\right) = \sum_{i=1}^{n}\left[64 - \left(1 + \frac{3i}{n}\right)^3\right]\left(\frac{3}{n}\right)$$

$$= \frac{3}{n}\sum_{i=1}^{n}\left[63 - \frac{27i^3}{n^3} - \frac{27i^2}{n^2} - \frac{9i}{n}\right]$$

$$= \frac{3}{n}\left[63n - \frac{27}{n^3}\frac{n^2(n+1)^2}{4} - \frac{27}{n^2}\frac{n(n+1)(2n+1)}{6} - \frac{9}{n}\frac{n(n+1)}{2}\right]$$

$$= 189 - \frac{81}{4n^2}(n+1)^2 - \frac{81}{6n^2}(n+1)(2n+1) - \frac{27}{2}\frac{n+1}{n}$$

$$\text{Area} = \lim_{n \to \infty} s(n) = 189 - \frac{81}{4} - 27 - \frac{27}{2} = \frac{513}{4} = 128.25$$

57. $y = x^2 - x^3$ on $[-1, 1]$. $\left(\text{Note: } \Delta x = \dfrac{1-(-1)}{n} = \dfrac{2}{n}\right)$

Again, $T(n)$ is neither an upper nor a lower sum.

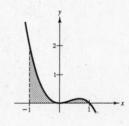

$$T(n) = \sum_{i=1}^{n} f\left(-1 + \frac{2i}{n}\right)\left(\frac{2}{n}\right) = \sum_{i=1}^{n}\left[\left(-1 + \frac{2i}{n}\right)^2 - \left(-1 + \frac{2i}{n}\right)^3\right]\left(\frac{2}{n}\right)$$

$$= \sum_{i=1}^{n}\left[\left(1 - \frac{4i}{n} + \frac{4i^2}{n^2}\right) - \left(-1 + \frac{6i}{n} - \frac{12i^2}{n^2} + \frac{8i^3}{n^3}\right)\right]\left(\frac{2}{n}\right)$$

$$= \sum_{i=1}^{n}\left[2 - \frac{10i}{n} + \frac{16i^2}{n^2} - \frac{8i^3}{n^3}\right]\left(\frac{2}{n}\right) = \frac{4}{n}\sum_{i=1}^{n}1 - \frac{20}{n^2}\sum_{i=1}^{n}i + \frac{32}{n^3}\sum_{i=1}^{n}i^2 - \frac{16}{n^4}\sum_{i=1}^{n}i^3$$

$$= \frac{4}{n}(n) - \frac{20}{n^2}\cdot\frac{n(n+1)}{2} + \frac{32}{n^3}\cdot\frac{n(n+1)(2n+1)}{6} - \frac{16}{n^4}\cdot\frac{n^2(n+1)^2}{4}$$

$$= 4 - 10\left(1 + \frac{1}{n}\right) + \frac{16}{3}\left(2 + \frac{3}{n} + \frac{1}{n^2}\right) - 4\left(1 + \frac{2}{n} + \frac{1}{n^2}\right)$$

$$\text{Area} = \lim_{n \to \infty} T(n) = 4 - 10 + \frac{32}{3} - 4 = \frac{2}{3}$$

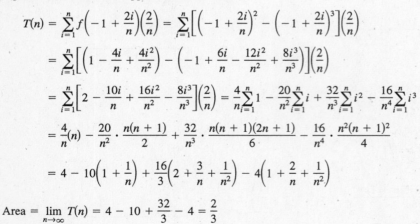

59. $f(y) = 3y, 0 \le y \le 2$ $\left(\textbf{Note:}\ \Delta y = \dfrac{2-0}{n} = \dfrac{2}{n}\right)$

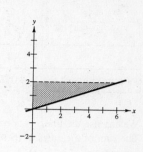

$$S(n) = \sum_{i=1}^{n} f(m_i)\, \Delta y = \sum_{i=1}^{n} f\left(\frac{2i}{n}\right)\left(\frac{2}{n}\right) = \sum_{i=1}^{n} 3\left(\frac{2i}{n}\right)\left(\frac{2}{n}\right)$$

$$= \frac{12}{n^2} \sum_{i=1}^{n} i = \left(\frac{12}{n^2}\right)\cdot\frac{n(n+1)}{2} = \frac{6(n+1)}{n} = 6 + \frac{6}{n}$$

$$\text{Area} = \lim_{n\to\infty} S(n) = \lim_{n\to\infty}\left(6 + \frac{6}{n}\right) = 6$$

61. $f(y) = y^2, 0 \le y \le 3$ $\left(\textbf{Note:}\ \Delta y = \dfrac{3-0}{n} = \dfrac{3}{n}\right)$

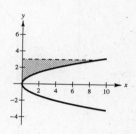

$$S(n) = \sum_{i=1}^{n} f\left(\frac{3i}{n}\right)\left(\frac{3}{n}\right) = \sum_{i=1}^{n}\left(\frac{3i}{n}\right)^2\left(\frac{3}{n}\right) = \frac{27}{n^3}\sum_{i=1}^{n} i^2$$

$$= \frac{27}{n^3}\cdot\frac{n(n+1)(2n+1)}{6} = \frac{9}{n^2}\left(\frac{2n^2+3n+1}{2}\right) = 9 + \frac{27}{2n} + \frac{9}{2n^2}$$

$$\text{Area} = \lim_{n\to\infty} S(n) = \lim_{n\to\infty}\left(9 + \frac{27}{2n} + \frac{9}{2n^2}\right) = 9$$

63. $g(y) = 4y^2 - y^3, 1 \le y \le 3.$ $\left(\textbf{Note:}\ \Delta y = \dfrac{3-1}{n} = \dfrac{2}{n}\right)$

$$S(n) = \sum_{i=1}^{n} g\left(1 + \frac{2i}{n}\right)\left(\frac{2}{n}\right)$$

$$= \sum_{i=1}^{n}\left[4\left(1 + \frac{2i}{n}\right)^2 - \left(1 + \frac{2i}{n}\right)^3\right]\frac{2}{n}$$

$$= \frac{2}{n}\sum_{i=1}^{n}\left[4\left[1 + \frac{4i}{n} + \frac{4i^2}{n^2}\right] - \left[1 + \frac{6i}{n} + \frac{12i^2}{n^2} + \frac{8i^3}{n^3}\right]\right]$$

$$= \frac{2}{n}\sum_{i=1}^{n}\left[3 + \frac{10i}{n} + \frac{4i^2}{n^2} - \frac{8i^3}{n^3}\right] = \frac{2}{n}\left[3n + \frac{10}{n}\frac{n(n+1)}{2} + \frac{4}{n^2}\frac{n(n+1)(2n+1)}{6} - \frac{8}{n^3}\frac{n^2(n+1)^2}{4}\right]$$

$$\text{Area} = \lim_{n\to\infty} S(n) = 6 + 10 + \frac{8}{3} - 4 = \frac{44}{3}$$

65. $f(x) = x^2 + 3, 0 \le x \le 2, n = 4$

Let $c_i = \dfrac{x_i + x_{i-1}}{2}$.

$$\Delta x = \frac{1}{2}, c_1 = \frac{1}{4}, c_2 = \frac{3}{4}, c_3 = \frac{5}{4}, c_4 = \frac{7}{4}$$

$$\text{Area} \approx \sum_{i=1}^{n} f(c_i)\,\Delta x = \sum_{i=1}^{4}[c_i^2 + 3]\left(\frac{1}{2}\right)$$

$$= \frac{1}{2}\left[\left(\frac{1}{16} + 3\right) + \left(\frac{9}{16} + 3\right) + \left(\frac{25}{16} + 3\right) + \left(\frac{49}{16} + 3\right)\right]$$

$$= \frac{69}{8}$$

67. $f(x) = \sqrt{x-1},\ [1, 2], n = 4$

Let $c_i = \dfrac{x_i + x_{i-1}}{2}$.

$$\Delta x = \frac{1}{4}, c_1 = \frac{9}{8}, c_2 = \frac{11}{8}, c_3 = \frac{13}{8}, c_4 = \frac{15}{8}$$

$$\text{Area} \approx \sum_{i=1}^{n} f(c_i)\,\Delta x = \sum_{i=1}^{4}\sqrt{c_i - 1}\left(\frac{1}{4}\right)$$

$$= \frac{1}{4}\left[\sqrt{1/8} + \sqrt{3/8} + \sqrt{5/8} + \sqrt{7/8}\right]$$

$$\approx 0.6730$$

69. $f(x) = \sqrt{x}$ on $[0, 4]$.

n	4	8	12	16	20
Approximate area	5.3838	5.3523	5.3439	5.3403	5.3384

(Exact value is 16/3.)

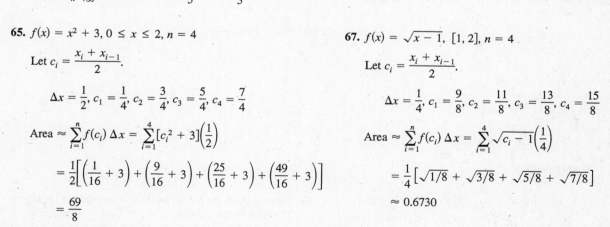

71. $f(x) = \dfrac{5x}{x^2 + 1}, [1, 3]$

n	4	8	12	16	20
Approximate Area	4.0272	4.0246	4.0241	4.0239	4.0238

73. We can use the line $y = x$ bounded by $x = a$ and $x = b$. The sum of the areas of these inscribed rectangles is the lower sum.

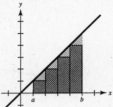

The sum of the areas of these circumscribed rectangles is the upper sum.

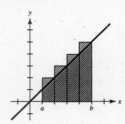

We can see that the rectangles do not contain all of the area in the first graph and the rectangles in the second graph cover more than the area of the region.

The exact value of the area lies between these two sums.

75. (a) In each case, $\Delta x = 4/n$. The lower sum uses left endpoints, $(i-1)(4/n)$. The upper sum uses right endpoints, $(i)(4/n)$. The Midpoint Rule uses midpoints, $\left(i - \frac{1}{2}\right)(4/n)$.

(b)

n	4	8	20	100	200
$s(n)$	15.333	17.368	18.459	18.995	19.06
$S(n)$	21.733	20.568	19.739	19.251	19.188
$M(n)$	19.403	19.201	19.137	19.125	19.125

(c) $s(n)$ increases because the lower sum approaches the exact value as n increases. $S(n)$ decreases because the upper sum approaches the exact value as n increases. Because of the shape of the graph, the lower sum is always smaller than the exact value, whereas the upper sum is always larger.

77.

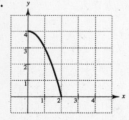

b. $A \approx 6$ square units

79. True.   (Theorem 6.2 (2))

81. (a) $y = (-4.09 \times 10^{-5})x^3 + 0.016x^2 - 2.67x + 452.9$

(b)

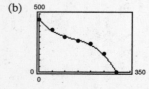

(c) Using the integration capability of a graphing utility, you obtain

$$A \approx 76{,}897.5 \text{ ft}^2.$$

83. Suppose there are n rows and $n + 1$ columns in the figure. The stars on the left total $1 + 2 + \cdots + n$, as do the stars on the right. There are $n(n + 1)$ stars in total, hence

$$2[1 + 2 + \cdots + n] = n(n + 1)$$

$$1 + 2 + \cdots + n = \tfrac{1}{2}(n)(n + 1).$$

Section 6.3   Riemann Sums and Definite Integrals

1. $f(x) = \sqrt{x}, y = 0, x = 0, x = 3, c_i = \dfrac{3i^2}{n^2}$

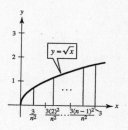

$\Delta x_i = \dfrac{3i^2}{n^2} - \dfrac{3(i-1)^2}{n^2} = \dfrac{3}{n^2}(2i - 1)$

$\displaystyle \lim_{n \to \infty} \sum_{i=1}^{n} f(c_i)\Delta x_i = \lim_{n \to \infty} \sum_{i=1}^{n} \sqrt{\dfrac{3i^2}{n^2}} \dfrac{3}{n^2}(2i - 1)$

$\displaystyle \qquad\qquad\qquad = \lim_{n \to \infty} \dfrac{3\sqrt{3}}{n^3} \sum_{i=1}^{n} (2i^2 - i)$

$\displaystyle \qquad\qquad\qquad = \lim_{n \to \infty} \dfrac{3\sqrt{3}}{n^3}\left[2\dfrac{n(n+1)(2n+1)}{6} - \dfrac{n(n+1)}{2} \right]$

$\displaystyle \qquad\qquad\qquad = \lim_{n \to \infty} 3\sqrt{3}\left[\dfrac{(n+1)(2n+1)}{3n^2} - \dfrac{n+1}{2n^2} \right]$

$\displaystyle \qquad\qquad\qquad = 3\sqrt{3}\left[\dfrac{2}{3} - 0 \right] = 2\sqrt{3} \approx 3.464$

3. $y = 6$ on $[4, 10]$. $\left(\textbf{Note: } \Delta x = \dfrac{10 - 4}{n} = \dfrac{6}{n}, \|\Delta\| \to 0 \text{ as } n \to \infty \right)$

$\displaystyle \sum_{i=1}^{n} f(c_i)\,\Delta x_i = \sum_{i=1}^{n} f\left(4 + \dfrac{6i}{n}\right)\left(\dfrac{6}{n}\right) = \sum_{i=1}^{n} 6\left(\dfrac{6}{n}\right) = \sum_{i=1}^{n} \dfrac{36}{n} = 36$

$\displaystyle \int_{4}^{10} 6\,dx = \lim_{n \to \infty} 36 = 36$

5. $y = x^3$ on $[-1, 1]$. $\left(\textbf{Note: } \Delta x = \dfrac{1 - (-1)}{n} = \dfrac{2}{n}, \|\Delta\| \to 0 \text{ as } n \to \infty \right)$

$\displaystyle \sum_{i=1}^{n} f(c_i)\,\Delta x_i = \sum_{i=1}^{n} f\left(-1 + \dfrac{2i}{n}\right)\left(\dfrac{2}{n}\right) = \sum_{i=1}^{n} \left(-1 + \dfrac{2i}{n}\right)^3\left(\dfrac{2}{n}\right) = \sum_{i=1}^{n}\left[-1 + \dfrac{6i}{n} - \dfrac{12i^2}{n^2} + \dfrac{8i^3}{n^3} \right]\left(\dfrac{2}{n}\right)$

$\displaystyle \qquad\qquad\qquad = -2 + \dfrac{12}{n^2}\sum_{i=1}^{n} i - \dfrac{24}{n^3}\sum_{i=1}^{n} i^2 + \dfrac{16}{n^4}\sum_{i=1}^{n} i^3$

$\displaystyle \qquad\qquad\qquad = -2 + 6\left(1 + \dfrac{1}{n}\right) - 4\left(2 + \dfrac{3}{n} + \dfrac{1}{n^2}\right) + 4\left(1 + \dfrac{2}{n} + \dfrac{1}{n^2}\right) = \dfrac{2}{n}$

$\displaystyle \int_{-1}^{1} x^3\,dx = \lim_{n \to \infty} \dfrac{2}{n} = 0$

7. $y = x^2 + 1$ on $[1, 2]$. $\left(\textbf{Note: } \Delta x = \dfrac{2 - 1}{n} = \dfrac{1}{n}, \|\Delta\| \to 0 \text{ as } n \to \infty \right)$

$\displaystyle \sum_{i=1}^{n} f(c_i)\,\Delta x_i = \sum_{i=1}^{n} f\left(1 + \dfrac{i}{n}\right)\left(\dfrac{1}{n}\right) = \sum_{i=1}^{n}\left[\left(1 + \dfrac{i}{n}\right)^2 + 1\right]\left(\dfrac{1}{n}\right) = \sum_{i=1}^{n}\left[1 + \dfrac{2i}{n} + \dfrac{i^2}{n^2} + 1 \right]\left(\dfrac{1}{n}\right)$

$\displaystyle \qquad\qquad\qquad = 2 + \dfrac{2}{n^2}\sum_{i=1}^{n} i + \dfrac{1}{n^3}\sum_{i=1}^{n} i^2 = 2 + \left(1 + \dfrac{1}{n}\right) + \dfrac{1}{6}\left(2 + \dfrac{3}{n} + \dfrac{1}{n^2}\right) = \dfrac{10}{3} + \dfrac{3}{2n} + \dfrac{1}{6n^2}$

$\displaystyle \int_{1}^{2} (x^2 + 1)\,dx = \lim_{n \to \infty}\left(\dfrac{10}{3} + \dfrac{3}{2n} + \dfrac{1}{6n^2} \right) = \dfrac{10}{3}$

9. $\lim\limits_{\|\Delta\|\to 0} \sum\limits_{i=1}^{n} (3c_i + 10)\, \Delta x_i = \int_{-1}^{5} (3x + 10)\, dx$

on the interval $[-1, 5]$.

11. $\lim\limits_{\|\Delta\|\to 0} \sum\limits_{i=1}^{n} \sqrt{c_i^2 + 4}\, \Delta x_i = \int_{0}^{3} \sqrt{x^2 + 4}\, dx$

on the interval $[0, 3]$.

13. $\displaystyle\int_{0}^{5} 3\, dx$

15. $\displaystyle\int_{-4}^{4} (4 - |x|)\, dx$

17. $\displaystyle\int_{-2}^{2} (4 - x^2)\, dx$

19. $\displaystyle\int_{0}^{2} \sqrt{x + 1}\, dx$

21. $\displaystyle\int_{0}^{2} y^3\, dy$

23. Rectangle

$A = bh = 3(4)$

$A = \displaystyle\int_{0}^{3} 4\, dx = 12$

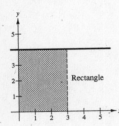

25. Triangle

$A = \dfrac{1}{2}bh = \dfrac{1}{2}(4)(4)$

$A = \displaystyle\int_{0}^{4} x\, dx = 8$

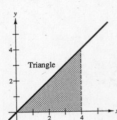

27. Trapezoid

$A = \dfrac{b_1 + b_2}{2}h = \left(\dfrac{5 + 9}{2}\right)2$

$A = \displaystyle\int_{0}^{2} (2x + 5)\, dx = 14$

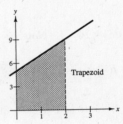

29. Triangle

$A = \dfrac{1}{2}bh = \dfrac{1}{2}(2)(1)$

$A = \displaystyle\int_{-1}^{1} (1 - |x|)\, dx = 1$

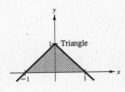

31. Semicircle

$A = \dfrac{1}{2}\pi r^2 = \dfrac{1}{2}\pi(3)^2$

$A = \displaystyle\int_{-3}^{3} \sqrt{9 - x^2}\, dx = \dfrac{9\pi}{2}$

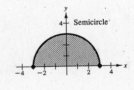

In Exercises 33–39, $\displaystyle\int_{2}^{4} x^3\, dx = 60,\ \int_{2}^{4} x\, dx = 6,\ \int_{2}^{4} dx = 2$

33. $\displaystyle\int_{4}^{2} x\, dx = -\int_{2}^{4} x\, dx = -6$

35. $\displaystyle\int_{2}^{4} 4x\, dx = 4\int_{2}^{4} x\, dx = 4(6) = 24$

37. $\displaystyle\int_{2}^{4} (x - 8)\, dx = \int_{2}^{4} x\, dx - 8\int_{2}^{4} dx = 6 - 8(2) = -10$

39. $\displaystyle\int_{2}^{4} \left(\dfrac{1}{2}x^3 - 3x + 2\right) dx = \dfrac{1}{2}\int_{2}^{4} x^3\, dx - 3\int_{2}^{4} x\, dx + 2\int_{2}^{4} dx$

$= \dfrac{1}{2}(60) - 3(6) + 2(2) = 16$

41. (a) $\int_0^7 f(x)\,dx = \int_0^5 f(x)\,dx + \int_5^7 f(x)\,dx = 10 + 3 = 13$

(b) $\int_5^0 f(x)\,dx = -\int_0^5 f(x)\,dx = -10$

(c) $\int_5^5 f(x)\,dx = 0$

(d) $\int_0^5 3f(x)\,dx = 3\int_0^5 f(x)\,dx = 3(10) = 30$

43. (a) $\int_2^6 [f(x) + g(x)]\,dx = \int_2^6 f(x)\,dx + \int_2^6 g(x)\,dx$

$= 10 + (-2) = 8$

(b) $\int_2^6 [g(x) - f(x)]\,dx = \int_2^6 g(x)\,dx - \int_2^6 f(x)\,dx$

$= -2 - 10 = -12$

(c) $\int_2^6 2g(x)\,dx = 2\int_2^6 g(x)\,dx = 2(-2) = -4$

(d) $\int_2^6 3f(x)\,dx = 3\int_2^6 f(x)\,dx = 3(10) = 30$

45. (a) Quarter circle below x-axis: $-\frac{1}{4}\pi r^2 = -\frac{1}{4}\pi(2)^2 = -\pi$

(b) Triangle: $\frac{1}{2}bh = \frac{1}{2}(4)(2) = 4$

(c) Triangle + Semicircle below x-axis: $-\frac{1}{2}(2)(1) - \frac{1}{2}\pi(2)^2 = -(1 + 2\pi)$

(d) Sum of parts (b) and (c): $4 - (1 + 2\pi) = 3 - 2\pi$

(e) Sum of absolute values of (b) and (c): $4 + (1 + 2\pi) = 5 + 2\pi$

(f) Answer to (d) plus $2(10) = 20$: $(3 - 2\pi) + 20 = 23 - 2\pi$

47. (a) $\int_0^5 [f(x) + 2]\,dx = \int_0^5 f(x)\,dx + \int_0^5 2\,dx = 4 + 10 = 14$ (b) $\int_{-2}^3 f(x + 2)\,dx = \int_0^5 f(x)\,dx = 4$ (Let $u = x + 2$.)

(c) $\int_{-5}^5 f(x)\,dx = 2\int_0^5 f(x)\,dx = 2(4) = 8$ (f even)   (d) $\int_{-5}^5 f(x)\,dx = 0$ (f odd)

49. The left endpoint approximation will be greater than the actual area: $>$

51. $f(x) = \dfrac{1}{x - 4}$

is not integrable on the interval $[3, 5]$ and f has a discontinuity at $x = 4$.

53.

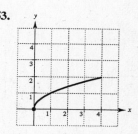

a. $A \approx 5$ square units

55. $\int_0^3 x\sqrt{3 - x}\,dx$

n	4	8	12	16	20
$L(n)$	3.6830	3.9956	4.0707	4.1016	4.1177
$M(n)$	4.3082	4.2076	4.1838	4.1740	4.1690
$R(n)$	3.6830	3.9956	4.0707	4.1016	4.1177

57. True

59. True

61. False

$\int_0^2 (-x)\,dx = -2$

63. $f(x) = x^2 + 3x$, $[0, 8]$

$x_0 = 0$, $x_1 = 1$, $x_2 = 3$, $x_3 = 7$, $x_4 = 8$

$\Delta x_1 = 1$, $\Delta x_2 = 2$, $\Delta x_3 = 4$, $\Delta x_4 = 1$

$c_1 = 1$, $c_2 = 2$, $c_3 = 5$, $c_4 = 8$

$$\sum_{i=1}^{4} f(c_i)\,\Delta x = f(1)\,\Delta x_1 + f(2)\,\Delta x_2 + f(5)\,\Delta x_3 + f(8)\,\Delta x_4$$

$$= (4)(1) + (10)(2) + (40)(4) + (88)(1) = 272$$

65. $f(x) = \begin{cases} 0, & x = 0 \\ \dfrac{1}{x}, & 0 < x \le 1 \end{cases}$

The limit

$$\lim_{\|\Delta\| \to 0} \sum_{i=1}^{n} f(c_i)\Delta x_i$$

does not exist. This does not contradict Theorem 6.4 because f is not continuous on $[0, 1]$.

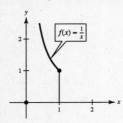

67. To find $\int_0^2 [\![x]\!]\,dx$, use a geometric approach.

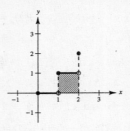

Thus,

$$\int_0^2 [\![x]\!]\,dx = 1(2 - 1) = 1.$$

Section 6.4   The Fundamental Theorem of Calculus

1. $f(x) = \dfrac{4}{x^2 + 1}$

$\displaystyle\int_0^{\pi} \dfrac{4}{x^2 + 1}\,dx$ is positive.

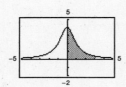

3. $f(x) = x\sqrt{x^2 + 1}$

$\displaystyle\int_{-2}^{2} x\sqrt{x^2 + 1}\,dx = 0$

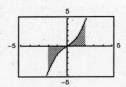

5. $\displaystyle\int_0^1 2x\,dx = \left[x^2\right]_0^1 = 1 - 0 = 1$

7. $\displaystyle\int_{-1}^{0} (x - 2)\,dx = \left[\dfrac{x^2}{2} - 2x\right]_{-1}^{0} = 0 - \left(\dfrac{1}{2} + 2\right) = -\dfrac{5}{2}$

9. $\displaystyle\int_{-1}^{1} (t^2 - 2)\,dt = \left[\dfrac{t^3}{3} - 2t\right]_{-1}^{1} = \left(\dfrac{1}{3} - 2\right) - \left(-\dfrac{1}{3} + 2\right) = -\dfrac{10}{3}$

11. $\displaystyle\int_0^1 (2t - 1)^2\,dt = \int_0^1 (4t^2 - 4t + 1)\,dt = \left[\dfrac{4}{3}t^3 - 2t^2 + t\right]_0^1 = \dfrac{4}{3} - 2 + 1 = \dfrac{1}{3}$

13. $\displaystyle\int_1^2 \left(\dfrac{3}{x^2} - 1\right)dx = \left[-\dfrac{3}{x} - x\right]_1^2 = \left(-\dfrac{3}{2} - 2\right) - (-3 - 1) = \dfrac{1}{2}$

15. $\displaystyle\int_1^4 \dfrac{u - 2}{\sqrt{u}}\,du = \int_1^4 (u^{1/2} - 2u^{-1/2})\,du = \left[\dfrac{2}{3}u^{3/2} - 4u^{1/2}\right]_1^4 = \left[\dfrac{2}{3}(\sqrt{4})^3 - 4\sqrt{4}\right] - \left[\dfrac{2}{3} - 4\right] = \dfrac{2}{3}$

17. $\int_{-1}^{1} (\sqrt[3]{t} - 2) \, dt = \left[\frac{3}{4} t^{4/3} - 2t \right]_{-1}^{1} = \left(\frac{3}{4} - 2 \right) - \left(\frac{3}{4} + 2 \right) = -4$

19. $\int_{0}^{1} \frac{x - \sqrt{x}}{3} \, dx = \frac{1}{3} \int_{0}^{1} (x - x^{1/2}) \, dx = \frac{1}{3} \left[\frac{x^2}{2} - \frac{2}{3} x^{3/2} \right]_{0}^{1} = \frac{1}{3} \left(\frac{1}{2} - \frac{2}{3} \right) = -\frac{1}{18}$

21. $\int_{-1}^{0} (t^{1/3} - t^{2/3}) \, dt = \left[\frac{3}{4} t^{4/3} - \frac{3}{5} t^{5/3} \right]_{-1}^{0} = 0 - \left(\frac{3}{4} + \frac{3}{5} \right) = -\frac{27}{20}$

23. $\int_{0}^{3} |2x - 3| \, dx = \int_{0}^{3/2} (3 - 2x) \, dx + \int_{3/2}^{3} (2x - 3) \, dx \ \left(\text{split up the integral at the zero } x = \frac{3}{2} \right)$

$\qquad = \left[3x - x^2 \right]_{0}^{3/2} + \left[x^2 - 3x \right]_{3/2}^{3} = \left(\frac{9}{2} - \frac{9}{4} \right) - 0 + (9 - 9) - \left(\frac{9}{4} - \frac{9}{2} \right) = 2\left(\frac{9}{2} - \frac{9}{4} \right) = \frac{9}{2}$

25. $\int_{0}^{3} |x^2 - 4| \, dx = \int_{0}^{2} (4 - x^2) \, dx + \int_{2}^{3} (x^2 - 4) \, dx \ = \left[4x - \frac{x^3}{3} \right]_{0}^{2} + \left[\frac{x^3}{3} - 4x \right]_{2}^{3} = \left(8 - \frac{8}{3} \right) + (9 - 12) - \left(\frac{8}{3} - 8 \right) = \frac{23}{3}$

27. $A = \int_{0}^{1} (x - x^2) \, dx = \left[\frac{x^2}{2} - \frac{x^3}{3} \right]_{0}^{1} = \frac{1}{6}$

29. $A = \int_{-1}^{1} (1 - x^4) \, dx = \left[x - \frac{1}{5} x^5 \right]_{-1}^{1} = \frac{8}{5}$

31. $A = \int_{0}^{4} (2x)^{1/3} \, dx = 2^{1/3} \int_{0}^{4} x^{1/3} \, dx$

$\qquad = 2^{1/3} \left[\frac{3}{4} x^{4/3} \right]_{0}^{4}$

$\qquad = 2^{1/3} \frac{3}{4} 4 \cdot 4^{1/3}$

$\qquad = 6$

33. Since $y \geq 0$ on $[0, 2]$,

$\qquad A = \int_{0}^{2} (3x^2 + 1) \, dx = \left[x^3 + x \right]_{0}^{2} = 8 + 2 = 10.$

35. Since $y \geq 0$ on $[0, 2]$,

$\qquad A = \int_{0}^{2} (x^3 + x) \, dx = \left[\frac{x^4}{4} + \frac{x^2}{2} \right]_{0}^{2} = 4 + 2 = 6.$

37. $\int_{0}^{2} x^3 \, dx = \frac{x^4}{4} \Big]_{0}^{2} = 4$

$\qquad f(c)(2 - 0) = 4$

$\qquad\quad c^3 = 2$

$\qquad\qquad c = 2^{1/3} \approx 1.2599$

39. $\int_{0}^{3} (-x^2 + 4x) \, dx = \left[-\frac{x^3}{3} + 2x^2 \right]_{0}^{3} = -9 + 18 = 9$

$\qquad f(c)(3 - 0) = 9$

$\qquad\quad -c^2 + 4c = 3$

$\qquad\ c^2 - 4c + 3 = 0$

$\qquad (c - 1)(c - 3) = 0$

$\qquad\qquad c = 1, 3$

41. $\dfrac{1}{2-(-2)}\displaystyle\int_{-2}^{2}(4-x^2)\,dx = \dfrac{1}{4}\Big[4x - \dfrac{1}{3}x^3\Big]_{-2}^{2}$

$$= \dfrac{1}{4}\Big[\Big(8-\dfrac{8}{3}\Big)-\Big(-8+\dfrac{8}{3}\Big)\Big]$$

$$= \dfrac{8}{3}$$

Average value $= \dfrac{8}{3}$

$4 - x^2 = \dfrac{8}{3}$ when $x^2 = 4 - \dfrac{8}{3}$

or $x = \pm\dfrac{2\sqrt{3}}{3} \approx \pm 1.155.$

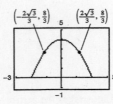

43. $\dfrac{1}{4-0}\displaystyle\int_{0}^{4}(x-2x^{1/2})\,dx = \dfrac{1}{4}\Big[\dfrac{x^2}{2}-\dfrac{4}{3}x^{3/2}\Big]_{0}^{4}$

$$= \dfrac{1}{4}\Big[8-\dfrac{4}{3}(8)\Big] = -\dfrac{2}{3}$$

$x - 2x^{1/2} = -\dfrac{2}{3}$

$3x - 6x^{1/2} + 2 = 0$

$\sqrt{x} = \dfrac{6 \pm \sqrt{36-24}}{6} = 1 \pm \dfrac{\sqrt{3}}{3}$

$x = \dfrac{4}{3} \pm \dfrac{2\sqrt{3}}{3} \quad (x \approx 0.179, 2.488)$

45. The distance traveled is $\int_{0}^{8} v(t)\,dt$. The area under the curve from $0 \le t \le 8$ is approximately (18 squares) (30) ≈ 540 ft.

47. If f is continuous on $[a,b]$ and $F'(x) = f(x)$ on $[a,b]$, then $\displaystyle\int_{a}^{b} f(x)\,dx = F(b) - F(a).$

49. $\displaystyle\int_{0}^{2} f(x)\,dx = -(\text{area of region } A) = -1.5$

51. $\displaystyle\int_{0}^{6}|f(x)|\,dx = -\int_{0}^{2} f(x)\,dx + \int_{2}^{6} f(x)\,dx = 1.5 + 5.0 = 6.5$

53. $\displaystyle\int_{0}^{6}[2+f(x)]\,dx = \int_{0}^{6} 2\,dx + \int_{0}^{6} f(x)\,dx$

$$= 12 + 3.5 = 15.5$$

55. $\dfrac{1}{5-0}\displaystyle\int_{0}^{5}(0.1729t + 0.1522t^2 - 0.0374t^3)\,dt \approx \dfrac{1}{5}\Big[0.08645t^2 + 0.05073t^3 - 0.00935t^4\Big]_{0}^{5} \approx 0.5318$ liter

57. (a) $v = -8.61 \times 10^{-4}t^3 + 0.0782t^2 - 0.208t + 0.0952$

(b)
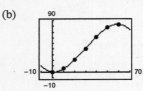

(c) $\displaystyle\int_{0}^{60} v(t)\,dt = \Big[\dfrac{-8.61\times10^{-4}t^4}{4} + \dfrac{0.0782t^3}{3} - \dfrac{0.208t^2}{2} + 0.0952t\Big]_{0}^{60} \approx 2472$ meters

59. $F(x) = \displaystyle\int_{0}^{x}(t-5)\,dt = \Big[\dfrac{t^2}{2}-5t\Big]_{0}^{x} = \dfrac{x^2}{2}-5x$

$F(2) = \dfrac{4}{2} - 5(2) = -8$

$F(5) = \dfrac{25}{2} - 5(5) = -\dfrac{25}{2}$

$F(8) = \dfrac{64}{2} - 5(8) = -8$

61. $F(x) = \displaystyle\int_{1}^{x}\dfrac{10}{v^2}\,dv = \int_{1}^{x} 10v^{-2}\,dv = \dfrac{-10}{v}\Big]_{1}^{x}$

$$= -\dfrac{10}{x} + 10 = 10\Big(1 - \dfrac{1}{x}\Big)$$

$F(2) = 10\Big(\dfrac{1}{2}\Big) = 5$

$F(5) = 10\Big(\dfrac{4}{5}\Big) = 8$

$F(8) = 10\Big(\dfrac{7}{8}\Big) = \dfrac{35}{4}$

63. $g(x) = \int_0^x f(t)dt$

(a) $g(0) = \int_0^0 f(t)dt = 0$

$g(2) = \int_0^2 f(t)dt \approx 4 + 2 + 1 = 7$

$g(4) = \int_0^4 f(t)dt \approx 7 + 2 = 9$

$g(6) = \int_0^6 f(t)dt \approx 9 + (-1) = 8$

$g(8) = \int_0^8 f(t)dt \approx 8 - 3 = 5$

(b) g increasing on $(0, 4)$ and decreasing on $(4, 8)$

(c) g is a maximum of 9 at $x = 4$.

(d)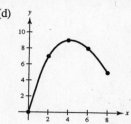

65. (a) $\int_0^x (t + 2)\, dt = \left[\dfrac{t^2}{2} + 2t\right]_0^x = \dfrac{1}{2}x^2 + 2x$

(b) $\dfrac{d}{dx}\left[\dfrac{1}{2}x^2 + 2x\right] = x + 2$

67. (a) $\int_8^x \sqrt[3]{t}\, dt = \left[\dfrac{3}{4}t^{4/3}\right]_8^x = \dfrac{3}{4}(x^{4/3} - 16) = \dfrac{3}{4}x^{4/3} - 12$

(b) $\dfrac{d}{dx}\left[\dfrac{3}{4}x^{4/3} - 12\right] = x^{1/3} = \sqrt[3]{x}$

69. (a) $F(x) = \int_1^x \dfrac{1}{t^2}\, dt = -\dfrac{1}{t}\Big]_1^x = 1 - \dfrac{1}{x}$

(b) $F'(x) = \dfrac{d}{dx}\left(1 - \dfrac{1}{x}\right) = \dfrac{1}{x^2}$

71. $F(x) = \int_{-2}^x (t^2 - 2t)\, dt$

$F'(x) = x^2 - 2x$

73. $F(x) = \int_{-1}^x \sqrt{t^4 + 1}\, dt$

$F'(x) = \sqrt{x^4 + 1}$

75. $F(x) = \int_x^{x+2} (4t + 1)\, dt = \left[2t^2 + t\right]_x^{x+2}$

$F'(x) = 8$

Alternate solution:

$F(x) = \int_x^{x+2} (4t + 1)\, dt$

$= \int_x^0 (4t + 1)\, dt + \int_0^{x+2} (4t + 1)\, dt$

$= -\int_0^x (4t + 1)\, dt + \int_0^{x+2} (4t + 1)\, dt$

$F'(x) = -(4x + 1) + 4(x + 2) + 1 = 8$

77. $F(x) = \int_2^{x^2} t^{-3}\, dt = \left[\dfrac{t^{-2}}{-2}\right]_2^{x^2} = \left[-\dfrac{1}{2t^2}\right]_2^{x^2} = \dfrac{-1}{2x^4} + \dfrac{1}{8} \implies F'(x) = 2x^{-5}$

Alternate solution: $F'(x) = (x^2)^{-3}(2x) = 2x^{-5}$

79. $g(x) = \int_0^x f(t)\, dt$

$g(0) = 0, g(1) \approx \frac{1}{2}, g(2) \approx 1, g(3) \approx \frac{1}{2}, g(4) = 0$

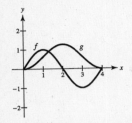

g has a relative maximum at $x = 2$.

83. True

87. $f(x) = \int_0^{1/x} \frac{1}{t^2 + 1}\, dt + \int_0^x \frac{1}{t^2 + 1}\, dt$

By the Second Fundamental Theorem of Calculus, we have

$$f'(x) = \frac{1}{(1/x)^2 + 1}\left(-\frac{1}{x^2}\right) + \frac{1}{x^2 + 1}$$

$$= -\frac{1}{1 + x^2} + \frac{1}{x^2 + 1} = 0.$$

Since $f'(x) = 0$, $f(x)$ must be constant.

89. $x(t) = t^3 - 6t^2 + 9t - 2$

$x'(t) = 3t^2 - 12t + 9$

$\quad\quad = 3(t^2 - 4t + 3)$

$\quad\quad = 3(t - 3)(t - 1)$

Total distance $= \int_0^5 |x'(t)|\, dt$

$\quad\quad\quad = \int_0^5 3|(t - 3)(t - 1)|\, dt$

$\quad\quad\quad = 3\int_0^1 (t^2 - 4t + 3)\, dt - 3\int_1^3 (t^2 - 4t + 3)\, dt + 3\int_3^5 (t^2 - 4t + 3)\, dt$

$\quad\quad\quad = 4 + 4 + 20$

$\quad\quad\quad = 28$ units

91. Total distance $= \int_1^4 |x'(t)|\, dt$

$\quad\quad\quad = \int_1^4 |v(t)|\, dt$

$\quad\quad\quad = \int_1^4 \frac{1}{\sqrt{t}}\, dt$

$\quad\quad\quad = 2t^{1/2} \Big]_1^4$

$\quad\quad\quad = 2(2 - 1) = 2$ units

81. (a) $C(x) = 5000\left(25 + 3\int_0^x t^{1/4}\, dt\right)$

$\quad\quad = 5000\left(25 + 3\left[\frac{4}{5}t^{5/4}\right]_0^x\right)$

$\quad\quad = 5000\left(25 + \frac{12}{5}x^{5/4}\right) = 1000(125 + 12x^{5/4})$

(b) $C(1) = 1000(125 + 12(1)) = \$137,000$

$\quad\quad C(5) = 1000(125 + 12(5)^{5/4}) \approx \$214,721$

$\quad\quad C(10) = 1000(125 + 12(10)^{5/4}) \approx \$338,394$

85. The function $f(x) = \frac{1}{x^2}$ is not continuous at $x = 0$.

You cannot simply integrate from -1 to 1.

Section 6.5   Integration by Substitution

$\int f(g(x))g'(x)\,dx$	$u = g(x)$	$du = g'(x)\,dx$
1. $\int (5x^2 + 1)^2(10x)\,dx$	$5x^2 + 1$	$10x\,dx$
3. $\int \dfrac{x}{\sqrt{x^2 + 1}}\,dx$	$x^2 + 1$	$2x\,dx$

5. $\int (1 + 2x)^4\,2\,dx = \dfrac{(1 + 2x)^5}{5} + C$

    Check: $\dfrac{d}{dx}\left[\dfrac{(1 + 2x)^5}{5} + C\right] = 2(1 + 2x)^4$

7. $\int (9 - x^2)^{1/2}(-2x)\,dx = \dfrac{(9 - x^2)^{3/2}}{3/2} + C = \dfrac{2}{3}(9 - x^2)^{3/2} + C$

    Check: $\dfrac{d}{dx}\left[\dfrac{2}{3}(9 - x^2)^{3/2} + C\right] = \dfrac{2}{3} \cdot \dfrac{3}{2}(9 - x^2)^{1/2}(-2x) = \sqrt{9 - x^2}(-2x)$

9. $\int x^3(x^4 + 3)^2\,dx = \dfrac{1}{4}\int (x^4 + 3)^2(4x^3)\,dx = \dfrac{1}{4}\dfrac{(x^4 + 3)^3}{3} + C = \dfrac{(x^4 + 3)^3}{12} + C$

    Check: $\dfrac{d}{dx}\left[\dfrac{(x^4 + 3)^3}{12} + C\right] = \dfrac{3(x^4 + 3)^2}{12}(4x^3) = (x^4 + 3)^2(x^3)$

11. $\int x^2(x^3 - 1)^4\,dx = \dfrac{1}{3}\int (x^3 - 1)^4(3x^2)\,dx = \dfrac{1}{3}\left[\dfrac{(x^3 - 1)^5}{5}\right] + C = \dfrac{(x^3 - 1)^5}{15} + C$

    Check: $\dfrac{d}{dx}\left[\dfrac{(x^3 - 1)^5}{15} + C\right] = \dfrac{5(x^3 - 1)^4(3x^2)}{15} = x^2(x^3 - 1)^4$

13. $\int t\sqrt{t^2 + 2}\,dt = \dfrac{1}{2}\int (t^2 + 2)^{1/2}(2t)\,dt = \dfrac{1}{2}\dfrac{(t^2 + 2)^{3/2}}{3/2} + C = \dfrac{(t^2 + 2)^{3/2}}{3} + C$

    Check: $\dfrac{d}{dt}\left[\dfrac{(t^2 + 2)^{3/2}}{3} + C\right] = \dfrac{3/2(t^2 + 2)^{1/2}(2t)}{3} = (t^2 + 2)^{1/2}t$

15. $\int 5x(1 - x^2)^{1/3}\,dx = -\dfrac{5}{2}\int (1 - x^2)^{1/3}(-2x)\,dx = -\dfrac{5}{2} \cdot \dfrac{(1 - x^2)^{4/3}}{4/3} + C = -\dfrac{15}{8}(1 - x^2)^{4/3} + C$

    Check: $\dfrac{d}{dx}\left[-\dfrac{15}{8}(1 - x^2)^{4/3} + C\right] = -\dfrac{15}{8} \cdot \dfrac{4}{3}(1 - x^2)^{1/3}(-2x) = 5x(1 - x^2)^{1/3} = 5x\sqrt[3]{1 - x^2}$

17. $\int \dfrac{x}{(1 - x^2)^3}\,dx = -\dfrac{1}{2}\int (1 - x^2)^{-3}(-2x)\,dx = -\dfrac{1}{2}\dfrac{(1 - x^2)^{-2}}{-2} + C = \dfrac{1}{4(1 - x^2)^2} + C$

    Check: $\dfrac{d}{dx}\left[\dfrac{1}{4(1 - x^2)^2} + C\right] = \dfrac{1}{4}(-2)(1 - x^2)^{-3}(-2x) = \dfrac{x}{(1 - x^2)^3}$

19. $\displaystyle\int \frac{x^2}{(1 + x^3)^2}\, dx = \frac{1}{3}\int (1 + x^3)^{-2}(3x^2)\, dx = \frac{1}{3}\left[\frac{(1 + x^3)^{-1}}{-1}\right] + C = -\frac{1}{3(1 + x^3)} + C$

Check: $\displaystyle\frac{d}{dx}\left[-\frac{1}{3(1 + x^3)} + C\right] = -\frac{1}{3}(-1)(1 + x^3)^{-2}(3x^2) = \frac{x^2}{(1 + x^3)^2}$

21. $\displaystyle\int \frac{x}{\sqrt{1 - x^2}}\, dx = -\frac{1}{2}\int (1 - x^2)^{-1/2}(-2x)\, dx = -\frac{1}{2}\frac{(1 - x^2)^{1/2}}{1/2} + C = -\sqrt{1 - x^2} + C$

Check: $\displaystyle\frac{d}{dx}[-(1 - x^2)^{1/2} + C] = -\frac{1}{2}(1 - x^2)^{-1/2}(-2x) = \frac{x}{\sqrt{1 - x^2}}$

23. $\displaystyle\int \left(1 + \frac{1}{t}\right)^3\left(\frac{1}{t^2}\right)dt = -\int \left(1 + \frac{1}{t}\right)^3\left(-\frac{1}{t^2}\right)dt = -\frac{[1 + (1/t)]^4}{4} + C$

Check: $\displaystyle\frac{d}{dt}\left[-\frac{[1 + (1/t)]^4}{4} + C\right] = -\frac{1}{4}(4)\left(1 + \frac{1}{t}\right)^3\left(-\frac{1}{t^2}\right) = \frac{1}{t^2}\left(1 + \frac{1}{t}\right)^3$

25. $\displaystyle\int \frac{1}{\sqrt{2x}}\, dx = \frac{1}{2}\int (2x)^{-1/2}\, 2\, dx = \frac{1}{2}\left[\frac{(2x)^{1/2}}{1/2}\right] + C = \sqrt{2x} + C$

Check: $\displaystyle\frac{d}{dx}\left[\sqrt{2x} + C\right] = \frac{1}{2}(2x)^{-1/2}(2) = \frac{1}{\sqrt{2x}}$

27. $\displaystyle\int \frac{x^2 + 3x + 7}{\sqrt{x}}\, dx = \int (x^{3/2} + 3x^{1/2} + 7x^{-1/2})\, dx = \frac{2}{5}x^{5/2} + 2x^{3/2} + 14x^{1/2} + C = \frac{2}{5}\sqrt{x}(x^2 + 5x + 35) + C$

Check: $\displaystyle\frac{d}{dx}\left[\frac{2}{5}x^{5/2} + 2x^{3/2} + 14x^{1/2} + C\right] = \frac{x^2 + 3x + 7}{\sqrt{x}}$

29. $\displaystyle\int t^2\left(t - \frac{2}{t}\right)dt = \int (t^3 - 2t)\, dt = \frac{1}{4}t^4 - t^2 + C$

Check: $\displaystyle\frac{d}{dt}\left[\frac{1}{4}t^4 - t^2 + C\right] = t^3 - 2t = t^2\left(t - \frac{2}{t}\right)$

31. $\displaystyle\int (9 - y)\sqrt{y}\, dy = \int (9y^{1/2} - y^{3/2})\, dy = 9\left(\frac{2}{3}y^{3/2}\right) - \frac{2}{5}y^{5/2} + C = \frac{2}{5}y^{3/2}(15 - y) + C$

Check: $\displaystyle\frac{d}{dy}\left[\frac{2}{5}y^{3/2}(15 - y) + C\right] = \frac{d}{dy}\left[6y^{3/2} - \frac{2}{5}y^{5/2} + C\right] = 9y^{1/2} - y^{3/2} = (9 - y)\sqrt{y}$

33. $\displaystyle y = \int \left[4x + \frac{4x}{\sqrt{16 - x^2}}\right]dx$

$\displaystyle = 4\int x\, dx - 2\int (16 - x^2)^{-1/2}(-2x)\, dx$

$\displaystyle = 4\left(\frac{x^2}{2}\right) - 2\left[\frac{(16 - x^2)^{1/2}}{1/2}\right] + C$

$\displaystyle = 2x^2 - 4\sqrt{16 - x^2} + C$

35. $\displaystyle y = \int \frac{x + 1}{(x^2 + 2x - 3)^2}\, dx$

$\displaystyle = \frac{1}{2}\int (x^2 + 2x - 3)^{-2}(2x + 2)\, dx$

$\displaystyle = \frac{1}{2}\left[\frac{(x^2 + 2x - 3)^{-1}}{-1}\right] + C$

$\displaystyle = -\frac{1}{2(x^2 + 2x - 3)} + C$

37. (a)

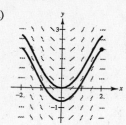

(b) $\dfrac{dy}{dx} = x\sqrt{4-x^2}, \ (2, 2)$

$$y = \int x\sqrt{4-x^2}\,dx = -\frac{1}{2}\int (4-x^2)^{1/2}(-2x\,dx)$$

$$= -\frac{1}{2}\cdot\frac{2}{3}(4-x^2)^{3/2} + C = -\frac{1}{3}(4-x^2)^{3/2} + C$$

$(2, 2)$: $2 = -\dfrac{1}{3}(4-2^2)^{3/2} + C \implies C = 2$

$$y = -\frac{1}{3}(4-x^2)^{3/2} + 2$$

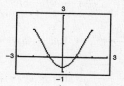

39. $u = x + 2, \ x = u - 2, \ dx = du$

$$\int x\sqrt{x+2}\,dx = \int (u-2)\sqrt{u}\,du$$

$$= \int (u^{3/2} - 2u^{1/2})\,du$$

$$= \frac{2}{5}u^{5/2} - \frac{4}{3}u^{3/2} + C$$

$$= \frac{2u^{3/2}}{15}(3u - 10) + C$$

$$= \frac{2}{15}(x+2)^{3/2}[3(x+2) - 10] + C$$

$$= \frac{2}{15}(x+2)^{3/2}(3x - 4) + C$$

41. $u = 1 - x, \ x = 1 - u, \ dx = -du$

$$\int x^2\sqrt{1-x}\,dx = -\int (1-u)^2\sqrt{u}\,du$$

$$= -\int (u^{1/2} - 2u^{3/2} + u^{5/2})\,du$$

$$= -\left(\frac{2}{3}u^{3/2} - \frac{4}{5}u^{5/2} + \frac{2}{7}u^{7/2}\right) + C$$

$$= -\frac{2u^{3/2}}{105}(35 - 42u + 15u^2) + C$$

$$= -\frac{2}{105}(1-x)^{3/2}[35 - 42(1-x) + 15(1-x)^2] + C$$

$$= -\frac{2}{105}(1-x)^{3/2}(15x^2 + 12x + 8) + C$$

43. $u = 2x - 1, x = \dfrac{1}{2}(u + 1), dx = \dfrac{1}{2}\,du$

$$\int \frac{x^2 - 1}{\sqrt{2x - 1}}\,dx = \int \frac{[(1/2)(u + 1)]^2 - 1}{\sqrt{u}}\frac{1}{2}\,du$$

$$= \frac{1}{8}\int u^{-1/2}[(u^2 + 2u + 1) - 4]\,du$$

$$= \frac{1}{8}\int (u^{3/2} + 2u^{1/2} - 3u^{-1/2})\,du$$

$$= \frac{1}{8}\left(\frac{2}{5}u^{5/2} + \frac{4}{3}u^{3/2} - 6u^{1/2}\right) + C$$

$$= \frac{u^{1/2}}{60}(3u^2 + 10u - 45) + C$$

$$= \frac{\sqrt{2x - 1}}{60}[3(2x - 1)^2 + 10(2x - 1) - 45] + C$$

$$= \frac{1}{60}\sqrt{2x - 1}(12x^2 + 8x - 52) + C$$

$$= \frac{1}{15}\sqrt{2x - 1}(3x^2 + 2x - 13) + C$$

45. $u = x + 1, x = u - 1, dx = du$

$$\int \frac{-x}{(x + 1) - \sqrt{x + 1}}\,dx = \int \frac{-(u - 1)}{u - \sqrt{u}}\,du$$

$$= -\int \frac{(\sqrt{u} + 1)(\sqrt{u} - 1)}{\sqrt{u}(\sqrt{u} - 1)}\,du$$

$$= -\int (1 + u^{-1/2})\,du$$

$$= -(u + 2u^{1/2}) + C$$

$$= -u - 2\sqrt{u} + C$$

$$= -(x + 1) - 2\sqrt{x + 1} + C$$

$$= -x - 2\sqrt{x + 1} - 1 + C$$

$$= -\left(x + 2\sqrt{x + 1}\right) + C_1$$

where $C_1 = -1 + C$.

47. Let $u = x^2 + 1, du = 2x\,dx.$

$$\int_{-1}^{1} x(x^2 + 1)^3\,dx = \frac{1}{2}\int_{-1}^{1} (x^2 + 1)^3(2x)\,dx$$

$$= \left[\frac{1}{8}(x^2 + 1)^4\right]_{-1}^{1} = 0$$

49. Let $u = x^3 + 1, du = 3x^2\,dx$

$$\int_{1}^{2} 2x^2\sqrt{x^3 + 1}\,dx = 2 \cdot \frac{1}{3}\int_{1}^{2} (x^3 + 1)^{1/2}(3x^2)\,dx$$

$$= \left[\frac{2}{3}\frac{(x^3 + 1)^{3/2}}{3/2}\right]_{1}^{2}$$

$$= \frac{4}{9}\left[(x^3 + 1)^{3/2}\right]_{1}^{2}$$

$$= \frac{4}{9}\left[27 - 2\sqrt{2}\right] = 12 - \frac{8}{9}\sqrt{2}$$

51. Let $u = 2x + 1, du = 2\,dx.$

$$\int_{0}^{4} \frac{1}{\sqrt{2x + 1}}\,dx = \frac{1}{2}\int_{0}^{4} (2x + 1)^{-1/2}(2)\,dx = \left[\sqrt{2x + 1}\right]_{0}^{4} = \sqrt{9} - \sqrt{1} = 2$$

53. Let $u = 1 + \sqrt{x}$, $du = \dfrac{1}{2\sqrt{x}}\,dx$.

$$\int_1^9 \frac{1}{\sqrt{x}\left(1 + \sqrt{x}\right)^2}\,dx = 2\int_1^9 \left(1 + \sqrt{x}\right)^{-2}\left(\frac{1}{2\sqrt{x}}\right)dx = \left[-\frac{2}{1 + \sqrt{x}}\right]_1^9 = -\frac{1}{2} + 1 = \frac{1}{2}$$

55. $u = 2 - x$, $x = 2 - u$, $dx = -du$

When $x = 1$, $u = 1$. When $x = 2$, $u = 0$.

$$\int_1^2 (x - 1)\sqrt{2 - x}\,dx = \int_1^0 -[(2 - u) - 1]\sqrt{u}\,du = \int_1^0 (u^{3/2} - u^{1/2})\,du = \left[\frac{2}{5}u^{5/2} - \frac{2}{3}u^{3/2}\right]_1^0 = -\left[\frac{2}{5} - \frac{2}{3}\right] = \frac{4}{15}$$

57. Let $u = \sqrt{x - 5}$, $x = u^2 + 5$, $dx = 2u\,du$.

$$\int_5^{14} x\sqrt{x - 5}\,dx = \int_0^3 (u^2 + 5)\,u\,(2u\,du)$$

$$= \int_0^3 (2u^4 + 10u^2)\,du$$

$$= \left[\frac{2u^5}{5} + \frac{10u^3}{3}\right]_0^3$$

$$= \frac{2(243)}{5} + 10(9) = \frac{936}{5}$$

59. $u = x + 1$, $x = u - 1$, $dx = du$

When $x = 0$, $u = 1$. When $x = 7$, $u = 8$.

$$\text{Area} = \int_0^7 x\sqrt[3]{x + 1}\,dx = \int_1^8 (u - 1)\sqrt[3]{u}\,du$$

$$= \int_1^8 (u^{4/3} - u^{1/3})\,du = \left[\frac{3}{7}u^{7/3} - \frac{3}{4}u^{4/3}\right]_1^8 = \left(\frac{384}{7} - 12\right) - \left(\frac{3}{7} - \frac{3}{4}\right) = \frac{1209}{28}$$

61. $\int_0^4 \dfrac{x}{\sqrt{2x + 1}}\,dx \approx 3.333 = \dfrac{10}{3}$

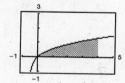

63. $\int_3^7 x\sqrt{x - 3}\,dx \approx 28.8 = \dfrac{144}{5}$

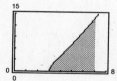

65. $\int (2x - 1)^2\,dx = \dfrac{1}{2}\int (2x - 1)^2\,2\,dx = \dfrac{1}{6}(2x - 1)^3 + C_1 = \dfrac{4}{3}x^3 - 2x^2 + x - \dfrac{1}{6} + C_1$

$\int (2x - 1)^2\,dx = \int (4x^2 - 4x + 1)\,dx = \dfrac{4}{3}x^3 - 2x^2 + x + C_2$

They differ by a constant: $C_2 = C_1 - \dfrac{1}{6}$.

67. $f(x) = x^2(x^2 + 1)$ is even.

$$\int_{-2}^2 x^2(x^2 + 1)\,dx = 2\int_0^2 (x^4 + x^2)\,dx = 2\left[\frac{x^5}{5} + \frac{x^3}{3}\right]_0^2$$

$$= 2\left[\frac{32}{5} + \frac{8}{3}\right] = \frac{272}{15}$$

69. $f(x) = x(x^2 + 1)^3$ is odd.

$$\int_{-2}^2 x(x^2 + 1)^3\,dx = 0$$

71. $\int_0^2 x^2 \, dx = \left[\dfrac{x^3}{3}\right]_0^2 = \dfrac{8}{3}$; the function x^2 is an even function.

(a) $\int_{-2}^0 x^2 \, dx = \int_0^2 x^2 \, dx = \dfrac{8}{3}$

(b) $\int_{-2}^2 x^2 \, dx = 2\int_0^2 x^2 \, dx = \dfrac{16}{3}$

(c) $\int_0^2 (-x^2) \, dx = -\int_0^2 x^2 \, dx = -\dfrac{8}{3}$

(d) $\int_{-2}^0 3x^2 \, dx = 3\int_0^2 x^2 \, dx = 8$

73. $\int_{-4}^4 (x^3 + 6x^2 - 2x - 3) \, dx = \int_{-4}^4 (x^3 - 2x) \, dx + \int_{-4}^4 (6x^2 - 3) \, dx = 0 + 2\int_0^4 (6x^2 - 3) \, dx = 2\Big[2x^3 - 3x\Big]_0^4 = 232$

75. If $u = 5 - x^2$, then $du = -2x \, dx$ and $\int x(5 - x^2)^3 \, dx = -\dfrac{1}{2}\int (5 - x^2)^3(-2x) \, dx = -\dfrac{1}{2}\int u^3 \, du.$

77. $\dfrac{dQ}{dt} = k(100 - t)^2$

$Q(t) = \int k(100 - t)^2 \, dt = -\dfrac{k}{3}(100 - t)^3 + C$

$Q(100) = C = 0$

$Q(t) = -\dfrac{k}{3}(100 - t)^3$

$Q(0) = -\dfrac{k}{3}(100)^3 = 2,000,000 \implies k = -6$

Thus, $Q(t) = 2(100 - t)^3$. When $t = 50$, $Q(50) = \$250,000$.

79. Let $u = x + h$, then $du = dx$. When $x = a$, $u = a + h$. When $x = b$, $u = b + h$. Thus,

$\int_a^b f(x + h) \, dx = \int_{a+h}^{b+h} f(u) \, du = \int_{a+h}^{b+h} f(x) \, dx.$

81. False

$\int (2x + 1)^2 \, dx = \dfrac{1}{2}\int (2x + 1)^2 \, 2 \, dx = \dfrac{1}{6}(2x + 1)^3 + C$

83. True

$\int_{-10}^{10} (ax^3 + bx^2 + cx + d) \, dx = \underbrace{\int_{-10}^{10} (ax^3 + cx) \, dx}_{\text{Odd}} + \underbrace{\int_{-10}^{10} (bx^2 + d) \, dx}_{\text{Even}} = 0 + 2\int_0^{10} (bx^2 + d) \, dx$

85. Let $f(x)$ be an odd function and let $u = -x$. Then,

$\int_{-a}^a f(x) \, dx = \int_a^{-a} f(-u)(-du)$

$= \int_a^{-a} f(u) \, du$

$= -\int_{-a}^a f(u) \, du$

Thus,

$2\int_{-a}^a f(x) \, dx = 0 \implies \int_{-a}^a f(x) \, dx = 0$

Section 6.6   Numerical Integration

1. Exact: $\displaystyle\int_0^2 x^2\,dx = \left[\frac{1}{3}x^3\right]_0^2 = \frac{8}{3} \approx 2.6667$

Trapezoidal: $\displaystyle\int_0^2 x^2\,dx \approx \frac{1}{4}\left[0 + 2\left(\frac{1}{2}\right)^2 + 2(1)^2 + 2\left(\frac{3}{2}\right)^2 + (2)^2\right] = \frac{11}{4} = 2.7500$

Simpson's: $\displaystyle\int_0^2 x^2\,dx \approx \frac{1}{6}\left[0 + 4\left(\frac{1}{2}\right)^2 + 2(1)^2 + 4\left(\frac{3}{2}\right)^2 + (2)^2\right] = \frac{8}{3} \approx 2.6667$

3. Exact: $\displaystyle\int_0^2 x^3\,dx = \left[\frac{x^4}{4}\right]_0^2 = 4.000$

Trapezoidal: $\displaystyle\int_0^2 x^3\,dx \approx \frac{1}{4}\left[0 + 2\left(\frac{1}{2}\right)^3 + 2(1)^3 + 2\left(\frac{3}{2}\right)^3 + (2)^3\right] = \frac{17}{4} = 4.2500$

Simpson's: $\displaystyle\int_0^2 x^3\,dx \approx \frac{1}{6}\left[0 + 4\left(\frac{1}{2}\right)^3 + 2(1)^3 + 4\left(\frac{3}{2}\right)^3 + (2)^3\right] = \frac{24}{6} = 4.0000$

5. Exact: $\displaystyle\int_0^2 x^3\,dx = \left[\frac{1}{4}x^4\right]_0^2 = 4.0000$

Trapezoidal: $\displaystyle\int_0^2 x^3\,dx \approx \frac{1}{8}\left[0 + 2\left(\frac{1}{4}\right)^3 + 2\left(\frac{2}{4}\right)^3 + 2\left(\frac{3}{4}\right)^3 + 2(1)^3 + 2\left(\frac{5}{4}\right)^3 + 2\left(\frac{6}{4}\right)^3 + 2\left(\frac{7}{4}\right)^3 + 8\right] = 4.0625$

Simpson's: $\displaystyle\int_0^2 x^3\,dx \approx \frac{1}{12}\left[0 + 4\left(\frac{1}{4}\right)^3 + 2\left(\frac{2}{4}\right)^3 + 4\left(\frac{3}{4}\right)^3 + 2(1)^3 + 4\left(\frac{5}{4}\right)^3 + 2\left(\frac{6}{4}\right)^3 + 4\left(\frac{7}{4}\right)^3 + 8\right] = 4.0000$

7. Exact: $\displaystyle\int_4^9 \sqrt{x}\,dx = \left[\frac{2}{3}x^{3/2}\right]_4^9 = 18 - \frac{16}{3} = \frac{38}{3} \approx 12.6667$

Trapezoidal: $\displaystyle\int_4^9 \sqrt{x}\,dx \approx \frac{5}{16}\left[2 + 2\sqrt{\frac{37}{8}} + 2\sqrt{\frac{21}{4}} + 2\sqrt{\frac{47}{8}} + 2\sqrt{\frac{26}{4}} + 2\sqrt{\frac{57}{8}} + 2\sqrt{\frac{31}{4}} + 2\sqrt{\frac{67}{8}} + 3\right]$

≈ 12.6640

Simpson's: $\displaystyle\int_4^9 \sqrt{x}\,dx \approx \frac{5}{24}\left[2 + 4\sqrt{\frac{37}{8}} + \sqrt{21} + 4\sqrt{\frac{47}{8}} + \sqrt{26} + 4\sqrt{\frac{57}{8}} + \sqrt{31} + 4\sqrt{\frac{67}{8}} + 3\right] \approx 12.6667$

9. Exact: $\displaystyle\int_1^2 \frac{1}{(x+1)^2}\,dx = \left[-\frac{1}{x+1}\right]_1^2 = -\frac{1}{3} + \frac{1}{2} = \frac{1}{6} \approx 0.1667$

Trapezoidal: $\displaystyle\int_1^2 \frac{1}{(x+1)^2}\,dx \approx \frac{1}{8}\left[\frac{1}{4} + 2\left(\frac{1}{((5/4)+1)^2}\right) + 2\left(\frac{1}{((3/2)+1)^2}\right) + 2\left(\frac{1}{((7/4)+1)^2}\right) + \frac{1}{9}\right]$

$= \frac{1}{8}\left(\frac{1}{4} + \frac{32}{81} + \frac{8}{25} + \frac{32}{121} + \frac{1}{9}\right) \approx 0.1676$

Simpson's: $\displaystyle\int_1^2 \frac{1}{(x+1)^2}\,dx \approx \frac{1}{12}\left[\frac{1}{4} + 4\left(\frac{1}{((5/4)+1)^2}\right) + 2\left(\frac{1}{((3/2)+1)^2}\right) + 4\left(\frac{1}{((7/4)+1)^2}\right) + \frac{1}{9}\right]$

$= \frac{1}{12}\left(\frac{1}{4} + \frac{64}{81} + \frac{8}{25} + \frac{64}{121} + \frac{1}{9}\right) \approx 0.1667$

11. $\displaystyle\int_0^4 \frac{1}{x+1}\,dx, \; n = 4$

Trapezoidal Rule: 1.6833

Simpson's Rule: 1.6222

Graphing utility: 1.6094

13. Trapezoidal: $\displaystyle\int_0^2 \sqrt{1+x^3}\,dx \approx \frac{1}{4}\left[1 + 2\sqrt{1+(1/8)} + 2\sqrt{2} + 2\sqrt{1+(27/8)} + 3\right] \approx 3.2833$

Simpson's: $\displaystyle\int_0^2 \sqrt{1+x^3}\,dx \approx \frac{1}{6}\left[1 + 4\sqrt{1+(1/8)} + 2\sqrt{2} + 4\sqrt{1+(27/8)} + 3\right] \approx 3.2396$

Graphing utility: 3.2413

15. $\displaystyle\int_0^1 \sqrt{x}\sqrt{1-x}\,dx = \int_0^1 \sqrt{x(1-x)}\,dx$

Trapezoidal: $\displaystyle\int_0^1 \sqrt{x(1-x)}\,dx \approx \frac{1}{8}\left[0 + 2\sqrt{\frac{1}{4}\left(1-\frac{1}{4}\right)} + 2\sqrt{\frac{1}{2}\left(1-\frac{1}{2}\right)} + 2\sqrt{\frac{3}{4}\left(1-\frac{3}{4}\right)}\right] \approx 0.3415$

Simpson's: $\displaystyle\int_0^1 \sqrt{x(1-x)}\,dx \approx \frac{1}{12}\left[0 + 4\sqrt{\frac{1}{4}\left(1-\frac{1}{4}\right)} + 2\sqrt{\frac{1}{2}\left(1-\frac{1}{2}\right)} + 4\sqrt{\frac{3}{4}\left(1-\frac{3}{4}\right)}\right] \approx 0.3720$

Graphing utility: 0.3927

17. $\displaystyle\int_{-2}^2 \frac{1}{x^2+1}\,dx \quad n=8$

Trapezoidal Rule: 2.2077

Simpson's Rule: 2.2103

Graphing utility: 2.2143

19. $\displaystyle\int_1^7 \frac{\sqrt{x-1}}{x}\,dx, \; n=6$

Trapezoidal Rule: 2.3521

Simpson's Rule: 2.4385

Graphing utility: 2.5326

21.

The Trapezoidal Rule overestimates the area if the graph of the integrand is concave up.

23. $f(x) = 2x + 3$

$f'(x) = 2$

$f''(x) = 0$

The error is 0 for both rules.

25. $f(x) = x^3$

$f'(x) = 3x^2$

$f''(x) = 6x$

$f'''(x) = 6$

$f^{(4)}(x) = 0$

(a) Trapezoidal: Error $\leq \dfrac{(2-0)^3}{12(4^2)}(12) = 0.5$ since

$f''(x)$ is maximum in $[0, 2]$ when $x = 2$.

(b) Simpson's: Error $\leq \dfrac{(2-0)^5}{180(4^4)}(0) = 0$ since

$f^{(4)}(x) = 0$.

27. $f''(x) = \dfrac{2}{x^3}$ in $[1, 3]$.

(a) $|f''(x)|$ is maximum when $x = 1$ and $|f''(1)| = 2$.

Trapezoidal: Error $\leq \dfrac{2^3}{12n^2}(2) < 0.00001$, $n^2 > 133{,}333.33$, $n > 365.15$; let $n = 366$.

$f^{(4)}(x) = \dfrac{24}{x^5}$ in $[1, 3]$

(b) $|f^{(4)}(x)|$ is maximum when $x = 1$ and when $|f^{(4)}(1)| = 24$.

Simpson's: Error $\leq \dfrac{2^5}{180n^4}(24) < 0.00001$, $n^4 > 426{,}666.67$, $n > 25.56$; let $n = 26$.

29. $f(x) = x^{-2}$, $[1, 3]$

    (a) $f'(x) = -2x^{-3}$, $f''(x) = 6x^{-4}$

        $|f''(x)|$ is maximum when $x = 1$ and $|f''(1)| = 6$.

        Trapezoidal: Error $\leq \dfrac{8}{12n^2}(6) < 0.00001$, $n^2 > 400{,}000$, $n > 632.5$; Let $n = 633$

    (b) $f'''(x) = -24x^{-5}$, $f^{(4)}(x) = 120x^{-6}$

        $|f^{(4)}(x)|$ is maximum when $x = 1$ and $|f^{(4)}(1)| = 120$.

        Simpson's Rule: Error $\leq \dfrac{32}{180n^4}(120) < 0.00001$, $n^4 > 2{,}133{,}333.3$, $n > 38.2$; Let $n = 40$ (must be even)

31. $f(x) = \sqrt{1 + x}$

    (a) $f''(x) = -\dfrac{1}{4(1 + x)^{3/2}}$ in $[0, 2]$.

        $|f''(x)|$ is maximum when $x = 0$ and $|f''(0)| = \dfrac{1}{4}$.

        Trapezoidal: Error $\leq \dfrac{8}{12n^2}\left(\dfrac{1}{4}\right) < 0.00001$, $n^2 > 16{,}666.67$, $n > 129.10$; let $n = 130$.

    (b) $f^{(4)}(x) = \dfrac{-15}{16(1 + x)^{7/2}}$ in $[0, 2]$

        $|f^{(4)}(x)|$ is maximum when $x = 0$ and $|f^{(4)}(0)| = \dfrac{15}{16}$.

        Simpson's: Error $\leq \dfrac{32}{180n^4}\left(\dfrac{15}{16}\right) < 0.00001$, $n^4 > 16{,}666.67$, $n > 11.36$; let $n = 12$.

33. Let $f(x) = Ax^3 + Bx^2 + Cx + D$. Then $f^{(4)}(x) = 0$.

    Simpson's: Error $\leq \dfrac{(b - a)^5}{180n^4}(0) = 0$

    Therefore, Simpson's Rule is exact when approximating the integral of a cubic polynomial.

    Example: $\displaystyle\int_0^1 x^3\, dx = \dfrac{1}{6}\left[0 + 4\left(\dfrac{1}{2}\right)^3 + 1\right] = \dfrac{1}{4}$

    This is the exact value of the integral.

35. $f(x) = \sqrt{2 + 3x^2}$ on $[0, 4]$.

n	$L(n)$	$M(n)$	$R(n)$	$T(n)$	$S(n)$
4	12.7771	15.3965	18.4340	15.6055	15.4845
8	14.0868	15.4480	16.9152	15.5010	15.4662
10	14.3569	15.4544	16.6197	15.4883	15.4658
12	14.5386	15.4578	16.4242	15.4814	15.4657
16	14.7674	15.4613	16.1816	15.4745	15.4657
20	14.9056	15.4628	16.0370	15.4713	15.4657

37. $W = \int_0^5 100x\sqrt{125 - x^3}\, dx$

Simpson's Rule: $n = 12$

$$\int_0^5 100x\sqrt{125 - x^3}\, dx \approx \frac{5}{3(12)}\left[0 + 400\left(\frac{5}{12}\right)\sqrt{125 - \left(\frac{5}{12}\right)^3} + 200\left(\frac{10}{12}\right)\sqrt{125 - \left(\frac{10}{12}\right)^3}\right.$$
$$\left. + 400\left(\frac{15}{12}\right)\sqrt{125 - \left(\frac{15}{12}\right)^3} + \cdots + 0\right] \approx 10{,}233.58 \text{ ft} \cdot \text{lb}$$

39. $\int_0^{1/2} \dfrac{6}{\sqrt{1 - x^2}}\, dx$ Simpson's Rule, $n = 6$

$$\pi \approx \frac{\left(\dfrac{1}{2} - 0\right)}{3(6)}[6 + 4(6.0209) + 2(6.0851) + 4(6.1968) + 2(6.3640) + 4(6.6002) + 6.9282]$$

$$\approx \frac{1}{36}[113.098] \approx 3.1416$$

41. Area $\approx \dfrac{1000}{2(10)}[125 + 2(125) + 2(120) + 2(112) + 2(90) + 2(90) + 2(95) + 2(88) + 2(75) + 2(35)] = 89{,}250 \text{ sq m}$

43. $f(x) = \dfrac{1}{x + 1}$, $n = 10$

Use a computer algebra system to determine t such that

$$\frac{t}{30}\left[\frac{1}{0 + 1} + 4\left(\frac{1}{\frac{t}{10} + 1}\right) + 2\left(\frac{1}{\frac{2t}{10} + 1}\right) + 4\left(\frac{1}{\frac{3t}{10} + 1}\right) + \cdots + 4\left(\frac{1}{\frac{9t}{10} + 1}\right) + \frac{1}{t + 1}\right] = 2$$

to find $t = 6.366$.

Review Exercises for Chapter 6

1.

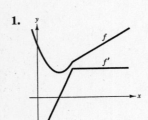

3. $\displaystyle\int (2x^2 + x - 1)\, dx = \frac{2}{3}x^3 + \frac{1}{2}x^2 - x + C$

5. $\displaystyle\int \frac{x^3 + 1}{x^2}\, dx = \int \left(x + \frac{1}{x^2}\right) dx = \frac{1}{2}x^2 - \frac{1}{x} + C$

7. $\displaystyle\int \sqrt[3]{x}\,(x + 3)\, dx = \int (x^{4/3} + 3x^{1/3})\, dx$
$$= \frac{3}{7}x^{7/3} + \frac{9}{4}x^{4/3} + C$$

9. $f'(x) = -2x$, $(-1, 1)$

$f(x) = \displaystyle\int -2x\, dx = -x^2 + C$

When $x = -1$:

$y = -1 + C = 1$

$C = 2$

$y = 2 - x^2$

11. (a)

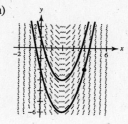

(b) $\dfrac{dy}{dx} = 2x - 4, \quad (4, -2)$

$$y = \int (2x - 4)dx = x^2 - 4x + C$$

$$-2 = 16 - 16 + C \Rightarrow C = -2$$

$$y = x^2 - 4x - 2$$

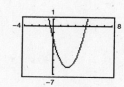

13. $a(t) = a$

$$v(t) = \int a\, dt = at + C_1$$

$$v(0) = 0 + C_1 = 0 \text{ when } C_1 = 0.$$

$$v(t) = at$$

$$s(t) = \int at\, dt = \frac{a}{2}t^2 + C_2$$

$$s(0) = 0 + C_2 = 0 \text{ when } C_2 = 0.$$

$$s(t) = \frac{a}{2}t^2$$

$$s(30) = \frac{a}{2}(30)^2 = 3600 \text{ or}$$

$$a = \frac{2(3600)}{(30)^2} = 8 \text{ ft/sec}^2.$$

$$v(30) = 8(30) = 240 \text{ ft/sec}$$

15. $a(t) = -32$

$$v(t) = -32t + 96$$

$$s(t) = -16t^2 + 96t$$

(a) $v(t) = -32t + 96 = 0$ when $t = 3$ sec.

(b) $s(3) = -144 + 288 = 144$ ft

(c) $v(t) = -32t + 96 = \dfrac{96}{2}$ when $t = \dfrac{3}{2}$ sec.

(d) $s\left(\dfrac{3}{2}\right) = -16\left(\dfrac{9}{4}\right) + 96\left(\dfrac{3}{2}\right) = 108$ ft

17. $\displaystyle\sum_{i=1}^{8} \frac{1}{4i} = \frac{1}{4(1)} + \frac{1}{4(2)} + \cdots + \frac{1}{4(8)}$

19. $\displaystyle\sum_{i=1}^{n} \left(\frac{3}{n}\right)\left(\frac{i+1}{n}\right)^2 = \frac{3}{n}\left(\frac{1+1}{n}\right)^2 + \frac{3}{n}\left(\frac{2+1}{n}\right)^2 + \cdots + \frac{3}{n}\left(\frac{n+1}{n}\right)^2$

21. $\displaystyle\sum_{i=1}^{10} 3i = 3\left(\frac{10(11)}{2}\right) = 165$

23. $\displaystyle\sum_{i=1}^{20} (i+1)^2 = \sum_{i=1}^{20} (i^2 + 2i + 1)$

$$= \frac{20(21)(41)}{6} + 2\frac{20(21)}{2} + 20$$

$$= 2870 + 420 + 20 = 3310$$

25. (a) $\displaystyle\sum_{i=1}^{10} (2i - 1)$ (b) $\displaystyle\sum_{i=1}^{n} i^3$ (c) $\displaystyle\sum_{i=1}^{10} (4i + 2)$

27. $y = \dfrac{10}{x^2 + 1}$, $\Delta x = \dfrac{1}{2}$, $n = 4$

$$S(n) = S(4) = \frac{1}{2}\left[\frac{10}{1} + \frac{10}{(1/2)^2 + 1} + \frac{10}{(1)^2 + 1} + \frac{10}{(3/2)^2 + 1}\right]$$

$$\approx 13.0385$$

$$s(n) = s(4) = \frac{1}{2}\left[\frac{10}{(1/2)^2 + 1} + \frac{10}{1 + 1} + \frac{10}{(3/2)^2 + 1} + \frac{10}{2^2 + 1}\right]$$

$$\approx 9.0385$$

$$9.0385 < \text{Area of Region} < 13.0385$$

29. $y = 6 - x$, $\Delta x = \dfrac{4}{n}$, right endpoints

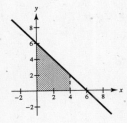

$$\text{Area} = \lim_{n \to \infty} \sum_{i=1}^{n} f(c_i)\,\Delta x$$

$$= \lim_{n \to \infty} \sum_{i=1}^{n} \left(6 - \frac{4i}{n}\right)\frac{4}{n}$$

$$= \lim_{n \to \infty} \frac{4}{n}\left[6n - \frac{4}{n}\frac{n(n + 1)}{2}\right]$$

$$= \lim_{n \to \infty} \left[24 - 8\frac{n + 1}{n}\right] = 24 - 8 = 16$$

31. $y = 5 - x^2$, $\Delta x = \dfrac{3}{n}$

$$\text{Area} = \lim_{n \to \infty} \sum_{i=1}^{n} f(c_i)\,\Delta x$$

$$= \lim_{n \to \infty} \sum_{i=1}^{n} \left[5 - \left(-2 + \frac{3i}{n}\right)^2\right]\left(\frac{3}{n}\right)$$

$$= \lim_{n\infty} \frac{3}{n} \sum_{i=1}^{n} \left[1 + \frac{12i}{n} - \frac{9i^2}{n^2}\right]$$

$$= \lim_{n \to \infty} \frac{3}{n}\left[n + \frac{12}{n}\frac{n(n + 1)}{2} - \frac{9}{n^2}\frac{n(n + 1)(2n + 1)}{6}\right]$$

$$= \lim_{n \to \infty} \left[3 + 18\frac{n + 1}{n} - \frac{9}{2}\frac{(n + 1)(2n + 1)}{n^2}\right]$$

$$= 3 + 18 - 9 = 12$$

33. $x = 5y - y^2$, $2 \le y \le 5$, $\Delta y = \dfrac{3}{n}$

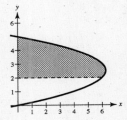

$$\text{Area} = \lim_{n \to \infty} \sum_{i=1}^{n}\left[5\left(2 + \frac{3i}{n}\right) - \left(2 + \frac{3i}{n}\right)^2\right]\left(\frac{3}{n}\right)$$

$$= \lim_{n \to \infty} \frac{3}{n} \sum_{i=1}^{n}\left[10 + \frac{15i}{n} - 4 - 12\frac{i}{n} - \frac{9i^2}{n^2}\right]$$

$$= \lim_{n \to \infty} \frac{3}{n} \sum_{i=1}^{n}\left[6 + \frac{3i}{n} - \frac{9i^2}{n^2}\right]$$

$$= \lim_{n \to \infty} \frac{3}{n}\left[6n + \frac{3}{n}\frac{n(n + 1)}{2} - \frac{9}{n^2}\frac{n(n + 1)(2n + 1)}{6}\right]$$

$$= \left[18 + \frac{9}{2} - 9\right] = \frac{27}{2}$$

35. $\displaystyle \lim_{\|\Delta\| \to \infty} \sum_{i=1}^{n} (2c_i - 3) \, \Delta xi = \int_{4}^{6} (2x - 3) \, dx$

37.

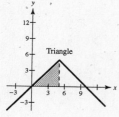

$$\text{Area} = \frac{1}{2}(\text{base})(\text{height})$$

$$= \frac{1}{2}(5)(5) = \frac{25}{2}$$

39. (a) $\displaystyle \int_{2}^{6} [f(x) + g(x)] \, dx = \int_{2}^{6} f(x) \, dx + \int_{2}^{6} g(x) \, dx = 10 + 3 = 13$

(b) $\displaystyle \int_{2}^{6} [f(x) - g(x)] \, dx = \int_{2}^{6} f(x) \, dx - \int_{2}^{6} g(x) \, dx = 10 - 3 = 7$

(c) $\displaystyle \int_{2}^{6} [2f(x) - 3g(x)] \, dx = 2\int_{2}^{6} f(x) \, dx - 3\int_{2}^{6} g(x) \, dx = 2(10) - 3(3) = 11$

(d) $\displaystyle \int_{2}^{6} 5f(x) \, dx = 5\int_{2}^{6} f(x) \, dx = 5(10) = 50$

41. $\displaystyle \int_{1}^{8} \left(\sqrt[3]{x} + 1 \right) dx = \left[\frac{3}{4}x^{4/3} + x \right]_{1}^{8} = \left[\frac{3}{4}(16) + 8 \right] - \left[\frac{3}{4} + 1 \right] = \frac{73}{4}$ (c)

43. $\displaystyle \int_{0}^{4} (2 + x) \, dx = \left[2x + \frac{x^2}{2} \right]_{0}^{4} = 8 + \frac{16}{2} = 16$ **45.** $\displaystyle \int_{-1}^{1} (4t^3 - 2t) \, dt = \left[t^4 - t^2 \right]_{-1}^{1} = 0$

47. $\displaystyle \int_{4}^{9} x\sqrt{x} \, dx = \int_{4}^{9} x^{3/2} \, dx = \left[\frac{2}{5}x^{5/2} \right]_{4}^{9} = \frac{2}{5}\left[\left(\sqrt{9} \right)^5 - \left(\sqrt{4} \right)^5 \right] = \frac{2}{5}(243 - 32) = \frac{422}{5}$

49. $\displaystyle \int_{1}^{3} (2x - 1) \, dx = \left[x^2 - x \right]_{1}^{3} = 6$ **51.** $\displaystyle \int_{3}^{4} (x^2 - 9) \, dx = \left[\frac{x^3}{3} - 9x \right]_{3}^{4}$

$$= \left(\frac{64}{3} - 36 \right) - (9 - 27)$$

$$= \frac{64}{3} - \frac{54}{3} = \frac{10}{3}$$

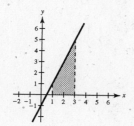

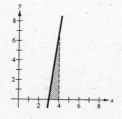

53. $\int_0^1 (x - x^3)\, dx = \left[\dfrac{x^2}{2} - \dfrac{x^4}{4}\right]_0^1 = \dfrac{1}{2} - \dfrac{1}{4} = \dfrac{1}{4}$

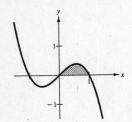

55. Area $= \int_1^9 \dfrac{4}{\sqrt{x}}\, dx = \left[\dfrac{4x^{1/2}}{(1/2)}\right]_1^9 = 8(3 - 1) = 16$

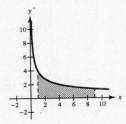

57. $\dfrac{1}{9 - 4}\int_4^9 \dfrac{1}{\sqrt{x}}\, dx = \left[\dfrac{1}{5} 2\sqrt{x}\right]_4^9 = \dfrac{2}{5}(3 - 2) = \dfrac{2}{5}$ Average value

$\dfrac{2}{5} = \dfrac{1}{\sqrt{x}}$

$\sqrt{x} = \dfrac{5}{2}$

$x = \dfrac{25}{4}$

59. $F'(x) = x^2\sqrt{1 + x^3}$

61. $\int (x^2 + 1)^3\, dx = \int (x^6 + 3x^4 + 3x^2 + 1)\, dx = \dfrac{x^7}{7} + \dfrac{3}{5}x^5 + x^3 + x + C$

63. $\int x(x^2 + 1)^3\, dx = \dfrac{1}{2}\int (x^2 + 1)^3\,(2x)\, dx = \dfrac{(x^2 + 1)^4}{8} + C$

65. $u = x^3 + 3, du = 3x^2\, dx$

$\int \dfrac{x^2}{\sqrt{x^3 + 3}}\, dx = \int (x^3 + 3)^{-1/2}\, x^2\, dx = \dfrac{1}{3}\int (x^3 + 3)^{-1/2}\, 3x^2\, dx = \dfrac{2}{3}(x^3 + 3)^{1/2} + C$

67. $u = 1 - 3x^2, du = -6x\, dx$

$\int x(1 - 3x^2)^4\, dx = -\dfrac{1}{6}\int (1 - 3x^2)^4(-6x\, dx) = -\dfrac{1}{30}(1 - 3x^2)^5 + C = \dfrac{1}{30}(3x^2 - 1)^5 + C$

69. $u = \sqrt{x + 5}, x = u^2 - 5, dx = 2u\, du$

$\int x^2\sqrt{x + 5}\, dx = \int (u^2 - 5)^2\, u(2u)\, du = \int (2u^6 - 20u^4 + 50u^2)\, du$

$= \dfrac{2(x + 5)^{7/2}}{7} - 4(x + 5)^{5/2} + \dfrac{50}{3}(x + 5)^{3/2} + C$

71. $\int_{-1}^2 x(x^2 - 4)\, dx = \dfrac{1}{2}\int_{-1}^2 (x^2 - 4)(2x)\, dx = \dfrac{1}{2}\dfrac{(x^2 - 4)^2}{2}\Big]_{-1}^2 = \dfrac{1}{4}[0 - 9] = -\dfrac{9}{4}$

73. $\int_0^3 \dfrac{1}{\sqrt{1 + x}}\, dx = \int_0^3 (1 + x)^{-1/2}\, dx = \left[2(1 + x)^{1/2}\right]_0^3 = 4 - 2 = 2$

75. $u = 1 - y, y = 1 - u, dy = -du$

When $y = 0, u = 1$. When $y = 1, u = 0$.

$$2\pi \int_0^1 (y + 1)\sqrt{1 - y}\, dy = 2\pi \int_1^0 -[(1 - u) + 1]\sqrt{u}\, du$$

$$= 2\pi \int_1^0 (u^{3/2} - 2u^{1/2})\, du = 2\pi \left[\frac{2}{5}u^{5/2} - \frac{4}{3}u^{3/2}\right]_1^0 = \frac{28\pi}{15}$$

77. $C = \dfrac{15,000}{M} \displaystyle\int_t^{t+1} p\, ds = \dfrac{15,000}{M} \displaystyle\int_t^{t+1} (1.20 + 0.04s)\, ds$

$$= \frac{15,000}{M}\left[1.20s + 0.02s^2\right]_t^{t+1}$$

(a) 2005 corresponds to $t = 15$.

$$C = \frac{15,000}{M}\left[1.20s + 0.02s^2\right]_{15}^{16} = \frac{27,300}{M} \text{ dollars}$$

(b) 2007 corresponds to $t = 17$.

$$C = \frac{15,000}{M}\left[1.20s + 0.02s^2\right]_{17}^{18} = \frac{28,500}{M} \text{ dollars}$$

79. Trapezoidal Rule ($n = 4$): $\displaystyle\int_0^1 \frac{x^{3/2}}{3 - x^2}\, dx \approx \frac{1}{8}\left[0 + \frac{2(1/4)^{3/2}}{3 - (1/4)^2} + \frac{2(1/2)^{3/2}}{3 - (1/2)^2} + \frac{2(3/4)^{3/2}}{3 - (3/4)^2} + \frac{1}{2}\right] \approx 0.172$

Simpson's Rule ($n = 4$): $\displaystyle\int_0^1 \frac{x^{3/2}}{3 - x^2}\, dx \approx \frac{1}{12}\left[0 + \frac{4(1/4)^{3/2}}{3 - (1/4)^2} + \frac{2(1/2)^{3/2}}{3 - (1/2)^2} + \frac{4(3/4)^{3/2}}{3 - (3/4)^2} + \frac{1}{2}\right] \approx 0.166$

Graphing utility: 0.166

Problem Solving for Chapter 6

1. (a) $L(1) = \displaystyle\int_1^1 \frac{1}{t}\, dt = 0$

(b) $L'(x) = \dfrac{1}{x}$ by the Second Fundamental Theorem of Calculus.

$L'(1) = 1$

(c) $L(x) = 1 = \displaystyle\int_1^x \frac{1}{t}\, dt$ for $x \approx 2.718$

$\displaystyle\int_1^{2.718} \frac{1}{t}\, dt = 0.999896$

(Note: The exact value of x is e, the base of the natural logarithm function.)

(d) We first show that $\displaystyle\int_1^{x_1} \frac{1}{t}\, dt = \int_{1/x_1}^1 \frac{1}{t}\, dt$.

To see this, let $u = \dfrac{t}{x_1}$ and $du = \dfrac{1}{x_1}\, dt$.

Then $\displaystyle\int_1^{x_1} \frac{1}{t}\, dt = \int_{1/x_1}^1 \frac{1}{ux_1}(x_1\, du) = \int_{1/x_1}^1 \frac{1}{u}\, du = \int_{1/x_1}^1 \frac{1}{t}\, dt$.

Now, $L(x_1 x_2) = \displaystyle\int_1^{x_1 x_2} \frac{1}{t}\, dt = \int_{1/x_1}^{x_2} \frac{1}{u}\, du \left(\text{using } u = \frac{t}{x_1}\right)$

$$= \int_{1/x_1}^1 \frac{1}{u}\, du + \int_1^{x_2} \frac{1}{u}\, du$$

$$= \int_1^{x_1} \frac{1}{u}\, du + \int_1^{x_2} \frac{1}{u}\, du$$

$$= L(x_1) + L(x_2).$$

3. (a) Slope $= \dfrac{9 - 0}{3 - 0} = 3$

(b) $f'(x) = 2x$

Average value $= \dfrac{1}{3 - 0} \displaystyle\int_0^3 f'(x)\, dx = \dfrac{1}{3} \int_0^3 2x\, dx$

$$= \frac{1}{3}\left[x^2\right]_0^3 = 3$$

(c) Average value $= \dfrac{1}{b - a} \displaystyle\int_a^b f'(x)\, dx = \dfrac{f(b) - f(a)}{b - a}$

5. (a)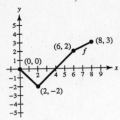

(b)

x	0	1	2	3	4	5	6	7	8
$F(x)$	0	$-\frac{1}{2}$	-2	$-\frac{7}{2}$	-4	$-\frac{7}{2}$	-2	$\frac{1}{4}$	3

(c) $f(x) = \begin{cases} -x, & 0 \le x < 2 \\ x - 4, & 2 \le x < 6 \\ \frac{1}{2}x - 1, & 6 \le x \le 8 \end{cases}$

$F(x) = \int_0^x f(t)\,dt = \begin{cases} (-x^2/2), & 0 \le x < 2 \\ (x^2/2) - 4x + 4, & 2 \le x < 6 \\ (1/4)x^2 - x - 5, & 6 \le x \le 8 \end{cases}$

$F'(x) = f(x)$. F is decreasing on $(0, 4)$ and increasing on $(4, 8)$. Therefore, the minimum is -4 at $x = 4$, and the maximum is 3 at $x = 8$.

(d) $F''(x) = f'(x) = \begin{cases} -1, & 0 < x < 2 \\ 1, & 2 < x < 6 \\ \frac{1}{2}, & 6 < x < 8 \end{cases}$

$x = 2$ is a point of inflection, whereas $x = 6$ is not.

7. (a) $\int_{-1}^1 \sqrt{x + 2}\,dx \approx f\left(-\frac{1}{\sqrt{3}}\right) + f\left(\frac{1}{\sqrt{3}}\right)$

$\approx 1.1927 + 1.6054 = 2.7981$

Exact: 2.7974

Error: ≈ 0.0007

(b) $\int_{-1}^1 \frac{1}{1 + x^2}\,dx \approx \frac{1}{1 + \left(\frac{1}{3}\right)} + \frac{1}{1 + \frac{1}{3}} = \frac{3}{2}$

$\left(\textbf{Note:}\ \text{Exact answer is } \frac{\pi}{2} \approx 1.5708.\right)$

(c) Let $p(x)\,dx = ax^3 + bx^2 + cx + d.$

$\int_{-1}^1 p(x)\,dx = \left[\frac{ax^4}{4} + \frac{bx^3}{3} + \frac{cx^2}{2} + dx\right]_{-1}^1 = \frac{2b}{3} + 2d$

$p\left(-\frac{1}{\sqrt{3}}\right) + p\left(\frac{1}{\sqrt{3}}\right) = \left(\frac{b}{3} + d\right) + \left(\frac{b}{3} + d\right) = \frac{2b}{3} + 2d$

9. Consider $F(x) = [f(x)]^2 \Rightarrow F'(x) = 2f(x)\,f'(x)$. Thus,

$\int_a^b f(x)\,f'(x)\,dx = \int_a^b \frac{1}{2}F'(x)\,dx$

$= \left[\frac{1}{2}F(x)\right]_a^b$

$= \frac{1}{2}[F(b) - F(a)]$

$= \frac{1}{2}[f(b)^2 - f(a)^2].$

11. Consider $\int_0^1 x^5\,dx = \left.\frac{x^6}{6}\right]_0^1 = \frac{1}{6}.$

The corresponding Riemann Sum using right endpoints is

$S(n) = \frac{1}{n}\left[\left(\frac{1}{n}\right)^5 + \left(\frac{2}{n}\right)^5 + \cdots + \left(\frac{n}{n}\right)^5\right]$

$= \frac{1}{n^6}[1^5 + 2^5 + \cdots + n^5].$

Thus, $\lim_{n \to \infty} S(n) = \lim_{n \to \infty} \frac{1^5 + 2^5 + \cdots + n^5}{n^6} = \frac{1}{6}.$

13. By Theorem 6.8, $0 < f(x) \le M \Rightarrow \int_a^b f(x)\,dx \le \int_a^b M\,dx = M(b - a).$

Similarly, $m \le f(x) \Rightarrow m(b - a) = \int_a^b m\,dx \le \int_a^b f(x)\,dx$. Thus, $m(b - a) \le \int_a^b f(x)\,dx \le M(b - a).$

On the interval $[0, 1]$, $1 \le \sqrt{1 + x^4} \le \sqrt{2}$ and $b - a = 1$. Thus, $1 \le \int_0^1 \sqrt{1 + x^4}\,dx \le \sqrt{2}.$

$\left(\textbf{Note:}\ \int_0^1 \sqrt{1 + x^4}\,dx \approx 1.0894\right)$

C H A P T E R 7
Exponential and Logarithmic Functions

CHAPTER 7
Exponential and Logarithmic Functions

Section 7.1   Exponential Functions and Their Graphs

1. $(3.4)^{5.6} \approx 946.852$ **3.** $(1.005)^{400} \approx 7.352$ **5.** $5^{-\pi} \approx 0.006$ **7.** $100^{\sqrt{2}} \approx 673.639$

9. $f(x) = 3^x$

    $g(x) = 3^{x-4}$

    Because $g(x) = f(x - 4)$, the graph of g can be obtained by shifting the graph of f four units to the right.

11. $f(x) = -2^x$

    $g(x) = 5 - 2^x$

    Because $g(x) = 5 + f(x)$, the graph of g can be obtained by shifting the graph of f five units upward.

13. $f(x) = \left(\frac{3}{5}\right)^x$

    $g(x) = -\left(\frac{3}{5}\right)^{x+4}$

    Because $g(x) = -f(x + 4)$, the graph of g can be obtained by reflecting the graph of f in the x-axis and shifting f four units to the left.

15. $f(x) = 0.3^x$

    $g(x) = -0.3^x + 5$

    Because $g(x) = -f(x) + 5$, the graph of g can be obtained by reflecting the graph of f in the x-axis and shifting the graph five units upward.

17. $f(x) = 2^x$

    Increasing

    Asymptote: $y = 0$

    Intercept: $(0, 1)$

    Matches graph (d)

19. $f(x) = 2^{-x}$

    Decreasing

    Asymptote: $y = 0$

    Intercept: $(0, 1)$

    Matches graph (a)

21. $f(x) = \left(\frac{1}{2}\right)^x$

x	-2	-1	0	1	2
$f(x)$	4	2	1	0.5	0.25

Asymptote: $y = 0$

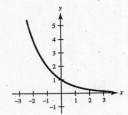

23. $f(x) = \left(\frac{1}{2}\right)^{-x} = 2^x$

x	-2	-1	0	1	2
$f(x)$	0.25	0.5	1	2	4

Asymptote: $y = 0$

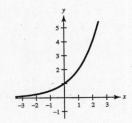

25. $f(x) = 2^{x-1}$

x	-2	-1	0	1	2
$f(x)$	0.125	0.25	0.5	1	2

Asymptote: $y = 0$

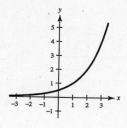

27. $f(x) = 4^{x-3} + 3$

x	-1	0	1	2	3
$f(x)$	3.004	3.016	3.063	3.25	4

Asymptote: $y = 3$

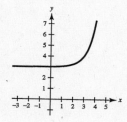

29. $g(x) = 5^x$

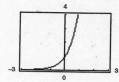

31. $f(x) = \left(\frac{1}{5}\right)^x = 5^{-x}$

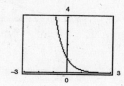

33. $h(x) = 5^{x-2}$

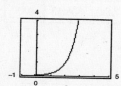

35. $g(x) = 5^{-x} - 3$

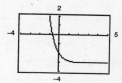

37. $y = 2^{-x^2}$

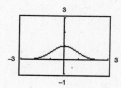

39. $y = 3^{x-2} + 1$

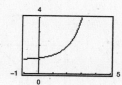

41. $y = 1.08^{-5x}$

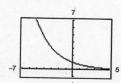

43. $e^{1/2} \approx 1.649$

45. $e^{-3/4} \approx 0.472$

47. $40e^{0.012} \approx 40.483$

49. $500e^{-1.250} \approx 143.252$

51. $f(x) = e^x$

x	-2	-1	0	1	2
$f(x)$	0.135	0.368	1	2.718	7.389

Asymptote: $y = 0$

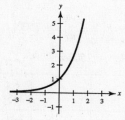

53. $f(x) = 3e^{x+4}$

x	-8	-7	-6	-5	-4
$f(x)$	0.055	0.149	0.406	1.104	3

Asymptote: $y = 0$

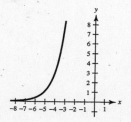

55. $f(x) = 2e^{x-2} + 4$

x	-2	-1	0	1	2
$f(x)$	4.037	4.100	4.271	4.736	6

Asymptote: $y = 4$

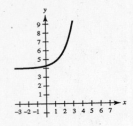

57. $s(t) = 2e^{0.12t}$

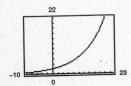

59. $g(x) = 1 + e^{-x}$

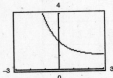

61. $f(x) = 1000e^{0.01x}$

x	-300	-200	-100	-50	0
$f(x)$	49.79	135.34	367.88	606.53	1000

63. $f(x) = 50e^{-0.0025x}$

x	-1000	-750	-500	-250	0
$f(x)$	609.12	326.04	174.52	93.41	50

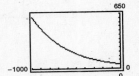

65. $f(x) = \dfrac{250}{1 + 2e^{-0.01x}}$

x	-400	0	200	400	600
$f(x)$	2.27	83.33	196.75	241.17	248.77

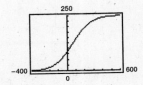

67. $P = \$2500$, $r = 8\%$, $t = 10$ years

Compounded n times per year: $A = P\left(1 + \dfrac{r}{n}\right)^{nt} = 2500\left(1 + \dfrac{0.08}{n}\right)^{10n}$

Compounded continuously: $A = Pe^{rt} = 2500e^{0.08(10)}$

n	1	2	4	12	365	Continuous Compounding
A	\$5397.31	\$5477.81	\$5520.10	\$5549.10	\$5563.36	\$5563.85

69. $P = \$2500$, $r = 8\%$, $t = 20$ years

Compounded n times per year: $A = P\left(1 + \dfrac{r}{n}\right)^{nt} = 2500\left(1 + \dfrac{0.08}{n}\right)^{20n}$

Compounded continuously: $A = Pe^{rt} = 2500e^{0.08(20)}$

n	1	2	4	12	365	Continuous Compounding
A	\$11,652.39	\$12,002.55	\$12,188.60	\$12,317.01	\$12,380.41	\$12,382.58

71. $A = Pe^{rt}$

$A = 12{,}000e^{0.08t}$

t	1	10	20	30	40	50
A	\$12,999.44	\$26,706.49	\$59,436.39	\$132,278.12	\$294,390.36	\$655,177.80

73. $A = Pe^{rt}$

$A = 12{,}000e^{0.065t}$

t	1	10	20	30	40	50
A	\$12,805.91	\$22,986.49	\$44,031.56	\$84,344.25	\$161,564.86	\$309,484.08

75. $f(x) = 3^{x-2}$

$\qquad = 3^x 3^{-2}$

$\qquad = 3^x\left(\dfrac{1}{3^2}\right)$

$\qquad = \dfrac{1}{9}(3^x)$

$\qquad = h(x)$

Thus, $f(x) \neq g(x)$, but $f(x) = h(x)$.

77. $f(x) = 16(4^{-x})$ $\qquad$ and $\qquad$ $f(x) = 16(4^{-x})$

$\qquad = 4^2(4^{-x})$ $\qquad\qquad\qquad\qquad = 16(2^2)^{-x}$

$\qquad = 4^{2-x}$ $\qquad\qquad\qquad\qquad\quad = 16(2^{-2x})$

$\qquad = \left(\dfrac{1}{4}\right)^{-(2-x)}$ $\qquad\qquad\qquad = h(x)$

$\qquad = \left(\dfrac{1}{4}\right)^{x-2}$

$\qquad = g(x)$

Thus, $f(x) = g(x) = h(x)$.

79. $y = 3^x$ and $y = 4^x$

x	-2	-1	0	1	2
3^x	$\frac{1}{9}$	$\frac{1}{3}$	1	3	9
4^x	$\frac{1}{16}$	$\frac{1}{4}$	1	4	16

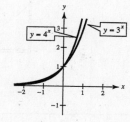

(a) $4^x < 3^x$ when $x < 0$.

(b) $4^x > 3^x$ when $x > 0$.

81.

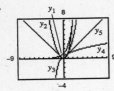

$$y_1 = e^x$$
$$y_2 = x^2$$
$$y_3 = x^3$$
$$y_4 = \sqrt{x}$$
$$y_5 = |x|$$

The function that increases at the fastest rate for "large" values of x is $y_1 = e^x$. (Note: One of the intersection points of $y = e^x$ and $y = x^3$ is approximately $(4.536, 93)$ and past this point $e^x > x^3$. This is not shown on the graph above.)

83. It usually implies rapid growth.

85. $A = 25,000e^{(0.0875)(25)} \approx \$222,822.57$

87. (a) The steeper curve represents the investment earning compound interest, because compound interest earns more than simple interest. With simple interest there is no compounding so the growth is linear.

89. $C(10) = 23.95(1.04)^{10} \approx \$35.45$

(b) Compound interest formula: $A = 500\left(1 + \dfrac{0.07}{1}\right)^{(1)t}$

$$= 500(1.07)^t$$

Simple interest formula: $A = Prt + P$

$$= 500(0.07)t + 500$$

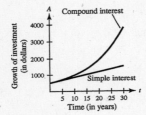

91. $P(t) = 100e^{0.2197t}$

(a) $P(0) = 100$

(b) $P(5) \approx 300$

(c) $P(10) \approx 900$

93. $Q = 25\left(\frac{1}{2}\right)^{t/1599}$

(a) $Q(0) = 25$ grams

(b) $Q(1000) = 25\left(\frac{1}{2}\right)^{1000/1599} \approx 16.21$ grams

(c)

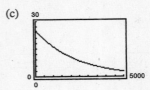

95. $P = 107,428e^{-0.15h}$

(a)

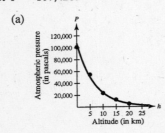

(b)

h	0	5	10	15	20
P (data)	101,293	54,735	23,294	12,157	5,069
P (model)	107,428	50,745	23,970	11,323	5,349

(c) $P(8) \approx 32,357$ pascals

(d) $P = 21,000$ when $h \approx 10.88$ kilometers.

97. $P = 115.5e^{0.0445t}$ where $t = 0$ corresponds to 1970 and P is in thousands.

(a)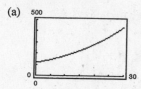

(b) For 1970 use $t = 0$: $P = 115,500$ people

For 1980 use $t = 10$: $P \approx 180,240$ people

For 1990 use $t = 20$: $P \approx 281,260$ people

For 2000 use $t = 30$: $P \approx 438,900$ people

The rate at which the population is increasing is increasing, therefore the population is growing exponentially and does not fit a linear model.

(c) $P = 750$ when $t \approx 42$ which corresponds to 2012.

99. $P = 1.879e^{0.01451t}$ where $t = 50$ corresponds to 1850.

(a) and (b)

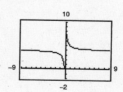

(c) $P = 40$ when $t \approx 211$ which corresponds to 2011.

101. $D = -182.08 + 677.14e^{-0.0142t}$, $t = 0 \leftrightarrow 1970$

(a) $D(0) \approx 495.06$ deaths per 100,000 people in 1970.

$D(10) \approx 405.42$ deaths per 100,000 people in 1980.

$D(20) \approx 327.65$ deaths per 100,000 people in 1990.

$D(30) \approx 260.17$ deaths per 100,000 people in 2000.

(b) No. The model decreases rapidly and eventually becomes negative as t increases.

103. True

105. False, $e \neq \dfrac{271,801}{99,990}$.

e is an irrational number.

107. (a) $f(x) = \dfrac{8}{1 + e^{-0.5x}}$

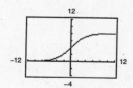

Horizontal asymptotes: $y = 0$ and $y = 8$

(b) $g(x) = \dfrac{8}{1 + e^{-0.5/x}}$

Horizontal asymptote: $y = 4$

Vertical asymptote: $x = 0$

Section 7.2   Logarithmic Functions and Their Graphs

1. $\log_4 64 = 3 \implies 4^3 = 64$

3. $\log_7 \frac{1}{49} = -2 \implies 7^{-2} = \frac{1}{49}$

5. $\log_{32} 4 = \frac{2}{5} \implies 32^{2/5} = 4$

7. $\log_{1/4} 64 = -3 \implies \left(\frac{1}{4}\right)^{-3} = 64$

9. $5^3 = 125 \implies \log_5 125 = 3$

11. $81^{1/4} = 3 \implies \log_{81} 3 = \frac{1}{4}$

13. $6^{-2} = \frac{1}{36} \implies \log_6 \frac{1}{36} = -2$

15. $\left(\frac{2}{3}\right)^{-2} = \frac{9}{4} \implies \log_{2/3} \frac{9}{4} = -2$

17. $a^k = b \implies \log_a b = k$

19. $\log_2 16 = \log_2 2^4 = 4$

21. $\log_{16} 4 = \log_{16} 16^{1/2} = \frac{1}{2}$

23. $\log_7 1 = \log_7 7^0 = 0$

25. $\log_{10} 0.01 = \log_{10} 10^{-2} = -2$

27. $\log_8 32 = \log_8 8^{5/3} = \frac{5}{3}$

29. $\log_a a^2 = 2$

31. $\log_{10} 345 \approx 2.538$

33. $\log_{10} \frac{4}{5} \approx -0.097$

35. $f(x) = \log_3 x + 2$

Asymptote: $x = 0$

Point on graph: $(1, 2)$

Matches graph (c)

37. $f(x) = -\log_3(x + 2)$

Asymptote: $x = -2$

Point on graph: $(-1, 0)$

Matches graph (d)

39. $f(x) = \log_3(1 - x)$

Asymptote: $x = 1$

Point on graph: $(0, 0)$

Matches graph (b)

41. $f(x) = \log_4 x$

Domain: $x > 0 \implies$ The domain is $(0, \infty)$.

x-intercept: $(1, 0)$

Vertical asymptote: $x = 0$

$y = \log_4 x \implies 4^y = x$

x	$\frac{1}{4}$	1	2	4
$f(x)$	-1	0	$\frac{1}{2}$	1

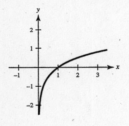

43. $y = -\log_3 x + 2$

Domain: $(0, \infty)$

x-intercept:

$$-\log_3 x + 2 = 0$$
$$2 = \log_3 x$$
$$3^2 = x$$
$$9 = x$$

The x-intercept is $(9, 0)$.

Vertical asymptote: $x = 0$

$y = -\log_3 x + 2$

$\log_3 x = 2 - y \implies 3^{2-y} = x$

x	27	9	3	1	$\frac{1}{3}$
y	-1	0	1	2	3

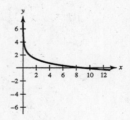

45. $f(x) = -\log_6(x + 2)$

Domain: $x + 2 > 0 \implies x > -2$

The domain is $(-2, \infty)$.

x-intercept:

$$0 = -\log_6(x + 2)$$
$$0 = \log_6(x + 2)$$
$$6^0 = x + 2$$
$$1 = x + 2$$
$$-1 = x$$

The x-intercept is $(-1, 0)$.

Vertical asymptote: $x + 2 = 0 \implies x = -2$

$$y = -\log_6(x + 2)$$
$$-y = \log_6(x + 2)$$
$$6^{-y} - 2 = x$$

x	4	-1	$-1\frac{5}{6}$	$-1\frac{35}{36}$
$f(x)$	-1	0	1	2

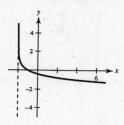

47. $y = \log_{10}\left(\dfrac{x}{5}\right)$

Domain: $\dfrac{x}{5} > 0 \implies x > 0$

The domain is $(0, \infty)$.

x-intercept:

$$\log_{10}\left(\frac{x}{5}\right) = 0$$
$$\frac{x}{5} = 10^0$$
$$\frac{x}{5} = 1 \implies x = 5$$

The x-intercept is $(5, 0)$.

Vertical asymptote: $\dfrac{x}{5} = 0 \implies x = 0$

The vertical asymptote is the y-axis.

x	1	2	3	4	5	6	7
y	-0.70	-0.40	-0.22	-0.10	0	0.08	0.15

49. $f(x) = \log_{10}(x + 1)$

Domain: $x + 1 > 0 \implies x > -1$

The domain is $(-1, \infty)$.

x-intercept:

$$\log_{10}(x + 1) = 0$$
$$x + 1 = 10^0$$
$$x + 1 = 1$$
$$x = 0$$

The x-intercept is $(0, 0)$.

Vertical asymptote: $x + 1 = 0 \implies x = -1$

$y = \log_{10}(x + 1) \implies 10^y = x + 1 \implies 10^y - 1 = x$

x	-0.9	0	9
y	-1	0	1

51. $\ln 4 = 1.386 \ldots \implies e^{1.386 \cdots} = 4$

53. $\ln 1 = 0 \implies e^0 = 1$

55. $\ln e^{-3} = -3 \implies e^{-3} = e^{-3}$

57. $e^3 = 20.0855 \ldots \implies \ln 20.0855 \ldots = 3$

59. $e^0 = 1 \implies \ln 1 = 0$

61. $\sqrt{e} = 1.6487 \ldots \implies e^{1/2} = 1.6487 \ldots$
$\implies \ln 1.6487 \ldots = \frac{1}{2}$

63. $\ln e^3 = 3$

65. $\ln e^{-2.1} = -2.1$

67. $\ln\left(\dfrac{1}{e^3}\right) = \ln e^{-3} = -3$

69. $\ln 18.42 \approx 2.913$

71. $3 \ln 0.32 \approx -3.418$

73. $\ln\left(1 + \sqrt{3}\right) \approx 1.005$

75. $\ln \frac{2}{3} \approx -0.405$

77. $f(x) = -\ln x$

The graph of f can be obtained by reflecting the graph of $y = \ln x$ about the x-axis.

Matches graph (d)

79. $f(x) = \ln x - 1$

The graph of f can be obtained by shifting the graph of $y = \ln x$ one unit downward.

Matches graph (a)

81. $f(x) = \ln(1 - x) = \ln(-x + 1)$

The graph of f can be obtained by shifting the graph of $y = \ln x$ one unit to the left and then reflecting it about the y-axis.

Matches graph (c)

83. $f(x) = \ln(x - 2)$

Domain: $x - 2 > 0 \implies x > 2$

The domain is $(2, \infty)$.

x-intercept: $0 = \ln(x - 2)$
$e^0 = x - 2$
$3 = x$

The x-intercept is $(3, 0)$.

Vertical asymptote:
$x - 2 = 0 \implies x = 2$

x	2.5	3	4	5
$f(x)$	-0.69	0	0.69	1.10

85. $g(x) = \ln(-x)$

Domain: $-x > 0 \implies x < 0$

The domain is $(-\infty, 0)$.

x-intercept: $0 = \ln(-x)$
$e^0 = -x$
$-1 = x$

The x-intercept is $(-1, 0)$.

Vertical asymptote: $-x = 0 \implies x = 0$

x	-0.5	-1	-2	-3
$g(x)$	-0.69	0	0.69	1.10

87. $f(x) = \ln(x^2 - 1)$

Domain: $x^2 - 1 > 0 \implies x^2 > 1 \implies x < -1$ or $x > 1$

The domain is $(-\infty, -1) \cup (1, \infty)$.

x-intercepts: $\ln(x^2 - 1) = 0 \implies x^2 - 1 = 1 \implies x^2 = 2 \implies x = \pm\sqrt{2}$

The x-intercepts are $\left(\pm\sqrt{2}, 0\right)$.

Vertical asymptotes: $x^2 - 1 = 0 \implies x^2 = 1 \implies x = \pm 1$

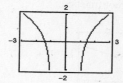

x	± 2	± 3	± 4
$f(x)$	1.099	2.079	2.708

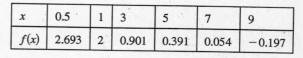

89. $f(x) = -\ln x + 2$

Domain: $x > 0$

The domain is $(0, \infty)$.

x-intercept: $-\ln x + 2 = 0 \implies \ln x = 2 \implies x = e^2$

The x-intercept is $(e^2, 0)$.

Vertical asymptote: $x = 0$

x	0.5	1	3	5	7	9
$f(x)$	2.693	2	0.901	0.391	0.054	-0.197

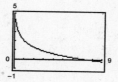

91. $f(x) = |\ln x|$

(a)

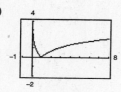

(b) Increasing on $(1, \infty)$

Decreasing on $(0, 1)$

(c) Relative minimum: $(1, 0)$

93. $f(x) = \dfrac{x}{2} - \ln\dfrac{x}{4}$

(a)

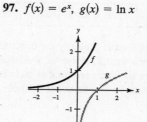

(b) Increasing on $(2, \infty)$.

Decreasing on $(0, 2)$.

(c) Relative minimum: $\left(2, 1 - \ln\frac{1}{2}\right)$

95. $f(x) = 3^x$, $g(x) = \log_3 x$

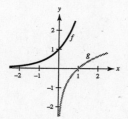

f and g are inverses. Their graphs are reflected about the line $y = x$.

97. $f(x) = e^x$, $g(x) = \ln x$

f and g are inverses. Their graphs are reflected about the line $y = x$.

99. (a) $f(x) = \ln x$

$g(x) = \sqrt{x}$

The natural log function grows at a slower rate than the square root function.

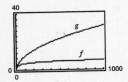

(b) $f(x) = \ln x$

$g(x) = \sqrt[4]{x}$

The natural log function grows at a slower rate than the fourth root function.

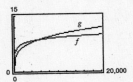

101. $f(x) = \log_{10} x$

(a) Domain: $(0, \infty)$

(b)
$$y = \log_{10} x$$
$$x = \log_{10} y$$
$$10^x = y$$
$$f^{-1}(x) = 10^x$$

(c) Since $\log_{10} 1000 = 3$ and $\log_{10} 10{,}000 = 4$, the interval in which $f(x)$ will be found is $(3, 4)$.

(d) When $f(x)$ is negative, x is in the interval $(0, 1)$.

(e)
$$0 = \log_{10} 1$$
$$1 = \log_{10} 10$$
$$2 = \log_{10} 100$$
$$3 = \log_{10} 1000$$

When $f(x)$ is increased by one unit, x is increased by a factor of 10.

(f) $f(x_1) = 3n$ $f(x_2) = n$

$\log_{10} x_1 = 3n$ $\log_{10} x_2 = n$

$x_1 = 10^{3n}$ $x_2 = 10^n$

$x_1 : x_2 = 10^{3n} : 10^n = 10^{2n} : 1$

103. $t = \dfrac{10 \ln 2}{\ln 67 - \ln 50} \approx 23.68$ years

105. $t = \dfrac{\ln K}{0.095}$

(a)

K	1	2	4	6	8	10	12
t	0	7.3	14.6	18.9	21.9	24.2	26.2

The number of years required to multiply the original investment by K increases with K. However, the larger the value of K, the fewer the years required to increase the value of the investment by an additional multiple of the original investment.

(b)

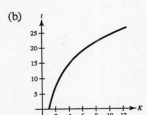

107. $y = 80.4 - 11 \ln x$

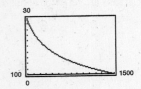

$y(300) = 80.4 - 11 \ln 300 \approx 17.66 \text{ ft}^3/\text{min}$

109. $f(x) = \dfrac{\ln x}{x}$

(a)

x	1	5	10	10^2	10^4	10^6
$f(x)$	0	0.322	0.230	0.046	0.00092	0.0000138

(b) As $x \to \infty$, $f(x) \to 0$.

(c)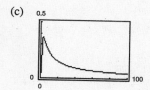

111. $t = 12.542 \ln\left(\dfrac{1100.65}{1100.65 - 1000}\right) \approx 30$ years

113. Total amount $= (1100.65)(12)(30) = \$396{,}234$

Interest $= 396{,}234 - 150{,}000 = \$246{,}234$

115. $y = 33.06 + 0.312\sqrt{t}\ln t$, $1 \leftrightarrow 1991$

(a)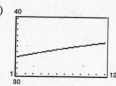

(b) $y(1) \approx 33.06$ years

   $y(12) \approx 35.75$ years

(c) The model would probably work for a few years after 2002 since the population of the United States is gradually aging.

(d) The model is undefined for $t \le 0$ so cannot be used for years prior to 1991.

117. False. You can determine the graph of $f(x) = \log_6 x$ by graphing $g(x) = 6^x$ and reflecting it about the line $y = x$.

119. False. To find the domain of $f(x) = \log_2(bx + c)$ where b and c are nonzero constants, set $bx + c > 0$ and solve.

$$bx + c > 0 \implies x > -\dfrac{c}{b}$$

The domain of $f(x) = \log_2(bx + c)$ is $\left(-\dfrac{c}{b}, \infty\right)$.

Section 7.3   Using Properties of Logarithms

1. $\log_3 7 = \dfrac{\log_{10} 7}{\log_{10} 3} = \dfrac{\ln 7}{\ln 3} \approx 1.771$

3. $\log_{1/2} 4 = \dfrac{\log_{10} 4}{\log_{10} \frac{1}{2}} = \dfrac{\ln 4}{\ln \frac{1}{2}} = -2.000$

5. $\log_9(0.4) = \dfrac{\log_{10} 0.4}{\log_{10} 9} = \dfrac{\ln 0.4}{\ln 9} \approx -0.417$

7. $\log_{15} 1250 = \dfrac{\log_{10} 1250}{\log_{10} 15} = \dfrac{\ln 1250}{\ln 15} \approx 2.633$

9. (a) $\log_5 x = \dfrac{\log_{10} x}{\log_{10} 5}$

(b) $\log_5 x = \dfrac{\ln x}{\ln 5}$

11. (a) $\log_{1/5} x = \dfrac{\log_{10} x}{\log_{10}\left(\frac{1}{5}\right)}$

(b) $\log_{1/5} x = \dfrac{\ln x}{\ln\left(\frac{1}{5}\right)}$

13. (a) $\log_x \dfrac{3}{10} = \dfrac{\log_{10}\left(\frac{3}{10}\right)}{\log_{10} x}$

(b) $\log_x \dfrac{3}{10} = \dfrac{\ln\left(\frac{3}{10}\right)}{\ln x}$

15. (a) $\log_{2.6} x = \dfrac{\log_{10} x}{\log_{10} 2.6}$

(b) $\log_{2.6} x = \dfrac{\ln x}{\ln 2.6}$

17. $f(x) = \log_2 x = \dfrac{\log_{10} x}{\log_{10} 2} = \dfrac{\ln x}{\ln 2}$

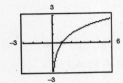

19. $f(x) = \log_{1/2} x = \dfrac{\log_{10} x}{\log_{10} \frac{1}{2}} = \dfrac{\ln x}{\ln\left(\frac{1}{2}\right)}$

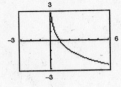

21. $f(x) = \log_{11.8} x = \dfrac{\log_{10} x}{\log_{10} 11.8} = \dfrac{\ln x}{\ln 11.8}$

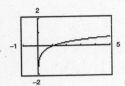

23. $\log_{10} 5x = \log_{10} 5 + \log_{10} x$

25. $\log_{10} \dfrac{5}{x} = \log_{10} 5 - \log_{10} x$

27. $\log_8 x^4 = 4 \log_8 x$

29. $\ln \sqrt{z} = \ln z^{1/2} = \frac{1}{2} \ln z$

31. $\ln xyz = \ln x + \ln y + \ln z$

33. $\ln \sqrt{a-1} = \frac{1}{2} \ln(a-1), \ a > 1$

35. $\ln z(z-1)^2 = \ln z + \ln(z-1)^2$
$= \ln z + 2 \ln(z-1), \ z > 1$

37. $\ln \sqrt[3]{\dfrac{x}{y}} = \dfrac{1}{3} \ln \dfrac{x}{y}$
$= \dfrac{1}{3} [\ln x - \ln y]$
$= \dfrac{1}{3} \ln x - \dfrac{1}{3} \ln y$

39. $\ln\left(\dfrac{x^4 \sqrt{y}}{z^5}\right) = \ln x^4 \sqrt{y} - \ln z^5$
$= \ln x^4 + \ln \sqrt{y} - \ln z^5$
$= 4 \ln x + \dfrac{1}{2} \ln y - 5 \ln z$

41. $\log_b\left(\dfrac{x^2}{y^2 z^3}\right) = \log_b x^2 - \log_b y^2 z^3$
$= \log_b x^2 - [\log_b y^2 + \log_b z^3]$
$= 2 \log_b x - 2 \log_b y - 3 \log_b z$

43. $\ln x + \ln 3 = \ln 3x$

45. $\log_4 z - \log_4 y = \log_4 \dfrac{z}{y}$

47. $2 \log_2(x+4) = \log_2(x+4)^2$

49. $\frac{1}{4} \log_3 5x = \log_3 (5x)^{1/4} = \log_3 \sqrt[4]{5x}$

51. $\ln x - 3 \ln(x+1) = \ln x - \ln(x+1)^3$
$= \ln \dfrac{x}{(x+1)^3}$

53. $\ln(x - 2) - \ln(x + 2) = \ln\left(\dfrac{x - 2}{x + 2}\right)$

55. $\ln x - 4[\ln(x + 2) + \ln(x - 2)] = \ln x - 4\ln(x + 2)(x - 2)$

$$= \ln x - 4\ln(x^2 - 4)$$

$$= \ln x - \ln(x^2 - 4)^4$$

$$= \ln \dfrac{x}{(x^2 - 4)^4}$$

57. $\dfrac{1}{3}[2\ln(x + 3) + \ln x - \ln(x^2 - 1)] = \dfrac{1}{3}[\ln(x + 3)^2 + \ln x - \ln(x^2 - 1)]$

$$= \dfrac{1}{3}[\ln x(x + 3)^2 - \ln(x^2 - 1)]$$

$$= \dfrac{1}{3}\ln\dfrac{x(x + 3)^2}{x^2 - 1}$$

$$= \ln \sqrt[3]{\dfrac{x(x + 3)^2}{x^2 - 1}}$$

59. $\dfrac{1}{3}[\ln y + 2\ln(y + 4)] - \ln(y - 1) = \dfrac{1}{3}[\ln y + \ln(y + 4)^2] - \ln(y - 1)$

$$= \dfrac{1}{3}\ln y(y + 4)^2 - \ln(y - 1)$$

$$= \ln\sqrt[3]{y(y + 4)^2} - \ln(y - 1)$$

$$= \ln\dfrac{\sqrt[3]{y(y + 4)^2}}{y - 1}$$

61. $\dfrac{1}{2}[\ln(x + 3) + \ln(x - 3)] - \ln(x^2 - 9) = \dfrac{1}{2}\ln[(x + 3)(x - 3)] - \ln[(x + 3)(x - 3)]$

$$= -\dfrac{1}{2}\ln[(x + 3)(x - 3)]$$

$$= \ln[(x^2 - 9)^{-1/2}]$$

$$= \ln\left[\dfrac{1}{\sqrt{x^2 - 9}}\right]$$

$$= \ln\left[1/\sqrt{(x + 3)(x - 3)}\right]$$

63. $\ln(x + 2) + \ln(x - 2) - \ln\sqrt{x^2 - 4} = \ln[(x + 2)(x - 2)] - \dfrac{1}{2}\ln(x^2 - 4)$

$$= \ln(x^2 - 4) - \dfrac{1}{2}\ln(x^2 - 4)$$

$$= \dfrac{1}{2}\ln(x^2 - 4)$$

$$= \ln\sqrt{x^2 - 4}$$

$$= \ln\sqrt{(x + 2)(x - 2)}$$

65. $\log_2 \dfrac{32}{4} = \log_2 32 - \log_2 4 \neq \dfrac{\log_2 32}{\log_2 4}$ **67.** $\log_3 9 = 2\log_3 3 = 2$

The first two expressions are equal by Property 2.

69. $\log_4 16^{1.2} = 1.2(\log_4 16) = 1.2 \log_4 4^2 = 1.2(2) = 2.4$

71. $\log_3(-9)$ is undefined. -9 is not in the domain of $\log_3 x$.

73. $\log_5 75 - \log_5 3 = \log_5 \frac{75}{3}$

$= \log_5 25$

$= \log_5 5^2$

$= 2 \log_5 5$

$= 2$

75. $\ln e^2 - \ln e^5 = 2 - 5 = -3$

77. $\log_{10} 0$ is undefined. 0 is not in the domain of $\log_{10} x$.

79. $\ln e^{4.5} = 4.5$

81. $\log_4 8 = \frac{\log_2 8}{\log_2 4} = \frac{\log_2 2^3}{\log_2 2^2} = \frac{3}{2}$

83. $\log_5 \frac{1}{250} = \log_5 \left(\frac{1}{125} \cdot \frac{1}{2} \right) = \log_5 \frac{1}{125} + \log_5 \frac{1}{2}$

$= \log_5 5^{-3} + \log_5 2^{-1}$

$= -3 - \log_5 2$

85. $\ln(5e^6) = \ln 5 + \ln e^6 = \ln 5 + 6 = 6 + \ln 5$

87. $\log_b 10 = \log_b(2 \cdot 5) = \log_b 2 + \log_b 5 = 0.2891 + 0.6712 = 0.9603$

89. $\log_b 2.5 = \log_b \frac{5}{2} = \log_b 5 - \log_b 2 = 0.6712 - 0.2891$

$= 0.3821$

91. $\log_b \sqrt{15b} = \frac{1}{2} \log_b(3 \cdot 5 \cdot b)$

$= \frac{1}{2}[\log_b 3 + \log_b 5 + \log_b b]$

$= \frac{1}{2}[0.4582 + 0.6712 + 1]$

$= 1.0647$

93. $\log_b \left(\frac{25}{3b} \right) = \log_b 25 - \log_b 3b$

$= \log_b 5^2 - [\log_b 3 + \log_b b]$

$= 2 \log_b 5 - \log_b 3 - \log_b b$

$= 2(0.6712) - 0.4582 - 1$

$= -0.1158$

95. $f(x) = \log_{10} x$

$g(x) = \frac{\ln x}{\ln 10}$

$f(x) = g(x)$

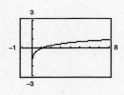

97. $f(x) = \ln \frac{x}{2}$, $g(x) = \frac{\ln x}{\ln 2}$, $h(x) = \ln x - \ln 2$

$f(x) = h(x)$ by Property 2.

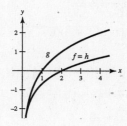

99. $\beta = 10 \log_{10} \left(\frac{I}{10^{-12}} \right)$

$= 10[\log_{10} I - \log_{10} 10^{-12}]$

$= 10[\log_{10} I - (-12)\log_{10} 10]$

$= 10[\log_{10} I + 12]$

For $I = 10^{-6}$, $\beta = 10[\log_{10} 10^{-6} + 12] = 60$ decibels

101. $f(x) = \ln x$

False, $f(0) \neq 0$ since 0 is not in the domain of $f(x)$.
$f(1) = \ln 1 = 0$

103. False. $f(x) - f(2) = \ln x - \ln 2 = \ln \dfrac{x}{2} \neq \ln(x - 2)$

105. False.

$$f(u) = 2f(v) \implies \ln u = 2 \ln v \implies \ln u$$
$$= \ln v^2 \implies u = v^2$$

107. Let $x = \log_b u$ and $y = \log_b v$, then $b^x = u$ and $b^y = v$.

$$\frac{u}{v} = \frac{b^x}{b^y} = b^{x-y}$$

Then $\log_b\left(\dfrac{u}{v}\right) = \log_b(b^{x-y}) = x - y = \log_b u - \log_b v.$

Section 7.4   Exponential and Logarithmic Equations

1. $4^{2x-7} = 64$

(a) $x = 5$

$4^{2(5)-7} = 4^3 = 64$

Yes, $x = 5$ is a solution.

(b) $x = 2$

$4^{2(2)-7} = 4^{-3} = \dfrac{1}{64} \neq 64$

No, $x = 2$ is not a solution.

3. $3e^{x+2} = 75$

(a) $x = -2 + e^{25}$

$3e^{(-2 + e^{25}) + 2} = 3e^{e^{25}} \neq 75$

No, $x = -2 + e^{25}$ is not a solution.

(b) $x = -2 + \ln 25$

$3e^{(-2 + \ln 25) + 2} = 3e^{\ln 25} = 3(25) = 75$

Yes, $x = -2 + \ln 25$ is a solution.

(c) $x \approx 1.219$

$3e^{1.219 + 2} = 3e^{3.219} \approx 75$

Yes, $x \approx 1.219$ is a solution.

5. $\log_4(3x) = 3 \implies 3x = 4^3 \implies 3x = 64$

(a) $x \approx 21.3560$

$3(21.3560) = 64.0680 \neq 64$

No, $x \approx 21.3560$ is not a solution.

(b) $x = -4$

$3(-4) = -12 \neq 64$

No, $x = -4$ is not a solution.

(c) $x = \dfrac{64}{3}$

$3\left(\dfrac{64}{3}\right) = 64$

Yes, $x = \dfrac{64}{3}$ is a solution.

7. $4^x = 16$

$4^x = 4^2$

$x = 2$

9. $5^x = \dfrac{1}{625}$

$5^x = \dfrac{1}{5^4}$

$5^x = 5^{-4}$

$x = -4$

11. $27^x = 9$

$(3^3)^x = 3^2$

$3^{3x} = 3^2$

$3x = 2$

$x = \dfrac{2}{3}$

13. $\left(\dfrac{1}{2}\right)^x = 32$

$2^{-x} = 2^5$

$-x = 5$

$x = -5$

15. $\left(\frac{3}{4}\right)^x = \frac{27}{64}$

$\left(\frac{3}{4}\right)^x = \left(\frac{3}{4}\right)^3$

$x = 3$

17. $3^{x-1} = 27$

$3^{x-1} = 3^3$

$x - 1 = 3$

$x = 4$

19. $\ln x - \ln 2 = 0$

$\ln x = \ln 2$

$x = 2$

21. $e^x = 2$

$\ln e^x = \ln 2$

$x = \ln 2$

$x \approx 0.693$

23. $\ln x = -1$

$e^{\ln x} = e^{-1}$

$x = e^{-1}$

$x \approx 0.368$

25. $\log_x 16 = 4$

$x^4 = 16$

$x = \sqrt[4]{16}$

$x = 2$

27. $\log_{10} x - 2 = 0$

$\log_{10} x = 2$

$10^{\log_{10} x} = 10^2$

$x = 10^2$

$x = 100$

29. $\log_{1/2} x = -3$

$\left(\frac{1}{2}\right)^{-3} = x$

$2^3 = x$

$8 = x$

31. $f(x) = g(x)$

$2^x = 8$

$2^x = 2^3$

$x = 3$

Point of intersection:
$(3, 8)$

33. $f(x) = g(x)$

$\log_3 x = 2$

$x = 3^2$

$x = 9$

Point of intersection:
$(9, 2)$

35. $\log_{10} 10^{x^2} = x^2$

37. $8^{\log_8(x-2)} = x - 2$

39. $\ln e^{7x+2} = 7x + 2$

41. $e^{\ln(5x+2)} = 5x + 2$

43. $-1 + \ln e^{2x} = -1 + 2x = 2x - 1$

45. $e^x = 10$

$x = \ln 10 \approx 2.303$

47. $7 - 2e^x = 5$

$-2e^x = -2$

$e^x = 1$

$x = \ln 1 = 0$

49. $e^{3x} = 12$

$3x = \ln 12$

$x = \dfrac{\ln 12}{3} \approx 0.828$

51. $500e^{-x} = 300$

$e^{-x} = \dfrac{3}{5}$

$-x = \ln \dfrac{3}{5}$

$x = -\ln \dfrac{3}{5} = \ln \dfrac{5}{3} \approx 0.511$

53. $e^{2x} - 4e^x - 5 = 0$

$(e^x + 1)(e^x - 5) = 0$

$e^x = -1$ or $e^x = 5$

(No solution) $x = \ln 5 \approx 1.609$

55. $20(100 - e^{x/2}) = 500$

$100 - e^{x/2} = 25$

$-e^{x/2} = -75$

$e^{x/2} = 75$

$\dfrac{x}{2} = \ln 75$

$x = 2 \ln 75 \approx 8.635$

57. $10^x = 42$

$x = \log_{10} 42 \approx 1.623$

59.
$$3^{2x} = 80$$
$$\ln 3^{2x} = \ln 80$$
$$2x \ln 3 = \ln 80$$
$$x = \frac{\ln 80}{2 \ln 3} \approx 1.994$$

61.
$$5^{-t/2} = 0.20$$
$$5^{-t/2} = \frac{1}{5}$$
$$5^{-t/2} = 5^{-1}$$
$$-\frac{t}{2} = -1$$
$$t = 2$$

63.
$$2^{3-x} = 565$$
$$\ln 2^{3-x} = \ln 565$$
$$(3 - x) \ln 2 = \ln 565$$
$$3 \ln 2 - x \ln 2 = \ln 565$$
$$-x \ln 2 = \ln 565 - \ln 2^3$$
$$x \ln 2 = \ln 8 - \ln 565$$
$$x = \frac{\ln 8 - \ln 565}{\ln 2} \approx -6.142$$

65. $g(x) = 6e^{1-x} - 25$

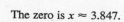

The zero is $x \approx -0.427$.

67. $f(x) = 3e^{3x/2} - 962$

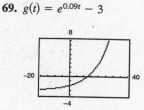

The zero is $x \approx 3.847$.

69. $g(t) = e^{0.09t} - 3$

The zero is $t \approx 12.207$.

71. $h(t) = e^{0.125t} - 8$

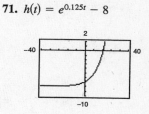

The zero is $t \approx 16.636$.

73.
$$8(10^{3x}) = 12$$
$$10^{3x} = \frac{12}{8}$$
$$\log_{10} 10^{3x} = \log_{10} \left(\frac{3}{2}\right)$$
$$3x = \log_{10}\left(\frac{3}{2}\right)$$
$$x = \tfrac{1}{3}\log_{10}\left(\frac{3}{2}\right) \approx 0.059$$

75.
$$3(5^{x-1}) = 21$$
$$5^{x-1} = 7$$
$$\ln 5^{x-1} = \ln 7$$
$$(x - 1) \ln 5 = \ln 7$$
$$x - 1 = \frac{\ln 7}{\ln 5}$$
$$x = 1 + \frac{\ln 7}{\ln 5} \approx 2.209$$

77.
$$\left(1 + \frac{0.065}{365}\right)^{365t} = 4$$
$$\ln\left(1 + \frac{0.065}{365}\right)^{365t} = \ln 4$$
$$365t \ln\left(1 + \frac{0.065}{365}\right) = \ln 4$$
$$t = \frac{\ln 4}{365 \ln\left(1 + \frac{0.065}{365}\right)} \approx 21.330$$

79.
$$\left(1 + \frac{0.10}{12}\right)^{12t} = 2$$
$$\ln\left(1 + \frac{0.10}{12}\right)^{12t} = \ln 2$$
$$12t \ln\left(1 + \frac{0.10}{12}\right) = \ln 2$$
$$t = \frac{\ln 2}{12 \ln\left(1 + \frac{0.10}{12}\right)} \approx 6.960$$

81. $\dfrac{3000}{2 + e^{2x}} = 2$

$3000 = 2(2 + e^{2x})$

$1500 = 2 + e^{2x}$

$1498 = e^{2x}$

$\ln 1498 = 2x$

$x = \dfrac{\ln 1498}{2} \approx 3.656$

83. $\ln x = -3$

$x = e^{-3} \approx 0.050$

85. $\ln 2x = 2.4$

$2x = e^{2.4}$

$x = \dfrac{e^{2.4}}{2} \approx 5.512$

87. $3 \ln 5x = 10$

$\ln 5x = \dfrac{10}{3}$

$5x = e^{10/3}$

$x = \dfrac{e^{10/3}}{5} \approx 5.606$

89. $\ln \sqrt{x + 2} = 1$

$\sqrt{x + 2} = e^1$

$x + 2 = e^2$

$x = e^2 - 2 \approx 5.389$

91. $\ln(x - 2) - \ln 3 = 2$

$\ln\left(\dfrac{x - 2}{3}\right) = 2$

$\dfrac{x - 2}{3} = e^2$

$x - 2 = 3e^2$

$x = 3e^2 + 2 \approx 24.167$

93. $\ln(2x - 1) + \ln 2 = -2$

$\ln[2(2x - 1)] = -2$

$2(2x - 1) = e^{-2}$

$4x - 2 = e^{-2}$

$4x = e^{-2} + 2$

$x = \dfrac{e^{-2} + 2}{4} \approx 0.534$

95. $\ln x + \ln (x - 1) = 1$

$\ln[x(x - 1)] = 1$

$x(x - 1) = e^1$

$x^2 - x - e = 0$

By the Quadratic Formula we have:

$x = \dfrac{1 \pm \sqrt{1 + 4e}}{2}$

$x = \dfrac{1 - \sqrt{1 + 4e}}{2} \approx -1.223$ is extraneous.

$x = \dfrac{1 + \sqrt{1 + 4e}}{2} \approx 2.223$ is the only solution.

97. $\ln x + \ln(x - 2) = 1$

$\ln[x(x - 2)] = 1$

$x(x - 2) = e^1$

$x^2 - 2x - e = 0$

$x = \dfrac{2 \pm \sqrt{4 + 4e}}{2}$

$\quad = \dfrac{2 \pm 2\sqrt{1 + e}}{2}$

$\quad = 1 \pm \sqrt{1 + e}$

The negative value is extraneous. The only solution is

$x = 1 + \sqrt{1 + e} \approx 2.928.$

99. $\ln(x + 5) = \ln(x - 1) - \ln(x + 1)$

$$\ln(x + 5) = \ln\left(\frac{x - 1}{x + 1}\right)$$

$$x + 5 = \frac{x - 1}{x + 1}$$

$$(x + 5)(x + 1) = x - 1$$

$$x^2 + 6x + 5 = x - 1$$

$$x^2 + 5x + 6 = 0$$

$$(x + 2)(x + 3) = 0$$

$$x = -2 \quad \text{or} \quad x = -3$$

Both of these solutions are extraneous, so the equation has no solution.

101. $\log_{10}(z - 3) = 2$

$$10^{\log_{10}(z-3)} = 10^2$$

$$z - 3 = 10^2$$

$$z = 10^2 + 3 = 103$$

103. $6 \log_3(0.5x) = 11$

$$\log_3(0.5x) = \frac{11}{6}$$

$$3^{\log_3(0.5x)} = 3^{11/6}$$

$$0.5x = 3^{11/6}$$

$$x = 2(3^{11/6}) \approx 14.988$$

105. $\log_{10}(x + 4) - \log_{10} x = \log_{10}(x + 2)$

$$\log_{10}\left(\frac{x + 4}{x}\right) = \log_{10}(x + 2)$$

$$\frac{x + 4}{x} = x + 2$$

$$x + 4 = x^2 + 2x$$

$$0 = x^2 + x - 4$$

$$x = \frac{-1 \pm \sqrt{17}}{2} \quad \text{Quadratic Formula}$$

Choosing the positive value of x (the negative value is extraneous), we have

$$x = \frac{-1 + \sqrt{17}}{2} \approx 1.562.$$

107. $\log_4 x - \log_4(x - 1) = \frac{1}{2}$

$$\log_4\left(\frac{x}{x - 1}\right) = \frac{1}{2}$$

$$4^{\log_4\left(\frac{x}{x-1}\right)} = 4^{1/2}$$

$$\frac{x}{x - 1} = 4^{1/2}$$

$$x = 2(x - 1)$$

$$x = 2x - 2$$

$$-x = -2$$

$$x = 2$$

109. $\log_{10} 8x - \log_{10}\left(1 + \sqrt{x}\right) = 2$

$$\log_{10}\frac{8x}{1 + \sqrt{x}} = 2$$

$$\frac{8x}{1 + \sqrt{x}} = 10^2$$

$$8x = 100\left(1 + \sqrt{x}\right)$$

$$2x = 25\left(1 + \sqrt{x}\right)$$

$$2x = 25 + 25\sqrt{x}$$

$$2x - 25 = 25\sqrt{x}$$

$$(2x - 25)^2 = \left(25\sqrt{x}\right)^2$$

$$4x^2 - 100x + 625 = 625x$$

$$4x^2 - 725x + 625 = 0$$

$$x = \frac{725 \pm \sqrt{725^2 - 4(4)(625)}}{2(4)}$$

$$x = \frac{725 \pm \sqrt{515625}}{8}$$

$$x = \frac{25\left(29 \pm 5\sqrt{33}\right)}{8}$$

$$x \approx 0.866 \text{ (extraneous)} \quad \text{or} \quad x \approx 180.384$$

The only solution is $x = \dfrac{25\left(29 + 5\sqrt{33}\right)}{8} \approx 180.384$.

111. $y_1 = 7$

$y_2 = 2^x$

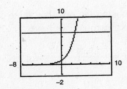

From the graph we have $x \approx 2.807$ when $y = 7$.

The point of intersection is approximately $(2.807, 7)$.

113. $y_1 = 3$

$y_2 = \ln x$

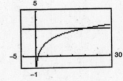

From the graph we have $x \approx 20.086$ when $y = 3$.

The point of intersection is approximately $(20.086, 3)$.

115. Rewrite the given equation in a form that allows use of the one-to-one properties of exponential functions, or rewrite the exponential equation in logarithmic form and use the properties of logarithms. Be sure to check for extraneous solutions.

117. $A = Pe^{rt}$

(a) $A = (2P)e^{rt} = 2(Pe^{rt})$ This doubles your money.

(b) $A = Pe^{(2r)t} = Pe^{rt}e^{rt} = e^{rt}(Pe^{rt})$

(c) $A = Pe^{r(2t)} = Pe^{rt}e^{rt} = e^{rt}(Pe^{rt})$

Doubling the interest rate yields the same result as doubling the number of years.

If $2 > e^{rt}$ (i.e., $rt < \ln 2$), then doubling your investment would yield the most money. If $rt > \ln 2$, then doubling either the interest rate or the number of years would yield more money.

119. (a) $A = Pe^{rt}$

$$2000 = 1000e^{0.085t}$$

$$2 = e^{0.085t}$$

$$\ln 2 = 0.085t$$

$$\frac{\ln 2}{0.085} = t$$

$$t \approx 8.2 \text{ years}$$

(b) Yes, it takes twice as long to quadruple.

(c) and (d)

Doubling Time	Quadrupling Time
$2P = Pe^{rt}$	$4P = Pe^{rt}$
$2 = e^{rt}$	$4 = e^{rt}$
$\ln 2 = rt$	$\ln 4 = rt$
$\dfrac{\ln 2}{r} = t$	$\dfrac{\ln 4}{r} = t$
	$\dfrac{\ln 2^2}{r} = t$
	$\dfrac{2\ln 2}{r} = t$
	$2\left(\dfrac{\ln 2}{r}\right) = t$

The time required for the investment to quadruple is twice the doubling time.

121. $A = Pe^{rt}$

$$3000 = 1000e^{0.085t}$$

$$3 = e^{0.085t}$$

$$\ln 3 = 0.085t$$

$$\frac{\ln 3}{0.085} = t$$

$$t \approx 12.9 \text{ years}$$

123. $p = 7000\left(1 - \dfrac{6}{6 + e^{-0.0012x}}\right)$

(a) When $p = \$350$:

$$350 = 7000\left(1 - \frac{6}{6 + e^{-0.0012x}}\right)$$

$$0.05 = 1 - \frac{6}{6 + e^{-0.0012x}}$$

$$\frac{6}{6 + e^{-0.0012x}} = 0.95$$

$$6 = 0.95(6 + e^{-0.0012x})$$

$$6 = 5.7 + 0.95e^{-0.0012x}$$

$$\frac{0.3}{0.95} = e^{-0.0012x}$$

$$\ln\left(\frac{6}{19}\right) = -0.0012x$$

$$x = \frac{\ln(6/19)}{-0.0012} \approx 961 \text{ units}$$

(b) When $p = \$300$

$$300 = 7000\left(1 - \frac{6}{6 + e^{-0.0012x}}\right)$$

$$\frac{3}{70} = 1 - \frac{6}{6 + e^{-0.0012x}}$$

$$\frac{6}{6 + e^{-0.0012x}} = \frac{67}{70}$$

$$6 = \frac{67}{70}(6 + e^{-0.0012x})$$

$$\frac{420}{67} = 6 + e^{-0.0012x}$$

$$\frac{18}{67} = e^{-0.0012x}$$

$$\ln\left(\frac{18}{67}\right) = -0.0012x$$

$$x = \frac{\ln(18/67)}{-0.0012} \approx 1095 \text{ units}$$

125. $V = 6.7e^{-48.1/t}$, $t \geq 0$

(a)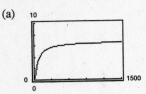

(b) As $t \to \infty$, $V \to 6.7$.

Horizontal asymptote: $V = 6.7$

The yield will approach 6.7 million cubic feet per acre.

(c) $1.3 = 6.7e^{-48.1/t}$

$$\frac{1.3}{6.7} = e^{-48.1/t}$$

$$\ln\left(\frac{13}{67}\right) = \frac{-48.1}{t}$$

$$t = \frac{-48.1}{\ln(13/67)} \approx 29.3 \text{ years}$$

127. (a) From the graph shown in the textbook, we see horizontal asymptotes at $y = 0$ and $y = 100$.
These represent the lower and upper percent bounds; the range falls between 0% and 100%.

(b) Males

$$50 = \frac{100}{1 + e^{-0.6114(x - 69.71)}}$$

$$1 + e^{-0.6114(x - 69.71)} = 2$$

$$e^{-0.6114(x - 69.71)} = 1$$

$$-0.6114(x - 69.71) = \ln 1$$

$$-0.6114(x - 69.71) = 0$$

$$x = 69.71 \text{ inches}$$

Females

$$50 = \frac{100}{1 + e^{-0.66607(x - 64.51)}}$$

$$1 + e^{-0.66607(x - 64.51)} = 2$$

$$e^{-0.66607(x - 64.51)} = 1$$

$$-0.66607(x - 64.51) = \ln 1$$

$$-0.66607(x - 64.51) = 0$$

$$x = 64.51 \text{ inches}$$

129. $T = 20[1 + 7(2^{-h})]$

(a) From the graph in the textbook we see a horizontal
asymptote at $T = 20$.
This represents the room temperature.

(b) $100 = 20[1 + 7(2^{-h})]$

$$5 = 1 + 7(2^{-h})$$

$$4 = 7(2^{-h})$$

$$\frac{4}{7} = 2^{-h}$$

$$\ln\left(\frac{4}{7}\right) = \ln 2^{-h}$$

$$\ln\left(\frac{4}{7}\right) = -h \ln 2$$

$$\frac{\ln\left(\frac{4}{7}\right)}{-\ln 2} = h$$

$$h \approx 0.81 \text{ hour}$$

131. $N = 17.02 - 3.096 \ln t$,

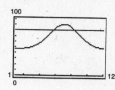

The high temperature is likely to be above 80° in the months of June, July, and August.

133. $P = \dfrac{8.617}{1 + 4.65e^{-0.0380t}} + 1.0005$ where $t = 0$ corresponds to 1950.

(a)

Year	Population	Model
1950	2.526	2.526
1960	3.058	3.062
1970	3.722	3.715
1980	4.454	4.465
1990	5.276	5.273
2000	6.079	6.083

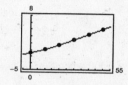

The model is a good fit.

(b) For 2010: $P(60) \approx 6.840$

For 2020: $P(70) \approx 7.503$

For 2030: $P(80) \approx 8.050$

(c) $\dfrac{8.617}{1 + 4.65e^{-0.0380t}} + 1.0005 = 8$

$8.617 = 6.9995(1 + 4.65e^{-0.0380t})$

$\dfrac{86,170}{69,995} = 1 + 4.65e^{-0.0380t}$

$\dfrac{1}{4.65}\left[\dfrac{86,170}{69,995} - 1\right] = e^{-0.0380t}$

$t = \dfrac{\ln\left\{\dfrac{1}{4.65}\left(\dfrac{86,170}{69,995} - 1\right)\right\}}{-0.0380} \approx 79$

According to the model, the world population would reach 8 billion in the year 2029.

(d)

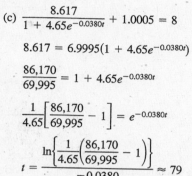

The horizontal asymptote is at $y = 8.617 + 1.0005 = 9.6175$. This would imply that the world population maximum would be 9.6175 billion.

(e) World population models can be used for short-range predictions, but there are too many unknown factors to use them for long-range predicitons.

135. $\log_a(uv) = \log_a u + \log_a v$

True by Property 1 in Section 7.3.

137. $\log_a(u - v) = \log_a u - \log_a v$

False. $1.95 \approx \log_{10}(100 - 10) \neq \log_{10} 100 - \log_{10} 10 = 1$

Section 7.5   Exponential and Logarithmic Models

1. $y = 2e^{x/4}$

This is an exponential growth model.

Matches graph (c)

3. $y = 6 + \log_{10}(x + 2)$

This is a logarithmic function shifted up 6 units and left 2 units.

Matches graph (b)

5. $y = \ln(x + 1)$

This is a logarithmic model.

Matches graph (d)

7.
$$y = ae^{bx}$$
$$1 = ae^{b(0)} \implies 1 = a$$
$$10 = e^{b(3)}$$
$$\ln 10 = 3b$$
$$\frac{\ln 10}{3} = b \implies b \approx 0.7675$$

Thus, $y = e^{0.7675x}$.

9.
$$y = ae^{bx}$$
$$5 = ae^{b(0)} \implies 5 = a$$
$$1 = 5e^{b(4)}$$
$$\frac{1}{5} = e^{4b}$$
$$\ln\left(\frac{1}{5}\right) = 4b$$
$$\frac{\ln\left(\frac{1}{5}\right)}{4} = b \implies b \approx -0.4024$$

Thus, $y = 5e^{-0.4024x}$.

11. (a) Logarithmic

(b) Logistic

(c) Exponential (decay)

(d) Linear

(e) None of the above (appears to be a combination of a linear and a quadratic)

(f) Exponential (growth)

13. $y = \dfrac{a}{1 + be^{-xt}}$

This curve does not have a vertical asymptote as long as $1 + be^{-xt} \neq 0$. This means $b \neq -e^{xt}$. Since $-e^{xt} < 0$ for all x and t, b cannot be negative. Thus, if $b > 0$, the curve has no vertical asymptotes. Also, if $b = 0$, $y = a$ is a horizontal line and the graph would not have a vertical asymptote.

15. Since $A = 1000e^{0.12t}$, the time to double is given by $2000 = 1000e^{0.12t}$ and we have
$$2000 = 1000e^{0.12t}$$
$$2 = e^{0.12t}$$
$$\ln 2 = \ln e^{0.12t}$$
$$\ln 2 = 0.12t$$
$$t = \frac{\ln 2}{0.12} \approx 5.78 \text{ years.}$$

Amount after 10 years: $A = 1000e^{1.2} \approx \$3320.12$

17. Since $A = 750e^{rt}$ and $A = 1500$ when $t = 7.75$, we have the following.
$$1500 = 750e^{7.75r}$$
$$2 = e^{7.75r}$$
$$\ln 2 = \ln e^{7.75r}$$
$$\ln 2 = 7.75r$$
$$r = \frac{\ln 2}{7.75} \approx 0.089438 = 8.9438\%$$

Amount after 10 years: $A = 750e^{0.089438(10)} \approx \$1834.37$

19. Since $A = 500e^{rt}$ and $A = \$1505.00$ when $t = 10$, we have the following.
$$1505.00 = 500e^{10r}$$
$$r = \frac{\ln(1505.00/500)}{10} \approx 0.110 = 11.0\%$$

The time to double is given by
$$1000 = 500e^{0.110t}$$
$$t = \frac{\ln 2}{0.110} \approx 6.3 \text{ years.}$$

21. Since $A = Pe^{0.045t}$ and $A = 10,000.00$ when $t = 10$, we have the following.
$$10,000.00 = Pe^{0.045(10)}$$
$$\frac{10,000.00}{e^{0.045(10)}} = P \approx \$6376.28$$

The time to double is given by
$$t = \frac{\ln 2}{0.045} \approx 15.40 \text{ years.}$$

23. $500,000 = P\left(1 + \dfrac{0.075}{12}\right)^{12(20)}$

$$P = \dfrac{500,000}{\left(1 + \dfrac{0.075}{12}\right)^{12(20)}} = \dfrac{500,000}{1.00625^{240}} \approx \$112,087.09$$

25. $P = 1000, r = 11\%$

  (a) $n = 1$

       $(1 + 0.11)^t = 2$

         $t \ln 1.11 = \ln 2$

            $t = \dfrac{\ln 2}{\ln 1.11} \approx 6.642$ years

  (c) $n = 365$

       $\left(1 + \dfrac{0.11}{365}\right)^{365t} = 2$

   $365t \ln\left(1 + \dfrac{0.11}{365}\right) = \ln 2$

          $t = \dfrac{\ln 2}{365 \ln\left(1 + \dfrac{0.11}{365}\right)} \approx 6.302$ years

  (b) $n = 12$

       $\left(1 + \dfrac{0.11}{12}\right)^{12t} = 2$

   $12t \ln\left(1 + \dfrac{0.11}{12}\right) = \ln 2$

          $t = \dfrac{\ln 2}{12 \ln\left(1 + \dfrac{0.11}{12}\right)} \approx 6.330$ years

  (d) Continuously

      $e^{0.11t} = 2$

      $0.11t = \ln 2$

         $t = \dfrac{\ln 2}{0.11} \approx 6.301$ years

27. $3P = Pe^{rt}$

       $3 = e^{rt}$

   $\ln 3 = rt$

   $\dfrac{\ln 3}{r} = t$

r	2%	4%	6%	8%	10%	12%
$t = \dfrac{\ln 3}{r}$ (years)	54.93	27.47	18.31	13.73	10.99	9.16

29. $3P = P(1 + r)^t$

          $3 = (1 + r)^t$

     $\ln 3 = \ln(1 + r)^t$

     $\ln 3 = t \ln(1 + r)$

  $\dfrac{\ln 3}{\ln(1 + r)} = t$

r	2%	4%	6%	8%	10%	12%
$t = \dfrac{\ln 3}{\ln(1 + r)}$ (years)	55.48	28.01	18.85	14.27	11.53	9.69

31. Continuous compounding results in faster growth.

  $A = 1 + 0.075[\![t]\!]$ and $A = e^{0.07t}$

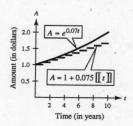

33. $P = 73,796e^{0.030t}$ where $t = 0$ corresponds to 1970.

$73,796e^{0.030t} = 350,000$

$0.030t = \ln\left(\dfrac{350,000}{73,796}\right)$

$t \approx 52$ which corresponds to the year 2022.

35. $P = 1,458,000e^{-0.016t}$ where $t = 0$ corresponds to 1970.

$1,458,000e^{-0.016t} = 800,000$

$-0.016t = \ln\left(\dfrac{800,000}{1,458,000}\right)$

$t \approx 38$ which corresponds to the year 2008.

37. $N = 100e^{kt}$

$300 = 100e^{5k}$

$3 = e^{5k}$

$\ln 3 = \ln e^{5k}$

$\ln 3 = 5k$

$k = \dfrac{\ln 3}{5} \approx 0.2197$

$N = 100e^{0.2197t}$

$200 = 100e^{0.2197t}$

$t = \dfrac{\ln 2}{0.2197} \approx 3.15$ hours

39. $y = Ce^{kt}$

$\dfrac{1}{2}C = Ce^{(1599)k}$

$\dfrac{1}{2} = e^{(1599)k}$

$\ln \dfrac{1}{2} = \ln e^{(1599)k}$

$\ln \dfrac{1}{2} = 1599k$

$k = \dfrac{\ln(1/2)}{1599}$

When $t = 100$, we have

$y = Ce^{[\ln(1/2)/1599](100)} \approx 0.957C = 95.7\%C.$

After 100 years, approximately 95.7% of the radioactive radium will remain.

41. $(0,\ 22{,}000),\ (2,\ 13{,}000)$

(a) $m = \dfrac{13{,}000 - 22{,}000}{2 - 0} = -4500$

$b = 22{,}000$

Thus, $V = -4500t + 22{,}000.$

(b) $\quad a = 22{,}000$

$13{,}000 = 22{,}000e^{k(2)}$

$\dfrac{13}{22} = e^{2k}$

$\ln\left(\dfrac{13}{22}\right) = \ln e^{2k}$

$\ln\left(\dfrac{13}{22}\right) = 2k \implies k \approx -0.263$

Thus, $V = 22{,}000e^{-0.263t}.$

(c) The exponential model depreciates faster in the first two years.

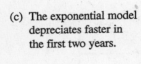

(d)

t	1	3
$V = -4500t + 22{,}000$	\$17,500	\$8500
$V = 22{,}000e^{-0.263t}$	\$16,912	\$9995

(e) The slope of the linear model means that the car depreciates \$4500 per year.

43. $S(t) = 100(1 - e^{kt})$

(a) $\quad 15 = 100(1 - e^{k(1)})$

$-85 = -100e^{k}$

$\dfrac{85}{100} = e^{k}$

$0.85 = e^{k}$

$\ln 0.85 = \ln e^{k}$

$k = \ln 0.85$

$k \approx -0.1625$

$S(t) = 100(1 - e^{-0.1625t})$

(b)

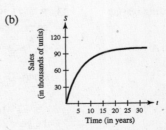

(c) $S(5) = 100(1 - e^{-0.1625(5)})$

$\approx 55.625 = 55{,}625$ units

45. $S = 10(1 - e^{kx})$

$x = 5$ (in hundreds), $S = 2.5$ (in thousands)

(a) $2.5 = 10(1 - e^{k(5)})$

$0.25 = 1 - e^{5k}$

$e^{5k} = 0.75$

$\ln e^{5k} = \ln 0.75$

$5k = \ln 0.75$

$k \approx -0.0575$

$S = 10(1 - e^{-0.0575x})$

(b) When $x = 7$, $S = 10(1 - e^{-0.0575(7)}) \approx 3.314$
which corresponds to 3314 units.

47. $N = 30(1 - e^{kt})$

(a) $N = 19$, $t = 20$

$19 = 30(1 - e^{20k})$

$30e^{20k} = 11$

$e^{20k} = \dfrac{11}{30}$

$\ln e^{20k} = \ln\left(\dfrac{11}{30}\right)$

$20k = \ln\dfrac{11}{30}$

$k \approx -0.050$

$N = 30(1 - e^{-0.050t})$

(b) $N = 25$

$25 = 30(1 - e^{-0.050t})$

$\dfrac{5}{30} = e^{-0.050t}$

$\ln\left(\dfrac{5}{30}\right) = \ln e^{-0.050t}$

$\ln\left(\dfrac{5}{30}\right) = -0.050t$

$t = -\dfrac{\ln(5/30)}{-0.050} \approx 36$ days

(c) No. It is not a linear function.

49. $\dfrac{1}{2}C = Ce^{k(1599)}$

$0.5 = e^{k(1599)}$

$\ln 0.5 = \ln e^{k(1599)}$

$\ln 0.5 = k(1599)$

$k = \dfrac{\ln 0.5}{1599}$

Given $C = 10$ grams after
1000, years we have

$y = 10e^{[(\ln 0.5)/1599](1000)}$

≈ 6.48 grams.

51. $\dfrac{1}{2}C = Ce^{k(5715)}$

$0.5 = e^{k(5715)}$

$\ln 0.5 = \ln e^{k(5715)}$

$\ln 0.5 = k(5715)$

$k = \dfrac{\ln 0.5}{5715}$

Given $y = 2$ grams after
1000 years, we have

$2 = Ce^{[(\ln 0.5)/5715](1000)}$

$C \approx 2.26$ grams.

53. $\dfrac{1}{2}C = Ce^{k(24,100)}$

$0.5 = e^{k(24,100)}$

$\ln 0.5 = \ln e^{k(24,100)}$

$\ln 0.5 = k(24,100)$

$k = \dfrac{\ln 0.5}{24,100}$

Given $y = 2.1$ grams after
1000 years, we have

$2.1 = Ce^{[(\ln 0.5)/24,100](1000)}$

$C \approx 2.16$ grams.

55. $R = \log_{10}\dfrac{I}{I_0} = \log_{10} I$ since $I_0 = 1$.

(a) $R = \log_{10} 80,500,000 \approx 7.91$

(b) $R = \log_{10} 48,275,000 \approx 7.68$

(c) $R = \log_{10} 251,200 \approx 5.40$

57. $\beta = 10 \log_{10} \dfrac{I}{I_0}$ where $I_0 = 10^{-12}$ watt/m^2

(a) $\beta = 10 \log_{10} \dfrac{10^{-10}}{10^{-12}} = 10 \log_{10} 10^2 = 20$ decibels

(c) $\beta = 10 \log_{10} \dfrac{10^{-2.5}}{10^{-12}} = 10 \log_{10} 10^{9.5} = 95$ decibels

(b) $\beta = 10 \log_{10} \dfrac{10^{-5}}{10^{-12}} = 10 \log_{10} 10^7 = 70$ decibels

(d) $\beta = 10 \log_{10} \dfrac{1}{10^{-12}} = 10 \log_{10} 10^{12} = 120$ decibels

59. $\beta = 10 \log_{10} \dfrac{I}{I_0}$

$\dfrac{\beta}{10} = \log_{10} \dfrac{I}{I_0}$

$10^{\beta/10} = 10^{\log_{10} I/I_0}$

$10^{\beta/10} = \dfrac{I}{I_0}$

$I = I_0 10^{\beta/10}$

% decrease $= \dfrac{I_0 10^{9.3} - I_0 10^{8.0}}{I_0 10^{9.3}} \times 100 \approx 95\%$

61. pH $= -\log_{10}[\text{H}^+] = -\log_{10}(2.3 \times 10^{-5}) \approx 4.64$

63. $5.8 = -\log_{10}[\text{H}^+]$

$-5.8 = \log_{10}[\text{H}^+]$

$10^{-5.8} = 10^{\log_{10}[\text{H}^+]}$

$10^{-5.8} = [\text{H}^+]$

$[\text{H}^+] \approx 1.58 \times 10^{-6}$ moles per liter

65. $2.5 = -\log_{10}[\text{H}^+]$

$-2.5 = \log_{10}[\text{H}^+]$

$10^{-2.5} = 10^{\log_{10}[\text{H}^+]}$

$10^{-2.5} = [\text{H}^+]$ for the fruit.

$9.5 = -\log_{10}[\text{H}^+]$

$10^{-9.5} = [\text{H}^+]$ for the antacid tablet.

$\dfrac{10^{-2.5}}{10^{-9.5}} = 10^7$

67. False. The domain can be the set of real numbers for a logistics growth function.

69. True. $e^{-x^2} > 0$ for all x, so $y \neq 0$ and there are no x-intercepts.

71. Answers will vary.

Review Exercises for Chapter 7

1. $(6.1)^{2.4} \approx 76.699$

3. $2^{-0.5\pi} \approx 0.337$

5. $60^{\sqrt{3}} \approx 1201.845$

7. $f(x) = 4^x$

Intercept: $(0, 1)$

Horizontal asymptote: x-axis

Increasing on: $(-\infty, \infty)$

Matches graph (c)

9. $f(x) = -4^x$

Intercept: $(0, -1)$

Horizontal asymptote: x-axis

Decreasing on: $(-\infty, \infty)$

Matches graph (a)

11. $f(x) = 4^{-x} + 4$

Horizontal asymptote: $y = 4$

x	-1	0	1	2	3
$f(x)$	8	5	4.25	4.0625	4.016

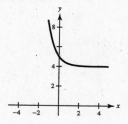

13. $f(x) = -2.65^{x+1}$

Horizontal asymptote: $y = 0$

x	-2	-1	0	1	2
$f(x)$	-0.377	-1	-2.65	-7.023	-18.61

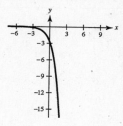

15. $f(x) = 5^{x-2} + 4$

Horizontal asymptote: $y = 4$

x	-1	0	1	2	3
$f(x)$	4.008	4.04	4.2	5	9

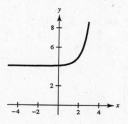

17. $f(x) = \left(\frac{1}{2}\right)^{-x} + 3 = 2^x + 3$

Horizontal asymptote: $y = 3$

x	-2	-1	0	1	2
$f(x)$	3.25	3.5	4	5	7

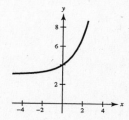

19. $e^8 \approx 2980.958$

21. $e^{-1.7} \approx 0.183$

23. $h(x) = e^{-x/2}$

x	-2	-1	0	1	2
$h(x)$	2.72	1.65	1	0.61	0.37

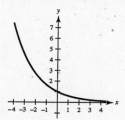

25. $f(x) = e^{x+2}$

x	-3	-2	-1	0	1
$f(x)$	0.37	1	2.72	7.39	20.09

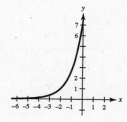

27. $A = 3500\left(1 + \dfrac{0.065}{n}\right)^{10n}$ or $A = 3500e^{(0.065)(10)}$

n	1	2	4	12	365	Continuous Compounding
A	\$6569.98	\$6635.43	\$6669.46	\$6692.64	\$6704.00	\$6704.39

29. $200,000 = Pe^{0.08t}$

$$P = \frac{200,000}{e^{0.08t}}$$

t	1	10	20	30	40	50
P	\$184,623.27	\$89,865.79	\$40,379.30	\$18,143.59	\$8,152.44	\$3,663.13

31. $F(t) = 1 - e^{-t/3}$

　(a) $F\left(\frac{1}{2}\right) \approx 0.154$

　(b) $F(2) \approx 0.487$

　(c) $F(5) \approx 0.811$

33. (a) $A = 50,000e^{(0.0875)(35)} \approx \$1,069,047.14$

　(b) The doubling time is

$$\frac{\ln 2}{0.0875} \approx 7.9 \text{ years.}$$

35. Order: $b < d < a < c$

37. 　　$4^3 = 64$

　　$\log_4 64 = 3$

39. $\log_{10} 1000 = \log_{10} 10^3 = 3$

41. $\log_2 \dfrac{1}{8} = \log_2 2^{-3} = -3$

43. $g(x) = \log_7 x \implies x = 7^y$

Vertical asymptote: $x = 0$

x	$\frac{1}{7}$	1	7	49
$g(x)$	-1	0	1	2

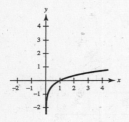

45. $f(x) = \log_{10}\left(\dfrac{x}{3}\right) \implies \dfrac{x}{3} = 10^y \implies x = 3(10^y)$

Vertical asymptote: $x = 0$

x	0.03	0.3	3	30
$f(x)$	-2	-1	0	1

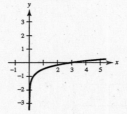

47. $f(x) = 4 - \log_{10}(x + 5)$

Vertical asymptote: $x = -5$

x	-4	-3	-2	-1	0	1
$f(x)$	4	3.70	3.52	3.40	3.30	3.22

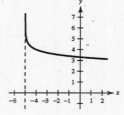

49. $\ln 22.6 \approx 3.118$

51. $\ln e^{-12} = -12$

53. $\ln\left(\sqrt{7} + 5\right) \approx 2.034$

55. $f(x) = \ln x + 3$

Domain: $(0, \infty)$

Vertical asymptote: $x = 0$

x	1	2	3	$\frac{1}{2}$	$\frac{1}{4}$
$f(x)$	3	3.69	4.10	2.31	1.61

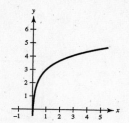

57. $h(x) = \ln(x^2) = 2\ln|x|$

Vertical asymptote: $x = 0$

x	± 0.5	± 1	± 2	± 3	± 4
y	-1.39	0	1.39	2.20	2.77

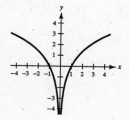

59. $s = 25 - \dfrac{13\ln(10/12)}{\ln 3} \approx 27.16$ miles

61. $\log_4 9 = \dfrac{\log_{10} 9}{\log_{10} 4} \approx 1.585$

$\log_4 9 = \dfrac{\ln 9}{\ln 4} \approx 1.585$

63. $\log_{1/2} 5 = \dfrac{\log_{10} 5}{\log_{10}(1/2)} \approx -2.322$

$\log_{1/2} 5 = \dfrac{\ln 5}{\ln(1/2)} \approx -2.322$

65. $\ln 8 + \ln 5 = \ln[(8)(5)] = \ln 40$

67. $\log_8\left(\dfrac{\sqrt{x}}{y^3}\right) = \log_8 \sqrt{x} - \log_8 y^3 = \dfrac{1}{2}\log_8 x - 3\log_8 y$

69. $\log_5 5x^2 = \log_5 5 + \log_5 x^2$

$= 1 + 2\log_5|x|$

71. $\log_{10} \dfrac{5\sqrt{y}}{x^2} = \log_{10} 5\sqrt{y} - \log_{10} x^2$

$= \log_{10} 5 + \log_{10} \sqrt{y} - \log_{10} x^2$

$= \log_{10} 5 + \dfrac{1}{2}\log_{10} y - 2\log_{10}|x|$

73. $\log_2 5 + \log_2 x = \log_2 5x$

75. $\dfrac{1}{2}\ln|2x - 1| - 2\ln|x + 1| = \ln\sqrt{|2x - 1|} - \ln|x + 1|^2$

$= \ln\dfrac{\sqrt{|2x - 1|}}{(x + 1)^2}$

77. $t = 50 \log_{10} \dfrac{18,000}{18,000 - h}$

(a) Domain: $0 \le h < 18,000$

(b)
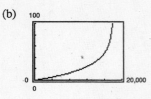

Vertical asymptote: $h = 18,000$

(c) As the plane approaches its absolute ceiling, it climbs at a slower rate, so the time required increases.

(d) $50 \log_{10} \dfrac{18,000}{18,000 - 4000} \approx 5.46$ minutes

79. $8^x = 512$

$8^x = 8^3$

$x = 3$

81. $6^x = \frac{1}{216}$

$6^x = 6^{-3}$

$x = -3$

83. $\log_7 x = 4$

$7^{\log_7 x} = 7^4$

$x = 7^4$

$x = 2401$

85. $e^x = 12$

$\ln e^x = \ln 12$

$x = \ln 12 \approx 2.485$

87. $3e^{-5x} = 132$

$e^{-5x} = 44$

$\ln e^{-5x} = \ln 44$

$-5x = \ln 44$

$x = \dfrac{\ln 44}{-5} \approx -0.757$

89. $e^x + 13 = 35$

$e^x = 22$

$\ln e^x = \ln 22$

$x = \ln 22 \approx 3.091$

91. $-4(5^x) = -68$

$5^x = 17$

$\ln 5^x = \ln 17$

$x \ln 5 = \ln 17$

$x = \dfrac{\ln 17}{\ln 5} \approx 1.760$

93. $e^{2x} - 7e^x + 10 = 0$

$(e^x - 2)(e^x - 5) = 0$

$e^x = 2 \qquad\qquad e^x = 5$

$\ln e^x = \ln 2 \qquad \ln e^x = \ln 5$

$x = \ln 2 \approx 0.693 \qquad x = \ln 5 \approx 1.609$

95. $2^{0.6x} - 3x = 0$

Graph $y_1 = 2^{0.6x} - 3x$.

The x-intercepts are at $x \approx 0.39$ and at $x \approx 7.48$.

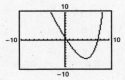

97. $25e^{-0.3x} = 12$

Graph $y_1 = 25e^{-0.3x}$ and $y_2 = 12$.

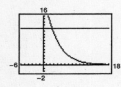

The graphs intersect at $x \approx 2.45$.

99. $2^{x-1} - x^2 = 0$

$2^{x-1} = x^2$

Graph $y_1 = 2^{x-1}$ and $y_2 = x^2$ and find the points of intersection.

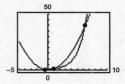

There are three points of intersection.

$x \approx -0.58, x = 1, x \approx 6.32$

101. $\ln 3x = 8.2$

$e^{\ln 3x} = e^{8.2}$

$3x = e^{8.2}$

$x = \dfrac{e^{8.2}}{3} \approx 1213.650$

103. $2 \ln 4x = 15$

$\ln 4x = \frac{15}{2}$

$e^{\ln 4x} = e^{7.5}$

$4x = e^{7.5}$

$x = \frac{1}{4}e^{7.5} \approx 452.011$

105. $\ln x - \ln 3 = 2$

$$\ln \frac{x}{3} = 2$$

$$e^{(\ln x/3)} = e^2$$

$$\frac{x}{3} = e^2$$

$$x = 3e^2 \approx 22.167$$

107. $\ln \sqrt{x + 1} = 2$

$$\frac{1}{2} \ln(x + 1) = 2$$

$$\ln(x + 1) = 4$$

$$e^{\ln(x+1)} = e^4$$

$$x + 1 = e^4$$

$$x = e^4 - 1 \approx 53.598$$

109. $\log_{10}(x - 1) = \log_{10}(x - 2) - \log_{10}(x + 2)$

$$\log_{10}(x - 1) = \log_{10}\left(\frac{x - 2}{x + 2}\right)$$

$$10^{\log_{10}(x-1)} = 10^{\log_{10}(x-2/x+2)}$$

$$x - 1 = \frac{x - 2}{x + 2}$$

$$(x - 1)(x + 2) = x - 2$$

$$x^2 + x - 2 = x - 2$$

$$x^2 = 0$$

$$x = 0$$

Since $x = 0$ is not in the domain of $\ln(x - 1)$ or of $\ln(x - 2)$, it is an extraneous solution. The equation has no solution.

111. $\log_3(1 - x) = -1$

$$1 - x = 3^{-1}$$

$$1 - \tfrac{1}{3} = x$$

$$x = \tfrac{2}{3} \approx 0.667$$

113. $2\ln(x + 3) + 3x = 8$

Graph $y_1 = 2\ln(x + 3) + 3x$ and $y_2 = 8$.

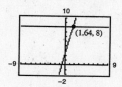

The graphs intersect at approximately $(1.64, 8)$. The solution of the equation is $x \approx 1.64$.

115. $4\ln(x + 5) - x = 10$

Graph $y_1 = 4\ln(x + 5) - x$ and $y_2 = 10$.

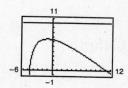

The graphs do not intersect. The equation has no solution.

117. $3(7550) = 7550e^{0.0725t}$

$$3 = e^{0.0725t}$$

$$\ln 3 = \ln e^{0.0725t}$$

$$\ln 3 = 0.0725t$$

$$t = \frac{\ln 3}{0.0725} \approx 15.2 \text{ years}$$

119. $y = e^{-2x/3}$

Exponential decay model

Matches graph (e)

121. $y = \ln(x + 3)$

Logarithmic model

Vertical asymptote: $x = -3$

Graph includes $(-2, 0)$

Matches graph (f)

123. $y = 2e^{-(x+4)^2/3}$

Gaussian model

Matches graph (a)

125. $P = 895,397e^{-0.011t}$ where $t = 0$ corresponds to 1970.

$$895,397e^{-0.011t} = 600,000$$

$$-0.011t = \ln\left(\frac{600,000}{895,397}\right)$$

$t \approx 36.4$ years which corresponds to the year 2006.

127. $20,000 = 10,000e^{r(5)}$

$$2 = e^{5r}$$

$$\ln 2 = 5r$$

$$\frac{\ln 2}{5} = r$$

$$r \approx 0.138629 = 13.8629\%$$

129.
$$y = ae^{bx}$$
$$2 = ae^{b(0)} \implies a = 2$$
$$3 = 2e^{b(4)}$$
$$1.5 = e^{4b}$$
$$\ln 1.5 = 4b \implies b \approx 0.1014$$
Thus, $y \approx 2e^{0.1014x}$

131. $y = 0.0499e^{-(x-71)^2/128}$

(a) Graph $y_1 = 0.0499e^{-(x-71)^2/128}$.

(b) The average test score is 71.

133.
$$\beta = 10 \log_{10}\left(\frac{I}{10^{-16}}\right)$$

$$125 = 10 \log_{10}\left(\frac{I}{10^{-16}}\right)$$

$$12.5 = \log_{10}\left(\frac{I}{10^{-16}}\right)$$

$$10^{12.5} = \frac{I}{10^{-16}}$$

$$I = 10^{-3.5} \text{ watt/cm}^2$$

Problem Solving for Chapter 7

1.

3. (a) $f(u + v) = a^{u+v}$
$$= a^u \cdot a^v$$
$$= f(u) \cdot f(v)$$

(b) $f(2x) = a^{2x}$
$$= (a^x)^2$$
$$= [f(x)]^2$$

5. $y_4 = (x - 1) - \frac{1}{2}(x-1)^2 + \frac{1}{3}(x-1)^3 - \frac{1}{4}(x-1)^4$

The pattern implies that

$$\ln x = (x-1) - \frac{1}{2}(x-1)^2 + \frac{1}{3}(x-1)^3 - \frac{1}{4}(x-1)^4 + \cdots.$$

7. $y = (1 + x)^{1/x}$

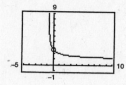

x	$(1 + x)^{1/x}$
-1	undefined
-0.1	2.867972
-0.01	2.731999
-0.001	2.719642
-0.0001	2.718418
-0.00001	2.718295
$\downarrow$	$\downarrow$
0^-	e

x	$(1 + x)^{1/x}$
1	2
0.1	2.593743
0.01	2.704814
0.001	2.716924
0.0001	2.718146
0.00001	2.718268
$\downarrow$	$\downarrow$
0^+	e

Near $x = 0$ the graph approaches e.

There is no y-intercept.

9. Interest: $u = M - \left(M - \dfrac{Pr}{12}\right)\left(1 + \dfrac{r}{12}\right)^{12t}$

Principal: $v = \left(M - \dfrac{Pr}{12}\right)\left(1 + \dfrac{r}{12}\right)^{12t}$

(a) $P = 120{,}000$, $t = 35$, $r = 0.075$, $M = 809.09$

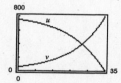

(b) In the early years of the mortgage, the majority of the monthly payment goes toward interest. The principal and interest are nearly equal when $t \approx 26$ years.

(c) $P = 120{,}000$, $t = 20$, $r = 0.075$, $M = 966.71$

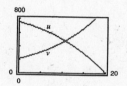

The interest is still the majority of the monthly payment in the early years. Now the principal and interest are nearly equal when $t \approx 10.729 \approx 11$ years.

11. $F(t) = 1 - e^{-t/3}$

    (a) $F(1) \approx 0.283$

    (b) $F(2.5) \approx 0.565$

    (c) $F(4) \approx 0.736$

13. (c)

The graph passes through $(0, 0)$. It is symmetric to the y-axis. $\lim\limits_{x \to \pm\infty} y = 6 \implies y = 6$ is a horizontal asymptote.

15. $y_1 = c_1\left(\dfrac{1}{2}\right)^{t/k_1}$ and $y_2 = c_2\left(\dfrac{1}{2}\right)^{t/k_2}$

$$c_1\left(\frac{1}{2}\right)^{t/k_1} = c_2\left(\frac{1}{2}\right)^{t/k_2}$$

$$\frac{c_1}{c_2} = \left(\frac{1}{2}\right)^{(t/k_2 - t/k_1)}$$

$$\ln\left(\frac{c_1}{c_2}\right) = \left(\frac{t}{k_2} - \frac{t}{k_1}\right)\ln\left(\frac{1}{2}\right)$$

$$\ln\left(\frac{c_1}{c_2}\right) = t\left(\frac{k_1 - k_2}{k_1 k_2}\right)(-\ln 2)$$

$$k_1 k_2 \ln\left(\frac{c_1}{c_2}\right) = t(k_2 - k_1)\ln 2$$

$$t = \frac{k_1 k_2 \ln(c_1/c_2)}{(k_2 - k_1)\ln 2}$$

17. $f(x) = e^x - e^{-x}$

$y = e^x - e^{-x}$

$x = e^y - e^{-y}$

$x = \dfrac{e^{2y} - 1}{e^y}$

$xe^y = e^{2y} - 1$

$e^{2y} - xe^y - 1 = 0$

$e^y = \dfrac{x \pm \sqrt{x^2 + 4}}{2}$ Quadratic Formula

Choosing the positive quantity for e^y we have

$y = \ln\left(\dfrac{x + \sqrt{x^2 + 4}}{2}\right)$. Thus,

$f^{-1}(x) = \ln\left(\dfrac{x + \sqrt{x^2 + 4}}{2}\right)$.

19. $f(x) = \dfrac{a^x + 1}{a^x - 1}, \ a > 0, \ a \neq 1$

$x = \dfrac{a^y + 1}{a^y - 1}$

$x(a^y - 1) = a^y + 1$

$xa^y - a^y = x + 1$

$a^y(x - 1) = x + 1$

$a^y = \dfrac{x + 1}{x - 1}$

$y = \log_a\left(\dfrac{x + 1}{x - 1}\right) = \dfrac{\ln\left(\dfrac{x + 1}{x - 1}\right)}{\ln a} = f^{-1}(x)$

C H A P T E R 8
Exponential and Logarithmic Functions
and Calculus

CHAPTER 8
Exponential and Logarithmic Functions and Calculus

Section 8.1   Exponential Functions: Differentiation and Integration

1. (a) $y = e^{3x}$

$y' = 3e^{3x}$

At $(0, 1)$, $y' = 3$.

(b) $y = e^{-3x}$

$y' = -3e^{-3x}$

At $(0, 1)$, $y' = -3$.

3. $f(x) = e^{2x}$

$f'(x) = 2e^{2x}$

5. $y = e^{-2x + x^2}$

$\dfrac{dy}{dx} = 2(x - 1)e^{-2x + x^2}$

7. $f(x) = e^{1/x} = e^{x^{-1}}$

$f'(x) = (-x^{-2})e^{x^{-1}} = -\dfrac{e^{1/x}}{x^2}$

9. $y = e^{\sqrt{x}}$

$\dfrac{dy}{dx} = \dfrac{e^{\sqrt{x}}}{2\sqrt{x}}$

11. $f(x) = (x + 1)e^{3x}$

$f'(x) = (x + 1)(3e^{3x}) + e^{3x}(1)$

$= e^{3x}(3x + 4)$

13. $f(x) = \dfrac{e^{x^2}}{x}$

$f'(x) = \dfrac{x(2xe^{x^2}) - e^{x^2}}{x^2}$

$= \dfrac{e^{x^2}(2x^2 - 1)}{x^2}$

15. $g(t) = (e^{-t} + e^{t})^3$

$g'(t) = 3(e^{-t} + e^{t})^2(e^{t} - e^{-t})$

17. $y = \dfrac{2}{e^x + e^{-x}} = 2(e^x + e^{-x})^{-1}$

$\dfrac{dy}{dx} = -2(e^x + e^{-x})^{-2}(e^x - e^{-x})$

$= \dfrac{-2(e^x - e^{-x})}{(e^x + e^{-x})^2}$

19. $y = x^2e^x - 2xe^x + 2e^x = e^x(x^2 - 2x + 2)$

$\dfrac{dy}{dx} = e^x(2x - 2) + e^x(x^2 - 2x + 2) = x^2e^x$

21. $xe^y - 10x + 3y = 0$

$xe^y\dfrac{dy}{dx} + e^y - 10 + 3\dfrac{dy}{dx} = 0$

$\dfrac{dy}{dx}(xe^y + 3) = 10 - e^y$

$\dfrac{dy}{dx} = \dfrac{10 - e^y}{xe^y + 3}$

23. $f(x) = 2e^{3x} + 3e^{-2x}$

$f'(x) = 6e^{3x} - 6e^{-2x}$

$f''(x) = 18e^{3x} + 12e^{-2x}$

25. $f(x) = (3 + 2x)e^{-3x}$

$f'(x) = (3 + 2x)(-3e^{-3x}) + e^{-3x}(2)$

$= e^{-3x}(-6x - 7)$

$f''(x) = e^{-3x}(-6) + (-6x - 7)(-3e^{-3x})$

$= 3e^{-3x}(6x + 5)$

27. $f(x) = \dfrac{e^x + e^{-x}}{2}$

$f'(x) = \dfrac{e^x - e^{-x}}{2} = 0$ when $x = 0$.

$f''(x) = \dfrac{e^x + e^{-x}}{2} > 0$

Relative minimum: $(0, 1)$

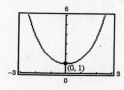

29. $g(x) = \dfrac{1}{\sqrt{2\pi}} e^{-(x-2)^2/2}$

$g'(x) = \dfrac{-1}{\sqrt{2\pi}}(x - 2)e^{-(x-2)^2/2}$

$g''(x) = \dfrac{1}{\sqrt{2\pi}}(x - 1)(x - 3)e^{-(x-2)^2/2}$

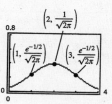

Relative maximum: $\left(2, \dfrac{1}{\sqrt{2\pi}}\right) \approx (2, 0.399)$

Points of inflection: $\left(1, \dfrac{1}{\sqrt{2\pi}} e^{-1/2}\right), \left(3, \dfrac{1}{\sqrt{2\pi}} e^{-1/2}\right) \approx (1, 0.242), (3, 0.242)$

31. $f(x) = x^2 e^{-x}$

$f'(x) = -x^2 e^{-x} + 2x e^{-x} = x e^{-x}(2 - x) = 0$ when $x = 0, 2$.

$f''(x) = -e^{-x}(2x - x^2) + e^{-x}(2 - 2x)$

$\quad = e^{-x}(x^2 - 4x + 2) = 0$ when $x = 2 \pm \sqrt{2}$.

Relative minimum: $(0, 0)$

Relative maximum: $(2, 4e^{-2})$

$\quad x = 2 \pm \sqrt{2}$

$\quad y = \left(2 \pm \sqrt{2}\right)^2 e^{-(2\pm\sqrt{2})}$

Point of inflection: $(3.414, 0.384), (0.586, 0.191)$

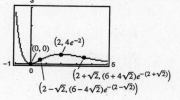

33. $g(t) = 1 + (2 + t)e^{-t}$

$g'(t) = -(1 + t)e^{-t}$

$g''(t) = t e^{-t}$

Relative maximum: $(-1, 1 + e) \approx (-1, 3.718)$

Point of inflection: $(0, 3)$

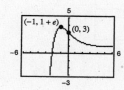

35. $A = 2xe^{-x^2}$

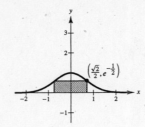

$$\frac{dA}{dx} = -4x^2 e^{-x^2} + 2e^{-x^2}$$

$$= 2e^{-x^2}(1 - 2x^2) = 0 \text{ when } x = \frac{\sqrt{2}}{2}.$$

$$A = \sqrt{2}e^{-1/2}$$

37. $y = \dfrac{L}{1 + ae^{-x/b}}, a > 0, b > 0, L > 0$

$$y' = \frac{-L\left(-\dfrac{a}{b}e^{-x/b}\right)}{(1 + ae^{-x/b})^2} = \frac{\dfrac{aL}{b}e^{-x/b}}{(1 + ae^{-x/b})^2}$$

$$y'' = \frac{(1 + ae^{-x/b})^2\left(\dfrac{-aL}{b^2}e^{-x/b}\right) - \left(\dfrac{aL}{b}e^{-x/b}\right)2(1 + ae^{-x/b})\left(\dfrac{-a}{b}e^{-x/b}\right)}{(1 + ae^{-x/b})^4}$$

$$= \frac{(1 + ae^{-x/b})\left(\dfrac{-aL}{b^2}e^{-x/b}\right) + 2\left(\dfrac{aL}{b}e^{-x/b}\right)\left(\dfrac{a}{b}e^{-x/b}\right)}{(1 + ae^{-x/b})^3}$$

$$= \frac{Lae^{-x/b}[ae^{-x/b} - 1]}{(1 + ae^{-x/b})^3 b^2}$$

$$y'' = 0 \text{ if } ae^{-x/b} = 1 \implies \frac{-x}{b} = \ln\left(\frac{1}{a}\right) \implies x = b \ln a$$

$$y(b \ln a) = \frac{L}{1 + ae^{-(b \ln a)/b}} = \frac{L}{1 + a(1/a)} = \frac{L}{2}$$

Therefore, the y-coordinate of the inflection point is $L/2$.

39. $y = e^{-x}$ Point: $(0, 1)$

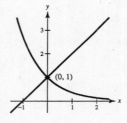

$y' = -e^{-x}$ (Slope of tangent line)

$-\dfrac{1}{y'} = e^x$ (Slope of normal line)

At $(0, 1)$, $e^x = 1$, $y = x + 1$.

41. (a)

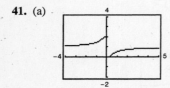

(b) When x increases without bound, $1/x$ approaches zero, and $e^{1/x}$ approaches 1. Therefore, $f(x)$ approaches $2/(1 + 1) = 1$. Thus, $f(x)$ has a horizontal asymptote at $y = 1$. As x approaches zero from the right, $1/x$ approaches ∞, $e^{1/x}$ approaches ∞ and $f(x)$ approaches zero. As x approaches zero from the left, $1/x$ approaches $-\infty$, $e^{1/x}$ approaches zero, and $f(x)$ approaches 2. The limit does not exist since the left limit does not equal the right limit. Therefore, $x = 0$ is a nonremovable discontinuity.

43. $f(x) = e^{x/2}, f(0) = 1$

$f'(x) = \frac{1}{2}e^{x/2}, f'(0) = \frac{1}{2}$

$f''(x) = \frac{1}{4}e^{x/2}, f''(0) = \frac{1}{4}$

$P_1(x) = 1 + \frac{1}{2}(x - 0) = \frac{x}{2} + 1, P_1(0) = 1$

$P_1'(x) = \frac{1}{2}, P_1'(0) = \frac{1}{2}$

$P_2(x) = 1 + \frac{1}{2}(x - 0) + \frac{1}{8}(x - 0)^2 = \frac{x^2}{8} + \frac{x}{2} + 1, P_2(0) = 1$

$P_2'(x) = \frac{1}{4}x + \frac{1}{2}, P_2'(0) = \frac{1}{2}$

$P_2''(x) = \frac{1}{4}, P_2''(0) = \frac{1}{4}$

The values of f, P_1, P_2 and their first derivatives agree at $x = 0$. The values of the second derivatives of f and P_2 agree at $x = 0$.

45. Let $u = 5x, du = 5\,dx$.

$\int e^{5x}\,5dx = e^{5x} + C$

47. Let $u = -2x, du = -2\,dx$.

$\int_0^1 e^{-2x}dx = -\frac{1}{2}\int_0^1 e^{-2x}(-2)\,dx = -\frac{1}{2}e^{-2x}\Big]_0^1$

$= \frac{1}{2}(1 - e^{-2}) = \frac{e^2 - 1}{2e^2}$

49. $\int xe^{-x^2}dx = -\frac{1}{2}\int e^{-x^2}(-2x)dx = -\frac{1}{2}e^{-x^2} + C$

51. $\int \frac{e^{\sqrt{x}}}{\sqrt{x}}dx = 2\int e^{\sqrt{x}}\left(\frac{1}{2\sqrt{x}}\right)dx = 2e^{\sqrt{x}} + C$

53. Let $u = \frac{3}{x}, du = -\frac{3}{x^2}\,dx$.

$\int_1^3 \frac{3e^{3/x}}{x^2}\,dx = -\frac{1}{3}\int_1^3 e^{3/x}\left(-\frac{3}{x^2}\right)dx = -\frac{1}{3}e^{3/x}\Big]_1^3 = \frac{e}{3}(e^2 - 1)$

55. $\int (1 + e^x)^2\,dx = \int (1 + 2e^x + e^{2x})\,dx = x + 2e^x + \frac{1}{2}e^{2x} + C$

57. $\int e^{-x}(1 + e^{-x})^2\,dx = -\int (1 + e^{-x})^2(-e^{-x})\,dx = -\frac{1}{3}(1 + e^{-x})^3 + C$

59. Let $u = 1 - e^x, du = -e^x\,dx$.

$\int e^x\sqrt{1 - e^x}\,dx = -\int (1 - e^x)^{1/2}(-e^x)\,dx = -\frac{2}{3}(1 - e^x)^{3/2} + C$

61. $\int \frac{e^{-x}}{(1 + e^{-x})^2}\,dx = -\int (1 + e^{-x})^{-2}(-e^{-x})\,dx = (1 + e^{-x})^{-1} + C = \frac{1}{1 + e^{-x}} + C$

63. $\int \frac{e^x + e^{-x}}{\sqrt{e^x - e^{-x}}}dx = \int (e^x - e^{-x})^{-1/2}(e^x + e^{-x})\,dx = 2(e^x - e^{-x})^{1/2} + C = 2\sqrt{e^x - e^{-x}} + C$

65. $\displaystyle\int \frac{5 - e^x}{e^{2x}} dx = \int 5e^{-2x} dx - \int e^{-x} dx = -\frac{5}{2}e^{-2x} + e^{-x} + C$

67. Let $u = ax^2, du = 2ax\, dx.$ (Assume $a \neq 0$)

$$y = \int xe^{ax^2} dx = \frac{1}{2a}\int e^{ax^2}(2ax)\, dx = \frac{1}{2a}e^{ax^2} + C$$

69. $f'(x) = \displaystyle\int \frac{1}{2}(e^x + e^{-x})\, dx = \frac{1}{2}(e^x - e^{-x}) + C_1$

$\quad f'(0) = C_1 = 0$

$\quad f(x) = \displaystyle\int \frac{1}{2}(e^x - e^{-x})\, dx = \frac{1}{2}(e^x + e^{-x}) + C_2$

$\quad f(0) = 1 + C_2 = 1 \implies C_2 = 0$

$\quad f(x) = \dfrac{1}{2}(e^x + e^{-x})$

71. (a)

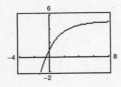

(b) $\dfrac{dy}{dx} = 2e^{-x/2}, \quad (0, 1)$

$$y = \int 2e^{-x/2}\, dx = -4\int e^{-x/2}\left(-\frac{1}{2}dx\right)$$

$$= -4e^{-x/2} + C$$

$(0, 1)$: $1 = -4e^0 + C = -4 + C \implies C = 5$

$y = -4e^{-x/2} + 5$

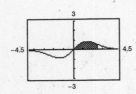

73. $\displaystyle\int_0^5 e^x\, dx = e^x\Big]_0^5 = e^5 - 1 \approx 147.413$

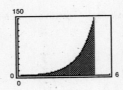

75. $\displaystyle\int_0^{\sqrt{6}} xe^{-(x^2/4)}\, dx = -2e^{-(x^2/4)}\Big]_0^{\sqrt{6}}$

$\qquad\qquad = 2 - 2e^{-3/2} = 2(1 - e^{-3/2}) \approx 1.554$

77. $\displaystyle\int_0^4 \sqrt{x}\, e^x\, dx, \ n = 12$

Midpoint Rule: 92.1898

Trapezoidal Rule: 93.8371

Simpson's Rule: 92.7385

79. $0.0665\displaystyle\int_{48}^{60} e^{-0.0139(t-48)^2}\, dt$

$n = 24$

Simpson's Rule: $0.4772 = 47.72\%$

81. $\displaystyle\int_0^x e^t\, dt \geq \int_0^x 1\, dt$

$\left[e^t\right]_0^x \geq \left[t\right]_0^x$

$e^x - 1 \geq x \implies e^x \geq 1 + x \text{ for } x \geq 0$

83. Yes. $f(x) = Ce^x$, C a constant.

85. $e^{-x} > 0 \implies \displaystyle\int_0^2 e^{-x}dx > 0.$

Section 8.2   Logarithmic Functions and Differentiation

1. $\displaystyle\lim_{x \to 3^+} \ln(x - 3) = -\infty$

3. $\displaystyle\lim_{x \to 2^-} \ln[x^2(3 - x)] = \ln 4 \approx 1.3863$

5. $y = \ln x^3 = 3 \ln x$

$y' = \dfrac{3}{x}$

At $(1, 0)$, $y' = 3$.

7. $y = \ln x^2 = 2 \ln x$

$y' = \dfrac{2}{x}$

At $(1, 0)$, $y' = 2$.

9. $g(x) = \ln x^2 = 2 \ln x$

$g'(x) = \dfrac{2}{x}$

11. $y = \ln\sqrt{x^4 - 4x} = \dfrac{1}{2}\ln(x^4 - 4x)$

$\dfrac{dy}{dx} = \dfrac{(4x^3 - 4)}{2(x^4 - 4x)} = \dfrac{2(x^3 - 1)}{x(x^3 - 4)}$

13. $y = (\ln x)^4$

$\dfrac{dy}{dx} = 4(\ln x)^3\left(\dfrac{1}{x}\right) = \dfrac{4(\ln x)^3}{x}$

15. $y = \ln x\sqrt{x^2 - 1} = \ln x + \dfrac{1}{2}\ln(x^2 - 1)$

$\dfrac{dy}{dx} = \dfrac{1}{x} + \dfrac{1}{2}\left(\dfrac{2x}{x^2 - 1}\right) = \dfrac{2x^2 - 1}{x(x^2 - 1)}$

17. $f(x) = \ln\dfrac{x}{x^2 + 1} = \ln x - \ln(x^2 + 1)$

$f'(x) = \dfrac{1}{x} - \dfrac{2x}{x^2 + 1} = \dfrac{1 - x^2}{x(x^2 + 1)}$

19. $g(t) = \dfrac{\ln t}{t^2}$

$g'(t) = \dfrac{t^2(1/t) - 2t \ln t}{t^4} = \dfrac{1 - 2 \ln t}{t^3}$

21. $y = \ln(\ln x^2)$

$\dfrac{dy}{dx} = \dfrac{(2x/x^2)}{\ln x^2} = \dfrac{2}{x \ln x^2} = \dfrac{1}{x \ln x}$

23. $y = \ln\sqrt{\dfrac{x + 1}{x - 1}} = \dfrac{1}{2}[\ln(x + 1) - \ln(x - 1)]$

$\dfrac{dy}{dx} = \dfrac{1}{2}\left[\dfrac{1}{x + 1} - \dfrac{1}{x - 1}\right] = \dfrac{1}{1 - x^2}$

25. $y = \ln e^{x^2} = x^2$

$\dfrac{dy}{dx} = 2x$

27. $y = \ln\left(\dfrac{1 + e^x}{1 - e^x}\right)$

$= \ln(1 + e^x) - \ln(1 - e^x)$

$\dfrac{dy}{dx} = \dfrac{e^x}{1 + e^x} + \dfrac{e^x}{1 - e^x} = \dfrac{2e^x}{1 - e^{2x}}$

29. $f(x) = e^{-x} \ln x$

$f'(x) = e^{-x}\left(\dfrac{1}{x}\right) - e^{-x} \ln x$

$= e^{-x}\left(\dfrac{1}{x} - \ln x\right)$

31. $f(x) = \ln\dfrac{\sqrt{4 + x^2}}{x} = \dfrac{1}{2}\ln(4 + x^2) - \ln x$

$f'(x) = \dfrac{x}{4 + x^2} - \dfrac{1}{x} = \dfrac{-4}{x(x^2 + 4)}$

33. $y = \dfrac{-\sqrt{x^2 + 1}}{x} + \ln\left(x + \sqrt{x^2 + 1}\right)$

$\dfrac{dy}{dx} = \dfrac{-x\left(x/\sqrt{x^2 + 1}\right) + \sqrt{x^2 + 1}}{x^2} + \left(\dfrac{1}{x + \sqrt{x^2 + 1}}\right)\left(1 + \dfrac{x}{\sqrt{x^2 + 1}}\right)$

$= \dfrac{1}{x^2\sqrt{x^2 + 1}} + \left(\dfrac{1\left(x - \sqrt{x^2 + 1}\right)}{-1}\right)\left(\dfrac{\sqrt{x^2 + 1} + x}{\sqrt{x^2 + 1}}\right) = \dfrac{\sqrt{x^2 + 1}}{x^2}$

35. $f(x) = \ln|x^2 - 1|$

$f'(x) = \dfrac{2x}{x^2 - 1}$

37. $f(x) = 4^x$

$f'(x) = (\ln 4)\, 4^x$

39. $y = 5^{x-2}$

$\dfrac{dy}{dx} = (\ln 5)\, 5^{x-2}$

41. $g(t) = t^2\, 2^t$

$g'(t) = t^2(\ln 2)2^t + (2t)\, 2^t$

$= t\, 2^t(t \ln 2 + 2)$

$= 2^t\, t(2 + t \ln 2)$

43. $y = \log_3 x$

$\dfrac{dy}{dx} = \dfrac{1}{x \ln 3}$

45. $f(x) = \log_2 \dfrac{x^2}{x - 1}$

$= 2 \log_2 x - \log_2 (x - 1)$

$f'(x) = \dfrac{2}{x \ln 2} - \dfrac{1}{(x - 1) \ln 2}$

$= \dfrac{x - 2}{(\ln 2)x(x - 1)}$

47. $y = \log_5 \sqrt{x^2 - 1} = \dfrac{1}{2} \log_5 (x^2 - 1)$

$\dfrac{dy}{dx} = \dfrac{1}{2} \cdot \dfrac{2x}{(x^2 - 1)\ln 5} = \dfrac{x}{(x^2 - 1)\ln 5}$

49. $g(t) = \dfrac{10 \log_4 t}{t} = \dfrac{10(\ln t)/(\ln 4)}{t} = \dfrac{10}{\ln 4}\left(\dfrac{\ln t}{t}\right)$

$g'(t) = \dfrac{10}{2 \ln 2}\left(\dfrac{t(1/t) - \ln t}{t^2}\right)$

$= \dfrac{5}{\ln 2}\left(\dfrac{1 - \ln t}{t^2}\right) = \dfrac{5(1 - \ln t)}{t^2 \ln 2}$

51. $y = x\sqrt{x^2 - 1}$

$\ln y = \ln x + \dfrac{1}{2} \ln(x^2 - 1)$

$\dfrac{1}{y}\left(\dfrac{dy}{dx}\right) = \dfrac{1}{x} + \dfrac{x}{x^2 - 1}$

$\dfrac{dy}{dx} = y\left[\dfrac{2x^2 - 1}{x(x^2 - 1)}\right] = \dfrac{2x^2 - 1}{\sqrt{x^2 - 1}}$

53. $y = \dfrac{x^2\sqrt{3x - 2}}{(x - 1)^2}$

$\ln y = 2 \ln x + \dfrac{1}{2} \ln(3x - 2) - 2 \ln(x - 1)$

$\dfrac{1}{y}\left(\dfrac{dy}{dx}\right) = \dfrac{2}{x} + \dfrac{3}{2(3x - 2)} - \dfrac{2}{x - 1}$

$\dfrac{dy}{dx} = y\left[\dfrac{3x^2 - 15x + 8}{2x(3x - 2)(x - 1)}\right]$

$= \dfrac{3x^3 - 15x^2 + 8x}{2(x - 1)^3\sqrt{3x - 2}}$

55. $y = \dfrac{x(x-1)^{3/2}}{\sqrt{x+1}}$

$\ln y = \ln x + \dfrac{3}{2}\ln(x-1) - \dfrac{1}{2}\ln(x+1)$

$\dfrac{1}{y}\left(\dfrac{dy}{dx}\right) = \dfrac{1}{x} + \dfrac{3}{2}\left(\dfrac{1}{x-1}\right) - \dfrac{1}{2}\left(\dfrac{1}{x+1}\right)$

$\dfrac{dy}{dx} = \dfrac{y}{2}\left[\dfrac{2}{x} + \dfrac{3}{x-1} - \dfrac{1}{x+1}\right]$

$= \dfrac{y}{2}\left[\dfrac{4x^2 + 4x - 2}{x(x^2-1)}\right]$

$= \dfrac{(2x^2 + 2x - 1)\sqrt{x-1}}{(x+1)^{3/2}}$

57. $y = x^{2/x}$

$\ln y = \dfrac{2}{x}\ln x$

$\dfrac{1}{y}\left(\dfrac{dy}{dx}\right) = \dfrac{2}{x}\left(\dfrac{1}{x}\right) + \ln x\left(-\dfrac{2}{x^2}\right)$

$\dfrac{dy}{dx} = \dfrac{2y}{x^2}(1 - \ln x) = 2x^{(2/x)-2}(1 - \ln x)$

59. $y = (x-2)^{x+1}$

$\ln y = (x+1)\ln(x-2)$

$\dfrac{1}{y}\left(\dfrac{dy}{dx}\right) = (x+1)\left(\dfrac{1}{x-2}\right) + \ln(x-2)$

$\dfrac{dy}{dx} = y\left[\dfrac{x+1}{x-2} + \ln(x-2)\right]$

$= (x-2)^{x+1}\left[\dfrac{x+1}{x-2} + \ln(x-2)\right]$

61. (a) $y = 3x^2 - \ln x,\ (1, 3)$

$\dfrac{dy}{dx} = 6x - \dfrac{1}{x}$

When $x = 1$, $\dfrac{dy}{dx} = 5$.

Tangent line: $y - 3 = 5(x - 1)$

$y = 5x - 2$

(b)
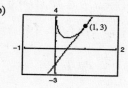

63. $x^2 - 3\ln y + y^2 = 10$

$2x - \dfrac{3}{y}\dfrac{dy}{dx} + 2y\dfrac{dy}{dx} = 0$

$2x = \dfrac{dy}{dx}\left(\dfrac{3}{y} - 2y\right)$

$\dfrac{dy}{dx} = \dfrac{2x}{(3/y) - 2y} = \dfrac{2xy}{3 - 2y^2}$

65. $y = 2(\ln x) + 3$

$y' = \dfrac{2}{x}$

$y'' = -\dfrac{2}{x^2}$

$xy'' + y' = x\left(-\dfrac{2}{x^2}\right) + \dfrac{2}{x} = 0$

67. $y = \dfrac{x^2}{2} - \ln x$, Domain: $(0, \infty)$

$y' = x - \dfrac{1}{x} = \dfrac{(x+1)(x-1)}{x} = 0$ when $x = \pm 1$.

$y'' = 1 + \dfrac{1}{x^2} > 0$

Relative minimum: $\left(1, \dfrac{1}{2}\right)$

69. $y = x\ln x$, Domain: $(0, \infty)$

$y' = x\left(\dfrac{1}{x}\right) + \ln x = 1 + \ln x = 0$ when $x = e^{-1}$.

$y'' = \dfrac{1}{x} > 0$

Relative minimum:
$(e^{-1}, -e^{-1})$

71. $y = \dfrac{x}{\ln x}$, Domain: $(0, 1) \cup (1, \infty)$

$y' = \dfrac{(\ln x)(1) - (x)(1/x)}{(\ln x)^2} = \dfrac{\ln x - 1}{(\ln x)^2} = 0$ when $x = e$.

$y'' = \dfrac{2 - \ln x}{x(\ln x)^3} = 0$ when $x = e^2$.

Relative minimum: (e, e)

Point of inflection: $(e^2, e^2/2)$

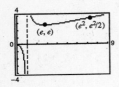

73. $y = x^2 - \ln x$, Domain: $x > 0$

$y' = 2x - \dfrac{1}{x} = 0$

$2x^2 = 1$

$x = \dfrac{1}{\sqrt{2}} = \dfrac{\sqrt{2}}{2}$

$y'' = 2 + \dfrac{1}{x^2} \neq 0$ No inflection points

Relative minimum: $\left(\dfrac{\sqrt{2}}{2}, \dfrac{1}{2} + \dfrac{1}{2}\ln 2 \right) \approx (0.7071, 0.8466)$

75. $f(x) = \ln x, \quad f(1) = 0$

$f'(x) = \dfrac{1}{x}, \quad f'(1) = 1$

$f''(x) = -\dfrac{1}{x^2}, \quad f''(1) = -1$

$P_1(x) = f(1) + f'(1)(x - 1) = x - 1, \quad P_1(1) = 0$

$P_2(x) = f(1) + f'(1)(x - 1) + \dfrac{1}{2}f''(1)(x - 1)^2$

$= (x - 1) - \dfrac{1}{2}(x - 1)^2, \quad P_2(1) = 0$

$P_1{}'(x) = 1, \quad P_1{}'(1) = 1$

$P_2{}'(x) = 1 - (x - 1) = 2 - x, \quad P_2{}'(1) = 1$

$P_2{}''(x) = -1, \quad P_2{}''(1) = -1$

The values of f, P_1, P_2, and their first derivatives agree at $x = 1$. The values of the second derivatives of f and P_2 agree at $x = 1$.

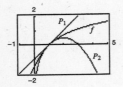

77. $g(x) = \ln f(x), f(x) > 0$

$g'(x) = \dfrac{f'(x)}{f(x)}$

(a) Yes. If the graph of g is increasing, then $g'(x) > 0$. Since $f(x) > 0$, you know that $f'(x) = g'(x)f(x)$ and thus, $f'(x) > 0$. Therefore, the graph of f is increasing.

(b) No. Let $f(x) = x^2 + 1$ (positive and concave up). $g(x) = \ln(x^2 + 1)$ is not concave up.

79. $t = \dfrac{5.315}{-6.7968 + \ln x}, \quad 1000 < x$

(a) $t(1167.41) \approx 20$ years

$T = (1167.41)(20)(12) = \$280{,}178.40$

(c) $\dfrac{dt}{dx} = -5.315(-6.7968 + \ln x)^{-2}\left(\dfrac{1}{x}\right)$

$= -\dfrac{5.315}{x(-6.7968 + \ln x)^2}$

When $x = 1167.41$, $dt/dx \approx -0.0645$.

When $x = 1068.45$, $dt/dx \approx -0.1585$.

(b) $t(1068.45) \approx 30$ years

$T = (1068.45)(30)(12) = \$384{,}642.00$

(d) There are two obvious benefits to paying a higher monthly payment:

1. The term is lower.

2. The total amount paid is lower.

81. (a)

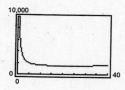

(b) $T'(p) = \dfrac{34.96}{p} + \dfrac{3.955}{\sqrt{p}}$

$T'(10) \approx 4.75 \text{ deg/lb/in}^2$

$T'(70) \approx 0.97 \text{ deg/lb/in}^2$

(c)

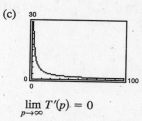

$\lim\limits_{p \to \infty} T'(p) = 0$

As the pressure increases, the rate of change of temperature approaches zero.

83. $C = 6000 + 300x + 300x \ln x$

Average cost: $\overline{C} = \dfrac{C}{x} = \dfrac{6000}{x} + 300 + 300 \ln x$

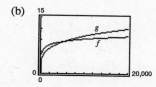

Minimum average cost: $1498.72

Analytically,

$\overline{C} = \dfrac{6000}{x} + 300 + 300 \ln x$

$\dfrac{d\overline{C}}{dx} = -\dfrac{6000}{x^2} + \dfrac{300}{x} = \dfrac{300x - 6000}{x^2}$

Critical Number: $x = 20$

On $(0, 20)$, $\dfrac{d\overline{C}}{dx} < 0 \implies \overline{C}$ is decreasing.

On $(20, \infty)$, $\dfrac{d\overline{C}}{dx} > 0 \implies \overline{C}$ is increasing.

Thus, $\overline{C}(20) \approx 1498.72$ is the minimum value.

85. (a) $\lim\limits_{t \to \infty} 6.7 e^{(-48.1)/t} = 6.7 e^0 = 6.7$ million ft³

(b) $V' = \dfrac{322.27}{t^2} e^{-(48.1)/t}$

$V'(20) \approx 0.073$ million ft³/yr

$V'(60) \approx 0.040$ million ft³/yr

87. (a)

(b)

The rate of growth of $\ln x$ is slower than $\sqrt{x}$ and $\sqrt[4]{x}$ for large x.

89. $y = \ln x$

$y' = \dfrac{1}{x} > 0$ for $x > 0$.

Since $\ln x$ is increasing on its entire domain $(0, \infty)$, it is a strictly monotonic function and therefore, is one-to-one.

91. False.

π is a constant.

$\dfrac{d}{dx}[\ln \pi] = 0$

93. Let $y = \ln x$

$e^y = x$

$e^y \dfrac{dy}{dx} = 1$

$\dfrac{dy}{dx} = \dfrac{1}{e^y} = \dfrac{1}{x}$.

Section 8.3   Logarithmic Functions and Integration

1. $\displaystyle\int \frac{3}{x}\,dx = 3\int \frac{1}{x}\,dx = 3\ln|x| + C$

3. $u = x + 1,\, du = dx$

$$\int \frac{1}{x+1}\,dx = \ln|x+1| + C$$

5. $u = 3 - 2x,\, du = -2\,dx$

$$\int \frac{1}{3-2x}\,dx = -\frac{1}{2}\int \frac{1}{3-2x}(-2)\,dx$$

$$= -\frac{1}{2}\ln|3 - 2x| + C$$

7. $u = x^2 + 1,\, du = 2x\,dx$

$$\int \frac{x}{x^2+1}\,dx = \frac{1}{2}\int \frac{1}{x^2+1}(2x)\,dx$$

$$= \frac{1}{2}\ln(x^2 + 1) + C$$

$$= \ln\sqrt{x^2 + 1} + C$$

9. $\displaystyle\int \frac{x^2 - 4}{x}\,dx = \int \left(x - \frac{4}{x}\right)dx$

$$= \frac{x^2}{2} - 4\ln|x| + C$$

11. $u = x^3 + 3x^2 + 9x,\, du = 3(x^2 + 2x + 3)\,dx$

$$\int \frac{x^2 + 2x + 3}{x^3 + 3x^2 + 9x}\,dx = \frac{1}{3}\int \frac{3(x^2 + 2x + 3)}{x^3 + 3x^2 + 9x}\,dx = \frac{1}{3}\ln|x^3 + 3x^2 + 9x| + C$$

13. $\displaystyle\int \frac{x^2 - 3x + 2}{x+1}\,dx = \int \left(x - 4 + \frac{6}{x+1}\right)dx$

$$= \frac{x^2}{2} - 4x + 6\ln|x+1| + C$$

15. $\displaystyle\int \frac{x^3 - 3x^2 + 5}{x-3}\,dx = \int \left(x^2 + \frac{5}{x-3}\right)dx$

$$= \frac{x^3}{3} + 5\ln|x - 3| + C$$

17. $\displaystyle\int \frac{x^4 + x - 4}{x^2 + 2}\,dx = \int \left(x^2 - 2 + \frac{x}{x^2+2}\right)dx$

$$= \frac{x^3}{3} - 2x + \frac{1}{2}\ln(x^2 + 2) + C$$

19. $u = \ln x,\, du = \dfrac{1}{x}\,dx$

$$\int \frac{(\ln x)^2}{x}\,dx = \frac{1}{3}(\ln x)^3 + C$$

21. Let $u = 1 + e^{-x},\, du = -e^{-x}\,dx.$

$$\int \frac{e^{-x}}{1 + e^{-x}}\,dx = -\int \frac{-e^{-x}}{1 + e^{-x}}\,dx = -\ln(1 + e^{-x}) + C$$

$$= \ln\left(\frac{e^x}{e^x + 1}\right) + C = x - \ln(e^x + 1) + C$$

23. Let $u = e^x - e^{-x},\, du = (e^x + e^{-x})\,dx.$

$$\int \frac{e^x + e^{-x}}{e^x - e^{-x}}\,dx = \ln|e^x - e^{-x}| + C$$

25. $u = x + 1,\, du = dx$

$$\int \frac{1}{\sqrt{x+1}}\,dx = \int (x+1)^{-1/2}\,dx$$

$$= 2(x+1)^{1/2} + C$$

$$= 2\sqrt{x+1} + C$$

27. $\displaystyle\int \frac{2x}{(x-1)^2}\,dx = \int \frac{2x - 2 + 2}{(x-1)^2}\,dx$

$$= \int \frac{2(x-1)}{(x-1)^2}\,dx + 2\int \frac{1}{(x-1)^2}\,dx$$

$$= 2\int \frac{1}{x-1}\,dx + 2\int \frac{1}{(x-1)^2}\,dx$$

$$= 2\ln|x-1| - \frac{2}{(x-1)} + C$$

29. $u = 1 + \sqrt{2x}$, $du = \dfrac{1}{\sqrt{2x}}\,dx \Rightarrow (u - 1)\,du = dx$

$$\int \frac{1}{1 + \sqrt{2x}}\,dx = \int \frac{(u-1)}{u}\,du = \int \left(1 - \frac{1}{u}\right)du$$

$$= u - \ln|u| + C_1$$

$$= \left(1 + \sqrt{2x}\right) - \ln\left|1 + \sqrt{2x}\right| + C_1$$

$$= \sqrt{2x} - \ln\left(1 + \sqrt{2x}\right) + C$$

where $C = C_1 + 1$.

31. $u = \sqrt{x} - 3$, $du = \dfrac{1}{2\sqrt{x}}\,dx \Rightarrow 2(u + 3)\,du = dx$

$$\int \frac{\sqrt{x}}{\sqrt{x} - 3}\,dx = 2\int \frac{(u+3)^2}{u}\,du = 2\int \frac{u^2 + 6u + 9}{u}\,du = 2\int \left(u + 6 + \frac{9}{u}\right)du$$

$$= 2\left[\frac{u^2}{2} + 6u + 9\ln|u|\right] + C_1 = u^2 + 12u + 18\ln|u| + C_1$$

$$= \left(\sqrt{x} - 3\right)^2 + 12\left(\sqrt{x} - 3\right) + 18\ln\left|\sqrt{x} - 3\right| + C_1$$

$$= x + 6\sqrt{x} + 18\ln\left|\sqrt{x} - 3\right| + C \text{ where } C = C_1 - 27.$$

33. $\displaystyle\int 3^x\,dx = \frac{3^x}{\ln 3} + C$

35. $\displaystyle\int x5^{-x^2}\,dx = -\frac{1}{2}\int 5^{-x^2}(-2x)\,dx$

$$= -\left(\frac{1}{2}\right)\frac{5^{-x^2}}{\ln 5} + C$$

$$= \frac{-1}{2\ln 5}\left(5^{-x^2}\right) + C$$

37. $\displaystyle\int \frac{3^{2x}}{1 + 3^{2x}}\,dx$, $u = 1 + 3^{2x}$, $du = 2(\ln 3)3^{2x}\,dx$

$$\frac{1}{2\ln 3}\int \frac{(2\ln 3)3^{2x}}{1 + 3^{2x}}\,dx = \frac{1}{2\ln 3}\ln(1 + 3^{2x}) + C$$

39. $\displaystyle\int_{-1}^{2} 2^x\,dx = \left[\left(\frac{1}{\ln 2}\right)2^x\right]_{-1}^{2}$

$$= \left(\frac{1}{\ln 2}\right)\left[4 - \frac{1}{2}\right]$$

$$= \frac{7}{2\ln 2} = \frac{7}{\ln 4}$$

41. $\displaystyle\int_{0}^{4} \frac{5}{3x + 1}\,dx = \left[\frac{5}{3}\ln|3x + 1|\right]_{0}^{4}$

$$= \frac{5}{3}\ln 13 \approx 4.275$$

43. $u = 1 + \ln x$, $du = \dfrac{1}{x}\,dx$

$$\int_{1}^{e} \frac{(1 + \ln x)^2}{x}\,dx = \left[\frac{1}{3}(1 + \ln x)^3\right]_{1}^{e} = \frac{7}{3}$$

45. $\displaystyle\int_{0}^{1} \frac{x^2 - 2}{x + 1}\,dx = \int_{0}^{1}\left(x - 1 - \frac{1}{x + 1}\right)dx$

$$= \left[\frac{1}{2}x^2 - x - \ln|x + 1|\right]_{0}^{1} = -\frac{1}{2} - \ln 2$$

$$\approx -1.193$$

47. $\displaystyle\int \frac{1}{1 + \sqrt{x}}\, dx = 2\left(1 + \sqrt{x}\right) - 2\ln\left(1 + \sqrt{x}\right) + C_1$

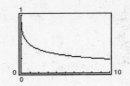

$\qquad\qquad = 2\left[\sqrt{x} - \ln\left(1 + \sqrt{x}\right)\right] + C$ where $C = C_1 + 2$.

49. $\displaystyle y = \int \frac{3}{2 - x}\, dx$

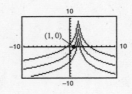

$\qquad = -3\displaystyle\int \frac{1}{x - 2}\, dx$

$\qquad = -3\ln|x - 2| + C$

$(1, 0):\ 0 = -3\ln|1 - 2| + C \implies C = 0$

$y = -3\ln|x - 2|$

51. $\dfrac{dy}{dx} = \dfrac{1}{x + 2},\ (0, 1)$

(a)

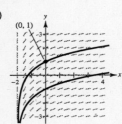

(b) $\quad y = \displaystyle\int \frac{1}{x + 2}\, dx = \ln|x + 2| + C$

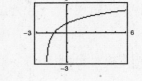

$\qquad y(0) = 1 \implies 1 = \ln 2 + C \implies C = 1 - \ln 2$

$\qquad$ Hence, $y = \ln|x + 2| + 1 - \ln 2 = \ln\left|\dfrac{x + 2}{2}\right| + 1.$

53. $\dfrac{dy}{dx} = 0.4^{x/3},\ \left(0, \dfrac{1}{2}\right)$

(a) 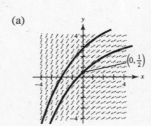 (b)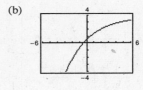

$\qquad y = \displaystyle\int 0.4^{x/3}\, dx = 3\int 0.4^{x/3}\left(\frac{1}{3}\, dx\right)$

$\qquad = \dfrac{3}{\ln 0.4}\, 0.4^{x/3} + C = \dfrac{-3}{\ln 2.5}(0.4)^{x/3} + C$

$\qquad \dfrac{1}{2} = -\dfrac{3}{\ln 2.5} + C \implies C = \dfrac{1}{2} + \dfrac{3}{\ln 2.5}$

$\qquad y = \dfrac{-3}{\ln 2.5}(0.4)^{x/3} + \dfrac{1}{2} + \dfrac{3}{\ln 2.5}$

$\qquad = \dfrac{3\left(1 - 0.4^{x/3}\right)}{\ln 2.5} + \dfrac{1}{2}$

55. $F(x) = \displaystyle\int_{1}^{x} \frac{1}{t}\, dt$

$\qquad F'(x) = \dfrac{1}{x}$

57. $F(x) = \displaystyle\int_{x}^{3x} \frac{1}{t}\, dt = \int_{1}^{3x} \frac{1}{t}\, dt - \int_{1}^{x} \frac{1}{t}\, dt$

$\qquad F'(x) = \dfrac{3}{3x} - \dfrac{1}{x} = 0$

59. $A = \displaystyle\int_{1}^{4} \frac{x^2 + 4}{x}\, dx = \int_{1}^{4}\left(x + \frac{4}{x}\right) dx$

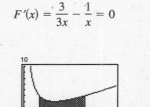

$\qquad = \left[\dfrac{x^2}{2} + 4\ln x\right]_{1}^{4} = (8 + 4\ln 4) - \dfrac{1}{2}$

$\qquad = \dfrac{15}{2} + 8\ln 2 \approx 13.045$ square units

61. $A = \int_0^3 3^x\,dx = \left[\dfrac{3^x}{\ln 3}\right]_0^3 = \dfrac{26}{\ln 3} \approx 23.666$

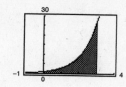

63. Power Rule

65. Substitution: $(u = x^2 + 4)$ and Log Rule

67. Divide the polynomials:

$$\frac{x^2}{x+1} = x - 1 + \frac{1}{x+1}$$

69. Average value $= \dfrac{1}{4-2}\displaystyle\int_2^4 \dfrac{8}{x^2}\,dx = 4\int_2^4 x^{-2}\,dx$

$$= \left[-4\frac{1}{x}\right]_2^4$$

$$= -4\left(\frac{1}{4} - \frac{1}{2}\right) = 1$$

71. Average value $= \dfrac{1}{e-1}\displaystyle\int_1^e \dfrac{\ln x}{x}\,dx = \dfrac{1}{e-1}\left[\dfrac{(\ln x)^2}{2}\right]_1^e = \dfrac{1}{e-1}\left(\dfrac{1}{2}\right) = \dfrac{1}{2e-2} \approx 0.291$

73. $P = \displaystyle\int \dfrac{3000}{1 + 0.25t}\,dt = (3000)(4)\int \dfrac{0.25}{1 + 0.25t}\,dt = 12{,}000\ln|1 + 0.25t| + C$

$P(0) = 12{,}000\ln|1 + 0.25(0)| + C = 1000$

$C = 1000$

$P = 12{,}000\ln|1 + 0.25t| + 1000 = 1000[12\ln|1 + 0.25t| + 1]$

$P(3) = 1000[12(\ln 1.75) + 1] \approx 7715$

75. $\dfrac{1}{50-40}\displaystyle\int_{40}^{50} \dfrac{90{,}000}{400 + 3x}\,dx = \Big[3000\ln|400 + 3x|\Big]_{40}^{50} \approx \$168.27$

77. (a) $\displaystyle\int_0^4 f(t)\,dt \approx 5.67$

$\displaystyle\int_0^4 g(t)\,dt \approx 5.67$

$\displaystyle\int_0^4 h(t)\,dt \approx 5.67$

(b)

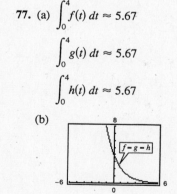

(c) The functions appear to be equal: $f(t) = g(t) = h(t)$

Analytically,

$$f(t) = 4\left(\frac{3}{8}\right)^{2t/3} = 4\left[\left(\frac{3}{8}\right)^{2/3}\right]^t = 4\left(\frac{9^{1/3}}{4}\right)^t = g(t)$$

$$h(t) = 4e^{-0.653886t} = 4\left[e^{-0.653886}\right]^t = 4(0.52002)^t$$

$$g(t) = 4\left(\frac{9^{1/3}}{4}\right)^t = 4(0.52002)^t.$$

No. The definite integrals over a given interval may be equal when the functions are not equal.

79. False

$$\frac{d}{dx}[\ln x] = \frac{1}{x}$$

81. False; the integrand has a nonremovable discontinuity at $x = 0$.

83. $F(x) = \displaystyle\int_x^{2x} \frac{1}{t}\, dt$

$$= \left[\ln|t| \right]_x^{2x}$$

$$= \ln(2x) - \ln(x) \text{ since } x > 0 \text{ on } (0, \infty).$$

$$= \ln\!\left(\frac{2x}{x}\right)$$

$$= \ln 2 \text{ which is constant on the interval } (0, \infty).$$

Section 8.4   Differential Equations: Growth and Decay

1. $\dfrac{dy}{dx} = x + 2$

$y = \displaystyle\int (x + 2)\, dx = \dfrac{x^2}{2} + 2x + C$

3. $\dfrac{dy}{dx} = y + 2$

$\dfrac{dy}{y + 2} = dx$

$\displaystyle\int \dfrac{1}{y + 2} dy = \int dx$

$\ln|y + 2| = x + C_1$

$y + 2 = e^{x + C_1} = Ce^x$

$y = Ce^x - 2$

5. $y' = \dfrac{5x}{y}$

$yy' = 5x$

$\displaystyle\int yy'\, dx = \int 5x\, dx$

$\displaystyle\int y\, dy = \int 5x\, dx$

$\dfrac{1}{2} y^2 = \dfrac{5}{2} x^2 + C_1$

$y^2 - 5x^2 = C$

7. $y' = \sqrt{x}\, y$

$\dfrac{y'}{y} = \sqrt{x}$

$\displaystyle\int \dfrac{y'}{y}\, dx = \int \sqrt{x}\, dx$

$\displaystyle\int \dfrac{dy}{y} = \int \sqrt{x}\, dx$

$\ln y = \dfrac{2}{3} x^{3/2} + C_1$

$y = e^{(2/3)x^{3/2} + C_1}$

$\quad = e^{C_1} e^{(2/3)x^{3/2}}$

$\quad = Ce^{(2/3)x^{3/2}}$

9. $(1 + x^2)y' - 2xy = 0$

$y' = \dfrac{2xy}{1 + x^2}$

$\dfrac{y'}{y} = \dfrac{2x}{1 + x^2}$

$\displaystyle\int \dfrac{y'}{y}\, dx = \int \dfrac{2x}{1 + x^2}\, dx$

$\displaystyle\int \dfrac{dy}{y} = \int \dfrac{2x}{1 + x^2}\, dx$

$\ln y = \ln(1 + x^2) + C_1$

$\ln y = \ln(1 + x^2) + \ln C$

$\ln y = \ln C(1 + x^2)$

$y = C(1 + x^2)$

11. $\dfrac{dQ}{dt} = \dfrac{k}{t^2}$

$\displaystyle\int \dfrac{dQ}{dt}\, dt = \int \dfrac{k}{t^2}\, dt$

$\displaystyle\int dQ = -\dfrac{k}{t} + C$

$Q = -\dfrac{k}{t} + C$

13. $\dfrac{dN}{ds} = k(250 - s)$

$\displaystyle\int \dfrac{dN}{ds}\, ds = \int k(250 - s)\, ds$

$\displaystyle\int dN = -\dfrac{k}{2}(250 - s)^2 + C$

$N = -\dfrac{k}{2}(250 - s)^2 + C$

15. (a)

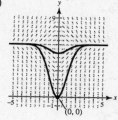

(b) $\dfrac{dy}{dx} = x(6 - y), \quad (0, 0)$

$\dfrac{dy}{y - 6} = -x$

$\ln|y - 6| = \dfrac{-x^2}{2} + C$

$y - 6 = e^{-x^2/2 + C} = C_1 e^{-x^2/2}$

$y = 6 + C_1 e^{-x^2/2}$

$(0, 0): \; 0 = 6 + C_1 \implies C_1 = -6 \implies y = 6 - 6e^{-x^2/2}$

17. $\dfrac{dy}{dt} = \dfrac{1}{2}t, \quad (0, 10)$

$\displaystyle\int dy = \int \dfrac{1}{2}t \, dt$

$y = \dfrac{1}{4}t^2 + C$

$10 = \dfrac{1}{4}(0)^2 + C \implies C = 10$

$y = \dfrac{1}{4}t^2 + 10$

19. $\dfrac{dy}{dt} = -\dfrac{1}{2}y, \quad (0, 10)$

$\displaystyle\int \dfrac{dy}{y} = \int -\dfrac{1}{2} \, dt$

$\ln y = -\dfrac{1}{2}t + C_1$

$y = e^{-(t/2) + C_1} = e^{C_1} e^{-t/2} = Ce^{-t/2}$

$10 = Ce^0 \implies C = 10$

$y = 10e^{-t/2}$

21. $\dfrac{dy}{dx} = ky$

$y = Ce^{kx}$ (Theorem 8.8)

$(0, 4): \; 4 = Ce^0 = C$

$(3, 10): \; 10 = 4e^{3k} \implies k = \dfrac{1}{3}\ln\left(\dfrac{5}{2}\right)$

When $x = 6, y = 4e^{1/3 \ln(5/2)(6)} = 4e^{\ln(5/2)^2}$

$\qquad = 4\left(\dfrac{5}{2}\right)^2 = 25.$

23. $\dfrac{dV}{dt} = kV$

$V = Ce^{kt}$ (Theorem 8.8)

$(0, 20{,}000): \; C = 20{,}000$

$(4, 12{,}500): \; 12{,}500 = 20{,}000e^{4k} \implies k = \dfrac{1}{4}\ln\left(\dfrac{5}{8}\right)$

When $t = 6, V = 20{,}000e^{1/4 \ln(5/8)(6)} = 20{,}000e^{\ln(5/8)^{3/2}}$

$\qquad = 20{,}000\left(\dfrac{5}{8}\right)^{3/2} \approx 9882.118.$

25. $y = Ce^{kt}, \; \left(0, \dfrac{1}{2}\right), (5, 5)$

$C = \dfrac{1}{2}$

$y = \dfrac{1}{2}e^{kt}$

$5 = \dfrac{1}{2}e^{5k}$

$k = \dfrac{\ln 10}{5} \approx 0.4605$

$y = \dfrac{1}{2}e^{0.4605t}$

27. $y = Ce^{kt}$, $(1, 1)$, $(5, 5)$

$$1 = Ce^k$$

$$5 = Ce^{5k}$$

$$5Ce^k = Ce^{5k}$$

$$5e^k = e^{5k}$$

$$5 = e^{4k}$$

$$k = \frac{\ln 5}{4} \approx 0.4024$$

$$y = Ce^{0.4024t}$$

$$1 = Ce^{0.4024}$$

$$C \approx 0.6687$$

$$y = 0.6687e^{0.4024t}$$

29. $\dfrac{dy}{dx} = \dfrac{1}{2}xy$

$\dfrac{dy}{dx} > 0$ when $xy > 0$. Quadrants I and III.

31. $\dfrac{1}{2}C = Ce^{k(1599)}$

$$0.5 = e^{k(1599)}$$

$$\ln 0.5 = k(1599)$$

$$k = \frac{\ln 0.5}{1599}$$

Given $C = 20$, after 1000 years we have:

$$y = 20e^{[(\ln 0.5)/1599](1000)} \approx 12.96 \text{ grams},$$

After 10,000 years we have:

$$y \approx 20e^{[(\ln 0.5)/1599](10,000)} \approx 0.26 \text{ grams}.$$

33. $\dfrac{1}{2}C = Ce^{k(5715)}$

$$0.5 = e^{k(5715)}$$

$$\ln 0.5 = k(5715)$$

$$k = \frac{\ln 0.5}{5715}$$

Given $y = 2$ grams after 10,000 years, we have:

$$2 = Ce^{[(\ln 0.5)/5715](10,000)} \implies C \approx 6.73 \text{ grams}.$$

After 1000 years we have:

$$y \approx 6.73e^{[(\ln 0.5)/5715](1000)} \approx 5.96 \text{ grams}.$$

35. $\dfrac{1}{2}C = Ce^{k(24,100)}$

$$0.5 = e^{k(24,100)}$$

$$\ln 0.5 = k(24,100)$$

$$k = \frac{\ln 0.5}{24,100}$$

Given $y = 2.5$ grams after 1000 years, we have:

$$2.5 = Ce^{[(\ln 0.5)/24,100](1000)} \implies C \approx 2.57 \text{ grams}.$$

After 10,000 years we have:

$$y \approx 2.57e^{[(\ln 0.5)/24,100](10,000)} \approx 1.93 \text{ grams}.$$

37. (a) $7.7 = Ce^{-0.009(1)}$

$$C = \frac{7.7}{e^{-0.009}} \approx 7.77$$

$$P = 7.77e^{-0.009t}$$

(b) $P = 7.77e^{-0.009(15)} \approx 6.79$ million people.

39. (a) $5.2 = Ce^{0.026(1)}$

$$C = \frac{5.2}{e^{0.026}} \approx 5.07$$

$$P = 5.07e^{0.026t}$$

(b) $P = 5.07e^{0.026(15)} \approx 7.49$ million people.

41. $S = Ce^{k/t}$

(a) $S = 5$ when $t = 1$.

$$5 = Ce^k$$

$$\lim_{t \to \infty} Ce^{k/t} = C = 30$$

$$5 = 30e^k$$

$$k = \ln \frac{1}{6} \approx -1.7918$$

$$S \approx 30e^{-1.7918/t}$$

(b) When $t = 5$, $S \approx 20.9646$ which is 20,965 units.

43. $R = \dfrac{\ln I - 0}{\ln 10}, I = e^{R \ln 10} = 10^R$

(a) $8.3 = \dfrac{\ln I - 0}{\ln 10}$

$\quad I = 10^{8.3} \approx 199{,}526{,}231.5$

(b) $2R = \dfrac{\ln I - 0}{\ln 10}$

$\quad I = e^{2R \ln 10} = e^{2R \ln 10} = (e^{R \ln 10})^2 = (10^R)^2$

Increases by a factor of $e^{R \ln 10}$ or 10^R.

(c) $\dfrac{dR}{dI} = \dfrac{1}{I \ln 10}$

45. $A(t) = V(t)e^{-0.10t} = 100{,}000e^{0.8\sqrt{t}}\, e^{-0.10t} = 100{,}000e^{0.8\sqrt{t}-0.10t}$

$\dfrac{dA}{dt} = 100{,}000\left(\dfrac{0.4}{\sqrt{t}} - 0.10\right)e^{0.8\sqrt{t}-0.10t} = 0$ when 16.

The timber should be harvested in the year 2014, (1998 + 16). **Note:** You could also use a graphing utility to graph $A(t)$ and find the maximum of $A(t)$. Use the viewing rectangle $0 \le x \le 30$ and $0 \le y \le 600{,}000$.

47. False. If $y = Ce^{kt}$, $y' = Cke^{kt} \ne$ constant.

49. True

Review Exercises for Chapter 8

1. $y = e^{-x/2}$

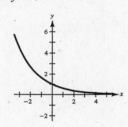

3. $f(x) = e^{-x^3}$

$\quad f'(x) = e^{-x^3}(-3x^2) = -3x^2 e^{-x^3}$

5. $g(t) = t^2 e^t$

$\quad g'(x) = t^2 e^t + 2t e^t = t e^t(t + 2)$

7. $y = \sqrt{e^{2x} + e^{-2x}}$

$\quad y' = \dfrac{1}{2}(e^{2x} + e^{-2x})^{-1/2}(2e^{2x} - 2e^{-2x})$

$\quad = \dfrac{e^{2x} - e^{-2x}}{\sqrt{e^{2x} + e^{-2x}}}$

9. $g(x) = \dfrac{x^2}{e^x}$

$\quad g'(x) = \dfrac{e^x(2x) - x^2 e^x}{e^{2x}} = \dfrac{x(2 - x)}{e^x}$

11. $e^x + y^2 = 0$

$\quad e^x + 2yy' = 0$

$\quad\quad y' = \dfrac{-e^x}{2y}$

13. Let $u = -3x^2$, $du = -6x\, dx$.

$\quad \displaystyle\int xe^{-3x^2}\, dx = -\dfrac{1}{6}\int e^{-3x^2}(-6x)\, dx = -\dfrac{1}{6}e^{-3x^2} + C$

15. $\displaystyle\int \dfrac{e^{4x} - e^{2x} + 1}{e^x}\, dx = \int (e^{3x} - e^x + e^{-x})\, dx$

$\quad = \dfrac{1}{3}e^{3x} - e^x - e^{-x} + C$

$\quad = \dfrac{e^{4x} - 3e^{2x} - 3}{3e^x} + C$

17. $\displaystyle\int xe^{1-x^2}\, dx = -\dfrac{1}{2}\int e^{1-x^2}(-2x)\, dx$

$\quad = -\dfrac{1}{2}e^{1-x^2} + C$

19. Let $u = e^x - 1$, $du = e^x \, dx$.

$$\int \frac{e^x}{(e^x - 1)^{3/2}} \, dx = \int (e^x - 1)^{-3/2} \, e^x \, dx$$

$$= -2(e^x - 1)^{-1/2} + C$$

$$= \frac{-2}{\sqrt{e^x - 1}} + C$$

21.
$$y = 5e^{2x} - 12e^{3x}$$
$$y' = 10e^{2x} - 36e^{3x}$$
$$y'' = 20e^{2x} - 108e^{3x}$$
$$y'' - 5y' + 6y = (20e^{2x} - 108e^{3x}) - 5(10e^{2x} - 36e^{3x}) + 6(5e^{2x} - 12e^{3x})$$
$$= 0$$

23. Area $= \displaystyle\int_0^4 xe^{-x^2} \, dx = \left[-\frac{1}{2}e^{-x^2} \right]_0^4$

$$= -\frac{1}{2}(e^{-16} - 1) \approx 0.500$$

25. $g(x) = \ln\sqrt{x} = \dfrac{1}{2}\ln x$

$$g'(x) = \frac{1}{2x}$$

27. $f(x) = x\sqrt{\ln x}$

$$f'(x) = \left(\frac{x}{2}\right)(\ln x)^{-1/2}\left(\frac{1}{x}\right) + \sqrt{\ln x}$$

$$= \frac{1}{2\sqrt{\ln x}} + \sqrt{\ln x} = \frac{1 + 2\ln x}{2\sqrt{\ln x}}$$

29. $y = \dfrac{1}{b^2}\left[\ln(a + bx) + \dfrac{a}{a + bx} \right]$

$$\frac{dy}{dx} = \frac{1}{b^2}\left[\frac{b}{a + bx} - \frac{ab}{(a + bx)^2} \right] = \frac{x}{(a + bx)^2}$$

31. $y = -\dfrac{1}{a}\ln\left(\dfrac{a + bx}{x}\right) = -\dfrac{1}{a}[\ln(a + bx) - \ln x]$

$$\frac{dy}{dx} = -\frac{1}{a}\left(\frac{b}{a + bx} - \frac{1}{x}\right) = \frac{1}{x(a + bx)}$$

33. $f(x) = 3^{x-1}$

$$f'(x) = 3^{x-1}\ln 3$$

35. $y = x^{2x+1}$

$$\ln y = (2x + 1)\ln x$$

$$\frac{y'}{y} = \frac{2x + 1}{x} + 2\ln x$$

$$y' = y\left(\frac{2x + 1}{x} + 2\ln x\right) = x^{2x+1}\left(\frac{2x + 1}{x} + 2\ln x\right)$$

37. $g(x) = \log_3\sqrt{1 - x} = \dfrac{1}{2}\log_3(1 - x)$

$$g'(x) = \frac{1}{2}\frac{-1}{(1 - x)\ln 3} = \frac{1}{2(x - 1)\ln 3}$$

39. $\ln x + y^2 = 0$

$$\frac{1}{x} + 2yy' = 0$$

$$2yy' = \frac{-1}{x}$$

$$y' = \frac{-1}{2xy}$$

41. $y = \left(\dfrac{6x}{x^2+1}\right)^{1/2}$

$\ln y = \dfrac{1}{2}[\ln(6x) - \ln(x^2+1)]$

$\dfrac{y'}{y} = \dfrac{1}{2}\left[\dfrac{1}{x} - \dfrac{2x}{x^2+1}\right]$

$y' = \dfrac{1}{2}\left(\dfrac{6x}{x^2+1}\right)^{1/2}\left(\dfrac{x^2+1-2x^2}{x(x^2+1)}\right)$

$\quad = \dfrac{\sqrt{6}(1-x^2)}{2\sqrt{x}(x^2+1)^{3/2}}$

43. $\qquad y\ln|1-x| = 1$

$y\left(\dfrac{-1}{1-x}\right) + y'\ln|1-x| = 0$

$\dfrac{-y}{1-x} + y'\left(\dfrac{1}{y}\right) = 0$

$y' = \dfrac{y^2}{1-x}$

45.

h	0	5	10	15	20
P	10,332	5583	2376	1240	517
$\ln P$	9.243	8.627	7.773	7.123	6.248

(a)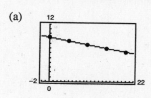

$y = -0.1499h + 9.3018$ is the regression line for data $(h, \ln P)$.

(c)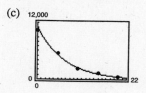

(b) $\ln P = ah + b$

$P = e^{ah+b} = e^b e^{ah}$

$P = Ce^{ah}, C = e^b$

For our data, $a = -0.1499$ and $C = e^{9.3018} = 10{,}957.7$.

$P = 10{,}957.7e^{-0.1499h}$

(d) $\dfrac{dP}{dh} = (10{,}957.7)(-0.1499)e^{-0.1499h}$

$\quad = -1642.56e^{-0.1499h}$

For $h=5$, $\dfrac{dP}{dh} = -776.3$. For $h=18$, $\dfrac{dP}{dh} \approx -110.6$.

47. $u = 7x - 2, du = 7dx$

$\displaystyle\int \dfrac{1}{7x-2}\,dx = \dfrac{1}{7}\int \dfrac{1}{7x-2}(7)\,dx = \dfrac{1}{7}\ln|7x-2| + C$

49. $\displaystyle\int_1^4 \dfrac{x+1}{x}\,dx = \int_1^4\left(1+\dfrac{1}{x}\right)dx = \left[x + \ln|x|\right]_1^4 = 3 + \ln 4$

51. Let $u = e^{2x} + e^{-2x}, du = (2e^{2x} - e^{-2x})\,dx$.

$\displaystyle\int \dfrac{e^{2x}-e^{-2x}}{e^{2x}+e^{-2x}}\,dx = \dfrac{1}{2}\int \dfrac{2e^{2x}-2e^{-2x}}{e^{2x}+e^{-2x}}\,dx$

$\quad = \dfrac{1}{2}\ln(e^{2x}+e^{-2x}) + C$

53. Let $u = e^x - 1, du = e^x\,dx$.

$\displaystyle\int \dfrac{e^x}{e^x-1}\,dx = \ln|e^x-1| + C$

55. $\displaystyle\int (x+1)5^{(x+1)^2}\,dx = \dfrac{1}{2}\dfrac{1}{\ln 5}5^{(x+1)^2} + C$

57. $\dfrac{dy}{dx} = \dfrac{x^2+3}{x}$

$\displaystyle\int dy = \int\left(x+\dfrac{3}{x}\right)dx$

$y = \dfrac{x^2}{2} + 3\ln|x| + C$

59. (a) $P = \dfrac{1}{100}\left(25 + \displaystyle\int_{2.5}^{10} \dfrac{25}{x}\,dx\right) = \dfrac{1}{100}\left(25 + \Big[25\ln x\Big]_{2.5}^{10}\right) = \dfrac{1}{4}(1 + \ln 4) \approx 0.60$

(b) $P = \dfrac{1}{100}\left(50 + \displaystyle\int_{5}^{10} \dfrac{50}{x}\,dx\right) = \dfrac{1}{100}\left(50 + \Big[50\ln x\Big]_{5}^{10}\right) = \dfrac{1}{2}(1 + \ln 2) \approx 0.85$

61. $P(h) = 30e^{kh}$

$P(18,000) = 30e^{18,000k} = 15$

$k = \dfrac{\ln(1/2)}{18,000} = \dfrac{-\ln 2}{18,000}$

$P(h) = 30e^{-(h\ln 2)/18,000}$

$P(35,000) = 30e^{-(35,000\ln 2)/18,000} \approx 7.79$ inches

63. $P = Ce^{0.015t}$

$2C = Ce^{0.015t}$

$2 = e^{0.015t}$

$\ln 2 = 0.015t$

$t = \dfrac{\ln 2}{0.015} \approx 46.21$ years

Problem Solving for Chapter 8

1. $y = e^{-x}, y' = -e^{-x} \Rightarrow y - b = -e^{-a}(x - a)$ is equation of tangent line.

Simplifying

$y = -e^{-a}x + b + ae^{-a}$

$y = -e^{-a}x + e^{-a} + ae^{-a}$

Let $x = 0$: $y = e^{-a}(1 + a)$. y-intercept is $R = (0, (1 + a)e^{-a})$.

Let $y = 0$: $-x + 1 + a = 0$. x-intercept is $Q = (a + 1, 0)$.

Area of $\triangle ORQ = A(a) = \dfrac{1}{2}(a + 1)(1 + a)e^{-a} = \dfrac{1}{2}e^{-a}(1 + a)^2$

$A'(a) = \dfrac{1}{2}e^{-a}2(1 + a) - \dfrac{1}{2}e^{-a}(1 + a)^2 = 0$

$\dfrac{1}{2}e^{-a}2(1 + a) = \dfrac{1}{2}e^{-a}(1 + a)^2$

$2 = 1 + a$

$a = 1, b = e^{-1}$

$P = (1, e^{-1})$ or $P = \left(1, \dfrac{1}{e}\right)$

Maximum area $= \dfrac{1}{2}e^{-1}(2)^2 = 2e^{-1} = \dfrac{2}{e} \approx 0.7358$ square units

3. (a) $\displaystyle\int_{0}^{2} f(x)\,dx = \int_{0}^{1} f(x)\,dx + \int_{1}^{2} f(x)\,dx.$

Use substitution for the integral on the right.

$x = u + 1, dx = du.$ Then

$\displaystyle\int_{0}^{2} f(x)\,dx = \int_{0}^{1} f(x)\,dx + \int_{0}^{1} f(u + 1)\,du$

$= \displaystyle\int_{0}^{1} [f(x) + f(x + 1)]\,dx.$

(b) $\displaystyle\int_{0}^{1}\left(\sqrt{x} + \sqrt{x + 1}\right)dx = \int_{0}^{2}\sqrt{x}\,dx = \dfrac{2}{3}x^{3/2}\Big]_{0}^{2} = \dfrac{4}{3}\sqrt{2}$

(c) $\displaystyle\int_{0}^{1}(e^x + e^{x+1})\,dx = \int_{0}^{2} e^x\,dx = e^x\Big]_{0}^{2} = e^2 - 1$

5. (a) $e^x > 1$

$$\int_0^x e^t \, dt > \int_0^x 1 \, dt$$

$$e^x - 1 > x$$

$$e^x > 1 + x$$

(b) $\displaystyle\int_0^x e^t \, dt > \int_0^x (1 + t) \, dt$

$$e^x - 1 > x + \frac{x^2}{2}$$

$$e^x > 1 + x + \frac{x^2}{2}$$

(c) Continuing this pattern,

$$e^x > 1 + x + \frac{x^2}{2!} + \frac{x^3}{3!}$$

$$\vdots$$

$$e^x > 1 + x + \frac{x^2}{2!} + \cdots + \frac{x^n}{n!}.$$

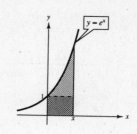

7. $y = \ln x$

$$y' = \frac{1}{x}$$

$$y - b = \frac{1}{a}(x - a)$$

$$y = \frac{1}{a}x + b - 1 \quad \text{Tangent line}$$

If $x = 0$, $c = b - 1$. Thus, the distance between b and c is

$$b - c = b - (b - 1) = 1.$$

9. (a) $\dfrac{dy}{dt} = y^{1.01}$

$$\int y^{-1.01} \, dy = \int dt$$

$$\frac{y^{-0.01}}{-0.01} = t + C_1$$

$$\frac{1}{y^{0.01}} = -0.01t + C$$

$$y^{0.01} = \frac{1}{C - 0.01t}$$

$$y = \frac{1}{(C - 0.01t)^{100}}$$

$y(0) = 1$: $1 = \dfrac{1}{C^{100}} \implies C = 1$

Hence, $y = \dfrac{1}{(1 - 0.01t)^{100}}.$

$\displaystyle\lim_{t \to T^-} y = \infty$ when $T = 100$.

(b) $\displaystyle\int y^{-(1+\varepsilon)} \, dy = \int k \, dt$

$$\frac{y^{-\varepsilon}}{-\varepsilon} = kt + C_1$$

$$y^{-\varepsilon} = -\varepsilon kt + C$$

$$y = \frac{1}{(C - \varepsilon kt)^{1/\varepsilon}}$$

$y(0) = y_0 = \dfrac{1}{C^{1/\varepsilon}} \implies C^{1/\varepsilon} = \dfrac{1}{y_0} \implies C = \left(\dfrac{1}{y_0}\right)^{\varepsilon}$

Hence, $y = \dfrac{1}{\left(\left(\dfrac{1}{y_0}\right)^{\varepsilon} - \varepsilon kt\right)^{1/\varepsilon}}.$

For $t \to \dfrac{1}{y_0^{\varepsilon} \varepsilon k}$, $y \to \infty$.

11. Let $u = 1 + \sqrt{x},\ \sqrt{x} = u - 1,\ x = u^2 - 2u + 1$,

$dx = (2u - 2)\,du$.

$$\text{Area} = \int_1^4 \frac{1}{\sqrt{x} + x}\,dx = \int_2^3 \frac{2u - 2}{(u - 1) + (u^2 - 2u + 1)}\,du$$

$$= \int_2^3 \frac{2(u - 1)}{u^2 - u}\,du$$

$$= \int_2^3 \frac{2}{u}\,du$$

$$= \Big[\, 2 \ln |u| \,\Big]_2^3$$

$$= 2 \ln 3 - 2 \ln 2 = 2 \ln\!\left(\frac{3}{2}\right)$$

$$\approx 0.8109$$

13. (a) $\dfrac{dS}{dt} = k_1 S(L - S)$

$S = \dfrac{L}{1 + Ce^{-kt}}$ is a solution because

$$\frac{dS}{dt} = -L(1 + Ce^{-kt})^{-2}(-C\,ke^{-kt})$$

$$= \frac{LC\,ke^{-kt}}{(1 + Ce^{-kt})^2}$$

$$= \left(\frac{k}{L}\right)\frac{L}{1 + Ce^{-kt}} \cdot \frac{C\,Le^{-kt}}{1 + Ce^{-kt}}$$

$$= \left(\frac{k}{L}\right)\frac{L}{1 + Ce^{-kt}} \cdot \left(L - \frac{L}{1 + Ce^{-kt}}\right)$$

$$= k_1 S(L - S),\ \text{where } k_1 = \frac{k}{L}.$$

$L = 100$. Also, $S = 10$ when $t = 0 \implies C = 9$. And,
$S = 20$ when $t = 1 \implies k = -\ln(4/9)$.

Particular Solution. $S = \dfrac{100}{1 + 9e^{\ln(4/9)t}}$

$$= \frac{100}{1 + 9e^{-0.8109t}}$$

(c)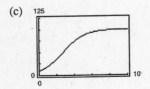

(b) $\dfrac{dS}{dt} = k_1 S(100 - S)$

$$\frac{d^2S}{dt^2} = k_1\!\left[S\!\left(-\frac{dS}{dt}\right) + (100 - S)\frac{dS}{dt} \right]$$

$$= k_1(100 - 2S)\frac{dS}{dt}$$

$$= 0 \text{ when } S = 50 \text{ or } \frac{dS}{dt} = 0.$$

Choosing $S = 50$, we have:

$$50 = \frac{100}{1 + 9e^{\ln(4/9)t}}$$

$$2 = 1 + 9e^{\ln(4/9)t}$$

$$\frac{\ln(1/9)}{\ln(4/9)} = t$$

$$t \approx 2.7 \text{ months (This is the inflection point)}$$

(d)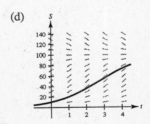

(e) Sales will decrease toward the line $S = L$.

CHAPTER 9
Trigonometric Functions

CHAPTER 9
Trigonometric Functions

Section 9.1   Radian and Degree Measure

1.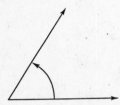

The angle shown is approximately 2 radians.

3.

The angle shown is approximately -3 radians.

5.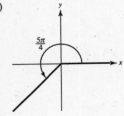

The angle shown is approximately 1 radian.

7. (a) Since $0 < \dfrac{\pi}{5} < \dfrac{\pi}{2}$; $\dfrac{\pi}{5}$ lies in Quadrant I.

   (b) Since $\pi < \dfrac{7\pi}{5} < \dfrac{3\pi}{2}$; $\dfrac{7\pi}{5}$ lies in Quadrant III.

9. (a) Since $-\dfrac{\pi}{2} < -\dfrac{\pi}{12} < 0$; $-\dfrac{\pi}{12}$ lies in Quadrant IV.

   (b) Since $-\dfrac{3\pi}{2} < -\dfrac{11\pi}{9} < -\pi$; $-\dfrac{11\pi}{9}$ lies in Quadrant II.

11. (a) Since $\pi < 3.5 < \dfrac{3\pi}{2}$; 3.5 lies in Quadrant III.

    (b) Since $\dfrac{\pi}{2} < 2.25 < \pi$; 2.25 lies in Quadrant II.

13. (a) 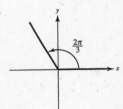   (b)

15. (a) 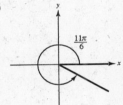   (b)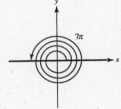

17. (a) Coterminal angles for $\dfrac{\pi}{6}$

$$\frac{\pi}{6} + 2\pi = \frac{13\pi}{6}$$

$$\frac{\pi}{6} - 2\pi = -\frac{11\pi}{6}$$

(b) Coterminal angles for $\dfrac{5\pi}{6}$

$$\frac{5\pi}{6} + 2\pi = \frac{17\pi}{6}$$

$$\frac{5\pi}{6} - 2\pi = -\frac{7\pi}{6}$$

19. (a) Coterminal angles for $\dfrac{2\pi}{3}$

$$\dfrac{2\pi}{3} + 2\pi = \dfrac{8\pi}{3}$$

$$\dfrac{2\pi}{3} - 2\pi = -\dfrac{4\pi}{3}$$

(b) Coterminal angles for $\dfrac{\pi}{12}$

$$\dfrac{\pi}{12} + 2\pi = \dfrac{25\pi}{12}$$

$$\dfrac{\pi}{12} - 2\pi = -\dfrac{23\pi}{12}$$

21. (a) Complement: $\dfrac{\pi}{2} - \dfrac{\pi}{3} = \dfrac{\pi}{6}$

Supplement: $\pi - \dfrac{\pi}{3} = \dfrac{2\pi}{3}$

(b) Complement: Not possible; $\dfrac{3\pi}{4}$ is greater than $\dfrac{\pi}{2}$.

Supplement: $\pi - \dfrac{3\pi}{4} = \dfrac{\pi}{4}$

23. (a) Complement: $\dfrac{\pi}{2} - 1 \approx 0.57$

Supplement: $\pi - 1 \approx 2.14$

(b) Complement: Not possible. 2 is greater than $\dfrac{\pi}{2}$.

Supplement: $\pi - 2 \approx 1.14$

25.

The angle shown is approximately $210°$.

27.

The angle shown is approximately $-60°$.

29.

The angle shown is approximately $165°$.

31. (a) Since $90° < 130° < 180°$; $130°$ lies in Quadrant II.

(b) Since $270° < 285° < 360°$; $285°$ lies in Quadrant IV.

33. (a) Since $-180° < -132°50' < -90°$; $-132°\ 50'$ lies in Quadrant III.

(b) Since $-360° < -336° < -270°$; $-336°$ lies in Quadrant I.

35. (a)   (b)

37. (a)   (b)

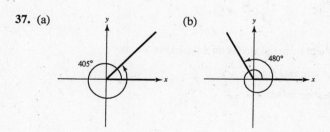

39. (a) Coterminal angles for $45°$

$$45° + 360° = 405°$$

$$45° - 360° = -315°$$

(b) Coterminal angles for $-36°$

$$-36° + 360° = 324°$$

$$-36° - 360° = -396°$$

41. (a) Coterminal angles for $120°$

$$120° + 360° = 480°$$

$$120° - 360° = -240°$$

(b) Coterminal angles for $-180°$

$$-180° + 360° = 180°$$

$$-180° - 360° = -540°$$

43. (a) Complement: $90° - 18° = 72°$

Supplement: $180° - 18° = 162°$

(b) Complement: Not possible; 115° is greater than 90°.

Supplement: $180° - 115° = 65°$

45. (a) Complement: $90° - 79° = 11°$

Supplement: $180° - 79° = 101°$

(b) Complement: Not possible. 150° is greater than 90°.

Supplement: $180° - 150° = 30°$

47. (a) $30° = 30°\left(\dfrac{\pi}{180°}\right) = \dfrac{\pi}{6}$

(b) $150° = 150°\left(\dfrac{\pi}{180°}\right) = \dfrac{5\pi}{6}$

49. (a) $-120° = -120°\left(\dfrac{\pi}{180°}\right) = -\dfrac{2\pi}{3}$

(b) $-240° = -240°\left(\dfrac{\pi}{180°}\right) = -\dfrac{4\pi}{3}$

51. $115° = 115°\left(\dfrac{\pi}{180°}\right) \approx 2.007$ radians

53. $-216.35° = -216.35°\left(\dfrac{\pi}{180°}\right) \approx -3.776$ radians

55. $532° = 532°\left(\dfrac{\pi}{180°}\right) \approx 9.285$ radians

57. $-0.83° = -0.83°\left(\dfrac{\pi}{180°}\right) \approx -0.014$ radian

59. (a) $\dfrac{3\pi}{2} = \dfrac{3\pi}{2}\left(\dfrac{180}{\pi}\right)^° = 270°$

(b) $\dfrac{7\pi}{6} = \dfrac{7\pi}{6}\left(\dfrac{180}{\pi}\right)^° = 210°$

61. (a) $-\dfrac{3\pi}{4} = -\dfrac{3\pi}{4}\left(\dfrac{180°}{\pi}\right) = -135°$

(b) $-\dfrac{7\pi}{6} = -\dfrac{7\pi}{6}\left(\dfrac{180°}{\pi}\right) = -210°$

63. $\dfrac{\pi}{7} = \dfrac{\pi}{7}\left(\dfrac{180°}{\pi}\right) \approx 25.714°$

65. $\dfrac{15\pi}{8} = \dfrac{15\pi}{8}\left(\dfrac{180°}{\pi}\right) = 337.5°$

67. $-4.2\pi = -4.2\pi\left(\dfrac{180°}{\pi}\right) = -756°$

69. $-2 = -2\left(\dfrac{180°}{\pi}\right) \approx -114.592°$

71. (a) $54° \, 45' = 54° + \left(\dfrac{45}{60}\right)^° = 54.75°$

(b) $-128° \, 30' = -128° - \left(\dfrac{30}{60}\right)^° = -128.5°$

73. (a) $85° \, 18' \, 30'' = \left(85 + \dfrac{18}{60} + \dfrac{30}{3600}\right)^° \approx 85.308°$

(b) $330° \, 25'' = \left(330 + \dfrac{25}{3600}\right)^° \approx 330.007°$

75. (a) $240.6° = 240° + 0.6(60)' = 240° \, 36'$

(b) $-145.8° = -[145° + 0.8(60')] = -145° \, 48'$

77. (a) $2.5° = 2° \, 30'$

(b) $-3.58° = -3° \, 34' \, 48''$

79. (a) An angle is in standard position if its vertex is at the origin and its initial side is on the positive *x*-axis.

(b) A negative angle is generated by a clockwise rotation of the terminal side.

(c) Two angles in standard position with the same terminal sides are coterminal.

(d) An obtuse angle measures between 90° and 180°.

81. 1 Radian $= \left(\dfrac{180}{\pi}\right)^° \approx 57.3°$, so one radian is much larger than one degree.

83. $s = r\theta$

$6 = 5\theta$

$\theta = \frac{6}{5}$ radians

85. $s = r\theta$

$32 = 7\theta$

$\theta = \frac{32}{7} = 4\frac{4}{7}$ radians

87. $s = r\theta$

$6 = 27\theta$

$\theta = \frac{6}{27} = \frac{2}{9}$ radian

89. $s = r\theta$

$25 = 14.5\theta$

$\theta = \frac{25}{14.5} = \frac{50}{29}$ radians

91. $s = r\theta$, θ in radians

$s = 15(180)\left(\frac{\pi}{180}\right) = 15\pi$ inches

≈ 47.12 inches

93. $s = r\theta$, θ in radians

$s = 3(1) = 3$ meters

95. $\theta = 41° \, 15' \, 31'' - 32° \, 47' \, 0'' = 8° \, 28' \, 31'' \approx 8.47528° \approx 0.14792$ radian

$s = r\theta = 4000(0.14797) \approx 591.08$ miles

97. $\theta = 42° \, 7' \, 45'' - 25° \, 46' \, 26'' = 16° \, 21' \, 19'' \approx 0.285453$ radian

$s = r\theta = 4000(0.285453) \approx 1141.81$ miles

99. (a) 65 miles per hour $= \frac{65(5280)}{60} = 5720$ feet per minute

The circumference of the tire is $C = 2.5\pi$ feet.

The number of revolutions per minute is $r = \frac{5720}{2.5\pi} \approx 728.3$ rev/min.

(b) The angular speed is $\frac{\theta}{t}$.

$\theta = \frac{5720}{2.5\pi}(2\pi) = 4576$ radians

Angular speed $= \frac{4576 \text{ radians}}{1 \text{ minute}} = 4576$ rad/min

101. Circumference: $C = 2\pi(1.68) = 3.36\pi$ inches

360 rev/min = 6 rev/sec

Linear speed: $(3.36\pi)(6) = 20.16\pi$ inches/sec $= 1209.6\pi$ inches/minute

103. False. A measurement of 4π radians corresponds to two complete revolutions from the initial to the terminal side of an angle.

105. False. $1° = \frac{\pi}{180}$ radian

Section 9.2   Trigonometric Functions: The Unit Circle

1. $x = -\frac{8}{17}, \quad y = \frac{15}{17}$

$\sin\theta = y = \frac{15}{17}$

$\cos\theta = x = -\frac{8}{17}$

$\tan\theta = \frac{y}{x} = -\frac{15}{8}$

$\csc\theta = \frac{1}{y} = \frac{17}{15}$

$\sec\theta = \frac{1}{x} = -\frac{17}{8}$

$\cot\theta = \frac{x}{y} = -\frac{8}{15}$

3. $x = \frac{12}{13}, \quad y = -\frac{5}{13}$

$\sin\theta = y = -\frac{5}{13}$

$\cos\theta = x = \frac{12}{13}$

$\tan\theta = \frac{y}{x} = -\frac{5}{12}$

$\csc\theta = \frac{1}{y} = -\frac{13}{5}$

$\sec\theta = \frac{1}{x} = \frac{13}{12}$

$\cot\theta = \frac{x}{y} = -\frac{12}{5}$

5. $t = \dfrac{\pi}{4}$ corresponds to $\left(\dfrac{\sqrt{2}}{2}, \dfrac{\sqrt{2}}{2}\right)$.

7. $t = \dfrac{7\pi}{6}$ corresponds to $\left(-\dfrac{\sqrt{3}}{2}, -\dfrac{1}{2}\right)$.

9. $t = \dfrac{4\pi}{3}$ corresponds to $\left(-\dfrac{1}{2}, -\dfrac{\sqrt{3}}{2}\right)$.

11. $t = \dfrac{3\pi}{2}$ corresponds to $(0, -1)$.

13. $t = \dfrac{\pi}{4}$ corresponds to $\left(\dfrac{\sqrt{2}}{2}, \dfrac{\sqrt{2}}{2}\right)$.

$\sin t = y = \dfrac{\sqrt{2}}{2}$

$\cos t = x = \dfrac{\sqrt{2}}{2}$

$\tan t = \dfrac{y}{x} = 1$

15. $t = -\dfrac{\pi}{6}$ corresponds to $\left(\dfrac{\sqrt{3}}{2}, -\dfrac{1}{2}\right)$.

$\sin t = y = -\dfrac{1}{2}$

$\cos t = x = \dfrac{\sqrt{3}}{2}$

$\tan t = \dfrac{y}{x} = -\dfrac{1}{\sqrt{3}} = -\dfrac{\sqrt{3}}{3}$

17. $t = -\dfrac{7\pi}{4}$ corresponds to $\left(\dfrac{\sqrt{2}}{2}, \dfrac{\sqrt{2}}{2}\right)$.

$\sin t = y = \dfrac{\sqrt{2}}{2}$

$\cos t = x = \dfrac{\sqrt{2}}{2}$

$\tan t = \dfrac{y}{x} = 1$

19. $t = \dfrac{11\pi}{6}$ corresponds to $\left(\dfrac{\sqrt{3}}{2}, -\dfrac{1}{2}\right)$.

$\sin t = y = -\dfrac{1}{2}$

$\cos t = x = \dfrac{\sqrt{3}}{2}$

$\tan t = \dfrac{y}{x} = -\dfrac{1}{\sqrt{3}} = -\dfrac{\sqrt{3}}{3}$

21. $t = -\dfrac{3\pi}{2}$ corresponds to $(0, 1)$.

$\sin t = y = 1$
$\cos t = x = 0$
$\tan t = \dfrac{y}{x}$ is undefined.

23. $t = \pi$ corresponds to the point $(x, y) = (-1, 0)$.

$\sin \pi = y = 0$

$\cos \pi = x = -1$

$\tan \pi = \dfrac{y}{x} = \dfrac{0}{-1} = 0$

25. $t = \dfrac{3\pi}{4}$ corresponds to $\left(-\dfrac{\sqrt{2}}{2}, \dfrac{\sqrt{2}}{2}\right)$.

$\sin t = y = \dfrac{\sqrt{2}}{2}$ $\csc t = \dfrac{1}{y} = \sqrt{2}$

$\cos t = x = -\dfrac{\sqrt{2}}{2}$ $\sec t = \dfrac{1}{x} = -\sqrt{2}$

$\tan t = \dfrac{y}{x} = -1$ $\cot t = \dfrac{x}{y} = -1$

27. $t = \dfrac{\pi}{2}$ corresponds to $(0, 1)$.

$\sin t = y = 1$ $\csc t = \dfrac{1}{y} = 1$

$\cos t = x = 0$ $\sec t = \dfrac{1}{x}$ is undefined.

$\tan t = \dfrac{y}{x}$ is undefined. $\cot t = \dfrac{x}{y} = 0$

29. $t = -\dfrac{\pi}{3}$ corresponds to $\left(\dfrac{1}{2}, -\dfrac{\sqrt{3}}{2}\right)$.

$\sin t = y = -\dfrac{\sqrt{3}}{2}$ $\qquad$ $\csc t = \dfrac{1}{y} = -\dfrac{2\sqrt{3}}{3}$

$\cos t = x = \dfrac{1}{2}$ $\qquad$ $\sec t = \dfrac{1}{x} = 2$

$\tan t = \dfrac{y}{x} = -\sqrt{3}$ $\qquad$ $\cot t = \dfrac{x}{y} = -\dfrac{\sqrt{3}}{3}$

31. $\sin 5\pi = \sin \pi = 0$

33. $\cos \dfrac{8\pi}{3} = \cos \dfrac{2\pi}{3} = -\dfrac{1}{2}$

35. $\cos(-3\pi) = \cos \pi = -1$

37. $\sin\left(-\dfrac{9\pi}{4}\right) = \sin\left(\dfrac{7\pi}{4}\right) = -\dfrac{\sqrt{2}}{2}$

39. $\sin t = \dfrac{1}{3}$

(a) $\sin(-t) = -\sin t = -\dfrac{1}{3}$

(b) $\csc(-t) = -\csc t = -3$

41. $\cos(-t) = -\dfrac{1}{5}$

(a) $\cos t = \cos(-t) = -\dfrac{1}{5}$

(b) $\sec(-t) = \dfrac{1}{\cos(-t)} = -5$

43. $\sin t = \dfrac{4}{5}$

(a) $\sin(\pi - t) = \sin t = \dfrac{4}{5}$

(b) $\sin(t + \pi) = -\sin t = -\dfrac{4}{5}$

45. $\sin \dfrac{\pi}{4} \approx 0.7071$

47. $\csc 1.3 = \dfrac{1}{\sin 1.3} \approx 1.0378$

49. $\cos(-1.7) \approx -0.1288$

51. $\csc 0.8 = \dfrac{1}{\sin 0.8} \approx 1.3940$

53. $\sec 22.8 = \dfrac{1}{\cos 22.8} \approx -1.4486$

55. (a) $\sin 5 \approx -1$

(b) $\cos 2 \approx -0.4$

57. (a) $\sin t = 0.25$

$t \approx 0.25$ or 2.89

(b) $\cos t = -0.25$

$t \approx 1.82$ or 4.46

59. $\cos 1.5 \approx 0.0707$

$2 \cos 0.75 \approx 1.4634$

$\cos 2t \neq 2 \cos t$

61. (a) The points have y-axis symmetry.

(b) $\sin t_1 = \sin(\pi - t_1)$ since they have the same y-value.

(c) $\cos(\pi - t_1) = -\cos t_1$ since the x-values have the opposite signs.

63. $\cos \theta = x = \cos(-\theta)$

$\sec \theta = \dfrac{1}{x} = \sec(-\theta)$

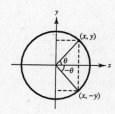

65. $h(t) = f(t)g(t)$ is an odd function, since

$h(-t) = f(-t)g(-t)$

$\qquad = -f(t)g(t)$

$\qquad = -h(t).$

67. $y(t) = \frac{1}{4}\cos 6t$

  (a) $y(0) = \frac{1}{4}\cos 0 = 0.2500$ feet

  (b) $y\left(\frac{1}{4}\right) = \frac{1}{4}\cos\frac{3}{2} \approx 0.0177$ feet

  (c) $y\left(\frac{1}{2}\right) = \frac{1}{4}\cos 3 \approx -0.2475$ feet

71. False. Because the sine function is odd, its graph is symmetric with respect to the origin, not the y-axis.

69. False. $\sin(-t) = -\sin t$ means the function is odd, not that the sine of a negative angle is a negative number.

For example: $\sin\left(-\frac{3\pi}{2}\right) = -\sin\left(\frac{3\pi}{2}\right) = -(-1) = 1.$

Even though the angle is negative, the sine value is positive.

Section 9.3   Right Triangle Trigonometry

1. hyp $= \sqrt{6^2 + 8^2} = \sqrt{36 + 64} = \sqrt{100} = 10$

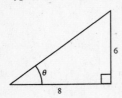

$$\sin\theta = \frac{\text{opp}}{\text{hyp}} = \frac{6}{10} = \frac{3}{5} \qquad \csc\theta = \frac{\text{hyp}}{\text{opp}} = \frac{10}{6} = \frac{5}{3}$$

$$\cos\theta = \frac{\text{adj}}{\text{hyp}} = \frac{8}{10} = \frac{4}{5} \qquad \sec\theta = \frac{\text{hyp}}{\text{adj}} = \frac{10}{8} = \frac{5}{4}$$

$$\tan\theta = \frac{\text{opp}}{\text{adj}} = \frac{6}{8} = \frac{3}{4} \qquad \cot\theta = \frac{\text{adj}}{\text{opp}} = \frac{8}{6} = \frac{4}{3}$$

3. adj $= \sqrt{41^2 - 9^2} = \sqrt{1681 - 81} = \sqrt{1600} = 40$

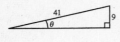

$$\sin\theta = \frac{\text{opp}}{\text{hyp}} = \frac{9}{41} \qquad \csc\theta = \frac{\text{hyp}}{\text{opp}} = \frac{41}{9}$$

$$\cos\theta = \frac{\text{adj}}{\text{hyp}} = \frac{40}{41} \qquad \sec\theta = \frac{\text{hyp}}{\text{adj}} = \frac{41}{40}$$

$$\tan\theta = \frac{\text{opp}}{\text{adj}} = \frac{9}{40} \qquad \cot\theta = \frac{\text{adj}}{\text{opp}} = \frac{40}{9}$$

5. adj $= \sqrt{3^2 - 1^2} = \sqrt{8} = 2\sqrt{2}$

$$\sin\theta = \frac{\text{opp}}{\text{hyp}} = \frac{1}{3} \qquad \csc\theta = \frac{\text{hyp}}{\text{opp}} = 3$$

$$\cos\theta = \frac{\text{adj}}{\text{hyp}} = \frac{2\sqrt{2}}{3} \qquad \sec\theta = \frac{\text{hyp}}{\text{adj}} = \frac{3}{2\sqrt{2}} = \frac{3\sqrt{2}}{4}$$

$$\tan\theta = \frac{\text{opp}}{\text{adj}} = \frac{1}{2\sqrt{2}} = \frac{\sqrt{2}}{4} \qquad \cot\theta = \frac{\text{adj}}{\text{opp}} = 2\sqrt{2}$$

adj $= \sqrt{6^2 - 2^2} = \sqrt{32} = 4\sqrt{2}$

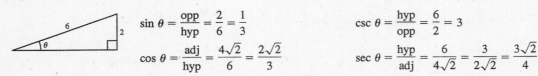

$$\sin\theta = \frac{\text{opp}}{\text{hyp}} = \frac{2}{6} = \frac{1}{3} \qquad \csc\theta = \frac{\text{hyp}}{\text{opp}} = \frac{6}{2} = 3$$

$$\cos\theta = \frac{\text{adj}}{\text{hyp}} = \frac{4\sqrt{2}}{6} = \frac{2\sqrt{2}}{3} \qquad \sec\theta = \frac{\text{hyp}}{\text{adj}} = \frac{6}{4\sqrt{2}} = \frac{3}{2\sqrt{2}} = \frac{3\sqrt{2}}{4}$$

$$\tan\theta = \frac{\text{opp}}{\text{adj}} = \frac{2}{4\sqrt{2}} = \frac{1}{2\sqrt{2}} = \frac{\sqrt{2}}{4} \qquad \cot\theta = \frac{\text{adj}}{\text{opp}} = \frac{4\sqrt{2}}{2} = 2\sqrt{2}$$

The function values are the same since the triangles are similar and the corresponding sides are proportional.

7. opp $= \sqrt{5^2 - 4^2} = 3$

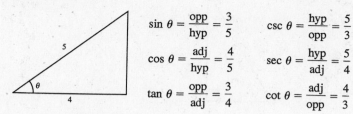

$$\sin \theta = \frac{\text{opp}}{\text{hyp}} = \frac{3}{5} \qquad \csc \theta = \frac{\text{hyp}}{\text{opp}} = \frac{5}{3}$$

$$\cos \theta = \frac{\text{adj}}{\text{hyp}} = \frac{4}{5} \qquad \sec \theta = \frac{\text{hyp}}{\text{adj}} = \frac{5}{4}$$

$$\tan \theta = \frac{\text{opp}}{\text{adj}} = \frac{3}{4} \qquad \cot \theta = \frac{\text{adj}}{\text{opp}} = \frac{4}{3}$$

opp $= \sqrt{1.25^2 - 1^2} = 0.75$

$$\sin \theta = \frac{\text{opp}}{\text{hyp}} = \frac{0.75}{1.25} = \frac{3}{5} \qquad \csc \theta = \frac{\text{hyp}}{\text{opp}} = \frac{1.25}{0.75} = \frac{5}{3}$$

$$\cos \theta = \frac{\text{adj}}{\text{hyp}} = \frac{1}{1.25} = \frac{4}{5} \qquad \sec \theta = \frac{\text{hyp}}{\text{adj}} = \frac{1.25}{1} = \frac{5}{4}$$

$$\tan \theta = \frac{\text{opp}}{\text{adj}} = \frac{0.75}{1} = \frac{3}{4} \qquad \cot \theta = \frac{\text{adj}}{\text{opp}} = \frac{1}{0.75} = \frac{4}{3}$$

The function values are the same since the triangles are similar and the corresponding sides are proportional.

9. Given: $\sin \theta = \dfrac{3}{4} = \dfrac{\text{opp}}{\text{hyp}}$

$3^2 + (\text{adj})^2 = 4^2$

$\qquad \text{adj} = \sqrt{7}$

$\cos \theta = \dfrac{\sqrt{7}}{4}$

$\tan \theta = \dfrac{3\sqrt{7}}{7}$

$\cot \theta = \dfrac{\sqrt{7}}{3}$

$\sec \theta = \dfrac{4\sqrt{7}}{7}$

$\csc \theta = \dfrac{4}{3}$

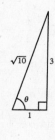

11. Given: $\sec \theta = 2 = \dfrac{2}{1} = \dfrac{\text{hyp}}{\text{adj}}$

$(\text{opp})^2 + 1^2 = 2^2$

$\qquad \text{opp} = \sqrt{3}$

$\sin \theta = \dfrac{\sqrt{3}}{2}$

$\cos \theta = \dfrac{1}{2}$

$\tan \theta = \sqrt{3}$

$\cot \theta = \dfrac{\sqrt{3}}{3}$

$\csc \theta = \dfrac{2\sqrt{3}}{3}$

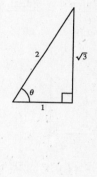

13. Given: $\tan \theta = 3 = \dfrac{3}{1} = \dfrac{\text{opp}}{\text{adj}}$

$3^2 + 1^2 = (\text{hyp})^2$

$\qquad \text{hyp} = \sqrt{10}$

$\sin \theta = \dfrac{3\sqrt{10}}{10}$

$\cos \theta = \dfrac{\sqrt{10}}{10}$

$\cot \theta = \dfrac{1}{3}$

$\sec \theta = \sqrt{10}$

$\csc \theta = \dfrac{\sqrt{10}}{3}$

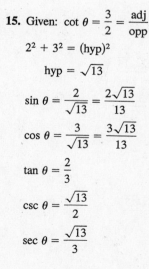

15. Given: $\cot \theta = \dfrac{3}{2} = \dfrac{\text{adj}}{\text{opp}}$

$2^2 + 3^2 = (\text{hyp})^2$

$\qquad \text{hyp} = \sqrt{13}$

$\sin \theta = \dfrac{2}{\sqrt{13}} = \dfrac{2\sqrt{13}}{13}$

$\cos \theta = \dfrac{3}{\sqrt{13}} = \dfrac{3\sqrt{13}}{13}$

$\tan \theta = \dfrac{2}{3}$

$\csc \theta = \dfrac{\sqrt{13}}{2}$

$\sec \theta = \dfrac{\sqrt{13}}{3}$

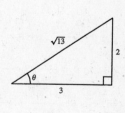

17.

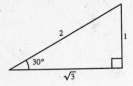

Function	θ(deg)	θ(rad)	Function Value
sin	30°	$\dfrac{\pi}{6}$	$\dfrac{1}{2}$

19.

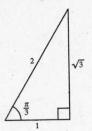

Function	θ(deg)	θ(rad)	Function Value
tan	60°	$\dfrac{\pi}{3}$	$\sqrt{3}$

21.

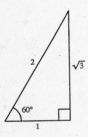

Function	θ(deg)	θ(rad)	Function Value
sec	60°	$\dfrac{\pi}{3}$	2

23.

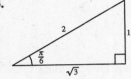

Function	θ(deg)	θ(rad)	Function Value
cos	30°	$\dfrac{\pi}{6}$	$\dfrac{\sqrt{3}}{2}$

25.

Function	θ(deg)	θ(rad)	Function Value
tan	45°	$\dfrac{\pi}{4}$	1

27. $\sin 60° = \dfrac{\sqrt{3}}{2}$, $\cos 60° = \dfrac{1}{2}$

(a) $\tan 60° = \dfrac{\sin 60°}{\cos 60°} = \sqrt{3}$

(b) $\sin 30° = \cos 60° = \dfrac{1}{2}$

(c) $\cos 30° = \sin 60° = \dfrac{\sqrt{3}}{2}$

(d) $\cot 60° = \dfrac{\cos 60°}{\sin 60°} = \dfrac{1}{\sqrt{3}} = \dfrac{\sqrt{3}}{3}$

29. $\csc \theta = \dfrac{\sqrt{13}}{2}$, $\sec \theta = \dfrac{\sqrt{13}}{3}$

(a) $\sin \theta = \dfrac{1}{\csc \theta} = \dfrac{2}{\sqrt{13}} = \dfrac{2\sqrt{13}}{13}$

(b) $\cos \theta = \dfrac{1}{\sec \theta} = \dfrac{3}{\sqrt{13}} = \dfrac{3\sqrt{13}}{13}$

(c) $\tan \theta = \dfrac{\sin \theta}{\cos \theta} = \dfrac{\dfrac{2\sqrt{13}}{13}}{\dfrac{3\sqrt{13}}{13}} = \dfrac{2}{3}$

(d) $\sec(90° - \theta) = \csc \theta = \dfrac{\sqrt{13}}{2}$

31. $\cos \alpha = \dfrac{1}{3}$

(a) $\sec \alpha = \dfrac{1}{\cos \alpha} = 3$

(b) $\sin^2 \alpha + \cos^2 \alpha = 1$

$\sin^2 \alpha + \left(\dfrac{1}{3}\right)^2 = 1$

$\sin^2 \alpha = \dfrac{8}{9}$

$\sin \alpha = \dfrac{2\sqrt{2}}{3}$

(c) $\cot \alpha = \dfrac{\cos \alpha}{\sin \alpha} = \dfrac{\dfrac{1}{3}}{2\sqrt{2}/3} = \dfrac{1}{2\sqrt{2}} = \dfrac{\sqrt{2}}{4}$

(d) $\sin(90° - \alpha) = \cos \alpha = \dfrac{1}{3}$

33. (a) $\sin 10° \approx 0.1736$

(b) $\cos 80° \approx 0.1736$

Note: $\cos 80° = \sin(90° - 80°) = \sin 10°$

35. (a) $\sin 16.35° \approx 0.2815$

(b) $\csc 16.35° = \dfrac{1}{\sin 16.35°} \approx 3.5523$

37. (a) $\sec 42°12' = \sec 42.2° = \dfrac{1}{\cos 42.2°} \approx 1.3499$

(b) $\csc 48°7' = \dfrac{1}{\sin \left(48 + \frac{7}{60}\right)°} \approx 1.3432$

39. (a) $\cot 11°15' = \dfrac{1}{\tan 11.25°} \approx 5.0273$

(b) $\tan 11° 15' = \tan 11.25° \approx 0.1989$

41. (a) $\csc 32° 40' 3'' = \dfrac{1}{\sin 32.6675°} \approx 1.8527$

(b) $\tan 44° 28' 16'' \approx \tan 44.4711° \approx 0.9817$

43. (a) $\sin \theta = \dfrac{1}{2} \implies \theta = 30° = \dfrac{\pi}{6}$

(b) $\csc \theta = 2 \implies \theta = 30° = \dfrac{\pi}{6}$

45. (a) $\sec \theta = 2 \implies \theta = 60° = \dfrac{\pi}{3}$

(b) $\cot \theta = 1 \implies \theta = 45° = \dfrac{\pi}{4}$

47. (a) $\csc \theta = \dfrac{2\sqrt{3}}{3} \implies \theta = 60° = \dfrac{\pi}{3}$

(b) $\sin \theta = \dfrac{\sqrt{2}}{2} \implies \theta = 45° = \dfrac{\pi}{4}$

49. (a) $\sin \theta = 0.0145 \implies \theta \approx 0.83° \approx 0.015$ radian

(b) $\sin \theta = 0.4565 \implies \theta \approx 27° \approx 0.474$ radian

51. (a) $\tan \theta = 0.0125 \implies \theta \approx 0.72° \approx 0.012$ radian

(b) $\tan \theta = 2.3545 \implies \theta \approx 67° \approx 1.169$ radians

53.

$\tan 30° = \dfrac{30}{x}$

$\dfrac{1}{\sqrt{3}} = \dfrac{30}{x}$

$x = 30\sqrt{3}$

55.

$\tan 60° = \dfrac{32}{x}$

$\sqrt{3} = \dfrac{32}{x}$

$\sqrt{3}\, x = 32$

$x = \dfrac{32}{\sqrt{3}} = \dfrac{32\sqrt{3}}{3}$

57. $\tan \theta \cot \theta = \tan \theta \left(\dfrac{1}{\tan \theta}\right) = 1$

59. $\tan \alpha \cos \alpha = \left(\dfrac{\sin \alpha}{\cos \alpha}\right) \cos \alpha = \sin \alpha$

61. $(1 + \cos \theta)(1 - \cos \theta) = 1 - \cos^2 \theta$

$\qquad\qquad = (\sin^2 \theta + \cos^2 \theta) - \cos^2 \theta$

$\qquad\qquad = \sin^2 \theta$

63. $(\sec \theta + \tan \theta)(\sec \theta - \tan \theta) = \sec^2 \theta - \tan^2 \theta$

$\qquad\qquad\qquad = (1 + \tan^2 \theta) - \tan^2 \theta$

$\qquad\qquad\qquad = 1$

65. $\dfrac{\sin \theta}{\cos \theta} + \dfrac{\cos \theta}{\sin \theta} = \dfrac{\sin^2 \theta + \cos^2 \theta}{\sin \theta \cos \theta}$

$\qquad\qquad = \dfrac{1}{\sin \theta \cos \theta}$

$\qquad\qquad = \dfrac{1}{\sin \theta} \cdot \dfrac{1}{\cos \theta}$

$\qquad\qquad = \csc \theta \sec \theta$

67. This is true because the corresponding sides of similar triangles are proportional.

69. (a)

θ	0.1	0.2	0.3	0.4	0.5
$\sin \theta$	0.0998	0.1987	0.2955	0.3894	0.4794

(b) As $\theta \rightarrow 0$, $\sin \theta \rightarrow 0$.

71. (a)

Not drawn to scale

(b) $\dfrac{6}{3} = \dfrac{h}{135}$

(c) $2(135) = h$

$\quad\quad h = 270$ feet

73. (a)

85°

Not drawn to scale

(b) $\sin 85° = \dfrac{h}{20}$

(c) $h = 20 \sin 85° \approx 19.9$ meters

(d) As θ decreases, the height decreases and the base increases. The hypotenuse is fixed at 20 meters.

(e) $h = 20 \sin \theta$

θ	h(meters)
80°	19.7
70°	18.8
60°	17.3
50°	15.3
40°	12.9
30°	10.0
20°	6.8
10°	3.5

(f) As θ approaches zero, h, the height, approaches zero.

75. $\tan 3.5° = \dfrac{h}{x + 13}$ and $\tan 9° = \dfrac{h}{x}$

$h = (x + 13) \tan 3.5°$ and $h = x \tan 9°$

$x \tan 3.5° + 13 \tan 3.5° = x \tan 9°$

$\quad\quad 13 \tan 3.5° = x (\tan 9° - \tan 3.5°)$

$\quad \dfrac{13 \tan 3.5°}{\tan 9° - \tan 3.5°} = x$

$\quad\quad\quad\quad x \approx 8.1783$ miles

$h = x \tan 9° \approx 1.3$ miles

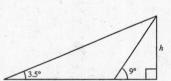

Not drawn to scale

77. (a)

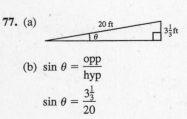

(b) $\sin \theta = \dfrac{\text{opp}}{\text{hyp}}$

$\sin \theta = \dfrac{3\frac{1}{3}}{20}$

(c) $\sin \theta = \dfrac{1}{6} \Rightarrow \theta = 9.59°$

79. $\tan 3° = \dfrac{x}{15}$

$x = 15 \tan 3°$

$d = 5 + 2x = 5 + 2(15 \tan 3°) \approx 6.57$ centimeters

81. $x \approx 2.588, \quad y \approx 9.659$

$\sin 75° = \dfrac{y}{10} \approx 0.97 \qquad \cot 75° = \dfrac{x}{y} \approx 0.27$

$\cos 75° = \dfrac{x}{10} \approx 0.26 \qquad \sec 75° = \dfrac{10}{x} \approx 3.86$

$\tan 75° = \dfrac{y}{x} \approx 3.73 \qquad \csc 75° = \dfrac{10}{y} \approx 1.04$

83. True, $\csc x = \dfrac{1}{\sin x} \implies \sin 60° \csc 60° = \sin 60°\left(\dfrac{1}{\sin 60°}\right) = 1.$

85. False, $\dfrac{\sqrt{2}}{2} + \dfrac{\sqrt{2}}{2} = \sqrt{2} \neq 1.$

87. False,

$\dfrac{\sin 60°}{\sin 30°} = \dfrac{\cos 30°}{\sin 30°} = \cot 30° \approx 1.7321; \sin 2° \approx 0.0349.$

89. False,

$2 \sin 30° = 1; \sin 60° = \dfrac{\sqrt{3}}{2}.$

91. False,

$\sin 30° = \dfrac{1}{2}; 1 - \cos 30° = 1 - \dfrac{\sqrt{3}}{2} = \dfrac{2 - \sqrt{3}}{2}.$

93. False,

$\dfrac{1}{2} \tan \dfrac{\pi}{3} = \dfrac{\sqrt{3}}{2}; \tan \dfrac{\pi}{6} = \dfrac{\sqrt{3}}{3} = \dfrac{1}{\sqrt{3}}.$

95. False,

$\sin \dfrac{6}{\pi} \approx 0.94; \csc \dfrac{\pi}{6} = 2.$

Section 9.4   Trigonometric Functions of Any Angle

1. (a) $(x, y) = (4, 3)$

$r = \sqrt{16 + 9} = 5$

$\sin \theta = \dfrac{y}{r} = \dfrac{3}{5} \qquad \csc \theta = \dfrac{r}{y} = \dfrac{5}{3}$

$\cos \theta = \dfrac{x}{r} = \dfrac{4}{5} \qquad \sec \theta = \dfrac{r}{x} = \dfrac{5}{4}$

$\tan \theta = \dfrac{y}{x} = \dfrac{3}{4} \qquad \cot \theta = \dfrac{x}{y} = \dfrac{4}{3}$

(b) $(x, y) = (8, -15)$

$r = \sqrt{64 + 225} = 17$

$\sin \theta = \dfrac{y}{r} = -\dfrac{15}{17} \qquad \csc \theta = \dfrac{r}{y} = -\dfrac{17}{15}$

$\cos \theta = \dfrac{x}{r} = \dfrac{8}{17} \qquad \sec \theta = \dfrac{r}{x} = \dfrac{17}{8}$

$\tan \theta = \dfrac{y}{x} = -\dfrac{15}{8} \qquad \cot \theta = \dfrac{x}{y} = -\dfrac{8}{15}$

3. (a) $(x, y) = \left(-\sqrt{3}, -1\right)$

$r = \sqrt{3 + 1} = 2$

$\sin \theta = \dfrac{y}{r} = -\dfrac{1}{2}$

$\cos \theta = \dfrac{x}{r} = -\dfrac{\sqrt{3}}{2}$

$\tan \theta = \dfrac{y}{x} = \dfrac{\sqrt{3}}{3}$

$\csc \theta = \dfrac{r}{y} = -2$

$\sec \theta = \dfrac{r}{x} = -\dfrac{2\sqrt{3}}{3}$

$\cot \theta = \dfrac{x}{y} = \sqrt{3}$

(b) $(x, y) = (-4, 1)$

$r = \sqrt{16 + 1} = \sqrt{17}$

$\sin \theta = \dfrac{y}{r} = \dfrac{\sqrt{17}}{17}$

$\cos \theta = \dfrac{x}{r} = -\dfrac{4\sqrt{17}}{17}$

$\tan \theta = \dfrac{y}{x} = -\dfrac{1}{4}$

$\csc \theta = \dfrac{r}{y} = \sqrt{17}$

$\sec \theta = \dfrac{r}{x} = -\dfrac{\sqrt{17}}{4}$

$\cot \theta = \dfrac{x}{y} = -4$

5. $(x, y) = (7, 24)$

$r = \sqrt{49 + 576} = 25$

$\sin \theta = \dfrac{y}{r} = \dfrac{24}{25}$ $\csc \theta = \dfrac{r}{y} = \dfrac{25}{24}$

$\cos \theta = \dfrac{x}{r} = \dfrac{7}{25}$ $\sec \theta = \dfrac{r}{x} = \dfrac{25}{7}$

$\tan \theta = \dfrac{y}{x} = \dfrac{24}{7}$ $\cot \theta = \dfrac{x}{y} = \dfrac{7}{24}$

7. $(x, y) = (-4, 10)$

$r = \sqrt{16 + 100} = 2\sqrt{29}$

$\sin \theta = \dfrac{y}{r} = \dfrac{5\sqrt{29}}{29}$ $\csc \theta = \dfrac{r}{y} = \dfrac{\sqrt{29}}{5}$

$\cos \theta = \dfrac{x}{r} = -\dfrac{2\sqrt{29}}{29}$ $\sec \theta = \dfrac{r}{x} = -\dfrac{\sqrt{29}}{2}$

$\tan \theta = \dfrac{y}{x} = -\dfrac{5}{2}$ $\cot \theta = \dfrac{x}{y} = -\dfrac{2}{5}$

9. $(x, y) = \left(-\dfrac{5}{12}, 1\right)$

$r = \sqrt{\dfrac{25}{144} + 1} = \sqrt{\dfrac{169}{144}} = \dfrac{13}{12}$

$\sin \theta = \dfrac{y}{r} = \dfrac{1}{13/12} = \dfrac{12}{13}$

$\cos \theta = \dfrac{x}{r} = \dfrac{-5/12}{13/12} = -\dfrac{5}{13}$

$\tan \theta = \dfrac{y}{x} = \dfrac{1}{-5/12} = -\dfrac{12}{5}$

$\csc \theta = \dfrac{r}{y} = \dfrac{13/12}{1} = \dfrac{13}{12}$

$\sec \theta = \dfrac{r}{x} = \dfrac{13/12}{-5/12} = -\dfrac{13}{5}$

$\cot \theta = \dfrac{x}{y} = \dfrac{-5/12}{1} = -\dfrac{5}{12}$

11. $\sin \theta < 0 \implies \theta$ lies in Quadrant III or in Quadrant IV.

$\cos \theta < 0 \implies \theta$ lies in Quadrant II or in Quadrant III.

$\sin \theta < 0$ *and* $\cos \theta < 0 \implies \theta$ lies in Quadrant III.

13. $\sin \theta > 0 \implies \theta$ lies in Quadrant I or in Quadrant II.

$\tan \theta < 0 \implies \theta$ lies in Quadrant II or in Quadrant IV.

$\sin \theta > 0$ *and* $\tan \theta < 0 \implies \theta$ lies in Quadrant II.

15. $\sin \theta = \dfrac{y}{r} = \dfrac{3}{5} \implies x^2 = 25 - 9 = 16$

θ in Quadrant II $\implies x = -4$

$\sin \theta = \dfrac{y}{r} = \dfrac{3}{5}$ $\csc \theta = \dfrac{r}{y} = \dfrac{5}{3}$

$\cos \theta = \dfrac{x}{r} = -\dfrac{4}{5}$ $\sec \theta = \dfrac{r}{x} = -\dfrac{5}{4}$

$\tan \theta = \dfrac{y}{x} = -\dfrac{3}{4}$ $\cot \theta = \dfrac{x}{y} = -\dfrac{4}{3}$

17. $\tan \theta = \dfrac{y}{x} = \dfrac{-15}{8}$

$\sin \theta < 0$ and $\tan \theta < 0 \implies \theta$ is in Quadrant IV $\implies$ $y < 0$ and $x > 0$

$x = 8, y = -15, r = 17$

$\sin \theta = \dfrac{y}{r} = -\dfrac{15}{17}$ $\csc \theta = \dfrac{r}{y} = -\dfrac{17}{15}$

$\cos \theta = \dfrac{x}{r} = \dfrac{8}{17}$ $\sec \theta = \dfrac{r}{x} = \dfrac{17}{8}$

$\tan \theta = \dfrac{y}{x} = -\dfrac{15}{8}$ $\cot \theta = \dfrac{x}{y} = -\dfrac{8}{15}$

19. $\cot \theta = \dfrac{x}{y} = -\dfrac{3}{1} = \dfrac{3}{-1}$

$\cos \theta > 0 \implies \theta$ is in Quadrant IV $\implies x$ is positive; $x = 3, y = -1, r = \sqrt{10}$

$\sin \theta = \dfrac{y}{r} = -\dfrac{\sqrt{10}}{10}$ $\csc \theta = \dfrac{r}{y} = -\sqrt{10}$

$\cos \theta = \dfrac{x}{r} = \dfrac{3\sqrt{10}}{10}$ $\sec \theta = \dfrac{r}{x} = \dfrac{\sqrt{10}}{3}$

$\tan \theta = \dfrac{y}{x} = -\dfrac{1}{3}$ $\cot \theta = \dfrac{x}{y} = -3$

21. $\sec \theta = \dfrac{r}{x} = \dfrac{2}{-1} \implies y^2 = 4 - 1 = 3$

$\sin \theta > 0 \implies \theta$ is in Quadrant II $\implies y = \sqrt{3}$

$$\sin \theta = \frac{y}{r} = \frac{\sqrt{3}}{2} \qquad\qquad \csc \theta = \frac{r}{y} = \frac{2\sqrt{3}}{3}$$

$$\cos \theta = \frac{x}{r} = -\frac{1}{2} \qquad\qquad \sec \theta = \frac{r}{x} = -2$$

$$\tan \theta = \frac{y}{x} = -\sqrt{3} \qquad\qquad \cot \theta = \frac{x}{y} = -\frac{\sqrt{3}}{3}$$

23. $\cot \theta$ is undefined, $\dfrac{\pi}{2} \le \theta \le \dfrac{3\pi}{2} \implies y = 0 \implies \theta = \pi$

$\sin \pi = 0$ $\qquad\qquad$ $\csc \pi$ is undefined.

$\cos \pi = -1$ $\qquad\qquad$ $\sec \pi = -1$

$\tan \pi = 0$ $\qquad\qquad$ $\cot \pi$ is undefined.

25. To find a point on the terminal side of θ, use any point on the line $y = -x$ that lies in Quadrant II. $(-1, 1)$ is one such point.

$x = -1, y = 1, r = \sqrt{2}$

$$\sin \theta = \frac{1}{\sqrt{2}} = \frac{\sqrt{2}}{2}$$

$$\cos \theta = -\frac{1}{\sqrt{2}} = -\frac{\sqrt{2}}{2}$$

$\tan \theta = -1$

$\csc \theta = \sqrt{2}$

$\sec \theta = -\sqrt{2}$

$\cot \theta = -1$

27. To find a point on the terminal side of θ, use any point on the line $y = 2x$ that lies in Quadrant III. $(-1, -2)$ is one such point.

$x = -1, y = -2, r = \sqrt{5}$

$$\sin \theta = -\frac{2}{\sqrt{5}} = -\frac{2\sqrt{5}}{5} \qquad \csc \theta = \frac{\sqrt{5}}{-2} = -\frac{\sqrt{5}}{2}$$

$$\cos \theta = -\frac{1}{\sqrt{5}} = -\frac{\sqrt{5}}{5} \qquad \sec \theta = \frac{\sqrt{5}}{-1} = -\sqrt{5}$$

$$\tan \theta = \frac{-2}{-1} = 2 \qquad\qquad \cot \theta = \frac{-1}{-2} = \frac{1}{2}$$

29. $(x, y) = (-1, 0), r = 1$

$$\cos \pi = \frac{x}{r} = \frac{-1}{1} = -1$$

31. $(x, y) = (-1, 0), r = 1$

$$\sec \pi = \frac{r}{x} = \frac{1}{-1} = -1$$

33. $(x, y) = (0, 1), r = 1$

$$\tan \frac{\pi}{2} = \frac{y}{x} = \frac{1}{0} \text{ undefined.}$$

35. $(x, y) = (0, 1)$

$$\cot \frac{\pi}{2} = \frac{x}{y} = \frac{0}{1} = 0$$

37. $\theta = \dfrac{2\pi}{3}$

$$\theta' = \pi - \frac{2\pi}{3} = \frac{\pi}{3}$$

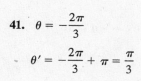

39. $\theta = \dfrac{5\pi}{6}$

$$\theta' = \pi - \frac{5\pi}{6} = \frac{\pi}{6}$$

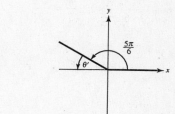

41. $\theta = -\dfrac{2\pi}{3}$

$$\theta' = -\frac{2\pi}{3} + \pi = \frac{\pi}{3}$$

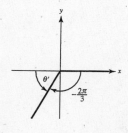

43. $\theta = 120°$

$$\theta' = 180° - 120° = 60°$$

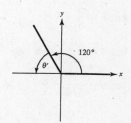

45. $\theta = 225°$

$\theta' = 225° - 180° = 45°$

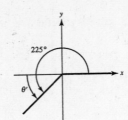

47. $\theta = -135°$

$\theta' = -135° + 180° = 45°$

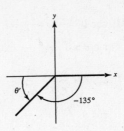

49. $\theta' = 45°$, Quadrant III

$\sin 225° = -\sin 45° = -\dfrac{\sqrt{2}}{2}$

$\cos 225° = -\cos 45° = -\dfrac{\sqrt{2}}{2}$

$\tan 225° = \tan 45° = 1$

51. $\theta' = 30°$, Quadrant I

$\sin 750° = \sin 30° = \dfrac{1}{2}$

$\cos 750° = \cos 30° = \dfrac{\sqrt{3}}{2}$

$\tan 750° = \tan 30° = \dfrac{\sqrt{3}}{3}$

53. $\theta' = 30°$, Quadrant III

$\sin(-150°) = -\sin 30° = -\dfrac{1}{2}$

$\cos(-150°) = -\cos 30° = -\dfrac{\sqrt{3}}{2}$

$\tan(-150°) = \tan 30° = \dfrac{\sqrt{3}}{3}$

55. $\theta' = \dfrac{\pi}{3}$, Quadrant III

$\sin \dfrac{4\pi}{3} = -\sin \dfrac{\pi}{3} = -\dfrac{\sqrt{3}}{2}$

$\cos \dfrac{4\pi}{3} = -\cos \dfrac{\pi}{3} = -\dfrac{1}{2}$

$\tan \dfrac{4\pi}{3} = \tan \dfrac{\pi}{3} = \sqrt{3}$

57. $\theta' = \dfrac{\pi}{6}$, Quadrant II

$\sin\left(-\dfrac{7\pi}{6}\right) = \sin \dfrac{\pi}{6} = \dfrac{1}{2}$

$\cos\left(-\dfrac{7\pi}{6}\right) = -\cos \dfrac{\pi}{6} = -\dfrac{\sqrt{3}}{2}$

$\tan\left(-\dfrac{7\pi}{6}\right) = -\tan \dfrac{\pi}{6} = -\dfrac{\sqrt{3}}{3}$

59. $\theta' = \dfrac{\pi}{4}$, Quadrant II

$\sin \dfrac{11\pi}{4} = \sin \dfrac{\pi}{4} = \dfrac{\sqrt{2}}{2}$

$\cos \dfrac{11\pi}{4} = -\cos \dfrac{\pi}{4} = -\dfrac{\sqrt{2}}{2}$

$\tan \dfrac{11\pi}{4} = -\tan \dfrac{\pi}{4} = -1$

61. $\theta' = \dfrac{\pi}{6}$, Quadrant III

$\sin\left(-\dfrac{17\pi}{6}\right) = -\sin \dfrac{\pi}{6} = -\dfrac{1}{2}$

$\cos\left(-\dfrac{17\pi}{6}\right) = -\cos \dfrac{\pi}{6} = -\dfrac{\sqrt{3}}{2}$

$\tan\left(-\dfrac{17\pi}{6}\right) = \tan \dfrac{\pi}{6} = \dfrac{\sqrt{3}}{3}$

63. $\sin \theta = -\dfrac{3}{5}$

$\sin^2 \theta + \cos^2 \theta = 1$

$\cos^2 \theta = 1 - \sin^2 \theta$

$\cos^2 \theta = 1 - \left(-\dfrac{3}{5}\right)^2$

$\cos^2 \theta = 1 - \dfrac{9}{25}$

$\cos^2 \theta = \dfrac{16}{25}$

$\cos \theta > 0$ in Quadrant IV.

$\cos \theta = \dfrac{4}{5}$

65. $\tan \theta = \dfrac{3}{2}$

$\sec^2 \theta = 1 + \tan^2 \theta$

$\sec^2 \theta = 1 + \left(\dfrac{3}{2}\right)^2$

$\sec^2 \theta = 1 + \dfrac{9}{4}$

$\sec^2 \theta = \dfrac{13}{4}$

$\sec \theta < 0$ in Quadrant III.

$\sec \theta = -\dfrac{\sqrt{13}}{2}$

67. $\cos \theta = \dfrac{5}{8}$

$\cos \theta = \dfrac{1}{\sec \theta} \implies \sec \theta = \dfrac{1}{\cos \theta}$

$\sec \theta = \dfrac{1}{5/8} = \dfrac{8}{5}$

69. $\sin 10° \approx 0.1736$

71. $\cos(-110°) \approx -0.3420$

73. $\tan 4.5 \approx 4.6373$

75. $\tan\left(\dfrac{5\pi}{9}\right) \approx -5.6713$

77. $\sin(-0.65) \approx -0.6052$

79. (a) $\sin\theta = \dfrac{1}{2} \implies$ reference angle is $30°$ or $\dfrac{\pi}{6}$ and θ is in Quadrant I or Quadrant II.

Values in degrees: $30°, 150°$

Values in radians: $\dfrac{\pi}{6}, \dfrac{5\pi}{6}$

(b) $\sin\theta = -\dfrac{1}{2} \implies$ reference angle is $30°$ or $\dfrac{\pi}{6}$ and θ is in Quadrant III or Quadrant IV.

Values in degrees: $210°, 330°$

Values in radians: $\dfrac{7\pi}{6}, \dfrac{11\pi}{6}$

81. (a) $\csc\theta = \dfrac{2\sqrt{3}}{3} \implies$ reference angle is $60°$ or $\dfrac{\pi}{3}$ and θ is in Quadrant I or Quadrant II.

Values in degrees: $60°, 120°$

Values in radians: $\dfrac{\pi}{3}, \dfrac{2\pi}{3}$

(b) $\cot\theta = -1 \implies$ reference angle is $45°$ or $\dfrac{\pi}{4}$ and θ is in Quadrant II or Quadrant IV.

Values in degrees: $135°, 315°$

Values in radians: $\dfrac{3\pi}{4}, \dfrac{7\pi}{4}$

83. (a) $\tan\theta = 1 \implies$ reference angle is $45°$ or $\dfrac{\pi}{4}$ and θ is in Quadrant I or Quadrant III.

Values in degrees: $45°, 225°$

Values in radians: $\dfrac{\pi}{4}, \dfrac{5\pi}{4}$

(b) $\cot\theta = -\sqrt{3} \implies$ reference angle is $30°$ or $\dfrac{\pi}{6}$ and θ is in Quadrant II or Quadrant IV.

Values in degrees: $150°, 330°$

Values in radians: $\dfrac{5\pi}{6}, \dfrac{11\pi}{6}$

85. $\sin\theta = 0.8191$

Quadrant I: $\theta = \sin^{-1} 0.8191 \approx 54.99°$

Quadrant II: $\theta = 180° - \sin^{-1} 0.8191 \approx 125.01°$

87. $\cos\theta = -0.4367 \implies \theta' \approx 64.11°$

Quadrant II: $\theta \approx 180° - 64.11° = 115.89°$

Quadrant III: $\theta \approx 180° + 64.11° = 244.11°$

89. $\tan\theta = 2.521$

Quadrant I: $\theta = \tan^{-1} 2.521 \approx 68.36°$

Quadrant III: $\theta = 180° + \tan^{-1} 2.521 \approx 248.36°$

91. $\sec\theta = -4.156 \implies \theta' = \cos^{-1}\left(\dfrac{1}{4.156}\right) \approx 76.08°$

Quadrant II: $\theta = 180° - \theta' \approx 103.92°$

Quadrant III: $\theta = 180° + \theta' \approx 256.08°$

93. $\cot\theta = 15.254 \implies \theta' = \tan^{-1}\left(\dfrac{1}{15.254}\right) \approx 3.75°$

Quadrant I: $\theta = \theta' \approx 3.75°$

Quadrant III: $\theta = 180° + \theta' \approx 183.75°$

95. $\cos\theta = 0.9848 \implies \theta' \approx 0.175$

Quadrant I: $\theta = \cos^{-1}(0.9848) \approx 0.175$

Quadrant IV: $\theta = 2\pi - \theta' \approx 6.109$

97. $\tan\theta = 1.192 \implies \theta' \approx 0.873$

Quadrant I: $\theta = \tan^{-1} 1.192 \approx 0.873$

Quadrant III: $\theta = \pi + \theta' \approx 4.014$

99. $\sec\theta = -2.6667 \implies \theta' = \cos^{-1}\left(\dfrac{1}{2.6667}\right) \approx 1.1864$

Quadrant II: $\theta = \pi - 1.1864 \approx 1.955$

Quadrant III: $\theta = \pi + 1.1864 \approx 4.328$

101. As θ increases from $0°$ to $90°$, x decreases from 12 cm to 0 cm and y increases from 0 cm to 12 cm.

Therefore, $\sin\theta = \dfrac{y}{12}$ increases from 0 to 1 and $\cos\theta = \dfrac{x}{12}$ decreases from 1 to 0. Thus,

$\tan\theta = \dfrac{y}{x}$ increases without bound, and when $\theta = 90°$ the tangent is undefined.

103. First, determine a positive coterminal angle. Then determine the trigonometric function of the reference angle and prefix the appropriate sign.

105. $S = 23.1 + 0.442t + 4.3\sin\dfrac{\pi t}{6}$

(a) February 2005 $\Rightarrow$ $t = 2$

$S = 23.1 + 0.442(2) + 4.3\sin\dfrac{2\pi}{6}$

≈ 27.7 thousand or 27,700 units

(b) February 2006 $\Rightarrow$ $t = 14$

$S = 23.1 + 0.442(14) + 4.3\sin\dfrac{14\pi}{6}$

≈ 33.0 thousand or 33,000 units

(c) September 2005 $\Rightarrow$ $t = 9$

$S = 23.1 + 0.442(9) + 4.3\sin\dfrac{9\pi}{6}$

≈ 22.8 thousand or 22,800 units

(d) September 2006 $\Rightarrow$ $t = 21$

$S = 23.1 + 0.442(21) + 4.3\sin\dfrac{21\pi}{6}$

≈ 28.1 thousand or 28,100 units

107. $y(t) = 2e^{-t}\cos 6t$

(a) $t = 0$

$y(0) = 2e^{-0}\cos 0 = 2$ centimeters

(b) $t = \dfrac{1}{4}$

$y\left(\dfrac{1}{4}\right) = 2e^{-1/4}\cos\left(6\cdot\dfrac{1}{4}\right) \approx 0.11$ centimeter

(c) $t = \dfrac{1}{2}$

$y\left(\dfrac{1}{2}\right) = 2e^{-1/2}\cos\left(6\cdot\dfrac{1}{2}\right) \approx -1.2$ centimeters

109. $\sin\theta = \dfrac{6}{d} \Rightarrow d = \dfrac{6}{\sin\theta}$

(a) $\theta = 30°$

$d = \dfrac{6}{\sin 30°} = \dfrac{6}{1/2} = 12$ miles

(b) $\theta = 90°$

$d = \dfrac{6}{\sin 90°} = \dfrac{6}{1} = 6$ miles

(c) $\theta = 120°$

$d = \dfrac{6}{\sin 120°} \approx 6.9$ miles

111. False. In each of the four quadrants, the sign of the secant function and the cosine function will be the same since they are reciprocals of each other.

113. True. If $\sin\theta = \dfrac{1}{4}$ and $\cos\theta < 0$, then θ lies in Quadrant II. Let $y = 1$ and $r = 4$, then $x = -\sqrt{4^2 - 1^2} = -\sqrt{15}$ and $\tan\theta = \dfrac{1}{-\sqrt{15}} = -\dfrac{\sqrt{15}}{15}$.

Section 9.5   Graphs of Sine and Cosine Functions

1. $y = 3\sin 2x$

Period: $\dfrac{2\pi}{2} = \pi$

Amplitude: $|3| = 3$

3. $y = \dfrac{5}{2}\cos\dfrac{x}{2}$

Period: $\dfrac{2\pi}{\frac{1}{2}} = 4\pi$

Amplitude: $\left|\dfrac{5}{2}\right| = \dfrac{5}{2}$

5. $y = \dfrac{1}{2}\sin\dfrac{\pi x}{3}$

Period: $\dfrac{2\pi}{\frac{\pi}{3}} = 6$

Amplitude: $\left|\dfrac{1}{2}\right| = \dfrac{1}{2}$

7. $y = -2 \sin x$

Period: $\dfrac{2\pi}{1} = 2\pi$

Amplitude: $|-2| = 2$

9. $y = 3 \sin 10x$

Period: $\dfrac{2\pi}{10} = \dfrac{\pi}{5}$

Amplitude: $|3| = 3$

11. $y = \dfrac{1}{2} \cos \dfrac{2x}{3}$

Period: $\dfrac{2\pi}{\frac{2}{3}} = 3\pi$

Amplitude: $\left|\dfrac{1}{2}\right| = \dfrac{1}{2}$

13. $y = \dfrac{1}{4} \sin 2\pi x$

Period: $\dfrac{2\pi}{2\pi} = 1$

Amplitude: $\left|\dfrac{1}{4}\right| = \dfrac{1}{4}$

15. $f(x) = \sin x$

$g(x) = \sin(x - \pi)$

The graph of g is a horizontal shift to the right π units of the graph of f (a phase shift).

17. $f(x) = \cos 2x$

$g(x) = -\cos 2x$

The graph of g is a reflection in the x-axis of the graph of f.

19. $f(x) = \cos x$

$g(x) = \cos 2x$

The period of f is twice that of g.

21. $f(x) = \sin 2x$

$g(x) = 3 + \sin 2x$

The graph of g is a vertical shift 3 units upward of the graph of f.

23. The graph of g has twice the amplitude as the graph of f. The period is the same.

25. The graph of g is a horizontal shift π units to the right of the graph of f.

27. $f(x) = -2 \sin x$

Period: 2π

Amplitude: 2

$g(x) = 4 \sin x$

Period: 2π

Amplitude: 4

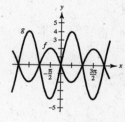

29. $f(x) = \cos x$

Period: 2π

Amplitude: 1

$g(x) = 1 + \cos x$
is a vertical shift of the graph of $f(x)$ one unit upward.

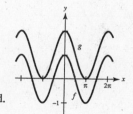

31. $f(x) = -\dfrac{1}{2} \sin \dfrac{x}{2}$

Period: 4π

Amplitude: $\dfrac{1}{2}$

$g(x) = 3 - \dfrac{1}{2} \sin \dfrac{x}{2}$ is the graph of $f(x)$ shifted vertically three units upward.

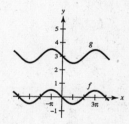

33. $f(x) = 2 \cos x$

Period: 2π

Amplitude: 2

$g(x) = 2 \cos(x + \pi)$ is the graph of $f(x)$ shifted π units to the left.

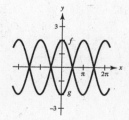

35. $y = -2 \sin 6x$; $a = -2$, $b = 6$, $c = 0$

Period: $\dfrac{2\pi}{6} = \dfrac{\pi}{3}$

Amplitude: $|-2| = 2$

Key points: $(0, 0), \left(\dfrac{\pi}{12}, -2\right), \left(\dfrac{\pi}{6}, 0\right), \left(\dfrac{\pi}{4}, 2\right), \left(\dfrac{\pi}{3}, 0\right)$

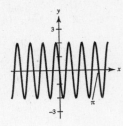

37. $y = \cos 2\pi x$

Period: $\dfrac{2\pi}{2\pi} = 1$

Amplitude: 1

Key points: $(0, 1), \left(\dfrac{1}{4}, 0\right), \left(\dfrac{1}{2}, -1\right), \left(\dfrac{3}{4}, 0\right)$

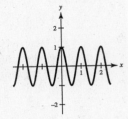

39. $y = -\sin \dfrac{2\pi x}{3}$; $a = -1, b = \dfrac{2\pi}{3}$, $c = 0$

Period: $\dfrac{2\pi}{\frac{2\pi}{3}} = 3$

Amplitude: 1

Key points: $(0, 0), \left(\dfrac{3}{4}, -1\right), \left(\dfrac{3}{2}, 0\right), \left(\dfrac{9}{4}, 1\right), (3, 0)$

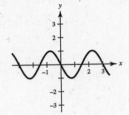

41. $y = \sin\left(x - \dfrac{\pi}{4}\right)$; $a = 1$, $b = 1$, $c = \dfrac{\pi}{4}$

Period: 2π

Amplitude: 1

Shift: Set $x - \dfrac{\pi}{4} = 0$ and $x - \dfrac{\pi}{4} = 2\pi$

$\qquad\quad x = \dfrac{\pi}{4} \qquad\qquad x = \dfrac{9\pi}{4}$

Key points: $\left(\dfrac{\pi}{4}, 0\right), \left(\dfrac{3\pi}{4}, 1\right), \left(\dfrac{5\pi}{4}, 0\right), \left(\dfrac{7\pi}{4}, -1\right), \left(\dfrac{9\pi}{4}, 0\right)$

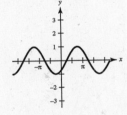

43. $y = 3 \cos (x + \pi)$

Period: 2π

Amplitude: 3

Shift: Set $x + \pi = 0$ and $x + \pi = 2\pi$

$\qquad\qquad x = -\pi \qquad\qquad x = \pi$

Key points: $(-\pi, 3), \left(-\dfrac{\pi}{2}, 0\right), (0, -3), \left(\dfrac{\pi}{2}, 0\right), (\pi, 3)$

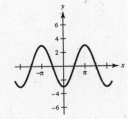

45. $y = 2 - \sin\dfrac{2\pi x}{3}$

Vertical shift 2 units upward of the graph in Exercise 39.

Key points: $(0, 2)$, $\left(\dfrac{3}{4}, 1\right)$, $\left(\dfrac{3}{2}, 2\right)$, $\left(\dfrac{9}{4}, 3\right)$, $(3, 2)$

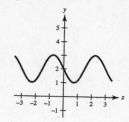

47. $y = 2 + \dfrac{1}{2}\cos(\pi x)$

Period: $\dfrac{2\pi}{\pi} = 2$

Amplitude: $\dfrac{1}{2}$

Vertical shift 2 units upward.

Key points: $(0, 2.5)$, $(0.5, 2)$, $(1, 1.5)$, $(1.5, 2)$, $(2, 2.5)$

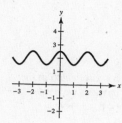

49. $y = 3\cos(x + \pi) - 3$

Vertical shift 3 units downward of the graph in Exercise 43.

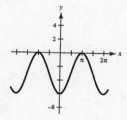

51. $y = \dfrac{2}{3}\cos\left(\dfrac{x}{2} - \dfrac{\pi}{4}\right)$; $a = \dfrac{2}{3}$, $b = \dfrac{1}{2}$, $c = \dfrac{\pi}{4}$

Period: 4π

Amplitude: $\dfrac{2}{3}$

Shift: $\dfrac{x}{2} - \dfrac{\pi}{4} = 0$ and $\dfrac{x}{2} - \dfrac{\pi}{4} = 2\pi$

$\qquad\quad x = \dfrac{\pi}{2} \qquad\qquad x = \dfrac{9\pi}{2}$

Key points: $\left(\dfrac{\pi}{2}, \dfrac{2}{3}\right)$, $\left(\dfrac{3\pi}{2}, 0\right)$, $\left(\dfrac{5\pi}{2}, \dfrac{-2}{3}\right)$, $\left(\dfrac{7\pi}{2}, 0\right)$, $\left(\dfrac{9\pi}{2}, \dfrac{2}{3}\right)$

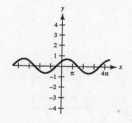

53. $y = -2\sin(4x + \pi)$

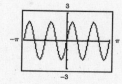

55. $y = \cos\left(2\pi x - \dfrac{\pi}{2}\right) + 1$

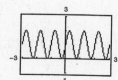

57. $y = 5\sin(\pi - 2x) + 10$

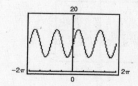

59. $y = -0.1 \sin\left(\dfrac{\pi x}{10} + \pi\right)$

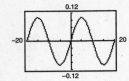

61. $f(x) = a \cos x + d$

Amplitude: $\dfrac{1}{2}[3 - (-1)] = 2 \implies a = 2$

Vertical shift 1 unit upward of $g(x) = 2\cos x \implies d = 1$.

Thus, $f(x) = 2 \cos x + 1$.

63. $f(x) = a \cos x + d$

Amplitude: $\dfrac{1}{2}[8 - 0] = 4$

Since $f(x)$ is the graph of $g(x) = 4 \cos x$ reflected in the x-axis and shifted vertically 4 units upward, we have $a = -4$ and $d = 4$. Thus, $f(x) = -4 \cos x + 4$.

65. $y = a \sin(bx - c)$

Amplitude: $|a| = |3|$ Since the graph is reflected in the x-axis, we have $a = -3$.

Period: $\dfrac{2\pi}{b} = \pi \implies b = 2$

Phase shift: $c = 0$

Thus, $y = -3 \sin 2x$.

67. $y = a \sin(bx - c)$

Amplitude: $a = 2$

Period: $2\pi \implies b = 1$

Phase shift: $bx - c = 0$ when $x = -\dfrac{\pi}{4}$

$$(1)\left(\dfrac{-\pi}{4}\right) - c = 0 \implies c = -\dfrac{\pi}{4}$$

Thus, $y = 2 \sin\left(x + \dfrac{\pi}{4}\right)$.

69. $y_1 = \sin x$

$y_2 = -\dfrac{1}{2}$

In the interval

$[-2\pi, 2\pi]$, $\sin x = -\dfrac{1}{2}$ when

$x = -\dfrac{5\pi}{6}, -\dfrac{\pi}{6}, \dfrac{7\pi}{6}, \dfrac{11\pi}{6}.$

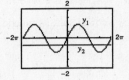

71. $y_1 = \cos x$

$y_2 = \dfrac{\sqrt{2}}{2}$

In the interval

$[-2\pi, 2\pi]$, $\cos x = \dfrac{\sqrt{2}}{2}$

when $x = \pm\dfrac{\pi}{4}, \pm\dfrac{7\pi}{4}.$

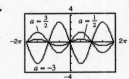

73.

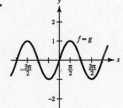

As a changes from $\frac{1}{2}$ to $\frac{3}{2}$, the amplitude increases. When $a = -3$, the amplitude again increases but the graph is reflected in the x-axis also.

75.

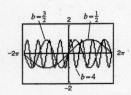

As b increases the period decreases.

77.

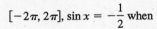
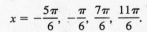

Since the graphs are the same, the conjecture is that

$$\sin(x) = \cos\left(x - \dfrac{\pi}{2}\right).$$

79.

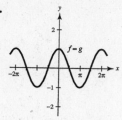

Since the graphs are the same, the conjecture is that

$$\cos x = -\sin\left(x - \frac{\pi}{2}\right).$$

81. (a)

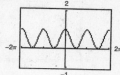

h is even

(b)

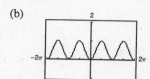

h is even

83. $y = 0.85 \sin \dfrac{\pi t}{3}$

(a) Time for one cycle = one period = $\dfrac{2\pi}{\pi/3} = 6$ sec

(b) Cycles per min = $\dfrac{60}{6} = 10$ cycles per min

(c) Amplitude: 0.85

Period: 6

Key points: $(0, 0)$, $\left(\dfrac{3}{2}, 0.85\right)$, $(3, 0)$, $\left(\dfrac{9}{2}, -0.85\right)$, $(6, 0)$

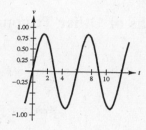

85. $y = 0.001 \sin 880\pi t$

(a) Period: $\dfrac{2\pi}{880\pi} = \dfrac{1}{440}$ seconds

(b) $f = \dfrac{1}{p} = 440$ cycles per second

87. (a) $A(t) = 19.7 \sin(0.47t - 1.74) + 70.24$

(b)

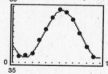

The model is a good fit for most months.

(d) Use the constant term of each model to estimate the average annual temperature.

Nantucket: 58°

Athens: 70.24°

(c)

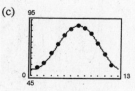

The model is a good fit.

(e) $N(t)$ has a period of 11 months, $A(t)$ has a period of about 13 months. The periods should both be 12 months to correspond with one year.

(f) Athens has slightly more variability in temperature since its amplitude is greater than the amplitude in the Nantucket model.

89. $S = 74.50 + 43.75 \sin \dfrac{\pi t}{6}$

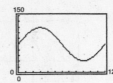

91. (a) and (c)

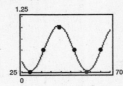

(b) $y = 0.51 \sin(0.216x - 1.62) + 0.54$

(d) For June 27, 2007, use $x = 543$.

$y(543) \approx 0.816$

93. False. $y = \frac{1}{2} \cos 2x$ has an amplitude that is **half** that of $y = \cos x$. For $y = a \cos bx$, the amplitude is $|a|$.

Section 9.6   Graphs of Other Trigonometric Functions

1. $y = \sec 3x$

Period: $\dfrac{2\pi}{3}$

Matches graph (e).

3. $y = \dfrac{1}{2} \cot \pi x$

Period: $\dfrac{\pi}{\pi} = 1$

Matches graph (a).

5. $y = -\csc x$

Period: 2π

Matches graph (d).

7. $y = \dfrac{1}{3} \tan x$

Period: π

Two consecutive asymptotes:

$x = -\dfrac{\pi}{2}$ and $x = \dfrac{\pi}{2}$

x	$-\dfrac{\pi}{4}$	0	$\dfrac{\pi}{4}$
y	$-\dfrac{1}{3}$	0	$\dfrac{1}{3}$

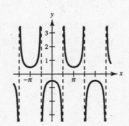

9. $y = \tan 3x$

Period: $\dfrac{\pi}{3}$

Two consecutive asymptotes:

$3x = -\dfrac{\pi}{2} \implies x = -\dfrac{\pi}{6}$

$3x = \dfrac{\pi}{2} \implies x = \dfrac{\pi}{6}$

x	$-\dfrac{\pi}{12}$	0	$\dfrac{\pi}{12}$
y	-1	0	1

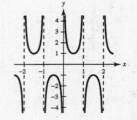

11. $y = -\dfrac{1}{2} \sec x$

Graph $y = -\dfrac{1}{2} \cos x$ first.

Period: 2π

One cycle: 0 to 2π

13. $y = \csc \pi x$

Graph $y = \sin \pi x$ first.

Period: $\dfrac{2\pi}{\pi} = 2$

One cycle: 0 to 2

15. $y = \sec \pi x - 1$

Graph $y = \cos \pi x$ first

Period: $\dfrac{2\pi}{\pi} = 2$

One cycle: 0 to 2

Vertical shift 1 unit downward.

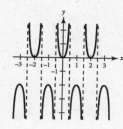

17. $y = \csc \dfrac{x}{2}$

Graph $y = \sin \dfrac{x}{2}$ first.

Period: $\dfrac{2\pi}{\frac{1}{2}} = 4\pi$

One cycle: 0 to 4π

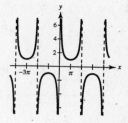

19. $y = \cot \dfrac{x}{2}$

Period: $\dfrac{\pi}{\frac{1}{2}} = 2\pi$

Two consecutive asymptotes:

$\dfrac{x}{2} = 0 \implies x = 0$

$\dfrac{x}{2} = \pi \implies x = 2\pi$

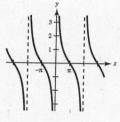

x	$\dfrac{\pi}{2}$	π	$\dfrac{3\pi}{2}$
y	1	0	-1

21. $y = \dfrac{1}{2} \sec 2x$

Graph $y = \dfrac{1}{2} \cos 2x$ first.

Period: $\dfrac{2\pi}{2} = \pi$

One cycle: 0 to π

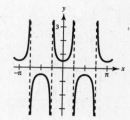

23. $y = \tan \dfrac{\pi x}{4}$

Period: $\dfrac{\pi}{\frac{\pi}{4}} = 4$

Two consecutive asymptotes:

$\dfrac{\pi x}{4} = -\dfrac{\pi}{2} \implies x = -2$

$\dfrac{\pi x}{4} = \dfrac{\pi}{2} \implies x = 2$

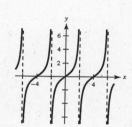

x	-1	0	1
y	-1	0	1

25. $y = \csc (\pi - x)$

Graph $y = \sin(\pi - x)$ first.

Period: 2π

Shift: Set $\pi - x = 0$ and $\pi - x = 2\pi$

$\qquad\qquad x = \pi \qquad\qquad x = -\pi$

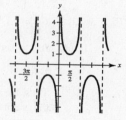

27. $y = \dfrac{1}{4} \csc\left(x + \dfrac{\pi}{4}\right)$

Graph $y = \dfrac{1}{4} \sin\left(x + \dfrac{\pi}{4}\right)$ first.

Period: 2π

Shift:

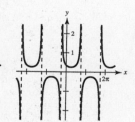

Set $x + \dfrac{\pi}{4} = 0 \qquad$ and $\qquad x + \dfrac{\pi}{4} = 2\pi$

$\qquad x = -\dfrac{\pi}{4} \quad$ to $\qquad\quad x = \dfrac{7\pi}{4}$

29. $y = \tan \dfrac{x}{3}$

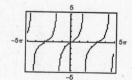

31. $y = -2 \sec 4x$

$= \dfrac{-2}{\cos 4x}$

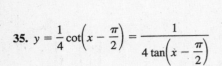

33. $y = \tan\left(x - \dfrac{\pi}{4}\right)$

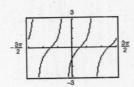

35. $y = \dfrac{1}{4}\cot\left(x - \dfrac{\pi}{2}\right) = \dfrac{1}{4\tan\left(x - \dfrac{\pi}{2}\right)}$

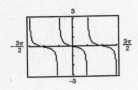

37. $y = 0.1\tan\left(\dfrac{\pi x}{4} + \dfrac{\pi}{4}\right)$

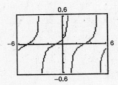

39. $\tan x = 1$

$x = -\dfrac{7\pi}{4},\ -\dfrac{3\pi}{4},\ \dfrac{\pi}{4},\ \dfrac{5\pi}{4}$

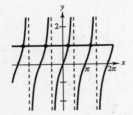

41. $\cot x = -\dfrac{\sqrt{3}}{3}$

$x = -\dfrac{4\pi}{3},\ -\dfrac{\pi}{3},\ \dfrac{2\pi}{3},\ \dfrac{5\pi}{3}$

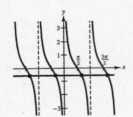

43. $\sec x = -2$

$x = \pm\dfrac{2\pi}{3},\ \pm\dfrac{4\pi}{3}$

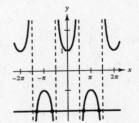

45. $\csc x = \sqrt{2}$

$x = -\dfrac{7\pi}{4},\ -\dfrac{5\pi}{4},\ \dfrac{\pi}{4},\ \dfrac{3\pi}{4}$

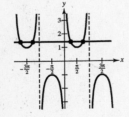

47. $f(x) = \sec x$

$\sec(-x) = \sec x$

Thus, the function is even and the graph of $y = \sec x$ has y-axis symmetry.

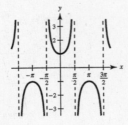

49. $g(x) = \sin(x + \pi)$

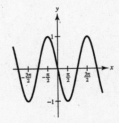

The graph has origin symmetry, so the function is odd.

51. $y_1 = \sin x \csc x$ and $y_2 = 1$

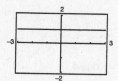

$$\sin x \csc x = \sin x \left(\frac{1}{\sin x} \right) = 1, \sin x \neq 0$$

The expressions are equivalent except when $\sin x = 0$ and y_1 is undefined.

53. $y_1 = \dfrac{\cos x}{\sin x}$ and $y_2 = \cot x = \dfrac{1}{\tan x}$

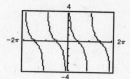

$$\cot x = \frac{\cos x}{\sin x}$$

The expressions are equivalent.

55. $f(x) = \sin x + \cos\left(x + \dfrac{\pi}{2}\right), g(x) = 0$

$f(x) = g(x)$ The graph is the line $y = 0$.

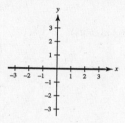

57. $f(x) = \sin^2 x, g(x) = \dfrac{1}{2}(1 - \cos 2x)$

$f(x) = g(x)$

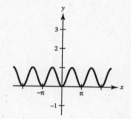

59. $f(x) = 2^{-x/4} \cos \pi x$

$-2^{-x/4} \leq f(x) \leq 2^{-x/4}$

The damping factor is $y = 2^{-x/4}$.

As $x \to \infty, f(x) \to 0$

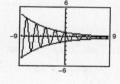

61. $g(x) = e^{-x^2/2} \sin x$

$-e^{-x^2/2} \leq g(x) \leq e^{-x^2/2}$

The damping factor is $y = e^{-x^2/2}$.

As $x \to \infty, g(x) \to 0$

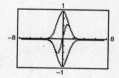

63. $y = \dfrac{6}{x} + \cos x, \ x > 0$

As $x \to 0, \ y \to \infty$.

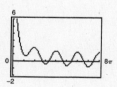

65. $g(x) = \dfrac{\sin x}{x}$

As $x \to 0, \ g(x) \to 1$.

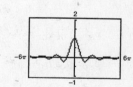

67. $f(x) = \sin \dfrac{1}{x}$

As $x \to 0, f(x)$ oscillates between -1 and 1.

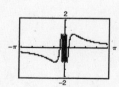

69. $f(x) = |x \cos x|$

As $x \to 0, f(x) \to 0$.

Matches graph (d).

71. $g(x) = |x| \sin x$

As $x \to 0, g(x) \to 0$.

Matches graph (b).

73. $f(x) = 2 \sin x$

$g(x) = \dfrac{1}{2} \csc x$

(a)

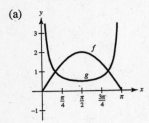

(b) $f > g$ on the interval, $\dfrac{\pi}{6} < x < \dfrac{5\pi}{6}$

(c) As $x \to \pi$, $f(x) = 2 \sin x \to 0$ and

$g(x) = \dfrac{1}{2} \csc x \to \infty$ if $x \to \pi^+$ and $-\infty$ if

$x \to \pi^-$ since $g(x)$ is the reciprocal of $f(x)$.

75. $\tan x = \dfrac{7}{d}$

$d = \dfrac{7}{\tan x} = 7 \cot x$

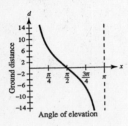

77. (a) $S = 74 + 3t + 40 \sin \dfrac{\pi t}{6}$

(b) Maximum sales: March

$(t \approx 3.27, S \approx 123.41 \text{ thousand units})$

Minimum sales: August

$(t \approx 8.73, S \approx 60.59 \text{ thousand units})$

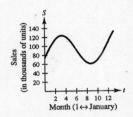

79. $H(t) = 54.33 - 20.38 \cos \dfrac{\pi t}{6} - 15.69 \sin \dfrac{\pi t}{6}$

$L(t) = 39.36 - 15.70 \cos \dfrac{\pi t}{6} - 14.16 \sin \dfrac{\pi t}{6}$

(a) Period of $\cos \dfrac{\pi t}{6}$: $\dfrac{2\pi}{\pi/6} = 12$

     Period of $\sin \dfrac{\pi t}{6}$: $\dfrac{2\pi}{\pi/6} = 12$

     Period of $H(t)$: 12

     Period of $L(t)$: 12

(b) From the graph, it appears that the greatest difference between high and low temperatures occurs in summer. The smallest difference occurs in winter.

(c) The highest high and low temperatures appear to occur around the middle of July, roughly one month after the time when the sun is northernmost in the sky.

81. True. Since $y = \csc x = \dfrac{1}{\sin x}$, for a given value of x, the y-coordinate of $\csc x$ is the reciprocal of the y-coordinate of $\sin x$.

83. True. $f(x) = \tan 2x$ has a period of $\frac{\pi}{2}$ and has consecutive asymptotes at:

$$2x = -\frac{\pi}{2} \quad \text{and} \quad 2x = \frac{\pi}{2}$$

$$x = -\frac{\pi}{4} \qquad\quad x = \frac{\pi}{4}$$

All the vertical asymptotes can be written as $x = \frac{\pi}{4} + n\left(\frac{\pi}{2}\right)$ where n is an integer.

85. As $x \to \frac{\pi}{2}$ from the left, $f(x) = \tan x \to \infty$.

As $x \to \frac{\pi}{2}$ from the right, $f(x) = \tan x \to -\infty$.

87.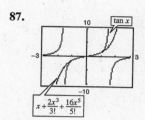

The polynomial approximation is best near the origin and becomes worse as x moves away from 0.

89. (a) $y_1 = \frac{4}{\pi}\left(\sin \pi x + \frac{1}{3}\sin 3\pi x\right)$

(b) $y_3 = \frac{4}{\pi}\left(\sin \pi x + \frac{1}{3}\sin 3\pi x + \frac{1}{5}\sin 5\pi x + \frac{1}{7}\sin 7\pi x\right)$

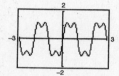

$$y_2 = \frac{4}{\pi}\left(\sin \pi x + \frac{1}{3}\sin 3\pi x + \frac{1}{5}\sin 5\pi x\right)$$

(c) $y_4 = \frac{4}{\pi}\left(\sin \pi x + \frac{1}{3}\sin 3\pi x + \frac{1}{5}\sin 5\pi x + \frac{1}{7}\sin 7\pi x + \frac{1}{9}\sin 9\pi x\right)$

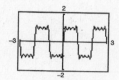

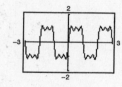

Section 9.7   Inverse Trigonometric Functions

1. $y = \arcsin \frac{1}{2} \implies \sin y = \frac{1}{2}$ for

$-\frac{\pi}{2} \le y \le \frac{\pi}{2} \implies y = \frac{\pi}{6}$

3. $y = \arccos \frac{1}{2} \implies \cos y = \frac{1}{2}$ for

$0 \le y \le \pi \implies y = \frac{\pi}{3}$

5. $y = \arctan \frac{\sqrt{3}}{3} \implies \tan y = \frac{\sqrt{3}}{3}$ for

$-\frac{\pi}{2} < y < \frac{\pi}{2} \implies y = \frac{\pi}{6}$

7. $y = \arccos\left(-\frac{\sqrt{3}}{2}\right) \implies \cos y = -\frac{\sqrt{3}}{2}$ for

$0 \le y \le \pi \implies y = \frac{5\pi}{6}$

9. $y = \arctan\left(-\sqrt{3}\right) \implies \tan y = -\sqrt{3}$ for

$-\frac{\pi}{2} < y < \frac{\pi}{2} \implies y = -\frac{\pi}{3}$

11. $y = \arccos\left(-\frac{1}{2}\right) \implies \cos y = -\frac{1}{2}$ for

$0 \le y \le \pi \implies y = \frac{2\pi}{3}$

13. $y = \arcsin \frac{\sqrt{3}}{2} \implies \sin y = \frac{\sqrt{3}}{2}$ for

$-\frac{\pi}{2} \le y \le \frac{\pi}{2} \implies y = \frac{\pi}{3}$

15. $y = \arctan 0 \Rightarrow \tan y = 0$ for

$$-\frac{\pi}{2} < y < \frac{\pi}{2} \Rightarrow y = 0$$

17. $\arccos 0.28 = \cos^{-1} 0.28 \approx 1.29$

19. $\arcsin(-0.75) = \sin^{-1}(-0.75) \approx -0.85$

21. $\arctan(-3) = \tan^{-1}(-3) \approx -1.25$

23. $\arcsin 0.31 = \sin^{-1} 0.31 \approx 0.32$

25. $\arccos(-0.41) = \cos^{-1}(-0.41)$
≈ 1.99

27. $\arctan 0.92 = \tan^{-1} 0.92 \approx 0.74$

29. $\arcsin\left(\frac{3}{4}\right) = \sin^{-1}(0.75) \approx 0.85$

31. $\arctan\left(\frac{7}{2}\right) = \tan^{-1}(3.5) \approx 1.29$

33. This is the graph of $y = \arctan x$. The coordinates are

$$\left(-\sqrt{3}, -\frac{\pi}{3}\right), \left(-\frac{\sqrt{3}}{3}, -\frac{\pi}{6}\right), \text{ and } \left(1, \frac{\pi}{4}\right).$$

35. $f(x) = \tan x$ and $g(x) = \arctan x$

Graph $y_1 = \tan x$

Graph $y_2 = \tan^{-1} x$

Graph $y_3 = x$

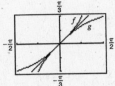

37. $\tan \theta = \dfrac{x}{4}$

$\theta = \arctan \dfrac{x}{4}$

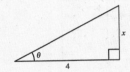

39. $\sin \theta = \dfrac{x+2}{5}$

$\theta = \arcsin\left(\dfrac{x+2}{5}\right)$

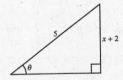

41. $\cos \theta = \dfrac{x+3}{2x}$

$\theta = \arccos\left(\dfrac{x+3}{2x}\right)$

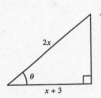

43. $\sin(\arcsin 0.3) = 0.3$

45. $\cos[\arccos(-0.1)] = -0.1$

47. $\arcsin(\sin 3\pi) = \arcsin(0) = 0$

Note: 3π is not in the range of the arcsine function.

49. Let $y = \arctan \dfrac{3}{4}$. Then,

$\tan y = \dfrac{3}{4}, \ 0 < y < \dfrac{\pi}{2}$

and $\sin y = \dfrac{3}{5}$.

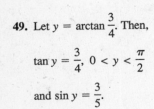

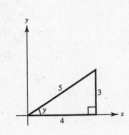

51. Let $y = \arctan 2$. Then,

$\tan y = 2 = \dfrac{2}{1}, \ 0 < y < \dfrac{\pi}{2}$

and $\cos y = \dfrac{1}{\sqrt{5}} = \dfrac{\sqrt{5}}{5}$.

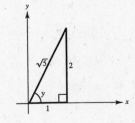

53. Let $y = \arcsin \dfrac{5}{13}$. Then,

$\sin y = \dfrac{5}{13},\ 0 < y < \dfrac{\pi}{2}$

and $\cos y = \dfrac{12}{13}$.

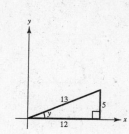

55. Let $y = \arctan\left(-\dfrac{3}{5}\right)$. Then,

$\tan y = -\dfrac{3}{5},\ -\dfrac{\pi}{2} < y < 0$

and $\sec y = \dfrac{\sqrt{34}}{5}$.

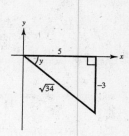

57. Let $y = \arccos\left(-\dfrac{2}{3}\right)$. Then,

$\cos y = -\dfrac{2}{3},\ \dfrac{\pi}{2} < y < \pi$

and $\sin y = \dfrac{\sqrt{5}}{3}$.

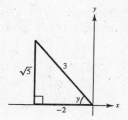

59. Let $y = \arctan x$. Then,

$\tan y = x = \dfrac{x}{1}$

and $\cot y = \dfrac{1}{x}$.

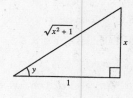

61. Let $y = \arcsin(2x)$. Then,

$\sin y = 2x = \dfrac{2x}{1}$

and $\cos y = \sqrt{1 - 4x^2}$.

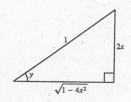

63. Let $y = \arccos x$. Then,

$\cos y = x = \dfrac{x}{1}$

and $\sin y = \sqrt{1 - x^2}$.

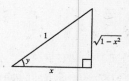

65. Let $y = \arccos\left(\dfrac{x}{3}\right)$. Then,

$\cos y = \dfrac{x}{3}$

and $\tan y = \dfrac{\sqrt{9 - x^2}}{x}$.

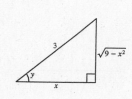

67. Let $y = \arctan \dfrac{x}{\sqrt{2}}$. Then,

$\tan y = \dfrac{x}{\sqrt{2}}$

and $\csc y = \dfrac{\sqrt{x^2 + 2}}{x}$.

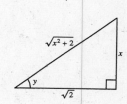

69. $f(x) = \sin(\arctan 2x)$, $g(x) = \dfrac{2x}{\sqrt{1 + 4x^2}}$

Let $y = \arctan 2x$. Then,

$\tan y = 2x = \dfrac{2x}{1}$

and $\sin y = \dfrac{2x}{\sqrt{1 + 4x^2}}$.

$g(x) = \dfrac{2x}{\sqrt{1 + 4x^2}} = f(x)$

The graph has horizontal asymptotes at $y = \pm 1$.

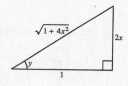

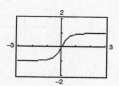

71. Let $y = \arctan \dfrac{9}{x}$. Then,

$\tan y = \dfrac{9}{x}$ and $\sin y = \dfrac{9}{\sqrt{x^2 + 81}},\ x > 0;\ \dfrac{-9}{\sqrt{x^2 + 81}},\ x < 0.$

Thus, $y = \arcsin \dfrac{9}{\sqrt{x^2 + 81}},\ x > 0;\ y = \arcsin \dfrac{-9}{\sqrt{x^2 + 81}},\ x < 0.$

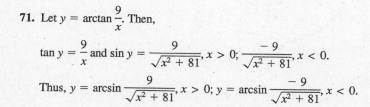

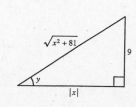

73. Let $y = \arccos \dfrac{3}{\sqrt{x^2 - 2x + 10}}$. Then,

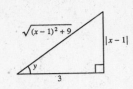

$$\cos y = \frac{3}{\sqrt{x^2 - 2x + 10}} = \frac{3}{\sqrt{(x - 1)^2 + 9}}$$

and $\sin y = \dfrac{|x - 1|}{\sqrt{(x - 1)^2 + 9}}$.

Thus, $y = \arcsin \dfrac{|x - 1|}{\sqrt{(x - 1)^2 + 9}} = \arcsin \dfrac{|x - 1|}{\sqrt{x^2 - 2x + 10}}$.

75. $y = 2 \arccos x$

Domain: $-1 \le x \le 1$

Range: $0 \le y \le 2\pi$

Vertical stretch of $f(x) = \arccos x$

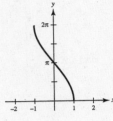

77. The graph of $f(x) = \arcsin(x - 1)$ is a horizontal translation of the graph of $y = \arcsin x$ by one unit.

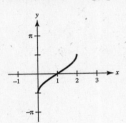

79. $f(x) = \arctan 2x$

Domain: all real numbers

Range: $-\dfrac{\pi}{2} < y < \dfrac{\pi}{2}$

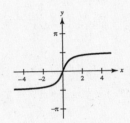

81. $h(v) = \tan(\arccos v) = \dfrac{\sqrt{1 - v^2}}{v}$

Domain: $-1 \le v \le 1, v \ne 0$

Range: all real numbers

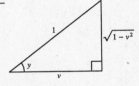

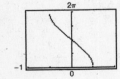

83. $f(x) = 2 \arccos(2x)$

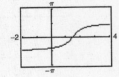

85. $f(x) = \arctan(2x - 3)$

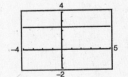

87. $f(x) = \pi - \arcsin\left(\dfrac{2}{3}\right) \approx 2.412$

89. $y = \operatorname{arccot} x$ if and only if $\cot y = x$.

Domain: $-\infty < x < \infty$

Range: $0 < x < \pi$

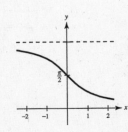

91. $y = \text{arccsc } x$ if and only if $\csc y = x$.

Domain: $(-\infty, -1] \cup [1, \infty)$

Range: $\left[-\dfrac{\pi}{2}, 0\right) \cup \left(0, \dfrac{\pi}{2}\right]$

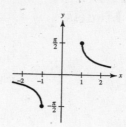

93. Let $y = \arcsin(-x)$. Then,

$$\sin y = -x$$
$$-\sin y = x$$
$$\sin(-y) = x$$
$$-y = \arcsin x$$
$$y = -\arcsin x.$$

Therefore, $\arcsin(-x) = -\arcsin x$.

95.
$$y = \pi - \arccos x$$
$$\cos y = \cos(\pi - \arccos x)$$
$$\cos y = \cos \pi \cos(\arccos x) + \sin \pi \sin(\arccos x)$$
$$\cos y = -x$$
$$y = \arccos(-x)$$

Therefore, $\arccos(-x) = \pi - \arccos x$.

97. $f(t) = 3\cos 2t + 3\sin 2t = \sqrt{3^2 + 3^2}\sin\left(2t + \arctan\dfrac{3}{3}\right)$

$$= 3\sqrt{2}\sin(2t + \arctan 1)$$

$$= 3\sqrt{2}\sin\left(2t + \dfrac{\pi}{4}\right)$$

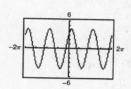

The graphs are the same.

99. (a) $\sin\theta = \dfrac{5}{s}$

$$\theta = \arcsin\dfrac{5}{s}$$

(b) $s = 40$: $\theta = \arcsin\dfrac{5}{40} \approx 0.13$

$s = 20$: $\theta = \arcsin\dfrac{5}{20} \approx 0.25$

101. $\beta = \arctan\dfrac{3x}{x^2 + 4}$

(a)

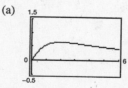

(b) β is maximum when $x = 2$.

(c) The graph has a horizontal asymptote at $\beta = 0$. As x increases, β decreases. This means the size of the picture in the photo gets smaller.

103. (a) $\tan\theta = \dfrac{x}{20}$

$$\theta = \arctan\dfrac{x}{20}$$

(b) $x = 5$: $\theta = \arctan\dfrac{5}{20} \approx 0.24$

$x = 12$: $\theta = \arctan\dfrac{12}{20} \approx 0.54$

105. False.

$\dfrac{5\pi}{6}$ is not in the range of $\arcsin(x)$.

$\arcsin\dfrac{1}{2} = \dfrac{\pi}{6}$

107. False.

$\sin\theta = 0.2$ has two solutions for $0 \le \theta < 2\pi$:

$\theta = \arcsin 0.2$ and $\theta = \pi - \arcsin 0.2$

Section 9.8 Applications and Models

1. Given: $A = 20°$, $b = 10$

$$\tan A = \frac{a}{b} \implies a = b \tan A = 10 \tan 20° \approx 3.64$$

$$\cos A = \frac{b}{c} \implies c = \frac{b}{\cos A} = \frac{10}{\cos 20°} \approx 10.64$$

$$B = 90° - 20° = 70°$$

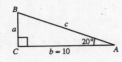

3. Given: $B = 71°$, $b = 24$

$$\tan B = \frac{b}{a} \implies a = \frac{b}{\tan B} = \frac{24}{\tan 71°} \approx 8.26$$

$$\sin B = \frac{b}{c} \implies c = \frac{b}{\sin B} = \frac{24}{\sin 71°} \approx 25.38$$

$$A = 90° - 71° = 19°$$

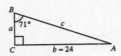

5. Given: $a = 6$, $b = 10$

$$c^2 = a^2 + b^2 \implies c = \sqrt{36 + 100}$$
$$= 2\sqrt{34} \approx 11.66$$

$$\tan A = \frac{a}{b} = \frac{6}{10} \implies A = \arctan \frac{3}{5} \approx 30.96°$$

$$B = 90° - 30.96° = 59.04°$$

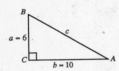

7. $b = 16$, $c = 52$

$$a = \sqrt{52^2 - 16^2}$$
$$= \sqrt{2448} = 12\sqrt{17} \approx 49.48$$

$$\cos A = \frac{16}{52}$$

$$A = \arccos \frac{16}{52} \approx 72.08°$$

$$B = 90° - 72.08° \approx 17.92°$$

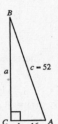

9. $A = 12°15'$, $c = 430.5$

$$B = 90° - 12°15' = 77°45'$$

$$\sin 12°15' = \frac{a}{430.5}$$
$$a = 430.5 \sin 12°15' \approx 91.34$$

$$\cos 12°15' = \frac{b}{430.5}$$
$$b = 430.5 \cos 12°15' \approx 420.70$$

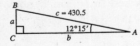

11. $\tan \theta = \dfrac{h}{\frac{1}{2}b} \implies h = \dfrac{1}{2} b \tan \theta$

$$h = \frac{1}{2}(4) \tan 52° \approx 2.56 \text{ inches}$$

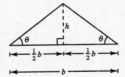

13. $\tan \theta = \dfrac{h}{\frac{1}{2}b} \implies h = \dfrac{1}{2} b \tan \theta$

$$h = \frac{1}{2}(46) \tan 41° \approx 19.99 \text{ inches}$$

15. Given A, C, a, we can solve for all the unknown parts.

$$B = 180° - A - C \quad \text{or} \quad B = 90° - A$$

$$\tan B = \frac{b}{a} \implies b = a \tan B$$

$$c = \sqrt{a^2 + b^2}$$

17. Given C, a, c, we can solve for all the remaining parts.

$$b = \sqrt{c^2 - a^2}$$

$$\sin A = \frac{a}{c} \implies A = \sin^{-1}\left(\frac{a}{c}\right)$$

$$B = 90° - A$$

19. N 45° W

45° west of north

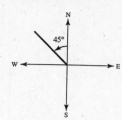

21. S 75° W

75° west of south

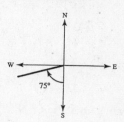

23. $d = a \sin \omega t$

Amplitude: $|a|$

25. $\tan 25° = \dfrac{50}{x}$

$x = \dfrac{50}{\tan 25°} \approx 107.2$ feet

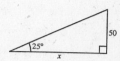

27. (a)

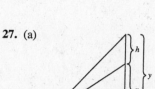
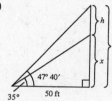

29. $\sin 80° = \dfrac{h}{20}$

$20 \sin 80° = h$

$h \approx 19.7$ feet

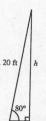

(b) Let the height of the church $= x$ and the height of the church and steeple $= y$. Then,

$$\tan 35° = \frac{x}{50} \quad \text{and} \quad \tan 47° 40' = \frac{y}{50}$$

$$x = 50 \tan 35° \text{ and } y = 50 \tan 47° 40'$$

$$h = y - x = 50 \left(\tan 47° 40' - \tan 35°\right).$$

(c) $h \approx 19.9$ feet

31. $\sin 34° = \dfrac{x}{4000}$

$x = 4000 \sin 34°$

≈ 2236.8 feet

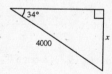

33. $\tan \theta = \dfrac{75}{50}$

$\theta = \arctan \dfrac{3}{2} \approx 56.3°$

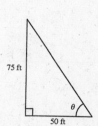

35.

Not drawn to scale

$5 \text{ miles} = 5 \text{ miles}\left(\dfrac{5280 \text{ ft}}{1 \text{ mile}}\right) = 26,400 \text{ feet}$

$\tan \theta = \dfrac{950}{26,400}$

$\theta = \text{artan}\left(\dfrac{950}{26,400}\right) \approx 2.06°$

37. Since the airplane speed is

$$\left(275\frac{\text{ft}}{\text{sec}}\right)\left(60\frac{\text{sec}}{\text{min}}\right) = 16,500\frac{\text{ft}}{\text{min}},$$

after one minute its distance travelled in 16,500 feet.

$\sin 18° = \dfrac{a}{16,500}$

$a = 16,500 \sin 18°$

≈ 5099 ft

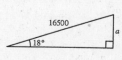

39. $\sin 10.5° = \dfrac{x}{4}$

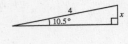

$\quad\quad x = 4 \sin 10.5°$

$\quad\quad\quad \approx 0.73$ mile

41. $\theta = 32°, \quad \phi = 68°$

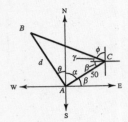

   (a) $\alpha = 90° - 32° = 58°$

      Bearing from A to C: N 58° E

   (b) $\quad \beta = \theta = 32°$

$\quad\quad\quad \gamma = 90° - \phi = 22°$

$\quad\quad\quad C = \beta + \gamma = 54°$

$\quad\quad \tan C = \dfrac{d}{50} \implies \tan 54° = \dfrac{d}{50} \implies d \approx 68.82$ meters

43. $\tan \theta = \dfrac{45}{30} \implies \theta \approx 56.3°$

    Bearing: N 56.3° W

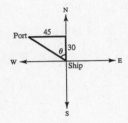

45. $\tan 6.5° = \dfrac{350}{d} \implies d \approx 3071.91$ ft

$\quad\quad \tan 4° = \dfrac{350}{D} \implies D \approx 5005.23$ ft

    Distance between ships: $D - d \approx 1933.32$ ft

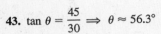

47. The plane has traveled $1.5 (600) = 900$ miles.

$\quad \sin 38° = \dfrac{a}{900} \implies a \approx 554$ miles north

$\quad \cos 38° = \dfrac{b}{900} \implies b \approx 709$ miles east

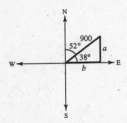

49. $\tan 57° = \dfrac{a}{x} \implies x = a \cot 57°$

$\quad \tan 16° = \dfrac{a}{x + \frac{55}{6}}$

$\quad \tan 16° = \dfrac{a}{a \cot 57° + \frac{55}{6}}$

$\quad \cot 16° = \dfrac{a \cot 57° + \frac{55}{6}}{a}$

$\quad a \cot 16° - a \cot 57° = \dfrac{55}{6} \implies a \approx 3.23$ miles

$\quad\quad\quad\quad\quad\quad\quad\quad\quad\quad\quad \approx 17,054$ ft

51. L_1: $3x - 2y = 5 \implies y = \dfrac{3}{2}x - \dfrac{5}{2} \implies m_1 = \dfrac{3}{2}$

$\quad L_2$: $\quad x + y = 1 \implies y = -x + 1 \implies m_2 = -1$

$\quad \tan \alpha = \left| \dfrac{-1 - \frac{3}{2}}{1 + (-1)\left(\frac{3}{2}\right)} \right| = \left| \dfrac{-\frac{5}{2}}{-\frac{1}{2}} \right| = 5$

$\quad\quad \alpha = \arctan 5 \approx 78.7°$

53. The diagonal of the base has a length of
$\sqrt{a^2 + a^2} = \sqrt{2}a.$

Now, we have $\tan\theta = \dfrac{a}{\sqrt{2}a} = \dfrac{1}{\sqrt{2}}$

$\theta = \arctan\dfrac{1}{\sqrt{2}}$

$\theta \approx 35.3°.$

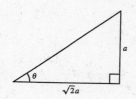

55. $\sin 36° = \dfrac{d}{25} \implies d \approx 14.69$

Length of side: $2d \approx 29.4$ inches

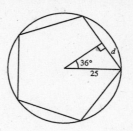

57. $\cos 30° = \dfrac{b}{r}$

$b = r\cos 30°$

$b = \dfrac{\sqrt{3}r}{2}$

$y = 2b = 2\left(\dfrac{\sqrt{3}r}{2}\right) = \sqrt{3}r$

59.

$\tan 35° = \dfrac{b}{10}$

$b = 10\tan 35° \approx 7$

$\cos 35° = \dfrac{10}{a}$

$a = \dfrac{10}{\cos 35°} \approx 12.2$

61. $d = 4\cos 8\pi t$

(a) Maximum displacement = amplitude = 4

(b) Frequency $= \dfrac{\omega}{2\pi} = \dfrac{8\pi}{2\pi}$

$\qquad\qquad$ = 4 cycles per unit of time

(c) $8\pi t = \dfrac{\pi}{2} \implies t = \dfrac{1}{16}$

63. $d = \dfrac{1}{16}\sin 120\pi t$

(a) Maximum displacement = amplitude $= \dfrac{1}{16}$

(b) Frequency $= \dfrac{\omega}{2\pi} = \dfrac{120\pi}{2\pi}$

$\qquad\qquad$ = 60 cycles per unit of time

(c) $120\pi t = \pi \implies t = \dfrac{1}{120}$

65. $d = 0$ when $t = 0$, $a = 4$, Period = 2

Use $d = a\sin\omega t$ since $d = 0$ when $t = 0$.

$\dfrac{2\pi}{\omega} = 2 \implies \omega = \pi$

Thus, $d = 4\sin\pi t.$

67. $d = 3$ when $t = 0$, $a = 3$, Period = 1.5

Use $d = a\cos\omega t$ since $d = 3$ when $t = 0$.

$\dfrac{2\pi}{\omega} = 1.5 \implies \omega = \dfrac{4\pi}{3}$

Thus, $d = 3\cos\left(\dfrac{4\pi}{3}t\right) = 3\cos\left(\dfrac{4\pi t}{3}\right).$

69.
$$d = a \sin \omega t$$

$$\text{Period} = \frac{2\pi}{\omega} = \frac{1}{\text{frequency}}$$

$$\frac{2\pi}{\omega} = \frac{1}{264}$$

$$\omega = 2\pi(264) = 528\pi$$

71. $y = \frac{1}{4} \cos 16t, \ t > 0$

(a)

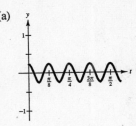

(b) Period: $\dfrac{2\pi}{16} = \dfrac{\pi}{8}$

(c) $\dfrac{1}{4} \cos 16t = 0$ when $16t = \dfrac{\pi}{2} \implies t = \dfrac{\pi}{32}$

73. False. Since the tower is not exactly vertical, a right triangle with sides 191 ft and d is not formed.

75. False. $\tan \beta = \dfrac{h}{d} \implies h = d \tan \beta$

Review Exercises for Chapter 9

1. $\theta \approx 1$ radian

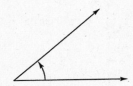

3. $\theta \approx 5$ radians

5. $\theta = \dfrac{11\pi}{4}$

Coterminal angles:

$$\frac{11\pi}{4} - 2\pi = \frac{3\pi}{4}$$

$$\frac{3\pi}{4} - 2\pi = -\frac{5\pi}{4}$$

7. $\theta = -\dfrac{4\pi}{3}$

Coterminal angles:

$$-\frac{4\pi}{3} + 2\pi = \frac{2\pi}{3}$$

$$-\frac{4\pi}{3} - 2\pi = -\frac{10\pi}{3}$$

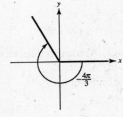

9. $\theta = 70°$

Coterminal angles:
$$70° + 360° = 430°$$
$$70° - 360° = -290°$$

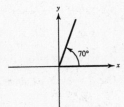

11. $\theta = -110°$

Coterminal angles:
$$-110° + 360° = 250°$$
$$-110° - 360° = -470°$$

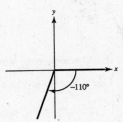

13. $\dfrac{5\pi \ \text{rad}}{7} = \dfrac{5\pi \ \text{rad}}{7} \cdot \dfrac{180°}{\pi \ \text{rad}} \approx 128.57°$

15. $-3.5 \ \text{rad} = -3.5 \ \text{rad} \cdot \dfrac{180°}{\pi \ \text{rad}} \approx -200.54°$

17. $480° = 480° \cdot \dfrac{\pi \ \text{rad}}{180°} = \dfrac{8\pi}{3} \ \text{radians} \approx 8.3776 \ \text{radians}$

19. $-33°45' = -33.75° = -33.75° \cdot \dfrac{\pi \text{ rad}}{180°} = -\dfrac{3\pi}{16} \text{ radian} \approx -0.5890 \text{ radian}$

21. (a) Angular speed $= \dfrac{\left(33\frac{1}{3}\right)(2\pi) \text{ radians}}{1 \text{ minute}}$

$\qquad\qquad\qquad\quad = 66\frac{2}{3}\pi \text{ radians per minute}$

(b) Linear speed $= \dfrac{6\left(66\frac{2}{3}\pi\right) \text{ inches}}{1 \text{ minute}}$

$\qquad\qquad\qquad = 400\pi \text{ inches per minute}$

23. $t = \dfrac{2\pi}{3}$ corresponds to the point $\left(-\dfrac{1}{2}, \dfrac{\sqrt{3}}{2}\right)$.

25. $t = \dfrac{5\pi}{6}$ corresponds to the point $\left(-\dfrac{\sqrt{3}}{2}, \dfrac{1}{2}\right)$.

27. $t = \dfrac{7\pi}{6}$ corresponds to the point $\left(-\dfrac{\sqrt{3}}{2}, -\dfrac{1}{2}\right)$.

$\sin \dfrac{7\pi}{6} = y = -\dfrac{1}{2}$ $\qquad$ $\csc \dfrac{7\pi}{6} = \dfrac{1}{y} = -2$

$\cos \dfrac{7\pi}{6} = x = -\dfrac{\sqrt{3}}{2}$ $\qquad$ $\sec \dfrac{7\pi}{6} = \dfrac{1}{x} = -\dfrac{2\sqrt{3}}{3}$

$\tan \dfrac{7\pi}{6} = \dfrac{y}{x} = \dfrac{1}{\sqrt{3}} = \dfrac{\sqrt{3}}{3}$ $\qquad$ $\cot \dfrac{7\pi}{6} = \dfrac{x}{y} = \sqrt{3}$

29. $t = -\dfrac{2\pi}{3}$ corresponds to the point $\left(-\dfrac{1}{2}, -\dfrac{\sqrt{3}}{2}\right)$.

$\sin\left(-\dfrac{2\pi}{3}\right) = y = -\dfrac{\sqrt{3}}{2}$ $\qquad$ $\csc\left(-\dfrac{2\pi}{3}\right) = \dfrac{1}{y} = -\dfrac{2\sqrt{3}}{3}$

$\cos\left(-\dfrac{2\pi}{3}\right) = x = -\dfrac{1}{2}$ $\qquad$ $\sec\left(-\dfrac{2\pi}{3}\right) = \dfrac{1}{x} = -2$

$\tan\left(-\dfrac{2\pi}{3}\right) = \dfrac{y}{x} = \sqrt{3}$ $\qquad$ $\cot\left(-\dfrac{2\pi}{3}\right) = \dfrac{x}{y} = \dfrac{\sqrt{3}}{3}$

31. $\sin \dfrac{11\pi}{4} = \sin \dfrac{3\pi}{4} = \dfrac{\sqrt{2}}{2}$

33. $\sin\left(-\dfrac{17\pi}{6}\right) = \sin\left(-\dfrac{5\pi}{6}\right) = -\dfrac{1}{2}$ $\qquad$ **35.** $\cos 84° \approx 0.10$ $\qquad$ **37.** $\sec \dfrac{12\pi}{5} = \dfrac{1}{\cos\left(\dfrac{12\pi}{5}\right)} \approx 3.24$

39. opp $= 4$, adj $= 5$, hyp $= \sqrt{4^2 + 5^2} = \sqrt{41}$

$\sin \theta = \dfrac{\text{opp}}{\text{hyp}} = \dfrac{4}{\sqrt{41}} = \dfrac{4\sqrt{41}}{41}$ $\qquad$ $\csc \theta = \dfrac{\text{hyp}}{\text{opp}} = \dfrac{\sqrt{41}}{4}$

$\cos \theta = \dfrac{\text{adj}}{\text{hyp}} = \dfrac{5}{\sqrt{41}} = \dfrac{5\sqrt{41}}{41}$ $\qquad$ $\sec \theta = \dfrac{\text{hyp}}{\text{adj}} = \dfrac{\sqrt{41}}{5}$

$\tan \theta = \dfrac{\text{opp}}{\text{adj}} = \dfrac{4}{5}$ $\qquad$ $\cot \theta = \dfrac{\text{adj}}{\text{opp}} = \dfrac{5}{4}$

41. opp $= 4$, hyp $= 8$, adj $= \sqrt{8^2 - 4^2} = \sqrt{48} = 4\sqrt{3}$

$\sin \theta = \dfrac{\text{opp}}{\text{hyp}} = \dfrac{4}{8} = \dfrac{1}{2}$ $\qquad$ $\csc \theta = \dfrac{\text{hyp}}{\text{opp}} = \dfrac{8}{4} = 2$

$\cos \theta = \dfrac{\text{adj}}{\text{hyp}} = \dfrac{4\sqrt{3}}{8} = \dfrac{\sqrt{3}}{2}$ $\qquad$ $\sec \theta = \dfrac{\text{hyp}}{\text{adj}} = \dfrac{8}{4\sqrt{3}} = \dfrac{2\sqrt{3}}{3}$

$\tan \theta = \dfrac{\text{opp}}{\text{adj}} = \dfrac{4}{4\sqrt{3}} = \dfrac{\sqrt{3}}{3}$ $\qquad$ $\cot \theta = \dfrac{\text{adj}}{\text{opp}} = \dfrac{4\sqrt{3}}{4} = \sqrt{3}$

43. $\sin \theta = \dfrac{1}{3}$

    (a) $\csc \theta = \dfrac{1}{\sin \theta} = 3$

    (b) $\sin^2 \theta + \cos^2 \theta = 1$

$$\left(\dfrac{1}{3}\right)^2 + \cos^2\theta = 1$$

$$\cos^2 \theta = 1 - \dfrac{1}{9}$$

$$\cos^2 \theta = \dfrac{8}{9}$$

$$\cos \theta = \sqrt{\dfrac{8}{9}}$$

$$\cos \theta = \dfrac{2\sqrt{2}}{3}$$

    (c) $\sec \theta = \dfrac{1}{\cos \theta} = \dfrac{3}{2\sqrt{2}} = \dfrac{3\sqrt{2}}{4}$

    (d) $\tan \theta = \dfrac{\sin \theta}{\cos \theta} = \dfrac{\frac{1}{3}}{\frac{2\sqrt{2}}{3}} = \dfrac{1}{2\sqrt{2}} = \dfrac{\sqrt{2}}{4}$

45. $\csc \theta = \dfrac{6}{5}$

    (a) $\sin \theta = \dfrac{1}{\csc \theta} = \dfrac{5}{6}$

    (b) $\sin^2 \theta + \cos^2 \theta = 1$

$$\left(\dfrac{5}{6}\right)^2 + \cos^2\theta = 1$$

$$\cos^2 \theta = 1 - \dfrac{25}{36}$$

$$\cos^2 \theta = \dfrac{11}{36}$$

$$\cos \theta = \sqrt{\dfrac{11}{36}}$$

$$\cos \theta = \dfrac{\sqrt{11}}{6}$$

    (c) $\sec \theta = \dfrac{1}{\cos \theta} = \dfrac{6}{\sqrt{11}} = \dfrac{6\sqrt{11}}{11}$

    (d) $\tan \theta = \dfrac{\sin \theta}{\cos \theta} = \dfrac{\frac{5}{6}}{\frac{\sqrt{11}}{6}} = \dfrac{5}{\sqrt{11}} = \dfrac{5\sqrt{11}}{11}$

47. $\tan 33° \approx 0.65$

49. $\sin 34.2° \approx 0.56$

51. $\cot 15°14' \approx \cot 15.2333° = \dfrac{1}{\tan 15.2333°} \approx 3.67$

53. $\sin 1°10' = \dfrac{x}{3.5}$

    $x = 3.5 \sin 1°10' \approx 0.07$ Kilometer

Not drawn to scale

55. $x = 12, y = 16, r = \sqrt{144 + 256} = \sqrt{400} = 20$

    $\sin \theta = \dfrac{y}{r} = \dfrac{4}{5}$ $\csc \theta = \dfrac{r}{y} = \dfrac{5}{4}$

    $\cos \theta = \dfrac{x}{r} = \dfrac{3}{5}$ $\sec \theta = \dfrac{r}{x} = \dfrac{5}{3}$

    $\tan \theta = \dfrac{y}{x} = \dfrac{4}{3}$ $\cot \theta = \dfrac{x}{y} = \dfrac{3}{4}$

57. $x = \dfrac{2}{3}, y = \dfrac{5}{2}$

$$r = \sqrt{\left(\dfrac{2}{3}\right)^2 + \left(\dfrac{5}{2}\right)^2} = \dfrac{\sqrt{241}}{6}$$

$$\sin\theta = \dfrac{y}{r} = \dfrac{\dfrac{5}{2}}{\dfrac{\sqrt{241}}{6}} = \dfrac{15}{\sqrt{241}} = \dfrac{15\sqrt{241}}{241}$$

$$\csc\theta = \dfrac{r}{y} = \dfrac{\dfrac{\sqrt{241}}{6}}{\dfrac{5}{2}} = \dfrac{2\sqrt{241}}{30} = \dfrac{\sqrt{241}}{15}$$

$$\cos\theta = \dfrac{x}{r} = \dfrac{\dfrac{2}{3}}{\dfrac{\sqrt{241}}{6}} = \dfrac{4}{\sqrt{241}} = \dfrac{4\sqrt{241}}{241}$$

$$\sec\theta = \dfrac{r}{x} = \dfrac{\dfrac{\sqrt{241}}{6}}{\dfrac{2}{3}} = \dfrac{\sqrt{241}}{4}$$

$$\tan\theta = \dfrac{y}{x} = \dfrac{\dfrac{5}{2}}{\dfrac{2}{3}} = \dfrac{15}{4}$$

$$\cot\theta = \dfrac{x}{y} = \dfrac{\dfrac{2}{3}}{\dfrac{5}{2}} = \dfrac{4}{15}$$

59. $x = -0.5, y = 4.5$

$$r = \sqrt{(-0.5)^2 + 4.5^2} = \sqrt{20.5}$$

$$\sin\theta = \dfrac{y}{r} = \dfrac{4.5}{\sqrt{20.5}} \approx 1$$

$$\cos\theta = \dfrac{x}{r} = \dfrac{-0.5}{\sqrt{20.5}} \approx -0.1$$

$$\tan\theta = \dfrac{y}{x} = -9$$

$$\csc\theta = \dfrac{r}{y} = \dfrac{\sqrt{20.5}}{4.5} \approx 1$$

$$\sec\theta = \dfrac{r}{x} = \dfrac{\sqrt{20.5}}{-0.5} \approx -9$$

$$\cot\theta = \dfrac{x}{y} = -\dfrac{1}{9} \approx -0.1$$

61. $(x, 4x), x > 0$

$$x = x, y = 4x$$

$$r = \sqrt{x^2 + (4x)^2} = \sqrt{17}x$$

$$\sin\theta = \dfrac{y}{r} = \dfrac{4x}{\sqrt{17}x} = \dfrac{4\sqrt{17}}{17}$$

$$\cos\theta = \dfrac{x}{r} = \dfrac{x}{\sqrt{17}x} = \dfrac{\sqrt{17}}{17}$$

$$\tan\theta = \dfrac{y}{x} = \dfrac{4x}{x} = 4$$

$$\csc\theta = \dfrac{r}{y} = \dfrac{\sqrt{17}x}{4x} = \dfrac{\sqrt{17}}{4}$$

$$\sec\theta = \dfrac{r}{x} = \dfrac{\sqrt{17}x}{x} = \sqrt{17}$$

$$\cot\theta = \dfrac{x}{y} = \dfrac{x}{4x} = \dfrac{1}{4}$$

63. $\sec\theta = \dfrac{6}{5}, \tan\theta < 0 \implies \theta$ is in Quadrant IV.

$$r = 6, x = 5, y = -\sqrt{36 - 25} = -\sqrt{11}$$

$$\sin\theta = \dfrac{y}{r} = -\dfrac{\sqrt{11}}{6}$$

$$\cos\theta = \dfrac{x}{r} = \dfrac{5}{6}$$

$$\tan\theta = \dfrac{y}{x} = -\dfrac{\sqrt{11}}{5}$$

$$\csc\theta = \dfrac{r}{y} = -\dfrac{6\sqrt{11}}{11}$$

$$\sec\theta = \dfrac{r}{x} = \dfrac{6}{5}$$

$$\cot\theta = \dfrac{x}{y} = -\dfrac{5\sqrt{11}}{11}$$

65. $\sin\theta = \dfrac{3}{8}, \cos\theta < 0 \implies \theta$ is in Quadrant II.

$$y = 3, r = 8, x = -\sqrt{55}$$

$$\sin\theta = \dfrac{y}{r} = \dfrac{3}{8}$$

$$\cos\theta = \dfrac{x}{r} = -\dfrac{\sqrt{55}}{8}$$

$$\tan\theta = \dfrac{y}{x} = -\dfrac{3}{\sqrt{55}} = -\dfrac{3\sqrt{55}}{55}$$

$$\csc\theta = \dfrac{8}{3}$$

$$\sec\theta = -\dfrac{8}{\sqrt{55}} = -\dfrac{8\sqrt{55}}{55}$$

$$\cot\theta = -\dfrac{\sqrt{55}}{3}$$

67. $\cos \theta = \dfrac{x}{r} = \dfrac{-2}{5} \implies y^2 = 21$

$\sin \theta > 0 \implies \theta$ is in Quadrant II $\implies y = \sqrt{21}$.

$\sin \theta = \dfrac{y}{r} = \dfrac{\sqrt{21}}{5}$

$\tan \theta = \dfrac{y}{x} = -\dfrac{\sqrt{21}}{2}$

$\csc \theta = \dfrac{r}{y} = \dfrac{5}{\sqrt{21}} = \dfrac{5\sqrt{21}}{21}$

$\sec \theta = \dfrac{r}{x} = \dfrac{5}{-2} = -\dfrac{5}{2}$

$\cot \theta = \dfrac{x}{y} = \dfrac{-2}{\sqrt{21}} = -\dfrac{2\sqrt{21}}{21}$

69. $\tan \dfrac{\pi}{3} = \sqrt{3}$

71. $\cos\left(-\dfrac{7\pi}{3}\right) = \cos \dfrac{\pi}{3} = \dfrac{1}{2}$

73. $\cos 495° = -\cos 45° = -\dfrac{\sqrt{2}}{2}$

75. $\sin 4 \approx -0.76$

77. $\sin(-3.2) \approx 0.06$

79. $\sin 3\pi = 0$

81. $\sec \dfrac{12\pi}{5} = \dfrac{1}{\cos\left(\dfrac{12\pi}{5}\right)} \approx 3.24$

83. $y = \sin x$

Amplitude: 1

Period: 2π

85. $f(x) = 5 \sin \dfrac{2x}{5}$

Amplitude: 5

Period: $\dfrac{2\pi}{\frac{2}{5}} = 5\pi$

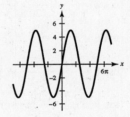

87. $y = 2 + \sin x$

Shift the graph of
$y = \sin x$ two units upward.

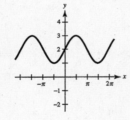

89. $g(t) = \dfrac{5}{2} \sin(t - \pi)$

Amplitude: $\dfrac{5}{2}$

Period: 2π

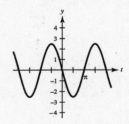

91. $y = 3 \sin x$

Amplitude: 3

Period: 2π

Matches graph (d).

93. $y = 2 \sin \pi x$

Amplitude: 2

Period: 2

Matches graph (b).

95. $y = a \sin bx$

(a) $a = 2, \dfrac{2\pi}{b} = \dfrac{1}{264} \implies b = 528\pi$

$y = 2 \sin(528\pi x)$

(b) $f = \dfrac{1}{\frac{1}{264}} = 264$ cycles per second.

97. $f(x) = \tan x$

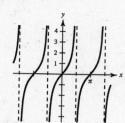

99. $f(x) = \cot x$

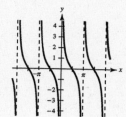

101. $f(x) = \sec x$

Graph $y = \cos x$ first.

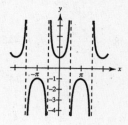

103. $f(x) = \csc x$

Graph $y = \sin x$ first.

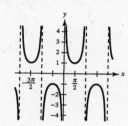

105. $f(x) = x \cos x$

Graph $y = x$ and $y = -x$ first.

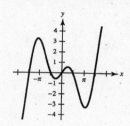

107. $\arcsin\left(-\dfrac{1}{2}\right) = -\arcsin\dfrac{1}{2} = -\dfrac{\pi}{6}$

109. $\arccos \dfrac{\sqrt{3}}{2} = \dfrac{\pi}{6}$

111. $\cos^{-1}(-1) = \pi$

113. $\arccos 0.324 \approx 1.24$ radians

115. $\arctan 0.123 \approx 0.12$ radian

117. $\arctan 5.783 \approx 1.40$ radians

119. $\tan^{-1}(-1.5) \approx -0.98$ radian

121. $\arcsin 0.4 \approx 0.41$ radian

123. $\sin^{-1}(-0.44) \approx -0.46$ radian

125. $\sin(\arcsin 0.72) = 0.72$

127. $\arctan(\tan \pi) = \arctan 0 = 0$

129. $\cos\left(\arctan \dfrac{3}{4}\right) = \dfrac{4}{5}$. Use a right triangle.

Let $\theta = \arctan \dfrac{3}{4}$

then $\tan \theta = \dfrac{3}{4}$

and $\cos \theta = \dfrac{4}{5}$.

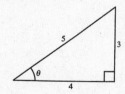

131. $\sec\left(\arctan \dfrac{12}{5}\right) = \dfrac{13}{5}$. Use a right triangle.

Let $\theta = \arctan \dfrac{12}{5}$

then $\tan \theta = \dfrac{12}{5}$

and $\sec \theta = \dfrac{13}{5}$.

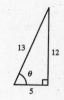

133. $\tan \theta = \dfrac{70}{30}$

$\theta = \arctan\left(\dfrac{70}{30}\right) \approx 66.8°$

135. $\sin 48° = \dfrac{d_1}{650} \implies d_1 \approx 483$

$\left. \cos 25° = \dfrac{d_2}{810} \implies d_2 \approx 734 \right\} \; d_1 + d_2 \approx 1217$

$\cos 48° = \dfrac{d_3}{650} \implies d_3 \approx 435$

$\left. \sin 25° = \dfrac{d_4}{810} \implies d_4 \approx 342 \right\} \; d_3 - d_4 \approx 93$

$\tan \theta \approx \dfrac{93}{1217} \implies \theta \approx 4.4°$

$\sec 4.4° \approx \dfrac{D}{1217} \implies D \approx 1217 \sec 4.4° \approx 1221$

The distance is 1221 miles and the bearing is N 85.6° E.

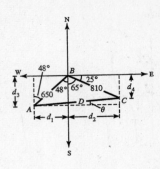

Problem Solving for Chapter 9

1. The area of a circle is $A = \pi r^2 \implies \pi = \dfrac{A}{r^2}$.

The circumference of a circle is $C = 2\pi r$.

$C = 2\left(\dfrac{A}{r^2}\right) r$

$C = \dfrac{2A}{r}$

$\dfrac{Cr}{2} = A$

For a sector, $C = s = r\theta$

Thus, $A = \dfrac{(r\theta)r}{2} = \dfrac{1}{2}\theta r^2$ for a sector.

3. $y_2 = \dfrac{\pi}{2} - y_1$

$\arctan x + \arctan \dfrac{1}{x} = y_1 + y_2$

$\qquad\qquad = y_1 + \left(\dfrac{\pi}{2} - y_1\right) = \dfrac{\pi}{2}$

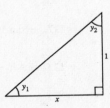

5. $\arcsin x = \arcsin \dfrac{x}{1} = \arctan \dfrac{x}{\sqrt{1-x^2}}, \; x > 0$

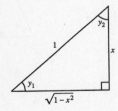

7. (a)

θ	L_1	L_2	$L_1 + L_2$
0.1	$\dfrac{2}{\sin 0.1}$	$\dfrac{3}{\cos 0.1}$	23.0
0.2	$\dfrac{2}{\sin 0.2}$	$\dfrac{3}{\cos 0.2}$	13.1
0.3	$\dfrac{2}{\sin 0.3}$	$\dfrac{3}{\cos 0.3}$	9.9
0.4	$\dfrac{2}{\sin 0.4}$	$\dfrac{3}{\cos 0.4}$	8.4

(b)

θ	L_1	L_2	$L_1 + L_2$
0.5	$\dfrac{2}{\sin 0.5}$	$\dfrac{3}{\cos 0.5}$	7.6
0.6	$\dfrac{2}{\sin 0.6}$	$\dfrac{3}{\cos 0.6}$	7.2
0.7	$\dfrac{2}{\sin 0.7}$	$\dfrac{3}{\cos 0.7}$	7.0
0.8	$\dfrac{2}{\sin 0.8}$	$\dfrac{3}{\cos 0.8}$	7.1

The minimum length of the elevator is 7.0 meters.

(c) $\quad L = L_1 + L_2 = \dfrac{2}{\sin \theta} + \dfrac{3}{\cos \theta}$

(d)

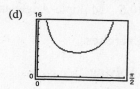

From the graph, it appears that the minimum length is 7.0 meters, which agrees with the estimate of part (b).

9. $f(x) = \sqrt{x}$

$\quad g(x) = 6 \arctan x$

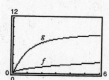

As x increases to infinity, g approaches 3π, but f has no maximum. Using the solve feature of the graphing utility, you find $a \approx 87.54$.

11. $d_1 + a_1 \sin(b_1 x + c_1) = d_2 + a_2 \cos(b_2 x + c_2)$. Assume that b_1 and b_2 are not both zero.

(a) $d_1 = d_2$ (b) $|a_1| = |a_2|$ (c) $|b_1| = |b_2|$

(d) $c_1 - c_2 = \dfrac{(2n + 1)\pi}{2}$, for some integer n

(e) Answers will vary. Examples:

$$1 + \sin x = 1 + \cos\left(x - \frac{\pi}{2}\right)$$

$$2 \sin(3x + \pi) = -2 \cos\left(-3x + \frac{\pi}{2}\right)$$

13. $d = \dfrac{(12\pi)(60)(24)(365)}{(5280)(12)} = \dfrac{1095\pi}{11} \approx 312.7$ miles

15. $y = Ae^{-kt} \cos bt = \frac{1}{5}e^{-t/10} \cos 6t$

(a) A is changed from $\frac{1}{5}$ to $\frac{1}{3}$: The displacement is increased.

(b) k is changed from $\frac{1}{10}$ to $\frac{1}{3}$: The friction damps the oscillations more rapidly.

(c) b is changed from 6 to 9: The frequency of oscillation is increased.

C H A P T E R 1 0
Analytic Trigonometry

CHAPTER 10
Analytic Trigonometry

Section 10.1   Using Fundamental Trigonometric Identities

1. $\sin x = \dfrac{\sqrt{3}}{2}$, $\cos x = -\dfrac{1}{2}$ $\Rightarrow$ x is in Quadrant II.

$\tan x = \dfrac{\sin x}{\cos x} = \dfrac{\sqrt{3}/2}{-1/2} = -\sqrt{3}$

$\cot x = \dfrac{1}{\tan x} = -\dfrac{1}{\sqrt{3}} = -\dfrac{\sqrt{3}}{3}$

$\sec x = \dfrac{1}{\cos x} = \dfrac{1}{-1/2} = -2$

$\csc x = \dfrac{1}{\sin x} = \dfrac{1}{\sqrt{3}/2} = \dfrac{2}{\sqrt{3}} = \dfrac{2\sqrt{3}}{3}$

3. $\sec \theta = \sqrt{2}$, $\sin \theta = -\dfrac{\sqrt{2}}{2}$ $\Rightarrow$ θ is in Quadrant IV.

$\cos \theta = \dfrac{1}{\sec \theta} = \dfrac{1}{\sqrt{2}} = \dfrac{\sqrt{2}}{2}$

$\tan \theta = \dfrac{\sin \theta}{\cos \theta} = \dfrac{-\sqrt{2}/2}{\sqrt{2}/2} = -1$

$\cot \theta = \dfrac{1}{\tan \theta} = -1$

$\csc \theta = \dfrac{1}{\sin \theta} = -\sqrt{2}$

5. $\tan x = \dfrac{5}{12}$, $\sec x = -\dfrac{13}{12}$ $\Rightarrow$ x is in

Quadrant III.

$\cos x = \dfrac{1}{\sec x} = -\dfrac{12}{13}$

$\sin x = -\sqrt{1 - \cos^2 x} = -\sqrt{1 - \dfrac{144}{169}} = -\dfrac{5}{13}$

$\cot x = \dfrac{1}{\tan x} = \dfrac{12}{5}$

$\csc x = \dfrac{1}{\sin x} = -\dfrac{13}{5}$

7. $\sec \phi = \dfrac{3}{2}$, $\csc \phi = -\dfrac{3\sqrt{5}}{5}$ $\Rightarrow$ ϕ is in Quadrant IV.

$\sin \phi = \dfrac{1}{\csc \phi} = \dfrac{1}{-3\sqrt{5}/5} = -\dfrac{\sqrt{5}}{3}$

$\cos \phi = \dfrac{1}{\sec \phi} = \dfrac{1}{3/2} = \dfrac{2}{3}$

$\tan \phi = \dfrac{\sin \phi}{\cos \phi} = \dfrac{-\sqrt{5}/3}{2/3} = -\dfrac{\sqrt{5}}{2}$

$\cot \phi = \dfrac{1}{\tan \phi} = \dfrac{1}{-\sqrt{5}/2} = -\dfrac{2}{\sqrt{5}} = -\dfrac{2\sqrt{5}}{5}$

9. $\sin(-x) = -\dfrac{1}{3}$ $\Rightarrow$ $\sin x = \dfrac{1}{3}$, $\tan x = -\dfrac{\sqrt{2}}{4}$ $\Rightarrow$ x is

in Quadrant II.

$\cos x = -\sqrt{1 - \sin^2 x} = -\sqrt{1 - \dfrac{1}{9}} = -\dfrac{2\sqrt{2}}{3}$

$\cot x = \dfrac{1}{\tan x} = \dfrac{1}{-\sqrt{2}/4} = -2\sqrt{2}$

$\sec x = \dfrac{1}{\cos x} = \dfrac{1}{-2\sqrt{2}/3} = -\dfrac{3\sqrt{2}}{4}$

$\csc x = \dfrac{1}{\sin x} = \dfrac{1}{1/3} = 3$

11. $\tan \theta = 2$, $\sin \theta < 0$ $\Rightarrow$ θ is in Quadrant III.

$\sec \theta = -\sqrt{\tan^2 \theta + 1} = -\sqrt{4 + 1} = -\sqrt{5}$

$\cos \theta = \dfrac{1}{\sec \theta} = -\dfrac{1}{\sqrt{5}} = -\dfrac{\sqrt{5}}{5}$

$\sin \theta = -\sqrt{1 - \cos^2 \theta}$

$= -\sqrt{1 - \dfrac{1}{5}} = -\dfrac{2}{\sqrt{5}} = -\dfrac{2\sqrt{5}}{5}$

$\csc \theta = \dfrac{1}{\sin \theta} = -\dfrac{\sqrt{5}}{2}$

$\cot \theta = \dfrac{1}{\tan \theta} = \dfrac{1}{2}$

13. $\sin\theta = -1$, $\cot\theta = 0 \implies \theta = \dfrac{3\pi}{2}$

$\cos\theta = \sqrt{1 - \sin^2\theta} = 0$

$\sec\theta$ is undefined.

$\tan\theta$ is undefined.

$\csc\theta = -1$

15. $\sec x \cos x = \sec x \cdot \dfrac{1}{\sec x} = 1$

The expression is matched with (d).

17. $\cot^2 x - \csc^2 x = \cot^2 x - (1 + \cot^2 x) = -1$

The expression is matched with (b).

19. $\dfrac{\sin(-x)}{\cos(-x)} = \dfrac{-\sin x}{\cos x} = -\tan x$

The expression is matched with (e).

21. $\sin x \sec x = \sin x \cdot \dfrac{1}{\cos x} = \tan x$

The expression is matched with (b).

23. $\sec^4 x - \tan^4 x = (\sec^2 x + \tan^2 x)(\sec^2 x - \tan^2 x)$

$\qquad = (\sec^2 x + \tan^2 x)(1) = \sec^2 x + \tan^2 x$

The expression is matched with (f).

25. $\dfrac{\sec^2 x - 1}{\sin^2 x} = \dfrac{\tan^2 x}{\sin^2 x} = \dfrac{\sin^2 x}{\cos^2 x} \cdot \dfrac{1}{\sin^2 x} = \sec^2 x$

The expression is matched with (e).

27. $\cot\theta \sec\theta = \dfrac{\cos\theta}{\sin\theta} \cdot \dfrac{1}{\cos\theta} = \dfrac{1}{\sin\theta} = \csc\theta$

29. $\sin\phi(\csc\phi - \sin\phi) = (\sin\phi)\dfrac{1}{\sin\phi} - \sin^2\phi$

$\qquad\qquad = 1 - \sin^2\phi = \cos^2\phi$

31. $\dfrac{\cot x}{\csc x} = \dfrac{\cos x/\sin x}{1/\sin x}$

$\qquad = \dfrac{\cos x}{\sin x} \cdot \dfrac{\sin x}{1} = \cos x$

33. $\dfrac{1 - \sin^2 x}{\csc^2 x - 1} = \dfrac{\cos^2 x}{\cot^2 x} = \cos^2 x \tan^2 x = (\cos^2 x)\dfrac{\sin^2 x}{\cos^2 x}$

$\qquad\qquad\qquad = \sin^2 x$

35. $\sec\alpha\dfrac{\sin\alpha}{\tan\alpha} = \dfrac{1}{\cos\alpha}(\sin\alpha)\cot\alpha = \dfrac{1}{\cos\alpha}(\sin\alpha)\left(\dfrac{\cos\alpha}{\sin\alpha}\right) = 1$

37. $\cos\left(\dfrac{\pi}{2} - x\right)\sec x = (\sin x)(\sec x) = (\sin x)\left(\dfrac{1}{\cos x}\right) = \dfrac{\sin x}{\cos x} = \tan x$

39. $\dfrac{\cos^2 y}{1 - \sin y} = \dfrac{1 - \sin^2 y}{1 - \sin y} = \dfrac{(1 + \sin y)(1 - \sin y)}{1 - \sin y} = 1 + \sin y$

41. $\sin\beta\tan\beta + \cos\beta = (\sin\beta)\dfrac{\sin\beta}{\cos\beta} + \cos\beta$

$\qquad\qquad = \dfrac{\sin^2\beta}{\cos\beta} + \dfrac{\cos^2\beta}{\cos\beta}$

$\qquad\qquad = \dfrac{\sin^2\beta + \cos^2\beta}{\cos\beta}$

$\qquad\qquad = \dfrac{1}{\cos\beta}$

$\qquad\qquad = \sec\beta$

43. $\cot u \sin u + \tan u \cos u = \dfrac{\cos u}{\sin u}(\sin u) + \dfrac{\sin u}{\cos u}(\cos u)$

$\qquad\qquad = \cos u + \sin u$

45. $\tan^2 x - \tan^2 x \sin^2 x = \tan^2 x(1 - \sin^2 x)$

$$= \tan^2 x \cos^2 x$$

$$= \frac{\sin^2 x}{\cos^2 x} \cdot \cos^2 x$$

$$= \sin^2 x$$

47. $\sin^2 x \sec^2 x - \sin^2 x = \sin^2 x(\sec^2 x - 1)$

$$= \sin^2 x \tan^2 x$$

49. $\dfrac{\sec^2 x - 1}{\sec x - 1} = \dfrac{(\sec x + 1)(\sec x - 1)}{\sec x - 1} = \sec x + 1$

51. $\tan^4 x + 2\tan^2 x + 1 = (\tan^2 x + 1)^2$

$$= (\sec^2 x)^2$$

$$= \sec^4 x$$

53. $\sin^4 x - \cos^4 x = (\sin^2 x + \cos^2 x)(\sin^2 x - \cos^2 x)$

$$= (1)(\sin^2 x - \cos^2 x)$$

$$= \sin^2 x - \cos^2 x$$

55. $\csc^3 x - \csc^2 x - \csc x + 1 = \csc^2 x(\csc x - 1) - 1(\csc x - 1)$

$$= (\csc^2 x - 1)(\csc x - 1)$$

$$= \cot^2 x(\csc x - 1)$$

57. $(\sin x + \cos x)^2 = \sin^2 x + 2\sin x \cos x + \cos^2 x$

$$= (\sin^2 x + \cos^2 x) + 2\sin x \cos x$$

$$= 1 + 2\sin x \cos x$$

59. $(2\csc x + 2)(2\csc x - 2) = 4\csc^2 x - 4 = 4(\csc^2 x - 1) = 4\cot^2 x$

61. $\dfrac{1}{1 + \cos x} + \dfrac{1}{1 - \cos x} = \dfrac{1 - \cos x + 1 + \cos x}{(1 + \cos x)(1 - \cos x)}$

$$= \dfrac{2}{1 - \cos^2 x}$$

$$= \dfrac{2}{\sin^2 x}$$

$$= 2\csc^2 x$$

63. $\dfrac{\cos x}{1 + \sin x} + \dfrac{1 + \sin x}{\cos x} = \dfrac{\cos^2 x + (1 + \sin x)^2}{\cos x(1 + \sin x)}$

$$= \dfrac{\cos^2 x + 1 + 2\sin x + \sin^2 x}{\cos x(1 + \sin x)}$$

$$= \dfrac{2 + 2\sin x}{\cos x(1 + \sin x)}$$

$$= \dfrac{2(1 + \sin x)}{\cos x(1 + \sin x)}$$

$$= \dfrac{2}{\cos x}$$

$$= 2\sec x$$

65. $\dfrac{\sin^2 y}{1 - \cos y} = \dfrac{1 - \cos^2 y}{1 - \cos y}$

$$= \dfrac{(1 + \cos y)(1 - \cos y)}{1 - \cos y}$$

$$= 1 + \cos y$$

67. $\dfrac{3}{\sec x - \tan x} \cdot \dfrac{\sec x + \tan x}{\sec x + \tan x} = \dfrac{3(\sec x + \tan x)}{\sec^2 x - \tan^2 x}$

$$= \dfrac{3(\sec x + \tan x)}{1}$$

$$= 3(\sec x + \tan x)$$

69. $\cos\theta = \sqrt{1 - \sin^2\theta}$ is **not** an identity.

$\cos^2\theta + \sin^2\theta = 1 \implies \cos\theta = \pm\sqrt{1 - \sin^2\theta}$

71. $\dfrac{\sin k\theta}{\cos k\theta} = \tan\theta$ is **not** an identity.

$\dfrac{\sin k\theta}{\cos k\theta} = \tan k\theta$

73. $\sin\theta \csc\theta = 1$ is an identity.

$\sin\theta \cdot \dfrac{1}{\sin\theta} = 1$, provided $\sin\theta \neq 0$.

75. Since $\sin^2\theta + \cos^2\theta = 1$ and $\cos^2\theta = 1 - \sin^2\theta$:

$\cos\theta = \pm\sqrt{1 - \sin^2\theta}$

$\tan\theta = \dfrac{\sin\theta}{\cos\theta} = \pm\dfrac{\sin\theta}{\sqrt{1 - \sin^2\theta}}$

$\cot\theta = \dfrac{1}{\tan\theta} = \pm\dfrac{\sqrt{1 - \sin^2\theta}}{\sin\theta}$

$\sec\theta = \dfrac{1}{\cos\theta} = \pm\dfrac{1}{\sqrt{1 - \sin^2\theta}}$

$\csc\theta = \dfrac{1}{\sin\theta}$

77. $y_1 = \cos\left(\dfrac{\pi}{2} - x\right)$, $y_2 = \sin x$

x	0.2	0.4	0.6	0.8	1.0	1.2	1.4
y_1	0.1987	0.3894	0.5646	0.7175	0.8414	0.9320	0.9854
y_2	0.1987	0.3894	0.5646	0.7175	0.8414	0.9320	0.9854

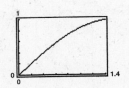

Conclusion: $y_1 = y_2$

79. $y_1 = \dfrac{\cos x}{1 - \sin x}$, $y_2 = \dfrac{1 + \sin x}{\cos x}$

x	0.2	0.4	0.6	0.8	1.0	1.2	1.4
y_1	1.2230	1.5085	1.8958	2.4650	3.4082	5.3319	11.6814
y_2	1.2230	1.5085	1.8958	2.4650	3.4082	5.3319	11.6814

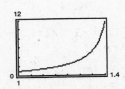

Conclusion: $y_1 = y_2$

81. $\cos x \cot x + \sin x = \cos x\left(\dfrac{\cos x}{\sin x}\right) + \sin x$

$\qquad = \dfrac{\cos^2 x}{\sin x} + \dfrac{\sin^2 x}{\sin x}$

$\qquad = \dfrac{\cos^2 x + \sin^2 x}{\sin x} = \dfrac{1}{\sin x} = \csc x$

$y_1 = \cos x \cot x + \sin x$ and $y_2 = \csc x = \dfrac{1}{\sin x}$

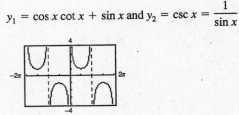

83. $\dfrac{1}{\sin x}\left(\dfrac{1}{\cos x} - \cos x\right) = \dfrac{1}{\sin x \cos x} - \dfrac{\cos x}{\sin x}$

$\qquad = \dfrac{1 - \cos^2 x}{\sin x \cos x} = \dfrac{\sin^2 x}{\sin x \cos x} = \dfrac{\sin x}{\cos x} = \tan x$

$y_1 = \dfrac{1}{\sin x}\left(\dfrac{1}{\cos x} - \cos x\right)$ and $y_2 = \tan x$

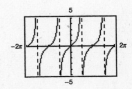

85. Let $x = 3 \cos \theta$, then

$$\sqrt{9 - x^2} = \sqrt{9 - (3 \cos \theta)^2} = \sqrt{9 - 9 \cos^2 \theta} = \sqrt{9(1 - \cos^2 \theta)}$$
$$= \sqrt{9 \sin^2 \theta} = 3 \sin \theta.$$

87. Let $x = 3 \sec \theta$, then

$$\sqrt{x^2 - 9} = \sqrt{(3 \sec \theta)^2 - 9}$$
$$= \sqrt{9 \sec^2 \theta - 9}$$
$$= \sqrt{9(\sec^2 \theta - 1)}$$
$$= \sqrt{9 \tan^2 \theta}$$
$$= 3 \tan \theta.$$

89. Let $x = 5 \tan \theta$, then

$$\sqrt{x^2 + 25} = \sqrt{(5 \tan \theta)^2 + 25}$$
$$= \sqrt{25 \tan^2 \theta + 25}$$
$$= \sqrt{25(\tan^2 \theta + 1)}$$
$$= \sqrt{25 \sec^2 \theta}$$
$$= 5 \sec \theta$$

91. Let $x = 3 \sin \theta$, then $\sqrt{9 - x^2} = 3$ becomes

$$\sqrt{9 - (3 \sin \theta)^2} = 3$$
$$\sqrt{9 - 9 \sin^2 \theta} = 3$$
$$\sqrt{9(1 - \sin^2 \theta)} = 3$$
$$\sqrt{9 \cos^2 \theta} = 3$$
$$3 \cos \theta = 3$$
$$\cos \theta = 1$$
$$\sin \theta = \sqrt{1 - \cos^2 \theta} = \sqrt{1 - (1)^2} = 0.$$

93. Let $x = 2 \cos \theta$, then $\sqrt{4 - x^2} = \sqrt{2}$ becomes

$$\sqrt{4 - (2 \cos \theta)^2} = \sqrt{2}$$
$$\sqrt{4 - 4 \cos^2 \theta} = \sqrt{2}$$
$$\sqrt{4(1 - \cos^2 \theta)} = \sqrt{2}$$
$$\sqrt{4 \sin^2 \theta} = \sqrt{2}$$
$$2 \sin \theta = \sqrt{2}, \text{ since } \sqrt{4 - x^2} > 0$$
$$\sin \theta = \frac{\sqrt{2}}{2}$$
$$\cos \theta = \sqrt{1 - \sin^2 \theta}$$
$$= \sqrt{1 - \frac{1}{2}}$$
$$= \sqrt{\frac{1}{2}}$$
$$= \frac{\sqrt{2}}{2}.$$

95. $\sin \theta = \sqrt{1 - \cos^2 \theta}$

Let $y_1 = \sin x$ and $y_2 = \sqrt{1 - \cos^2 x}$, $0 \le x < 2\pi$.

$y_1 = y_2$ for $0 \le x \le \pi$, so we have

$\sin \theta = \sqrt{1 - \cos^2 \theta}$ for $0 \le \theta \le \pi$.

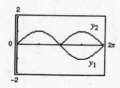

97. $\sec \theta = \sqrt{1 + \tan^2 \theta}$

Let $y_1 = \dfrac{1}{\cos x}$ and $y_2 = \sqrt{1 + \tan^2 x}$, $0 \le x < 2\pi$.

$y_1 = y_2$ for $0 \le x < \dfrac{\pi}{2}$ and $\dfrac{3\pi}{2} < x < 2\pi$, so we have

$\sec \theta = \sqrt{1 + \tan^2 \theta}$ for $0 \le \theta < \dfrac{\pi}{2}$ and $\dfrac{3\pi}{2} < \theta < 2\pi$.

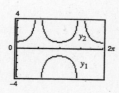

99. $\ln|\cos x| - \ln|\sin x| = \ln\dfrac{|\cos x|}{|\sin x|} = \ln|\cot x|$

101. $\ln|\cot t| + \ln(1 + \tan^2 t) = \ln\left[|\cot t|(1 + \tan^2 t)\right]$

$$= \ln|\cot t \sec^2 t| = \ln\left|\dfrac{\cos t}{\sin t}\cdot\dfrac{1}{\cos^2 t}\right|$$

$$= \ln\left|\dfrac{1}{\sin t \cos t}\right| = \ln|\csc t \sec t|$$

103. (a) $\csc^2 132° - \cot^2 132° \approx 1.8107 - 0.8107 = 1$

   (b) $\csc^2\dfrac{2\pi}{7} - \cot^2\dfrac{2\pi}{7} \approx 1.6360 - 0.6360 = 1$

105. $\cos\left(\dfrac{\pi}{2} - \theta\right) = \sin\theta$

   (a) $\theta = 80°$

   $\cos(90° - 80°) = \sin 80°$

   $0.9848 = 0.9848$

   (b) $\theta = 0.8$

   $\cos\left(\dfrac{\pi}{2} - 0.8\right) = \sin 0.8$

   $0.7174 = 0.7174$

107. $\mu W \cos\theta = W\sin\theta$

$$\mu = \dfrac{W\sin\theta}{W\cos\theta} = \tan\theta$$

109. True.

For example, $\sin(-x) = -\sin(x)$ means the graph of $\sin(x)$ is symmetric about the origin.

111. True.

$1 + \tan^2\theta = \sec^2\theta$

$\tan^2\theta = \sec^2\theta - 1$

For $\dfrac{\pi}{2} < \theta \le \pi$, $\tan\theta \le 0$ so

$\tan\theta = -\sqrt{\sec^2\theta - 1}.$

113. (a) $\displaystyle\lim_{x\to(\pi/2)^-}\sin x = 1$

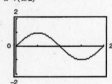

   (b) $\displaystyle\lim_{x\to(\pi/2)^-}\csc x = 1$

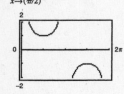

115. (a) $\displaystyle\lim_{x\to(\pi/2)^-}\tan x = \infty$

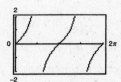

   (b) $\displaystyle\lim_{x\to(\pi/2)^-}\cot x = 0$

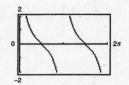

Section 10.2   Verifying Trigonometric Identities

1. $\sin t \csc t = \sin t\left(\dfrac{1}{\sin t}\right) = 1$

3. $(1 + \sin \alpha)(1 - \sin \alpha) = 1 - \sin^2 \alpha = \cos^2 \alpha$

5. $\cos^2 \beta - \sin^2 \beta = (1 - \sin^2 \beta) - \sin^2 \beta$
$\qquad\qquad\qquad = 1 - 2 \sin^2 \beta$

7. $\tan^2 \theta + 4 = (\sec^2 \theta - 1) + 4$
$\qquad\qquad = \sec^2 \theta + 3$

9. $\sin^2 \alpha - \sin^4 \alpha = \sin^2 \alpha(1 - \sin^2 \alpha)$
$\qquad\qquad\qquad = (1 - \cos^2 \alpha)(\cos^2 \alpha)$
$\qquad\qquad\qquad = \cos^2 \alpha - \cos^4 \alpha$

11. $\dfrac{\csc^2 \theta}{\cot \theta} = \csc^2 \theta\left(\dfrac{1}{\cot \theta}\right) = \csc^2 \theta \tan \theta$
$\qquad = \left(\dfrac{1}{\sin^2 \theta}\right)\left(\dfrac{\sin \theta}{\cos \theta}\right) = \left(\dfrac{1}{\sin \theta}\right)\left(\dfrac{1}{\cos \theta}\right)$
$\qquad = \csc \theta \sec \theta$

13. $\dfrac{\cot^2 t}{\csc t} = \dfrac{\cos^2 t}{\sin^2 t} \cdot \sin t$
$\quad = \dfrac{\cos^2 t}{\sin t}$
$\quad = \dfrac{1 - \sin^2 t}{\sin t} = \dfrac{1}{\sin t} - \dfrac{\sin^2 t}{\sin t}$
$\quad = \csc t - \sin t$

15. $\sin^{1/2} x \cos x - \sin^{5/2} x \cos x = \sin^{1/2} x \cos x(1 - \sin^2 x) = \sin^{1/2} x \cos x \cdot \cos^2 x = \cos^3 x\sqrt{\sin x}$

17. $\dfrac{1}{\sec x \tan x} = \cos x \cot x = \cos x \cdot \dfrac{\cos x}{\sin x}$
$\qquad\qquad\qquad = \dfrac{\cos^2 x}{\sin x}$
$\qquad\qquad\qquad = \dfrac{1 - \sin^2 x}{\sin x}$
$\qquad\qquad\qquad = \dfrac{1}{\sin x} - \sin x$
$\qquad\qquad\qquad = \csc x - \sin x$

19. $\dfrac{1}{\sec x} - \dfrac{1}{\cos x} = \cos x - \sec x$ by the reciprocal identities.

21. $\sin x \cos x + \sin^3 x \sec x = \sin x\left[\cos x + \sin^2 x\left(\dfrac{1}{\cos x}\right)\right]$
$\qquad\qquad\qquad = \sin x\left[\dfrac{\cos^2 x + \sin^2 x}{\cos x}\right]$
$\qquad\qquad\qquad = \sin x\left(\dfrac{1}{\cos x}\right)$
$\qquad\qquad\qquad = \dfrac{\sin x}{\cos x}$
$\qquad\qquad\qquad = \tan x$

23. $\dfrac{1}{\tan x} + \dfrac{1}{\cot x} = \dfrac{\cot x + \tan x}{\tan x \cot x}$
$\qquad\qquad\qquad = \dfrac{\cot x + \tan x}{1}$
$\qquad\qquad\qquad = \tan x + \cot x$

25. $\dfrac{\cos\theta\cot\theta}{1-\sin\theta}-1=\dfrac{\cos\theta\cot\theta-(1-\sin\theta)}{1-\sin\theta}$

$$=\dfrac{\cos\theta\left(\dfrac{\cos\theta}{\sin\theta}\right)-1+\sin\theta}{1-\sin\theta}\cdot\dfrac{\sin\theta}{\sin\theta}$$

$$=\dfrac{\cos^2\theta-\sin\theta+\sin^2\theta}{\sin\theta(1-\sin\theta)}$$

$$=\dfrac{1-\sin\theta}{\sin\theta(1-\sin\theta)}$$

$$=\dfrac{1}{\sin\theta}$$

$$=\csc\theta$$

27. $\dfrac{1}{\sin x+1}+\dfrac{1}{\csc x+1}=\dfrac{\csc x+1+\sin x+1}{(\sin x+1)(\csc x+1)}$

$$=\dfrac{\sin x+\csc x+2}{\sin x\csc x+\sin x+\csc x+1}$$

$$=\dfrac{\sin x+\csc x+2}{1+\sin x+\csc x+1}$$

$$=\dfrac{\sin x+\csc x+2}{\sin x+\csc x+2}$$

$$=1$$

29. $\tan\left(\dfrac{\pi}{2}-\theta\right)\tan\theta=\cot\theta\tan\theta=\left(\dfrac{1}{\tan\theta}\right)\tan\theta=1$

31. $\dfrac{\csc(-x)}{\sec(-x)}=\dfrac{\dfrac{1}{\sin(-x)}}{\dfrac{1}{\cos(-x)}}=\dfrac{\cos(-x)}{\sin(-x)}=\dfrac{\cos x}{-\sin x}=-\cot x$

33. $\dfrac{\cos(-\theta)}{1+\sin(-\theta)}=\dfrac{\cos\theta}{1-\sin\theta}\cdot\dfrac{1+\sin\theta}{1+\sin\theta}$

$$=\dfrac{\cos\theta(1+\sin\theta)}{1-\sin^2\theta}$$

$$=\dfrac{\cos\theta(1+\sin\theta)}{\cos^2\theta}$$

$$=\dfrac{1+\sin\theta}{\cos\theta}$$

$$=\dfrac{1}{\cos\theta}+\dfrac{\sin\theta}{\cos\theta}$$

$$=\sec\theta+\tan\theta$$

35. $\dfrac{\sin x\cos y+\cos x\sin y}{\cos x\cos y-\sin x\sin y}=\dfrac{\dfrac{\sin x\cos y}{\cos x\cos y}+\dfrac{\cos x\sin y}{\cos x\cos y}}{\dfrac{\cos x\cos y}{\cos x\cos y}-\dfrac{\sin x\sin y}{\cos x\cos y}}$

$$=\dfrac{\tan x+\tan y}{1-\tan x\tan y}$$

37. $\dfrac{\tan x+\cot y}{\tan x\cot y}=\dfrac{\dfrac{1}{\cot x}+\dfrac{1}{\tan y}}{\dfrac{1}{\cot x}\cdot\dfrac{1}{\tan y}}\cdot\dfrac{\cot x\tan y}{\cot x\tan y}=\tan y+\cot x$

39. $\sqrt{\dfrac{1+\sin\theta}{1-\sin\theta}}=\sqrt{\dfrac{1+\sin\theta}{1-\sin\theta}\cdot\dfrac{1+\sin\theta}{1+\sin\theta}}$

$$=\sqrt{\dfrac{(1+\sin\theta)^2}{1-\sin^2\theta}}$$

$$=\sqrt{\dfrac{(1+\sin\theta)^2}{\cos^2\theta}}$$

$$=\dfrac{1+\sin\theta}{|\cos\theta|}$$

41. $\cos^2\beta+\cos^2\left(\dfrac{\pi}{2}-\beta\right)=\cos^2\beta+\sin^2\beta=1$

43. $\sin t\csc\left(\dfrac{\pi}{2}-t\right)=\sin t\sec t=\sin t\left(\dfrac{1}{\cos t}\right)$

$$=\dfrac{\sin t}{\cos t}=\tan t$$

45. $2 \sec^2 x - 2 \sec^2 x \sin^2 x - \sin^2 x - \cos^2 x = 2 \sec^2 x(1 - \sin^2 x) - (\sin^2 x + \cos^2 x)$

$$= 2 \sec^2 x(\cos^2 x) - 1$$

$$= 2 \cdot \frac{1}{\cos^2 x} \cdot \cos^2 x - 1$$

$$= 2 - 1$$

$$= 1$$

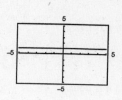

47. $2 + \cos^2 x - 3 \cos^4 x = (1 - \cos^2 x)(2 + 3 \cos^2 x)$

$$= \sin^2 x(2 + 3 \cos^2 x)$$

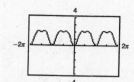

49. $\csc^4 x - 2 \csc^2 x + 1 = (\csc^2 x - 1)^2$

$$= (\cot^2 x)^2 = \cot^4 x$$

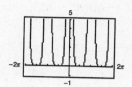

51. $\sec^4 \theta - \tan^4 \theta = (\sec^2 \theta + \tan^2 \theta)(\sec^2 \theta - \tan^2 \theta)$

$$= (1 + \tan^2 \theta + \tan^2 \theta)(1)$$

$$= 1 + 2 \tan^2 \theta$$

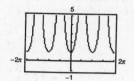

53. $\dfrac{\cos x}{1 + \sin x} = \dfrac{\cos x}{1 + \sin x} \cdot \dfrac{1 - \sin x}{1 - \sin x}$

$$= \frac{\cos x(1 - \sin x)}{1 - \sin^2 x}$$

$$= \frac{\cos x(1 - \sin x)}{\cos^2 x}$$

$$= \frac{1 - \sin x}{\cos x}$$

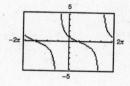

55. $\dfrac{\tan^3 \alpha - 1}{\tan \alpha - 1} = \dfrac{(\tan \alpha - 1)(\tan^2 \alpha + \tan \alpha + 1)}{\tan \alpha - 1} = \tan^2 \alpha + \tan \alpha + 1$

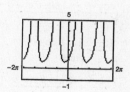

57. Since $\sin^2 \theta = 1 - \cos^2 \theta$, then

$\sin \theta = \pm\sqrt{1 - \cos^2 \theta}$; $\sin \theta \neq \sqrt{1 - \cos^2 \theta}$ if θ lies in Quadrant III or IV.

One such angle is $\theta = \dfrac{7\pi}{4}$.

59. $\sqrt{\tan^2 x} = |\tan x|$

$\sqrt{\tan^2 x} \neq \tan x$ if x lies in Quadrant II or IV.

One such angle is $x = \dfrac{3\pi}{4}$.

61. When n is even,

$$\cos\left[\frac{(2n+1)\pi}{2}\right] = \cos\frac{\pi}{2} = 0.$$

When n is odd,

$$\cos\left[\frac{(2n+1)\pi}{2}\right] = \cos\frac{3\pi}{2} = 0.$$

Thus, $\cos\left[\frac{(2n+1)\pi}{2}\right] = 0$ for all integers n.

63. $\ln|\sec\theta| = \ln\left|\frac{1}{\cos\theta}\right|$

$$= \ln 1 - \ln|\cos\theta|$$

$$= -\ln|\cos\theta|$$

65. $\sin^2 25° + \sin^2 65° = \sin^2 25° + \cos^2(90° - 65°) = \sin^2 25° + \cos^2 25° = 1$

67. $\cos^2 20° + \cos^2 52° + \cos^2 38° + \cos^2 70° = \cos^2 20° + \cos^2 52° + \sin^2(90° - 38°) + \sin^2(90° - 70°)$

$$= \cos^2 20° + \cos^2 52° + \sin^2 52° + \sin^2 20°$$

$$= (\cos^2 20° + \sin^2 20°) + (\cos^2 52° + \sin^2 52°)$$

$$= 1 + 1$$

$$= 2$$

69. $\cos x - \csc x \cot x = \cos x - \left(\frac{1}{\sin x}\right)\left(\frac{\cos x}{\sin x}\right)$

$$= \cos x\left(1 - \frac{1}{\sin^2 x}\right)$$

$$= \cos x(1 - \csc^2 x)$$

$$= -\cos x(\csc^2 x - 1)$$

$$= -\cos x \cot^2 x$$

71. False. For the equation to be an identity, it must be true for all values of θ in the domain.

73. False

$$\frac{(\cos\theta + \sin\theta)^2}{\sin\theta} = \frac{\cos^2\theta + 2\sin\theta\cos\theta + \sin^2\theta}{\sin\theta}$$

$$= \frac{1 + 2\sin\theta\cos\theta}{\sin\theta}$$

$$= \frac{1}{\sin\theta} + \frac{2\sin\theta\cos\theta}{\sin\theta}$$

$$= \csc\theta + 2\cos\theta$$

Section 10.3   Solving Trigonometric Equations

1. $2\cos x - 1 = 0$

(a) $2\cos\frac{\pi}{3} - 1 = 2\left(\frac{1}{2}\right) - 1 = 0$

(b) $2\cos\frac{5\pi}{3} - 1 = 2\left(\frac{1}{2}\right) - 1 = 0$

3. $3 \tan^2 2x - 1 = 0$

(a) $3\left[\tan 2\left(\dfrac{\pi}{12}\right)\right]^2 - 1 = 3 \tan^2 \dfrac{\pi}{6} - 1$

$\qquad\qquad\qquad = 3\left(\dfrac{1}{\sqrt{3}}\right)^2 - 1$

$\qquad\qquad\qquad = 0$

(b) $3\left[\tan 2\left(\dfrac{5\pi}{12}\right)\right]^2 - 1 = 3 \tan^2 \dfrac{5\pi}{6} - 1$

$\qquad\qquad\qquad = 3\left(-\dfrac{1}{\sqrt{3}}\right)^2 - 1$

$\qquad\qquad\qquad = 0$

5. $2 \sin^2 x - \sin x - 1 = 0$

(a) $2 \sin^2 \dfrac{\pi}{2} - \sin \dfrac{\pi}{2} - 1 = 2(1)^2 - 1 - 1$

$\qquad\qquad\qquad\qquad = 0$

(b) $2 \sin^2 \dfrac{7\pi}{6} - \sin \dfrac{7\pi}{6} - 1 = 2\left(-\dfrac{1}{2}\right)^2 - \left(-\dfrac{1}{2}\right) - 1$

$\qquad\qquad\qquad\qquad = \dfrac{1}{2} + \dfrac{1}{2} - 1 = 0$

7. $2 \cos x + 1 = 0$

$\qquad 2 \cos x = -1$

$\qquad \cos x = -\dfrac{1}{2}$

$\qquad x = \dfrac{2\pi}{3} + 2n\pi$

$\qquad$ or $\quad x = \dfrac{4\pi}{3} + 2n\pi$

9. $\sqrt{3} \csc x - 2 = 0$

$\qquad \sqrt{3} \csc x = 2$

$\qquad \csc x = \dfrac{2}{\sqrt{3}}$

$\qquad x = \dfrac{\pi}{3} + 2n\pi$

$\qquad$ or $\quad x = \dfrac{2\pi}{3} + 2n\pi$

11. $3 \sec^2 x - 4 = 0$

$\qquad \sec^2 x = \dfrac{4}{3}$

$\qquad \sec x = \pm\dfrac{2}{\sqrt{3}}$

$\qquad x = \dfrac{\pi}{6} + n\pi$

$\qquad$ or $\quad x = \dfrac{5\pi}{6} + n\pi$

13. $\sin x(\sin x + 1) = 0$

$\quad \sin x = 0 \quad$ or $\quad \sin x = -1$

$\quad x = n\pi \qquad\qquad x = \dfrac{3\pi}{2} + 2n\pi$

15. $4 \cos^2 x - 1 = 0$

$\qquad \cos^2 x = \dfrac{1}{4}$

$\qquad \cos^2 x = \pm\dfrac{1}{2}$

$\qquad x = \dfrac{\pi}{3} + n\pi$

$\qquad$ or $\quad x = \dfrac{2\pi}{3} + n\pi$

17. $\qquad\qquad \cos^3 x = \cos x$

$\qquad \cos^3 x - \cos x = 0$

$\quad \cos x(\cos^2 x - 1) = 0$

$\quad \cos x = 0 \qquad$ or $\quad \cos^2 x - 1 = 0$

$\qquad x = \dfrac{\pi}{2}, \dfrac{3\pi}{2} \qquad\qquad \cos x = \pm 1$

$\qquad\qquad\qquad\qquad\qquad x = 0, \pi$

19. $\qquad 3 \tan^3 x - \tan x = 0$

$\qquad \tan x(3 \tan^2 x - 1) = 0$

$\qquad \tan x = 0 \quad$ or $\quad 3 \tan^2 x - 1 = 0$

$\qquad x = 0, \pi \qquad\qquad \tan x = \pm\dfrac{\sqrt{3}}{3}$

$\qquad\qquad\qquad\qquad x = \dfrac{\pi}{6}, \dfrac{5\pi}{6}, \dfrac{7\pi}{6}, \dfrac{11\pi}{6}$

21. $\sec^2 x - \sec x - 2 = 0$

$(\sec x - 2)(\sec x + 1) = 0$

$\sec x - 2 = 0 \quad$ or $\quad \sec x + 1 = 0$

$\qquad \sec x = 2 \qquad\qquad\qquad \sec x = -1$

$$x = \frac{\pi}{3}, \frac{5\pi}{3} \qquad\qquad\qquad x = \pi$$

23. $2 \sin x + \csc x = 0$

$2 \sin x + \dfrac{1}{\sin x} = 0$

$2 \sin^2 x + 1 = 0$

$\sin^2 x = -\dfrac{1}{2} \implies$ No solution

25. $2 \cos^2 x + \cos x - 1 = 0$

$(2 \cos x - 1)(\cos x + 1) = 0$

$2 \cos x - 1 = 0 \quad$ or $\quad \cos x + 1 = 0$

$\qquad \cos x = \dfrac{1}{2} \qquad\qquad\qquad \cos x = -1$

$$x = \frac{\pi}{3}, \frac{5\pi}{3} \qquad\qquad\qquad x = \pi$$

27. $2 \sec^2 x + \tan^2 x - 3 = 0$

$2(\tan^2 x + 1) + \tan^2 x - 3 = 0$

$3 \tan^2 x - 1 = 0$

$\tan x = \pm \dfrac{\sqrt{3}}{3}$

$$x = \frac{\pi}{6}, \frac{5\pi}{6}, \frac{7\pi}{6}, \frac{11\pi}{6}$$

29. $\cos 2x = \dfrac{1}{2}$

$2x = \dfrac{\pi}{3} + 2n\pi \quad$ or $\quad 2x = \dfrac{5\pi}{3} + 2n\pi$

$x = \dfrac{\pi}{6} + n\pi \qquad\qquad x = \dfrac{5\pi}{6} + n\pi$

31. $\tan 3x = 1$

$3x = \dfrac{\pi}{4} + 2n\pi \quad$ or $\quad 3x = \dfrac{5\pi}{4} + 2n\pi$

$x = \dfrac{\pi}{12} + \dfrac{2n\pi}{3} \qquad\qquad x = \dfrac{5\pi}{12} + \dfrac{2n\pi}{3}$

These can be combined as $x = \dfrac{\pi}{12} + \dfrac{n\pi}{3}$.

33. $\cos\left(\dfrac{x}{2}\right) = \dfrac{\sqrt{2}}{2}$

$\dfrac{x}{2} = \dfrac{\pi}{4} + 2n\pi \quad$ or $\quad \dfrac{x}{2} = \dfrac{7\pi}{4} + 2n\pi$

$x = \dfrac{\pi}{2} + 4n\pi \qquad\qquad x = \dfrac{7\pi}{2} + 4n\pi$

35. $2 \sin^2 2x = 1$

$\sin 2x = \pm \dfrac{1}{\sqrt{2}} = \pm \dfrac{\sqrt{2}}{2}$

$2x = \dfrac{\pi}{4} + \dfrac{n\pi}{2}$

Thus, $x = \dfrac{\pi}{8} + \dfrac{n\pi}{4}$

37. $\tan 3x (\tan x - 1) = 0$

$\tan 3x = 0 \quad$ or $\quad \tan x - 1 = 0$

$\qquad 3x = n\pi \qquad\qquad\qquad \tan x = 1$

$\qquad x = \dfrac{n\pi}{3} \qquad\qquad\qquad x = \dfrac{\pi}{4} + n\pi$

39. $y = \sin \dfrac{\pi x}{2} + 1$

$\sin \dfrac{\pi x}{2} + 1 = 0$

$\sin \dfrac{\pi x}{2} = -1$

$\dfrac{\pi x}{2} = \dfrac{3\pi}{2} + 2n\pi$

$x = 3 + 4n$

For $-2 < x < 4$ the intercepts are -1 and 3.

41. $y = \tan^2\left(\dfrac{\pi x}{6}\right) - 3$

$\tan^2\left(\dfrac{\pi x}{6}\right) - 3 = 0$

$\tan^2\left(\dfrac{\pi x}{6}\right) = 3$

$\tan\left(\dfrac{\pi x}{6}\right) = \pm\sqrt{3}$

$\dfrac{\pi x}{6} = \dfrac{\pi}{3} + n\pi$ or $\dfrac{\pi x}{6} = \dfrac{2\pi}{3} + n\pi$

$x = 2 + 6n$ or $x = 4 + 6n$

For $-3 < x < 3$ the intercepts are -2 and 2.

43. $6y^2 - 13y + 6 = 0$

$(3y - 2)(2y - 3) = 0$

$3y - 2 = 0$ or $2y - 3 = 0$

$y = \dfrac{2}{3}$ $y = \dfrac{3}{2}$

$6\cos^2 x - 13\cos x + 6 = 0$

$(3\cos x - 2)(2\cos x - 3) = 0$

$3\cos x - 2 = 0$ or $2\cos x - 3 = 0$

$\cos x = \dfrac{2}{3}$ $\cos x = \dfrac{3}{2}$ (No solution)

$x \approx 0.8411 + 2n\pi, 5.4421 + 2n\pi$

45. $f(x) = \cos\dfrac{1}{x}$

(a) Domain: All real numbers except $x = 0$.

(b) Symmetry: y-axis; Asymptote: $y = 1$

(c) The function values oscillate between -1 and 1 more rapidly.

(d) Infinitely many

(e) $\cos\dfrac{1}{x} = 0$

$\dfrac{1}{x} = n\pi + \dfrac{\pi}{2} \implies x = \dfrac{1}{n\pi + \pi/2}$ x is the greatest when $n = 0$. Then $x = \dfrac{2}{\pi} \approx 0.6366$

47. $2\sin x + \cos x = 0$

$2\sin x = -\cos x$

$2 = -\dfrac{\cos x}{\sin x}$

$2 = -\cot x$

$-2 = \cot x$

$-\dfrac{1}{2} = \tan x$

$x = \arctan\left(-\dfrac{1}{2}\right)$

$x = \pi - \arctan\left(\dfrac{1}{2}\right) \approx 2.6779$ or $x = 2\pi - \arctan\left(\dfrac{1}{2}\right) \approx 5.8195$

Graph $y_1 = 2\sin x + \cos x$.

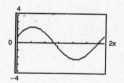

The x-intercepts occur at $x \approx 2.6779$ and $x \approx 5.8195$.

49.
$$\frac{1 + \sin x}{\cos x} + \frac{\cos x}{1 + \sin x} = 4$$

$$\frac{(1 + \sin x)^2 + \cos^2 x}{\cos x(1 + \sin x)} = 4$$

$$\frac{1 + 2\sin x + \sin^2 x + \cos^2 x}{\cos x(1 + \sin x)} = 4$$

$$\frac{2 + 2\sin x}{\cos x(1 + \sin x)} = 4$$

$$\frac{2}{\cos x} = 4$$

$$\cos x = \frac{1}{2}$$

$$x = \frac{\pi}{3}, \frac{5\pi}{3}$$

Graph $y_1 = \dfrac{1 + \sin x}{\cos x} + \dfrac{\cos x}{1 + \sin x} - 4.$

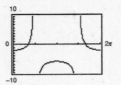

The x-intercepts occur at $x = \dfrac{\pi}{3} \approx 1.0472$ and

$x = \dfrac{5\pi}{3} \approx 5.2360.$

51. $x \tan x - 1 = 0$

Graph $y_1 = x \tan x - 1.$

The x-intercepts occur at $x \approx 0.8603$ and $x \approx 3.4256$

53. $\sec^2 x + 0.5 \tan x - 1 = 0$

Graph $y_1 = \dfrac{1}{(\cos x)^2} + 0.5 \tan x - 1.$

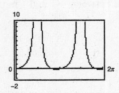

The x-intercepts occur at $x = 0$, $x \approx 2.6779$,
$x = \pi \approx 3.1416$, and $x \approx 5.8195.$

55. $2 \tan^2 x + 7 \tan x - 15 = 0$

$(2 \tan x - 3)(\tan x + 5) = 0$

$2 \tan x - 3 = 0 \qquad$ or $\quad \tan x + 5 = 0$

$\quad \tan x = 1.5 \qquad\qquad\qquad \tan x = -5$

$\quad x \approx 0.9828, 4.1244 \qquad\qquad x \approx 1.7682, 4.9098$

Graph $y_1 = 2 \tan^2 x + 7 \tan x - 15.$

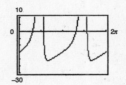

The x-intercepts occur at $x \approx 0.9828$, $x \approx 1.7682$,
$x \approx 4.1244$, and $x \approx 4.9098.$

57. $12 \sin^2 x - 13 \sin x + 3 = 0$

$$\sin x = \frac{-(-13) \pm \sqrt{(-13)^2 - 4(12)(3)}}{2(12)}$$

$$= \frac{13 \pm 5}{24}$$

$\sin x = \dfrac{1}{3} \qquad$ or $\quad \sin x = \dfrac{3}{4}$

$x \approx 0.3398, \ 2.8018 \qquad\qquad x \approx 0.8481, \ 2.2935$

Graph $y_1 = 12 \sin^2 x - 13 \sin x + 3.$

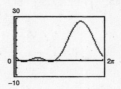

The x-intercepts occur at $x \approx 0.3398$, $x \approx 0.8481$,
$x \approx 2.2935,$ and $x \approx 2.8018.$

59. $\tan^2 x + 3\tan x + 1 = 0$

$$\tan x = \frac{-3 \pm \sqrt{3^2 - 4(1)(1)}}{2(1)} = \frac{-3 \pm \sqrt{5}}{2}$$

$\tan x = \dfrac{-3 - \sqrt{5}}{2}$ or $\tan x = \dfrac{-3 + \sqrt{5}}{2}$

$x \approx 1.9357, \ 5.0773$ $x \approx 2.7767, \ 5.9183$

Graph $y_1 = \tan^2 x + 3\tan x + 1$.

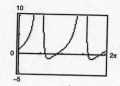

The x-intercepts occur at $x \approx 1.9357$, $x \approx 2.7767$, $x \approx 5.0773$, and $x \approx 5.9183$.

61. $\tan^2 x - 6\tan x + 5 = 0$

$(\tan x - 1)(\tan x - 5) = 0$

$\tan x - 1 = 0$ or $\tan x - 5 = 0$

$\tan x = 1$ $\tan x = 5$

$x = \dfrac{\pi}{4}, \dfrac{5\pi}{4}$ $x = \arctan 5, \ \arctan 5 + \pi$

62. $\sec^2 x + \tan x - 3 = 0$

$1 + \tan^2 x + \tan x - 3 = 0$

$\tan^2 x + \tan x - 2 = 0$

$(\tan x + 2)(\tan x - 1) = 0$

$\tan x + 2 = 0$ $\tan x - 1 = 0$

$\tan x = -2$ $\tan x = 1$

$x = \arctan(-2) + n\pi$ $x = \arctan(1) + n\pi$

$\approx -1.1071 + n\pi$ $= \dfrac{\pi}{4} + n\pi$

Solutions in $[0, 2\pi)$ are $\arctan(-2) + \pi, \ \arctan(-2) + 2\pi, \ \dfrac{\pi}{4}, \dfrac{5\pi}{4}$.

63. $2\cos^2 x - 5\cos x + 2 = 0$

$(2\cos x - 1)(\cos x - 2) = 0$

$2\cos x - 1 = 0$ or $\cos x - 2 = 0$

$\cos x = \dfrac{1}{2}$ $\cos x = 2$

$x = \dfrac{\pi}{3}, \dfrac{5\pi}{3}$ $x = \arccos 2,$

$2\pi - \arccos 2$

65. (a) $f(x) = \sin x + \cos x$

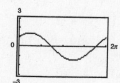

Maximum point: $(0.7854, 1.4142)$

Minimum point: $(3.9270, -1.4142)$

—CONTINUED—

65. —CONTINUED—

(b) $\cos x - \sin x = 0$

$$\cos x = \sin x$$

$$1 = \frac{\sin x}{\cos x}$$

$$\tan x = 1$$

$$x = \frac{\pi}{4}, \frac{5\pi}{4} \approx 0.7854, 3.9270$$

$$f\left(\frac{\pi}{4}\right) = \sin\frac{\pi}{4} + \cos\frac{\pi}{4} = \frac{\sqrt{2}}{2} + \frac{\sqrt{2}}{2} = \sqrt{2}$$

$$f\left(\frac{5\pi}{4}\right) = \sin\frac{5\pi}{4} + \cos\frac{5\pi}{4} = -\sin\frac{\pi}{4} + \left(-\cos\frac{\pi}{4}\right) = -\frac{\sqrt{2}}{2} - \frac{\sqrt{2}}{2} = -\sqrt{2}$$

Therefore, the maximum point in the interval $[0, 2\pi)$ is $\left(\pi/4, \sqrt{2}\right)$ and the minimum point is $\left(5\pi/4, -\sqrt{2}\right)$.

67. (a) $f(x) = 2\sin x + \cos 2x$

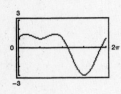

Maximum points: $(0.5240, 1.5), (2.6180, 1.5)$

Minimum points: $(1.5708, 1.0), (4.7124, -3.0)$

(b) $2\cos x - 4\sin x \cos x = 0$

$$2\cos x(1 - 2\sin x) = 0$$

$2\cos x = 0$ $\qquad\qquad 1 - 2\sin x = 0$

$x = \dfrac{\pi}{2}, \dfrac{3\pi}{2}$ $\qquad\qquad \sin x = \dfrac{1}{2}$

$\approx 1.5708, 4.7124$ $\qquad\qquad x = \dfrac{\pi}{6}, \dfrac{5\pi}{6}$

$f\left(\dfrac{\pi}{2}\right) = 1$ $\qquad\qquad\qquad \approx 0.5240, 2.6180$

$f\left(\dfrac{3\pi}{2}\right) = -3$ $\qquad\qquad f\left(\dfrac{\pi}{6}\right) = f\left(\dfrac{5\pi}{6}\right) = 1.5$

Maximum points: $\left(\dfrac{\pi}{6}, 1.5\right), \left(\dfrac{5\pi}{6}, 1.5\right)$

Minimum points: $\left(\dfrac{\pi}{2}, 1\right), \left(\dfrac{3\pi}{2}, -3\right)$

69.
$$y = \frac{1}{12}(\cos 8t - 3 \sin 8t)$$

$$\frac{1}{12}(\cos 8t - 3 \sin 8t) = 0$$

$$\cos 8t = 3 \sin 8t$$

$$\frac{1}{3} = \tan 8t$$

$$8t \approx 0.32175 + n\pi$$

$$t \approx 0.04 + \frac{n\pi}{8}$$

In the interval $0 \le t \le 1$, $t \approx 0.04$, 0.43, and 0.83.

71. $S = 74.50 + 43.75 \sin \dfrac{\pi t}{6}$

t	1	2	3	4	5	6	7	8	9	10	11	12
S	96.4	112.4	118.3	112.4	96.4	74.5	52.6	36.6	30.8	36.6	52.6	74.5

Sales exceed 100,000 units during February, March, and April.

73. Range = 1000 yards = 3000 feet

$$v_0 = 1200 \text{ feet per second}$$

$$r = \frac{1}{32} v_0{}^2 \sin 2\theta$$

$$3000 = \frac{1}{32}(1200)^2 \sin 2\theta$$

$$\sin 2\theta \approx 0.066667$$

$$2\theta \approx 3.8°$$

$$\theta \approx 1.9°$$

75. True. The period of $2 \sin 4t - 1$ is $\dfrac{\pi}{2}$ and the period of $2 \sin t - 1$ is 2π.

In the interval $[0, 2\pi)$ the first equation has four cycles whereas the second equation has only one cycle, thus the first equation has four times the x-intercepts (solutions) as the second equation.

77. True

$$\tan 4x = 0$$

In general, $4x = 0 + n\pi$ so

$$x = \frac{n\pi}{4}, \text{ where } n \text{ is an integer.}$$

79. $y_1 = 2 \sin x$

$y_2 = 3x + 1$

From the graph we see that there is only one point of intersection.

Section 10.4 Sum and Difference Formulas

1. (a) $\cos\left(\dfrac{\pi}{4} + \dfrac{\pi}{3}\right) = \cos\dfrac{\pi}{4}\cos\dfrac{\pi}{3} - \sin\dfrac{\pi}{4}\sin\dfrac{\pi}{3}$

$\qquad\qquad\qquad = \dfrac{\sqrt{2}}{2}\cdot\dfrac{1}{2} - \dfrac{\sqrt{2}}{2}\cdot\dfrac{\sqrt{3}}{2}$

$\qquad\qquad\qquad = \dfrac{\sqrt{2} - \sqrt{6}}{4}$

(b) $\cos\dfrac{\pi}{4} + \cos\dfrac{\pi}{3} = \dfrac{\sqrt{2}}{2} + \dfrac{1}{2} = \dfrac{\sqrt{2} + 1}{2}$

3. (a) $\sin\left(\dfrac{7\pi}{6} - \dfrac{\pi}{3}\right) = \sin\dfrac{5\pi}{6} = \sin\dfrac{\pi}{6} = \dfrac{1}{2}$

(b) $\sin\dfrac{7\pi}{6} - \sin\dfrac{\pi}{3} = -\dfrac{1}{2} - \dfrac{\sqrt{3}}{2} = \dfrac{-1 - \sqrt{3}}{2}$

5. (a) $\cos(120° + 45°) = \cos 120°\cos 45° - \sin 120°\sin 45°$

$\qquad\qquad\qquad = \left(-\dfrac{1}{2}\right)\left(\dfrac{\sqrt{2}}{2}\right) - \left(\dfrac{\sqrt{3}}{2}\right)\left(\dfrac{\sqrt{2}}{2}\right)$

$\qquad\qquad\qquad = \dfrac{-\sqrt{2} - \sqrt{6}}{4}$

(b) $\cos 120° + \cos 45° = -\dfrac{1}{2} + \dfrac{\sqrt{2}}{2} = \dfrac{-1 + \sqrt{2}}{2}$

7. $\sin 105° = \sin(60° + 45°)$

$\qquad = \sin 60°\cos 45° + \cos 60°\sin 45°$

$\qquad = \dfrac{\sqrt{3}}{2}\cdot\dfrac{\sqrt{2}}{2} + \dfrac{1}{2}\cdot\dfrac{\sqrt{2}}{2}$

$\qquad = \dfrac{\sqrt{2}}{4}\left(\sqrt{3} + 1\right)$

$\cos 105° = \cos(60° + 45°)$

$\qquad = \cos 60°\cos 45° - \sin 60°\sin 45°$

$\qquad = \dfrac{1}{2}\cdot\dfrac{\sqrt{2}}{2} - \dfrac{\sqrt{3}}{2}\cdot\dfrac{\sqrt{2}}{2}$

$\qquad = \dfrac{\sqrt{2}}{4}\left(1 - \sqrt{3}\right)$

$\tan 105° = \tan(60° + 45°)$

$\qquad = \dfrac{\tan 60° + \tan 45°}{1 - \tan 60°\tan 45°}$

$\qquad = \dfrac{\sqrt{3} + 1}{1 - \sqrt{3}} = \dfrac{\sqrt{3} + 1}{1 - \sqrt{3}}\cdot\dfrac{1 + \sqrt{3}}{1 + \sqrt{3}}$

$\qquad = \dfrac{4 + 2\sqrt{3}}{-2} = -2 - \sqrt{3}$

9. $\sin 195° = \sin(225° - 30°)$

$\qquad = \sin 225°\cos 30° - \cos 225°\sin 30°$

$\qquad = -\sin 45°\cos 30° + \cos 45°\sin 30°$

$\qquad = -\dfrac{\sqrt{2}}{2}\cdot\dfrac{\sqrt{3}}{2} + \dfrac{\sqrt{2}}{2}\cdot\dfrac{1}{2}$

$\qquad = \dfrac{\sqrt{2}}{4}\left(1 - \sqrt{3}\right)$

$\cos 195° = \cos(225° - 30°)$

$\qquad = \cos 225°\cos 30° + \sin 225°\sin 30°$

$\qquad = -\cos 45°\cos 30° - \sin 45°\sin 30°$

$\qquad = -\dfrac{\sqrt{2}}{2}\cdot\dfrac{\sqrt{3}}{2} - \dfrac{\sqrt{2}}{2}\cdot\dfrac{1}{2}$

$\qquad = -\dfrac{\sqrt{2}}{4}\left(\sqrt{3} + 1\right)$

$\tan 195° = \tan(225° - 30°)$

$\qquad = \dfrac{\tan 225° - \tan 30°}{1 + \tan 225°\tan 30°}$

$\qquad = \dfrac{\tan 45° - \tan 30°}{1 + \tan 45°\tan 30°}$

$\qquad = \dfrac{1 - \left(\dfrac{\sqrt{3}}{3}\right)}{1 + \left(\dfrac{\sqrt{3}}{3}\right)} = \dfrac{3 - \sqrt{3}}{3 + \sqrt{3}}\cdot\dfrac{3 - \sqrt{3}}{3 - \sqrt{3}}$

$\qquad = \dfrac{12 - 6\sqrt{3}}{6} = 2 - \sqrt{3}$

11. $\sin\dfrac{11\pi}{12} = \sin\left(\dfrac{3\pi}{4} + \dfrac{\pi}{6}\right)$

$\qquad = \sin\dfrac{3\pi}{4}\cos\dfrac{\pi}{6} + \cos\dfrac{3\pi}{4}\sin\dfrac{\pi}{6}$

$\qquad = \dfrac{\sqrt{2}}{2}\cdot\dfrac{\sqrt{3}}{2} + \left(-\dfrac{\sqrt{2}}{2}\right)\dfrac{1}{2}$

$\qquad = \dfrac{\sqrt{2}}{4}\left(\sqrt{3} - 1\right)$

$\cos\dfrac{11\pi}{12} = \cos\left(\dfrac{3\pi}{4} + \dfrac{\pi}{6}\right)$

$\qquad = \cos\dfrac{3\pi}{4}\cos\dfrac{\pi}{6} - \sin\dfrac{3\pi}{4}\sin\dfrac{\pi}{6}$

$\qquad = -\dfrac{\sqrt{2}}{2}\cdot\dfrac{\sqrt{3}}{2} - \dfrac{\sqrt{2}}{2}\cdot\dfrac{1}{2}$

$\qquad = -\dfrac{\sqrt{2}}{4}\left(\sqrt{3} + 1\right)$

$\tan\dfrac{11\pi}{12} = \tan\left(\dfrac{3\pi}{4} + \dfrac{\pi}{6}\right)$

$\qquad = \dfrac{\tan\dfrac{3\pi}{4} + \tan\dfrac{\pi}{6}}{1 - \tan\dfrac{3\pi}{4}\tan\dfrac{\pi}{6}}$

$\qquad = \dfrac{-1 + \dfrac{\sqrt{3}}{3}}{1 - (-1)\dfrac{\sqrt{3}}{3}}$

$\qquad = \dfrac{-3 + \sqrt{3}}{3 + \sqrt{3}}\cdot\dfrac{3 - \sqrt{3}}{3 - \sqrt{3}}$

$\qquad = \dfrac{-12 + 6\sqrt{3}}{6} = -2 + \sqrt{3}$

13. $\sin\dfrac{17\pi}{12} = \sin\left(\dfrac{9\pi}{4} - \dfrac{5\pi}{6}\right)$

$\qquad = \sin\dfrac{9\pi}{4}\cos\dfrac{5\pi}{6} - \cos\dfrac{9\pi}{4}\sin\dfrac{5\pi}{6}$

$\qquad = \dfrac{\sqrt{2}}{2}\left(-\dfrac{\sqrt{3}}{2}\right) - \left(\dfrac{\sqrt{2}}{2}\right)\left(\dfrac{1}{2}\right)$

$\qquad = -\dfrac{\sqrt{2}}{4}\left(\sqrt{3} + 1\right)$

$\cos\dfrac{17\pi}{12} = \cos\left(\dfrac{9\pi}{4} - \dfrac{5\pi}{6}\right)$

$\qquad = \cos\dfrac{9\pi}{4}\cos\dfrac{5\pi}{6} + \sin\dfrac{9\pi}{4}\sin\dfrac{5\pi}{6}$

$\qquad = \dfrac{\sqrt{2}}{2}\left(-\dfrac{\sqrt{3}}{2}\right) + \dfrac{\sqrt{2}}{2}\left(\dfrac{1}{2}\right)$

$\qquad = \dfrac{\sqrt{2}}{4}\left(1 - \sqrt{3}\right)$

$\tan\dfrac{17\pi}{12} = \tan\left(\dfrac{9\pi}{4} - \dfrac{5\pi}{6}\right)$

$\qquad = \dfrac{\tan(9\pi/4) - \tan(5\pi/6)}{1 + \tan(9\pi/4)\tan(5\pi/6)}$

$\qquad = \dfrac{1 - (-\sqrt{3}/3)}{1 + (-\sqrt{3}/3)}$

$\qquad = \dfrac{3 + \sqrt{3}}{3 - \sqrt{3}}\cdot\dfrac{3 + \sqrt{3}}{3 + \sqrt{3}}$

$\qquad = \dfrac{12 + 6\sqrt{3}}{6} = 2 + \sqrt{3}$

15. $\qquad 285° = 225° + 60°$

$\sin 285° = \sin(225° + 60°)$

$\qquad = \sin 225°\cos 60° + \cos 225°\sin 60°$

$\qquad = -\dfrac{\sqrt{2}}{2}\left(\dfrac{1}{2}\right) - \dfrac{\sqrt{2}}{2}\left(\dfrac{\sqrt{3}}{2}\right) = -\dfrac{\sqrt{2}}{4}\left(\sqrt{3} + 1\right)$

$\cos 285° = \cos(225° + 60°)$

$\qquad = \cos 225°\cos 60° - \sin 225°\sin 60°$

$\qquad = -\dfrac{\sqrt{2}}{2}\left(\dfrac{1}{2}\right) - \left(-\dfrac{\sqrt{2}}{2}\right)\left(\dfrac{\sqrt{3}}{2}\right) = \dfrac{\sqrt{2}}{4}\left(\sqrt{3} - 1\right)$

$\tan 285° = \tan(225° + 60°)$

$\qquad = \dfrac{\tan 225° + \tan 60°}{1 - \tan 225°\tan 60°} = \dfrac{1 + \sqrt{3}}{1 - \sqrt{3}}\cdot\dfrac{1 + \sqrt{3}}{1 + \sqrt{3}}$

$\qquad = \dfrac{4 + 2\sqrt{3}}{-2} = -2 - \sqrt{3} = -\left(2 + \sqrt{3}\right)$

17. $\quad -165° = -(120° + 45°)$

$$\sin(-165°) = \sin[-(120° + 45°)]$$
$$= -\sin(120° + 45°)$$
$$= -[\sin 120° \cos 45° + \cos 120° \sin 45°]$$
$$= -\left[\frac{\sqrt{3}}{2} \cdot \frac{\sqrt{2}}{2} - \frac{1}{2} \cdot \frac{\sqrt{2}}{2}\right]$$
$$= -\frac{\sqrt{2}}{4}(\sqrt{3} - 1)$$

$$\cos(-165°) = \cos[-(120° + 45°)]$$
$$= \cos(120° + 45°)$$
$$= \cos 120° \cos 45° - \sin 120° \sin 45°$$
$$= -\frac{1}{2} \cdot \frac{\sqrt{2}}{2} - \frac{\sqrt{3}}{2} \cdot \frac{\sqrt{2}}{2}$$
$$= -\frac{\sqrt{2}}{4}(1 + \sqrt{3})$$

$$\tan(-165°) = \tan[-(120° + 45°)]$$
$$= -\tan(120° + \tan 45°)$$
$$= -\frac{\tan 120° + \tan 45°}{1 - \tan 120° \tan 45°}$$
$$= -\frac{-\sqrt{3} + 1}{1 - (-\sqrt{3})(1)}$$
$$= -\frac{1 - \sqrt{3}}{1 + \sqrt{3}} \cdot \frac{1 - \sqrt{3}}{1 - \sqrt{3}}$$
$$= -\frac{4 - 2\sqrt{3}}{-2}$$
$$= 2 - \sqrt{3}$$

19. $\quad \dfrac{13\pi}{12} = \dfrac{3\pi}{4} + \dfrac{\pi}{3}$

$$\sin \frac{13\pi}{12} = \sin\left(\frac{3\pi}{4} + \frac{\pi}{3}\right)$$
$$= \sin \frac{3\pi}{4} \cos \frac{\pi}{3} + \cos \frac{3\pi}{4} \sin \frac{\pi}{3}$$
$$= \frac{\sqrt{2}}{2} \cdot \frac{1}{2} + \left(-\frac{\sqrt{2}}{2}\right)\left(\frac{\sqrt{3}}{2}\right)$$
$$= \frac{\sqrt{2}}{4}(1 - \sqrt{3})$$

$$\cos \frac{13\pi}{12} = \cos\left(\frac{3\pi}{4} + \frac{\pi}{3}\right)$$
$$= \cos \frac{3\pi}{4} \cos \frac{\pi}{3} - \sin \frac{3\pi}{4} \sin \frac{\pi}{3}$$
$$= -\frac{\sqrt{2}}{2} \cdot \frac{1}{2} - \frac{\sqrt{2}}{2} \cdot \frac{\sqrt{3}}{2} = -\frac{\sqrt{2}}{4}(1 + \sqrt{3})$$

$$\tan \frac{13\pi}{12} = \tan\left(\frac{3\pi}{4} + \frac{\pi}{3}\right)$$
$$= \frac{\tan\left(\frac{3\pi}{4}\right) + \tan\left(\frac{\pi}{3}\right)}{1 - \tan\left(\frac{3\pi}{4}\right)\tan\left(\frac{\pi}{3}\right)}$$
$$= \frac{-1 + \sqrt{3}}{1 - (-1)(\sqrt{3})}$$
$$= -\frac{1 - \sqrt{3}}{1 + \sqrt{3}} \cdot \frac{1 - \sqrt{3}}{1 - \sqrt{3}}$$
$$= -\frac{4 - 2\sqrt{3}}{-2}$$
$$= 2 - \sqrt{3}$$

21. $\quad -\dfrac{13\pi}{12} = -\left(\dfrac{3\pi}{4} + \dfrac{\pi}{3}\right)$

$$\sin\left[-\left(\frac{3\pi}{4} + \frac{\pi}{3}\right)\right] = -\sin\left(\frac{3\pi}{4} + \frac{\pi}{3}\right)$$
$$= -\left[\sin \frac{3\pi}{4} \cos \frac{\pi}{3} + \cos \frac{3\pi}{4} \sin \frac{\pi}{3}\right]$$
$$= -\left[\frac{\sqrt{2}}{2}\left(\frac{1}{2}\right) + \left(-\frac{\sqrt{2}}{2}\right)\left(\frac{\sqrt{3}}{2}\right)\right]$$
$$= -\frac{\sqrt{2}}{4}(1 - \sqrt{3}) = \frac{\sqrt{2}}{4}(\sqrt{3} - 1)$$

$$\cos\left[-\left(\frac{3\pi}{4} + \frac{\pi}{3}\right)\right] = \cos\left(\frac{3\pi}{4} + \frac{\pi}{3}\right)$$
$$= \cos \frac{3\pi}{4} \cos \frac{\pi}{3} - \sin \frac{3\pi}{4} \sin \frac{\pi}{3}$$
$$= -\frac{\sqrt{2}}{2}\left(\frac{1}{2}\right) - \frac{\sqrt{2}}{2}\left(\frac{\sqrt{3}}{2}\right)$$
$$= -\frac{\sqrt{2}}{4}(\sqrt{3} + 1)$$

$$\tan\left[-\left(\frac{3\pi}{4} + \frac{\pi}{3}\right)\right] = -\tan\left(\frac{3\pi}{4} + \frac{\pi}{3}\right)$$
$$= -\frac{\tan(3\pi/4) + \tan(\pi/3)}{1 - \tan(3\pi/4) \tan(\pi/3)}$$
$$= -\frac{-1 + \sqrt{3}}{1 - (-\sqrt{3})}$$
$$= \frac{1 - \sqrt{3}}{1 + \sqrt{3}} \cdot \frac{1 - \sqrt{3}}{1 - \sqrt{3}}$$
$$= \frac{4 - 2\sqrt{3}}{-2}$$
$$= -2 + \sqrt{3}$$

23. $\cos 25° \cos 15° - \sin 25° \sin 15° = \cos(25° + 15°) = \cos 40°$

25. $\dfrac{\tan 325° - \tan 86°}{1 + \tan 325° \tan 86°} = \tan(325° - 86°) = \tan 239°$

27. $\sin 3 \cos 1.2 - \cos 3 \sin 1.2 = \sin(3 - 1.2) = \sin 1.8$

29. $\dfrac{\tan 2x + \tan x}{1 - \tan 2x \tan x} = \tan(2x + x) = \tan 3x$

31. $\sin 330° \cos 30° - \cos 330° \sin 30° = \sin(330° - 30°)$
$$= \sin 300°$$
$$= -\frac{\sqrt{3}}{2}$$

33. $\sin \dfrac{\pi}{12} \cos \dfrac{\pi}{4} + \cos \dfrac{\pi}{12} \sin \dfrac{\pi}{4} = \sin\left(\dfrac{\pi}{12} + \dfrac{\pi}{4}\right)$
$$= \sin \frac{\pi}{3}$$
$$= \frac{\sqrt{3}}{2}$$

35. $\dfrac{\tan 25° + \tan 110°}{1 - \tan 25° \tan 110°} = \tan(25° + 110°)$
$$= \tan 135°$$
$$= -1$$

For Exercises 37–43, we have:

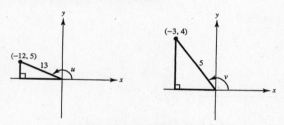

Figures for Exercises 37–43

$\sin u = \frac{5}{13}$, u in Quadrant II $\implies$ $\cos u = -\frac{12}{13}$, $\tan u = -\frac{5}{12}$
$\cos v = -\frac{3}{5}$, v in Quadrant II $\implies$ $\sin v = \frac{4}{5}$, $\tan v = -\frac{4}{3}$.

37. $\sin(u + v) = \sin u \cos v + \cos u \sin v$
$$= \left(\frac{5}{13}\right)\left(-\frac{3}{5}\right) + \left(-\frac{12}{13}\right)\left(\frac{4}{5}\right)$$
$$= -\frac{63}{65}$$

39. $\cos(u + v) = \cos u \cos v - \sin u \sin v$
$$= \left(-\frac{12}{13}\right)\left(-\frac{3}{5}\right) - \left(\frac{5}{13}\right)\left(\frac{4}{5}\right)$$
$$= \frac{16}{65}$$

41. $\tan(u + v) = \dfrac{\tan u + \tan v}{1 - \tan u \tan v} = \dfrac{-\frac{5}{12} + \left(-\frac{4}{3}\right)}{1 - \left(-\frac{5}{12}\right)\left(-\frac{4}{3}\right)} = \dfrac{-\frac{21}{12}}{1 - \frac{5}{9}}$
$$= \left(-\frac{7}{4}\right)\left(\frac{9}{4}\right) = -\frac{63}{16}$$

43. $\sec(v - u) = \dfrac{1}{\cos(v - u)} = \dfrac{1}{\cos v \cos u + \sin v \sin u}$
$$= \dfrac{1}{\left(-\frac{3}{5}\right)\left(-\frac{12}{13}\right) + \left(\frac{4}{5}\right)\left(\frac{5}{13}\right)} = \dfrac{1}{\left(\frac{36}{65}\right) + \left(\frac{20}{65}\right)} = \dfrac{1}{\frac{56}{65}}$$
$$= \frac{65}{56}$$

For Exercises 45–49, we have:

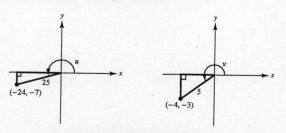

Figures for Exercises 45–49

$\sin u = -\frac{7}{25}$, u in Quadrant III $\implies$ $\cos u = -\frac{24}{25}$, $\tan u = \frac{7}{24}$
$\cos v = -\frac{4}{5}$, v in Quadrant III $\implies$ $\sin v = -\frac{3}{5}$, $\tan v = \frac{3}{4}$.

45. $\cos(u + v) = \cos u \cos v - \sin u \sin v$

$$= \left(-\frac{24}{25}\right)\left(-\frac{4}{5}\right) - \left(-\frac{7}{25}\right)\left(-\frac{3}{5}\right)$$

$$= \frac{3}{5}$$

47. $\tan(u - v) = \dfrac{\tan u - \tan v}{1 + \tan u \tan v}$

$$= \frac{\frac{7}{24} - \frac{3}{4}}{1 + \left(\frac{7}{24}\right)\left(\frac{3}{4}\right)} = \frac{-\frac{11}{24}}{\frac{39}{32}} = -\frac{44}{117}$$

49. $\sec(u + v) = \dfrac{1}{\cos(u + v)} = \dfrac{1}{\frac{3}{5}} = \dfrac{5}{3}$

Use Exercise 45 for $\cos(u + v)$.

51. $\sin(3\pi - x) = \sin 3\pi \cos x - \sin x \cos 3\pi$

$$= (0)(\cos x) - (-1)(\sin x)$$

$$= \sin x$$

53. $\sin\left(\dfrac{\pi}{6} + x\right) = \sin\dfrac{\pi}{6}\cos x + \cos\dfrac{\pi}{6}\sin x$

$$= \frac{1}{2}\left(\cos x + \sqrt{3}\sin x\right)$$

55. $\cos(\pi - \theta) + \sin\left(\dfrac{\pi}{2} + \theta\right) = \cos\pi\cos\theta + \sin\pi\sin\theta + \sin\dfrac{\pi}{2}\cos\theta + \cos\dfrac{\pi}{2}\sin\theta$

$$= (-1)(\cos\theta) + (0)(\sin\theta) + (1)(\cos\theta) + (\sin\theta)(0)$$

$$= -\cos\theta + \cos\theta$$

$$= 0$$

57. $\cos\left(\dfrac{3\pi}{2} - x\right) = \cos\dfrac{3\pi}{2}\cos x + \sin\dfrac{3\pi}{2}\sin x$

$$= (0)(\cos x) + (-1)(\sin x)$$

$$= -\sin x$$

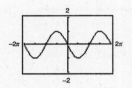

59. $\sin\left(\dfrac{3\pi}{2} + \theta\right) = \sin\dfrac{3\pi}{2}\cos\theta + \cos\dfrac{3\pi}{2}\sin\theta$

$$= (-1)(\cos\theta) + (0)(\sin\theta)$$

$$= -\cos\theta$$

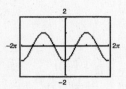

61. $\cos(x + y)\cos(x - y) = (\cos x \cos y - \sin x \sin y)(\cos x \cos y + \sin x \sin y)$

$$= \cos^2 x \cos^2 y - \sin^2 x \sin^2 y$$

$$= \cos^2 x(1 - \sin^2 y) - \sin^2 x \sin^2 y$$

$$= \cos^2 x - \cos^2 x \sin^2 y - \sin^2 x \sin^2 y$$

$$= \cos^2 x - \sin^2 y(\cos^2 x + \sin^2 x)$$

$$= \cos^2 x - \sin^2 y$$

63. $\sin(x + y) + \sin(x - y) = \sin x \cos y + \cos x \sin y + \sin x \cos y - \cos x \sin y$

$$= 2\sin x \cos y$$

65. $\sin(\arcsin x + \arccos x) = \sin(\arcsin x)\cos(\arccos x) + \sin(\arccos x)\cos(\arcsin x)$

$$= x \cdot x + \sqrt{1 - x^2} \cdot \sqrt{1 - x^2}$$

$$= x^2 + 1 - x^2$$

$$= 1$$

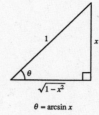
$\theta = \arcsin x$

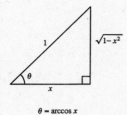
$\theta = \arccos x$

67. $\cos(\arccos x + \arcsin x) = \cos(\arccos x)\cos(\arcsin x) - \sin(\arccos x)\sin(\arcsin x)$

$$= x \cdot \sqrt{1 - x^2} - \sqrt{1 - x^2} \cdot x$$

$$= 0$$

(Use the triangles in Exercise 65.)

69.
$$\sin\left(x + \frac{\pi}{3}\right) + \sin\left(x - \frac{\pi}{3}\right) = 1$$

$$\sin x \cos \frac{\pi}{3} + \cos x \sin \frac{\pi}{3} + \sin x \cos \frac{\pi}{3} - \cos x \sin \frac{\pi}{3} = 1$$

$$2 \sin x(0.5) = 1$$

$$\sin x = 1$$

$$x = \frac{\pi}{2}$$

71.
$$\cos\left(x + \frac{\pi}{4}\right) - \cos\left(x - \frac{\pi}{4}\right) = 1$$

$$\cos x \cos \frac{\pi}{4} - \sin x \sin \frac{\pi}{4} - \left(\cos x \cos \frac{\pi}{4} + \sin x \sin \frac{\pi}{4}\right) = 1$$

$$-2 \sin x\left(\frac{\sqrt{2}}{2}\right) = 1$$

$$-\sqrt{2} \sin x = 1$$

$$\sin x = -\frac{1}{\sqrt{2}}$$

$$\sin x = -\frac{\sqrt{2}}{2}$$

$$x = \frac{5\pi}{4}, \frac{7\pi}{4}$$

73.
$$\tan(x + \pi) + 2\sin(x + \pi) = 0$$

$$\frac{\tan x + \tan \pi}{1 - \tan x \tan \pi} + 2(\sin x \cos \pi + \cos x \sin \pi) = 0$$

$$\frac{\tan x + 0}{1 - \tan x(0)} + 2[\sin x(-1) + \cos x(0)] = 0$$

$$\frac{\tan x}{1} - 2\sin x = 0$$

$$\frac{\sin x}{\cos x} = 2\sin x$$

$$\sin x = 2\sin x \cos x$$

$$\sin x(1 - 2\cos x) = 0$$

$$\sin x = 0 \quad \text{or} \quad \cos x = \frac{1}{2}$$

$$x = 0, \pi \qquad x = \frac{\pi}{3}, \frac{5\pi}{3}$$

75. $\sin(x + \Delta x) = \sin x \cos \Delta x + \cos x \sin \Delta x$

77. $\sin(\theta + \Delta \theta) = \sin \theta \cos \Delta \theta + \cos \theta \sin \Delta \theta$

79. False. $\sin(u \pm v) = \sin u \cos v \pm \cos u \sin v$.
In Exercises 2–3, parts (a) and (b) are unequal.

81. False. $\cos\left(x - \dfrac{\pi}{2}\right) = \cos x \cos \dfrac{\pi}{2} + \sin x \sin \dfrac{\pi}{2}$

$$= (\cos x)(0) + (\sin x)(1)$$

$$= \sin x$$

83. $C = \arctan \dfrac{b}{a} \implies \sin C = \dfrac{b}{\sqrt{a^2 + b^2}}, \cos C = \dfrac{a}{\sqrt{a^2 + b^2}}$

$$\sqrt{a^2 + b^2} \sin(B\theta + C) = \sqrt{a^2 + b^2}\left(\sin B\theta \cdot \frac{a}{\sqrt{a^2 + b^2}} + \frac{b}{\sqrt{a^2 + b^2}} \cdot \cos B\theta\right) = a \sin B\theta + b \cos B\theta$$

85. $\sin \theta + \cos \theta$

$a = 1, b = 1, B = 1$

(a) $C = \arctan \dfrac{b}{a} = \arctan 1 = \dfrac{\pi}{4}$

$\sin \theta + \cos \theta = \sqrt{a^2 + b^2} \sin(B\theta + C)$

$$= \sqrt{2} \sin\left(\theta + \frac{\pi}{4}\right)$$

(b) $C = \arctan \dfrac{a}{b} = \arctan 1 = \dfrac{\pi}{4}$

$\sin \theta + \cos \theta = \sqrt{a^2 + b^2} \cos(B\theta - C)$

$$= \sqrt{2} \cos\left(\theta - \frac{\pi}{4}\right)$$

87. $12 \sin 3\theta + 5 \cos 3\theta$

$a = 12, b = 5, B = 3$

(a) $C = \arctan \dfrac{b}{a} = \arctan \dfrac{5}{12} \approx 0.3948$

$12 \sin 3\theta + 5 \cos 3\theta = \sqrt{a^2 + b^2} \sin(B\theta + C)$

$$\approx 13 \sin(3\theta + 0.3948)$$

(b) $C = \arctan \dfrac{a}{b} = \arctan \dfrac{12}{5} \approx 1.1760$

$12 \sin 3\theta + 5 \cos 3\theta = \sqrt{a^2 + b^2} \cos(B\theta - C)$

$$\approx 13 \cos(3\theta - 1.1760)$$

89. $C = \arctan \dfrac{b}{a} = \dfrac{\pi}{2} \implies a = 0$

$\sqrt{a^2 + b^2} = 2 \implies b = 2$

$B = 1$

$2 \sin\left(\theta + \dfrac{\pi}{2}\right) = (0)(\sin\theta) + (2)(\cos\theta) = 2\cos\theta$

91. $\dfrac{\cos(x + \Delta x) - \cos x}{\Delta x} = \dfrac{\cos x \cos \Delta x - \sin x \sin \Delta x - \cos x}{\Delta x}$

$= \dfrac{-\sin x \sin \Delta x - \cos x + \cos x \cos \Delta x}{\Delta x}$

$= \dfrac{-\sin x \sin \Delta x}{\Delta x} - \dfrac{\cos x(1 - \cos \Delta x)}{\Delta x}$

$= -\sin x \left(\dfrac{\sin \Delta x}{\Delta x}\right) - \cos x\left(\dfrac{1 - \cos \Delta x}{\Delta x}\right)$

Section 10.5   Multiple-Angle and Product-to-Sum Formulas

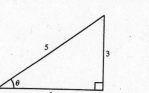

Figure for Exercises 1–7

1. $\sin \theta = \dfrac{3}{5}$

3. $\cos 2\theta = 2\cos^2 \theta - 1$

$= 2\left(\dfrac{4}{5}\right)^2 - 1$

$= \dfrac{32}{25} - 1$

$= \dfrac{7}{25}$

5. $\tan 2\theta = \dfrac{2\tan\theta}{1 - \tan^2\theta}$

$= \dfrac{2(3/4)}{1 - (3/4)^2}$

$= \dfrac{3/2}{1 - 9/16}$

$= \dfrac{3}{2} \cdot \dfrac{16}{7}$

$= \dfrac{24}{7}$

7. $\csc 2\theta = \dfrac{1}{\sin 2\theta}$

$= \dfrac{1}{24/25}$ From Exercise #4

$= \dfrac{25}{24}$

9.
$$\sin 2x - \sin x = 0$$
$$2 \sin x \cos x - \sin x = 0$$
$$\sin x(2 \cos x - 1) = 0$$
$$\sin x = 0 \quad \text{or} \quad 2 \cos x - 1 = 0$$
$$x = 0, \ \pi \qquad \cos x = \frac{1}{2}$$
$$x = \frac{\pi}{3}, \frac{5\pi}{3}$$
$$x = 0, \ \frac{\pi}{3}, \ \pi, \ \frac{5\pi}{3}$$

11. $4 \sin x \cos x = 1$
$$2 \sin 2x = 1$$
$$\sin 2x = \frac{1}{2}$$
$$2x = \frac{\pi}{6} + 2n\pi \quad \text{or} \quad 2x = \frac{5\pi}{6} + 2n\pi$$
$$x = \frac{\pi}{12} + n\pi \qquad x = \frac{5\pi}{12} + n\pi$$
$$x = \frac{\pi}{12}, \frac{13\pi}{12} \qquad x = \frac{5\pi}{12}, \frac{17\pi}{12}$$

13.
$$\cos 2x = \cos x$$
$$\cos^2 x - \sin^2 x = \cos x$$
$$\cos^2 x - (1 - \cos^2 x) - \cos x = 0$$
$$2 \cos^2 x - \cos x - 1 = 0$$
$$(2 \cos x + 1)(\cos x - 1) = 0$$
$$2 \cos x + 1 = 0 \quad \text{or} \quad \cos x - 1 = 0$$
$$\cos x = -\frac{1}{2} \qquad \cos x = 1$$
$$x = \frac{2\pi}{3}, \frac{4\pi}{3} \qquad x = 0$$

15.
$$\tan 2x - \cot x = 0$$
$$\frac{2 \tan x}{1 - \tan^2 x} = \cot x$$
$$2 \tan x = \cot x(1 - \tan^2 x)$$
$$2 \tan x = \cot x - \cot x \tan^2 x$$
$$2 \tan x = \cot x - \tan x$$
$$3 \tan x = \cot x$$
$$3 \tan x - \cot x = 0$$
$$3 \tan x - \frac{1}{\tan x} = 0$$
$$\frac{3 \tan^2 x - 1}{\tan x} = 0$$
$$\frac{1}{\tan x}(3 \tan^2 x - 1) = 0$$
$$\cot x(3 \tan^2 x - 1) = 0$$
$$\cot x = 0 \quad \text{or} \quad 3 \tan^2 x - 1 = 0$$
$$x = \frac{\pi}{2}, \frac{3\pi}{2} \qquad \tan^2 x = \frac{1}{3}$$
$$\tan x = \pm \frac{\sqrt{3}}{3}$$
$$x = \frac{\pi}{6}, \frac{5\pi}{6}, \frac{7\pi}{6}, \frac{11\pi}{6}$$
$$x = \frac{\pi}{6}, \frac{\pi}{2}, \frac{5\pi}{6}, \frac{7\pi}{6}, \frac{3\pi}{2}, \frac{11\pi}{6}$$

17.
$$\sin 4x = -2 \sin 2x$$
$$\sin 4x + 2 \sin 2x = 0$$
$$2 \sin 2x \cos 2x + 2 \sin 2x = 0$$
$$2 \sin 2x(\cos 2x + 1) = 0$$
$$2 \sin 2x = 0 \quad \text{or} \quad \cos 2x + 1 = 0$$
$$\sin 2x = 0 \qquad \cos 2x = -1$$
$$2x = n\pi \qquad 2x = \pi + 2n\pi$$
$$x = \frac{n}{2}\pi \qquad x = \frac{\pi}{2} + n\pi$$
$$x = 0, \ \frac{\pi}{2}, \ \pi, \ \frac{3\pi}{2} \qquad x = \frac{\pi}{2}, \frac{3\pi}{2}$$

19. $6 \sin x \cos x = 3(2 \sin x \cos x)$

$\qquad\qquad\quad = 3 \sin 2x$

21. $4 - 8 \sin^2 x = 4(1 - 2 \sin^2 x)$

$\qquad\qquad\quad = 4 \cos 2x$

23. $\sin u = -\dfrac{4}{5}, \; \pi < u < \dfrac{3\pi}{2} \implies \cos u = -\dfrac{3}{5}$

$\sin 2u = 2 \sin u \cos u = 2\left(-\dfrac{4}{5}\right)\left(-\dfrac{3}{5}\right) = \dfrac{24}{25}$

$\cos 2u = \cos^2 u - \sin^2 u = \dfrac{9}{25} - \dfrac{16}{25} = -\dfrac{7}{25}$

$\tan 2u = \dfrac{2 \tan u}{1 - \tan^2 u} = \dfrac{2\left(\frac{4}{3}\right)}{1 - \frac{16}{9}} = \dfrac{8}{3}\left(-\dfrac{9}{7}\right) = -\dfrac{24}{7}$

25. $\tan u = \dfrac{3}{4}, \; 0 < u < \dfrac{\pi}{2} \implies \sin u = \dfrac{3}{5}$ and $\cos u = \dfrac{4}{5}$

$\sin 2u = 2 \sin u \cos u = 2\left(\dfrac{3}{5}\right)\left(\dfrac{4}{5}\right) = \dfrac{24}{25}$

$\cos 2u = \cos^2 u - \sin^2 u = \dfrac{16}{25} - \dfrac{9}{25} = \dfrac{7}{25}$

$\tan 2u = \dfrac{2 \tan u}{1 - \tan^2 u} = \dfrac{2\left(\frac{3}{4}\right)}{1 - \frac{9}{16}} = \dfrac{3}{2}\left(\dfrac{16}{7}\right) = \dfrac{24}{7}$

27. $\sec u = -\dfrac{5}{2}, \; \dfrac{\pi}{2} < u < \pi \implies \sin u = \dfrac{\sqrt{21}}{5}$ and $\cos u = -\dfrac{2}{5}$

$\sin 2u = 2 \sin u \cos u = 2\left(\dfrac{\sqrt{21}}{5}\right)\left(-\dfrac{2}{5}\right) = -\dfrac{4\sqrt{21}}{25}$

$\cos 2u = \cos^2 u - \sin^2 u = \left(-\dfrac{2}{5}\right)^2 - \left(\dfrac{\sqrt{21}}{5}\right)^2 = -\dfrac{17}{25}$

$\tan 2u = \dfrac{2 \tan u}{1 - \tan^2 u} = \dfrac{2\left(-\dfrac{\sqrt{21}}{2}\right)}{1 - \left(-\dfrac{\sqrt{21}}{2}\right)^2}$

$\qquad = \dfrac{-\sqrt{21}}{1 - \frac{21}{4}} = \dfrac{4\sqrt{21}}{17}$

29. $\cos^4 x = (\cos^2 x)(\cos^2 x) = \left(\dfrac{1 + \cos 2x}{2}\right)\left(\dfrac{1 + \cos 2x}{2}\right) = \dfrac{1 + 2\cos 2x + \cos^2 2x}{4}$

$\qquad\qquad = \dfrac{1 + 2\cos 2x + \dfrac{1 + \cos 4x}{2}}{4}$

$\qquad\qquad = \dfrac{2 + 4\cos 2x + 1 + \cos 4x}{8}$

$\qquad\qquad = \dfrac{3 + 4\cos 2x + \cos 4x}{8}$

$\qquad\qquad = \dfrac{1}{8}(3 + 4\cos 2x + \cos 4x)$

31. $(\sin^2 x)(\cos^2 x) = \left(\dfrac{1 - \cos 2x}{2}\right)\left(\dfrac{1 + \cos 2x}{2}\right)$

$\qquad\qquad = \dfrac{1 - \cos^2 2x}{4}$

$\qquad\qquad = \dfrac{1}{4}\left(1 - \dfrac{1 + \cos 4x}{2}\right)$

$\qquad\qquad = \dfrac{1}{8}(2 - 1 - \cos 4x)$

$\qquad\qquad = \dfrac{1}{8}(1 - \cos 4x)$

33. $\sin^2 x \cos^4 x = \sin^2 x \cos^2 x \cos^2 x = \left(\dfrac{1 - \cos 2x}{2}\right)\left(\dfrac{1 + \cos 2x}{2}\right)\left(\dfrac{1 + \cos 2x}{2}\right)$

$$= \frac{1}{8}(1 - \cos 2x)(1 + \cos 2x)(1 + \cos 2x)$$

$$= \frac{1}{8}(1 - \cos^2 2x)(1 + \cos 2x)$$

$$= \frac{1}{8}(1 + \cos 2x - \cos^2 2x - \cos^3 2x)$$

$$= \frac{1}{8}\left[1 + \cos 2x - \left(\frac{1 + \cos 4x}{2}\right) - \cos 2x \left(\frac{1 + \cos 4x}{2}\right)\right]$$

$$= \frac{1}{16}[2 + 2\cos 2x - 1 - \cos 4x - \cos 2x - \cos 2x \cos 4x]$$

$$= \frac{1}{16}\left[1 + \cos 2x - \cos 4x - \left(\frac{1}{2}\cos 2x + \frac{1}{2}\cos 6x\right)\right]$$

$$= \frac{1}{32}(2 + 2\cos 2x - 2\cos 4x - \cos 2x - \cos 6x)$$

$$= \frac{1}{32}(2 + \cos 2x - 2\cos 4x - \cos 6x)$$

$\sin \theta = \frac{8}{17}$

$\cos \theta = \frac{15}{17}$

Figure for Exercises 35–39

35. $\cos \dfrac{\theta}{2} = \sqrt{\dfrac{1 + \cos \theta}{2}} = \sqrt{\dfrac{1 + \frac{15}{17}}{2}} = \sqrt{\dfrac{32}{34}} = \sqrt{\dfrac{16}{17}} = \dfrac{4\sqrt{17}}{17}$

37. $\tan \dfrac{\theta}{2} = \dfrac{\sin \theta}{1 + \cos \theta} = \dfrac{\frac{8}{17}}{1 + \frac{15}{17}} = \dfrac{8}{17} \cdot \dfrac{17}{32} = \dfrac{1}{4}$

39. $\csc \dfrac{\theta}{2} = \dfrac{1}{\sin \frac{\theta}{2}} = \dfrac{1}{\sqrt{\dfrac{(1 - \cos \theta)}{2}}} = \dfrac{1}{\sqrt{\dfrac{1 - \frac{15}{17}}{2}}}$

$$= \dfrac{1}{\sqrt{\dfrac{1}{17}}} = \sqrt{17}$$

41. $\sin 75° = \sin\left(\dfrac{1}{2} \cdot 150°\right) = \sqrt{\dfrac{1 - \cos 150°}{2}} = \sqrt{\dfrac{1 + \frac{\sqrt{3}}{2}}{2}} = \dfrac{1}{2}\sqrt{2 + \sqrt{3}}$

$\cos 75° = \cos\left(\dfrac{1}{2} \cdot 150°\right) = \sqrt{\dfrac{1 + \cos 150°}{2}} = \sqrt{\dfrac{1 - \frac{\sqrt{3}}{2}}{2}} = \dfrac{1}{2}\sqrt{2 - \sqrt{3}}$

$\tan 75° = \tan\left(\dfrac{1}{2} \cdot 150°\right) = \dfrac{\sin 150°}{1 + \cos 150°} = \dfrac{\frac{1}{2}}{1 - \frac{\sqrt{3}}{2}} = \dfrac{1}{2 - \sqrt{3}} \cdot \dfrac{2 + \sqrt{3}}{2 + \sqrt{3}} = \dfrac{2 + \sqrt{3}}{4 - 3} = 2 + \sqrt{3}$

43. $\sin 112° 30' = \sin\left(\frac{1}{2} \cdot 225°\right) = \sqrt{\frac{1 - \cos 225°}{2}} = \sqrt{\frac{1 + \frac{\sqrt{2}}{2}}{2}} = \frac{1}{2}\sqrt{2 + \sqrt{2}}$

$\cos 112° 30' = \cos\left(\frac{1}{2} \cdot 225°\right) = -\sqrt{\frac{1 + \cos 225°}{2}} = -\sqrt{\frac{1 - \frac{\sqrt{2}}{2}}{2}} = -\frac{1}{2}\sqrt{2 - \sqrt{2}}$

$\tan 112° 30' = \tan\left(\frac{1}{2} \cdot 225°\right) = \frac{\sin 225°}{1 + \cos 225°} = \frac{-\frac{\sqrt{2}}{2}}{1 - \frac{\sqrt{2}}{2}} = -1 - \sqrt{2}$

45. $\sin\frac{\pi}{8} = \sin\left[\frac{1}{2}\left(\frac{\pi}{4}\right)\right] = \sqrt{\frac{1 - \cos\frac{\pi}{4}}{2}} = \frac{1}{2}\sqrt{2 - \sqrt{2}}$

$\cos\frac{\pi}{8} = \cos\left[\frac{1}{2}\left(\frac{\pi}{4}\right)\right] = \sqrt{\frac{1 + \cos\frac{\pi}{4}}{2}} = \frac{1}{2}\sqrt{2 + \sqrt{2}}$

$\tan\frac{\pi}{8} = \tan\left[\frac{1}{2}\left(\frac{\pi}{4}\right)\right] = \frac{\sin\frac{\pi}{4}}{1 + \cos\frac{\pi}{4}} = \frac{\frac{\sqrt{2}}{2}}{1 + \frac{\sqrt{2}}{2}} = \sqrt{2} - 1$

47. $\sin\frac{3\pi}{8} = \sin\left(\frac{1}{2} \cdot \frac{3\pi}{4}\right) = \sqrt{\frac{1 - \cos\frac{3\pi}{4}}{2}} = \sqrt{\frac{1 + \frac{\sqrt{2}}{2}}{2}} = \frac{1}{2}\sqrt{2 + \sqrt{2}}$

$\cos\frac{3\pi}{8} = \cos\left(\frac{1}{2} \cdot \frac{3\pi}{4}\right) = \sqrt{\frac{1 + \cos\frac{3\pi}{4}}{2}} = \sqrt{\frac{1 - \frac{\sqrt{2}}{2}}{2}} = \frac{1}{2}\sqrt{2 - \sqrt{2}}$

$\tan\frac{3\pi}{8} = \tan\left(\frac{1}{2} \cdot \frac{3\pi}{4}\right) = \frac{\sin\frac{3\pi}{4}}{1 + \cos\frac{3\pi}{4}} = \frac{\frac{\sqrt{2}}{2}}{1 - \frac{\sqrt{2}}{2}} = \frac{\frac{\sqrt{2}}{2}}{\frac{(2 - \sqrt{2})}{2}} = \frac{\sqrt{2}}{2 - \sqrt{2}} \cdot \frac{2 + \sqrt{2}}{2 + \sqrt{2}} = \sqrt{2} + 1$

49. $\sin u = \frac{5}{13}, \frac{\pi}{2} < u < \pi \implies \cos u = -\frac{12}{13}$

$\sin\left(\frac{u}{2}\right) = \sqrt{\frac{1 - \cos u}{2}} = \sqrt{\frac{1 + \frac{12}{13}}{2}} = \frac{5\sqrt{26}}{26}$

$\cos\left(\frac{u}{2}\right) = \sqrt{\frac{1 + \cos u}{2}} = \sqrt{\frac{1 - \frac{12}{13}}{2}} = \frac{\sqrt{26}}{26}$

$\tan\left(\frac{u}{2}\right) = \frac{\sin u}{1 + \cos u} = \frac{\frac{5}{13}}{1 - \frac{12}{13}} = 5$

51. $\tan u = -\dfrac{5}{8}, \ \dfrac{3\pi}{2} < u < 2\pi \ \Rightarrow \ \sin u = -\dfrac{5}{\sqrt{89}}$ and $\cos u = \dfrac{8}{\sqrt{89}}$

$$\sin\left(\frac{u}{2}\right) = \sqrt{\frac{1 - \cos u}{2}} = \sqrt{\frac{1 - \dfrac{8}{\sqrt{89}}}{2}}\sqrt{\frac{\sqrt{89} - 8}{2\sqrt{89}}} = \sqrt{\frac{89 - 8\sqrt{89}}{178}}$$

$$\cos\left(\frac{u}{2}\right) = -\sqrt{\frac{1 + \cos u}{2}} = -\sqrt{\frac{1 + \dfrac{8}{\sqrt{89}}}{2}} = -\sqrt{\frac{\sqrt{89} + 8}{2\sqrt{89}}} = -\sqrt{\frac{89 + 8\sqrt{89}}{178}}$$

$$\tan\left(\frac{u}{2}\right) = \frac{1 - \cos u}{\sin u} = \frac{1 - \dfrac{8}{\sqrt{89}}}{-\dfrac{5}{\sqrt{89}}} = \frac{8 - \sqrt{89}}{5}$$

53. $\csc u = -\dfrac{5}{3}, \ \pi < u < \dfrac{3\pi}{2} \ \Rightarrow \ \sin u = -\dfrac{3}{5}$ and $\cos u = -\dfrac{4}{5}$

$$\sin\left(\frac{u}{2}\right) = \sqrt{\frac{1 - \cos u}{2}} = \sqrt{\frac{1 + \frac{4}{5}}{2}} = \frac{3\sqrt{10}}{10}$$

$$\cos\left(\frac{u}{2}\right) = -\sqrt{\frac{1 + \cos u}{2}} = -\sqrt{\frac{1 - \frac{4}{5}}{2}} = -\frac{\sqrt{10}}{10}$$

$$\tan\left(\frac{u}{2}\right) = \frac{1 - \cos u}{\sin u} = \frac{1 + \frac{4}{5}}{-\frac{3}{5}} = -3$$

55. $\sqrt{\dfrac{1 - \cos 6x}{2}} = |\sin 3x|$

57. $-\sqrt{\dfrac{1 - \cos 8x}{1 + \cos 8x}} = -\dfrac{\sqrt{\dfrac{1 - \cos 8x}{2}}}{\sqrt{\dfrac{1 + \cos 8x}{2}}}$

$$= -\left|\frac{\sin 4x}{\cos 4x}\right|$$

$$= -|\tan 4x|$$

59. $\sin\dfrac{x}{2} + \cos x = 0$

$$\pm\sqrt{\frac{1 - \cos x}{2}} = -\cos x$$

$$\frac{1 - \cos x}{2} = \cos^2 x$$

$$0 = 2\cos^2 x + \cos x - 1$$

$$= (2\cos x - 1)(\cos x + 1)$$

$$\cos x = \frac{1}{2} \quad \text{or} \quad \cos x = -1$$

$$x = \frac{\pi}{3}, \frac{5\pi}{3} \qquad x = \pi$$

By checking these values in the original equation, we see that $x = \pi/3$ and $x = 5\pi/3$ are extraneous, and $x = \pi$ is the only solution.

61.
$$\cos \frac{x}{2} - \sin x = 0$$

$$\pm \sqrt{\frac{1 + \cos x}{2}} = \sin x$$

$$\frac{1 + \cos x}{2} = \sin^2 x$$

$$1 + \cos x = 2 \sin^2 x$$

$$1 + \cos x = 2 - 2\cos^2 x$$

$$2 \cos^2 x + \cos x - 1 = 0$$

$$(2 \cos x - 1)(\cos x + 1) = 0$$

$$2 \cos x - 1 = 0 \quad \text{or} \ \cos x + 1 = 0$$

$$\cos x = \frac{1}{2} \qquad\qquad \cos x = -1$$

$$x = \frac{\pi}{3}, \frac{5\pi}{3} \qquad\qquad x = \pi$$

$$x = \frac{\pi}{3}, \ \pi, \ \frac{5\pi}{3}$$

$\pi/3$, π, and $5\pi/3$
are all solutions
to the equation.

63. $6 \sin \dfrac{\pi}{4} \cos \dfrac{\pi}{4} = 6 \cdot \dfrac{1}{2}\left[\sin\left(\dfrac{\pi}{4} + \dfrac{\pi}{4}\right) + \sin\left(\dfrac{\pi}{4} - \dfrac{\pi}{4}\right) \right]$

$$= 3\left(\sin \frac{\pi}{2} + \sin 0 \right)$$

65. $\cos 4\theta \sin 6\theta = \frac{1}{2}[\sin(4\theta + 6\theta) - \sin(4\theta - 6\theta)]$

$$= \tfrac{1}{2}[\sin 10\theta - \sin(-2\theta)]$$

$$= \tfrac{1}{2}(\sin 10\theta + \sin 2\theta)$$

67. $5 \cos(-5\beta) \cos 3\beta = 5 \cdot \tfrac{1}{2}\left[\cos(-5\beta - 3\beta) + \cos(-5\beta + 3\beta) \right] = \tfrac{5}{2}\left[\cos(-8\beta) + \cos(-2\beta) \right]$

$$= \tfrac{5}{2}(\cos 8\beta + \cos 2\beta)$$

69. $\sin(x + y) \sin (x - y) = \frac{1}{2}(\cos 2y - \cos 2x)$

71. $\cos(\theta - \pi) \sin(\theta + \pi) = \frac{1}{2}(\sin 2\theta + \sin 2\pi)$

73. $10 \cos 75° \cos 15° = 10\left(\tfrac{1}{2}\right)[\cos(75° - 15°) + \cos(75° + 15°)] = 5[\cos 60° + \cos 90°]$

75. $\sin 60° + \sin 30° = 2 \sin\left(\dfrac{60° + 30°}{2}\right) \cos\left(\dfrac{60° - 30°}{2}\right)$

$$= 2 \sin 45° \cos 15°$$

77. $\cos \dfrac{3\pi}{4} - \cos \dfrac{\pi}{4} = -2 \sin\left(\dfrac{\frac{3\pi}{4} + \frac{\pi}{4}}{2}\right) \sin\left(\dfrac{\frac{3\pi}{4} - \frac{\pi}{4}}{2}\right)$

$$= -2 \sin \frac{\pi}{2} \sin \frac{\pi}{4}$$

79. $\sin 5\theta - \sin 3\theta = 2 \cos\left(\dfrac{5\theta + 3\theta}{2}\right) \sin\left(\dfrac{5\theta - 3\theta}{2}\right)$

$$= 2 \cos 4\theta \sin \theta$$

81. $\cos 6x + \cos 2x = 2 \cos\left(\dfrac{6x + 2x}{2}\right) \cos\left(\dfrac{6x - 2x}{2}\right)$

$$= 2 \cos 4x \cos 2x$$

83. $\sin(\alpha + \beta) - \sin(\alpha - \beta) = 2 \cos\left(\dfrac{\alpha + \beta + \alpha - \beta}{2}\right) \sin\left(\dfrac{\alpha + \beta - \alpha + \beta}{2}\right) = 2 \cos \alpha \sin \beta$

85. $\cos\left(\theta + \dfrac{\pi}{2}\right) - \cos\left(\theta - \dfrac{\pi}{2}\right) = -2\sin\left[\dfrac{\left(\theta + \frac{\pi}{2}\right) + \left(\theta - \frac{\pi}{2}\right)}{2}\right]\sin\left[\dfrac{\left(\theta + \frac{\pi}{2}\right) - \left(\theta - \frac{\pi}{2}\right)}{2}\right]$

$$= -2\sin\theta\sin\dfrac{\pi}{2}$$

87.
$$\sin 6x + \sin 2x = 0$$
$$2\sin\left(\dfrac{6x + 2x}{2}\right)\cos\left(\dfrac{6x - 2x}{2}\right) = 0$$
$$2(\sin 4x)\cos 2x = 0$$
$$\sin 4x = 0 \quad \text{or} \quad \cos 2x = 0$$
$$4x = n\pi \qquad 2x = \dfrac{\pi}{2} + n\pi$$
$$x = \dfrac{n\pi}{4} \qquad x = \dfrac{\pi}{4} + \dfrac{n\pi}{2}$$

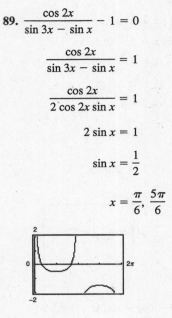

In the interval $[0, 2\pi)$ we have

$$x = 0,\ \dfrac{\pi}{4},\ \dfrac{\pi}{2},\ \dfrac{3\pi}{4},\ \pi,\ \dfrac{5\pi}{4},\ \dfrac{3\pi}{2},\ \dfrac{7\pi}{4}.$$

89. $\dfrac{\cos 2x}{\sin 3x - \sin x} - 1 = 0$

$$\dfrac{\cos 2x}{\sin 3x - \sin x} = 1$$

$$\dfrac{\cos 2x}{2\cos 2x \sin x} = 1$$

$$2\sin x = 1$$

$$\sin x = \dfrac{1}{2}$$

$$x = \dfrac{\pi}{6},\ \dfrac{5\pi}{6}$$

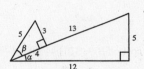

Figure for Exercises 91 and 93

91. $\sin^2\alpha = \left(\dfrac{5}{13}\right)^2 = \dfrac{25}{169}$

$$\sin^2\alpha = 1 - \cos^2\alpha = 1 - \left(\dfrac{12}{13}\right)^2$$

$$= 1 - \dfrac{144}{169} = \dfrac{25}{169}$$

93. $\sin\alpha\cos\beta = \left(\dfrac{5}{13}\right)\left(\dfrac{4}{5}\right) = \dfrac{4}{13}$

$$\sin\alpha\cos\beta = \cos\left(\dfrac{\pi}{2} - \alpha\right)\sin\left(\dfrac{\pi}{2} - \beta\right)$$

$$= \left(\dfrac{5}{13}\right)\left(\dfrac{4}{5}\right) = \dfrac{4}{13}$$

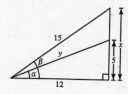

Figure for Exercises 95-101

$$x = \sqrt{15^2 - 12^2} = 9$$

$$y = \sqrt{12^2 + 5^2} = 13$$

$$\sin \alpha = \tfrac{5}{13} \qquad \sin \beta = \tfrac{9}{15} = \tfrac{3}{5}$$

$$\cos \alpha = \tfrac{12}{13} \qquad \cos \beta = \tfrac{12}{15} = \tfrac{4}{5}$$

$$\tan \alpha = \tfrac{5}{12} \qquad \tan \beta = \tfrac{9}{12} = \tfrac{3}{4}$$

95. $\sin \beta = \tfrac{9}{15} = \tfrac{3}{5}$

97. $\cos 2\beta = 2\cos^2 \beta - 1$
$$= 2\left(\tfrac{4}{5}\right)^2 - 1$$
$$= \tfrac{32}{25} - 1$$
$$= \tfrac{7}{25}$$

99. $\sin \dfrac{\beta}{2} = \sqrt{\dfrac{1 - \cos \beta}{2}}$
$$= \sqrt{\dfrac{1 - \tfrac{4}{5}}{2}}$$
$$= \sqrt{\dfrac{1}{10}}$$
$$= \dfrac{\sqrt{10}}{10}$$

101. $\tan 2\alpha = \dfrac{2 \tan \alpha}{1 - \tan^2 \alpha}$
$$= \dfrac{2\left(\tfrac{5}{12}\right)}{1 - \left(\tfrac{5}{12}\right)^2}$$
$$= \dfrac{\tfrac{5}{6}}{1 - \tfrac{25}{144}}$$
$$= \left(\tfrac{5}{6}\right)\left(\tfrac{144}{119}\right)$$
$$= \dfrac{120}{119}$$

103. $\csc 2\theta = \dfrac{1}{\sin 2\theta}$
$$= \dfrac{1}{2 \sin \theta \cos \theta}$$
$$= \dfrac{1}{\sin \theta} \cdot \dfrac{1}{2 \cos \theta}$$
$$= \dfrac{\csc \theta}{2 \cos \theta}$$

105. $\cos^2 2\alpha - \sin^2 2\alpha = \cos\left[2(2\alpha)\right]$
$$= \cos 4\alpha$$

107. $(\sin x + \cos x)^2 = \sin^2 x + 2 \sin x \cos x + \cos^2 x$
$$= (\sin^2 x + \cos^2 x) + 2 \sin x \cos x$$
$$= 1 + \sin 2x$$

109. $1 + \cos 10y = 1 + \cos^2 5y - \sin^2 5y$
$$= 1 + \cos^2 5y - (1 - \cos^2 5y)$$
$$= 2\cos^2 5y$$

111. $\sec \dfrac{u}{2} = \dfrac{1}{\cos \dfrac{u}{2}}$

$$= \pm \sqrt{\dfrac{2}{1 + \cos u}}$$

$$= \pm \sqrt{\dfrac{2 \sin u}{\sin u (1 + \cos u)}}$$

$$= \pm \sqrt{\dfrac{2 \sin u}{\sin u + \sin u \cos u}}$$

$$= \pm \sqrt{\dfrac{\dfrac{2 \sin u}{\cos u}}{\dfrac{\sin u}{\cos u} + \dfrac{\sin u \cos u}{\cos u}}}$$

$$= \pm \sqrt{\dfrac{2 \tan u}{\tan u + \sin u}}$$

113. $\dfrac{\sin x \pm \sin y}{\cos x + \cos y} = \dfrac{2 \sin\left(\dfrac{x \pm y}{2}\right) \cos\left(\dfrac{x \mp y}{2}\right)}{2 \cos\left(\dfrac{x + y}{2}\right) \cos\left(\dfrac{x - y}{2}\right)}$

$$= \tan\left(\dfrac{x \pm y}{2}\right)$$

115. $\dfrac{\cos 4x + \cos 2x}{\sin 4x + \sin 2x} = \dfrac{2 \cos\left(\dfrac{4x + 2x}{2}\right) \cos\left(\dfrac{4x - 2x}{2}\right)}{2 \sin\left(\dfrac{4x + 2x}{2}\right) \cos\left(\dfrac{4x - 2x}{2}\right)}$

$$= \dfrac{2 \cos 3x \cos x}{2 \sin 3x \cos x}$$

$$= \cot 3x$$

117. $\sin\left(\dfrac{\pi}{6} + x\right) + \sin\left(\dfrac{\pi}{6} - x\right) = 2 \sin \dfrac{\pi}{6} \cos x$

$$= 2 \cdot \dfrac{1}{2} \cos x$$

$$= \cos x$$

119. $\cos 3\beta = \cos(2\beta + \beta)$

$$= \cos 2\beta \cos\beta - \sin 2\beta \sin \beta$$

$$= (\cos^2 \beta - \sin^2 \beta)\cos \beta - 2 \sin \beta \cos \beta \sin \beta$$

$$= \cos^3 \beta - \sin^2 \beta \cos \beta - 2 \sin^2 \beta \cos \beta$$

$$= \cos^3 \beta - 3 \sin^2 \beta \cos \beta$$

121. $\dfrac{\cos 4x - \cos 2x}{2 \sin 3x} = \dfrac{-2 \sin\left(\dfrac{4x + 2x}{2}\right) \sin\left(\dfrac{4x - 2x}{2}\right)}{2 \sin 3x}$

$$= \dfrac{-2 \sin 3x \sin x}{2 \sin 3x}$$

$$= -\sin x$$

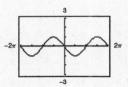

123. $\sin^2 x = \dfrac{1 - \cos 2x}{2} = \dfrac{1}{2} - \dfrac{\cos 2x}{2}$

125. $f(x) = \cos^2 x = \dfrac{1 + \cos 2x}{2} = \dfrac{1}{2} + \dfrac{\cos 2x}{2}$

Shifted upward by $\dfrac{1}{2}$ unit.

Amplitude: $|a| = \dfrac{1}{2}$

Period: $\dfrac{2\pi}{2} = \pi$

127. $\sin(2\arcsin x) = 2\sin(\arcsin x)\cos(\arcsin x)$

$$= 2x\sqrt{1-x^2}$$

129. $\cos(2\arcsin x) = 1 - 2\sin^2(\arcsin x)$

$$= 1 - 2x^2$$

131. $\cos(2\arctan x) = \cos^2(\arctan x) - \sin^2(\arctan x)$

$$= \left(\frac{1}{\sqrt{1+x^2}}\right)^2 - \left(\frac{x}{\sqrt{1+x^2}}\right)^2$$

$$= \frac{1-x^2}{1+x^2}$$

133. $f(x) = \sin^4 x + \cos^4 x$

(a) $\sin^4 x + \cos^4 x = (\sin^2 x)^2 + (\cos^2 x)^2$

$$= \left(\frac{1-\cos 2x}{2}\right)^2 + \left(\frac{1+\cos 2x}{2}\right)^2$$

$$= \frac{1}{4}[(1-\cos 2x)^2 + (1+\cos 2x)^2]$$

$$= \frac{1}{4}(1 - 2\cos 2x + \cos^2 2x + 1 + 2\cos 2x + \cos^2 2x)$$

$$= \frac{1}{4}(2 + 2\cos^2 2x)$$

$$= \frac{1}{4}\left[2 + 2\left(\frac{1+\cos 2(2x)}{2}\right)\right]$$

$$= \frac{1}{4}(3 + \cos 4x)$$

(b) $\sin^4 x + \cos^4 x = (\sin^2 x)^2 + \cos^4 x$

$$= (1 - \cos^2 x)^2 + \cos^4 x$$

$$= 1 - 2\cos^2 x + \cos^4 x + \cos^4 x$$

$$= 2\cos^4 x - 2\cos^2 x + 1$$

(c) $\sin^4 x + \cos^4 x = \sin^4 x + 2\sin^2 x \cos^2 x + \cos^4 x - 2\sin^2 x \cos^2 x$

$$= (\sin^2 x + \cos^2 x)^2 - 2\sin^2 x \cos^2 x$$

$$= 1 - 2\sin^2 x \cos^2 x$$

(d) $1 - 2\sin^2 x \cos^2 x = 1 - (2\sin x \cos x)(\sin x \cos x)$

$$= 1 - (\sin 2x)\left(\frac{1}{2}\sin 2x\right)$$

$$= 1 - \frac{1}{2}\sin^2 2x$$

(e) No, it does not mean that one of you is wrong. There is often more than one way to rewrite a trigonometric expression.

135. $r = \frac{1}{32}v_0{}^2 \sin 2\theta$

$$= \frac{1}{32}v_0{}^2(2\sin\theta\cos\theta)$$

$$= \frac{1}{16}v_0{}^2\sin\theta\cos\theta$$

137. False. For $u < 0$,

$$\sin 2u = -\sin(-2u)$$

$$= -2\sin(-u)\cos(-u)$$

$$= -2(-\sin u)\cos u$$

$$= 2\sin u \cos u.$$

139. True

$$\sqrt{\frac{1 - \cos(\pi/6)}{2}} = \sin\frac{(\pi/6)}{2}$$

$$= \sin\frac{\pi}{12} = \sin 2\left(\frac{\pi}{24}\right)$$

$$= 2\sin\frac{\pi}{24}\cos\frac{\pi}{24}$$

141. (a) $y = 4\sin\frac{x}{2} + \cos x$

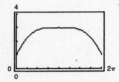

Maximum point: (3.1416, 3.0)

(b) $\qquad 2\cos\frac{x}{2} - \sin x = 0$

$$2\left(\pm\sqrt{\frac{1 + \cos x}{2}}\right) = \sin x$$

$$4\left(\frac{1 + \cos x}{2}\right) = \sin^2 x$$

$$2(1 + \cos x) = 1 - \cos^2 x$$

$$\cos^2 x + 2\cos x + 1 = 0$$

$$(\cos x + 1)^2 = 0$$

$$\cos x = -1$$

$$x = \pi \approx 3.1416$$

143. $f(x) = \cos 2x - 2\sin x$

(a)

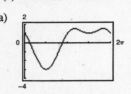

Maximum points: (3.6652, 1.5), (5.7596, 1.5)

Minimum points: (1.5708, −3), (4.7124, 1)

(b) $-2\cos x(2\sin x + 1) = 0$

$\qquad -2\cos x = 0 \qquad$ or $\quad 2\sin x + 1 = 0$

$\qquad\quad \cos x = 0 \qquad\qquad\qquad \sin x = -\dfrac{1}{2}$

$$x = \frac{\pi}{2}, \frac{3\pi}{2} \qquad\qquad x = \frac{7\pi}{6}, \frac{11\pi}{6}$$

$$\frac{\pi}{2} \approx 1.5708 \qquad\qquad \frac{7\pi}{6} \approx 3.6652$$

$$\frac{3\pi}{2} \approx 4.7124 \qquad\qquad \frac{11\pi}{6} \approx 5.7596$$

145. $f(x) = \dfrac{1}{2}\cos 2x + \cos x$

(a)

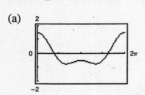

Maximum points: (0, 1.5), (3.1416, −0.5)

Minimum points: (2.0944, −0.75), (4.1888, −0.75)

(b) $\qquad\qquad -\sin 2x - \sin x = 0$

$$-2\sin x\cos x - \sin x = 0$$

$$-\sin x(2\cos x + 1) = 0$$

$\qquad -\sin x = 0 \qquad$ or $\quad 2\cos x + 1 = 0$

$\qquad\quad \sin x = 0 \qquad\qquad\qquad \cos x = -\dfrac{1}{2}$

$$x = 0, \pi \approx 3.1416 \qquad x = \frac{2\pi}{3}, \frac{4\pi}{3}$$

$$\approx 2.0944, 4.1888$$

Review Exercises for Chapter 10

1. $\dfrac{1}{\cos x} = \sec x$

3. $\dfrac{1}{\sec x} = \cos x$

5. $\dfrac{\cos x}{\sin x} = \cot x$

7. $\sin x = \dfrac{3}{5},\ \cos x = \dfrac{4}{5}$

$\tan x = \dfrac{\sin x}{\cos x} = \dfrac{\frac{3}{5}}{\frac{4}{5}} = \dfrac{3}{4}$

$\cot x = \dfrac{1}{\tan x} = \dfrac{4}{3}$

$\sec x = \dfrac{1}{\cos x} = \dfrac{5}{4}$

$\csc x = \dfrac{1}{\sin x} = \dfrac{5}{3}$

9. $\sin\left(\dfrac{\pi}{2} - x\right) = \dfrac{\sqrt{2}}{2} \Rightarrow \cos x = \dfrac{1}{\sqrt{2}} = \dfrac{\sqrt{2}}{2},\ \sin x = -\dfrac{\sqrt{2}}{2}$

$\tan x = \dfrac{\sin x}{\cos x} = \dfrac{-\dfrac{1}{\sqrt{2}}}{\dfrac{1}{\sqrt{2}}} = -1$

$\cot x = \dfrac{1}{\tan x} = -1$

$\sec x = \dfrac{1}{\cos x} = \sqrt{2}$

$\csc x = \dfrac{1}{\sin x} = -\sqrt{2}$

11. $\dfrac{1}{\cot^2 x + 1} = \dfrac{1}{\csc^2 x} = \sin^2 x$

13. $\tan^2 x(\csc^2 x - 1) = \tan^2 x(\cot^2 x) = \tan^2 x\left(\dfrac{1}{\tan^2 x}\right) = 1$

15. $\dfrac{\sin\left(\dfrac{\pi}{2} - \theta\right)}{\sin \theta} = \dfrac{\cos \theta}{\sin \theta} = \cot \theta$

17. $\cos^2 x + \cos^2 x \cot^2 x = \cos^2 x(1 + \cot^2 x) = \cos^2 x(\csc^2 x)$

$= \cos^2 x\left(\dfrac{1}{\sin^2 x}\right) = \dfrac{\cos^2 x}{\sin^2 x} = \cot^2 x$

19. $(\tan x + 1)^2 \cos x = (\tan^2 x + 2\tan x + 1)\cos x$

$= (\sec^2 x + 2\tan x)\cos x$

$= \sec^2 x \cos x + 2\left(\dfrac{\sin x}{\cos x}\right)\cos x$

$= \sec x + 2\sin x$

21. $\dfrac{1}{\csc \theta + 1} - \dfrac{1}{\csc \theta - 1} = \dfrac{(\csc \theta - 1) - (\csc \theta + 1)}{(\csc \theta + 1)(\csc \theta - 1)}$

$= \dfrac{-2}{\csc^2 \theta - 1}$

$= \dfrac{-2}{\cot^2 \theta}$

$= -2\tan^2 \theta$

23. $\cos x(\tan^2 x + 1) = \cos x \sec^2 x$

$= \dfrac{1}{\sec x} \sec^2 x$

$= \sec x$

25. $\cos\left(x + \dfrac{\pi}{2}\right) = \cos x \cos \dfrac{\pi}{2} - \sin x \sin \dfrac{\pi}{2}$

$= (\cos x)(0) - (\sin x)(1)$

$= -\sin x$

27. $\dfrac{1}{\tan \theta \csc \theta} = \dfrac{1}{\dfrac{\sin \theta}{\cos \theta} \cdot \dfrac{1}{\sin \theta}} = \cos \theta$

29. $\sin^5 x \cos^2 x = \sin^4 x \cos^2 x \sin x$

$= (1 - \cos^2 x)^2 \cos^2 x \sin x$

$= (1 - 2\cos^2 x + \cos^4 x)\cos^2 x \sin x$

$= (\cos^2 x - 2\cos^4 x + \cos^6 x)\sin x$

31. $\sin x = \sqrt{3} - \sin x$

$\sin x = \dfrac{\sqrt{3}}{2}$

$x = \dfrac{\pi}{3} + 2\pi n, \dfrac{2\pi}{3} + 2\pi n$

33. $3\sqrt{3}\,\tan u = 3$

$\tan u = \dfrac{1}{\sqrt{3}}$

$u = \dfrac{\pi}{6} + n\pi$

35. $3\csc^2 x = 4$

$\csc^2 x = \dfrac{4}{3}$

$\sin x = \pm\dfrac{\sqrt{3}}{2}$

$x = \dfrac{\pi}{3} + 2\pi n, \dfrac{2\pi}{3} + 2\pi n, \dfrac{4\pi}{3} + 2\pi n, \dfrac{5\pi}{3} + 2\pi n$

37. $\qquad 2\cos^2 x - \cos x = 1$

$2\cos^2 x - \cos x - 1 = 0$

$(2\cos x + 1)(\cos x - 1) = 0$

$2\cos x + 1 = 0 \qquad\qquad \cos x - 1 = 0$

$\cos x = -\dfrac{1}{2} \qquad\qquad \cos x = 1$

$x = \dfrac{2\pi}{3}, \dfrac{4\pi}{3} \qquad\qquad x = 0$

39. $\qquad \cos^2 x + \sin x = 1$

$1 - \sin^2 x + \sin x - 1 = 0$

$-\sin x(\sin x - 1) = 0$

$\sin x = 0 \qquad \sin x - 1 = 0$

$x = 0, \pi \qquad \sin x = 1$

$\qquad\qquad x = \dfrac{\pi}{2}$

41. $2\sin 2x - \sqrt{2} = 0$

$\sin 2x = \dfrac{\sqrt{2}}{2}$

$2x = \dfrac{\pi}{4} + 2\pi n, \dfrac{3\pi}{4} + 2\pi n$

$x = \dfrac{\pi}{8} + \pi n, \dfrac{3\pi}{8} + \pi n$

$x = \dfrac{\pi}{8}, \dfrac{3\pi}{8}, \dfrac{9\pi}{8}, \dfrac{11\pi}{8}$

43. $\cos 4x(\cos x - 1) = 0$

$\cos 4x = 0 \qquad\qquad\qquad \cos x - 1 = 0$

$4x = \dfrac{\pi}{2} + 2\pi n, \dfrac{3\pi}{2} + 2\pi n \qquad \cos x = 1$

$x = \dfrac{\pi}{8} + \dfrac{\pi}{2}n, \dfrac{3\pi}{8} + \dfrac{\pi}{2}n \qquad\qquad x = 0$

$x = 0, \dfrac{\pi}{8}, \dfrac{3\pi}{8}, \dfrac{5\pi}{8}, \dfrac{7\pi}{8}, \dfrac{9\pi}{8}, \dfrac{11\pi}{8}, \dfrac{13\pi}{8}, \dfrac{15\pi}{8}$

45. $\sin^2 x - 2\sin x = 0$

$\sin x(\sin x - 2) = 0$

$\sin x = 0 \qquad \sin x - 2 = 0$

$x = 0, \pi \qquad$ No solution

47. $\qquad \tan^2 \theta + \tan \theta - 12 = 0$

$(\tan \theta + 4)(\tan \theta - 3) = 0$

$\tan \theta + 4 = 0 \qquad\qquad \tan \theta - 3 = 0$

$\theta = \arctan(-4) + n\pi \qquad\qquad \theta = \arctan 3 + n\pi$

$\theta = \arctan(-4) + \pi, \arctan(-4) + 2\pi, \arctan 3, \arctan 3 + \pi$

49. $\sin 285° = \sin(315° - 30°)$

$$= \sin 315° \cos 30° - \cos 315° \sin 30°$$

$$= \left(-\frac{\sqrt{2}}{2}\right)\left(\frac{\sqrt{3}}{2}\right) - \left(\frac{\sqrt{2}}{2}\right)\left(\frac{1}{2}\right)$$

$$= -\frac{\sqrt{2}}{4}(\sqrt{3} + 1)$$

$\cos 285° = \cos(315° - 30°)$

$$= \cos 315° \cos 30° + \sin 315° \sin 30°$$

$$= \left(\frac{\sqrt{2}}{2}\right)\left(\frac{\sqrt{3}}{2}\right) + \left(-\frac{\sqrt{2}}{2}\right)\left(\frac{1}{2}\right)$$

$$= \frac{\sqrt{2}}{4}(\sqrt{3} - 1)$$

$\tan 285° = \tan(315° - 30°) = \dfrac{\tan 315° - \tan 30°}{1 + \tan 315° \tan 30°}$

$$= \frac{(-1) - \left(\frac{\sqrt{3}}{3}\right)}{1 + (-1)\left(\frac{\sqrt{3}}{3}\right)} = -2 - \sqrt{3}$$

51. $\sin\dfrac{25\pi}{12} = \sin\left(\dfrac{11\pi}{6} + \dfrac{\pi}{4}\right) = \sin\dfrac{11\pi}{6}\cos\dfrac{\pi}{4} + \cos\dfrac{11\pi}{6}\sin\dfrac{\pi}{4}$

$$= \left(-\frac{1}{2}\right)\left(\frac{\sqrt{2}}{2}\right) + \left(\frac{\sqrt{3}}{2}\right)\left(\frac{\sqrt{2}}{2}\right) = \frac{\sqrt{2}}{4}(\sqrt{3} - 1)$$

$\cos\dfrac{25\pi}{12} = \cos\left(\dfrac{11\pi}{6} + \dfrac{\pi}{4}\right) = \cos\dfrac{11\pi}{6}\cos\dfrac{\pi}{4} - \sin\dfrac{11\pi}{6}\sin\dfrac{\pi}{4}$

$$= \left(\frac{\sqrt{3}}{2}\right)\left(\frac{\sqrt{2}}{2}\right) - \left(-\frac{1}{2}\right)\left(\frac{\sqrt{2}}{2}\right) = \frac{\sqrt{2}}{4}(\sqrt{3} + 1)$$

$\tan\dfrac{25\pi}{12} = \tan\left(\dfrac{11\pi}{6} + \dfrac{\pi}{4}\right) = \dfrac{\tan\dfrac{11\pi}{6} + \tan\dfrac{\pi}{4}}{1 - \tan\dfrac{11\pi}{6}\tan\dfrac{\pi}{4}}$

$$= \frac{\left(-\frac{\sqrt{3}}{3}\right) + 1}{1 - \left(-\frac{\sqrt{3}}{3}\right)(1)} = 2 - \sqrt{3}$$

53. $\sin 60° \cos 45° - \cos 60° \sin 45° = \sin(60° - 45°)$

$$= \sin 15°$$

55. $\dfrac{\tan 25° + \tan 10°}{1 - \tan 25° \tan 10°} = \tan(25° + 10°)$

$$= \tan 35°$$

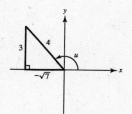

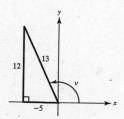

Figures for Exercises 57–61

57. $\sin(u + v) = \sin u \cos v + \cos u \sin v$

$$= \left(\frac{3}{4}\right)\left(-\frac{5}{13}\right) + \left(-\frac{\sqrt{7}}{4}\right)\left(\frac{12}{13}\right)$$

$$= -\frac{3}{52}(5 + 4\sqrt{7})$$

59. $\cos(u - v) = \cos u \cos v + \sin u \sin v$

$$= \left(-\frac{\sqrt{7}}{4}\right)\left(-\frac{5}{13}\right) + \left(\frac{3}{4}\right)\left(\frac{12}{13}\right)$$

$$= \frac{1}{52}(5\sqrt{7} + 36)$$

61. $\cos(u + v) = \cos u \cos v - \sin u \sin v$

$$= \left(-\frac{\sqrt{7}}{4}\right)\left(-\frac{5}{13}\right) - \left(\frac{3}{4}\right)\left(\frac{12}{13}\right)$$

$$= \frac{1}{52}(5\sqrt{7} - 36)$$

63. $\sin\left(x + \dfrac{\pi}{4}\right) - \sin\left(x - \dfrac{\pi}{4}\right) = 1$

$$2 \cos x \sin\frac{\pi}{4} = 1$$

$$\cos x = \frac{\sqrt{2}}{2}$$

$$x = \frac{\pi}{4}, \frac{7\pi}{4}$$

65. $\sin 4x = 2 \sin 2x \cos 2x$

$\qquad = 2[2 \sin x \cos x(\cos^2 x - \sin^2 x)]$

$\qquad = 4 \sin x \cos x(2 \cos^2 x - 1)$

$\qquad = 8 \cos^3 x \sin x - 4 \cos x \sin x$

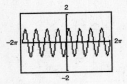

67. $\sin u = -\dfrac{4}{5}, \ \pi < u < \dfrac{3\pi}{2}$

$\cos u = -\sqrt{1 - \sin^2 u} = -\dfrac{3}{5}$

$\tan u = \dfrac{\sin u}{\cos u} = \dfrac{4}{3}$

$\sin 2u = 2 \sin u \cos u = 2\left(-\dfrac{4}{5}\right)\left(-\dfrac{3}{5}\right) = \dfrac{24}{25}$

$\cos 2u = \cos^2 u - \sin^2 u = \left(-\dfrac{3}{5}\right)^2 - \left(-\dfrac{4}{5}\right)^2 = -\dfrac{7}{25}$

$\tan 2u = \dfrac{2 \tan u}{1 - \tan^2 u} = \dfrac{2\left(\dfrac{4}{3}\right)}{1 - \left(\dfrac{4}{3}\right)^2} = -\dfrac{24}{7}$

69. $\tan^2 2x = \dfrac{\sin^2 2x}{\cos^2 2x} = \dfrac{\dfrac{1 - \cos 4x}{2}}{\dfrac{1 + \cos 4x}{2}} = \dfrac{1 - \cos 4x}{1 + \cos 4x}$

71. $\sin^2 x \tan^2 x = \sin^2 x\left(\dfrac{\sin^2 x}{\cos^2 x}\right) = \dfrac{\sin^4 x}{\cos^2 x}$

$\qquad = \dfrac{\left(\dfrac{1 - \cos 2x}{2}\right)^2}{\dfrac{1 + \cos 2x}{2}} = \dfrac{\dfrac{1 - 2\cos 2x + \cos^2 2x}{4}}{\dfrac{1 + \cos 2x}{2}}$

$\qquad = \dfrac{1 - 2\cos 2x + \dfrac{1 + \cos 4x}{2}}{2(1 + \cos 2x)}$

$\qquad = \dfrac{2 - 4\cos 2x + 1 + \cos 4x}{4(1 + \cos 2x)}$

$\qquad = \dfrac{3 - 4\cos 2x + \cos 4x}{4(1 + \cos 2x)}$

73. $\sin(-75°) = -\sqrt{\dfrac{1 - \cos 150°}{2}} = -\sqrt{\dfrac{1 - \left(-\dfrac{\sqrt{3}}{2}\right)}{2}} = -\dfrac{\sqrt{2 + \sqrt{3}}}{2}$

$\qquad = -\dfrac{1}{2}\sqrt{2 + \sqrt{3}}$

$\cos(-75°) = \sqrt{\dfrac{1 + \cos 150°}{2}} = \sqrt{\dfrac{1 + \left(-\dfrac{\sqrt{3}}{2}\right)}{2}} = \dfrac{\sqrt{2 - \sqrt{3}}}{2}$

$\qquad = \dfrac{1}{2}\sqrt{2 - \sqrt{3}}$

$\tan(-75°) = -\left(\dfrac{1 - \cos 150°}{\sin 150°}\right) = -\left(\dfrac{1 - \left(-\dfrac{\sqrt{3}}{2}\right)}{\dfrac{1}{2}}\right) = -\left(2 + \sqrt{3}\right)$

$\qquad = -2 - \sqrt{3}$

75. $\sin\left(\dfrac{19\pi}{12}\right) = -\sqrt{\dfrac{1 - \cos\dfrac{19\pi}{6}}{2}} = -\sqrt{\dfrac{1 - \left(-\dfrac{\sqrt{3}}{2}\right)}{2}} = -\dfrac{\sqrt{2 + \sqrt{3}}}{2}$

$\qquad = -\dfrac{1}{2}\sqrt{2 + \sqrt{3}}$

$\cos\left(\dfrac{19\pi}{12}\right) = \sqrt{\dfrac{1 + \cos\dfrac{19\pi}{6}}{2}} = \sqrt{\dfrac{1 + \left(-\dfrac{\sqrt{3}}{2}\right)}{2}} = \dfrac{\sqrt{2 - \sqrt{3}}}{2}$

$\qquad = \dfrac{1}{2}\sqrt{2 - \sqrt{3}}$

$\tan\left(\dfrac{19\pi}{12}\right) = \dfrac{1 - \cos\dfrac{19\pi}{6}}{\sin\dfrac{19\pi}{6}} = \dfrac{1 - \left(-\dfrac{\sqrt{3}}{2}\right)}{-\dfrac{1}{2}} = -2 - \sqrt{3}$

77. $-\sqrt{\dfrac{1 + \cos 10x}{2}} = -\left|\cos\dfrac{10x}{2}\right| = -|\cos 5x|$

79. $\qquad r = \dfrac{1}{32} v_0{}^2 \sin 2\theta$

$\text{range} = 100 \text{ feet}$

$v_0 = 80 \text{ feet per second}$

$r = \dfrac{1}{32}(80)^2 \sin 2\theta = 100$

$\sin 2\theta = 0.5$

$2\theta = 30°, 150°$

$\theta = 15° \text{ or } 75° = \dfrac{\pi}{12} \text{ or } \dfrac{5\pi}{12}$

81. $\cos\dfrac{\pi}{6}\sin\dfrac{\pi}{6} = \dfrac{1}{2}\left[\sin\dfrac{\pi}{3} - \sin 0\right]$

83. $\cos 5\theta \cos 3\theta = \dfrac{1}{2}[\cos 2\theta + \cos 8\theta]$

85. $\sin 60° + \sin 90° = 2\sin 75° \cos 15°$

87. $\cos\left(x + \dfrac{\pi}{6}\right) - \cos\left(x - \dfrac{\pi}{6}\right) = -2\sin(x)\sin\left(\dfrac{\pi}{6}\right)$

89. $y = \sqrt{x + 3} + 4\cos x$

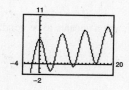

Zeros: $x \approx -1.8431, 2.1758, 3.9903, 8.8935, 9.8820$

91. (a) $y = 1.5 \sin 8t - 0.5 \cos 8t = \frac{1}{2}(3 \sin 8t - 1 \cos 8t)$

Using the identity

$a \sin B\theta + b \cos B\theta = \sqrt{a^2 + b^2} \sin(B\theta + C)$,

$C = \arctan \frac{b}{a}, a > 0$

(Exercise 83, Section 10.4), we have

$y = \frac{1}{2}\sqrt{(3)^2 + (-1)^2} \sin\left(8t + \arctan\left(-\frac{1}{3}\right)\right)$

$= \frac{\sqrt{10}}{2} \sin\left(8t - \arctan\left(\frac{1}{3}\right)\right)$.

(b) Amplitude $= \dfrac{\sqrt{10}}{2}$ feet

(c) Frequency $= \dfrac{1}{\dfrac{2\pi}{8}} = \dfrac{4}{\pi}$ cycles per second

Problem Solving for Chapter 10

1. $f(x) = \tan \dfrac{\pi x}{4}$

Since $\tan \pi/4 = 1$, $x = 1$ is the smallest nonnegative fixed point.

3. $A = 2x \cos x$, $0 < x < \dfrac{\pi}{2}$

(a)

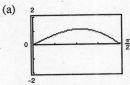

The maximum area of $A \approx 1.12$ occurs when $x \approx 0.86$.

(b) $A \geq 1$ for $0.6 < x < 1.1$

5. $f(x) = 3 \sin(0.6x - 2)$

(a) Zero: $\sin(0.6x - 2) = 0$

$0.6x - 2 = 0$

$0.6x = 2$

$x = \dfrac{2}{0.6} = \dfrac{10}{3}$

(c) $-0.45x^2 + 5.52x - 13.70 = 0$

$x = \dfrac{-5.52 \pm \sqrt{(5.52)^2 - 4(-0.45)(-13.70)}}{2(-0.45)}$

$x \approx 3.46, 8.81$

The zero of g on $[0, 6]$ is 3.46. The zero is close to the zero $\frac{10}{3} \approx 3.33$ of f.

(b) $g(x) = -0.45x^2 + 5.52x - 13.70$

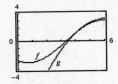

For $3.5 \leq x \leq 6$ the approximation appears to be good.

7. $\cos(n\pi + \theta) = \cos n\pi \cos \theta - \sin n\pi \sin \theta$

$= (-1)^n (\cos \theta) - (0)(\sin \theta)$

$= (-1)^n (\cos \theta)$, where n is an integer.

9.

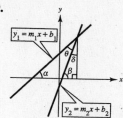

Let $m_2 > m_1 > 0$.

$m_1 = \tan \alpha$ and $m_2 = \tan \beta$

$\beta + \delta = 90° \implies \delta = 90° - \beta$

$\alpha + \theta + \delta = 90° \implies \alpha + \theta + (90° - \beta) = 90° \implies \theta = \beta - \alpha$

Therefore, $\theta = \arctan m_2 - \arctan m_1$, $m_2 > m_1 > 0$.

11. $\sin^2\left(\theta + \dfrac{\pi}{4}\right) + \sin^2\left(\theta - \dfrac{\pi}{4}\right) = \left[\sin\theta\cos\dfrac{\pi}{4} + \cos\theta\sin\dfrac{\pi}{4}\right]^2 + \left[\sin\theta\cos\dfrac{\pi}{4} - \cos\theta\sin\dfrac{\pi}{4}\right]^2$

$$= \left[\frac{\sin\theta}{\sqrt{2}} + \frac{\cos\theta}{\sqrt{2}}\right]^2 + \left[\frac{\sin\theta}{\sqrt{2}} - \frac{\cos\theta}{\sqrt{2}}\right]^2$$

$$= \frac{\sin^2\theta}{2} + \sin\theta\cos\theta + \frac{\cos^2\theta}{2} + \frac{\sin^2\theta}{2} - \sin\theta\cos\theta + \frac{\cos^2\theta}{2}$$

$$= \sin^2\theta + \cos^2\theta$$

$$= 1$$

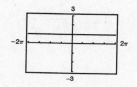

13. To prove the identity for $\sin(u + v)$ we first need to prove the identity for $\cos(u - v)$.
Assume $0 < v < u < 2\pi$ and locate u, v, and $u - v$ on the unit circle.

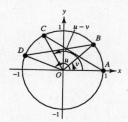

The coordinates of the points on the circle are:

$A = (1, 0)$, $B = (\cos v, \sin v)$, $C = (\cos(u - v), \sin(u - v))$, and $D = (\cos u, \sin u)$.

Since $\angle DOB = \angle COA$, chords AC and BD are equal. By the distance formula we have:

$$\sqrt{[\cos(u - v) - 1]^2 + [\sin(u - v) - 0]^2} = \sqrt{(\cos u - \cos v)^2 + (\sin u - \sin v)^2}$$

$$\cos^2(u - v) - 2\cos(u - v) + 1 + \sin^2(u - v) = \cos^2 u - 2\cos u\cos v + \cos^2 v + \sin^2 u - 2\sin u\sin v + \sin^2 v$$

$$[\cos^2(u - v) + \sin^2(u - v)] + 1 - 2\cos(u - v) = (\cos^2 u + \sin^2 u) + (\cos^2 v + \sin^2 v) - 2\cos u\cos v - 2\sin u\sin v$$

$$2 - 2\cos(u - v) = 2 - 2\cos u\cos v - 2\sin u\sin v$$

$$-2\cos(u - v) = -2(\cos u\cos v + \sin u\sin v)$$

$$\cos(u - v) = \cos u\cos v + \sin u\sin v$$

Now, to prove the identity for $\sin(u + v)$, use cofunction identities.

$$\sin(u + v) = \cos\left[\frac{\pi}{2} - (u + v)\right] = \cos\left[\left(\frac{\pi}{2} - u\right) - v\right] = \cos\left(\frac{\pi}{2} - u\right)\cos v + \sin\left(\frac{\pi}{2} - u\right)\sin v = \sin u\cos v + \cos u\sin v$$

15. (a) $A = \dfrac{1}{2}bh$

$\cos\dfrac{\theta}{2} = \dfrac{h}{10} \implies h = 10\cos\dfrac{\theta}{2}$

$\sin\dfrac{\theta}{2} = \dfrac{(1/2)b}{10} \implies \dfrac{1}{2}b = 10\sin\dfrac{\theta}{2}$

$A = 10\sin\dfrac{\theta}{2}10\cos\dfrac{\theta}{2} \implies A = 100\sin\dfrac{\theta}{2}\cos\dfrac{\theta}{2}$

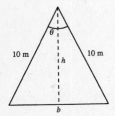

(b) $A = 100\sin\dfrac{\theta}{2}\cos\dfrac{\theta}{2}$

$A = 50\left(2\sin\dfrac{\theta}{2}\cos\dfrac{\theta}{2}\right)$

$A = 50\sin\theta$

When $\theta = \pi/2$, $\sin\theta = 1 \implies$ the area is a maximum.

$A = 50\sin\dfrac{\pi}{2} = 50(1) = 50$ square feet

17. $\displaystyle\lim_{x\to 0^-} f(x) = \lim_{x\to 0^-}(a^2 - 2) = a^2 - 2$

$\displaystyle\lim_{x\to 0^+} f(x) = \lim_{x\to 0^+}\dfrac{ax}{\tan x} = a\left(\text{because } \lim_{x\to 0}\dfrac{\tan x}{x} = 1\right)$

Thus,

$$a^2 - 2 = a$$
$$a^2 - a - 2 = 0$$
$$(a - 2)(a + 1) = 0$$
$$a = -1, 2$$

CHAPTER 11
Trigonometric Functions and Calculus

CHAPTER 11
Trigonometric Functions and Calculus

Section 11.1   Limits of Trigonometric Functions

1.

x	-0.1	-0.01	-0.001	0.001	0.01	0.1
$f(x)$	0.9983	0.99998	1.0000	1.0000	0.99998	0.9983

$$\lim_{x \to 0} \frac{\sin x}{x} \approx 1.0000 \quad \text{(Actual limit is 1.) (Make sure you use radian mode.)}$$

3. $\lim\limits_{x \to \pi/2} \tan x$ does not exist since the function increases and decreases without bound as x approaches $\pi/2$.

5. $\lim\limits_{x \to 0} \cos(1/x)$ does not exist since the function oscillates between -1 and 1 as x approaches 0.

7. $\lim\limits_{x \to \pi/2} \sin x = \sin \dfrac{\pi}{2} = 1$

9. $\lim\limits_{x \to 2} \cos \dfrac{\pi x}{3} = \cos \dfrac{\pi 2}{3} = -\dfrac{1}{2}$

11. $\lim\limits_{x \to 0} \sec 2x = \sec 0 = 1$

13. $\lim\limits_{x \to 5\pi/6} \sin x = \sin \dfrac{5\pi}{6} = \dfrac{1}{2}$

15. $\lim\limits_{x \to 3} \tan\left(\dfrac{\pi x}{4}\right) = \tan \dfrac{3\pi}{4} = -1$

17. $f(x) = 3x - \cos x$ is continuous for all real x.

19. $f(x) = \begin{cases} \tan \frac{\pi x}{4}, & |x| < 1 \\ x, & |x| \geq 1 \end{cases} = \begin{cases} \tan \frac{\pi x}{4}, & -1 < x < 1 \\ x, & x \leq -1 \text{ or } x \geq 1 \end{cases}$

has **possible** discontinuities at $x = -1$, $x = 1$.

1. $f(-1) = -1 \qquad f(1) = 1$

2. $\lim\limits_{x \to -1} f(x) = -1 \qquad \lim\limits_{x \to 1} f(x) = 1$

3. $f(-1) = \lim\limits_{x \to -1} f(x) \qquad f(1) = \lim\limits_{x \to 1} f(x)$

f is continuous at $x = \pm 1$, therefore, f is continuous for all real x.

21. $f(x) = \csc 2x$ has nonremovable discontinuities at integer multiples of $\pi/2$.

23. $f(x) = \dfrac{\sin x}{x}$

Although the graph appears continuous on $[-1, 1]$, there is a discontinuity at $x = 0$. Examining a function graphically and analytically ensures that you will find all points where the function is not defined.

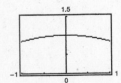

25. $f(x) = \tan 2x = \dfrac{\sin 2x}{\cos 2x}$ has vertical asymptotes at

$$x = \frac{(2n + 1)\pi}{4} = \frac{\pi}{4} + \frac{n\pi}{2}, \, n \text{ any integer.}$$

27. $s(t) = \dfrac{t}{\sin t}$ has vertical asymptotes at $t = n\pi$, n a nonzero integer. There is no vertical asymptote at $t = 0$ since

$$\lim_{t \to 0} \frac{t}{\sin t} = 1.$$

29. $f(t) = \dfrac{\sin 3t}{t}$

t	-0.1	-0.01	-0.001	0	0.001	0.01	0.1
$f(t)$	2.96	2.9996	3	?	3	2.9996	2.96

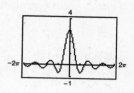

Analytically, $\displaystyle\lim_{t\to 0} \frac{\sin 3t}{t} = \lim_{t\to 0}\left[3\left(\frac{\sin 3t}{3t}\right)\right] = 3(1) = 3.$

31. $f(x) = \dfrac{\sin x^2}{x}$

x	-0.1	-0.01	-0.001	0	0.001	0.01	0.1
$f(x)$	-0.099998	-0.01	-0.001	?	0.001	0.01	0.099998

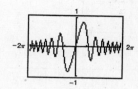

Analytically, $\displaystyle\lim_{x\to 0} \frac{\sin x^2}{x} = \lim_{x\to 0} x\left(\frac{\sin x^2}{x^2}\right) = 0(1) = 0.$

33. $f(x) = x \cos x$

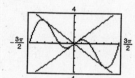

$\displaystyle\lim_{x\to 0} x \cos x = 0$

35. $f(x) = |x| \sin x$

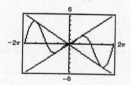

$\displaystyle\lim_{x\to 0} |x| \sin x = 0$

37. $f(x) = x \sin \dfrac{1}{x}$

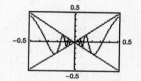

$\displaystyle\lim_{x\to 0} \left(x \sin \frac{1}{x}\right) = 0$

39. $\displaystyle\lim_{x\to 0} \frac{\sin x}{5x} = \lim_{x\to 0}\left[\left(\frac{\sin x}{x}\right)\left(\frac{1}{5}\right)\right] = (1)\left(\frac{1}{5}\right) = \frac{1}{5}$

41. $\displaystyle\lim_{x\to 0} \frac{\sin x(1-\cos x)}{2x^2} = \lim_{x\to 0}\left[\frac{1}{2}\cdot\frac{\sin x}{x}\cdot\frac{1-\cos x}{x}\right]$

$\qquad\qquad = \dfrac{1}{2}(1)(0) = 0$

43. $\displaystyle\lim_{\theta\to 0} \frac{\sin^2 x}{x} = \lim_{\theta\to 0}\left[\frac{\sin x}{x}\sin x\right] = (1)\sin 0 = 0$

45. $\displaystyle\lim_{h\to 0} \frac{(1-\cos h)^2}{h} = \lim_{h\to 0}\left[\frac{1-\cos h}{h}(1-\cos h)\right]$

$\qquad\qquad = (0)(0) = 0$

47. $\displaystyle\lim_{x\to (\pi/2)} \frac{\cos x}{\cot x} = \lim_{x\to (\pi/2)} \sin x = 1$

49. $\displaystyle\lim_{t\to 0} \frac{\sin 3t}{2t} = \lim_{t\to 0}\left(\frac{\sin 3t}{3t}\right)\left(\frac{3}{2}\right) = (1)\left(\frac{3}{2}\right) = \frac{3}{2}$

51. $\displaystyle\lim_{x\to \pi} \cot x$

Does not exist.

53. $\displaystyle\lim_{x\to 0^+} \frac{2}{\sin x} = \infty$

55. $\displaystyle\lim_{x\to \pi} \frac{\sqrt{x}}{\csc x} = \lim_{x\to \pi}\left(\sqrt{x}\sin x\right) = 0$

57. $\displaystyle\lim_{x\to (1/2)^-} x\sec(\pi x) = \infty$ and $\displaystyle\lim_{x\to (1/2)^+} x\sec(\pi x) = -\infty.$

Therefore, $\displaystyle\lim_{x\to (1/2)} x\sec(\pi x)$ does not exist.

59. $\lim\limits_{x \to \pi/6} \sin x = \dfrac{1}{2}$

means that as x gets closer to $\dfrac{\pi}{6}$, $\sin x$ gets closer to $\dfrac{1}{2}$.

61. No. $f(\pi)$ could equal any number, or not exist at all.

63. $f(x) = x$, $g(x) = \sin x$, $h(x) = \dfrac{\sin x}{x}$

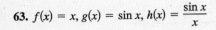

When you are "close to" 0 the magnitude of f is approximately equal to the magnitude of g. Thus, $|g|/|f| \approx 1$ when x is "close to" 0.

65. (a)

x	1	0.5	0.2	0.1	0.01	0.001	0.0001
$f(x)$	0.1585	0.0411	0.0067	0.0017	≈ 0	≈ 0	≈ 0

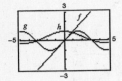

$\lim\limits_{x \to 0^+} \dfrac{x - \sin x}{x} = 0$

(b)

x	1	0.5	0.2	0.1	0.01	0.001	0.0001
$f(x)$	0.1585	0.0823	0.0333	0.0167	0.0017	≈ 0	≈ 0

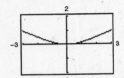

$\lim\limits_{x \to 0^+} \dfrac{x - \sin x}{x^2} = 0$

(c)

x	1	0.5	0.2	0.1	0.01	0.001	0.0001
$f(x)$	0.1585	0.1646	0.1663	0.1666	0.1667	0.1667	0.1667

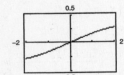

$\lim\limits_{x \to 0^+} \dfrac{x - \sin x}{x^3} = 0.1167 \ (1/6)$

(d)

x	1	0.5	0.2	0.1	0.01	0.001	0.0001
$f(x)$	0.1585	0.3292	0.8317	1.6658	16.67	166.7	1667.0

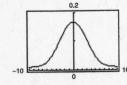

$\lim\limits_{x \to 0^+} \dfrac{x - \sin x}{x^4} = \infty$

For $n \geq 3$, $\lim\limits_{x \to 0^+} \dfrac{x - \sin x}{x^n} = \infty$.

67. (a) $\displaystyle\lim_{x\to0}\frac{1-\cos x}{x^2}=\lim_{x\to0}\frac{1-\cos x}{x^2}\cdot\frac{1+\cos x}{1+\cos x}$

$\qquad\qquad = \displaystyle\lim_{x\to0}\frac{1-\cos^2 x}{x^2(1+\cos x)}$

$\qquad\qquad = \displaystyle\lim_{x\to0}\frac{\sin^2 x}{x^2}\cdot\frac{1}{1+\cos x}$

$\qquad\qquad = (1)\left(\dfrac{1}{2}\right)=\dfrac{1}{2}$

(b) Thus, $\dfrac{1-\cos x}{x^2}\approx\dfrac{1}{2}\Rightarrow 1-\cos x\approx\dfrac{1}{2}x^2$

$\qquad\qquad\qquad\qquad\Rightarrow \cos x\approx1-\dfrac{1}{2}x^2$ for $x\approx0$.

(c) $\cos(0.1)\approx1-\dfrac{1}{2}(0.1)^2=0.995$

(d) $\cos(0.1)\approx0.9950$, which agrees with part (c).

69. (a) $A=\dfrac{1}{2}bh-\dfrac{1}{2}r^2\theta=\dfrac{1}{2}(10)(10\tan\theta)-\dfrac{1}{2}(10)^2\theta$

$\qquad\qquad = 50\tan\theta-50\theta$

$\qquad$ Domain: $\left(0,\dfrac{\pi}{2}\right)$

(b)

ϕ	0.3	0.6	0.9	1.2	1.5
$f(\theta)$	0.47	4.21	18.0	68.6	630.1

(c)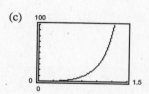

(d) $\displaystyle\lim_{\theta\to\pi/2^-}A=\infty$

71. The calculator was set in degree mode instead of radian mode.

73. $\displaystyle\lim_{x\to0}\frac{1-\cos x}{x}=\lim_{x\to0}\frac{1-\cos x}{x}\cdot\frac{1+\cos x}{1+\cos x}$

$\qquad\qquad = \displaystyle\lim_{x\to0}\frac{1-\cos^2 x}{x(1+\cos x)}=\lim_{x\to0}\frac{\sin^2 x}{x(1+\cos x)}$

$\qquad\qquad = \displaystyle\lim_{x\to0}\frac{\sin x}{x}\cdot\frac{\sin x}{1+\cos x}$

$\qquad\qquad = \left[\displaystyle\lim_{x\to0}\frac{\sin x}{x}\right]\left[\lim_{x\to0}\frac{\sin x}{1+\cos x}\right]$

$\qquad\qquad = (1)(0)=0$

75. False. Tangent, cotangent, secant and cosecant are not continuous at all real numbers.

Section 11.2   Trigonometric Functions: Differentiation

1. $f(x)=2\sin x+3\cos x$

$\quad f'(x)=2\cos x-3\sin x$

3. $f(x)=6\sqrt{x}+5\cos x=6x^{1/2}+5\cos x$

$\quad f'(x)=3x^{-1/2}-5\sin x=\dfrac{3}{\sqrt{x}}-5\sin x$

5. $f(x)=-x+\tan x$

$\quad f'(x)=-1+\sec^2 x=\tan^2 x$

7. $y=\dfrac{1}{x}-3\sin x$

$\quad y'=\dfrac{-1}{x^2}-3\cos x$

9. $f(x)=x^3\cos x$

$\quad f'(x)=x^3(-\sin x)+\cos x(3x^2)$

$\qquad\ = 3x^2\cos x-x^3\sin x$

11. $f(t)=t^2\sin t$

$\quad f'(t)=t^2\cos t+2t\sin t$

$\qquad\ = t(t\cos t+2\sin t)$

13. $f(t)=\dfrac{\cos t}{t}$

$\quad f'(t)=\dfrac{-t\sin t-\cos t}{t^2}=-\dfrac{t\sin t+\cos t}{t^2}$

15. $g(x) = \dfrac{\sin x}{x^2}$

$g'(x) = \dfrac{x^2 \cos x - \sin x (2x)}{x^4} = \dfrac{x \cos x - 2 \sin x}{x^3}$

17. $g(\theta) = \dfrac{\theta}{1 - \sin \theta}$

$g'(\theta) = \dfrac{1 - \sin \theta + \theta \cos \theta}{(\sin \theta - 1)^2}$ (form of answer may vary)

19. $y = \dfrac{3(1 - \sin x)}{2 \cos x} = \dfrac{3}{2}(\sec x - \tan x)$

$y' = \dfrac{3}{2}(\sec x \tan x - \sec^2 x) = \dfrac{3}{2}\sec x(\tan x - \sec x)$

$\quad = \dfrac{3}{2}(\sec x \tan x - \tan^2 x - 1)$

21. $y = -\csc x - \sin x$

$y' = \csc x \cot x - \cos x$

$\quad = \dfrac{\cos x}{\sin^2 x} - \cos x$

$\quad = \cos x(\csc^2 x - 1)$

$\quad = \cos x \cot^2 x$

23. $y = 2x \sin x + x^2 \cos x$

$y' = 2x \cos x + 2 \sin x + x^2(-\sin x) + 2x \cos x$

$\quad = 4x \cos x + 2 \sin x - x^2 \sin x$

25. (a) $y = \sin x$

$\qquad y' = \cos x$

$\qquad y'(0) = 1$

$\qquad$ 1 cycle in $[0, 2\pi]$.

(b) $y = \sin 2x$

$\qquad y' = 2 \cos 2x$

$\qquad y'(0) = 2$

$\qquad$ 2 cycles in $[0, 2\pi]$.

27. $f(x) = \sin x \cos x, \left(\dfrac{\pi}{2}, 0\right)$

$f'(x) = \cos^2 x - \sin^2 x$

$f'\left(\dfrac{\pi}{2}\right) = -1$

Tangent line: $y - 0 = -1\left(x - \dfrac{\pi}{2}\right)$

$\qquad\qquad y = -x + \dfrac{\pi}{2}$

29. $f(x) = e^{\sin x}, (0, 1)$

$f'(x) = e^{\sin x} \cdot \cos x$

$f'(0) = 1$

Tangent line: $y - 1 = 1(x - 0)$

$\qquad\qquad y = x + 1$

31. $y = \cos 3x$

$\dfrac{dy}{dx} = -3 \sin 3x$

33. $g(x) = 3 \tan 4x$

$g'(x) = 12 \sec^2 4x$

35. $y = \sin(\pi x)^2 = \sin(\pi^2 x^2)$

$y' = \cos(\pi^2 x^2)[2\pi^2 x] = 2\pi^2 x \cos(\pi^2 x^2)$

37. $h(x) = \sin 2x \cos 2x$

$h'(x) = \sin 2x(-2 \sin 2x) + \cos 2x(2 \cos 2x)$

$\quad = 2 \cos^2 2x - 2 \sin^2 2x$

$\quad = 2 \cos 4x$

Alternate solution: $h(x) = \dfrac{1}{2}\sin 4x$

$\qquad\qquad h'(x) = \dfrac{1}{2}\cos 4x(4) = 2 \cos 4x$

39. $f(x) = \dfrac{\cot x}{\sin x} = \dfrac{\cos x}{\sin^2 x}$

$f'(x) = \dfrac{\sin^2 x(-\sin x) - \cos x(2 \sin x \cos x)}{\sin^4 x}$

$\quad = \dfrac{-\sin^2 x - 2 \cos^2 x}{\sin^3 x} = \dfrac{-1 - \cos^2 x}{\sin^3 x}$

41. $y = 4 \sec^2 x$

$y' = 8 \sec x \cdot \sec x \tan x = 8 \sec^2 x \tan x$

43. $f(\theta) = \frac{1}{4}\sin^2 2\theta = \frac{1}{4}(\sin 2\theta)^2$

$f'(\theta) = 2\left(\frac{1}{4}\right)(\sin 2\theta)(\cos 2\theta)(2)$

$= \sin 2\theta \cos 2\theta = \frac{1}{4}\sin 4\theta$

45. $f(x) = 3\sec^2(\pi t - 1)$

$f'(t) = 6\sec(\pi t - 1)\sec(\pi t - 1)\tan(\pi t - 1)(\pi)$

$= 6\pi \sec^2(\pi t - 1)\tan(\pi t - 1) = \dfrac{6\pi \sin(\pi t - 1)}{\cos^3(\pi t - 1)}$

47. $y = \sqrt{x} + \frac{1}{4}\sin(2x)^2$

$= \sqrt{x} + \frac{1}{4}\sin(4x^2)$

$\dfrac{dy}{dx} = \frac{1}{2}x^{-1/2} + \frac{1}{4}\cos(4x^2)(8x)$

$= \dfrac{1}{2\sqrt{x}} + 2x\cos(2x)^2$

49. $y = \sin x^{1/3} + (\sin x)^{1/3}$

$y' = \cos x^{1/3}\left(\frac{1}{3}x^{-2/3}\right) + \frac{1}{3}(\sin x)^{-2/3}\cos x$

$= \frac{1}{3}\left[\dfrac{\cos x^{1/3}}{x^{2/3}} + \dfrac{\cos x}{(\sin x)^{2/3}}\right]$

51. $y = \ln|\sin x|$

$\dfrac{dy}{dx} = \dfrac{\cos x}{\sin x} = \cot x$

53. $y = \ln|\csc x - \cot x|$

$y' = \dfrac{-\csc x \cot x + \csc^2 x}{\csc x - \cot x}$

$= \dfrac{\csc x(\csc x - \cot x)}{\csc x - \cot x} = \csc x$

55. $y = \ln\left|\dfrac{\cos x}{1 - \sin x}\right| = \ln|\cos x| - \ln|1 - \sin x|$

$y' = \dfrac{1}{\cos x}(-\sin x) - \dfrac{1}{1 - \sin x}(-\cos x)$

$= \dfrac{-\sin x}{\cos x} + \dfrac{\cos x}{1 - \sin x} = \dfrac{-\sin x + \sin^2 x + \cos^2 x}{\cos x(1 - \sin x)}$

$= \dfrac{-\sin x + 1}{\cos x(1 - \sin x)}$

$= \dfrac{1}{\cos x} = \sec x$

57. Yes. f is continuous on $[0, \pi]$ and differentiable on $(0, \pi)$.

$f'(c) = \dfrac{f(b) - f(a)}{b - a}$

$\cos(c) = \dfrac{0 - 0}{\pi - 0} = 0$

$c = \dfrac{\pi}{2}$

59. $y = \dfrac{1 + \csc x}{1 - \csc x}$

$y' = \dfrac{(1 - \csc x)(-\csc x \cot x) - (1 + \csc x)(\csc x \cot x)}{(1 - \csc x)^2}$

$= \dfrac{-2\csc x \cot x}{(1 - \csc x)^2}$

$y'\left(\dfrac{\pi}{6}\right) = \dfrac{-2(2)(\sqrt{3})}{(1 - 2)^2} = -4\sqrt{3}$

61. $h(t) = \dfrac{\sec t}{t}$

$h'(t) = \dfrac{t(\sec t \tan t) - (\sec t)(1)}{t^2}$

$= \dfrac{\sec t(t \tan t - 1)}{t^2}$

$h'(\pi) = \dfrac{\sec \pi(\pi \tan \pi - 1)}{\pi^2} = \dfrac{1}{\pi^2}$

63. $y = e^x(\cos\sqrt{2}\,x + \sin\sqrt{2}\,x)$

$y' = e^x(-\sqrt{2}\sin\sqrt{2}\,x + \sqrt{2}\cos\sqrt{2}\,x) + e^x(\cos\sqrt{2}x + \sin\sqrt{2}\,x)$

$y'' = e^x(-2\cos\sqrt{2}\,x - 2\sin\sqrt{2}\,x) + 2e^x(-\sqrt{2}\sin\sqrt{2}\,x + \sqrt{2}\cos\sqrt{2}\,x) + e^x(\cos\sqrt{2}x + \sin\sqrt{2}\,x)$

$\quad = -e^x(\cos\sqrt{2}\,x + \sin\sqrt{2}\,x) + 2e^x(-\sqrt{2}\sin\sqrt{2}\,x + \sqrt{2}\cos\sqrt{2}\,x)$

Therefore, $y'' - 2y' + 3y = 0$.

65. $f(x) = x^n \sin x$

$f'(x) = x^n \cos x + nx^{n-1}\sin x$

$\quad = x^{n-1}(x\cos x + n\sin x)$

When $n = 1$: $f'(x) = x\cos x + \sin x$.

When $n = 2$: $f'(x) = x(x\cos x + 2\sin x)$.

When $n = 3$: $f'(x) = x^2(x\cos x + 3\sin x)$.

When $n = 4$: $f'(x) = x^3(x\cos x + 4\sin x)$.

For general n, $f'(x) = x^{n-1}(x\cos x + n\sin x)$.

67. $\sin x + 2\cos 2y = 1$

$\cos x - 4(\sin 2y)y' = 0$

$$y' = \frac{\cos x}{4\sin 2y}$$

69. $\sin x = x(1 + \tan y)$

$\cos x = x(\sec^2 y)y' + (1 + \tan y)(1)$

$$y' = \frac{\cos x - \tan y - 1}{x\sec^2 y}$$

71. $y = \sin(xy)$

$y' = [xy' + y]\cos(xy)$

$y' - x\cos(xy)y' = y\cos(xy)$

$$y' = \frac{y\cos(xy)}{1 - x\cos(xy)}$$

73. (a) $f(x) = \cos x \qquad f\left(\dfrac{\pi}{3}\right) = \dfrac{1}{2}$

$f'(x) = -\sin x \qquad f'\left(\dfrac{\pi}{3}\right) = -\dfrac{\sqrt{3}}{2}$

$f''(x) = -\cos x \qquad f''\left(\dfrac{\pi}{3}\right) = -\dfrac{1}{2}$

$P_1(x) = -\dfrac{\sqrt{3}}{2}\left(x - \dfrac{\pi}{3}\right) + \dfrac{1}{2}$

$P_2(x) = -\dfrac{1}{4}\left(x - \dfrac{\pi}{3}\right)^2 - \dfrac{\sqrt{3}}{2}\left(x - \dfrac{\pi}{3}\right) + \dfrac{1}{2}$

(b)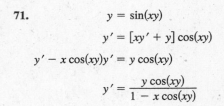

(c) P_2 is a better approximation than P_1.

(d) The accuracy worsens as you move away from $x = a = \dfrac{\pi}{3}$.

75. (a) $f(x) = \tan\dfrac{\pi x}{4}$ $\qquad\qquad f(1) = 1$

$f'(x) = \dfrac{\pi}{4}\sec^2\dfrac{\pi x}{4}$ $\qquad\qquad f'(1) = \dfrac{\pi}{4}(2) = \dfrac{\pi}{2}$

$f''(x) = \dfrac{\pi}{2}\sec^2\dfrac{\pi x}{4}\cdot\tan\dfrac{\pi x}{4}\left(\dfrac{\pi}{4}\right)$ $\qquad f''(1) = \dfrac{\pi^2}{8}(2)(1) = \dfrac{\pi^2}{4}$

(b)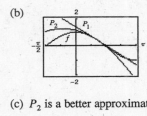

$P_1(x) = f'(1)(x - 1) + f(1) = \dfrac{\pi}{2}(x - 1) + 1$

$P_2(x) = \dfrac{1}{2}\left(\dfrac{\pi^2}{4}\right)(x - 1)^2 + f'(1)(x - 1) + f(1) = \dfrac{\pi^2}{8}(x - 1)^2 + \dfrac{\pi}{2}(x - 1) + 1$

(c) P_2 is a better approximation than P_1.

(d) The accuracy worsens as you move away from $x = a = 1$.

77. $y = \tan x$

$$\frac{dx}{dt} = 2$$

$$\frac{dy}{dt} = \sec^2 x \frac{dx}{dt}$$

(a) When $x = -\pi/3$,

$$\frac{dy}{dt} = (2)^2(2) = 8 \text{ cm/sec.}$$

(b) When $x = -\pi/4$,

$$\frac{dy}{dt} = \left(\sqrt{2}\right)^2(2) = 4 \text{ cm/sec.}$$

(c) When $x = 0$,

$$\frac{dy}{dt} = (1)^2(2) = 2 \text{ cm/sec.}$$

79. $f(x) = \cos \pi x, \left[0, \frac{1}{6}\right]$

$$f'(x) = -\pi \sin \pi x$$

Left endpoint: $(0, 1)$ Maximum

Right endpoint: $\left(\frac{1}{6}, \frac{\sqrt{3}}{2}\right)$ Minimum

81. $y = \frac{4}{x} + \tan \frac{\pi x}{8}, \left[1, \frac{7}{2}\right]$

$$y' = \frac{-4}{x^2} + \frac{\pi}{8} \sec^2 \frac{\pi x}{8} = 0$$

$$\frac{\pi}{8} \sec^2 \frac{\pi x}{8} = \frac{4}{x^2}$$

$$\cos^2 \frac{\pi x}{8} = \frac{\pi x^2}{32}$$

Using a graphing utility, $x \approx 2.1345$ and $f(x) \approx 2.9856$.

Endpoints: $\left(1, \sqrt{2} + 3\right) \approx (1, 4.4142)$

$$\left(\frac{7}{2}, 6.1702\right)$$

Minimum: $(2.1345, 2.9856)$

Maximum: $(3.5, 6.1702)$

83. $f(x) = \sin x, [0, 2\pi]$

$$f(0) = f(2\pi) = 0$$

f is continuous on $[0, 2\pi]$. f is differentiable on $(0, 2\pi)$. Rolle's Theorem applies.

$$f'(x) = \cos x$$

c values: $\frac{\pi}{2}, \frac{3\pi}{2}$

85. $f(x) = \cos 2x, \left[-\frac{\pi}{12}, \frac{\pi}{6}\right]$

$$f\left(-\frac{\pi}{12}\right) = \frac{\sqrt{3}}{2}$$

$$f\left(\frac{\pi}{6}\right) = \frac{1}{2}$$

$$f\left(-\frac{\pi}{12}\right) \neq f\left(\frac{\pi}{6}\right)$$

Rolle's Theorem does not apply.

87. $f(x) = \tan x, [0, \pi]$

$$f(0) = f(\pi) = 0$$

f is not continuous on $[0, \pi]$ since $f(\pi/2)$ does not exist. Rolle's Theorem does not apply.

89. $f(x) = 2 \sin x + \sin 2x$

$f'(x) = 2 \cos x + 2 \cos 2x = 0$

$2(2 \cos^2 x + \cos x - 1) = 0$

$2(\cos x + 1)(2 \cos x - 1) = 0$ when $\cos x = \dfrac{1}{2}, -1$ or $x = \dfrac{\pi}{3}, \pi, \dfrac{5\pi}{3}$.

$f''(x) = -2 \sin x - 4 \sin 2x$

$\quad = -2 \sin x(1 + 4 \cos x) = 0$ when $x = 0, 1.823, \pi, 4.460$.

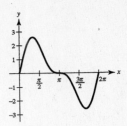

Relative maximum: $\left(\dfrac{\pi}{3}, 2.598 \right)$

Relative minimum: $\left(\dfrac{5\pi}{3}, -2.598 \right)$

Points of inflection: $(1.823, 1.452), (\pi, 0,) (4.46, -1.452)$

91. $y = \sin x - \dfrac{1}{18} \sin 3x, 0 \le x \le 2\pi$

$y' = \cos x - \dfrac{1}{6} \cos 3x = 0$ when $x = \dfrac{\pi}{2}, \dfrac{3\pi}{2}$.

$y'' = -\sin x + \dfrac{1}{2} \sin 3x = 0$ when $x = 0, \dfrac{\pi}{6}, \dfrac{5\pi}{6}, \pi, \dfrac{7\pi}{6}, \dfrac{11\pi}{6}$.

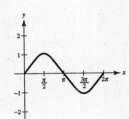

Relative maximum: $\left(\dfrac{\pi}{2}, \dfrac{19}{18} \right)$

Relative minimum: $\left(\dfrac{3\pi}{2}, -\dfrac{19}{18} \right)$

Inflection points: $\left(\dfrac{\pi}{6}, \dfrac{4}{9} \right), \left(\dfrac{5\pi}{6}, \dfrac{4}{9} \right), (\pi, 0), \left(\dfrac{7\pi}{6}, -\dfrac{4}{9} \right), \left(\dfrac{11\pi}{6}, -\dfrac{4}{9} \right)$

93. $f(x) = x - \sin x$

$f'(x) = 1 - \cos x \ge 0$

Therefore, f is nondecreasing, and there are no relative extrema.

$f''(x) = \sin x = 0$ when $x = 0, \pi, 2\pi, 3\pi, 4\pi$.

Points of inflection: $(\pi, \pi), (2\pi, 2\pi), (3\pi, 3\pi)$

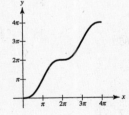

95. $y = 2(\csc x + \sec x), 0 < x < \dfrac{\pi}{2}$

$y' = 2(\sec x \tan x - \csc x \cot x) = 0 \implies x = \pi/4$

Relative minimum: $\left(\dfrac{\pi}{4}, 4\sqrt{2} \right)$

Vertical asymptotes: $x = 0, x = \dfrac{\pi}{2}$

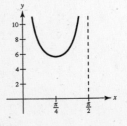

97. Since $\dfrac{\pi}{2} < 2 < 4 < \dfrac{3\pi}{2}$ and $f(x) = \sin x$ is decreasing

on $\left(\dfrac{\pi}{2}, \dfrac{3\pi}{2} \right), f(2) > f(4)$.

99. Answers will vary.
One example is

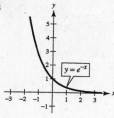

101. $y = A \cos \omega t$

  (a) Amplitude: $A = \dfrac{3.5}{2} = 1.75$

$$y = 1.75 \cos \omega t$$

     Period: $10 \implies \omega = \dfrac{2\pi}{10} = \dfrac{\pi}{5}$

$$y = 1.75 \cos \frac{\pi t}{5}$$

  (b) $v = y' = 1.75\left[-\dfrac{\pi}{5} \sin \dfrac{\pi t}{5} \right]$

$$= -0.35\pi \sin \frac{\pi t}{5}$$

103. $\tan \theta = \dfrac{y}{x}$

$$\frac{dx}{dt} = -600 \text{ mi/hr}$$

$$(\sec^2 \theta)\frac{d\theta}{dt} = -\frac{y}{x^2} \cdot \frac{dx}{dt}$$

$$\frac{d\theta}{dt} = \cos^2\theta\left(-\frac{y}{x^2}\right)\frac{dx}{dt} = \frac{x^2}{L^2}\left(-\frac{y}{x^2}\right)\frac{dx}{dt}$$

$$= \left(-\frac{y^2}{L^2}\right)\left(\frac{1}{y}\right)\frac{dx}{dt} = (-\sin^2\theta)\left(\frac{1}{5}\right)(-600) = 120\sin^2\theta$$

  (a) When $\theta = 30°$, $\dfrac{d\theta}{dt} = \dfrac{120}{4} = 30 \text{ rad/hr} = \dfrac{1}{2} \text{ rad/min.}$

  (b) When $\theta = 60°$, $\dfrac{d\theta}{dt} = 120\left(\dfrac{3}{4}\right) = 90 \text{ rad/hr} = \dfrac{3}{2} \text{ rad/min.}$

  (c) When $\theta = 75°$, $\dfrac{d\theta}{dt} = 120\sin^2 75° \approx 111.96 \text{ rad/hr} \approx 1.87 \text{ rad/min.}$

105. $r = \dfrac{v_0^{\,2}}{32}(\sin 2\theta)$

$v_0 = 2200 \text{ ft/sec}$

θ changes from $10°$ to $11°$.

$$dr = \frac{(2200)^2}{16}(\cos 2\theta)\,d\theta$$

$$\theta = 10\left(\frac{\pi}{180}\right)$$

$$d\theta = (11 - 10)\frac{\pi}{180}$$

$$\Delta r \approx dr$$

$$= \frac{(2200)^2}{16}\cos\left(\frac{20\pi}{180}\right)\left(\frac{\pi}{180}\right) \approx 4961 \text{ feet}$$

$$\approx 4961 \text{ feet}$$

107. $F \cos \theta = k(W - F \sin \theta)$

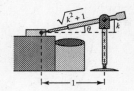

$$F = \frac{kW}{\cos \theta + k \sin \theta},$$

$$\frac{dF}{d\theta} = \frac{-kW(k \cos \theta - \sin \theta)}{(\cos \theta + k \sin \theta)^2} = 0$$

$k \cos \theta = \sin \theta \implies k = \tan \theta \implies \theta = \arctan k$

Since $\cos \theta + k \sin \theta = \dfrac{1}{\sqrt{k^2 + 1}} + \dfrac{k^2}{\sqrt{k^2 + 1}} = \sqrt{k^2 + 1}$,

the minimum force is $F = \dfrac{kW}{\cos \theta + k \sin \theta} = \dfrac{kW}{\sqrt{k^2 + 1}}.$

109. (a)

Base 1	Base 2	Altitude	Area
8	$8 + 16 \cos 10°$	$8 \sin 10°$	≈ 22.1
8	$8 + 16 \cos 20°$	$8 \sin 20°$	≈ 42.5
8	$8 + 16 \cos 30°$	$8 \sin 30°$	≈ 59.7
8	$8 + 16 \cos 40°$	$8 \sin 40°$	≈ 72.7
8	$8 + 16 \cos 50°$	$8 \sin 50°$	≈ 80.5
8	$8 + 16 \cos 60°$	$8 \sin 60°$	≈ 83.1

(b)

Base 1	Base 2	Altitude	Area
8	$8 + 16 \cos 10°$	$8 \sin 10°$	≈ 22.1
8	$8 + 16 \cos 20°$	$8 \sin 20°$	≈ 42.5
8	$8 + 16 \cos 30°$	$8 \sin 30°$	≈ 59.7
8	$8 + 16 \cos 40°$	$8 \sin 40°$	≈ 72.7
8	$8 + 16 \cos 50°$	$8 \sin 50°$	≈ 80.5
8	$8 + 16 \cos 60°$	$8 \sin 60°$	≈ 83.1
8	$8 + 16 \cos 70°$	$8 \sin 70°$	≈ 80.7
8	$8 + 16 \cos 80°$	$8 \sin 80°$	≈ 74.0
8	$8 + 16 \cos 90°$	$8 \sin 90°$	≈ 64.0

The maximum cross-sectional area is approximately 83.1 square feet.

(c) $A = (a + b)\dfrac{h}{2}$

$\quad = [8 + (8 + 16 \cos \theta)]\dfrac{8 \sin \theta}{2}$

$\quad = 64(1 + \cos \theta)\sin \theta, \, 0° < \theta < 90°$

(e)

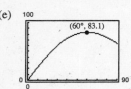

(d) $\dfrac{dA}{d\theta} = 64(1 + \cos \theta)\cos \theta + (-64 \sin \theta)\sin \theta$

$\quad = 64(\cos \theta + \cos^2 \theta - \sin^2 \theta)$

$\quad = 64(2 \cos^2 \theta + \cos \theta - 1)$

$\quad = 64(2 \cos \theta - 1)(\cos \theta + 1)$

$\quad = 0$ when $\theta = 60°, 180°, 300°$.

The maximum occurs when $\theta = 60°$.

111. $f(x) = \dfrac{\cos^2 \pi x}{\sqrt{x^2 + 1}}, \quad 0 < x < 4$

(a)

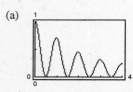

Critical numbers: 0.5, 0.9747, 1.5, 1.9796, 2.5, 2.9848, 3.5

(b) $f'(x) = \dfrac{-\cos \pi x[x \cos \pi x + 2\pi(x^2 + 1)\sin \pi x]}{(x^2 + 1)^{3/2}}$

$f'(x) = 0$ when $x \approx 0.5, 0.9747, 1.5, 1.9796, 2.5, 2.9848, 3.5$

They are the same values as in part (a).

113. False. If $y = (1 - x)^{1/2}$, then $y' = \frac{1}{2}(1 - x)^{-1/2}(-1)$.

115. False. First apply the Product Rule.

117. True. $y = \sin(bx)$

Slope: $y' = b\cos(bx)$

$$-b \le y' \le b \quad \text{(Assume } b > 0\text{)}$$

Section 11.3   Trigonometric Functions: Integration

1. $\displaystyle\int (2\sin x + 3\cos x)\, dx = -2\cos x + 3\sin x + C$

3. $\displaystyle\int (1 - \csc t \cot t)\, dt = t + \csc t + C$

5. $\displaystyle\int (\sec^2 \theta - \sin \theta)\, d\theta = \tan \theta + \cos \theta + C$

7. $\displaystyle\int (\tan^2 y + 1)\, dy = \int \sec^2 y\, dy = \tan y + C$

Check: $\dfrac{d}{dy}(\tan y + C) = \sec^2 y = \tan^2 y + 1$

9. $\displaystyle\int \pi \sin \pi x\, dx = -\cos \pi x + C$

11. $\displaystyle\int \sin 2x\, dx = \frac{1}{2}\int (\sin 2x)(2x)\, dx = -\frac{1}{2}\cos 2x + C$

13. $\displaystyle\int \frac{1}{\theta^2} \cos \frac{1}{\theta}\, d\theta = -\int \cos \frac{1}{\theta}\left(-\frac{1}{\theta^2}\right) d\theta = -\sin \frac{1}{\theta} + C$

15. $\displaystyle\int \sin 2x \cos 2x\, dx = \frac{1}{2}\int (\sin 2x)(2\cos 2x)\, dx = \frac{1}{2}\frac{(\sin 2x)^2}{2} + C = \frac{1}{4}\sin^2 2x + C$ OR

$\displaystyle\int \sin 2x \cos 2x\, dx = -\frac{1}{2}\int (\cos 2x)(-2\sin 2x)\, dx = -\frac{1}{2}\frac{(\cos 2x)^2}{2} + C_1 = -\frac{1}{4}\cos^2 2x + C_1$ OR

$\displaystyle\int \sin 2x \cos 2x\, dx = \frac{1}{2}\int 2\sin 2x \cos 2x\, dx = \frac{1}{2}\int \sin 4x\, dx = -\frac{1}{8}\cos 4x + C_2$

17. $\displaystyle\int \tan^4 x \sec^2 x\, dx = \frac{\tan^5 x}{5} + C = \frac{1}{5}\tan^5 x + C$

19. $\displaystyle\int \frac{\csc^2 x}{\cot^3 x}\, dx = -\int (\cot x)^{-3}(-\csc^2 x)\, dx$

$\displaystyle = -\frac{(\cot x)^{-2}}{-2} + C = \frac{1}{2\cot^2 x} + C = \frac{1}{2}\tan^2 x + C = \frac{1}{2}(\sec^2 x - 1) + C = \frac{1}{2}\sec^2 x + C_1$

21. $\displaystyle\int \cot^2 x\, dx = \int (\csc^2 x - 1)\, dx = -\cot x - x + C$

23. Let $u = e^x$, $du = e^x dx$.

$\displaystyle\int e^x \cos(e^x)\, dx = \int \cos u\, du = \sin u + C = \sin(e^x) + C$

25. $\displaystyle\int \frac{\cos \theta}{\sin \theta}\, d\theta = \ln|\sin \theta| + C$

$(u = \sin \theta, du = \cos \theta\, d\theta)$

27. $\displaystyle\int \csc 2x\, dx = \frac{1}{2}\int (\csc 2x)(2)\, dx$

$\displaystyle = -\frac{1}{2}\ln|\csc 2x + \cot 2x| + C$

29. $\displaystyle\int \frac{\cos t}{1 + \sin t}\, dt = \ln|1 + \sin t| + C$

31. $\displaystyle\int \frac{\sec x \tan x}{\sec x - 1}\, dx = \ln|\sec x - 1| + C$

33. $\displaystyle\int_0^{\pi} (1 + \sin x)\, dx = \left[x - \cos x \right]_0^{\pi}$

$$= (\pi + 1) - (0 - 1)$$

$$= 2 + \pi$$

35. $\displaystyle\int_{-\pi/6}^{\pi/6} \sec^2 x\, dx = \left[\tan x \right]_{-\pi/6}^{\pi/6} = \frac{\sqrt{3}}{3} - \left(-\frac{\sqrt{3}}{3} \right) = \frac{2\sqrt{3}}{3}$

37. $\displaystyle\int_{-\pi/3}^{\pi/3} 4\sec\theta \tan\theta\, d\theta = \left[4\sec\theta \right]_{-\pi/3}^{\pi/3} = 4(2) - 4(2) = 0$

39.

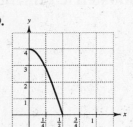

b. $A \approx \frac{4}{3}$ square units

41. (a)

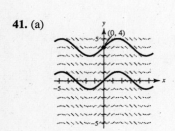

(b) $y = \displaystyle\int \cos x\, dx = \sin x + C$

$$4 = \sin 0 + C \implies C = 4$$

$$y = \sin x + 4$$

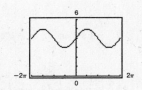

43. $s = \displaystyle\int \tan(2\theta)\, d\theta$

$$= \frac{1}{2}\int \tan(2\theta)(2\, d\theta)$$

$$= -\frac{1}{2}\ln|\cos 2\theta| + C$$

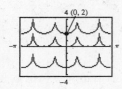

$(0, 2)$: $2 = -\dfrac{1}{2}\ln|\cos (0)| + C \implies C = 2$

$$s = -\frac{1}{2}\ln|\cos 2\theta| + 2$$

45. $\displaystyle\int \cos(1 - x)\, dx = -\sin(1 - x) + C$

47. $\displaystyle\int_{\pi/4}^{\pi/2} (\csc x - \sin x)\, dx = \left[-\ln|\csc x + \cot x| + \cos x \right]_{\pi/4}^{\pi/2}$

$$= \ln\left(\sqrt{2} + 1 \right) - \frac{\sqrt{2}}{2} \approx 0.174$$

49. $\displaystyle\int_0^{\pi/2} \sin^2 x\, dx \approx 0.7854 \left(= \frac{\pi}{4} \right)$

51. $\displaystyle\int_0^4 \sin \sqrt{x}\, dx \approx 3.4832$

53. Trapezoidal: $\displaystyle\int_0^{\sqrt{\pi/2}} \cos(x^2)\, dx \approx \frac{\sqrt{\pi/2}}{8} \cos 0 + 2\cos\left(\frac{\sqrt{\pi/2}}{4} \right)^2 + 2\cos\left(\frac{\sqrt{\pi/2}}{2} \right)^2 + 2\cos\left(\frac{3\sqrt{\pi/2}}{4} \right)^2 + \cos\left(\sqrt{\frac{\pi}{2}} \right)^2$

$$\approx 0.957$$

Simpson's: $\displaystyle\int_0^{\sqrt{\pi/2}} \cos(x^2)\, dx \approx \frac{\sqrt{\pi/2}}{12} \left[\cos 0 + 4\cos\left(\frac{\sqrt{\pi/2}}{4} \right)^2 + 2\cos\left(\frac{\sqrt{\pi/2}}{2} \right)^2 + 4\cos\left(\frac{3\sqrt{\pi/2}}{4} \right)^2 + \cos\left(\sqrt{\frac{\pi}{2}} \right)^2 \right]$

$$\approx 0.978$$

Graphing utility: 0.977

55. Trapezoidal: $\displaystyle\int_1^{1.1} \sin x^2\,dx \approx \frac{1}{80}[\sin(1) + 2\sin(1.025)^2 + 2\sin(1.05)^2 + 2\sin(1.075)^2 + \sin(1.1)^2] \approx 0.089$

Simpson's: $\displaystyle\int_1^{1.1} \sin x^2\,dx \approx \frac{1}{120}[\sin(1) + 4\sin(1.025)^2 + 2\sin(1.05)^2 + 4\sin(1.075)^2 + \sin(1.1)^2] \approx 0.089$

Graphing utility: 0.089

57. Trapezoidal: $\displaystyle\int_0^{\pi/4} x\tan x\,dx \approx \frac{\pi}{32}\left[0 + 2\left(\frac{\pi}{16}\right)\tan\left(\frac{\pi}{16}\right) + 2\left(\frac{2\pi}{16}\right)\tan\left(\frac{2\pi}{16}\right) + 2\left(\frac{3\pi}{16}\right)\tan\left(\frac{3\pi}{16}\right) + \frac{\pi}{4}\right] \approx 0.194$

Simpson's: $\displaystyle\int_0^{\pi/4} x\tan x\,dx \approx \frac{\pi}{48}\left[0 + 4\left(\frac{\pi}{16}\right)\tan\left(\frac{\pi}{16}\right) + 2\left(\frac{2\pi}{16}\right)\tan\left(\frac{2\pi}{16}\right) + 4\left(\frac{3\pi}{16}\right)\tan\left(\frac{3\pi}{16}\right) + \frac{\pi}{4}\right] \approx 0.186$

Graphing utility: 0.186

59. $A = \displaystyle\int_0^{\pi/2} \cos x\,dx = \left[\sin x\right]_0^{\pi/2} = 1$

61. $A = \displaystyle\int_0^{\pi} (2\sin x + \sin 2x)\,dx$

$= -\left[2\cos x + \dfrac{1}{2}\cos 2x\right]_0^{\pi} = 4$

63. $\text{Area} = \displaystyle\int_{\pi/2}^{2\pi/3} \sec^2\left(\frac{x}{2}\right)dx = 2\int_{\pi/2}^{2\pi/3} \sec^2\left(\frac{x}{2}\right)\left(\frac{1}{2}\right)dx = \left[2\tan\left(\frac{x}{2}\right)\right]_{\pi/2}^{2\pi/3} = 2\left(\sqrt{3} - 1\right)$

65. $\displaystyle\int_{-\pi/4}^{\pi/4} \sin x\,dx = 0$ since $\sin x$ is symmetric to the origin.

67. $f(x) = \sin^2 x\cos x$ is even.

$\displaystyle\int_{-\pi/2}^{\pi/2} \sin^2 x\cos x\,dx = 2\int_0^{\pi/2}\sin^2 x(\cos x)\,dx = 2\left[\frac{\sin^3 x}{3}\right]_0^{\pi/2} = \frac{2}{3}$

69. $\displaystyle\int_{-\pi/4}^{\pi/4} 2\sec^2 x\,dx = \left[2\tan x\right]_{-\pi/4}^{\pi/4} = 2(1) - 2(-1) = 4$

$f(c)\left[\dfrac{\pi}{4} - \left(-\dfrac{\pi}{4}\right)\right] = 4$

$2\sec^2 c = \dfrac{8}{\pi}$

$\sec^2 c = \dfrac{4}{\pi}$

$\sec c = \pm\dfrac{2}{\sqrt{\pi}}$

$c = \pm\operatorname{arcsec}\left(\dfrac{2}{\sqrt{\pi}}\right)$

$= \pm\arccos\dfrac{\sqrt{\pi}}{2} \approx \pm0.4817$

71. $\dfrac{1}{\pi - 0}\displaystyle\int_0^{\pi} \sin x\,dx = \left[-\dfrac{1}{\pi}\cos x\right]_0^{\pi} = \dfrac{2}{\pi}$

Average value $= \dfrac{2}{\pi}$

$\sin x = \dfrac{2}{\pi}$

$x \approx 0.690,\ 2.451$

73. $F(x) = \displaystyle\int_0^x t\cos t\,dt$

$F'(x) = x\cos x$

75. $F(x) = \displaystyle\int_{\pi/4}^x \sec^2 t\,dt$

$F'(x) = \sec^2 x$

77. No, $\tan x$ is not continuous at $x = \dfrac{\pi}{2}$.

79. $f(x) = x \cos x$ is odd and $[-\pi, \pi]$ is centered at 0.

81. (a)

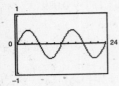

(b)

The area above the x-axis equals the area below the x-axis. Thus, the average value is zero.

The average value of S appears to be g.

83. $\dfrac{1}{b-a}\displaystyle\int_a^b \left[74.50 + 43.75 \sin \dfrac{\pi t}{6} \right] dt = \dfrac{1}{b-a}\left[74.50t - \dfrac{262.5}{\pi} \cos \dfrac{\pi t}{6} \right]_a^b$

(a) $\dfrac{1}{3}\left[74.50t - \dfrac{262.5}{\pi} \cos \dfrac{\pi t}{6} \right]_0^3 = \dfrac{1}{3}\left(223.5 + \dfrac{262.5}{\pi} \right) \approx 102.352$ thousand units

(b) $\dfrac{1}{3}\left[74.50t - \dfrac{262.5}{\pi} \cos \dfrac{\pi t}{6} \right]_3^6 = \dfrac{1}{3}\left(447 + \dfrac{262.5}{\pi} - 223.5 \right) \approx 102.352$ thousand units

(c) $\dfrac{1}{12}\left[74.50t - \dfrac{262.5}{\pi} \cos \dfrac{\pi t}{6} \right]_0^{12} = \dfrac{1}{12}\left(894 - \dfrac{262.5}{\pi} + \dfrac{262.5}{\pi} \right) = 74.5$ thousand units

85. $\dfrac{1}{b-a}\displaystyle\int_a^b \left[2 \sin(60\pi t) + \cos(120\pi t) \right] dt = \dfrac{1}{b-a}\left[-\dfrac{1}{30\pi} \cos(60\pi t) + \dfrac{1}{120\pi} \sin(120\pi t) \right]_a^b$

(a) $\dfrac{1}{(1/60) - 0}\left[-\dfrac{1}{30\pi} \cos(60\pi t) + \dfrac{1}{120\pi} \sin(120\pi t) \right]_0^{1/60} = 60\left[\left(\dfrac{1}{30\pi} + 0 \right) - \left(-\dfrac{1}{30\pi} \right) \right] = \dfrac{4}{\pi} \approx 1.273$ amps

(b) $\dfrac{1}{(1/240) - 0}\left[-\dfrac{1}{30\pi} \cos(60\pi t) + \dfrac{1}{120\pi} \sin(120\pi t) \right]_0^{1/240} = 240\left[\left(-\dfrac{1}{30\sqrt{2}\pi} + \dfrac{1}{120\pi} \right) - \left(-\dfrac{1}{30\pi} \right) \right]$

$= \dfrac{2}{\pi}\left(5 - 2\sqrt{2} \right) \approx 1.382$ amps

(c) $\dfrac{1}{(1/30) - 0}\left[-\dfrac{1}{30\pi} \cos(60\pi t) + \dfrac{1}{120\pi} \sin(120\pi t) \right]_0^{1/30} = 30\left[\left(\dfrac{-1}{30\pi} \right) - \left(-\dfrac{1}{30\pi} \right) \right] = 0$ amps

87. $-\ln|\cos x| + C = \ln\left| \dfrac{1}{\cos x} \right| + C = \ln|\sec x| + C$

89. $\ln|\sec x + \tan x| + C = \ln\left| \dfrac{(\sec x + \tan x)(\sec x - \tan x)}{(\sec x - \tan x)} \right| + C$

$= \ln\left| \dfrac{\sec^2 x - \tan^2 x}{\sec x - \tan x} \right| + C = \ln\left| \dfrac{1}{\sec x - \tan x} \right| + C = -\ln|\sec x - \tan x| + C$

91. True

$\displaystyle\int_a^b \sin x \, dx = \left[-\cos x \right]_a^b = -\cos b + \cos a = -\cos(b + 2\pi) + \cos a = \int_a^{b + 2\pi} \sin x \, dx$

93. False

$\displaystyle\int \sin^2 2x \cos 2x \, dx = \dfrac{1}{2}\int (\sin 2x)^2 (2 \cos 2x) \, dx = \dfrac{1}{2} \dfrac{(\sin 2x)^3}{3} + C = \dfrac{1}{6} \sin^3 2x + C$

Section 11.4   Inverse Trigonometric Functions: Differentiation

1. $y = \arcsin x$

$$y' = \frac{1}{\sqrt{1 - x^2}}$$

At $x = 0, y' = 1$.

3. $y = \arcsin \dfrac{x}{2}$

$$y' = \frac{1/2}{\sqrt{1 - x^2/4}}$$

At $x = 0, y' = \dfrac{1}{2}$.

5. $f(x) = 2 \arcsin(x - 1)$

$$f'(x) = \frac{2}{\sqrt{1 - (x - 1)^2}} = \frac{2}{\sqrt{2x - x^2}}$$

7. $g(x) = 3 \arccos \dfrac{x}{2}$

$$g'(x) = \frac{-3(1/2)}{\sqrt{1 - (x^2/4)}} = \frac{-3}{\sqrt{4 - x^2}}$$

9. $f(x) = \arctan \dfrac{x}{a}$

$$f'(x) = \frac{1/a}{1 + (x^2/a^2)} = \frac{a}{a^2 + x^2}$$

11. $g(x) = \dfrac{\arcsin 3x}{x}$

$$g'(x) = \frac{x\left(3/\sqrt{1 - 9x^2}\right) - \arcsin 3x}{x^2}$$

$$= \frac{3x - \sqrt{1 - 9x^2} \, \arcsin 3x}{x^2\sqrt{1 - 9x^2}}$$

13. $h(t) = \sin(\arccos t) = \sqrt{1 - t^2}$

$$h'(t) = \frac{1}{2}(1 - t^2)^{-1/2}(-2t) = \frac{-t}{\sqrt{1 - t^2}}$$

15. $y = x \arccos x - \sqrt{1 - x^2}$

$$y' = \arccos x - \frac{x}{\sqrt{1 - x^2}} - \frac{1}{2}(1 - x^2)^{-1/2}(-2x)$$

$$= \arccos x$$

17. $y = \dfrac{1}{2}\left(\dfrac{1}{2} \ln \dfrac{x + 1}{x - 1} + \arctan x\right)$

$$= \frac{1}{4}[\ln(x + 1) - \ln(x - 1)] + \frac{1}{2}\arctan x$$

$$\frac{dy}{dx} = \frac{1}{4}\left(\frac{1}{x + 1} - \frac{1}{x - 1}\right) + \frac{1/2}{1 + x^2} = \frac{1}{1 - x^4}$$

19. $y = x \arcsin x + \sqrt{1 - x^2}$

$$\frac{dy}{dx} = x\left(\frac{1}{\sqrt{1 - x^2}}\right) + \arcsin x - \frac{x}{\sqrt{1 - x^2}} = \arcsin x$$

21. $y = 8 \arcsin \dfrac{x}{4} - \dfrac{x\sqrt{16 - x^2}}{2}$

$$y' = 2\frac{1}{\sqrt{1 - (x/4)^2}} - \frac{\sqrt{16 - x^2}}{2} - \frac{x}{4}(16 - x^2)^{-1/2}(-2x)$$

$$= \frac{8}{\sqrt{16 - x^2}} - \frac{\sqrt{16 - x^2}}{2} + \frac{x^2}{2\sqrt{16 - x^2}}$$

$$= \frac{16 - (16 - x^2) + x^2}{2\sqrt{16 - x^2}} = \frac{x^2}{\sqrt{16 - x^2}}$$

23. $y = \arctan x + \dfrac{x}{1 + x^2}$

$$y' = \frac{1}{1 + x^2} + \frac{(1 + x^2) - x(2x)}{(1 + x^2)^2}$$

$$= \frac{(1 + x^2) + (1 - x^2)}{(1 + x^2)^2}$$

$$= \frac{2}{(1 + x^2)^2}$$

25. $f(x) = 2 \arcsin x, \left(\frac{1}{2}, \frac{\pi}{3}\right)$

$$f'(x) = \frac{2}{\sqrt{1 - x^2}}$$

$$f'\left(\frac{1}{2}\right) = \frac{2}{\sqrt{1 - 1/4}} = \frac{4}{\sqrt{3}} = \frac{4\sqrt{3}}{3}$$

Tangent line: $y - \frac{\pi}{3} = \frac{4\sqrt{3}}{3}\left(x - \frac{1}{2}\right)$

$$y = \frac{4\sqrt{3}}{3}x + \frac{\pi}{3} - \frac{2\sqrt{3}}{3}$$

27. $f(x) = \arcsin x, \, a = \frac{1}{2}$

$$f'(x) = \frac{1}{\sqrt{1 - x^2}}$$

$$f''(x) = \frac{x}{(1 - x^2)^{3/2}}$$

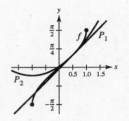

$$P_1(x) = f\left(\frac{1}{2}\right) + f'\left(\frac{1}{2}\right)\left(x - \frac{1}{2}\right) = \frac{\pi}{6} + \frac{2\sqrt{3}}{3}\left(x - \frac{1}{2}\right)$$

$$P_2(x) = f\left(\frac{1}{2}\right) + f'\left(\frac{1}{2}\right)\left(x - \frac{1}{2}\right) + \frac{1}{2}f''\left(\frac{1}{2}\right)\left(x - \frac{1}{2}\right)^2 = \frac{\pi}{6} + \frac{2\sqrt{3}}{3}\left(x - \frac{1}{2}\right) + \frac{2\sqrt{3}}{9}\left(x - \frac{1}{2}\right)^2$$

29. $\qquad f(x) = \operatorname{arcsec} x - x$

$$f'(x) = \frac{1}{|x|\sqrt{x^2 - 1}} - 1$$

$$= 0 \text{ when } |x|\sqrt{x^2 - 1} = 1.$$

$$x^2(x^2 - 1) = 1$$

$$x^4 - x^2 - 1 = 0 \text{ when } x^2 = \frac{1 + \sqrt{5}}{2} \text{ or}$$

$$x = \pm\sqrt{\frac{1 + \sqrt{5}}{2}} = \pm 1.272.$$

Relative maximum: $(1.272, -0.606)$

Relative minimum: $(-1.272, 3.747)$

31. $\quad f(x) = \arctan x - \arctan(x - 4)$

$$f'(x) = \frac{1}{1 + x^2} - \frac{1}{1 + (x - 4)^2} = 0$$

$$1 + x^2 = 1 + (x - 4)^2$$

$$0 = -8x + 16$$

$$x = 2$$

By the First Derivative Test, $(2, 2.214)$ is a relative maximum.

33. $f(x) = \arcsin x$ is an increasing function.

35. Since $\tan(\arctan x) = x$, $\dfrac{d}{dx}[\tan(\arctan x)] = \dfrac{d}{dx}(x) = 1$.

37. (a) $\cot \theta = \dfrac{x}{5}$

$$\theta = \operatorname{arccot}\left(\frac{x}{5}\right)$$

(b) $\dfrac{d\theta}{dt} = \dfrac{-\dfrac{1}{5}}{1 + \left(\dfrac{x}{5}\right)^2}\dfrac{dx}{dt} = \dfrac{-5}{x^2 + 25}\dfrac{dx}{dt}$

If $\dfrac{dx}{dt} = -400$ and $x = 10$, $\dfrac{d\theta}{dt} = 16$ rad/hr.

If $\dfrac{dx}{dt} = -400$ and $x = 3$, $\dfrac{d\theta}{dt} \approx 58.824$ rad/hr.

39. (a) $h(t) = -16t^2 + 256$

$-16t^2 + 256 = 0$ when $t = 4$ sec.

(b) $\tan \theta = \dfrac{h}{500} = \dfrac{-16t^2 + 256}{500}$

$\theta = \arctan\left[\dfrac{16}{500}(-t^2 + 16)\right]$

$\dfrac{d\theta}{dt} = \dfrac{-8t/125}{1 + [(4/125)(-t^2 + 16)]^2} = \dfrac{-1000t}{15{,}625 + 16(16 - t^2)^2}$

When $t = 1$, $d\theta/dt \approx -0.0520$ rad/sec.

When $t = 2$, $d\theta/dt \approx -0.1116$ rad/sec.

41. $\tan(\arctan x + \arctan y) = \dfrac{\tan(\arctan x) + \tan(\arctan y)}{1 - \tan(\arctan x)\tan(\arctan y)} = \dfrac{x + y}{1 - xy}, xy \neq 1$

Therefore,

$\arctan x + \arctan y = \arctan\left(\dfrac{x + y}{1 - xy}\right), xy \neq 1.$

Let $x = \frac{1}{2}$ and $y = \frac{1}{3}$.

$\arctan\left(\dfrac{1}{2}\right) + \arctan\left(\dfrac{1}{3}\right) = \arctan\dfrac{(1/2) + (1/3)}{1 - [(1/2) \cdot (1/3)]} = \arctan\dfrac{5/6}{1 - (1/6)} = \arctan\dfrac{5/6}{5/6} = \arctan 1 = \dfrac{\pi}{4}$

43. $f(x) = \arcsin\left(\dfrac{x - 2}{2}\right) - 2\arcsin\dfrac{\sqrt{x}}{2}$

$f'(x) = \dfrac{\dfrac{1}{2}}{\sqrt{1 - \left(\dfrac{x - 2}{2}\right)^2}} - 2\dfrac{\dfrac{1}{4}x^{-1/2}}{\sqrt{1 - \dfrac{x}{4}}}$

$= \dfrac{\dfrac{1}{2}}{\sqrt{1 - \left(\dfrac{x^2 - 4x + 4}{4}\right)}} - \dfrac{\dfrac{1}{2}}{\sqrt{x}\sqrt{1 - \dfrac{x}{4}}}$

$= \dfrac{1}{\sqrt{4x - x^2}} - \dfrac{1}{\sqrt{4x - x^2}} = 0$

Since $f'(x) = 0, f$ is constant.

45. True.

$\dfrac{d}{dx}[\arctan x] = \dfrac{1}{1 + x^2} > 0$ for all x.

47. False.

$\arccos\left(\dfrac{1}{2}\right) = \dfrac{\pi}{3}$

49. $y = (7 - 4x)^2$

$y' = 2(7 - 4x)(-4) = -8(7 - 4x) = -56 + 32x$

51. $y = \dfrac{x}{\sqrt{x + 5}}$

$y' = \dfrac{\sqrt{x + 5} - x\dfrac{1}{2}(x + 5)^{-1/2}}{x + 5}$

$= \dfrac{x + 5 - \dfrac{1}{2}x}{(x + 5)^{3/2}}$

$= \dfrac{10 + x}{2(x + 5)^{3/2}}$

53. $f(t) = 4\sin^2(2t)$

$f'(t) = 8\sin(2t)\cos(2t)(2)$

$= 16\sin(2t)\cos(2t)$

$= 8\sin(4t)$

55. $f(x) = \dfrac{\cos(4x)}{x}$

$f'(x) = \dfrac{-4\sin(4x)x - \cos(4x)}{x^2}$

57. $g(x) = \ln\dfrac{\sqrt{x+5}}{x^2+1} = \dfrac{1}{2}\ln(x+5) - \ln(x^2+1)$

$g'(x) = \dfrac{1}{2(x+5)} - \dfrac{2x}{x^2+1}$

$= \dfrac{-3x^2 - 20x + 1}{2(x+5)(x^2+1)}$

59. $s = t\ln(t^2+2)$

$s' = t\dfrac{2t}{t^2+2} + \ln(t^2+2)$

$= \dfrac{2t^2}{t^2+2} + \ln(t^2+2)$

61. $f(x) = e^{x/2}\sin x$

$f'(x) = e^{x/2}\cos x + \dfrac{1}{2}e^{x/2}\sin x$

$= \dfrac{1}{2}e^{x/2}[2\cos x + \sin x]$

63. $y = e^{-x}\ln x^2 = 2e^{-x}\ln x$

$y' = -2e^{-x}\ln x + \dfrac{2}{x}e^{-x}$

$= 2e^{-x}\left[\dfrac{1}{x} - \ln x\right]$

$= e^{-x}\left[\dfrac{2}{x} - \ln x^2\right]$

65. $y = 5^x$

$y' = 5^x \ln 5$

Section 11.5   Inverse Trigonometric Functions: Integration

1. $\displaystyle\int \dfrac{5}{\sqrt{9-x^2}}\,dx = 5\arcsin\left(\dfrac{x}{3}\right) + C$

3. Let $u = 3x$, $du = 3\,dx$.

$\displaystyle\int_0^{1/6} \dfrac{1}{\sqrt{1-9x^2}}\,dx = \dfrac{1}{3}\int_0^{1/6}\dfrac{1}{\sqrt{1-(3x)^2}}(3)\,dx = \left[\dfrac{1}{3}\arcsin(3x)\right]_0^{1/6} = \dfrac{\pi}{18}$

5. $\displaystyle\int \dfrac{7}{16+x^2}\,dx = \dfrac{7}{4}\arctan\left(\dfrac{x}{4}\right) + C$

7. Let $u = 2x$, $du = 2\,dx$.

$\displaystyle\int_0^{\sqrt{3}/2} \dfrac{1}{1+4x^2}\,dx = \dfrac{1}{2}\int_0^{\sqrt{3}/2}\dfrac{2}{1+(2x)^2}\,dx = \left[\dfrac{1}{2}\arctan(2x)\right]_0^{\sqrt{3}/2} = \dfrac{\pi}{6}$

9. $\displaystyle\int \dfrac{1}{x\sqrt{4x^2-1}}\,dx = \int \dfrac{2}{2x\sqrt{(2x)^2-1}}\,dx = \operatorname{arcsec}|2x| + C$

11. $\displaystyle\int \dfrac{x^3}{x^2+1}\,dx = \int\left[x - \dfrac{x}{x^2+1}\right]dx = \int x\,dx - \dfrac{1}{2}\int\dfrac{2x}{x^2+1}\,dx = \dfrac{1}{2}x^2 - \dfrac{1}{2}\ln(x^2+1) + C$ (Use long division.)

13. Let $u = t^2$, $du = 2t\,dt$.

$\displaystyle\int \dfrac{t}{\sqrt{1-t^4}}\,dt = \dfrac{1}{2}\int\dfrac{1}{\sqrt{1-(t^2)^2}}(2t)\,dt = \dfrac{1}{2}\arcsin(t^2) + C$

15. Let $u = \arcsin x$, $du = \dfrac{1}{\sqrt{1-x^2}}\,dx$.

$\displaystyle\int_0^{1/\sqrt{2}} \dfrac{\arcsin x}{\sqrt{1-x^2}}\,dx = \left[\dfrac{1}{2}\arcsin^2 x\right]_0^{1/\sqrt{2}} = \dfrac{\pi^2}{32} \approx 0.308$

17. Let $u = 1 - x^2$, $du = -2x\,dx$.

$$\int_{-1/2}^{0} \frac{x}{\sqrt{1 - x^2}}\,dx = -\frac{1}{2}\int_{-1/2}^{0} (1 - x^2)^{-1/2}(-2x)\,dx$$

$$= \left[-\sqrt{1 - x^2}\, \right]_{-1/2}^{0} = \frac{\sqrt{3} - 2}{2}$$

$$\approx -0.134$$

19. Let $u = e^{2x}$, $du = 2e^{2x}\,dx$.

$$\int \frac{e^{2x}}{4 + e^{4x}}\,dx = \frac{1}{2}\int \frac{2e^{2x}}{4 + (e^{2x})^2}\,dx = \frac{1}{4}\arctan\frac{e^{2x}}{2} + C$$

21. Let $u = \cos x$, $du = -\sin x\,dx$.

$$\int_{\pi/2}^{\pi} \frac{\sin x}{1 + \cos^2 x}\,dx = -\int_{\pi/2}^{\pi} \frac{-\sin x}{1 + \cos^2 x}\,dx$$

$$= \left[-\arctan(\cos x) \right]_{\pi/2}^{\pi} = \frac{\pi}{4}$$

23. $\displaystyle\int \frac{1}{\sqrt{x}\,\sqrt{1 - x}}\,dx$. $u = \sqrt{x}$, $x = u^2$, $dx = 2u\,du$

$$\int \frac{1}{u\sqrt{1 - u^2}}(2u\,du) = 2\int \frac{du}{\sqrt{1 - u^2}} = 2\arcsin u + C$$

$$= 2\arcsin\sqrt{x} + C$$

25. $\displaystyle\int \frac{x - 3}{x^2 + 1}\,dx = \frac{1}{2}\int \frac{2x}{x^2 + 1}\,dx - 3\int \frac{1}{x^2 + 1}\,dx$

$$= \frac{1}{2}\ln(x^2 + 1) - 3\arctan x + C$$

27. $\displaystyle\int \frac{x + 5}{\sqrt{9 - (x - 3)^2}}\,dx = \int \frac{(x - 3)}{\sqrt{9 - (x - 3)^2}}\,dx + \int \frac{8}{\sqrt{9 - (x - 3)^2}}\,dx$

$$= -\sqrt{9 - (x - 3)^2} + 8\arcsin\left(\frac{x - 3}{3}\right) + C$$

$$= -\sqrt{6x - x^2} + 8\arcsin\left(\frac{x}{3} - 1\right) + C$$

29. $\displaystyle\int_{0}^{2} \frac{1}{x^2 - 2x + 2}\,dx = \int_{0}^{2} \frac{1}{1 + (x - 1)^2}\,dx = \left[\arctan(x - 1) \right]_{0}^{2} = \frac{\pi}{2}$

31. $\displaystyle\int \frac{2x}{x^2 + 6x + 13}\,dx = \int \frac{2x + 6}{x^2 + 6x + 13}\,dx - 6\int \frac{1}{x^2 + 6x + 13}\,dx = \int \frac{2x + 6}{x^2 + 6x + 13}\,dx - 6\int \frac{1}{4 + (x + 3)^2}\,dx$

$$= \ln|x^2 + 6x + 13| - 3\arctan\left(\frac{x + 3}{2}\right) + C$$

33. $\displaystyle\int \frac{1}{\sqrt{-x^2 - 4x}}\,dx = \int \frac{1}{\sqrt{4 - (x + 2)^2}}\,dx$

$$= \arcsin\left(\frac{x + 2}{2}\right) + C$$

35. Let $u = -x^2 - 4x$, $du = (-2x - 4)\,dx$.

$$\int \frac{x + 2}{\sqrt{-x^2 - 4x}}\,dx = -\frac{1}{2}\int (-x^2 - 4x)^{-1/2}(-2x - 4)\,dx$$

$$= -\sqrt{-x^2 - 4x} + C$$

37. $\displaystyle\int_{2}^{3} \frac{2x - 3}{\sqrt{4x - x^2}}\,dx = \int_{2}^{3} \frac{2x - 4}{\sqrt{4x - x^2}}\,dx + \int_{2}^{3} \frac{1}{\sqrt{4x - x^2}}\,dx = -\int_{2}^{3} (4x - x^2)^{-1/2}(4 - 2x)\,dx + \int_{2}^{3} \frac{1}{\sqrt{4 - (x - 2)^2}}\,dx$

$$= \left[-2\sqrt{4x - x^2} + \arcsin\left(\frac{x - 2}{2}\right) \right]_{2}^{3} = 4 - 2\sqrt{3} + \frac{\pi}{6} \approx 1.059$$

39. Let $u = x^2 + 1$, $du = 2x\,dx$.

$$\int \frac{x}{x^4 + 2x^2 + 2}\,dx = \frac{1}{2}\int \frac{2x}{(x^2 + 1)^2 + 1}\,dx$$

$$= \frac{1}{2}\arctan(x^2 + 1) + C$$

41. Let $u = \sqrt{e^t - 3}$. Then $u^2 + 3 = e^t$, $2u\,du = e^t\,dt$, and $\dfrac{2u\,du}{u^2 + 3} = dt$.

$$\int \sqrt{e^t - 3}\,dt = \int \frac{2u^2}{u^2 + 3}\,du = \int 2\,du - \int 6\frac{1}{u^2 + 3}\,du$$

$$= 2u - 2\sqrt{3}\arctan \frac{u}{\sqrt{3}} + C = 2\sqrt{e^t - 3} - 2\sqrt{3}\arctan \sqrt{\frac{e^t - 3}{3}} + C$$

43. A perfect square trinomial is an expression in x with three terms that factors as a perfect square.

Example: $x^2 + 6x + 9 = (x + 3)^2$

45. (a) $\displaystyle\int \frac{1}{\sqrt{1 - x^2}}\,dx = \arcsin x + C$, $u = x$

(b) $\displaystyle\int \frac{x}{\sqrt{1 - x^2}}\,dx = -\sqrt{1 - x^2} + C$, $u = 1 - x^2$

(c) $\displaystyle\int \frac{1}{x\sqrt{1 - x^2}}\,dx$ cannot be evaluated using the basic integration rules.

47. (a)

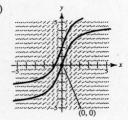

49. $\dfrac{dy}{dx} = \dfrac{10}{x\sqrt{x^2 - 1}}$, $y(3) = 0$

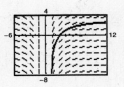

(b) $\dfrac{dy}{dx} = \dfrac{3}{1 + x^2}$, $(0, 0)$

$$y = 3\int \frac{dx}{1 + x^2} = 3\arctan x + C$$

$(0, 0)$: $0 = 3\arctan(0) + C \implies C = 0$

$$y = 3\arctan x$$

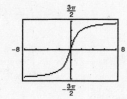

51. $A = \int_1^3 \frac{1}{x^2 - 2x + 1 + 4}\, dx = \int_1^3 \frac{1}{(x-1)^2 + 2^2}\, dx$

$= \left[\frac{1}{2} \arctan\left(\frac{x-1}{2}\right) \right]_1^3 = \frac{1}{2} \arctan(1) = \frac{\pi}{8} \approx 0.3927$

53. (a) $\int_0^1 \frac{4}{1 + x^2}\, dx = \left[4 \arctan x \right]_0^1 = 4 \arctan 1 - 4 \arctan 0 = 4\left(\frac{\pi}{4}\right) - 4(0) = \pi$

(b) Let $n = 6$.

$4\int_0^1 \frac{1}{1 + x^2}\, dx \approx 4\left(\frac{1}{18}\right)\left[1 + \frac{4}{1 + (1/36)} + \frac{2}{1 + (1/9)} + \frac{4}{1 + (1/4)} + \frac{2}{1 + (4/9)} + \frac{4}{1 + (25/36)} + \frac{1}{2} \right] \approx 3.1415918$

(c) 3.1415927

55. (a) $\frac{d}{dx}\left[\arcsin\left(\frac{u}{a}\right) + C \right] = \frac{1}{\sqrt{1 - (u^2/a^2)}}\left(\frac{u'}{a}\right) = \frac{u'}{\sqrt{a^2 - u^2}}$

Thus, $\int \frac{du}{\sqrt{a^2 - u^2}} = \arcsin\left(\frac{u}{a}\right) + C.$

(b) $\frac{d}{dx}\left[\frac{1}{a} \arctan \frac{u}{a} + C \right] = \frac{1}{a}\left[\frac{u'/a}{1 + (u/a)^2} \right] = \frac{1}{a^2}\left[\frac{u'}{(a^2 + u^2)/a^2} \right] = \frac{u'}{a^2 + u^2}$

Thus, $\int \frac{du}{a^2 + u^2} = \int \frac{u'}{a^2 + u^2}\, dx = \frac{1}{a} \arctan \frac{u}{a} + C.$

(c) Assume $u > 0$.

$\frac{d}{dx}\left[\frac{1}{a} \operatorname{arcsec} \frac{u}{a} + C \right] = \frac{1}{a}\left[\frac{u'/a}{(u/a)\sqrt{(u/a)^2 - 1}} \right] = \frac{1}{a}\left[\frac{u'}{u\sqrt{(u^2 - a^2)/a^2}} \right] = \frac{u'}{u\sqrt{u^2 - a^2}}.$

The case $u < 0$ is handled in a similar manner.

Thus, $\int \frac{du}{u\sqrt{u^2 - a^2}} = \int \frac{u'}{u\sqrt{u^2 - a^2}}\, dx = \frac{1}{a} \operatorname{arcsec} \frac{|u|}{a} + C.$

57. $\int \frac{1}{\sqrt{6x - x^2}}\, dx$

(a) $6x - x^2 = 9 - (x^2 - 6x + 9) = 9 - (x - 3)^2$

$\int \frac{1}{\sqrt{6x - x^2}}\, dx = \int \frac{dx}{\sqrt{9 - (x - 3)^2}} = \arcsin\left(\frac{x - 3}{3}\right) + C$

(b) $u = \sqrt{x},\ u^2 = x,\ 2u\, du = dx$

$\int \frac{1}{\sqrt{6u^2 - u^4}}(2u\, du) = \int \frac{2}{\sqrt{6 - u^2}}\, du = 2 \arcsin\left(\frac{u}{\sqrt{6}}\right) + C$

$= 2 \arcsin\left(\frac{\sqrt{x}}{\sqrt{6}}\right) + C$

(c)

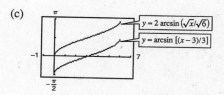

$y = 2 \arcsin\left(\sqrt{x}/\sqrt{6}\right)$

$y = \arcsin\left[(x - 3)/3\right]$

The antiderivatives differ by a constant, $\pi/2$.

Domain: $[0, 6]$

Section 11.6   Hyperbolic Functions

1. (a) $\sinh 3 = \frac{e^3 - e^{-3}}{2} \approx 10.018$

(b) $\tanh(-2) = \frac{\sinh(-2)}{\cosh(-2)} = \frac{e^{-2} - e^2}{e^{-2} + e^2} \approx -0.964$

3. (a) $\cosh^{-1}(2) = \ln\left(2 + \sqrt{3}\right) \approx 1.317$

(b) $\operatorname{sech}^{-1}\left(\frac{2}{3}\right) = \ln\left(\frac{1 + \sqrt{1 - (4/9)}}{2/3}\right) \approx 0.962$

5. $\tanh^2 x + \operatorname{sech}^2 x = \left(\dfrac{e^x - e^{-x}}{e^x + e^{-x}}\right)^2 + \left(\dfrac{2}{e^x + e^{-x}}\right)^2 = \dfrac{e^{2x} - 2 + e^{-2x} + 4}{(e^x + e^{-x})^2} = \dfrac{e^{2x} + 2 + e^{-2x}}{e^{2x} + 2 + e^{-2x}} = 1$

7. $\sinh x \cosh y + \cosh x \sinh y = \left(\dfrac{e^x - e^{-x}}{2}\right)\left(\dfrac{e^y + e^{-y}}{2}\right) + \left(\dfrac{e^x + e^{-x}}{2}\right)\left(\dfrac{e^y - e^{-y}}{2}\right)$

$$= \frac{1}{4}[e^{x+y} - e^{-x+y} + e^{x-y} - e^{-(x+y)} + e^{x+y} + e^{-x+y} - e^{x-y} - e^{-(x+y)}]$$

$$= \frac{1}{4}[2(e^{x+y} - e^{-(x+y)})] = \frac{e^{(x+y)} - e^{-(x+y)}}{2} = \sinh(x + y)$$

9. $\qquad \sinh x = \dfrac{3}{2}$

$\cosh^2 x - \left(\dfrac{3}{2}\right)^2 = 1 \implies \cosh^2 x = \dfrac{13}{4} \implies \cosh x = \dfrac{\sqrt{13}}{2}$

$\qquad \tanh x = \dfrac{3/2}{\sqrt{13}/2} = \dfrac{3\sqrt{13}}{13}$

$\qquad \operatorname{csch} x = \dfrac{1}{3/2} = \dfrac{2}{3}$

$\qquad \operatorname{sech} x = \dfrac{1}{\sqrt{13}/2} = \dfrac{2\sqrt{13}}{13}$

$\qquad \coth x = \dfrac{1}{3/\sqrt{13}} = \dfrac{\sqrt{13}}{3}$

11. $y = \sinh(1 - x^2)$

$\quad y' = -2x \cosh(1 - x^2)$

13. $f(x) = \ln(\sinh x)$

$\quad f'(x) = \dfrac{1}{\sinh x}(\cosh x) = \coth x$

15. $y = x \cosh x - \sinh x$

$\quad y' = x \sinh x + \cosh x - \cosh x = x \sinh x$

17. $f(t) = \arctan(\sinh t)$

$\quad f'(t) = \dfrac{1}{1 + \sinh^2 t}(\cosh t)$

$\qquad = \dfrac{\cosh t}{\cosh^2 t} = \operatorname{sech} t$

19. Let $y = g(x)$.

$\qquad y = x^{\cosh x}$

$\qquad \ln y = \cosh x \ln x$

$\qquad \dfrac{1}{y}\left(\dfrac{dy}{dx}\right) = \dfrac{\cosh x}{x} + \sinh x \ln x$

$\qquad \dfrac{dy}{dx} = \dfrac{y}{x}[\cosh x + x(\sinh x) \ln x]$

$\qquad = \dfrac{x^{\cosh x}}{x}[\cosh x + x(\sinh x) \ln x]$

21. $y = (\cosh x - \sinh x)^2$

$\quad y' = 2(\cosh x - \sinh x)(\sinh x - \cosh x)$

$\quad = -2(\cosh x - \sinh x)^2 = -2e^{-2x}$

23. $f(x) = \sin x \sinh x - \cos x \cosh x, \; -4 \le x \le 4$

$f'(x) = \sin x \cosh x + \cos x \sinh x - \cos x \sinh x + \sin x \cosh x$

$\quad\quad = 2 \sin x \cosh x = 0 \text{ when } x = 0, \pm \pi.$

Relative maxima: $(\pm \pi, \cosh \pi)$

Relative minimum: $(0, -1)$

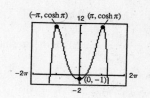

25. $g(x) = x \operatorname{sech} x = \dfrac{x}{\cosh x}$

Relative maximum:
$(1.20, 0.66)$

Relative minimum:
$(-1.20, -0.66)$

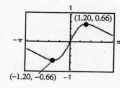

27. $y = a \sinh x$

$\quad y' = a \cosh x$

$\quad y'' = a \sinh x$

$\quad y''' = a \cosh x$

Therefore, $y''' - y' = 0.$

29. $f(x) = \tanh x \quad\quad\quad f(0) = 0$

$\quad f'(x) = \operatorname{sech}^2 x \quad\quad\quad f'(0) = 1$

$\quad f''(x) = -2 \operatorname{sech}^2 x \tanh x \quad f''(0) = 0$

$\quad P_1(x) = P_2(x) = x$

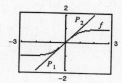

31. (a) $y = 10 + 15 \cosh \dfrac{x}{15}, \; -15 \le x \le 15$

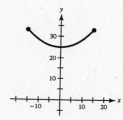

(b) At $x = \pm 15, y = 10 + 15 \cosh(1) \approx 33.146.$

At $x = 0, y = 10 + 15 \cosh(0) = 25.$

(c) $y' = \sinh \dfrac{x}{15}.$ At $x = 15, y' = \sinh(1) \approx 1.175$

33. Let $u = 1 - 2x, du = -2\, dx.$

$$\int \sinh(1 - 2x)\, dx = -\frac{1}{2} \int \sinh(1 - 2x)(-2)\, dx$$

$$= -\frac{1}{2} \cosh(1 - 2x) + C$$

35. Let $u = \sinh x, du = \cosh x\, dx.$

$$\int \frac{\cosh x}{\sinh x}\, dx = \ln|\sinh x| + C$$

37. Let $u = \dfrac{x^2}{2}, du = x\, dx.$

$$\int x \operatorname{csch}^2 \frac{x^2}{2}\, dx = \int \left(\operatorname{csch}^2 \frac{x^2}{2} \right) x\, dx = -\coth \frac{x^2}{2} + C$$

39. Let $u = \dfrac{1}{x}, du = -\dfrac{1}{x^2}\, dx.$

$$\int \frac{\operatorname{csch}(1/x) \coth(1/x)}{x^2}\, dx = -\int \operatorname{csch}\frac{1}{x} \coth \frac{1}{x}\left(-\frac{1}{x^2}\right) dx$$

$$= \operatorname{csch}\frac{1}{x} + C$$

41. $\displaystyle\int_0^4 \frac{1}{25 - x^2}\, dx = \left[\frac{1}{10} \ln \left| \frac{5 + x}{5 - x} \right| \right]_0^4 = \frac{1}{10} \ln 9 = \frac{1}{5} \ln 3$

43. Let $u = 2x, du = 2\, dx$.

$$\int_0^{\sqrt{2}/4} \frac{1}{\sqrt{1 - (2x)^2}}(2)\, dx = \left[\arcsin(2x) \right]_0^{\sqrt{2}/4} = \frac{\pi}{4}$$

45. Let $u = x^2, du = 2x\, dx$.

$$\int \frac{x}{x^4 + 1}\, dx = \frac{1}{2}\int \frac{2x}{(x^2)^2 + 1}\, dx = \frac{1}{2}\arctan(x^2) + C$$

47. $y = \cosh^{-1}(3x)$

$$y' = \frac{3}{\sqrt{9x^2 - 1}}$$

49. $y = \sinh^{-1}(\tan x)$

$$y' = \frac{1}{\sqrt{\tan^2 x + 1}}(\sec^2 x) = |\sec x|$$

51. $y = \coth^{-1}(\sin 2x)$

$$y' = \frac{1}{1 - \sin^2 2x}(2\cos 2x) = 2\sec 2x$$

53. $y = 2x\sinh^{-1}(2x) - \sqrt{1 + 4x^2}$

$$y' = 2x\left(\frac{2}{\sqrt{1 + 4x^2}}\right) + 2\sinh^{-1}(2x) - \frac{4x}{\sqrt{1 + 4x^2}}$$

$$= 2\sinh^{-1}(2x)$$

55. $y = a\,\text{sech}^{-1}\left(\dfrac{x}{a}\right) - \sqrt{a^2 - x^2}$

$$\frac{dy}{dx} = \frac{-1}{(x/a)\sqrt{1 - (x^2/a^2)}} + \frac{x}{\sqrt{a^2 - x^2}} = \frac{-a^2}{x\sqrt{a^2 - x^2}} + \frac{x}{\sqrt{a^2 - x^2}} = \frac{x^2 - a^2}{x\sqrt{a^2 - x^2}} = \frac{-\sqrt{a^2 - x^2}}{x}$$

57. See "Definition of the Hyperbolic Functions" on page 709.

59. $\displaystyle \int \frac{1}{\sqrt{1 + e^{2x}}}\, dx = \int \frac{e^x}{e^x\sqrt{1 + (e^x)^2}}\, dx = -\text{csch}^{-1}(e^x) + C = -\ln\left(\frac{1 + \sqrt{1 + e^{2x}}}{e^x}\right) + C$

61. Let $u = \sqrt{x}, du = \dfrac{1}{2\sqrt{x}}\, dx$.

$$\int \frac{1}{\sqrt{x}\sqrt{1 + x}}\, dx = 2\int \frac{1}{\sqrt{1 + (\sqrt{x})^2}}\left(\frac{1}{2\sqrt{x}}\right) dx = 2\sinh^{-1}\sqrt{x} + C = 2\ln\left(\sqrt{x} + \sqrt{1 + x}\right) + C$$

63. $\displaystyle \int \frac{-1}{4x - x^2}\, dx = \int \frac{1}{(x - 2)^2 - 4}\, dx = \frac{1}{4}\ln\left|\frac{(x - 2) - 2}{(x - 2) + 2}\right| = \frac{1}{4}\ln\left|\frac{x - 4}{x}\right| + C$

65. $\displaystyle \int \frac{1}{1 - 4x - 2x^2}\, dx = \int \frac{1}{3 - 2(x + 1)^2}\, dx = \frac{-1}{\sqrt{2}}\int \frac{\sqrt{2}}{\left[\sqrt{2}(x + 1)\right]^2 - (\sqrt{3})^2}\, dx$

$$= \frac{-1}{2\sqrt{6}}\ln\left|\frac{\sqrt{2}(x + 1) - \sqrt{3}}{\sqrt{2}(x + 1) + \sqrt{3}}\right| + C = \frac{1}{2\sqrt{6}}\ln\left|\frac{\sqrt{2}(x + 1) + \sqrt{3}}{\sqrt{2}(x + 1) - \sqrt{3}}\right| + C$$

67. $y = \displaystyle \int \frac{1}{\sqrt{16x^2 - 8x + 80}}\, dx$

$$= \int \frac{1}{\sqrt{(4x - 1)^2 + 79}}\, dx \quad (u = 4x - 1, a^2 = 79)$$

$$= \frac{1}{4}\ln\left[4x - 1 + \sqrt{(4x - 1)^2 + 79}\right] + C$$

$$= \frac{1}{4}\ln\left[4x - 1 + \sqrt{16x^2 - 8x + 80}\right] + C$$

$$= \frac{\ln[2\sqrt{2}\sqrt{2x^2 - x + 10} + 4x - 1]}{4} + C$$

69. $y = \int \dfrac{x^3 - 21x}{5 + 4x - x^2}\,dx$

$\qquad = \int \left(-x - 4 + \dfrac{20}{5 + 4x - x^2}\right) dx$

$\qquad = \int (-x - 4)\,dx + 20 \int \dfrac{1}{3^2 - (x - 2)^2}\,dx$

$\qquad = -\dfrac{x^2}{2} - 4x + \dfrac{20}{6} \ln\left|\dfrac{(x - 2) + 3}{(x - 2) - 3}\right| + C$

$\qquad = -\dfrac{x^2}{2} - 4x + \dfrac{10}{3} \ln\left|\dfrac{x + 1}{x - 5}\right| + C$

$\qquad = \dfrac{-x^2}{2} - 4x - \dfrac{10}{3} \ln\left|\dfrac{x - 5}{x + 1}\right| + C$

71. $A = \displaystyle\int_0^4 \operatorname{sech}\dfrac{x}{2} \tanh\dfrac{x}{2}\,dx$

$\qquad = -2 \operatorname{sech}\dfrac{x}{2}\Big]_0^4$

$\qquad = -2\,[\operatorname{sech} 2 - \operatorname{sech} 0]$

$\qquad = \dfrac{-4}{e^2 + e^{-2}} + 2$

73. $A = \displaystyle\int_0^2 \dfrac{5x}{\sqrt{x^4 + 1}}\,dx$

$\qquad = \dfrac{5}{2}\displaystyle\int_0^2 \dfrac{2x}{\sqrt{(x^2)^2 + 1}}\,dx$

$\qquad = \left[\dfrac{5}{2} \ln\left(x^2 + \sqrt{x^4 + 1}\right)\right]_0^2$

$\qquad = \dfrac{5}{2} \ln\left(4 + \sqrt{17}\right) \approx 5.237$

75. $y = \cosh x = \dfrac{e^x + e^{-x}}{2}$

$\qquad y' = \dfrac{e^x - e^{-x}}{2} = \sinh x$

77.

$\qquad\qquad y = \cosh^{-1} x$

$\qquad\quad \cosh y = x$

$\qquad (\sinh y)(y') = 1$

$\qquad\qquad y' = \dfrac{1}{\sinh y} = \dfrac{1}{\sqrt{\cosh^2 y - 1}} = \dfrac{1}{\sqrt{x^2 - 1}}$

79. $\displaystyle\int \dfrac{3k}{16}\,dt = \int \dfrac{1}{x^2 - 12x + 32}\,dx$

$\qquad \dfrac{3kt}{16} = \displaystyle\int \dfrac{1}{(x - 6)^2 - 4}\,dx = \dfrac{1}{2(2)} \ln\left|\dfrac{(x - 6) - 2}{(x - 6) + 2}\right| + C = \dfrac{1}{4} \ln\left|\dfrac{x - 8}{x - 4}\right| + C$

When $x = 0$: $\qquad t = 0$

$\qquad\qquad\qquad C = -\dfrac{1}{4} \ln(2)$

When $x = 1$: $\qquad t = 10$

$\qquad\qquad\qquad \dfrac{30k}{16} = \dfrac{1}{4} \ln\left|\dfrac{-7}{-3}\right| - \dfrac{1}{4} \ln(2) = \dfrac{1}{4} \ln\left(\dfrac{7}{6}\right)$

$\qquad\qquad\qquad k = \dfrac{2}{15} \ln\left(\dfrac{7}{6}\right)$

When $t = 20$: $\left(\dfrac{3}{16}\right)\left(\dfrac{2}{15}\right) \ln\left(\dfrac{7}{6}\right)(20) = \dfrac{1}{4} \ln\dfrac{x - 8}{2x - 8}$

$\qquad\qquad\qquad\qquad \ln\left(\dfrac{7}{6}\right)^2 = \ln\dfrac{x - 8}{2x - 8}$

$\qquad\qquad\qquad\qquad \dfrac{49}{36} = \dfrac{x - 8}{2x - 8}$

$\qquad\qquad\qquad\qquad 62x = 104$

$\qquad\qquad\qquad\qquad x = \dfrac{104}{62} = \dfrac{52}{31} \approx 1.677 \text{ kg}$

81. As k increases, the time required for the object to reach the ground increases.

Review Exercises for Chapter 11

1. $\displaystyle\lim_{x\to 3} \sec(\pi x) = \sec(3\pi) = -1$

3. $\displaystyle\lim_{x\to \pi/2} \cot x = \cot\left(\frac{\pi}{2}\right) = 0$

5. $\displaystyle\lim_{\alpha\to 0} \frac{\sin 5\alpha}{3\alpha} = \lim_{\alpha\to 0}\frac{\sin 5\alpha}{5\alpha}\cdot\frac{5}{3} = \frac{5}{3}$

7. $\displaystyle\lim_{\Delta x\to 0}\frac{\sin[(\pi/6)+\Delta x]-(1/2)}{\Delta x} = \lim_{\Delta x\to 0}\frac{\sin(\pi/6)\cos\Delta x+\cos(\pi/6)\sin\Delta x-(1/2)}{\Delta x}$

$$= \lim_{\Delta x\to 0}\frac{1}{2}\cdot\frac{(\cos\Delta x - 1)}{\Delta x} + \lim_{\Delta x\to 0}\frac{\sqrt{3}}{2}\cdot\frac{\sin\Delta x}{\Delta x}$$

$$= 0 + \frac{\sqrt{3}}{2}(1) = \frac{\sqrt{3}}{2}$$

9. $\displaystyle\lim_{x\to 0^+}\frac{\sin 4x}{5x} = \lim_{x\to 0^+}\left[\frac{4}{5}\left(\frac{\sin 4x}{4x}\right)\right] = \frac{4}{5}$

11. $\displaystyle\lim_{x\to 0^+}\frac{\csc 2x}{x} = \lim_{x\to 0^+}\frac{1}{x\sin 2x} = \infty$

13. $f(x) = \csc\dfrac{\pi x}{2}$

Nonremovable discontinuities at each even integer.

Continuous on

$$(2k, 2k+2)$$

for all integers k.

15. $f(x) = \dfrac{-4}{x} + \tan\dfrac{\pi x}{8}$ is continuous on $[1, 3]$.

$$f(1) = -4 + \tan\frac{\pi}{8} < 0 \text{ and } f(3) = -\frac{4}{3} + \tan\frac{3\pi}{8} > 0.$$

By the Intermediate Value Theorem, $f(1) = 0$ for at least one value of c between 1 and 3.

17. $f(\theta) = 2\theta - 3\sin\theta$

$f'(\theta) = 2 - 3\cos\theta$

19. $h(x) = \sqrt{x}\sin x = x^{1/2}\sin x$

$$h'(x) = \frac{1}{2\sqrt{x}}\sin x + \sqrt{x}\cos x$$

21. $y = \dfrac{x^2}{\cos x}$

$$y' = \frac{\cos x\,(2x) - x^2(-\sin x)}{\cos^2 x} = \frac{2x\cos x + x^2\sin x}{\cos^2 x}$$

23. $y = 3\sec x$

$y' = 3\sec x\tan x$

25. $y = \dfrac{1}{2}\csc 2x$

$$y' = \frac{1}{2}(-\csc 2x\cot 2x)(2)$$

$$= -\csc 2x\cot 2x$$

27. $y = \dfrac{x}{2} - \dfrac{\sin 2x}{4}$

$$y' = \frac{1}{2} - \frac{1}{4}\cos 2x(2)$$

$$= \frac{1}{2}(1 - \cos 2x) = \sin^2 x$$

29. $y = \dfrac{2}{3}\sin^{3/2}x - \dfrac{2}{7}\sin^{7/2}x$

$$y' = \sin^{1/2}x\cos x - \sin^{5/2}x\cos x$$

$$= (\cos x)\sqrt{\sin x}(1 - \sin^2 x)$$

$$= (\cos^3 x)\sqrt{\sin x}$$

31. $y = -x\tan x$

$$y' = -x\sec^2 x - \tan x$$

33. $y = \dfrac{\sin \pi x}{x + 2}$

$$y' = \dfrac{(x + 2)\pi \cos \pi x - \sin \pi x}{(x + 2)^2}$$

35. $y = \ln|\tan \theta|$

$$y' = \dfrac{1}{\tan \theta} \cdot \sec^2 \theta$$

$$= \dfrac{\cos \theta}{\sin \theta} \cdot \dfrac{1}{\cos^2 \theta}$$

$$= \csc \theta \cdot \sec \theta$$

$$\text{or, } y' = \dfrac{1}{\tan \theta}(\tan^2\theta + 1)$$

$$= \tan \theta + \cot \theta$$

37. $f(x) = \cot x$

$$f'(x) = -\csc^2 x$$

$$f'' = -2 \csc x(-\csc x \cdot \cot x)$$

$$= 2 \csc^2 x \cot x$$

39. $f(\theta) = 3 \tan \theta$

$$f'(\theta) = 3 \sec^2 \theta$$

$$f''(\theta) = 6 \sec \theta \,(\sec \theta \tan \theta)$$

$$= 6 \sec^2 \theta \tan \theta$$

41. $g(x) = \csc x, \left[\dfrac{\pi}{6}, \dfrac{\pi}{3}\right]$

$$g'(x) = -\csc x \cot x$$

Left endpoint: $\left(\dfrac{\pi}{6}, 2\right)$ Maximum

Right endpoint: $\left(\dfrac{\pi}{3}, \dfrac{2\sqrt{3}}{3}\right)$ Minimum

43.

$$y = 2 \sin x + 3 \cos x$$

$$y' = 2 \cos x - 3 \sin x$$

$$y'' = -2 \sin x - 3 \cos x$$

$$y'' + y = -(2 \sin x + 3 \cos x) + (2 \sin x + 3 \cos x)$$

$$= 0$$

45.

$$\tan(x + y) = x$$

$$(1 + y')\sec^2(x + y) = 1$$

$$y' = \dfrac{1 - \sec^2(x + y)}{\sec^2(x + y)}$$

$$= \dfrac{-\tan^2(x + y)}{\tan^2(x + y) + 1}$$

$$= -\dfrac{x^2}{x^2 + 1}$$

At $(0, 0)$: $y' = 0$.

47.
$$\csc \theta = \frac{L_1}{6} \text{ or } L_1 = 6 \csc \theta \qquad (\text{see figure})$$

$$\csc\left(\frac{\pi}{2} - \theta\right) = \frac{L_2}{9} \text{ or } L_2 = 9 \csc\left(\frac{\pi}{2} - \theta\right)$$

$$L = L_1 + L_2 = 6 \csc \theta + 9 \csc\left(\frac{\pi}{2} - \theta\right) = 6 \csc \theta + 9 \sec \theta$$

$$\frac{dL}{d\theta} = -6 \csc \theta \cot \theta + 9 \sec \theta \tan \theta = 0$$

$$\tan^3 \theta = \frac{2}{3} \Rightarrow \tan \theta = \frac{\sqrt[3]{2}}{\sqrt[3]{3}}$$

$$\sec \theta = \sqrt{1 + \tan^2 \theta} = \sqrt{1 + \left(\frac{2}{3}\right)^{2/3}} = \frac{\sqrt{3^{2/3} + 2^{2/3}}}{3^{1/3}}$$

$$\csc \theta = \frac{\sec \theta}{\tan \theta} = \frac{\sqrt{3^{2/3} + 2^{2/3}}}{2^{1/3}}$$

$$L = 6\frac{(3^{2/3} + 2^{2/3})^{1/2}}{2^{1/3}} + 9\frac{(3^{2/3} + 2^{2/3})^{1/2}}{3^{1/3}} = 3(3^{2/3} + 2^{2/3})^{3/2} \text{ ft} \approx 21.07 \text{ ft (Compare to Exercise 48 using } a = 6 \text{ and } b = 9.)$$

49. $\displaystyle\int (4x - 3 \sin x)\, dx = 2x^2 + 3 \cos x + C$

51. $\displaystyle\int \sin^3 x \cos x\, dx = \frac{1}{4} \sin^4 x + C$

53. $\displaystyle\int \frac{\sin \theta}{\sqrt{1 - \cos \theta}}\, dx = \int (1 - \cos \theta)^{-1/2} \sin \theta\, d\theta = 2(1 - \cos \theta)^{1/2} + C = 2\sqrt{1 - \cos \theta} + C$

55. $\displaystyle\int \tan^n x \sec^2 x\, dx = \frac{\tan^{n+1} x}{n + 1} + C, n \neq -1$

57. $\displaystyle\int \sec 2x \tan 2x\, dx = \frac{1}{2}\int (\sec 2x \tan 2x)(2)\, dx = \frac{1}{2} \sec 2x + C$

59. $\displaystyle\int \frac{\sin x}{1 + \cos x}\, dx = -\int \frac{-\sin x}{1 + \cos x}\, dx$

$$= -\ln|1 + \cos x| + C$$

61. $f(x) = \displaystyle\int \cos \frac{x}{2}\, dx = 2 \sin \frac{x}{2} + C$

Since $f(0) = 3 = 2 \sin 0 + C, C = 3$. Thus,

$$f(x) = 2 \sin \frac{x}{2} + 3.$$

63. $\displaystyle\int_0^\pi \cos\left(\frac{x}{2}\right) dx = 2\int_0^\pi \cos\left(\frac{x}{2}\right)\frac{1}{2}\, dx = \left[2 \sin\left(\frac{x}{2}\right)\right]_0^\pi = 2$

65. $\displaystyle\int_0^{\pi/3} \sec \theta\, d\theta = \left[\ln|\sec \theta + \tan \theta|\right]_0^{\pi/3} = \ln\left(2 + \sqrt{3}\right)$

67.

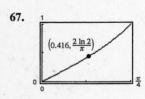

$$\text{Average value} = \frac{1}{\frac{\pi}{4} - 0}\int_0^{\pi/4} \tan x\, dx$$

$$= \frac{-4}{\pi}\ln|\cos x|\Big]_0^{\pi/4}$$

$$= \frac{-4}{\pi}\ln\left(\frac{\sqrt{2}}{2}\right)$$

$$= \frac{2}{\pi}\ln 2 \approx 0.4413$$

$$\tan x = \frac{2}{\pi}\ln 2 \Rightarrow x \approx 0.4156$$

69. Area $= \displaystyle\int_0^{\pi/3} \sec^2 x \, dx = \tan x \Big]_0^{\pi/3} = \sqrt{3}$

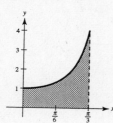

71. $F(x) = \displaystyle\int_0^x \tan^4 t \, dt$

$F'(x) = \tan^4 x$

73. (a) $C = 0.1 \displaystyle\int_8^{20} \left[12 \sin \frac{\pi(t-8)}{12} \right] dt = \left[-\frac{14.4}{\pi} \cos \frac{\pi(t-8)}{12} \right]_8^{20} = \frac{-14.4}{\pi}(-1-1) \approx \$9.17$

75. $\displaystyle\int_0^\pi \sqrt{x} \sin x \, dx, \; n = 4$

Trapezoidal Rule: 2.3290

Simpson's Rule: 2.4491

77. $y = \tan(\arcsin x) = \dfrac{x}{\sqrt{1-x^2}}$

$y' = \dfrac{(1-x^2)^{1/2} + x^2(1-x^2)^{-1/2}}{1-x^2} = (1-x^2)^{-3/2}$

79. $y = x \operatorname{arcsec} x$

$y' = \dfrac{x}{|x|\sqrt{x^2-1}} + \operatorname{arcsec} x$

81. $y = x(\arcsin x)^2 - 2x + 2\sqrt{1-x^2} \arcsin x$

$y' = \dfrac{2x \arcsin x}{\sqrt{1-x^2}} + (\arcsin x)^2 - 2 + \dfrac{2\sqrt{1-x^2}}{\sqrt{1-x^2}} - \dfrac{2x}{\sqrt{1-x^2}} \arcsin x = (\arcsin x)^2$

83. Let $u = e^{2x}, \; du = 2e^{2x} \, dx$.

$\displaystyle\int \frac{1}{e^{2x} + e^{-2x}} \, dx = \int \frac{e^{2x}}{1 + e^{4x}} \, dx = \frac{1}{2} \int \frac{1}{1 + (e^{2x})^2} (2e^{2x}) \, dx = \frac{1}{2} \arctan(e^{2x}) + C$

85. Let $u = 16 + x^2, \; du = 2x \, dx$.

$\displaystyle\int \frac{x}{16 + x^2} \, dx = \frac{1}{2} \int \frac{1}{16 + x^2} (2x) \, dx = \frac{1}{2} \ln(16 + x^2) + C$

87. Let $u = \arctan\left(\dfrac{x}{2}\right), \; du = \dfrac{2}{4 + x^2} \, dx$.

$\displaystyle\int \frac{\arctan(x/2)}{4 + x^2} \, dx = \frac{1}{2} \int \left(\arctan \frac{x}{2} \right)\left(\frac{2}{4 + x^2} \right) dx$

$= \dfrac{1}{4}\left(\arctan \dfrac{x}{2} \right)^2 + C$

89. $y' = \sqrt{1-y^2}$

(a)

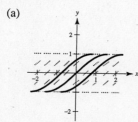

(b) Rate of change is greatest when $y = 0$. (slope lines are slope 1)

Rate of change is least when $y = \pm 1$. (slope lines are horizontal)

(c) $\displaystyle\int \frac{dy}{\sqrt{1-y^2}} = \int dx$

$\arcsin y = x + C$

$y = \sin(x + C), \; -\dfrac{\pi}{2} \le x + C \le \dfrac{\pi}{2}$

91. $y = 2x - \cosh\sqrt{x}$

$y' = 2 - \dfrac{1}{2\sqrt{x}}\left(\sinh\sqrt{x}\right) = 2 - \dfrac{\sinh\sqrt{x}}{2\sqrt{x}}$

93. Let $u = x^2$, $du = 2x\,dx$.

$\displaystyle\int \frac{x}{\sqrt{x^4 - 1}}\,dx = \frac{1}{2}\int \frac{1}{\sqrt{(x^2)^2 - 1}}(2x)\,dx$

$\displaystyle = \frac{1}{2}\ln\left(x^2 + \sqrt{x^4 - 1}\right) + C$

Problem Solving for Chapter 11

1. $f(x) = a + b\cos cx$

$f'(x) = -bc\sin cx$

At $(0, 1)$: $a + b = 1$ ⎯⎯ Equation 1

At $\left(\dfrac{\pi}{4}, \dfrac{3}{2}\right)$: $a + b\cos\left(\dfrac{c\pi}{4}\right) = \dfrac{3}{2}$ ⎯⎯ Equation 2

$-bc\sin\left(\dfrac{c\pi}{4}\right) = 1$ ⎯⎯ Equation 3

From Equation 1, $a = 1 - b$. Equation 2 becomes $(1 - b) + b\cos\left(\dfrac{c\pi}{4}\right) = \dfrac{3}{2} \Rightarrow -b + b\cos\dfrac{c\pi}{4} = \dfrac{1}{2}$.

From Equation 3, $b = \dfrac{-1}{c\sin\left(\dfrac{c\pi}{4}\right)}$. Thus $\dfrac{1}{c\sin\left(\dfrac{c\pi}{4}\right)} + \dfrac{-1}{c\sin\left(\dfrac{c\pi}{4}\right)}\cos\left(\dfrac{c\pi}{4}\right) = \dfrac{1}{2}$

$1 - \cos\left(\dfrac{c\pi}{4}\right) = \dfrac{1}{2}c\sin\left(\dfrac{c\pi}{4}\right)$.

Graphing the equation $g(c) = \dfrac{1}{2}c\sin\left(\dfrac{c\pi}{4}\right) + \cos\left(\dfrac{c\pi}{4}\right) - 1$, you see that many values of c will work.

One answer: $c = 2$, $b = -\dfrac{1}{2}$, $a = \dfrac{3}{2} \Rightarrow f(x) = \dfrac{3}{2} - \dfrac{1}{2}\cos 2x$

3. $E(\phi) = \dfrac{\tan\phi(1 - 0.1\tan\phi)}{0.1 + \tan\phi} = \dfrac{10\tan\phi - \tan^2\phi}{1 + 10\tan\phi}$

$E'(\phi) = \dfrac{(1 + 10\tan\phi)(10\sec^2\phi - 2\tan\phi\sec^2\phi) - (10\tan\phi - \tan^2\phi)10\sec^2\phi}{(1 + 10\tan\phi)^2} = 0$

$\Rightarrow (1 + 10\tan\phi)(10\sec^2\phi - 2\tan\phi\sec^2\phi) = (10\tan\phi - \tan^2\phi)10\sec^2\phi$

$\Rightarrow 10\sec^2\phi - 2\tan\phi\sec^2\phi + 100\tan\phi\sec^2\phi - 20\tan^2\phi\sec^2\phi = 100\tan\phi\sec^2\phi - 10\tan^2\phi\sec^2\phi$

$\Rightarrow 10 - 2\tan\phi = 10\tan^2\phi$

$\Rightarrow 10\tan^2\phi + 2\tan\phi - 10 = 0$

$\tan\phi = \dfrac{-2 \pm \sqrt{4 + 400}}{20} \approx 0.90499, -1.10499$

Using the positive value, $\phi \approx 0.736$, or $42.1°$.

5. $d = \sqrt{13^2 + x^2}$, $\sin \theta = \dfrac{x}{d}$

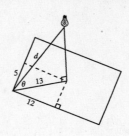

Let A be the amount of illumination at one of the corners, as indicated in the figure. Then

$$A = \frac{kI}{(13^2 + x^2)} \sin \theta = \frac{kIx}{(13^2 + x^2)^{3/2}}$$

$$A'(x) = kI \frac{(x^2 + 169)^{3/2}(1) - x\left(\dfrac{3}{2}\right)(x^2 + 169)^{1/2}(2x)}{(169 + x^2)^3} = 0$$

$$\Rightarrow\ (x^2 + 169)^{3/2} = 3x^2(x^2 + 169)^{1/2}$$

$$x^2 + 169 = 3x^2$$

$$2x^2 = 169$$

$$x = \frac{13}{\sqrt{2}} \approx 9.19 \text{ feet.}$$

By the First Derivative Test, this is a maximum.

7. Since $-|f(x)| \le f(x) \le |f(x)|$,

$$-\int_a^b |f(x)|\, dx \le \int_a^b f(x)\, dx \le \int_a^b |f(x)|\, dx \ \Rightarrow\ \left| \int_a^b f(x)\, dx \right| \le \int_a^b |f(x)|\, dx.$$

9. (a) The area of the sector of a circle of radius r is $A = \dfrac{tr^2}{2} = \dfrac{1}{2}r^2 t$. Since $r = 1$, $A = \dfrac{1}{2}(1)^2 t = \dfrac{1}{2}t$ or $t = 2A$.

   (b) Let $A = $ Area of AOP.

$$A = \frac{1}{2}\text{Base} \cdot \text{Height} - \int_1^{\cosh t} \sqrt{x^2 - 1}\, dx$$

$$= \frac{1}{2}\cosh t \cdot \sinh t - \int_1^{\cosh t} \sqrt{x^2 - 1}\, dx$$

$$= \frac{1}{2}\cosh t \cdot \sinh t - \left(\frac{-\ln|\sqrt{x^2 - 1} + x|}{2} + \frac{x\sqrt{x^2 - 1}}{2} \right)_1^{\cosh t}$$

$$= \frac{1}{2}\cosh t \cdot \sinh t - \frac{1}{2}\left[x\sqrt{x^2 - 1} - \ln\left|\sqrt{x^2 - 1} + x\right| \right]_1^{\cosh t}$$

$$= \frac{1}{2}\cosh t \cdot \sinh t - \frac{1}{2}\left(\cosh t \sinh t - \ln|\sinh t + \cosh t|\right)$$

$$= \frac{1}{2}\ln e^t$$

$$= \frac{1}{2}t$$

Therefore, $t = 2A$.

11. $\tan\theta_1 = \dfrac{3}{x}$

$\tan\theta_2 = \dfrac{6}{10-x}$

Minimize $\theta_1 + \theta_2$:

$f(x) = \theta_1 + \theta_2 = \arctan\left(\dfrac{3}{x}\right) + \arctan\left(\dfrac{6}{10-x}\right)$

$f'(x) = \dfrac{1}{1+\dfrac{9}{x^2}}\left(\dfrac{-3}{x^2}\right) + \dfrac{1}{1+\dfrac{36}{(10-x)^2}}\left(\dfrac{6}{(10-x)^2}\right) = 0$

$$\dfrac{3}{x^2+9} = \dfrac{6}{(10-x)^2+36}$$

$$(10-x)^2 + 36 = 2(x^2+9)$$

$$100 - 20x + x^2 + 36 = 2x^2 + 18$$

$$x^2 + 20x - 118 = 0$$

$$x = \dfrac{-20 \pm \sqrt{20^2 - 4(-118)}}{2} = -10 \pm \sqrt{218}$$

$a = -10 + \sqrt{218} \approx 4.7648 \quad f(a) \approx 1.4153$

$\theta = \pi - (\theta_1 + \theta_2) \approx 1.7263 \quad \text{or} \quad 98.9°$

Endpoints: $a = 0$: $\theta \approx 1.0304$

$\qquad\qquad\quad a = 10$: $\theta \approx 1.2793$

Maximum is 1.7263 at $a = -10 + \sqrt{218} \approx 4.7648$.

13. (a)

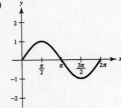

$$\int_0^\pi \sin x\, dx = -\int_\pi^{2\pi} \sin x\, dx \implies \int_0^{2\pi} \sin x\, dx = 0$$

(b)

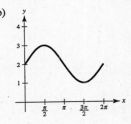

$$\int_0^{2\pi} (\sin x + 2)\, dx = 2(2\pi) = 4\pi$$

(c)

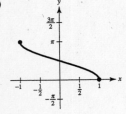

$$\int_{-1}^1 \arccos x\, dx = \dfrac{1}{2}(2\pi) = \pi$$

(d)

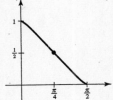

$y = \dfrac{1}{1 + (\tan x)^{\sqrt{2}}}$ is symmetric with respect to the

point $\left(\dfrac{\pi}{4}, \dfrac{1}{2}\right)$. $\displaystyle\int_0^{\pi/2} \dfrac{1}{1 + (\tan x)^{\sqrt{2}}}\, dx = \dfrac{\pi}{2}\left(\dfrac{1}{2}\right) = \dfrac{\pi}{4}$

CHAPTER 12
Topics in Analytic Geometry

CHAPTER 12
Topics in Analytic Geometry

Section 12.1   Introduction to Conics: Parabolas

1. $y^2 = -4x$

Vertex: $(0, 0)$

Opens to the left since
p is negative.

Matches graph (e).

3. $x^2 = -8y$

Vertex: $(0, 0)$

Opens downward since
p is negative.

Matches graph (d).

5. $(y - 1)^2 = 4(x - 3)$

Vertex: $(3, 1)$

Opens to the right since
p is positive.

Matches graph (a).

7. $y = \frac{1}{2}x^2$

$x^2 = 2y$

$x^2 = 4\left(\frac{1}{2}\right)y \implies h = 0, k = 0, p = \frac{1}{2}$

Vertex: $(0, 0)$

Focus: $\left(0, \frac{1}{2}\right)$

Directrix: $y = -\frac{1}{2}$

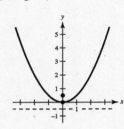

9. $y^2 = -6x$

$y^2 = 4\left(-\frac{3}{2}\right)x \implies h = 0, k = 0, p = -\frac{3}{2}$

Vertex: $(0, 0)$

Focus: $\left(-\frac{3}{2}, 0\right)$

Directrix: $x = \frac{3}{2}$

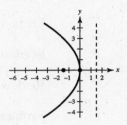

11. $x^2 + 6y = 0$

$x^2 = -6y = 4\left(-\frac{3}{2}\right)y \implies h = 0, k = 0, p = -\frac{3}{2}$

Vertex: $(0, 0)$

Focus: $\left(0, -\frac{3}{2}\right)$

Directrix: $y = \frac{3}{2}$

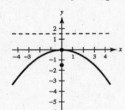

13. $(x - 1)^2 + 8(y + 2) = 0$

$(x - 1)^2 = 4(-2)(y + 2)$

$h = 1, k = -2, p = -2$

Vertex: $(1, -2)$

Focus: $(1, -4)$

Directrix: $y = 0$

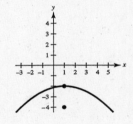

15. $\left(x + \frac{3}{2}\right)^2 = 4(y - 2)$

$\left(x + \frac{3}{2}\right)^2 = 4(1)(y - 2)$

$h = -\frac{3}{2}, k = 2, p = 1$

Vertex: $\left(-\frac{3}{2}, 2\right)$

Focus: $\left(-\frac{3}{2}, 3\right)$

Directrix: $y = 1$

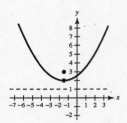

17.
$$y = \frac{1}{4}(x^2 - 2x + 5)$$
$$4y = x^2 - 2x + 5$$
$$4y - 5 + 1 = x^2 - 2x + 1$$
$$4y - 4 = (x - 1)^2$$
$$(x - 1)^2 = 4(1)(y - 1)$$

$h = 1, k = 1, p = 1$

Vertex: $(1, 1)$

Focus: $(1, 2)$

Directrix: $y = 0$

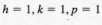

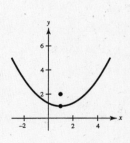

19.
$$y^2 + 6y + 8x + 25 = 0$$
$$y^2 + 6y + 9 = -8x - 25 + 9$$
$$(y + 3)^2 = 4(-2)(x + 2)$$
$h = -2, k = -3, p = -2$

Vertex: $(-2, -3)$

Focus: $(-4, -3)$

Directrix: $x = 0$

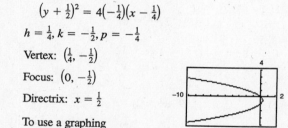

21. $x^2 + 4x + 6y - 2 = 0$
$$x^2 + 4x = -6y + 2$$
$$x^2 + 4x + 4 = -6y + 2 + 4$$
$$(x + 2)^2 = -6(y - 1)$$
$$(x + 2)^2 = 4\left(-\tfrac{3}{2}\right)(y - 1)$$
$h = -2, k = 1, p = -\tfrac{3}{2}$

Vertex: $(-2, 1)$

Focus: $\left(-2, -\tfrac{1}{2}\right)$

Directrix: $y = \tfrac{5}{2}$

On the graphing calculator, enter:

$y_1 = -\tfrac{1}{6}(x^2 + 4x - 2)$

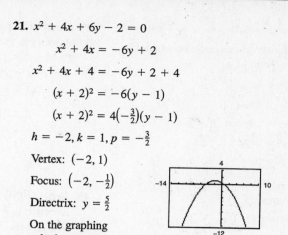

23. $y^2 + x + y = 0$
$$y^2 + y + \tfrac{1}{4} = -x + \tfrac{1}{4}$$
$$\left(y + \tfrac{1}{2}\right)^2 = 4\left(-\tfrac{1}{4}\right)\left(x - \tfrac{1}{4}\right)$$
$h = \tfrac{1}{4}, k = -\tfrac{1}{2}, p = -\tfrac{1}{4}$

Vertex: $\left(\tfrac{1}{4}, -\tfrac{1}{2}\right)$

Focus: $\left(0, -\tfrac{1}{2}\right)$

Directrix: $x = \tfrac{1}{2}$

To use a graphing calculator, enter:

$y_1 = -\tfrac{1}{2} + \sqrt{\tfrac{1}{4} - x}$

$y_2 = -\tfrac{1}{2} - \sqrt{\tfrac{1}{4} - x}$

25. $y^2 - 8x = 0 \implies y = \pm\sqrt{8x}$
$$x - y + 2 = 0 \implies y = x + 2$$
The point of tangency is $(2, 4)$.

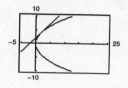

27. Vertex: $(0, 0) \implies h = 0, k = 0$

Graph opens upward.

$x^2 = 4py$

Point on graph: $(3, 6)$

$3^2 = 4p(6)$

$9 = 24p$

$\tfrac{3}{8} = p$

Thus, $x^2 = 4\left(\tfrac{3}{8}\right)y \implies x^2 = \tfrac{3}{2}y$.

29. Vertex: $(0, 0) \implies h = 0, k = 0$

Focus: $\left(0, -\tfrac{3}{2}\right) \implies p = -\tfrac{3}{2}$

$x^2 = 4py$

$x^2 = 4\left(-\tfrac{3}{2}\right)y$

$x^2 = -6y$

31. Vertex: $(0, 0) \implies h = 0, k = 0$

Focus: $(-2, 0) \implies p = -2$

$y^2 = 4px$

$y^2 = 4(-2)x$

$y^2 = -8x$

33. Vertex: $(0, 0) \implies h = 0, k = 0$

Directrix: $y = -1 \implies p = 1$

$x^2 = 4py$

$x^2 = 4(1)y$

$x^2 = 4y$

35. Vertex: $(0, 0) \implies h = 0, k = 0$

Directrix: $x = 2 \implies p = -2$

$y^2 = 4px$

$y^2 = 4(-2)x$

$y^2 = -8x$

37. Vertex: $(0, 0) \implies h = 0, k = 0$

Horizontal axis and passes through the point $(4, 6)$

$y^2 = 4px$

$6^2 = 4p(4)$

$36 = 16p \implies p = \frac{9}{4}$

$y^2 = 4\left(\frac{9}{4}\right)x$

$y^2 = 9x$

39. Vertex: $(3, 1)$ and opens downward. Passes through $(2, 0)$ and $(4, 0)$.

$y = -(x - 2)(x - 4)$

$= -x^2 + 6x - 8$

$= -(x - 3)^2 + 1$

$(x - 3)^2 = -(y - 1)$

41. Vertex: $(-4, 0)$ and opens to the right. Passes through $(0, 4)$.

$(y - 0)^2 = 4p(x + 4)$

$4^2 = 4p(0 + 4)$

$16 = 16p$

$1 = p$

$y^2 = 4(x + 4)$

43. Vertex: $(5, 2)$

Focus: $(3, 2)$

Horizontal axis

$p = 3 - 5 = -2$

$(y - 2)^2 = 4(-2)(x - 5)$

$(y - 2)^2 = -8(x - 5)$

45. Vertex: $(0, 4)$

Directrix: $y = 2$

Vertical axis

$p = 4 - 2 = 2$

$(x - 0)^2 = 4(2)(y - 4)$

$x^2 = 8(y - 4)$

47. Focus: $(2, 2)$

Directrix: $x = -2$

Horizontal axis

Vertex: $(0, 2)$

$p = 2 - 0 = 2$

$(y - 2)^2 = 4(2)(x - 0)$

$(y - 2)^2 = 8x$

49. $(y - 3)^2 = 6(x + 1)$

For the upper half of the parabola:

$y - 3 = +\sqrt{6(x + 1)}$

$y = \sqrt{6(x + 1)} + 3$

51. $x^2 = 4y$

$2x = 4\dfrac{dy}{dx}$

$\dfrac{dy}{dx} = \dfrac{x}{2}$

53. $y^2 = 6x$

$2y\dfrac{dy}{dx} = 6$

$\dfrac{dy}{dx} = \dfrac{3}{y}$

55. $(x - 2)^2 = 6(y + 3)$

$2(x - 2) = 6\dfrac{dy}{dx}$

$\dfrac{dy}{dx} = \dfrac{x - 2}{3}$

57. $(y + 3)^2 = -8(x - 2)$

$2(y + 3)\dfrac{dy}{dx} = -8$

$\dfrac{dy}{dx} = -\dfrac{4}{y + 3}$

59. $x^2 = 2y$

$2x = 2\dfrac{dy}{dx}$

$\dfrac{dy}{dx} = x$

At $(4, 8)$, $m = \dfrac{dy}{dx}\bigg]_{(4,\, 8)} = 4.$

Tangent line: $y - 8 = 4(x - 4)$

$4x - y - 8 = 0$

61. $y = -2x^2$

$\dfrac{dy}{dx} = -4x$

At $(-1, -2)$, $m = \dfrac{dy}{dx}\bigg]_{(-1,\, -2)} = 4.$

Tangent line: $y + 2 = 4(x + 1)$

$4x - y + 2 = 0$

63. $y^2 = 2(x - 3)$

$2y\dfrac{dy}{dx} = 2$

$\dfrac{dy}{dx} = \dfrac{1}{y}$

At $(5, 2)$, $m = \dfrac{dy}{dx}\bigg]_{(5,\, 2)} = \dfrac{1}{2}.$

Tangent line: $y - 2 = \dfrac{1}{2}(x - 5)$

$2y - 4 = x - 5$

$x - 2y - 1 = 0$

65. $(x - 1)^2 = 6(y + 2)$

$2(x - 1) = 6\dfrac{dy}{dx}$

$\dfrac{dy}{dx} = \dfrac{x - 1}{3}$

At $(-5, 4)$, $m = \dfrac{dy}{dx}\Big]_{(-5,\,4)} = -2$.

Tangent line: $y - 4 = -2(x + 5)$

$2x + y + 6 = 0$

67. $R = 265x - \dfrac{5}{4}x^2$

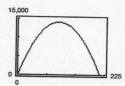

The revenue is maximum when $x = 106$ units.

69. A circle is formed when a plane intersects the top or bottom half of a double-napped cone and is perpendicular to the axis through the center of the cone.

71. A parabola is formed when a plane intersects the top or bottom half of a double-napped cone, is parallel to the side of the cone, and does not intersect the vertex.

73. $x^2 = 4py$

(a) As p increases, the graph becomes wider.

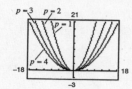

75. Vertex: $(0, 0) \implies h = 0, k = 0$

Focus: $(0, 4.5) \implies p = 4.5$

$(x - h)^2 = 4p(y - k)$

$(x - 0)^2 = 4(4.5)(y - 0)$

$x^2 = 18y$ or $y = \dfrac{1}{18}x^2$

(b)

Value of p	Focus
$p = 1$	$(0, 1)$
$p = 2$	$(0, 2)$
$p = 3$	$(0, 3)$
$p = 4$	$(0, 4)$

(c) The length of the focal chord is $4|p|$.

When $p = 1$, $4|p| = 4$.

When $p = 2$, $4|p| = 8$.

When $p = 3$, $4|p| = 12$.

When $p = 4$, $4|p| = 16$.

(d) The endpoints of the focal chord give us two easily determined additional points on the graph.

77. (a) Vertex: $(0, 0) \implies h = 0, k = 0$

Points on the parabola: $(\pm 16, -0.4)$

$x^2 = 4py$

$(\pm 16)^2 = 4p(-0.4)$

$256 = -1.6p$

$-160 = p$

$x^2 = 4(-160y)$

$x^2 = -640y$

$y = -\dfrac{1}{640}x^2$

(b) When $y = -0.1$ we have, $-0.1 = -\dfrac{1}{640}x^2$

$64 = x^2$

$\pm 8 = x$.

Thus, 8 feet away from the center of the road, the road surface is 0.1 foot lower than in the middle.

79. The position of the target is on the x-axis, so first, let $y = 0$.

$$0 = 30,000 - \frac{x^2}{39,204}$$

$$\frac{x^2}{39,204} = 30,000$$

$$x^2 = 30,000(39,204)$$

$$x = \sqrt{30,000(39,204)}$$

$$x \approx 34,294.606 \text{ feet}$$

Since the bomber is flying at 792 feet per second,

$$\frac{34,294.606 \text{ feet}}{792 \text{ feet per second}} \approx 43.3 \text{ seconds.}$$

The bomb should be released 43.3 seconds prior to being over the target.

81. $x^2 = 2y, y = 3$

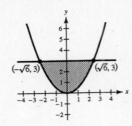

$$A = 2 \int_0^{\sqrt{6}} \left(3 - \frac{x^2}{2} \right) dx$$

$$= 2 \left[3x - \frac{x^3}{6} \right]_0^{\sqrt{6}}$$

$$= 4\sqrt{6} \text{ square units}$$

83. $y^2 = 4x, x = 5$

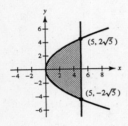

$$A = 2 \int_0^{2\sqrt{5}} \left(5 - \frac{y^2}{4} \right) dy$$

$$= 2 \left[5y - \frac{y^3}{12} \right]_0^{2\sqrt{5}}$$

$$= \frac{40\sqrt{5}}{3} \text{ square units}$$

85. $(x - 2)^2 = 4y, x = 0, x = 4, y = 0$

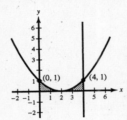

By symmetry $A = 2 \int_0^2 \frac{1}{4}(x - 2)^2 dx$

$$= \frac{1}{2} \frac{(x - 2)^3}{3} \Big]_0^2$$

$$= \frac{4}{3} \text{ square units}$$

87. False. It is not possible for a parabola to intersect its directrix. If the graph crossed the directrix there would exist points nearer the directrix than the focus.

Section 12.2   Ellipses and Implicit Differentiation

1. $\dfrac{x^2}{4} + \dfrac{y^2}{9} = 1$

Center: $(0, 0)$

$a = 3, b = 2$

Vertical major axis

Matches graph (b).

3. $\dfrac{x^2}{4} + \dfrac{y^2}{25} = 1$

Center: $(0, 0)$

$a = 5, b = 2$

Vertical major axis

Matches graph (d).

5. $\dfrac{(x - 2)^2}{16} + (y + 1)^2 = 1$

Center: $(2, -1)$

$a = 4, b = 1$

Horizontal major axis
Matches graph (a).

7. $\dfrac{x^2}{25} + \dfrac{y^2}{16} = 1$

Center: $(0, 0)$

$a = 5, b = 4, c = 3$

Foci: $(\pm 3, 0)$

Vertices: $(\pm 5, 0)$

$e = \dfrac{3}{5}$

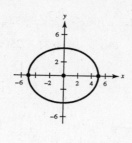

9. $\dfrac{x^2}{5} + \dfrac{y^2}{9} = 1$

$a = 3, b = \sqrt{5}, c = 2$

Center: $(0, 0)$

Foci: $(0, \pm 2)$

Vertices: $(0, \pm 3)$

$e = \dfrac{2}{3}$

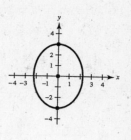

11. $\dfrac{(x + 3)^2}{16} + \dfrac{(y - 5)^2}{25} = 1$

Center: $(-3, 5)$

$a = 5, b = 4, c = 3$

Foci: $(-3, 8)(-3, 2)$

Vertices: $(-3, 10)(-3, 0)$

$e = \dfrac{3}{5}$

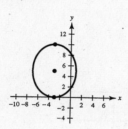

13. $\dfrac{(x + 5)^2}{\frac{9}{4}} + (y - 1)^2 = 1$

Center: $(-5, 1)$

$a = \dfrac{3}{2}, b = 1, c = \dfrac{\sqrt{5}}{2}$

Foci: $\left(-5 + \dfrac{\sqrt{5}}{2}, 1\right), \left(-5 - \dfrac{\sqrt{5}}{2}, 1\right)$

Vertices: $\left(-\dfrac{7}{2}, 1\right), \left(-\dfrac{13}{2}, 1\right)$

$e = \dfrac{\frac{\sqrt{5}}{2}}{\frac{3}{2}} = \dfrac{\sqrt{5}}{3}$

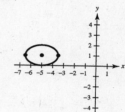

15. $9x^2 + 4y^2 + 36x - 24y + 36 = 0$

$9(x^2 + 4x + 4) + 4(y^2 - 6y + 9) = -36 + 36 + 36$

$9(x + 2)^2 + 4(y - 3) = 36$

$\dfrac{(x + 2)^2}{4} + \dfrac{(y - 3)^2}{9} = 1$

$a = 3, b = 2, c = \sqrt{5}$

Center: $(-2, 3)$

Foci: $\left(-2, 3 \pm \sqrt{5}\right)$

Vertices: $(-2, 6), (-2, 0)$

$e = \dfrac{\sqrt{5}}{3}$

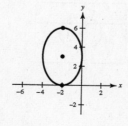

17. $x^2 + 5y^2 - 8x - 30y - 39 = 0$

$(x^2 - 8x + 16) + 5(y^2 - 6y + 9) = 39 + 16 + 45$

$(x - 4)^2 + 5(y - 3)^2 = 100$

$\dfrac{(x - 4)^2}{100} + \dfrac{(y - 3)^2}{20} = 1$

Center: $(4, 3)$

$a = 10, b = 2\sqrt{5}, c = 4\sqrt{5}$

Foci: $\left(4 \pm 4\sqrt{5}, 3\right)$

Vertices: $(14, 3), (-6, 3)$

$e = \dfrac{4\sqrt{5}}{10} = \dfrac{2\sqrt{5}}{5}$

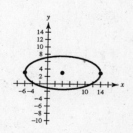

19. $3x^2 + y^2 + 9x - 5y + 1 = 0$

$3(x^2 + 3x) + (y^2 - 5y) = -1$

$3\left(x^2 + 3x + \frac{9}{4}\right) + \left(y^2 - 5y + \frac{25}{4}\right) = -1 + \frac{27}{4} + \frac{25}{4}$

$3\left(x + \frac{3}{2}\right)^2 + \left(y - \frac{5}{2}\right)^2 = 12$

$\dfrac{\left(x + \frac{3}{2}\right)^2}{4} + \dfrac{\left(y - \frac{5}{2}\right)^2}{12} = 1$

$a = \sqrt{12} = 2\sqrt{3}, b = 2, c = \sqrt{a^2 - b^2} = \sqrt{8} = 2\sqrt{2}$

Vertical major axis

Center: $\left(-\dfrac{3}{2}, \dfrac{5}{2}\right)$

Vertices: $\left(-\dfrac{3}{2}, \dfrac{5}{2} \pm 2\sqrt{3}\right)$

Foci: $\left(-\dfrac{3}{2}, \dfrac{5}{2} \pm 2\sqrt{2}\right)$

Eccentricity: $e = \dfrac{2\sqrt{2}}{2\sqrt{3}} = \dfrac{\sqrt{6}}{3}$

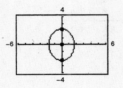

21. $16x^2 + 25y^2 - 32x + 50y + 16 = 0$

$16(x^2 - 2x + 1) + 25(y^2 + 2y + 1) = -16 + 16 + 25$

$16(x - 1)^2 + 25(y + 1)^2 = 25$

$\dfrac{(x - 1)^2}{\frac{25}{16}} + (y + 1)^2 = 1$

$a = \dfrac{5}{4}, b = 1, c = \dfrac{3}{4}$

Center: $(1, -1)$

Foci: $\left(\dfrac{7}{4}, -1\right), \left(\dfrac{1}{4}, -1\right)$

Vertices: $\left(\dfrac{9}{4}, -1\right), \left(-\dfrac{1}{4}, -1\right)$

$e = \dfrac{3}{5}$

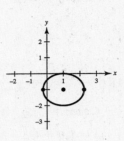

23. $5x^2 + 3y^2 = 15$

$\dfrac{x^2}{3} + \dfrac{y^2}{5} = 1$

Center: $(0, 0)$

$a = \sqrt{5}, b = \sqrt{3}, c = \sqrt{2}$

Foci: $\left(0, \pm\sqrt{2}\right)$

Vertices: $\left(0, \pm\sqrt{5}\right)$

To graph, solve for y.

$y^2 = \dfrac{15 - 5x^2}{3}$

$y_1 = \sqrt{\dfrac{15 - 5x^2}{3}}$

$y_2 = -\sqrt{\dfrac{15 - 5x^2}{3}}$

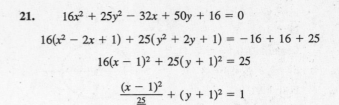

25. $12x^2 + 20y^2 - 12x + 40y - 37 = 0$

$12\left(x^2 - x + \dfrac{1}{4}\right) + 20(y^2 + 2y + 1) = 37 + 3 + 20$

$12\left(x - \dfrac{1}{2}\right)^2 + 20(y + 1)^2 = 60$

$\dfrac{\left(x - \frac{1}{2}\right)^2}{5} + \dfrac{(y + 1)^2}{3} = 1$

$a = \sqrt{5}, b = \sqrt{3}, c = \sqrt{2}$

Center: $\left(\dfrac{1}{2}, -1\right)$

Foci: $\left(\dfrac{1}{2} \pm \sqrt{2}, -1\right)$

Vertices: $\left(\dfrac{1}{2} \pm \sqrt{5}, -1\right)$

To graph, solve for y.

$(y + 1)^2 = 3\left[1 - \dfrac{(x - 0.5)^2}{5}\right]$

$y_1 = -1 + \sqrt{3\left[1 - \dfrac{(x - 0.5)^2}{5}\right]}$

$y_2 = -1 - \sqrt{3\left[1 - \dfrac{(x - 0.5)^2}{5}\right]}$

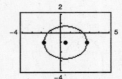

27. Center: $(0, 0)$

$a = 4, b = 2$

Vertical major axis

$$\frac{(x-h)^2}{b^2} + \frac{(y-k)^2}{a^2} = 1$$

$$\frac{x^2}{4} + \frac{y^2}{16} = 1$$

29. Vertices: $(\pm 6, 0)$

$a = 6, c = 2 \implies b = \sqrt{32} = 4\sqrt{2}$

Foci: $(\pm 2, 0)$

Horizontal major axis

Center: $(0, 0)$

$$\frac{(x-h)^2}{a^2} + \frac{(y-k)^2}{b^2} = 1$$

$$\frac{x^2}{36} + \frac{y^2}{32} = 1$$

31. Foci: $(\pm 5, 0) \implies c = 5$

Center: $(0, 0)$

Horizontal major axis

Major axis of length 12 $\implies 2a = 12$

$$a = 6$$

$$6^2 - b^2 = 5^2 \implies b^2 = 11$$

$$\frac{(x-h)^2}{a^2} + \frac{(y-k)^2}{b^2} = 1$$

$$\frac{x^2}{36} + \frac{y^2}{11} = 1$$

33. Vertices: $(0, \pm 5) \implies a = 5$

Center: $(0, 0)$

Vertical major axis

$$\frac{(x-h)^2}{b^2} + \frac{(y-k)^2}{a^2} = 1$$

$$\frac{x^2}{b^2} + \frac{y^2}{25} = 1$$

Point: $(4, 2)$

$$\frac{4^2}{b^2} + \frac{2^2}{25} = 1$$

$$\frac{16}{b^2} = 1 - \frac{4}{25} = \frac{21}{25}$$

$$400 = 21b^2$$

$$\frac{400}{21} = b^2$$

$$\frac{x^2}{\frac{400}{21}} + \frac{y^2}{25} = 1$$

$$\frac{21x^2}{400} + \frac{y^2}{25} = 1$$

35. Center: $(2, 3)$

$a = 3, \quad b = 1$

Vertical major axis

$$\frac{(x-h)^2}{b^2} + \frac{(y-k)^2}{a^2} = 1$$

$$\frac{(x-2)^2}{1} + \frac{(y-3)^2}{9} = 1$$

37. Center: $(-2, 3)$

$a = 4, \quad b = 3$

Horizontal major axis

$$\frac{(x-h)^2}{a^2} + \frac{(y-k)^2}{b^2} = 1$$

$$\frac{(x+2)^2}{16} + \frac{(y-3)^2}{9} = 1$$

39. Vertices: $(0, 4), (4, 4) \implies a = 2$

Minor axis of length 2 $\implies b = 1$

Center: $(2, 4) = (h, k)$

$$\frac{(x-h)^2}{a^2} + \frac{(y-k)^2}{b^2} = 1$$

$$\frac{(x-2)^2}{4} + \frac{(y-4)^2}{1} = 1$$

41. Foci: $(0, 0), (0, 8) \implies c = 4$

Major axis of length 16 $\implies a = 8$

$b^2 = a^2 - c^2 = 64 - 16 = 48$

Center: $(0, 4) = (h, k)$

$$\frac{(x-h)^2}{b^2} + \frac{(y-k)^2}{a^2} = 1$$

$$\frac{x^2}{48} + \frac{(y-4)^2}{64} = 1$$

43. Vertices: $(3, 1), (3, 9) \implies a = 4$

Center: $(3, 5)$

Minor axis of length 6 $\implies b = 3$

Vertical major axis

$$\frac{(x-h)^2}{b^2} + \frac{(y-k)^2}{a^2} = 1$$

$$\frac{(x-3)^2}{9} + \frac{(y-5)^2}{16} = 1$$

45. Center: (0, 4)

Vertices:

$(-4, 4), (4, 4) \Rightarrow a = 4$

$a = 2c \Rightarrow 4 = 2c \Rightarrow c = 2$

$2^2 = 4^2 - b^2 \Rightarrow b^2 = 12$

Horizontal major axis

$$\frac{(x - h)^2}{a^2} + \frac{(y - k)^2}{b^2} = 1$$

$$\frac{x^2}{16} + \frac{(y - 4)^2}{12} = 1$$

47. Vertices: $(\pm 5, 0) \Rightarrow a = 5$

Eccentricity: $\frac{3}{5} \Rightarrow c = \frac{3}{5}a = 3$

$b^2 = a^2 - c^2 = 25 - 9 = 16$

Center: $(0, 0) = (h, k)$

$$\frac{(x - h)^2}{a^2} + \frac{(y - k)^2}{b^2} = 1$$

$$\frac{x^2}{25} + \frac{y^2}{16} = 1$$

49. (a) The length of the string is $2a$.

(b) The path is an ellipse since the sum of the distances from the two fixed points is constant.

51. $\frac{x^2}{9} + \frac{y^2}{4} = 1$

$$\frac{2}{9}x + \frac{1}{2}y\frac{dy}{dx} = 0$$

$$\frac{dy}{dx} = \frac{-\frac{2}{9}x}{\frac{1}{2}y} = -\frac{4x}{9y}$$

53. $\frac{(x - 4)^2}{4} + \frac{(y + 2)^2}{16} = 1$

$4(x - 4)^2 + (y + 2)^2 = 16$

$8(x - 4) + 2(y + 2)\frac{dy}{dx} = 0$

$$\frac{dy}{dx} = -\frac{4(x - 4)}{y + 2}$$

55. $9x^2 + 4y^2 - 36x + 8y + 31 = 0$

$18x + 8y\frac{dy}{dx} - 36 + 8\frac{dy}{dx} = 0$

$\frac{dy}{dx}(8y + 8) = -18x + 36$

$$\frac{dy}{dx} = \frac{-18(x - 2)}{8(y + 1)} = -\frac{9(x - 2)}{4(y + 1)}$$

57. $\frac{(x - 2)^2}{16} + \frac{y^2}{12} = 1$

(a) $\frac{(x - 2)}{8} + \frac{y}{6}\frac{dy}{dx} = 0$

$$\frac{dy}{dx} = -\frac{3(x - 2)}{4y}$$

At $(0, 3)$, $\frac{dy}{dx} = \frac{1}{2}$.

Tangent line: $y - 3 = \frac{1}{2}(x - 0)$

$$\frac{1}{2}x - y + 3 = 0$$

(b) The center of the ellipse is $(2, 0)$. The point $(0, 3)$ is two units to the left and three units above the center. By symmetry, the point two units to the right and three unit below the center, $(4, -3)$, also has a slope of $\frac{1}{2}$.

$$y + 3 = \frac{1}{2}(x - 4)$$

$$0 = \frac{1}{2}x - y - 5 \text{ or } x - 2y - 10 = 0$$

(c)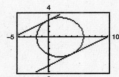

Graph: $y_1 = \sqrt{12 - \frac{3}{4}(x - 2)^2}$

$y_2 = -\sqrt{12 - \frac{3}{4}(x - 2)^2}$

$y_3 = \frac{1}{2}x + 3$

$y_4 = \frac{1}{2}x - 5$

59. $x^2 + 4y^2 + 6x - 16y + 9 = 0$

$$2x + 8y\frac{dy}{dx} + 6 - 16\frac{dy}{dx} = 0$$

$$\frac{dy}{dx}(8y - 16) = -2x - 6$$

$$\frac{dy}{dx} = \frac{-2(x + 3)}{8(y - 2)} = -\frac{x + 3}{4(y - 2)}$$

Horizontal tangents when $\frac{dy}{dx} = 0$: $x = -3$

$$(-3)^2 + 4y^2 + 6(-3) - 16y + 9 = 0$$

$$4y^2 - 16y = 0$$

$$4y(y - 4) = 0$$

$$y = 0 \text{ or } y = 4$$

The endpoints of the minor axis are $(-3, 0)$ and $(-3, 4)$.

Vertical tangents when $\frac{dy}{dx}$ does not exist: $y = 2$

$$x^2 + 4(2)^2 + 6x - 16(2) + 9 = 0$$

$$x^2 + 6x - 7 = 0$$

$$(x + 7)(x - 1) = 0$$

$$x = -7 \text{ or } x = 1$$

The endpoints of the major axis are $(-7, 2)$ and $(1, 2)$.

61. $\dfrac{x^2}{4} + \dfrac{y^2}{1} = 1$

$a = 2, b = 1$

Area $= \pi ab = 2\pi$ square units.

63. $3x^2 + 2y^2 = 6$

$$\frac{x^2}{2} + \frac{y^2}{3} = 1$$

$a = \sqrt{3}, b = \sqrt{2}$

Area $= \pi ab = \sqrt{6}\pi$ square units.

65. The tacks should be placed at the foci and the length of the string is the length of the major axis, $2a$.

Center: $(0, 0)$

$a = 3, b = 2, c = \sqrt{5}$

Foci (Positions of the tacks): $\left(\pm\sqrt{5}, 0\right)$

Length of string: 6 feet

67. Area of circle: $\pi r^2 = 100\pi$

Area of ellipse: $\pi(a)(10)$

$$10a\pi = 2(100\pi)$$

$$10a\pi = 200\pi$$

$$a = 20$$

Length of major axis: $2a = 40$

69. $a - c = 0.34$

$a + c = 4.08$

$2a = 4.42 \implies a = 2.21$

$c = 4.08 - a \implies c = 1.87$

$b^2 = a^2 - c^2 = 2.21^2 - 1.87^2 = 1.3872 \implies b \approx 1.18$

$$\frac{x^2}{a^2} + \frac{y^2}{b^2} = 1$$

$$\frac{x^2}{4.88} + \frac{y^2}{1.39} = 1$$

71. $\dfrac{x^2}{25} + \dfrac{y^2}{16} = 1 \implies y = \pm\dfrac{4}{5}\sqrt{25 - x^2}$

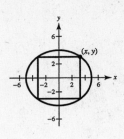

Let (x, y) be the corner of the rectangle in Quadrant I. Then the area to be maximized is:

$$A = (2x)(2y) = 4x\left(\dfrac{4}{5}\sqrt{25 - x^2}\right) = \dfrac{16}{5}x\sqrt{25 - x^2}$$

$$\dfrac{dA}{dx} = \dfrac{16}{5}\left[\dfrac{-x^2}{\sqrt{25 - x^2}} + \sqrt{25 - x^2}\right] = \dfrac{16}{5}\left[\dfrac{25 - 2x^2}{\sqrt{25 - x^2}}\right]$$

The critical numbers are $x = \pm5$ and $x = \pm\dfrac{5\sqrt{2}}{2}$.

Choosing x in Quadrant I yields $x = \dfrac{5\sqrt{2}}{2}$ and $y = \dfrac{4}{5}\sqrt{25 - \left(\dfrac{5\sqrt{2}}{2}\right)^2} = 2\sqrt{2}$.

The length is $2x = 5\sqrt{2}$ units.

The width is $2y = 4\sqrt{2}$ units.

73. False. The graph of $\dfrac{x^2}{4} + y^4 = 1$ is not an ellipse. The degree on y is 4, not 2.

75. True. The area of the circle is 16π. The area of the ellipse is $4\pi b$, where $b < 4$ since $2a = 8$ is the major axis length.

77. Let $A = C = 0$ and $D = E = F = 1$. Then $x + y + 1 = 0$ is a line, not an ellipse.

Section 12.3   Hyperbolas and Implicit Differentiation

1. $\dfrac{y^2}{9} - \dfrac{x^2}{25} = 1$

Center: $(0, 0)$

$a = 3, b = 5$

Vertical transverse axis

Matches graph (b).

3. $\dfrac{(x - 1)^2}{16} - \dfrac{y^2}{4} = 1$

Center: $(1, 0)$

$a = 4, b = 2$

Horizontal transverse axis

Matches graph (a).

5. $x^2 - y^2 = 1$

$a = 1, b = 1, c = \sqrt{2}$

Center: $(0, 0)$

Vertices: $(\pm1, 0)$

Foci: $\left(\pm\sqrt{2}, 0\right)$

Asymptotes: $y = \pm x$

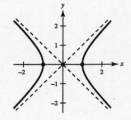

7. $\dfrac{y^2}{25} - \dfrac{x^2}{81} = 1$

$a = 5, b = 9, c = \sqrt{106}$

Center: $(0, 0)$

Vertices: $(0, \pm5)$

Foci: $\left(0, \pm\sqrt{106}\right)$

Asymptotes: $y = \pm\dfrac{5}{9}x$

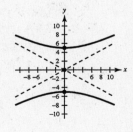

9. $\dfrac{(x-1)^2}{4} - \dfrac{(y+2)^2}{1} = 1$

$a = 2, b = 1, c = \sqrt{5}$

Center: $(1, -2)$

Vertices: $(-1, -2), (3, -2)$

Foci: $\left(1 \pm \sqrt{5}, -2\right)$

Asymptotes: $y = -2 \pm \dfrac{1}{2}(x - 1)$

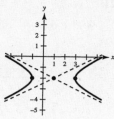

11. $\dfrac{(y+6)^2}{\frac{1}{9}} - \dfrac{(x-2)^2}{\frac{1}{4}} = 1$

$a = \dfrac{1}{3}, b = \dfrac{1}{2}, c = \dfrac{\sqrt{13}}{6}$

Center: $(2, -6)$

Vertices: $\left(2, -\dfrac{17}{3}\right), \left(2, -\dfrac{19}{3}\right)$

Foci: $\left(2, -6 \pm \dfrac{\sqrt{13}}{6}\right)$

Asymptotes: $y = -6 \pm \dfrac{2}{3}(x - 2)$

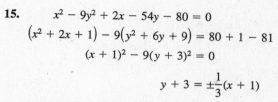

13. $\quad 9x^2 - y^2 - 36x - 6y + 18 = 0$

$9(x^2 - 4x + 4) - (y^2 + 6y + 9) = -18 + 36 - 9$

$9(x - 2)^2 - (y + 3)^2 = 9$

$\dfrac{(x-2)^2}{1} - \dfrac{(y+3)^2}{9} = 1$

$a = 1, b = 3, c = \sqrt{10}$

Center: $(2, -3)$

Vertices: $(1, -3), (3, -3)$

Foci: $\left(2 \pm \sqrt{10}, -3\right)$

Asymptotes: $y = -3 \pm 3(x - 2)$

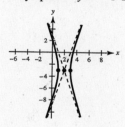

15. $\quad x^2 - 9y^2 + 2x - 54y - 80 = 0$

$\left(x^2 + 2x + 1\right) - 9\left(y^2 + 6y + 9\right) = 80 + 1 - 81$

$(x + 1)^2 - 9(y + 3)^2 = 0$

$\qquad\qquad y + 3 = \pm\dfrac{1}{3}(x + 1)$

Degenerate hyperbola is two lines intersecting at $(-1, -3)$.

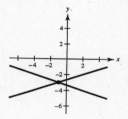

17. $\qquad 16y^2 - 9x^2 - 18x - 32y = 137$

$16(y^2 - 2y + 1) - 9(x^2 + 2x + 1) = 137 + 16 - 9$

$16(y - 1)^2 - 9(x + 1)^2 = 144$

$\dfrac{(y-1)^2}{9} - \dfrac{(x+1)^2}{16} = 1$

Center: $(-1, 1)$

$a = 3, b = 4 \Rightarrow c = 5$

Vertices: $(-1, 4), (-1, -2)$

Foci: $(-1, 6), (-1, -4)$

Asymptotes: $y = k \pm \dfrac{a}{b}(x - h) = 1 \pm \dfrac{3}{4}(x + 1)$

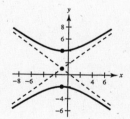

19. $2x^2 - 3y^2 = 6$

$$\frac{x^2}{3} - \frac{y^2}{2} = 1$$

$a = \sqrt{3}, b = \sqrt{2}, c = \sqrt{5}$

Center: $(0, 0)$

Vertices: $(\pm\sqrt{3}, 0)$

Foci: $(\pm\sqrt{5}, 0)$

Asymptotes: $y = \pm\sqrt{\dfrac{2}{3}}x = \pm\dfrac{\sqrt{6}}{3}x$

To use a graphing calculator, solve for y first.

$$y^2 = \frac{2x^2 - 6}{3}$$

$\left.\begin{array}{l} y_1 = \sqrt{\dfrac{2x^2 - 6}{3}} \\[18pt] y_2 = -\sqrt{\dfrac{2x^2 - 6}{3}} \end{array}\right\}$ Hyperbola

$\left.\begin{array}{l} y_3 = \dfrac{\sqrt{6}}{3}x \\[14pt] y_4 = -\dfrac{\sqrt{6}}{3}x \end{array}\right\}$ Asymptotes

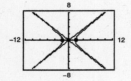

21. $9y^2 - x^2 + 2x + 54y + 62 = 0$

$9(y^2 + 6y + 9) - (x^2 - 2x + 1) = -62 - 1 + 81$

$9(y + 3)^2 - (x - 1)^2 = 18$

$$\frac{(y + 3)^2}{2} - \frac{(x - 1)^2}{18} = 1$$

$a = \sqrt{2}, b = 3\sqrt{2}, c = 2\sqrt{5}$

Center: $(1, -3)$

Vertices: $(1, -3 \pm \sqrt{2})$

Foci: $(1, -3 \pm 2\sqrt{5})$

Asymptotes: $y = -3 \pm \dfrac{1}{3}(x - 1)$

To use a graphing calculator, solve for y first.

$9(y + 3)^2 = 18 + (x - 1)^2$

$$y = -3 \pm \sqrt{\frac{18 + (x - 1)^2}{9}}$$

$\left.\begin{array}{l} y_1 = -3 + \dfrac{1}{3}\sqrt{18 + (x - 1)^2} \\[14pt] y_2 = -3 - \dfrac{1}{3}\sqrt{18 + (x - 1)^2} \end{array}\right\}$ Hyperbola

$\left.\begin{array}{l} y_3 = -3 + \dfrac{1}{3}(x - 1) \\[14pt] y_4 = -3 - \dfrac{1}{3}(x - 1) \end{array}\right\}$ Asymptotes

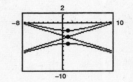

23. Vertices: $(0, \pm 2) \Rightarrow a = 2$

Foci: $(0, \pm 4) \Rightarrow c = 4$

$b^2 = c^2 - a^2 = 16 - 4 = 12$

Center: $(0, 0) = (h, k)$

$$\frac{(y - k)^2}{a^2} - \frac{(x - h)^2}{b^2} = 1$$

$$\frac{y^2}{4} - \frac{x^2}{12} = 1$$

25. Vertices: $(\pm 1, 0) \Rightarrow a = 1$

Asymptotes: $y = \pm 5x \Rightarrow \dfrac{b}{a} = 5, b = 5$

Center: $(0, 0) = (h, k)$

$$\frac{(x - h)^2}{a^2} - \frac{(y - k)^2}{b^2} = 1$$

$$\frac{x^2}{1} - \frac{y^2}{25} = 1$$

27. Foci: $(0, \pm 8) \implies c = 8$

Asymptotes: $y = \pm 4x \implies \dfrac{a}{b} = 4 \implies a = 4b$

Center: $(0, 0) = (h, k)$

$c^2 = a^2 + b^2 \implies 64 = 16b^2 + b^2$

$$\dfrac{64}{17} = b^2 \implies a^2 = \dfrac{1024}{17}$$

$$\dfrac{(y - k)^2}{a^2} - \dfrac{(x - h)^2}{b^2} = 1$$

$$\dfrac{y^2}{\frac{1024}{17}} - \dfrac{x^2}{\frac{64}{17}} = 1$$

$$\dfrac{17y^2}{1024} - \dfrac{17x^2}{64} = 1$$

29. Vertices: $(2, 0), (6, 0) \implies a = 2$

Foci: $(0, 0), (8, 0) \implies c = 4$

$b^2 = c^2 - a^2 = 16 - 4 = 12$

Center: $(4, 0) = (h, k)$

$$\dfrac{(x - h)^2}{a^2} - \dfrac{(y - k)^2}{b^2} = 1$$

$$\dfrac{(x - 4)^2}{4} - \dfrac{y^2}{12} = 1$$

31. Vertices: $(4, 1), (4, 9) \implies a = 4$

Foci: $(4, 0), (4, 10) \implies c = 5$

$b^2 = c^2 - a^2 = 25 - 16 = 9$

Center: $(4, 5) = (h, k)$

$$\dfrac{(y - k)^2}{a^2} - \dfrac{(x - h)^2}{b^2} = 1$$

$$\dfrac{(y - 5)^2}{16} - \dfrac{(x - 4)^2}{9} = 1$$

33. Vertices: $(2, 3), (2, -3) \implies a = 3$

Passes through the point: $(0, 5)$

Center: $(2, 0) = (h, k)$

$$\dfrac{(y - k)^2}{a^2} - \dfrac{(x - h)^2}{b^2} = 1$$

$$\dfrac{y^2}{9} - \dfrac{(x - 2)^2}{b^2} = 1 \implies \dfrac{(x - 2)^2}{b^2} = \dfrac{y^2}{9} - 1 = \dfrac{y^2 - 9}{9} \implies b^2 = \dfrac{9(x - 2)^2}{y^2 - 9} = \dfrac{9(-2)^2}{25 - 9} = \dfrac{36}{16} = \dfrac{9}{4}$$

$$\dfrac{y^2}{9} - \dfrac{(x - 2)^2}{9/4} = 1$$

$$\dfrac{y^2}{9} - \dfrac{4(x - 2)^2}{9} = 1$$

35. Vertices: $(0, 4), (0, 0) \implies a = 2$

Passes through the point $\left(\sqrt{5}, -1 \right)$

Center: $(0, 2) = (h, k)$

$$\dfrac{(y - k)^2}{a^2} - \dfrac{(x - h)^2}{b^2} = 1$$

$$\dfrac{(y - 2)^2}{4} - \dfrac{x^2}{b^2} = 1 \implies \dfrac{x^2}{b^2} = \dfrac{(y - 2)^2}{4} - 1 = \dfrac{(y - 2)^2 - 4}{4}$$

$$\implies b^2 = \dfrac{4x^2}{(y - 2)^2 - 4} = \dfrac{4\left(\sqrt{5} \right)^2}{(-1 - 2)^2 - 4} = \dfrac{20}{5} = 4$$

$$\dfrac{(y - 2)^2}{4} - \dfrac{x^2}{4} = 1$$

37. Vertices: $(1, 2), (3, 2) \implies a = 1$

Asymptotes: $y = x, y = 4 - x$

$\dfrac{b}{a} = 1 \implies \dfrac{b}{1} = 1 \implies b = 1$

Center: $(2, 2) = (h, k)$

$\dfrac{(x - h)^2}{a^2} - \dfrac{(y - k)^2}{b^2} = 1$

$\dfrac{(x - 2)^2}{1} - \dfrac{(y - 2)^2}{1} = 1$

39. Vertices: $(0, 2), (6, 2) \implies a = 3$

Asymptotes: $y = \dfrac{2}{3}x, y = 4 - \dfrac{2}{3}x$

$\dfrac{b}{a} = \dfrac{2}{3} \implies b = 2$

Center: $(3, 2) = (h, k)$

$\dfrac{(x - h)^2}{a^2} - \dfrac{(y - k)^2}{b^2} = 1$

$\dfrac{(x - 3)^2}{9} - \dfrac{(y - 2)^2}{4} = 1$

41. Foci: $(2, 2)$ and $(10, 2)$

From the foci we see that the transverse axis is horizontal, the center is $(6, 2)$ and $c = 4$.

Also, $2a = 6 \implies a = 3$.

$c^2 = a^2 + b^2 \implies 16 = 9 + b^2 \implies b = \sqrt{7}$

$c^2 = a^2 + b^2 \implies 16 = 9 + b^2 \implies b = \sqrt{7}$

$\dfrac{(x - 6)^2}{9} - \dfrac{(y - 2)^2}{7} = 1$

43. Center $(0, 0)$

Horizontal transverse axis.

Foci: $(\pm c, 0)$

$2a = d_1 - d_2$

$$2a = \left| \sqrt{(x + c)^2 + y^2} - \sqrt{(x - c)^2 + y^2} \right|$$

$$2a + \sqrt{(x - c)^2 + y^2} = \sqrt{(x + c)^2 + y^2}$$

$$4a^2 + 4a\sqrt{(x - c)^2 + y^2} + (x - c)^2 + y^2 = (x + c)^2 + y^2$$

$$4a\sqrt{(x - c)^2 + y^2} = 4cx - 4a^2$$

$$a\sqrt{(x - c)^2 + y^2} = cx - 2a$$

$$a^2(x^2 - acx + c^2 + y^2) = c^2x^2 - 2a^2cx + a^4$$

$$a^2(c^2 - a^2) = (c^2 - a^2)x^2 - a^2y^2$$

Let $b^2 = c^2 - a^2$. Then, $a^2b^2 = b^2x^2 - a^2y^2 \implies 1 = \dfrac{x^2}{a^2} - \dfrac{y^2}{b^2}$.

45. $\dfrac{x^2}{64} - \dfrac{y^2}{36} = 1$

$\dfrac{x}{32} - \dfrac{y}{18}\dfrac{dy}{dx} = 0$

$\dfrac{dy}{dx} = \dfrac{9x}{16y}$

47. $\dfrac{(y - 3)^2}{9} - \dfrac{(x + 1)^2}{9} = 1$

$\dfrac{2(y - 3)}{9}\dfrac{dy}{dx} - \dfrac{2(x + 1)}{9} = 0$

$\dfrac{dy}{dx} = \dfrac{x + 1}{y - 3}$

49. $x^2 - 2y^2 + 8y - 17 = 0$

$$2x - 4y\frac{dy}{dx} + 8\frac{dy}{dx} = 0$$

$$\frac{dy}{dx}(-4y + 8) = -2x$$

$$\frac{dy}{dx} = \frac{x}{2(y - 2)}$$

51. $x^2 - 4y^2 + 2x + 16y - 19 = 0$

$$2x - 8y\frac{dy}{dx} + 2 + 16\frac{dy}{dx} = 0$$

$$(16 - 8y)\frac{dy}{dx} = -2x - 2$$

$$\frac{dy}{dx} = \frac{2x + 2}{8y - 16} = \frac{x + 1}{4y - 8}$$

53. $\dfrac{(x - 2)^2}{16} - \dfrac{y^2}{12} = 1$

(a) $\dfrac{(x - 2)}{8} - \dfrac{y}{6}\dfrac{dy}{dx} = 0$

$$\frac{dy}{dx} = \frac{3(x - 2)}{4y}$$

At $(10, 6)$, $\dfrac{dy}{dx} = 1$.

Tangent line: $y = x - 4$ or $x - y - 4 = 0$

(b) The center of the hyperbola is $(2, 0)$. The point $(10, 6)$ is eight units to the right and six units above the center. By symmetry, the point eight units to the left and six units below the center, $(-6, -6)$ also has a slope of 1.

$y = x$ or $x - y = 0$

(c)

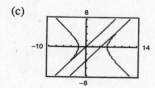

Graph: $y_1 = \sqrt{\dfrac{3}{4}(x - 2(^2 - 12}$

$$y_2 = -\sqrt{\frac{3}{4}(x - 2)^2 - 12}$$

$$y_3 = x - 4$$

$$y_4 = x$$

55. $4y^2 - x^2 + 6x + 40y + 75 = 0$

$$8y\frac{dy}{dx} - 2x + 6 + 40\frac{dy}{dx} = 0$$

$$\frac{dy}{dx}(8y + 40) = 2x - 6$$

$$\frac{dy}{dx} = \frac{x - 3}{4(y + 5)}$$

$$\frac{dy}{dx} = 0 \text{ when } x = 3.$$

$\dfrac{dy}{dx}$ does not exist when $y = -5$.

When $x = 3$: $4y^2 - 3^2 + 6(3) + 40y + 75 = 0$

$$4y^2 + 40y + 84 = 0$$

$$4(y + 7)(y + 3) = 0$$

$$y = -7 \text{ or } y = -3$$

When $y = -5$: $4(-5)^2 - x^2 + 6x + 40(-5) + 75 = 0$

$$x^2 - 6x + 25 = 0$$

No real solutions

The vertices are $(3, -7)$ and $(3, -3)$.

57. $x^2 + y^2 - 6x + 4y + 9 = 0$

$A = 1, C = 1$

$A = C \implies$ Circle

59. $4x^2 - y^2 - 4x - 3 = 0$

$A = 4, C = -1$

$AC = (4)(-1) = -4 < 0 \implies$ Hyperbola

61. $4x^2 + 3y^2 + 8x - 24y + 51 = 0$

$A = 4, C = 3$

$AC = 4(3) = 12 > 0 \implies$ Ellipse

63. $25x^2 - 10x - 200y - 119 = 0$

$A = 25, C = 0$

$AC = 25(0) = 0 \implies$ Parabola

65. Since $\overline{AB} = 1100$ feet and the sound takes one second longer to reach B than A, the explosion must occur on the vertical line through A and B below A.

Foci: $(\pm 3300, 0) \implies c = 3300$

Center: $(0, 0) = (h, k)$

$\dfrac{\overline{CE}}{1100} - \dfrac{\overline{AE}}{1100} = 4 \implies 2a = 4400, a = \dfrac{4400}{2} = 2200$

$b^2 = c^2 - a^2 = (3300)^2 - (2200)^2 = 6{,}050{,}000$

$\dfrac{x^2}{(2200)^2} - \dfrac{y^2}{6{,}050{,}000} = 1$

$$y^2 = 6{,}050{,}000\left(\dfrac{x^2}{(2200)^2} - 1\right)$$

$$y^2 = 6{,}050{,}000\left(\dfrac{(3300)^2}{(2200)^2} - 1\right) = 7{,}562{,}500$$

$$y = -2750$$

The explosion occurs at $(3300, -2750)$.

67. True. For a hyperbola, $c^2 = a^2 + b^2$ or $e^2 = \dfrac{c^2}{a^2} = 1 + \dfrac{b^2}{a^2}$. The larger the ratio of b to a, the larger the eccentricity $e = \dfrac{c}{a}$ of the hyperbola.

Section 12.4   Parametric Equations and Calculus

1. $x = \sqrt{t}, y = 3 - t$

(a)

t	0	1	2	3	4
x	0	1	$\sqrt{2}$	$\sqrt{3}$	2
y	3	2	1	0	-1

(b)

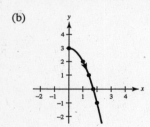

(c) $x^2 = t$
    $y = 3 - x^2$

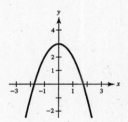

The graph of the rectangular equation shows the entire parabola rather than just the right half.

3. $x = 3t - 3 \implies t = \dfrac{x + 3}{3}$

$y = 2t + 1 \implies y = \dfrac{2}{3}(x + 3) + 1 = \dfrac{2}{3}x + 3$

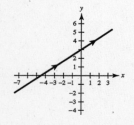

5. $x = \dfrac{1}{4}t \implies t = 4x$

$y = t^2 \implies y = 16x^2$

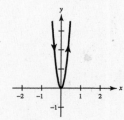

7. $x = t + 2 \implies t = x - 2$

$y = t^2 \qquad \implies y = (x - 2)^2$

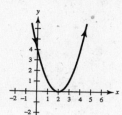

9. $x = t + 1 \implies t = x - 1$

$y = \dfrac{t}{t + 1} \implies y = \dfrac{x - 1}{x} = 1 - \dfrac{1}{x}$

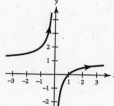

11. $x = 2(t + 1) \implies \dfrac{x}{2} - 1 = t \quad \text{or} \quad t = \dfrac{x - 2}{2}$

$y = |t - 2| \implies \qquad y = \left|\dfrac{x}{2} - 1 - 2\right| = \left|\dfrac{x}{2} - 3\right| = \left|\dfrac{x - 6}{2}\right|$

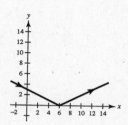

13. $x = 3 \cos \theta \implies \left(\dfrac{x}{3}\right)^2 = \cos^2 \theta$

$y = 3 \sin \theta \implies \left(\dfrac{y}{3}\right)^2 = \sin^2 \theta$

$\left(\dfrac{x}{3}\right)^2 + \left(\dfrac{y}{3}\right)^2 = 1$

$\quad x^2 + y^2 = 9$

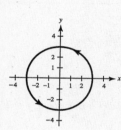

15. $x = 4 \sin 2\theta \implies \left(\dfrac{x}{4}\right)^2 = \sin^2 2\theta$

$y = 2 \cos 2\theta \implies \left(\dfrac{y}{2}\right)^2 = \cos^2 2\theta$

$\left(\dfrac{x}{4}\right)^2 + \left(\dfrac{y}{2}\right)^2 = 1$

$\dfrac{x^2}{16} + \dfrac{y^2}{4} = 1$

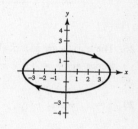

17. $x = 4 + 2 \cos \theta \implies \left(\dfrac{x - 4}{2}\right)^2 = \cos^2 \theta$

$y = -1 + \sin \theta \implies (y + 1)^2 = \sin^2 \theta$

$\dfrac{(x - 4)^2}{4} + \dfrac{(y + 1)^2}{1} = 1$

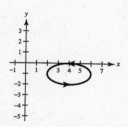

19. $x = e^{-t} \implies \dfrac{1}{x} = e^t$

$y = e^{3t} \implies y = (e^t)^3$

$y = \left(\dfrac{1}{x}\right)^3$

$y = \dfrac{1}{x^3}, \; x > 0, y > 0$

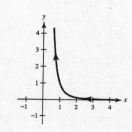

21. $x = t^3 \qquad \implies x^{1/3} = t$

$y = 3 \ln t \implies y = \ln t^3$

$y = \ln(x^{1/3})^3$

$y = \ln x$

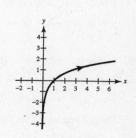

23. By eliminating the parameters in (a) to (d), we get $y = 2x + 1$. They differ from each other in orientation and in restricted domains. These curves are all smooth except for (b).

(a) $x = t, y = 2t + 1$

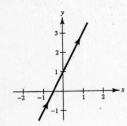

(b) $x = \cos\theta \qquad y = 2\cos\theta + 1$

$-1 \le x \le 1 \quad -1 \le y \le 3$

$\dfrac{dx}{d\theta} = \dfrac{dy}{d\theta} = 0$ when $\theta = 0, \pm\pi, \pm 2\pi, \ldots$

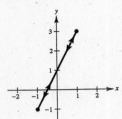

(c) $x = e^{-t} \qquad y = 2e^{-t} + 1$

$\quad x > 0 \qquad y > 1$

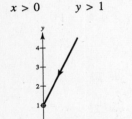

(d) $x = e^{t} \qquad y = 2e^{t} + 1$

$\quad x > 0 \qquad y > 1$

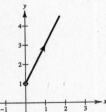

25. The curves are identical on $0 < \theta < \pi$.

Represent $y = 2(1 - x^2)$.

27. $x = x_1 + t(x_2 - x_1), y = y_1 + t(y_2 - y_1)$

$\dfrac{x - x_1}{x_2 - x_1} = t$

$y = y_1 + \left(\dfrac{x - x_1}{x_2 - x_1}\right)(y_2 - y_1)$

$y - y_1 = \dfrac{y_2 - y_1}{x_2 - x_1}(x - x_1) = m(x - x_1)$

29. $x = h + a\cos\theta, y = k + b\sin\theta$

$\dfrac{x - h}{a} = \cos\theta, \dfrac{y - k}{b} = \sin\theta$

$\dfrac{(x - h)^2}{a^2} + \dfrac{(y - k)^2}{b^2} = 1$

31. From Exercise 27 we have:

$x = 0 + t(6 - 0) = 6t$

$y = 0 + t(-3 - 0) = -3t$

33. From Exercise 28 we have:

$x = 3 + 4\cos\theta$

$y = 2 + 4\sin\theta$

35. Vertices: $(\pm 4, 0) \implies (h, k) = (0, 0)$ and $a = 4$

Foci: $(\pm 3, 0) \implies c = 3$

$c^2 = a^2 - b^2 \implies 9 = 16 - b^2 \implies b = \sqrt{7}$

From Exercise 29 we have:

$x = 4\cos\theta$

$y = \sqrt{7}\sin\theta$

37. Vertices: $(\pm 4, 0) \implies (h, k) = (0, 0)$ and $a = 4$

Foci: $(\pm 5, 0) \implies c = 5$

$c^2 = a^2 + b^2 \implies 25 = 16 + b^2 \implies b = 3$

From Exercise 30 we have:

$x = 4\sec\theta$

$y = 3\tan\theta$

39. $y = 3x - 2$

    (a) $t = x \implies x = t$ and $y = 3t - 2$

    (b) $t = 2 - x \implies x = -t + 2$ and $y = 3(-t + 2) - 2 = -3t + 4$

41. $y = x^2$

    (a) $t = x \implies x = t$ and $y = t^2$

    (b) $t = 2 - x \implies x = -t + 2$ and $y = (-t + 2)^2 = t^2 - 4t + 4$

43. $y = x^2 + 1$

    (a) $t = x \implies x = t$ and $y = t^2 + 1$

    (b) $t = 2 - x \implies x = -t + 2$ and $y = (-t + 2)^2 + 1 = t^2 - 4t + 5$

45. $y = \dfrac{1}{x}$

    (a) $t = x \implies x = t$ and $y = \dfrac{1}{t}$

    (b) $t = 2 - x \implies x = -t + 2$ and $y = \dfrac{1}{-t + 2} = \dfrac{-1}{t - 2}$

47. $x = 2 \sec t,\ y = 3 \tan t$

$$\sec^2 t - \tan^2 t = 1$$

$$\frac{x^2}{4} - \frac{y^2}{9} = 1$$

$$x = 2 \sec(-t) = 2 \sec t$$

$$y = 3 \tan(-t) = -3 \tan t$$

$$\sec^2 t - \tan^2 t = 1$$

$$\frac{x^2}{4} - \frac{y^2}{9} = 1$$

The graph would still be the same except the orientation of the curve would be reversed.

49. $x = 2 \cos \theta \implies -2 \le x \le 2$

    $y = \sin 2\theta \implies -1 \le y \le 1$

    Matches graph (b).

    Domain: $[-2, 2]$

    Range: $[-1, 1]$

51. $x = \frac{1}{2}(\cos \theta + \theta \sin \theta)$

    $y = \frac{1}{2}(\sin \theta - \theta \cos \theta)$

    Matches graph (d).

    Domain: $(-\infty, \infty)$

    Range: $(-\infty, \infty)$

53. $x = 4(\theta - \sin \theta)$

    $y = 4(1 - \cos \theta)$

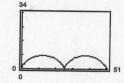

55. $x = \theta - \frac{3}{2} \sin \theta$

    $y = 1 - \frac{3}{2} \cos \theta$

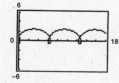

57. $x = 3 \cos^3 \theta$

    $y = 3 \sin^3 \theta$

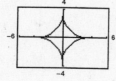

59. $x = 2 \cot \theta$

    $y = 2 \sin^2 \theta$

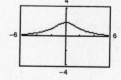

61. (a)

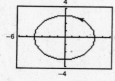

    (b) The orientation of the second curve is the reverse of the first curve.

    (c) The orientation will be reversed.

    (d) Many answers possible. For example, $x = 1 + t$, $y = 1 + 2t$, and $x = 1 - t, y = 1 - 2t$.

63. See definition on page 749.

65. $x = (v_0 \cos \theta)t$ and $y = h + (v_0 \sin \theta)t - 16t^2$

(a) $\theta = 60°$, $v_0 = 88$ ft/sec

$x = (88 \cos 60°)t$ and $y = (88 \sin 60°)t - 16t^2$

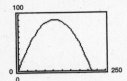

Maximum height: 90.8 feet

Range: 209.6 feet

(c) $\theta = 45°$, $v_0 = 88$ ft/sec

$x = (88 \cos 45°)t$ and $y = (88 \sin 45°)t - 16t^2$

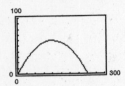

Maximum height: 60.5 ft

Range: 242.0 ft

(b) $\theta = 60°$, $v_0 = 132$ ft/sec

$x = (132 \cos 60°)t$ and $y = (132 \sin 60°)t - 16t^2$

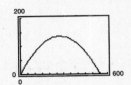

Maximum height: 204.2 feet

Range: 471.6 feet

(d) $\theta = 45°$, $v_0 = 132$ ft/sec

$x = (132 \cos 45°)t$ and $y = (132 \sin 45°)t - 16t^2$

Maximum height: 136.1 ft

Range: 544.5 ft

67. $x = 2t$, $y = 3t - 1$

(a) $\dfrac{dy}{dx} = \dfrac{3}{2}$

(b) $t = \dfrac{x}{2}$

$y = \dfrac{3x}{2} - 1$

$\dfrac{dy}{dx} = \dfrac{3}{2}$

69. $x = t + 1$, $y = t^2 + 3t$

(a) $\dfrac{dy}{dx} = 2t + 3$

(b) $x - 1 = t \implies y = (x - 1)^2 + 3(x - 1)$

$= x^2 + x - 2$

$\dfrac{dy}{dx} = 2x + 1$

$= 2(t + 1) + 1$

$= 2t + 3$

71. $x = 2 \cos t$, $y = 2 \sin t$

(a) $\dfrac{dy}{dx} = \dfrac{2 \cos t}{-2 \sin t} = -\cot t$

(b) $\cos^2 t + \sin^2 t = 1$

$\left(\dfrac{x}{2}\right)^2 + \left(\dfrac{y}{2}\right)^2 = 1$

$x^2 + y^2 = 4$

$2x + 2y \dfrac{dy}{dx} = 0$

$\dfrac{dy}{dx} = \dfrac{-x}{y} = \dfrac{-2 \cos t}{2 \sin t} = -\cot t$

73. $x = 2 + \sec t$, $y = 1 + 2 \tan t$

(a) $\dfrac{dy}{dx} = \dfrac{2 \sec^2 t}{\sec t \tan t} = 2 \sec t \cot t$

$= 2\left(\dfrac{1}{\cos t}\right)\left(\dfrac{\cos t}{\sin t}\right) = 2 \csc t$

(b) $\sec^2 t - \tan^2 t = 1$

$(x - 2)^2 - \dfrac{(y - 1)^2}{4} = 1$

$2(x - 2) - \dfrac{1}{2}(y - 1)\dfrac{dy}{dx} = 0$

$\dfrac{dy}{dx} = \dfrac{4(x - 2)}{y - 1} = \dfrac{4 \sec t}{2 \tan t} = 2 \csc t$

75. $x = \cos^3 t, \ y = \sin^3 t$

(a) $\dfrac{dy}{dx} = \dfrac{3 \sin^2 t \cos t}{-3 \cos^2 t \sin t} = -\tan t$

(b) $\qquad \cos^2 t + \sin^2 t = 1$

$x^{2/3} + y^{2/3} = 1$

$\dfrac{2}{3}x^{-1/3} + \dfrac{2}{3}y^{-1/3}\dfrac{dy}{dx} = 0$

$\dfrac{dy}{dx} = -\dfrac{y^{1/3}}{x^{1/3}} = -\dfrac{\sin t}{\cos t} = -\tan t$

77. $\dfrac{dy}{dx} = \dfrac{dy/dt}{dx/dt} = \dfrac{-4}{2t} = \dfrac{-2}{t}$

79. $\dfrac{dy}{dx} = \dfrac{dy/d\theta}{dx/d\theta} = \dfrac{-2\cos\theta\sin\theta}{2\sin\theta\cos\theta} = -1$

$\left[\textbf{Note: } x + y = 1 \implies y = 1 - x \text{ and } \dfrac{dy}{dx} = -1\right]$

81. $x = 2t, \ y = 3t - 1$

$\dfrac{dy}{dx} = \dfrac{dy/dt}{dx/dt} = \dfrac{3}{2}$

$\dfrac{d^2y}{dx^2} = 0$ Line

83. $x = t + 1, \ y = t^2 + 3t$

$\dfrac{dy}{dx} = \dfrac{2t + 3}{1} = 1$ when $t = -1$.

$\dfrac{d^2y}{dx^2} = 2$ concave upwards

85. $x = 2\cos\theta, \ y = 2\sin\theta$

$\dfrac{dy}{dx} = \dfrac{2\cos\theta}{-2\sin\theta} = -\cot\theta = -1$ when $\theta = \dfrac{\pi}{4}$.

$\dfrac{d^2y}{dx^2} = \dfrac{\csc^2\theta}{-2\sin\theta} = \dfrac{-\csc^3\theta}{2} = -\sqrt{2}$ when $\theta = \dfrac{\pi}{4}$.

concave downward

87. $x = 2 + \sec\theta, \ y = 1 + 2\tan\theta$

$\dfrac{dy}{dx} = \dfrac{2\sec^2\theta}{\sec\theta\tan\theta}$

$= \dfrac{2\sec\theta}{\tan\theta} = 2\csc\theta = 4$ when $\theta = \dfrac{\pi}{6}$.

$\dfrac{d^2y}{dx^2} = \dfrac{d\left[\dfrac{dy}{dx}\right]}{\dfrac{dx}{d\theta}} = \dfrac{-2\csc\theta\cot\theta}{\sec\theta\tan\theta}$

$= -2\cot^3\theta = -6\sqrt{3}$ when $\theta = \dfrac{\pi}{6}$.

concave downward

89. $x = \cos^3\theta, \ y = \sin^3\theta$

$\dfrac{dy}{dx} = \dfrac{3\sin^2\theta\cos\theta}{-3\cos^2\theta\sin\theta}$

$= -\tan\theta = -1$ when $\theta = \dfrac{\pi}{4}$.

$\dfrac{d^2y}{dx^2} = \dfrac{-\sec^2\theta}{-3\cos^2\theta\sin\theta} = \dfrac{1}{3\cos^4\theta\sin\theta}$

$= \dfrac{\sec^4\theta\csc\theta}{3} = \dfrac{4\sqrt{2}}{3}$ when $\theta = \dfrac{\pi}{4}$.

concave upward

91. $x = 1 - t, \ y = t^2$

Horizontal tangents: $\dfrac{dy}{dt} = 2t = 0$ when $t = 0$.

Point: $(1, 0)$

Vertical tangents: $\dfrac{dx}{dt} = -1 \neq 0$; none

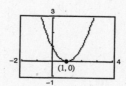

93. $x = 1 - t, \; y = t^3 - 3t$

Horizontal tangents: $\dfrac{dy}{dt} = 3t^2 - 3 = 0$ when $t = \pm 1$.

Points: $(0, -2), (2, 2)$

Vertical tangents: $\dfrac{dx}{dt} = -1 \neq 0$; none

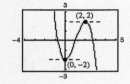

95. $x = 3 \cos \theta, \; y = 3 \sin \theta$

Horizontal tangents: $\dfrac{dy}{d\theta} = 3 \cos \theta = 0$ when $\theta = \dfrac{\pi}{2}, \dfrac{3\pi}{2}$.

Points: $(0, 3), (0, -3)$

Vertical tangents: $\dfrac{dx}{d\theta} = -3 \sin \theta = 0$ when $\theta = 0, \pi$.

Points: $(3, 0), (-3, 0)$

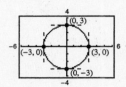

97. $x = 4 + 2 \cos \theta, \; y = -1 + \sin \theta$

Horizontal tangents: $\dfrac{dy}{d\theta} = \cos \theta = 0$ when $\theta = \dfrac{\pi}{2}, \dfrac{3\pi}{2}$.

Points: $(4, 0), (4, -2)$

Vertical tangents: $\dfrac{dx}{d\theta} = -2 \sin \theta = 0$ when $\theta = 0, \pi$.

Points: $(6, -1), (2, -1)$

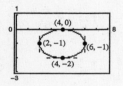

99. $x = \sec \theta, \; y = \tan \theta$

Horizontal tangents: $\dfrac{dy}{d\theta} = \sec^2 \theta \neq 0$; none

Vertical tangents: $\dfrac{dx}{d\theta} = \sec \theta \tan \theta = 0$ when $x = 0, \pi$.

Points: $(1, 0), (-1, 0)$

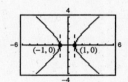

101. (a) $x = (v_0 \cos \theta)t$

$y = h + (v_0 \sin \theta)t - 16t^2$

$h = 5, v_0 = 240, \theta = 10°$

$x = (240 \cos 10°)t$

$y = 5 + (240 \sin 10°)t - 16t^2$

(b) $5 + (240 \sin 10°)t - 16t^2 = 0 \implies t = \dfrac{240 \sin 10° + \sqrt{(240 \sin 10°)^2 + 320}}{32}$

≈ 2.7196 seconds

Distance traveled $= x = (240 \cos 10°)(2.7196) \approx 643$ ft

(c)

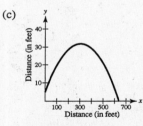

Maximum height: 32.1 feet

(d) From (b), $t \approx 2.72$ seconds

103. $x = (v_0 \cos \theta)t \implies t = \dfrac{x}{v_0 \cos \theta}$

$y = h + (v_0 \sin \theta)t - 16t^2$

$= h + (v_0 \sin \theta)\left(\dfrac{x}{v_0 \cos \theta}\right) - 16\left(\dfrac{x}{v_0 \cos \theta}\right)^2$

$= h + (\tan \theta)x - \dfrac{16x^2}{v_0^2 \cos^2 \theta}$

$= -\dfrac{16 \sec^2 \theta}{v_0^2}x^2 + (\tan \theta)x + h$

105. True.

$x = t$

$y = t^2 + 1 \implies y = x^2 + 1$

$x = 3t$

$y = 9t^2 + 1 \implies y = x^2 + 1$

107. False.

A single rectangular equation may have many different parametric representations.

$y = 3 - 2x$

Let $x = t$ and $y = 3 - 2t$.

Let $x = t + 1$ and $y = 3 - 2(t + 1) = 1 - 2t$.

Section 12.5   Polar Coordinates and Calculus

1. Polar Coordinates: $\left(4, -\frac{\pi}{3}\right)$

Additional representations

$\left(4, -\frac{\pi}{3} + 2\pi\right) = \left(4, \frac{5\pi}{3}\right)$

$\left(-4, -\frac{\pi}{3} - \pi\right) = \left(-4, -\frac{4\pi}{3}\right)$

3. Polar Coordinates: $\left(0, -\frac{7\pi}{6}\right)$

Additional representations

$\left(0, -\frac{7\pi}{6} + 2\pi\right) = \left(0, \frac{5\pi}{6}\right)$

$\left(0, -\frac{7\pi}{6} + \pi\right) = \left(0, -\frac{\pi}{6}\right)$

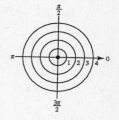

5. Polar Coordinates: $\left(\sqrt{2}, 2.36\right)$

Additional representations

$\left(\sqrt{2}, 2.36 + 2\pi\right) \approx \left(\sqrt{2}, 8.64\right)$

$\left(-\sqrt{2}, 2.36 - \pi\right) \approx \left(-\sqrt{2}, -0.78\right)$

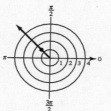

7. Polar Coordinates: $\left(3, \frac{\pi}{2}\right)$

$x = 3 \cos \frac{\pi}{2} = 0$

$y = 3 \sin \frac{\pi}{2} = 3$

Rectangular Coordinates: $(0, 3)$

9. Polar Coordinates: $\left(-1, \frac{5\pi}{4}\right)$

$x = -1 \cos\left(\frac{5\pi}{4}\right) = \frac{\sqrt{2}}{2}, y = -1 \sin\left(\frac{5\pi}{4}\right) = \frac{\sqrt{2}}{2}$

Rectangular Coordinates: $\left(\frac{\sqrt{2}}{2}, \frac{\sqrt{2}}{2}\right)$

11. Polar Coordinates: $\left(2, \frac{3\pi}{4}\right)$

$x = 2 \cos \frac{3\pi}{4} = -\sqrt{2}$

$y = 2 \sin \frac{3\pi}{4} = \sqrt{2}$

Rectangular Coordinates: $\left(-\sqrt{2}, \sqrt{2}\right)$

13. Polar Coordinates: $(-2.5, 1.1)$

$x = -2.5 \cos 1.1 \approx -1.134$

$y = -2.5 \sin 1.1 \approx -2.228$

Rectangular Coordinates: $(-1.134, -2.228)$

15. Rectangular Coordinates: $(1, 1)$

$r = \pm\sqrt{2}, \tan \theta = 1, \theta = \frac{\pi}{4} \text{ or } \frac{5\pi}{4}$

Polar Coordinates: $\left(\sqrt{2}, \frac{\pi}{4}\right), \left(-\sqrt{2}, \frac{5\pi}{4}\right)$

17. Rectangular Coordinates: $(-6, 0)$

$r = \pm 6, \tan \theta = 0, \theta = 0 \text{ or } \pi$

Polar Coordinates: $(6, \pi), (-6, 0)$

19. Rectangular Coordinates: $(-3, 4)$

$r = \pm\sqrt{9 + 16} = \pm 5, \tan \theta = -\frac{4}{3}, \theta \approx 2.2143, 5.3559$

Polar Coordinates: $(5, 2.2143), (-5, 5.3559)$

21. Rectangular Coordinates: $\left(-\sqrt{3}, -\sqrt{3}\right)$

$r = \pm\sqrt{3+3} = \pm\sqrt{6}$, $\tan\theta = 1$, $\theta = \dfrac{\pi}{4}$ or $\dfrac{5\pi}{4}$

Polar Coordinates: $\left(\sqrt{6}, \dfrac{5\pi}{4}\right), \left(-\sqrt{6}, \dfrac{\pi}{4}\right)$

23. Rectangular: $(3, -2)$

$(3, -2) \blacktriangleright$ Pol

$\approx (3.606, -0.5880)$

or $\left(\sqrt{13}, -0.5880\right)$

25. Rectangular: $\left(\sqrt{3}, 2\right)$

$\left(\sqrt{3}, 2\right) \blacktriangleright$ Pol

$\approx (2.646, 0.8571)$

or $\left(\sqrt{7}, 0.8571\right)$

27. Rectangular: $\left(\dfrac{5}{2}, \dfrac{4}{3}\right)$

$\left(\dfrac{5}{2}, \dfrac{4}{3}\right) \blacktriangleright$ Pol

$\approx (2.833, 0.4900)$ or

$\left(\dfrac{17}{6}, 0.4900\right)$

29. $x^2 + y^2 = 9$

$r = 3$

31. $y = 4$

$r\sin\theta = 4$

$r = 4\csc\theta$

33. $3x - y + 2 = 0$

$3r\cos\theta - r\sin\theta + 2 = 0$

$r(3\cos\theta - \sin\theta) = -2$

$r = \dfrac{-2}{3\cos\theta - \sin\theta}$

35. $xy = 16$

$(r\cos\theta)(r\sin\theta) = 16$

$r^2 = 16\sec\theta\csc\theta = 32\csc 2\theta$

37. $y^2 - 8x - 16 = 0$

$r^2\sin^2\theta - 8r\cos\theta - 16 = 0$

By the Quadratic Formula, we have:

$r = \dfrac{-(-8\cos\theta) \pm \sqrt{(-8\cos\theta)^2 - 4(\sin^2\theta)(-16)}}{2\sin^2\theta}$

$= \dfrac{8\cos\theta \pm \sqrt{64\cos^2\theta + 64\sin^2\theta}}{2\sin^2\theta}$

$= \dfrac{8\cos\theta \pm \sqrt{64(\cos^2\theta + \sin^2\theta)}}{2\sin^2\theta}$

$= \dfrac{8\cos\theta \pm 8}{2\sin^2\theta}$

$= \dfrac{4(\cos\theta \pm 1)}{1 - \cos^2\theta}$

$r = \dfrac{4(\cos\theta + 1)}{(1 + \cos\theta)(1 - \cos\theta)} = \dfrac{4}{1 - \cos\theta}$ or $r = \dfrac{4(\cos\theta - 1)}{(1 + \cos\theta)(1 - \cos\theta)} = \dfrac{-4}{1 + \cos\theta}$

39. $x^2 + y^2 = a^2$

$r^2 = a^2$

$r = a$

41. $y = b$

$r\sin\theta = b$

$r = b\csc\theta$

43. $x^2 + y^2 - 2ax = 0$

$r^2 - 2ar\cos\theta = 0$

$r(r - 2a\cos\theta) = 0$

$r - 2a\cos\theta = 0$

$r = 2a\cos\theta$

45.
$$r = 4 \sin \theta$$
$$r^2 = 4r \sin \theta$$
$$x^2 + y^2 = 4y$$
$$x^2 + y^2 - 4y = 0$$

47.
$$\theta = \frac{2\pi}{3}$$
$$\tan \theta = \tan \frac{2\pi}{3}$$
$$\frac{y}{x} = -\sqrt{3}$$
$$y = -\sqrt{3}x$$
$$\sqrt{3}x + y = 0$$

49.
$$r = 2 \csc \theta$$
$$r \sin \theta = 2$$
$$y = 2$$

51.
$$r = 2 \sin 3\theta$$
$$r = 2 \sin(\theta + 2\theta)$$
$$r = 2[\sin \theta \cos 2\theta + \cos \theta \sin 2\theta]$$
$$r = 2[\sin \theta(1 - 2\sin^2 \theta) + \cos \theta(2 \sin \theta \cos \theta)]$$
$$r = 2[\sin \theta - 2\sin^3 \theta + 2 \sin \theta \cos^2 \theta]$$
$$r = 2[\sin \theta - 2\sin^3 \theta + 2 \sin \theta(1 - \sin^2 \theta)]$$
$$r = 2(3 \sin \theta - 4 \sin^3 \theta)$$
$$r^4 = 6r^3 \sin \theta - 8r^3 \sin^3 \theta$$
$$(x^2 + y^2)^2 = 6(x^2 + y^2)y - 8y^3$$
$$(x^2 + y^2)^2 = 6x^2y - 2y^3$$

53.
$$r = \frac{6}{2 - 3 \sin \theta}$$
$$r(2 - 3 \sin \theta) = 6$$
$$2r = 6 + 3r \sin \theta$$
$$2(\pm\sqrt{x^2 + y^2}) = 6 + 3y$$
$$4(x^2 + y^2) = (6 + 3y)^2$$
$$4x^2 + 4y^2 = 36 + 36y + 9y^2$$
$$4x^2 - 5y^2 - 36y - 36 = 0$$

55.
$$r = 6$$
$$r^2 = 36$$
$$x^2 + y^2 = 36$$

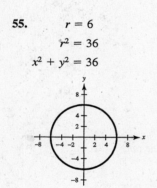

57.
$$\theta = \frac{\pi}{6}$$
$$\tan \theta = \tan \frac{\pi}{6}$$
$$\frac{y}{x} = \frac{\sqrt{3}}{3}$$
$$y = \frac{\sqrt{3}}{3}x$$
$$3y = \sqrt{3}x$$
$$-\sqrt{3}x + 3y = 0$$

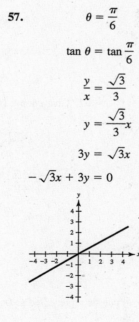

59.
$$r = 3 \sec \theta$$
$$r \cos \theta = 3$$
$$x = 3$$
$$x - 3 = 0$$

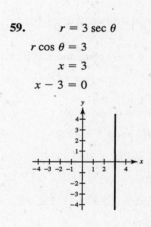

61. $r = 1 + \sin\theta, 0 \le \theta < 2\pi$

$x = r\cos\theta = (1 + \sin\theta)\cos\theta$

$y = r\sin\theta = (1 + \sin\theta)\sin\theta$

$\dfrac{dx}{d\theta} = (1 + \sin\theta)(-\sin\theta) + \cos^2\theta = 0$

$\quad -\sin\theta - \sin^2\theta + (1 - \sin^2\theta) = 0$

$\quad 2\sin^2\theta + \sin\theta - 1 = 0$

$\quad (2\sin\theta - 1)(\sin\theta + 1) = 0$

$\quad \sin\theta = \dfrac{1}{2} \implies \theta = \dfrac{\pi}{6}, \dfrac{5\pi}{6}$

$\quad \sin\theta = -1 \implies \theta = \dfrac{3\pi}{2}$

$\dfrac{dy}{d\theta} = (1 + \sin\theta)\cos\theta + \cos\theta\sin\theta = 0$

$\quad \cos\theta(2\sin\theta + 1) = 0$

$\quad \cos\theta = 0 \implies \theta = \dfrac{\pi}{2}, \dfrac{3\pi}{2}$

$\quad \sin\theta = -\dfrac{1}{2} \implies \theta = \dfrac{7\pi}{6}, \dfrac{11\pi}{6}$

Vertical tangents: $\left(\dfrac{3}{2}, \dfrac{\pi}{6}\right), \left(\dfrac{3}{2}, \dfrac{5\pi}{6}\right)$

Horizontal tangents: $\left(2, \dfrac{\pi}{2}\right), \left(\dfrac{1}{2}, \dfrac{7\pi}{6}\right), \left(\dfrac{1}{2}, \dfrac{11\pi}{6}\right)$

(**Note:** $\theta = \dfrac{3\pi}{2}$ corresponds to $(0, 0)$, which is a cusp.)

63. $r = 2\csc\theta + 3$

$x = r\cos\theta = (2\csc\theta + 3)\cos\theta = 2\cot\theta + 3\cos\theta$

$y = r\sin\theta = (2\csc\theta + 3)\sin\theta = 2 + 3\sin\theta$

$\dfrac{dx}{d\theta} = -2\csc^2\theta - 3\sin\theta = 0$

$\quad \dfrac{2}{\sin^2\theta} = -3\sin\theta$

$\quad \sin^3\theta = \dfrac{-2}{3}$

$\quad \sin\theta \approx -0.87358$

$\quad \theta \approx 4.2041, 5.2207$

$\dfrac{dy}{d\theta} = 3\cos\theta = 0 \implies \theta = \dfrac{\pi}{2}, \dfrac{3\pi}{2}$

Vertical tangents: $(0.7106, 4.2041), (0.7106, 5.2207)$

Horizontal tangents: $\left(5, \dfrac{\pi}{2}\right), \left(1, \dfrac{3\pi}{2}\right)$

65. $r = 2(h\cos\theta + k\sin\theta)$

$r^2 = 2r(h\cos\theta + k\sin\theta)$

$x^2 + y^2 = 2hx + 2ky$

$(x^2 - 2hx) + (y^2 - 2ky) = 0$

$(x^2 - 2hx + h^2) + (y^2 - 2ky + k^2) = h^2 + k^2$

$(x - h)^2 + (y - k)^2 = h^2 + k^2$

Radius: $\sqrt{h^2 + k^2}$

Center: (h, k)

67. Given point: $\left(4, \dfrac{\pi}{6}\right)$

(a) $\left(-4, \dfrac{\pi}{6}\right)$: Symmetric to the pole

(b) $\left(4, -\dfrac{\pi}{6}\right)$: Symmetric to the polar axis

(c) $\left(-4, -\dfrac{\pi}{6}\right)$: Symmetric to the line $\theta = \dfrac{\pi}{2}$

69. True. Because r is a directed distance, then the point (r, θ) can be represented as $(r, \theta + 2n\pi)$.

71. False.

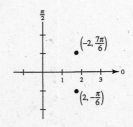

Section 12.6   Graphs of Polar Equations

1. $r = 3 \cos 2\theta$

Rose curve with 4 petals

3. $r = 3(1 - 2 \cos \theta)$

Limaçon with inner loop

5. $r = 6 \sin 2\theta$

Rose curve with 4 petals

7. $r = 5 + 4 \cos \theta$

$\theta = \dfrac{\pi}{2}$: $-r = 5 + 4 \cos(-\theta)$

$\qquad\qquad -r = 5 + 4 \cos \theta$

Not an equivalent equation

Polar axis: $r = 5 + 4 \cos(-\theta)$

$\qquad\qquad r = 5 + 4 \cos \theta$

Equivalent equation

Pole: $-r = 5 + 4 \cos \theta$

Not an equivalent equation

Answer: Symmetric with respect to polar axis

9. $r = \dfrac{2}{1 + \sin \theta}$

$\theta = \dfrac{\pi}{2}$: $r = \dfrac{2}{1 + \sin(\pi - \theta)}$

$\qquad\qquad r = \dfrac{2}{1 + \sin \pi \cos \theta - \cos \pi \sin \theta}$

$\qquad\qquad r = \dfrac{2}{1 + \sin \theta}$

Equivalent equation

Polar axis: $r = \dfrac{2}{1 + \sin(-\theta)}$

$\qquad\qquad r = \dfrac{2}{1 - \sin \theta}$

Not an equivalent equation

Pole: $-r = \dfrac{2}{1 + \sin \theta}$

Answer: Symmetric with respect to $\theta = \pi/2$

11. $r^2 = 16 \cos 2\theta$

$\theta = \dfrac{\pi}{2}$: $(-r)^2 = 16 \cos 2(-\theta)$

$\qquad\qquad r^2 = 16 \cos 2\theta$

Equivalent equation

Polar axis: $r^2 = 16 \cos 2(-\theta)$

$\qquad\qquad r^2 = 16 \cos 2\theta$

Equivalent equation

Pole: $(-r)^2 = 16 \cos 2\theta$

$\qquad\qquad r^2 = 16 \cos 2\theta$

Equivalent equation

Answer: Symmetric with respect to $\theta = \dfrac{\pi}{2}$, the

polar axis, and the pole

13. $|r| = |10(1 - \sin \theta)| = 10|1 - \sin \theta| \le 10(2) = 20$

$|1 - \sin \theta| = 2$

$1 - \sin \theta = 2 \qquad\qquad$ or $1 - \sin \theta = -2$

$\quad \sin \theta = -1 \qquad\qquad\qquad\qquad \sin \theta = 3$

$\qquad \theta = \dfrac{3\pi}{2} \qquad\qquad\qquad$ Not possible

Maximum: $|r| = 20$ when $\theta = \dfrac{3\pi}{2}$.

$0 = 10(1 - \sin \theta)$

$\sin \theta = 1$

$\theta = \dfrac{\pi}{2}$

Zero: $r = 0$ when $\theta = \dfrac{\pi}{2}$.

15. $|r| = |4 \cos 3\theta| = 4|\cos 3\theta| \leq 4$

$|\cos 3\theta| = 1$

$\cos 3\theta = \pm 1$

$\theta = 0, \dfrac{\pi}{3}, \dfrac{2\pi}{3}$

Maximum: $|r| = 4$ when $\theta = 0, \dfrac{\pi}{3}, \dfrac{2\pi}{3}$.

$0 = 4 \cos 3\theta$

$\cos 3\theta = 0$

$\theta = \dfrac{\pi}{6}, \dfrac{\pi}{2}, \dfrac{5\pi}{6}$

Zero: $r = 0$ when $\theta = \dfrac{\pi}{6}, \dfrac{\pi}{2}, \dfrac{5\pi}{6}$.

17. Circle: $r = 5$

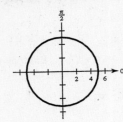

19. Circle: $r = \dfrac{\pi}{6}$

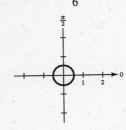

21. $r = 3 \sin \theta$

Symmetric with respect to $\theta = \dfrac{\pi}{2}$.

Circle with a radius of $\dfrac{3}{2}$.

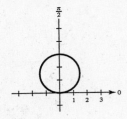

23. $r = 3(1 - \cos \theta)$

Symmetric with respect to the polar axis.

$\dfrac{a}{b} = \dfrac{3}{3} = 1 \implies$ Cardioid

$|r| = 6$ when $\theta = \pi$.

$r = 0$ when $\theta = 0$.

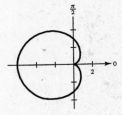

25. $r = 4(1 + \sin \theta)$

Symmetric with respect to $\theta = \dfrac{\pi}{2}$.

$\dfrac{a}{b} = \dfrac{4}{4} = 1 \implies$ Cardioid

$|r| = 8$ when $\theta = \dfrac{\pi}{2}$.

$r = 0$ when $\theta = \dfrac{3\pi}{2}$.

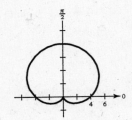

27. $r = 3 + 6 \sin \theta$

Symmetric with respect to $\theta = \dfrac{\pi}{2}$.

$\dfrac{a}{b} = \dfrac{3}{6} < 1 \implies$ Limaçon with inner loop

$|r| = 9$ when $\theta = \dfrac{\pi}{2}$.

$r = 0$ when $\theta = \dfrac{7\pi}{6}, \dfrac{11\pi}{6}$.

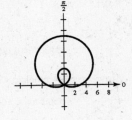

29. $r = 1 - 2\sin\theta$

Symmetric with respect to $\theta = \dfrac{\pi}{2}$.

$\dfrac{a}{b} = \dfrac{1}{2} < 1 \implies$ Limaçon with inner loop

$|r| = 3$ when $\theta = \dfrac{3\pi}{2}$.

$r = 0$ when $\theta = \dfrac{\pi}{6}, \dfrac{5\pi}{6}$.

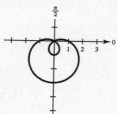

31. $r = 3 - 4\cos\theta$

Symmetric with respect to the polar axis.

$\dfrac{a}{b} = \dfrac{3}{4} < 1 \implies$ Limaçon with inner loop

$|r| = 7$ when $\theta = \pi$.

$r = 0$ when $\cos\theta = \dfrac{3}{4}$ or

$\theta \approx 0.723, 5.560$.

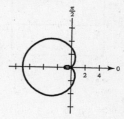

33. $r = 5\sin 2\theta$

Symmetric with respect to $\theta = \dfrac{\pi}{2}$, the polar axis and the pole.

Rose curve ($n = 2$) with 4 petals

$|r| = 5$ when $\theta = \dfrac{\pi}{4}, \dfrac{3\pi}{4}, \dfrac{5\pi}{4}, \dfrac{7\pi}{4}$.

$r = 0$ when $\theta = 0, \dfrac{\pi}{2}, \pi$.

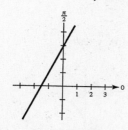

35.
$$r = 2\sec\theta$$
$$r = \dfrac{2}{\cos\theta}$$
$$r\cos\theta = 2$$
$$x = 2 \implies \text{Line}$$

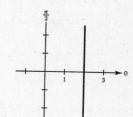

37.
$$r = \dfrac{3}{\sin\theta - 2\cos\theta}$$
$$r(\sin\theta - 2\cos\theta) = 3$$
$$y - 2x = 3$$
$$y = 2x + 3 \implies \text{Line}$$

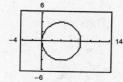

39. $r^2 = 9\cos 2\theta$

Symmetric with respect to the polar axis, $\theta = \dfrac{\pi}{2}$, and the pole.

Lemniscate

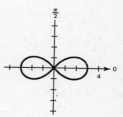

41. $r = 8\cos\theta$

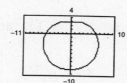

43. $r = 3(2 - \sin\theta)$

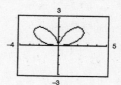

45. $r = 8\sin\theta\cos^2\theta$

47. $r = 1 + 2 \cos \theta$

Limaçon with inner loop

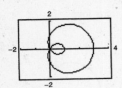

49. $r = 1 + \sin \theta$

Cardioid

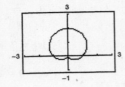

51. $r = 2 \sin 4\theta$

Rose curve

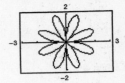

53. $r = \sin 5\theta$

Rose curve

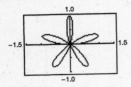

55. $r = 3 - 4 \cos \theta$

$0 \le \theta < 2\pi$

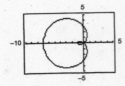

57. $r = 2 \cos\left(\dfrac{3\theta}{2}\right)$

$0 \le \theta < 4\pi$

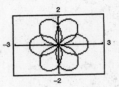

59. $r^2 = 9 \sin 2\theta$

Graph as $r_1 = 3\sqrt{\sin 2\theta}$ and $r_2 = -3\sqrt{\sin 2\theta}$.

It is traced out once for $0 \le \theta < \dfrac{\pi}{2}$.

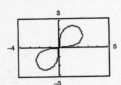

61. $r^2 = \dfrac{1}{\theta}$

$0 < \theta < \infty$

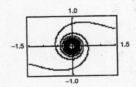

63. $r = 6 \cos \theta$

(a) $0 \le \theta \le \dfrac{\pi}{2}$

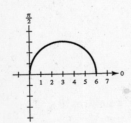

Upper half of circle

(b) $\dfrac{\pi}{2} \le \theta \le \pi$

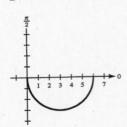

Lower half of circle

(c) $-\dfrac{\pi}{2} \le \theta \le \dfrac{\pi}{2}$

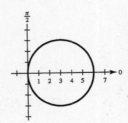

Full Circle

(d) $\dfrac{\pi}{4} \le \theta \le \dfrac{3\pi}{4}$

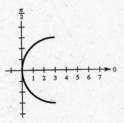

Left half of circle

65.
$$r = 2 - \sec\theta = 2 - \frac{1}{\cos\theta}$$

$$r\cos\theta = 2\cos\theta - 1$$

$$r(r\cos\theta) = 2r\cos\theta - r$$

$$\left(\pm\sqrt{x^2+y^2}\right)x = 2x - \left(\pm\sqrt{x^2+y^2}\right)$$

$$\left(\pm\sqrt{x^2+y^2}\right)(x+1) = 2x$$

$$\left(\pm\sqrt{x^2+y^2}\right) = \frac{2x}{x+1}$$

$$x^2 + y^2 = \frac{4x^2}{(x+1)^2}$$

$$y^2 = \frac{4x^2}{(x+1)^2} - x^2$$

$$= \frac{4x^2 - x^2(x+1)^2}{(x+1)^2} = \frac{4x^2 - x^2(x^2+2x+1)}{(x+1)^2}$$

$$= \frac{-x^4 - 2x^3 + 3x^2}{(x+1)^2} = \frac{-x^2(x^2+2x-3)}{(x+1)^2}$$

$$y = \pm\sqrt{\frac{x^2(3-2x-x^2)}{(x+1)^2}} = \pm\left|\frac{x}{x+1}\right|\sqrt{3-2x-x^2}$$

The graph has an asymptote at $x = -1$.

67. $r = \dfrac{3}{\theta}$

$$\theta = \frac{3}{r} = \frac{3\sin\theta}{r\sin\theta} = \frac{3\sin\theta}{y}$$

$$y = \frac{3\sin\theta}{\theta}$$

As $\theta \to 0, y \to 3$.

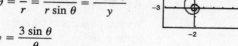

69. True. For a graph to have polar axis symmetry, replace (r, θ) by $(r, -\theta)$ or $(-r, \pi - \theta)$.

71. True. Since $\dfrac{a}{b} = \dfrac{1}{2} < 1$, the graph is a limaçon with an inner loop.

73. True

75. (a) $r = 1 - \sin\theta$

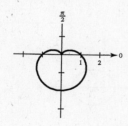

(b) $r = 1 - \sin\left(\theta - \dfrac{\pi}{4}\right)$

Rotate the graph of $r = 1 - \sin\theta$ through the angle $\dfrac{\pi}{4}$.

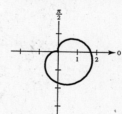

Section 12.7   Polar Equations of Conics

1. $r = \dfrac{4e}{1 + e \cos \theta}$

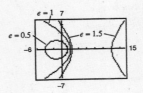

(a) $e = 1, r = \dfrac{4}{1 + \cos \theta}$, parabola

(b) $e = 0.5, r = \dfrac{2}{1 + 0.5 \cos \theta} = \dfrac{4}{2 + \cos \theta}$, ellipse

(c) $e = 1.5, r = \dfrac{6}{1 + 1.5 \cos \theta} = \dfrac{12}{2 + 3 \cos \theta}$, hyperbola

3. $r = \dfrac{4e}{1 - e \sin \theta}$

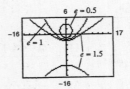

(a) $e = 1, r = \dfrac{4}{1 - \sin \theta}$, parabola

(b) $e = 0.5, r = \dfrac{2}{1 - 0.5 \sin \theta} = \dfrac{4}{2 - \sin \theta}$, ellipse

(c) $e = 1.5, r = \dfrac{6}{1 - 1.5 \sin \theta} = \dfrac{12}{2 - 3 \sin \theta}$, hyperbola

5. $r = \dfrac{2}{1 + \cos \theta}$

$e = 1 \implies$ Parabola

Vertical directrix to the right
of the pole

Matches graph (f).

7. $r = \dfrac{3}{1 + 2 \sin \theta}$

$e = 2 \implies$ Hyperbola

Matches graph (d).

9. $r = \dfrac{4}{2 + \cos \theta}$

$= \dfrac{2}{1 + 0.5 \cos \theta}$

$e = 0.5 \implies$ Ellipse

Vertical directrix to the right
of the pole

Matches graph (a).

11. $r = \dfrac{2}{1 - \cos \theta}$

$e = 1$, the graph is a parabola.

Vertex: $(1, \pi)$

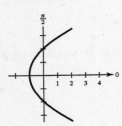

13. $r = \dfrac{5}{1 + \sin \theta}$

$e = 1$, the graph is a parabola.

Vertex: $\left(\dfrac{5}{2}, \dfrac{\pi}{2} \right)$

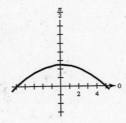

15. $r = \dfrac{2}{2 - \cos \theta} = \dfrac{1}{1 - (1/2) \cos \theta}$

$e = \dfrac{1}{2}1$, the graph is an ellipse.

Vertices: $(2, 0), \left(\dfrac{2}{3}, \pi \right)$

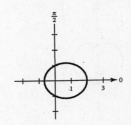

17. $r = \dfrac{6}{2 + \sin \theta} = \dfrac{3}{1 + (1/2)\sin \theta}$

$e = \dfrac{1}{2} < 1$, the graph is an ellipse.

Vertices: $\left(2, \dfrac{\pi}{2}\right), \left(6, \dfrac{3\pi}{2}\right)$

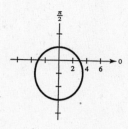

19. $r = \dfrac{3}{2 + 4\sin \theta} = \dfrac{3/2}{1 + 2\sin \theta}$

$e = 2 > 1$, the graph is a hyperbola.

Vertices: $\left(\dfrac{1}{2}, \dfrac{\pi}{2}\right), \left(-\dfrac{3}{2}, \dfrac{3\pi}{2}\right)$

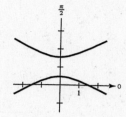

21. $r = \dfrac{3}{2 - 6\cos \theta} = \dfrac{3/2}{1 - 3\cos \theta}$

$e = 3 > 1$, the graph is a hyperbola.

Vertices: $\left(-\dfrac{3}{4}, 0\right), \left(\dfrac{3}{8}, \pi\right)$

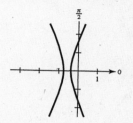

23. $r = \dfrac{4}{2 - \cos \theta} = \dfrac{2}{1 - (1/2)\cos \theta}$

$e = \dfrac{1}{2} < 1$, the graph is an ellipse.

Vertices: $(4, 0), \left(\dfrac{4}{3}, \pi\right)$

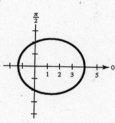

25. $r = \dfrac{-1}{1 - \sin \theta}$

$e = 1 \Rightarrow$ Parabola

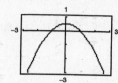

27. $r = \dfrac{-5}{2 + 4\sin \theta} = \dfrac{-(5/2)}{1 + 2\sin \theta}$

$e = 2 \Rightarrow$ Hyperbola

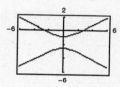

29. $r = \dfrac{3}{-4 + 2\cos \theta}$

$e = \dfrac{1}{2} \Rightarrow$ Ellipse

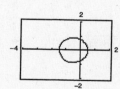

31. $r = \dfrac{4}{1 - 2\cos \theta}$

$e = 2 \Rightarrow$ Hyperbola

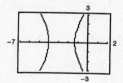

33. Parabola: $e = 1$

Directrix: $x = -1$

Vertical directrix to the left of the pole

$r = \dfrac{1(1)}{1 - 1\cos \theta} = \dfrac{1}{1 - \cos \theta}$

35. Ellipse: $e = \dfrac{1}{2}$

Directrix: $y = 1$

$p = 1$

Horizontal directrix above the pole

$r = \dfrac{(1/2)(1)}{1 + (1/2)\sin \theta} = \dfrac{1}{2 + \sin \theta}$

37. Hyperbola: $e = 2$

Directrix: $x = 1$

$p = 1$

Vertical directrix to the right of the pole

$r = \dfrac{2(1)}{1 + 2\cos \theta} = \dfrac{2}{1 + 2\cos \theta}$

39. Parabola

Vertex: $\left(1, -\dfrac{\pi}{2}\right) \Rightarrow e = 1, p = 2$

Horizontal directrix below the pole

$r = \dfrac{1(2)}{1 - 1\sin \theta} = \dfrac{2}{1 - \sin \theta}$

41. Parabola

Vertex: $(5, \pi) \implies e = 1, p = 10$

Vertical directrix to the left of the pole

$$r = \frac{1(10)}{1 - 1\cos\theta} = \frac{10}{1 - \cos\theta}$$

43. Ellipse: Vertices $(2, 0)$, $(10, \pi)$

Center: $(4, \pi)$; $c = 4$, $a = 6$, $e = \dfrac{2}{3}$

Vertical directrix to the right of the pole

$$r = \frac{(2/3)p}{1 + (2/3)\cos\theta} = \frac{2p}{3 + 2\cos\theta}$$

$$2 = \frac{2p}{3 + 2\cos 0}$$

$$p = 5$$

$$r = \frac{2(5)}{3 + 2\cos\theta} = \frac{10}{3 + 2\cos\theta}$$

45. Ellipse: Vertices $(20, 0)$, $(4, \pi)$

Center: $(8, 0)$; $c = 8$, $a = 12$, $e = \dfrac{2}{3}$

Vertical directrix to the left of the pole

$$r = \frac{(2/3)p}{1 - (2/3)\cos\theta} = \frac{2p}{3 - 2\cos\theta}$$

$$20 = \frac{2p}{3 - 2\cos 0}$$

$$p = 10$$

$$r = \frac{2(10)}{3 - 2\cos\theta} = \frac{20}{3 - 2\cos\theta}$$

47. Hyperbola: Vertices $\left(1, \dfrac{3\pi}{2}\right)$, $\left(9, \dfrac{3\pi}{2}\right)$

Center: $\left(5, \dfrac{3\pi}{2}\right)$; $c = 5$, $a = 4$, $e = \dfrac{5}{4}$

Horizontal directrix below the pole

$$r = \frac{(5/4)p}{1 - (5/4)\sin\theta} = \frac{5p}{4 - 5\sin\theta}$$

$$1 = \frac{5p}{4 - 5\sin(3\pi/2)}$$

$$p = \frac{9}{5}$$

$$r = \frac{5(9/5)}{4 - 5\sin\theta} = \frac{9}{4 - 5\sin\theta}$$

49.

$$\frac{x^2}{a^2} + \frac{y^2}{b^2} = 1$$

$$x^2 b^2 + y^2 a^2 = a^2 b^2$$

$$b^2 r^2 \cos^2\theta + a^2 r^2 \sin^2\theta = a^2 b^2$$

$$r^2[b^2 \cos^2\theta + a^2(1 - \cos^2\theta)] = a^2 b^2$$

$$r^2[a^2 + \cos^2\theta(b^2 - a^2)] = a^2 b^2$$

$$r^2 = \frac{a^2 b^2}{a^2 + (b^2 - a^2)\cos^2\theta} = \frac{a^2 b^2}{a^2 - c^2\cos^2\theta}$$

$$= \frac{b^2}{1 - (c/a)^2\cos^2\theta} = \frac{b^2}{1 - e^2\cos^2\theta}$$

51. $\dfrac{x^2}{169} + \dfrac{y^2}{144} = 1$

$a = 13$, $b = 12$, $c = 5$, $e = \dfrac{5}{13}$

$$r^2 = \frac{144}{1 - \frac{25}{169}\cos^2\theta} = \frac{24{,}336}{169 - 25\cos^2\theta}$$

53. $\dfrac{x^2}{9} - \dfrac{y^2}{16} = 1$

$a = 3$, $b = 4$, $c = 5$, $e = \dfrac{5}{3}$

$$r^2 = \frac{-16}{1 - \frac{25}{9}\cos^2\theta} = \frac{144}{25\cos^2\theta - 9}$$

55. Hyperbola

One focus: $\left(5, \dfrac{\pi}{2}\right)$

Vertices: $\left(4, \dfrac{\pi}{2}\right), \left(4, -\dfrac{\pi}{2}\right)$

$c = 5, a = 4, b = 3, e = \dfrac{5}{4}$

If the vertices were $(4, 0)$ and $(4, \pi)$, we would have:

$$r^2 = \dfrac{-9}{1 - \frac{25}{16}\cos^2\theta} = \dfrac{144}{25\cos^2\theta - 16}$$

The equation for the given hyperbola can be found by rotating through an angle of $\pi/2$, thus:

$$r^2 = \dfrac{144}{25\cos^2\left(\dfrac{\pi}{2} - \theta\right) - 16} = \dfrac{144}{25\sin^2\theta - 16}$$

57. Vertex: $\left(4100, \dfrac{\pi}{2}\right)$

Focus: $(0, 0)$

$e = 1, p = 8200$

$$r = \dfrac{ep}{1 + e\sin\theta} = \dfrac{8200}{1 + \sin\theta}$$

When $\theta = 30°, r = 8200/1.5 \approx 5466.67$.

Distance between the surface of the earth and the satellite is $r - 4000 \approx 1467$ miles.

59. False. If e remains fixed, and p changes, then the lengths of the major axis and minor axis change.

For example, graph $r = \dfrac{5}{1 - \frac{2}{3}\sin\theta}$, with $e = \dfrac{2}{3}$ and $p = \dfrac{15}{2}$ and $r = \dfrac{6}{1 - \frac{2}{3}\sin\theta}$, with $e = \dfrac{2}{3}$ and $p = 9$, on the same set of coordinate axes.

61. True.

$$r = \dfrac{10}{3 + 2\cos\theta} = \dfrac{\frac{10}{3}}{1 + \frac{2}{3}\cos\theta} \implies e = \dfrac{2}{3} \text{ and } p = 5$$

Ellipse with a horizontal major axis.

Center: $(-4, 0)$

$a = 6, c = 4, b = \sqrt{20}$

$$\dfrac{(x + 4)^2}{36} + \dfrac{y^2}{20} = 1$$

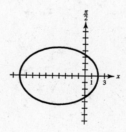

Review Exercises for Chapter 12

1. Hyperbola

3. Vertex: $(4, 2) = (h, k)$

Focus: $(4, 0) \implies p = -2$

$(x - h)^2 = 4p(y - k)$

$(x - 4)^2 = -8(y - 2)$

5. Vertex: $(0, 2) = (h, k)$

Directrix: $x = -3 \implies p = 3$

$(y - k)^2 = 4p(x - h)$

$(y - 2)^2 = 12x$

7. $x^2 = -2y \implies p = -\dfrac{1}{2}$

Focus: $\left(0, -\dfrac{1}{2}\right)$

$d_1 = b + \dfrac{1}{2}$

$d_2 = \sqrt{(2-0)^2 + \left(-2 + \dfrac{1}{2}\right)^2}$

$\quad = \sqrt{4 + \dfrac{9}{4}} = \dfrac{5}{2}$

$d_1 = d_2$

$b + \dfrac{1}{2} = \dfrac{5}{2}$

$\qquad b = 2$

The slope of the line is $m = \dfrac{-2-2}{2-0} = -2$

Tangent line: $y = -2x + 2$

x-intercept: $(1, 0)$

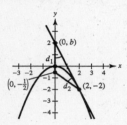

11. Vertices: $(-3, 0), (7, 0) \implies a = 5$
$\qquad\qquad\quad (h, k) = (2, 0)$

Foci: $(0, 0), (4, 0) \implies c = 2$

$b^2 = a^2 - c^2 = 25 - 4 = 21$

$\dfrac{(x-h)^2}{a^2} + \dfrac{(y-k)^2}{b^2} = 1$

$\qquad \dfrac{(x-2)^2}{25} + \dfrac{y^2}{21} = 1$

15. $2a = 10 \implies a = 5$

$b = 4$

$c^2 = a^2 - b^2 = 25 - 16 = 9 \implies c = 3$

The foci occur 3 feet from the center of the arch on a line connecting the tops of the pillars.

9. Parabola

Opens downward

Vertex: $(0, 12)$

$(x - h)^2 = 4p(y - k)$

$\qquad x^2 = 4p(y - 12)$

Solution points: $(\pm 4, 10)$

$16 = 4p(10 - 12)$

$16 = -8p$

$-2 = p$

$\quad x^2 = -8(y - 12)$

To find the x-intercepts, let $y = 0$.

$x^2 = 96$

$x = \pm\sqrt{96} = \pm 4\sqrt{6}$

At the base z, the archway is $2(4\sqrt{6}) = 8\sqrt{6}$ meters wide.

13. Vertices: $(0, \pm 6) \implies a = 6, (h, k) = (0, 0)$

Passes through $(2, 2)$

$\dfrac{(x-h)^2}{b^2} + \dfrac{(y-k)^2}{a^2} = 1$

$\qquad \dfrac{x^2}{b^2} + \dfrac{y^2}{36} = 1 \implies b^2 = \dfrac{36x^2}{36 - y^2} = \dfrac{36(4)}{36 - 4} = \dfrac{9}{2}$

$\qquad \dfrac{x^2}{9/2} + \dfrac{y^2}{36} = 1$

$\qquad \dfrac{2x^2}{9} + \dfrac{y^2}{36} = 1$

17. $\qquad 16x^2 + 9y^2 - 32x + 72y + 16 = 0$

$16(x^2 - 2x + 1) + 9(y^2 + 8y + 16) = -16 + 16 + 144$

$16(x - 1)^2 + 9(y + 4)^2 = 144$

$\qquad\qquad \dfrac{(x-1)^2}{9} + \dfrac{(y+4)^2}{16} = 1$

$a = 4, b = 3, c = \sqrt{7}$

Center: $(1, -4)$

Vertices: $(1, 0)$ and $(1, -8)$

Foci: $\left(1, -4 \pm \sqrt{7}\right)$

Eccentricity: $e = \dfrac{\sqrt{7}}{4}$

19. $\dfrac{(x+2)^2}{81} + \dfrac{(y-1)^2}{100} = 1$

$a = 10, b = 9, c = \sqrt{19}$

Center: $(-2, 1)$

Vertices: $(-2, 11)$ and $(-2, -9)$

Foci: $\left(-2, 1 \pm \sqrt{19}\right)$

Eccentricity: $e = \dfrac{\sqrt{19}}{10}$

21. $\dfrac{(x-1)^2}{9} + \dfrac{(y-5)^2}{25} = 1$

$\dfrac{2(x-1)}{9} + \dfrac{2(y-5)}{25}\dfrac{dy}{dx} = 0$

$\dfrac{dy}{dx} = -\dfrac{25(x-1)}{9(y-5)}$

At $(4, 5)$, $\dfrac{dy}{dx}$ is undefined. The tangent line is vertical.

$x = 4$

$x - 4 = 0$

23. $x^2 + 5y^2 = 10$

$\dfrac{x^2}{10} + \dfrac{y^2}{2} = 1$

$a = \sqrt{10}, b = \sqrt{2}$

Area $= \pi ab = \sqrt{20}\,\pi = 2\sqrt{5}\,\pi$ square units.

25. $(x^2 - 2x + 1) + 4(y^2 - 4y + 4) = -13 + 1 + 16$

$(x-1)^2 + 4(y-2)^2 = 4$

$\dfrac{(x-1)^2}{4} + \dfrac{(y-2)^2}{1} = 1$

Area $= \pi ab = \pi(2)(1) = 2\pi$

27. Vertices: $(0, \pm 1) \implies a = 1, (h, k) = (0, 0)$

Foci: $(0, \pm 3) \implies c = 3$

$b^2 = c^2 - a^2 = 9 - 1 = 8$

$\dfrac{(y-k)^2}{a^2} - \dfrac{(x-h)^2}{b^2} = 1$

$\qquad y^2 - \dfrac{x^2}{8} = 1$

29. Foci: $(0, 0), (8, 0) \implies c = 4, (h, k) = (4, 0)$

Asymptotes: $y = \pm 2(x - 4) \implies \dfrac{b}{a} = 2, b = 2a$

$b^2 = c^2 - a^2 \implies 4a^2 = 16 - a^2 \; 130 \implies a^2 = \dfrac{16}{5},$

$b^2 = \dfrac{64}{5}$

$\dfrac{(x-h)^2}{a^2} - \dfrac{(y-k)^2}{b^2} = 1$

$\dfrac{(x-4)^2}{\frac{16}{5}} - \dfrac{y^2}{\frac{64}{5}} = 1$

$\dfrac{5(x-4)^2}{16} - \dfrac{5y^2}{64} = 1$

31. $9x^2 - 16y^2 - 18x - 32y - 151 = 0$

$9(x^2 - 2x + 1) - 16(y^2 + 2y + 1) = 151 + 9 - 16$

$9(x-1)^2 - 16(y+1)^2 = 144$

$\dfrac{(x-1)^2}{16} - \dfrac{(y+1)^2}{9} = 1$

$a = 4, b = 3, c = 5$

Center: $(1, -1)$

Vertices: $(5, -1)$ and $(-3, -1)$

Foci: $(6, -1)$ and $(-4, -1)$

Asymptotes: $y = -1 \pm \dfrac{3}{4}(x - 1)$

$\qquad y = \dfrac{3}{4}x - \dfrac{7}{4} \quad \text{or} \quad y = -\dfrac{3}{4}x - \dfrac{1}{4}$

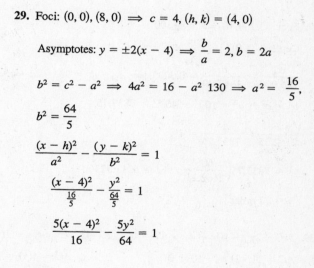

33. $\dfrac{(x-3)^2}{16} - \dfrac{(y+5)^2}{4} = 1$

$a = 4, b = 2, c = \sqrt{20} = 2\sqrt{5}$

Center: $(3, -5)$

Vertices: $(7, -5)$ and $(-1, -5)$

Foci: $\left(3 \pm 2\sqrt{5}, -5\right)$

Asymptotes: $y = -5 \pm \dfrac{1}{2}(x - 3)$

$$y = \dfrac{1}{2}x - \dfrac{13}{2} \quad \text{or} \quad y = -\dfrac{1}{2}x - \dfrac{7}{2}$$

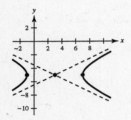

35. $\dfrac{x^2}{9} - \dfrac{y^2}{3} = 1$

$\dfrac{2x}{9} - \dfrac{2y}{3}\dfrac{dy}{dx} = 0$

$\dfrac{dy}{dx} = \dfrac{x}{3y}$

At $(6, 3)$, $\dfrac{dy}{dx} = \dfrac{6}{9} = \dfrac{2}{3}$.

Tangent line: $\quad y - 3 = \dfrac{2}{3}(x - 6)$

$3y - 9 = 2x - 12$

$2x - 3y - 3 = 0$

37. Foci: $(\pm 100, 0) \implies c = 100$

Center: $(0, 0)$

$\dfrac{d_2}{186,000} - \dfrac{d_1}{186,000} = 0.0005 \implies d_2 - d_1 = 93 = 2a \implies a = 46.5$

$b^2 = c^2 - a^2 = 100^2 - 46.5^2 = 7837.75$

$\dfrac{x^2}{2162.25} - \dfrac{y^2}{7837.75} = 1$

$y^2 = 7837.75\left(\dfrac{60^2}{2162.25} - 1\right) \approx 5211.5736$

$y \approx 72$ miles

39. $5x^2 - 2y^2 + 10x - 4y + 17 = 0$

$AC = 5(-2) = -10 < 0$

The graph is a hyperbola.

41. $4x^2 + y^2 - 16x = 0$

$AC = 4(1) = 4 > 0$

Ellipse

43. $x = 3\cos 0 = 3$

$y = 2\sin^2 0 = 0$

45. $x = 3\cos\dfrac{\pi}{6} = \dfrac{3\sqrt{3}}{2}$

$y = 2\sin^2\dfrac{\pi}{6} = \dfrac{1}{2}$

47. $x = 2t \implies \dfrac{x}{2} = t$

$y = 4t \implies y = 4\left(\dfrac{x}{2}\right) = 2x$

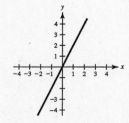

49. $x = t^2, \ x \geq 0$

$y = \sqrt{t} \implies y^2 = t$

$x = (y^2)^2 \implies x = y^4 \implies y = \sqrt[4]{x}$

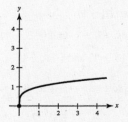

51. $x = 6\cos\theta$, $y = 6\sin\theta$

$\cos\theta = \dfrac{x}{6}$, $\sin\theta = \dfrac{y}{6}$

$\dfrac{x^2}{36} + \dfrac{y^2}{36} = 1$

$x^2 + y^2 = 36$

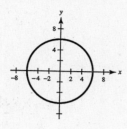

53. $x = t^2 - 4t$

$y = t^3$

$\dfrac{dx}{dt} = 2t - 4 = 0 \implies t = 2$

$\dfrac{dy}{dt} = 3t^2 \implies t = 0$

Vertical tangent: $(-4, 8)$, $(t = 2)$

Horizontal tangent: $(0, 0)$ $(t = 0)$

55. $x = 2\sin t$

$y = -\cos t$

$\dfrac{dx}{dt} = 2\cos t = 0 \implies t = \dfrac{\pi}{2}, \dfrac{3\pi}{2}$

$\dfrac{dy}{dt} = \sin t = 0 \implies t = 0, \pi$

Vertical tangents: $(2, 0)$ $(-2, 0)$

Horizontal tangents: $(0, -1)$, $(0, 1)$

57. $(h, k) = (-3, 4)$

$2a = 8 \implies a = 4$

$2b = 6 \implies b = 3$

$\dfrac{(x + 3)^2}{16} + \dfrac{(y - 4)^2}{9} = 1$

$x = -3 + 4\cos\theta$

$y = 4 + 3\sin\theta$

This solution is not unique.

59. Hyperbola

Asymptotes: $y = 3 \pm \dfrac{1}{2}(x + 1)$

Center: $(-1, 3)$

If the transverse axis is *horizontal*, then $\pm\dfrac{b}{a} = \pm\dfrac{1}{2}$.

Let $a = 2$ and $b = 1$.

$\dfrac{(x + 1)^2}{4} - (y - 3)^2 = 1$

Using the identity $\sec^2\theta - \tan^2\theta = 1$, we have

$\sec\theta = \dfrac{x + 1}{2} \implies x = 2\sec\theta - 1$

$\tan\theta = y - 3 \implies y = \tan\theta + 3$.

If the transverse axis is *vertical*, then $\pm\dfrac{a}{b} = \pm\dfrac{1}{2}$.

Let $a = 1$ and $b = 2$.

$(y - 3)^2 - \dfrac{(x + 1)^2}{4} = 1$

This time, $\sec\theta = y - 3 \implies y = \sec\theta + 3$

and $\tan\theta = \dfrac{x + 1}{2} \implies x = 2\tan\theta - 1$.

61. Polar coordinates: $\left(2, \dfrac{\pi}{4}\right)$

$x = 2\cos\dfrac{\pi}{4} = \sqrt{2}$

$y = 2\sin\dfrac{\pi}{4} = \sqrt{2}$

Rectangular coordinates:
$\left(\sqrt{2}, \sqrt{2}\right)$

63. Polar coordinates: $(-7, 4.19)$

$x = -7 \cos 4.19 \approx 3.4927$

$y = -7 \sin 4.19 \approx 6.0664$

Rectangular coordinates: $(3.4927, 6.0664)$

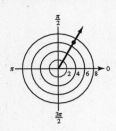

65. Rectangular coordinates: $(0, 2)$

$r = \pm\sqrt{0^2 + 2^2} = \pm 2$

$\tan \theta$ is undefined $\implies \theta = \dfrac{\pi}{2}, \dfrac{3\pi}{2}$

Polar coordinates: $\left(2, \dfrac{\pi}{2}\right), \left(-2, \dfrac{3\pi}{2}\right), \left(2, \dfrac{5\pi}{2}\right)$

67. Rectangular coordinates: $(4, 6)$

$r = \pm\sqrt{4^2 + 6^2} = \pm\sqrt{52} \approx \pm 7.2111$

$\tan \theta = \dfrac{6}{4} \implies \theta \approx 0.9828, 4.1244$

Polar coordinates: $(7.2111, 0.9828), (-7.2111, 4.1244),$

$(7.2111, 7.2660)$

69. $r = 3 \cos \theta$

$r^2 = 3r \cos \theta$

$x^2 + y^2 = 3x$

71. $r = \dfrac{2}{1 + \sin \theta}$

$r(1 + \sin \theta) = 2$

$r + r \sin \theta = 2$

$\pm\sqrt{x^2 + y^2} + y = 2$

$\pm\sqrt{x^2 + y^2} = 2 - y$

$x^2 + y^2 = (2 - y)^2$

$x^2 + y^2 = 4 - 4y + y^2$

$x^2 + 4y - 4 = 0$

73. $(x^2 + y^2)^2 = ax^2 y$

$(r^2)^2 = ar^2 \cos^2 \theta \, r \sin \theta$

$r = a \cos^2 \theta \sin \theta$

75. $xy = 5$

$(r \cos \theta)(r \sin \theta) = 5$

$r^2 = \dfrac{5}{\cos \theta \sin \theta}$

77. $r = 4$

Circle of radius 4 centered at the pole

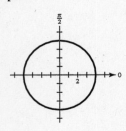

79. $r = 4 \sin 2\theta$

Symmetric with respect to $\theta = \pi/2$, the polar axis, and the pole.

Rose curve ($n = 2$) with 4 petals

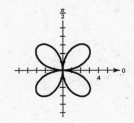

81. $r = -2(1 + \cos \theta)$

Symmetric with respect to the polar axis

$\dfrac{a}{b} = \dfrac{2}{2} = 1 \implies$ Cardioid

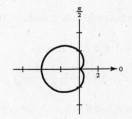

83. $r = 2 + 6 \sin \theta$

Limaçon with inner loop

$r = f(\sin \theta) \implies \theta = \dfrac{\pi}{2}$ symmetry

Maximum value: $|r| = 8$ when $\theta = \dfrac{\pi}{2}$.

Zeros: $2 + 6 \sin \theta = 0 \implies \sin \theta = -\dfrac{1}{3} \implies \theta \approx 3.4814, 5.9433$

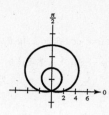

85. $r = -3 \cos 2\theta$

Rose curve with 4 petals

$r = f(\cos \theta) \implies$ polar axis symmetry

$\theta = \dfrac{\pi}{2}$: $r = -3 \cos 2(\pi - \theta) = -3 \cos(2\pi - 2\theta) = -3 \cos 2\theta$

Equivalent equation $\implies \theta = \dfrac{\pi}{2}$ symmetry

Pole: $r = -3 \cos 2(\pi + \theta) = -3 \cos(2\pi + 2\theta) = -3 \cos 2\theta$

Equivalent equation $\implies$ pole symmetry

Maximum value: $|r| = 3$ when $\theta = 0, \dfrac{\pi}{2}, \pi, \dfrac{3\pi}{2}$.

Zeros: $-3 \cos 2\theta = 0$ when $\cos 2\theta = 0 \implies \theta = \dfrac{\pi}{4}, \dfrac{3\pi}{4}, \dfrac{5\pi}{4}, \dfrac{7\pi}{4}$.

87. $r = 3(2 - \cos \theta)$

$\quad = 6 - 3 \cos \theta$

$\dfrac{a}{b} = \dfrac{6}{3} = 2$

The graph is a convex limaçon.

89. $r = 4 \cos 3\theta$

The graph is a rose curve with 3 petals.

91. $r = 2(1 + \sin \theta), \theta = \dfrac{\pi}{2}$

$\dfrac{dy}{dx} = \dfrac{2(1 + \sin \theta) \cos \theta + (2 \cos \theta) \sin \theta}{-2(1 + \sin \theta) \sin \theta + (2 \cos \theta) \cos \theta}$

$\quad = \dfrac{2 \sin \theta \cos \theta + \cos \theta}{\cos^2 \theta - \sin^2 \theta - \sin \theta}$

At $\theta = \dfrac{\pi}{2}, \dfrac{dy}{dx} = 0$.

93. $r = 3 + 2 \cos \theta, \theta = \pi$

$\dfrac{dy}{dx} = \dfrac{(3 + 2 \cos \theta) \cos \theta + (-2 \sin \theta) \sin \theta}{-(3 + 2 \cos \theta) \sin \theta + (-2 \sin \theta) \cos \theta}$

$\quad = \dfrac{2 \cos^2 \theta - 2 \sin^2 \theta + 3 \cos \theta}{-4 \sin \theta \cos \theta - 3 \sin \theta}$

At $\theta = \pi, \dfrac{dy}{dx}$ is undefined.

95. $r = 1 - 2 \sin \theta$

$x = r \cos \theta = (1 - 2 \sin \theta) \cos \theta = \cos \theta - 2 \sin \theta \cos \theta$

$y = r \sin \theta = (1 - 2 \sin \theta) \sin \theta = \sin \theta - 2 \sin^2 \theta$

$\dfrac{dx}{d\theta} = -\sin \theta + 2 \sin^2 \theta - 2 \cos^2 \theta$

$\qquad = -\sin \theta + 2 \sin^2 \theta - 2(1 - \sin^2 \theta)$

$\qquad = 4 \sin^2 \theta - \sin \theta - 2 = 0$

$\sin \theta = \dfrac{1 \pm \sqrt{33}}{8} \approx 0.84307, -0.59307$

$\theta \approx 1.0030, 2.1386, 3.7765, 5.6483$

$\dfrac{dy}{d\theta} = \cos \theta - 4 \sin \theta \cos \theta = \cos \theta (1 - 4 \sin \theta) = 0$

$\cos \theta = 0 \implies \theta = \dfrac{\pi}{2}, \dfrac{3\pi}{2}$

$\sin \theta = \dfrac{1}{4} \implies \theta = 0.2527, 2.8889$

Vertical tangents: $(-0.6861, 1.0030), (-0.6861, 2.1386),$

$\qquad (2.1861, 3.7765), (2.1861, 5.6483)$

Horizontal tangents: $\left(-1, \dfrac{\pi}{2} \right), \left(3, \dfrac{3\pi}{2} \right),$

$\qquad \left(\dfrac{1}{2}, 0.2527 \right), \left(\dfrac{1}{2}, 2.8889 \right)$

97. $r = \cos \theta, 0 \le \theta \le \pi$

$x = r \cos \theta = \cos^2 \theta$

$y = r \sin \theta = \cos \theta \sin \theta$

$\dfrac{dx}{d\theta} = -2 \cos \theta \sin \theta = 0 \implies \theta = 0, \dfrac{\pi}{2}$

$\dfrac{dy}{d\theta} = \cos^2 \theta - \sin^2 \theta = 0 \implies \theta = \dfrac{\pi}{4}, \dfrac{3\pi}{4}$

Vertical tangents: $(1, 0), \left(0, \dfrac{\pi}{2} \right)$

Horizontal tangents: $\left(\dfrac{\sqrt{2}}{2}, \dfrac{\pi}{4} \right) \left(-\dfrac{\sqrt{2}}{2}, \dfrac{3\pi}{4} \right)$

99. $r = \dfrac{1}{1 + 2 \sin \theta}, e = 2$

Hyperbola symmetric with respect to $\theta = \dfrac{\pi}{2}$ and having

vertices at $\left(\dfrac{1}{3}, \dfrac{\pi}{2} \right)$ and $\left(-1, \dfrac{3\pi}{2} \right)$.

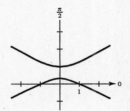

101. $r = \dfrac{4}{5 - 3 \cos \theta}$

$r = \dfrac{\frac{4}{5}}{1 - \left(\frac{3}{5} \right) \cos \theta}, e = \dfrac{3}{5}$

Ellipse symmetric with respect to the polar axis and

having vertices at $(2, 0)$ and $\left(\dfrac{1}{2}, \pi \right)$.

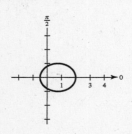

103. Parabola: $r = \dfrac{ep}{1 - e \cos \theta}, e = 1$

Vertex: $(2, \pi)$

Focus: $(0, 0) \implies p = 4$

$r = \dfrac{4}{1 - \cos \theta}$

105. Ellipse: $r = \dfrac{ep}{1 - e\cos\theta}$

Vertices: $(5, 0), (1, \pi) \implies a = 3$

One focus: $(0, 0) \implies c = 2$

$e = \dfrac{c}{a} = \dfrac{2}{3}, p = \dfrac{5}{2}$

$r = \dfrac{(2/3)(5/2)}{1 - (2/3)\cos\theta} = \dfrac{5/3}{1 - (2/3)\cos\theta} = \dfrac{5}{3 - 2\cos\theta}$

107. $a + c = 122{,}800 + 4000 = 126{,}800$

$a - c = 119 + 4000 = 4{,}119$

$2a = 130{,}919$

$a = 65{,}459.5$

$c = 61{,}340.5$

$e = \dfrac{c}{a} = \dfrac{61{,}340.5}{65{,}459.5} \approx 0.937$

$r = \dfrac{ep}{1 - e\cos\theta} \approx \dfrac{0.937p}{1 - 0.937\cos\theta}$

$r = 126{,}800$ when $\theta = 0$

$126{,}800 = \dfrac{ep}{1 - e\cos 0} = \dfrac{ep}{1 - e}$

$ep = 126{,}800(1 - e) \approx 7978.81$

Thus, $r = \dfrac{7978.81}{1 - 0.937\cos\theta}$

When $\theta = \dfrac{\pi}{3}, r \approx 15{,}011.9$

The distance from the surface of the earth and the satellite is $15{,}011.9 - 4000 \approx 11{,}011.9$ miles.

Problem Solving for Chapter 12

1. (a) $A = 2\displaystyle\int_0^b 2\sqrt{py}\,dy = 4\sqrt{p}\left(\dfrac{2}{3}y^{3/2}\right)\Big]_0^b = \dfrac{8}{3}p^{1/2}b^{3/2}$

(b) If $p = 2$ and $b = 4$, then $A = \dfrac{8}{3}\sqrt{2}(4)^{3/2} = \dfrac{64\sqrt{2}}{3}$ square units.

(c) As p approaches zero, the parabola becomes narrower and narrower, thus the area becomes smaller and smaller.

3. $x = \dfrac{4t}{1 + t^3}, \; y = \dfrac{4t^2}{1 + t^3}$

(a) $x^3 + y^3 = 4xy$

(b) $\dfrac{dy}{dt} = \dfrac{(1 + t^3)(8t) - 4t^2(3t^2)}{(1 + t^3)^2}$

$= \dfrac{4t(2 - t^3)}{(1 + t^3)^2} = 0$ when $t = 0$ or $t = \sqrt[3]{2}$.

Points: $(0, 0), \left(\dfrac{4\sqrt[3]{2}}{3}, \dfrac{4\sqrt[3]{4}}{3}\right) \approx (1.6799, 2.1165)$

(c) Since $\dfrac{dy}{dx} = \dfrac{dy/dt}{dx/dt}$, the points of horizontal tangency occur when $\dfrac{dy}{dt} = 0$ (but $\dfrac{dx}{dt} \neq 0$).

$\dfrac{dx}{dt} = \dfrac{4(1 - 2t^3)}{(1 + t^3)^2}$

$\dfrac{dy}{dt} = \dfrac{-4t(t^3 - 2)}{(t^3 + 1)^2} = 0$ when $t = 0$ or $t = \sqrt[3]{2}$.

When $t = 0$: $x = 0, y = 0$, and $\dfrac{dx}{dt} = 4$.

When $t = \sqrt[3]{2}$: $x = \dfrac{4\sqrt[3]{2}}{3}, y = \dfrac{4\sqrt[3]{4}}{3}$, and $\dfrac{dx}{dt} = -\dfrac{4}{3}$.

5. The coordinates of point (x, y) can be thought of as the sum of two vectors:

From origin to center of small circle:

$\langle 3 \cos \theta, 3 \sin \theta \rangle$

From center of small circle to point (x, y):

$\langle \cos \beta, \sin \beta \rangle$

Because the small circle rotates by 2θ when its center has rotated by θ, we have $\beta = \pi + 3\theta$.

$x = 3 \cos \theta + \cos(\pi + 3\theta) = 3 \cos \theta - \cos 3\theta$

$y = 3 \sin \theta + \sin(\pi + 3\theta) = 3 \sin \theta - \sin 3\theta$

7. (a) $(r_1, \theta_1) = (x_1, y_1)$ where $x_1 = r_1 \cos \theta_1$ and $y_1 = r_1 \sin \theta_1$.

$(r_2, \theta_2) = (x_2, y_2)$ where $x_2 = r_2 \cos \theta_2$ and $y_2 = r_2 \sin \theta_2$.

$$d = \sqrt{(x_1 - x_2)^2 + (y_1 - y_2)^2}$$
$$= \sqrt{x_1^2 - 2x_1x_2 + x_2^2 + y_1^2 - 2y_1y_2 + y_2^2}$$
$$= \sqrt{(x_1^2 + y_1^2) + (x_2^2 + y_2^2) - 2(x_1x_2 + y_1y_2)}$$
$$= \sqrt{r_1^2 + r_2^2 - 2(r_1r_2 \cos \theta_1 \cos \theta_2 + r_1r_2 \sin \theta_1 \sin \theta_2)}$$
$$= \sqrt{r_1^2 + r_2^2 - 2r_1r_2 \cos(\theta_1 - \theta_2)}$$

(b) If $\theta_1 = \theta_2$, then

$$d = \sqrt{r_1^2 + r_2^2 - 2r_1r_2}$$
$$= \sqrt{(r_1 - r_2)^2}$$
$$= |r_1 - r_2|.$$

This represents the distance between two points on the line $\theta = \theta_1 = \theta_2$.

(c) If $\theta_1 - \theta_2 = 90°$, then

$$d = \sqrt{r_1^2 + r_2^2}.$$

This is the result of the Pythagorean Theorem.

(d) The results should be the same. For example, use the points

$$\left(3, \frac{\pi}{6}\right) \text{ and } \left(4, \frac{\pi}{3}\right).$$

The distance is $d \approx 2.053$.
Now use the representations

$$\left(-3, \frac{7\pi}{6}\right) \text{ and } \left(-4, \frac{4\pi}{3}\right).$$

The distance is still $d \approx 2.053$.

9. $r = 3 \sin k\theta$

(a) $r = 3 \sin 1.5\theta$

$0 \le \theta < 4\pi$

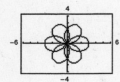

(b) $r = 3 \sin 2.5\theta$

$0 \le \theta < 4\pi$

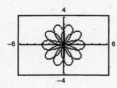

(c) Yes. $r = 3 \sin(k\theta)$.

Find the minimum value of $\theta(\theta > 0)$ that is a multiple of 2π that makes $k\theta$ a multiple of 2π.

11. $r = e^{\cos \theta} - 2 \cos 4\theta + \sin^5\left(\dfrac{\theta}{12}\right)$

(a) No, the graph appears to have a period of 2π but does not. For example, $r(\pi) \ne r(3\pi)$.

(b) By using the table feature of the calculator we have

$r \approx 4.077$ when $\theta \approx 5.54$ for $0 \le \theta \le 2\pi$ and

$r \approx 4.46$ when $\theta \approx 11.83$ for $0 \le \theta \le 4\pi$.

The graph is not periodic. As θ increases the maximum value of r changes.

13. Since the axis of symmetry is the x-axis, the vertex is $(h, 0)$ and $y^2 = 4p(x - h)$.

Also, since the focus is $(0, 0)$, $0 - h = p \implies h = -p$ and $y^2 = 4p(x + p)$.

CHAPTER 13
Additional Topics in Trigonometry

C H A P T E R   1 3
Additional Topics in Trigonometry

Section 13.1   Law of Sines

1. Given: $A = 30°$, $B = 45°$, $a = 20$

$C = 180° - A - B = 105°$

$b = \dfrac{a}{\sin A}(\sin B) = \dfrac{20 \sin 45°}{\sin 30°} = 20\sqrt{2} \approx 28.28$

$c = \dfrac{a}{\sin A}(\sin C) = \dfrac{20 \sin 105°}{\sin 30°} \approx 38.64$

3. Given: $A = 25°$, $B = 35°$, $a = 3.5$

$C = 180° - A - B = 120°$

$b = \dfrac{a}{\sin A}(\sin B) = \dfrac{3.5}{\sin 25°}(\sin 35°) \approx 4.8$

$c = \dfrac{a}{\sin A}(\sin C) = \dfrac{3.5}{\sin 25°}(\sin 120°) \approx 7.2$

5. Given: $A = 36°$, $a = 8$, $b = 5$

$\sin B = \dfrac{b \sin A}{a} = \dfrac{5 \sin 36°}{8} \approx 0.36737 \implies B \approx 21.55°$

$C = 180° - A - B \approx 180° - 36° - 21.55° = 122.45°$

$c = \dfrac{a}{\sin A}(\sin C) = \dfrac{8}{\sin 36°}(\sin 122.45°) \approx 11.49$

7. Given: $A = 102.4°$, $C = 16.7°$, $a = 21.6$

$B = 180° - A - C = 60.9°$

$b = \dfrac{a}{\sin A}(\sin B) = \dfrac{21.6}{\sin 102.4°}(\sin 60.9°) \approx 19.3$

$c = \dfrac{a}{\sin A}(\sin C) = \dfrac{21.6}{\sin 102.4°}(\sin 16.7°) \approx 6.4$

9. Given: $A = 83° \, 20'$, $C = 54.6°$, $c = 18.1$

$B = 180° - A - C = 180° - 83° \, 20' - 54° \, 36' = 42° \, 4'$

$a = \dfrac{c}{\sin C}(\sin A) = \dfrac{18.1}{\sin 54.6°}(\sin 83° \, 20') \approx 22.05$

$b = \dfrac{c}{\sin C}(\sin B) = \dfrac{18.1}{\sin 54.6°}(\sin 42° \, 4') \approx 14.88$

11. Given: $B = 15° \, 30'$, $a = 4.5$, $b = 6.8$

$\sin A = \dfrac{a \sin B}{b} = \dfrac{4.5 \sin 15° \, 30'}{6.8} \approx 0.17685 \implies A \approx 10° \, 11'$

$C = 180° - A - B \approx 180° - 10° \, 11' - 15° \, 30' = 154° \, 19'$

$c = \dfrac{b}{\sin B}(\sin C) = \dfrac{6.8}{\sin 15° \, 30'}(\sin 154° \, 19') \approx 11.03$

13. Given: $C = 145°$, $b = 4$, $c = 14$

$\sin B = \dfrac{b \sin C}{c} = \dfrac{4 \sin 145°}{14} \approx 0.16387 \implies B \approx 9.43°$

$A = 180° - B - C \approx 180° - 9.43° - 145° = 25.57°$

$a = \dfrac{c}{\sin C}(\sin A) \approx \dfrac{14}{\sin 145°}(\sin 25.57°) \approx 10.53$

15. Given: $A = 110°\ 15'$, $a = 48$, $b = 16$

$$\sin B = \frac{b \sin A}{a} = \frac{16 \sin 110°\ 15'}{48} \approx 0.31273 \implies B \approx 18°\ 13'$$

$$C = 180° - A - B \approx 180° - 110°\ 15' - 18°\ 13' = 51°\ 32'$$

$$c = \frac{a}{\sin A}(\sin C) = \frac{48}{\sin 110°\ 15'}(\sin 51°\ 32') \approx 40.06$$

17. Given: $A = 55°, B = 42°, c = \dfrac{3}{4}$

$$C = 180° - A - B = 83°$$

$$a = \frac{c}{\sin C}(\sin A) = \frac{0.75}{\sin 83°}(\sin 55°) \approx 0.62$$

$$b = \frac{c}{\sin C}(\sin B) = \frac{0.75}{\sin 83°}(\sin 42°) \approx 0.51$$

19. Given: $A = 58°, a = 11.4, b = 12.8$

$$\sin B = \frac{b \sin A}{a} = \frac{12.8 \sin 58°}{11.4} \approx 0.9522 \implies B \approx 72.2° \text{ or } B \approx 107.8°$$

Case 1

$B \approx 72.2°$

$C = 180° - A - B \approx 49.8°$

$$c = \frac{a}{\sin A}(\sin C) \approx \frac{11.4 \sin 49.8°}{\sin 58°} \approx 10.27$$

Case 2

$B \approx 107.8°$

$C = 180° - A - B \approx 14.2°$

$$c = \frac{a}{\sin A}(\sin C) \approx \frac{11.4 \sin 14.2°}{\sin 58°} \approx 3.30$$

21. Given: $a = 18, b = 20, A = 76°$

$h = 20 \sin 76° \approx 19.41$

Since $a < h$, no triangle is formed.

23. Given: $a = 125, b = 200, A = 110°$

No triangle is formed because A is obtuse and $a < b$.

25. Given: $A = 22°, a = \dfrac{5}{7}, b = \dfrac{5}{7}$

$$\sin B = \frac{b \sin A}{a} = \frac{\frac{5}{7} \sin 22°}{\frac{5}{7}} = \sin 22° \implies B = 22°$$

$$C = 180° - A - B = 136°$$

$$c = \frac{a \sin C}{\sin A} = \frac{\frac{5}{7} \sin 136°}{\sin 22°} \approx 1.32$$

27. Area $= \dfrac{1}{2}ab \sin C = \dfrac{1}{2}(4)(6) \sin 120° \approx 10.4$

29. Area $= \frac{1}{2}bc \sin A = \frac{1}{2}(57)(85) \sin 43°\ 45' \approx 1675.2$

31. Area $= \frac{1}{2}ac \sin B = \frac{1}{2}(105)(64) \sin(72°30') \approx 3204.5$

33. Given: $A = 36°$, $a = 5$

(a) One solution if $b \le 5$ or $b = \dfrac{5}{\sin 36°}$

(b) Two solutions if $5 < b < \dfrac{5}{\sin 36°}$

(c) No solution if $b > \dfrac{5}{\sin 36°}$

35. Given: $A = 10°$, $a = 10.8$

(a) One solution if $b \le 10.8$ or $b = \dfrac{10.8}{\sin 10°}$

(b) Two solutions if $10.8 < b < \dfrac{10.8}{\sin 10°}$

(c) No solution if $b > \dfrac{10.8}{\sin 10°}$

37. Law of Sines

If ABC is a triangle with sides a, b, and c, then

$$\frac{a}{\sin A} = \frac{b}{\sin B} = \frac{c}{\sin C}.$$

41. $\dfrac{\sin(42° - \theta)}{10} = \dfrac{\sin 48°}{17}$

$\sin(42° - \theta) \approx 0.43714$

$42° - \theta \approx 25.9°$

$\theta \approx 16.1°$

45. (a)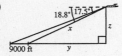

(b) $\dfrac{x}{\sin 17.5°} = \dfrac{9000}{\sin 1.3°}$

$x \approx 119{,}289.1261$ feet ≈ 22.6 miles

(c) $\dfrac{y}{\sin 71.2°} = \dfrac{x}{\sin 90°}$

$y = x \sin 71.2° \approx 119{,}289.1261 \sin 71.2°$

$\approx 112{,}924.963$ feet ≈ 21.4 miles

(d) $z = x \sin 18.8° \approx 119{,}289.1261 \sin 18.8° \approx 38{,}443$ feet

49. $\alpha = 180° - (180° - \phi + \theta) = \phi - \theta$

$\dfrac{d}{\sin \theta} = \dfrac{2}{\sin \alpha}$

$d = \dfrac{2 \sin \theta}{\sin(\phi - \theta)}$

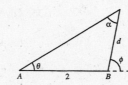

39. $C = 180° - 23° - 94° = 63°$

$h = \dfrac{35}{\sin 63°}(\sin 23°) \approx 15.3$ meters

43. Given: $c = 100$

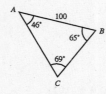

$A = 74° - 28° = 46°,$

$B = 180° - 41° - 74° = 65°,$

$C = 180° - 46° - 65° = 69°$

$a = \dfrac{c}{\sin C}(\sin A) = \dfrac{100}{\sin 69°}(\sin 46°) \approx 77$ meters

47.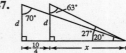

In 15 minutes the boat has traveled

$(10 \text{ mph})\left(\dfrac{1}{4} \text{ hr}\right) = \dfrac{10}{4}$ miles.

$\tan 63° = \dfrac{x}{d} \quad\Rightarrow\quad d \tan 63° = x$

$\tan 70° = \dfrac{x + (10/4)}{d} \Rightarrow d \tan 70° = x + \dfrac{10}{4}$

$\Rightarrow d \tan 70° - \dfrac{10}{4} = x$

$d \tan 70° - \dfrac{10}{4} = d \tan 63°$

$d \tan 70° - d \tan 63° = \dfrac{10}{4}$

$d(\tan 70° - \tan 63°) = 2.5$

$d = \dfrac{2.5}{\tan 70° - \tan 63°} \approx 3.2$ miles

51. True. The longest side of a triangle is always opposite the largest angle.

53. True.

Assume $C = 90°$. Then $\sin C = 1$

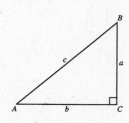

$$\sin A = \frac{a}{c} \implies c = \frac{a}{\sin A}$$

$$\implies \frac{a}{\sin A} = \frac{c}{\sin C}$$

$$\sin B = \frac{b}{c} \implies c = \frac{b}{\sin B}$$

$$\implies \frac{b}{\sin B} = \frac{c}{\sin C}.$$

Thus, $\dfrac{a}{\sin A} = \dfrac{b}{\sin B} = \dfrac{c}{\sin C}$.

55. (a) $A = 30°, a = 15, c = 25$

$$\sin C = \frac{25 \sin 30°}{15} \implies C \approx 56.4427°$$

$$B = 180° - A - C \implies B \approx 93.5573°$$

$$\text{Area} = \frac{1}{2} ac \sin B \approx \frac{1}{2}(15)(25) \sin 93.5573°$$

$$\approx 187.14 \text{ square units}$$

(b) $b = \dfrac{a \sin B}{\sin A} \approx \dfrac{15 \sin 93.5573°}{\sin 30°} \approx 29.94$

The coordinates for point C are as follows:

$x = b \cos A \approx 29.94 \cos 30° \approx 25.93$

$y = b \sin A \approx 29.94 \sin 30° \approx 14.97$

The line between $(0, 0)$ and C is: $y \approx 0.577x$

The line between $(25, 0)$ and C is: $y \approx 16.097(x - 25)$

Solving for x in each equation yields
$x \approx 1.733y$ and $x \approx 25 + 0.062y$.

$$\text{Area} \approx \int_0^{14.97} [(25 + 0.062y) - 1.733y]\, dy$$

$$= \int_0^{14.97} (25 - 1.671y)\,dy$$

$$= (25y - 0.835y^2) \Big]_0^{14.97}$$

$$\approx 187.13 \text{ square units}$$

Note: The answers are slightly different due to rounding errors.

Section 13.2   Law of Cosines

1. Given: $a = 7, b = 10, c = 15$

$$\cos C = \frac{a^2 + b^2 - c^2}{2ab} = \frac{49 + 100 - 225}{2(7)(10)} \approx -0.5429 \implies C \approx 122.88°$$

$$\sin B = \frac{b \sin C}{c} = \frac{10 \sin 122.88°}{15} \approx 0.5599 \implies B \approx 34.05°$$

$$A \approx 180° - 34.05° - 122.88° \approx 23.07°$$

3. Given: $A = 30°$, $b = 15$, $c = 30$

$$a^2 = b^2 + c^2 - 2bc \cos A = 225 + 900 - 2(15)(30) \cos 30° \approx 345.5771$$

$$a \approx 18.6$$

$$\cos B = \frac{a^2 + c^2 - b^2}{2ac} \approx \frac{(18.6)^2 + 900 - 225}{2(18.6)(30)} \approx 0.9148$$

$$B \approx 23.8°$$

$$C \approx 180° - 30° - 23.8° = 126.2°$$

5. $a = 11$, $b = 14$, $c = 20$

$$\cos C = \frac{a^2 + b^2 - c^2}{2ab} = \frac{121 + 196 - 400}{2(11)(14)} \approx -0.2695 \implies C \approx 105.63°$$

$$\sin B = \frac{b \sin C}{c} = \frac{14 \sin 105.63°}{20} \approx 0.6741 \implies B \approx 42.39°$$

$$A \approx 180° - 42.39° - 105.63° \approx 31.98°$$

7. Given: $a = 75.4$, $b = 52$, $c = 52$

$$\cos A = \frac{b^2 + c^2 - a^2}{2bc} = \frac{52^2 + 52^2 - 75.4^2}{2(52)(52)} = -0.05125 \implies A \approx 92.94°$$

$$\sin B = \frac{b \sin A}{a} \approx \frac{52(0.9987)}{75.4} \approx 0.68875 \implies B \approx 43.53°$$

$$C = B \approx 43.53°$$

9. Given: $A = 135°$, $b = 4$, $c = 9$

$$a^2 = b^2 + c^2 - 2bc \cos A = 16 + 81 - 2(4)(9)\cos 135° \approx 147.9117 \implies a \approx 12.16$$

$$\sin B = \frac{b \sin A}{a} = \frac{4 \sin 135°}{12.16} \approx 0.2326 \implies B \approx 13.45°$$

$$C \approx 180° - 135° - 13.45° \approx 31.55°$$

11. Given: $B = 10° \, 35'$, $a = 40$, $c = 30$

$$b^2 = a^2 + c^2 - 2ac \cos B = 1600 + 900 - 2(40)(30)\cos 10° \, 35' \approx 140.8268 \implies b \approx 11.9$$

$$\sin C = \frac{c \sin B}{b} = \frac{30 \sin 10° \, 35'}{11.9} \approx 0.4630 \implies C \approx 27.58° \approx 27° \, 35'$$

$$A \approx 180° - 10° \, 35' - 27° \, 35' = 141° \, 50'$$

13. Given: $B = 125° \, 40'$, $a = 32$, $c = 32$

$$b^2 = a^2 + c^2 - 2ac \cos B \approx 32^2 + 32^2 - 2(32)(32)(-0.5831) \approx 3242.1 \implies b \approx 56.9$$

$$A = C \implies 2A = 180° - 125° \, 40' = 54° \, 20' \implies A = C = 27° \, 10'$$

15. $C = 43°$, $a = \dfrac{4}{9}$, $b = \dfrac{7}{9}$

$$c^2 = a^2 + b^2 - 2ab \cos C = \left(\frac{4}{9}\right)^2 + \left(\frac{7}{9}\right)^2 - 2\left(\frac{4}{9}\right)\left(\frac{7}{9}\right)\cos 43° \approx 0.2968 \implies c \approx 0.5448$$

$$\sin A = \frac{a \sin C}{c} = \frac{(4/9)\sin 43°}{0.5448} \approx 0.5564 \implies A \approx 33.8°$$

$$B \approx 180° - 43° - 33.8° \approx 103.2°$$

17. $a = 5, \ b = 7, \ c = 10 \implies s = \dfrac{a + b + c}{2} = 11$

Area $= \sqrt{s(s - a)(s - b)(s - c)} = \sqrt{11(6)(4)(1)} \approx 16.25$

19. $a = 2.5, \ b = 10.2, \ c = 9 \implies s = \dfrac{a + b + c}{2} = 10.85$

Area $= \sqrt{s(s - a)(s - b)(s - c)} = \sqrt{10.85(8.35)(0.65)(1.85)} \approx 10.44$

21. $a = 12.32, \ b = 8.46, \ c = 15.05 \implies s = \dfrac{a + b + c}{2} = 17.915$

Area $= \sqrt{s(s - a)(s - b)(s - c)} = \sqrt{17.915(5.595)(9.455)(2.865)} \approx 52.11$

23. Law of Cosines

If ABC is a triangle with sides a, b, and c, then

Standard Form	*Alternate Form*

$a^2 = b^2 + c^2 - 2bc \cos A$ $\cos A = \dfrac{b^2 + c^2 - a^2}{2\,bc}$

$b^2 = a^2 + c^2 - 2ac \cos B$ $\cos B = \dfrac{a^2 + c^2 - b^2}{2\,ac}$

$c^2 = a^2 + b^2 - 2ab \cos C$ $\cos C = \dfrac{a^2 + b^2 - c^2}{2\,ab}$

25.

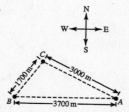

$\cos B = \dfrac{1700^2 + 3700^2 - 3000^2}{2(1700)(3700)} \implies B \approx 52.9°$

Bearing: $90° - 52.9° = \text{N } 37.1° \text{ E}$

$\cos C = \dfrac{1700^2 + 3000^2 - 3700^2}{2(1700)(3000)} \implies C \approx 100.2°$

Bearing: $A = 180° - 52.9° - 100.2° = 26.9° \implies \text{S } 63.1° \text{ E}$

27.

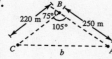

$b^2 = 220^2 + 250^2 - 2(220)(250)\cos 105° \implies b \approx 373.3$ meters

29.

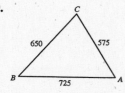

The largest angle is across from the largest side.

$\cos C = \dfrac{650^2 + 575^2 - 725^2}{2(650)(575)}$

$C \approx 72.3°$

31. $C = 180° - 53° - 67° = 60°$

$c^2 = a^2 + b^2 - 2ab \cos C$

$\quad = 36^2 + 48^2 - 2(36)(48)(0.5)$

$\quad = 1872$

$c \approx 43.3$ mi

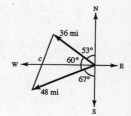

33. The angles at the base of the tower are 96° and 84°. The longer guy wire g_1 is given by:

$g_1{}^2 = 75^2 + 100^2 - 2(75)(100) \cos 96° \approx 17{,}192.9 \implies g_1 \approx 131.1$ feet

The shorter guy wire g_2 is given by:

$g_2{}^2 = 75^2 + 100^2 - 2(75)(100) \cos 84° \approx 14{,}057.1 \implies g_2 \approx 118.6$ feet

35. (a) $\cos \theta = \dfrac{165^2 + 216^2 - 368^2}{2(165)(216)}$

$\qquad \theta \approx 149.7°$

$\qquad \theta - 90° \approx 59.7°$

Bearing: N 59.7° E

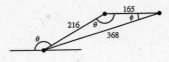

(b) $\cos \phi = \dfrac{368^2 + 165^2 - 216^2}{2(368)(165)}$

$\qquad \phi \approx 17.2°$

$\qquad 90° - \phi \approx 72.8°$

Bearing: N 72.8° E

37. (a) $x^2 = 330^2 + 420^2 - 2(330)(420) \cos 8°$

$\qquad \approx 10{,}797.69134$

$\qquad x \approx 103.9119 \approx 103.9$ feet

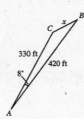

(b) Angle C does not change as θ changes.
To find C, use the Law of Sines.

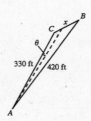

$\dfrac{\sin C}{420} = \dfrac{\sin 8°}{103.9119}$

$\sin C = \dfrac{420 \sin 8°}{103.9119} \approx 0.56252$

Since C is obtuse, $C \approx 180° - \sin^{-1} 0.5652$

$\qquad\qquad \approx 145.77°$

$B = 180° - C - \theta \approx 34.23° - \theta$

$\dfrac{\sin(34.23° - \theta)}{330} = \dfrac{\sin \theta}{x}$

$x = \dfrac{330 \sin \theta}{\sin(34.23° - \theta)}$

By the Quotient Rule,

$\dfrac{dx}{dt} = \dfrac{330\left[\sin(34.23° - \theta) \cos \theta \dfrac{d\theta}{dt} - \sin \theta \cos(34.23° - \theta)\left(-\dfrac{d\theta}{dt}\right)\right]}{\sin^2(34.23° - \theta)}$

$\qquad = \dfrac{330\dfrac{d\theta}{dt} \sin 34.23°}{\sin^2(34.23° - \theta)} \quad$ (by the sum and difference formula)

When $\theta = 3°$, $\dfrac{d\theta}{dt} = \dfrac{1.4\pi}{180}$ rad/sec, and we have

$\dfrac{dx}{dt} = \dfrac{330\left(\dfrac{1.4\pi}{180}\right) \sin 34.23°}{\sin^2 31.23°} \approx 16.87$ feet per second

39. $a^2 = 20^2 + 20^2 - 2(20)(20) \cos 11° \implies a \approx 3.8$ miles

41. $d^2 = 10^2 + 7^2 - 2(10)(7) \cos \theta$

$$\theta = \arccos\left[\frac{10^2 + 7^2 - d^2}{2(10)(7)}\right]$$

$$s = \frac{360° - \theta}{360°}(2\pi r) = \frac{(360° - \theta)\pi}{45°}$$

d (inches)	9	10	12	13	14	15	16
θ (degrees)	60.9°	69.5°	88.0°	98.2°	109.6°	122.9°	139.8°
s (inches)	20.88	20.28	18.99	18.28	17.48	16.55	15.37

43. $a = 200, b = 500, c = 600 \implies s = \dfrac{200 + 500 + 600}{2} = 650$

Area $= \sqrt{650(450)(150)(50)} \approx 46,837.5$ square feet

45. False. The average of the three sides of a triangle is $\dfrac{a + b + c}{3}$, not $\dfrac{a + b + c}{2}$.

47. True. Assume $C = 90°$, then $\cos C = 0$ and $c^2 = a^2 + b^2 - 2ab \cos C$ becomes $c^2 = a^2 + b^2$.
The Law of Cosines simplifies to the Pythagorean Theorem.

Section 13.3   Vectors in the Plane

1. Initial point: $(0, 0)$

Terminal point: $(3, 2)$

$\mathbf{v} = \langle 3 - 0, 2 - 0 \rangle = \langle 3, 2 \rangle$

$\|\mathbf{v}\| = \sqrt{3^2 + 2^2} = \sqrt{13}$

3. Initial point: $(2, 2)$

Terminal point: $(-1, 4)$

$\mathbf{v} = \langle -1 - 2, 4 - 2 \rangle = \langle -3, 2 \rangle$

$\|\mathbf{v}\| = \sqrt{(-3)^2 + 2^2} = \sqrt{13}$

5. Initial point: $(3, -2)$

Terminal point: $(3, 3)$

$\mathbf{v} = \langle 3 - 3, 3 - (-2) \rangle = \langle 0, 5 \rangle$

$\|\mathbf{v}\| = \sqrt{0^2 + 5^2} = \sqrt{25} = 5$

7. Initial point: $(-1, 5)$

Terminal point: $(15, 12)$

$\mathbf{v} = \langle 15 - (-1), 12 - 5 \rangle = \langle 16, 7 \rangle$

$\|\mathbf{v}\| = \sqrt{16^2 + 7^2} = \sqrt{305}$

9. Initial point: $(-3, -5)$

Terminal point: $(5, 1)$

$\mathbf{v} = \langle 5 - (-3), 1 - (-5) \rangle = \langle 8, 6 \rangle$

$\|\mathbf{v}\| = \sqrt{8^2 + 6^2} = \sqrt{100} = 10$

11. Initial point: $(1, 3)$

Terminal point: $(-8, -9)$

$\mathbf{v} = \langle -8 - 1, -9 - 3 \rangle = \langle -9, -12 \rangle$

$\|\mathbf{v}\| = \sqrt{(-9)^2 + (-12)^2} = \sqrt{225} = 15$

13.

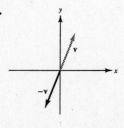

15.

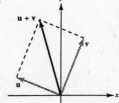

17. u + 2v

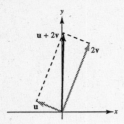

19. u = ⟨2, 1⟩, v = ⟨1, 3⟩

(a) **u + v = ⟨3, 4⟩**

(b) **u − v = ⟨1, −2⟩**

(c) **2u − 3v = ⟨4, 2⟩ − ⟨3, 9⟩**

$$= ⟨1, −7⟩$$

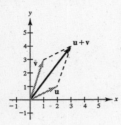

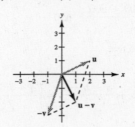

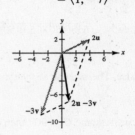

21. u = ⟨−5, 3⟩, v = ⟨0, 0⟩

(a) **u + v = ⟨−5, 3⟩ = u**

(b) **u − v = ⟨−5, 3⟩ = u**

(c) **2u − 3v = 2u = ⟨−10, 6⟩**

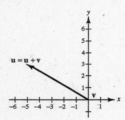

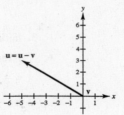

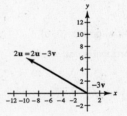

23. u = i + j, v = 2i − 3j

(a) **u + v = 3i − 2j**

(b) **u − v = −i + 4j**

(c) **2u − 3v = (2i + 2j) − (6i − 9j)**

$$= −4i + 11j$$

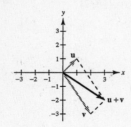

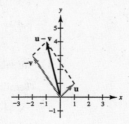

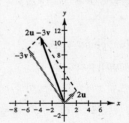

25. $\mathbf{u} = 2\mathbf{i}$, $\mathbf{v} = \mathbf{j}$

(a) $\mathbf{u} + \mathbf{v} = 2\mathbf{i} + \mathbf{j}$

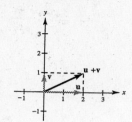

(b) $\mathbf{u} - \mathbf{v} = 2\mathbf{i} - \mathbf{j}$

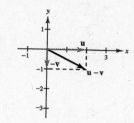

(c) $2\mathbf{u} - 3\mathbf{v} = 4\mathbf{i} - 3\mathbf{j}$

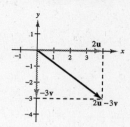

27. $\mathbf{v} = \dfrac{1}{\|\mathbf{u}\|}\mathbf{u} = \dfrac{1}{\sqrt{3^2 + 0^2}}\langle 3, 0 \rangle = \dfrac{1}{3}\langle 3, 0 \rangle = \langle 1, 0 \rangle$

29. $\mathbf{u} = \dfrac{1}{\|\mathbf{v}\|}\mathbf{v} = \dfrac{1}{\sqrt{(-2)^2 + 2^2}}\langle -2, 2 \rangle = \dfrac{1}{2\sqrt{2}}\langle -2, 2 \rangle$

$\qquad = \left\langle -\dfrac{1}{\sqrt{2}}, \dfrac{1}{\sqrt{2}} \right\rangle$

31. $\mathbf{u} = \dfrac{1}{\|\mathbf{v}\|}\mathbf{v} = \dfrac{1}{\sqrt{6^2 + (-2)^2}}(6\mathbf{i} - 2\mathbf{j}) = \dfrac{1}{\sqrt{40}}(6\mathbf{i} - 2\mathbf{j})$

$\qquad = \dfrac{1}{2\sqrt{10}}(6\mathbf{i} - 2\mathbf{j}) = \dfrac{3}{\sqrt{10}}\mathbf{i} - \dfrac{1}{\sqrt{10}}\mathbf{j}$

33. $\mathbf{u} = \dfrac{1}{\|\mathbf{w}\|}\mathbf{w} = \dfrac{1}{4}(4\mathbf{j}) = \mathbf{j}$

35. $\mathbf{u} = \dfrac{1}{\|\mathbf{w}\|}\mathbf{w} = \dfrac{1}{\sqrt{1^2 + (-2)^2}}(\mathbf{i} - 2\mathbf{j}) = \dfrac{1}{\sqrt{5}}(\mathbf{i} - 2\mathbf{j})$

$\qquad = \dfrac{1}{\sqrt{5}}\mathbf{i} - \dfrac{2}{\sqrt{5}}\mathbf{j}$

37. $5\left(\dfrac{1}{\|\mathbf{u}\|}\mathbf{u}\right) = 5\left(\dfrac{1}{\sqrt{3^2 + 3^2}}\langle 3, 3 \rangle\right) = \dfrac{5}{3\sqrt{2}}\langle 3, 3 \rangle$

$\qquad = \left\langle \dfrac{5}{\sqrt{2}}, \dfrac{5}{\sqrt{2}} \right\rangle$

39. $9\left(\dfrac{1}{\|\mathbf{u}\|}\mathbf{u}\right) = 9\left(\dfrac{1}{\sqrt{2^2 + 5^2}}\langle 2, 5 \rangle\right) = \dfrac{9}{\sqrt{29}}\langle 2, 5 \rangle$

$\qquad = \left\langle \dfrac{18}{\sqrt{29}}, \dfrac{45}{\sqrt{29}} \right\rangle$

41. $\mathbf{v} = \dfrac{3}{2}\mathbf{u} = \dfrac{3}{2}(2\mathbf{i} - \mathbf{j})$

$\qquad = 3\mathbf{i} - \dfrac{3}{2}\mathbf{j} = \left\langle 3, -\dfrac{3}{2} \right\rangle$

43. $\mathbf{v} = \mathbf{u} + 2\mathbf{w}$

$\qquad = (2\mathbf{i} - \mathbf{j}) + 2(\mathbf{i} + 2\mathbf{j})$

$\qquad = 4\mathbf{i} + 3\mathbf{j} = \langle 4, 3 \rangle$

45. $\mathbf{v} = \dfrac{1}{2}(3\mathbf{u} + \mathbf{w})$

$\qquad = \dfrac{1}{2}(6\mathbf{i} - 3\mathbf{j} + \mathbf{i} + 2\mathbf{j})$

$\qquad = \dfrac{7}{2}\mathbf{i} - \dfrac{1}{2}\mathbf{j} = \left\langle \dfrac{7}{2}, -\dfrac{1}{2} \right\rangle$

47. $\mathbf{v} = 3(\cos 60°\mathbf{i} + \sin 60°\mathbf{j})$

$\qquad \|\mathbf{v}\| = 3$, $\theta = 60°$

49. $\mathbf{v} = 6\mathbf{i} - 6\mathbf{j}$

$\qquad \|\mathbf{v}\| = \sqrt{6^2 + (-6)^2} = \sqrt{72}$

$\qquad\qquad = 6\sqrt{2}$

$\qquad \tan \theta = \dfrac{-6}{6} = -1$

Since $\mathbf{v}$ lies in Quadrant IV, $\theta = 315°$.

51. $\|\mathbf{v}\| = 5 \quad \theta = 0°$

$\mathbf{v} = \langle 5\cos 0°, 5\sin 0° \rangle$

$= \langle 5, 0 \rangle$

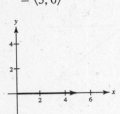

53. $\mathbf{v} = \left\langle \dfrac{7}{2}\cos 150°, \dfrac{7}{2}\sin 150° \right\rangle$

$= \left\langle -\dfrac{7\sqrt{3}}{4}, \dfrac{7}{4} \right\rangle$

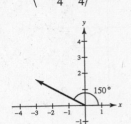

55. $\mathbf{v} = \left\langle 3\sqrt{2}\cos 150°, 3\sqrt{2}\sin 150° \right\rangle$

$= \left\langle -\dfrac{3\sqrt{6}}{2}, \dfrac{3\sqrt{2}}{2} \right\rangle$

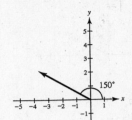

57. $\mathbf{v} = 2\left(\dfrac{1}{\sqrt{1^2+3^2}}\right)(\mathbf{i}+3\mathbf{j})$

$= \dfrac{2}{\sqrt{10}}(\mathbf{i}+3\mathbf{j})$

$= \dfrac{\sqrt{10}}{5}\mathbf{i} + \dfrac{3\sqrt{10}}{5}\mathbf{j}$

$= \left\langle \dfrac{\sqrt{10}}{5}, \dfrac{3\sqrt{10}}{5} \right\rangle$

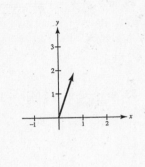

59. $\mathbf{u} = \langle 5\cos 0°, 5\sin 0° \rangle = \langle 5, 0 \rangle$

$\mathbf{v} = \langle 5\cos 90°, 5\sin 90° \rangle = \langle 0, 5 \rangle$

$\mathbf{u} + \mathbf{v} = \langle 5, 5 \rangle$

61. $\mathbf{u} = \langle 20\cos 45°, 20\sin 45° \rangle = \langle 10\sqrt{2}, 10\sqrt{2} \rangle$

$\mathbf{v} = \langle 50\cos 180°, 50\sin 180° \rangle = \langle -50, 0 \rangle$

$\mathbf{u} + \mathbf{v} = \langle 10\sqrt{2} - 50, 10\sqrt{2} \rangle$

63. $\mathbf{v} = \mathbf{i} + \mathbf{j}$

$\mathbf{w} = 2\mathbf{i} - 2\mathbf{j}$

$\mathbf{u} = \mathbf{v} - \mathbf{w} = -\mathbf{i} + 3\mathbf{j}$

$\|\mathbf{v}\| = \sqrt{2}$

$\|\mathbf{w}\| = 2\sqrt{2}$

$\|\mathbf{v} - \mathbf{w}\| = \sqrt{10}$

$\cos\alpha = \dfrac{\|\mathbf{v}\|^2 + \|\mathbf{w}\|^2 - \|\mathbf{v}-\mathbf{w}\|^2}{2\|\mathbf{v}\|\,\|\mathbf{w}\|} = \dfrac{2+8-10}{2\sqrt{2}\cdot 2\sqrt{2}} = 0$

$\alpha = 90°$

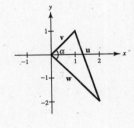

65. $\mathbf{v} = \mathbf{i} + \mathbf{j}$

$\mathbf{w} = 3\mathbf{i} - \mathbf{j}$

$\mathbf{u} = \mathbf{v} - \mathbf{w} = -2\mathbf{i} + 2\mathbf{j}$

$\cos\alpha = \dfrac{\|\mathbf{v}\|^2 + \|\mathbf{w}\|^2 - \|\mathbf{v}-\mathbf{w}\|^2}{2\|\mathbf{v}\|\,\|\mathbf{w}\|} = \dfrac{2+10-8}{2\sqrt{2}\,\sqrt{10}} \approx 0.4472$

$\alpha = 63.4°$

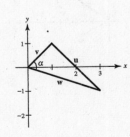

67. Force One: $\mathbf{u} = 45\mathbf{i}$

Force Two: $\mathbf{v} = 60\cos\theta\mathbf{i} + 60\sin\theta\mathbf{j}$

Resultant Force: $\mathbf{u} + \mathbf{v} = (45 + 60\cos\theta)\mathbf{i} + 60\sin\theta\mathbf{j}$

$\|\mathbf{u} + \mathbf{v}\| = \sqrt{(45 + 60\cos\theta)^2 + (60\sin\theta)^2} = 90$

$$2025 + 5400\cos\theta + 3600 = 8100$$

$$5400\cos\theta = 2475$$

$$\cos\theta = \frac{2475}{5400} \approx 0.4583$$

$$\theta \approx 62.7°$$

69. The difference $\mathbf{u} - \mathbf{v}$ is the vector from the terminal point of $\mathbf{v}$ to the terminal point of $\mathbf{u}$.

71. The sum of $\mathbf{u}$ and $\mathbf{v}$ will be greater in figure (a). The angle between $\mathbf{u}$ and $\mathbf{v}$ is acute.

73.
$$\mathbf{u} = (2000\cos 30°)\mathbf{i} + (2000\sin 30°)\mathbf{j}$$
$$\approx 1732.05\mathbf{i} + 1000\mathbf{j}$$
$$\mathbf{v} = (900\cos(-45°))\mathbf{i} + (900\sin(-45°))\mathbf{j}$$
$$\approx 636.4\mathbf{i} + -636.4\mathbf{j}$$
$$\mathbf{u} + \mathbf{v} \approx 2368.45\mathbf{i} + 363.6\mathbf{j}$$
$$\|\mathbf{u} + \mathbf{v}\| \approx \sqrt{(2368.45)^2 + (363.6)^2} \approx 2396.20 \text{ newtons}$$
$$\tan\theta = \frac{363.6}{2368.45} \approx 0.1535 \implies \theta \approx 8.7°$$

75.
$$\mathbf{u} = (70\cos 60°)\mathbf{i} - (70\sin 60°)\mathbf{j} \approx 35\mathbf{i} - 60.62\mathbf{j}$$
$$\mathbf{v} = (40\cos 45°)\mathbf{i} + (40\sin 45°)\mathbf{j} \approx 28.28\mathbf{i} + 28.28\mathbf{j}$$
$$\mathbf{w} = (60\cos 120°)\mathbf{i} + (60\sin 120°)\mathbf{j} \approx -30\mathbf{i} + 51.96\mathbf{j}$$
$$\mathbf{u} + \mathbf{v} + \mathbf{w} = 33.28\mathbf{i} + 19.62\mathbf{j}$$
$$\|\mathbf{u} + \mathbf{v} + \mathbf{w}\| \approx 38.6 \text{ pounds}$$
$$\tan\theta \approx \frac{19.62}{33.28} \approx 0.5895$$
$$\theta \approx 30.5°$$

77. Horizontal component of velocity: $1100\cos 5° \approx 1095.8$ feet per second

Vertical component of velocity: $1100\sin 5° \approx 95.9$ feet per second

79. Cable $\overrightarrow{AC}$: $\mathbf{u} = \|\mathbf{u}\|(\cos 45°\mathbf{i} - \sin 45°\mathbf{j})$

Cable $\overrightarrow{BC}$: $\mathbf{v} = \|\mathbf{v}\|(-\cos 30°\mathbf{i} - \sin 30°\mathbf{j})$

Resultant: $\mathbf{u} + \mathbf{v} = -3000\mathbf{j}$

$\|\mathbf{u}\| \cos 45° - \|\mathbf{v}\| \cos 30° = 0$

$-\|\mathbf{u}\| \sin 45° - \|\mathbf{v}\|\sin 30° = -3000$

Solving this system yields:

$T_{AC} = \|\mathbf{u}\| \approx 2689.7$ pounds

$T_{BC} = \|\mathbf{v}\| \approx 2196.2$ pounds

81. Towline 1: $\mathbf{u} = \|\mathbf{u}\|(\cos 18°\mathbf{i} + \sin 18°\mathbf{j})$

Towline 2: $\mathbf{v} = \|\mathbf{u}\|(\cos 18°\mathbf{i} - \sin 18°\mathbf{j})$

Resultant: $\mathbf{u} + \mathbf{v} = 6000\mathbf{i}$

$\|\mathbf{u}\| \cos 18° + \|\mathbf{u}\| \cos 18° = 6000$

$\|\mathbf{u}\| \approx 3154.4$

Therefore, the tension on each towline is

$\|\mathbf{u}\| \approx 3154.4$ pounds.

83. (a)

(b) Plane: $\mathbf{u} = (580 \cos 118°)\mathbf{i} + (580 \sin 118°)\mathbf{j} \approx -272.3\mathbf{i} + 512.1\mathbf{j}$

Wind: $\mathbf{v} = (60 \cos 45°)\mathbf{i} + (60 \sin 45°)\mathbf{j} \approx 42.4\mathbf{i} + 42.4\mathbf{j}$

$\mathbf{u} + \mathbf{v} \approx -229.9\mathbf{i} + 554.5\mathbf{j}$

$\|\mathbf{u} + \mathbf{v}\| \approx \sqrt{(-229.9)^2 + (554.5)^2} \approx 600.3$

$\tan \theta \approx -\dfrac{554.5}{229.9} \approx -2.4119 \implies \theta \approx 112.5$

The ground speed is 600.3 miles per hour and the heading is N 22.5° W.

85. Horizontal force: $\mathbf{u} = \|\mathbf{u}\|\mathbf{i}$

Weight: $\mathbf{w} = -\mathbf{j}$

Rope: $\mathbf{t} = \|\mathbf{t}\| (\cos 135°\mathbf{i} + \sin 135°\mathbf{j})$

$\mathbf{u} + \mathbf{w} + \mathbf{t} = 0 \implies \|\mathbf{u}\| + \|\mathbf{t}\| \cos 135° = 0$

$-1 + \|\mathbf{t}\| \sin 135° = 0$

$\|\mathbf{t}\| \approx \sqrt{2}$ pounds

$\|\mathbf{u}\| \approx 1$ pound

87. $\mathbf{u} = \langle 5 - 1, 2 - 6 \rangle = \langle 4, -4 \rangle$

$\mathbf{v} = \langle 9 - 4, 4 - 5 \rangle = \langle 5, -1 \rangle$

$\mathbf{u} - \mathbf{v} = \langle -1, -3 \rangle$ or $\mathbf{v} - \mathbf{u} = \langle 1, 3 \rangle$

89. True. See Example 1.

91. False, $a = b = 0$.

Section 13.4 Vectors and Dot Products

1. $\mathbf{u} = \langle 6, 1 \rangle$, $\mathbf{v} = \langle -2, 3 \rangle$

$\mathbf{u} \cdot \mathbf{v} = 6(-2) + 1(3) = -9$

3. $\mathbf{u} = 4\mathbf{i} - 2\mathbf{j}$, $\mathbf{v} = \mathbf{i} - \mathbf{j}$

$\mathbf{u} \cdot \mathbf{v} = 4(1) + (-2)(-1) = 6$

5. $\mathbf{u} = \langle 2, 2 \rangle$

$\mathbf{u} \cdot \mathbf{u} = 2(2) + 2(2) = 8$

The result is a scalar.

7. $\mathbf{u} = \langle 2, 2 \rangle$, $\mathbf{v} = \langle -3, 4 \rangle$

$(\mathbf{u} \cdot \mathbf{v})\mathbf{v} = [(2)(-3) + 2(4)]\langle -3, 4 \rangle$

$= 2\langle -3, 4 \rangle = \langle -6, 8 \rangle$

The result is a vector.

9. $\mathbf{u} = \langle -5, 12 \rangle$

$\|\mathbf{u}\| = \sqrt{\mathbf{u} \cdot \mathbf{u}} = \sqrt{(-5)^2 + 12^2} = 13$

11. $\mathbf{u} = 20\mathbf{i} + 25\mathbf{j}$

$\|\mathbf{u}\| = \sqrt{(20)^2 + (25)^2} = \sqrt{1025} = 5\sqrt{41}$

13. $\mathbf{u} = 6\mathbf{j}$

$\|\mathbf{u}\| = \sqrt{(0)^2 + (6)^2} = \sqrt{36} = 6$

15. $\mathbf{u} = \langle 1, 0 \rangle$, $\mathbf{v} = \langle 0, -2 \rangle$

$$\cos \theta = \frac{\mathbf{u} \cdot \mathbf{v}}{\|\mathbf{u}\| \|\mathbf{v}\|} = \frac{0}{(1)(2)} = 0$$

$$\theta = 90°$$

17. $\mathbf{u} = 3\mathbf{i} + 4\mathbf{j}$, $\mathbf{v} = -2\mathbf{j}$

$$\cos \theta = \frac{\mathbf{u} \cdot \mathbf{v}}{\|\mathbf{u}\| \|\mathbf{v}\|} = -\frac{8}{(5)(2)}$$

$$\theta = \arccos\left(-\frac{4}{5}\right)$$

$$\theta \approx 143.13°$$

19. $\mathbf{u} = 2\mathbf{i} - \mathbf{j}$, $\mathbf{v} = 6\mathbf{i} + 4\mathbf{j}$

$$\cos \theta = \frac{\mathbf{u} \cdot \mathbf{v}}{\|\mathbf{u}\| \|\mathbf{v}\|} = \frac{8}{\sqrt{5}\sqrt{52}} \implies \theta \approx 60.26°$$

21. $\mathbf{u} = 5\mathbf{i} + 5\mathbf{j}$, $\mathbf{v} = -6\mathbf{i} + 6\mathbf{j}$

$$\cos \theta = \frac{\mathbf{u} \cdot \mathbf{v}}{\|\mathbf{u}\| \|\mathbf{v}\|} = 0 \implies \theta = 90°$$

23. $\mathbf{u} = \left(\cos \frac{\pi}{3}\right)\mathbf{i} + \left(\sin \frac{\pi}{3}\right)\mathbf{j} = \frac{1}{2}\mathbf{i} + \frac{\sqrt{3}}{2}\mathbf{j}$

$\mathbf{v} = \left(\cos \frac{3\pi}{4}\right)\mathbf{i} + \left(\sin \frac{3\pi}{4}\right)\mathbf{j} = -\frac{\sqrt{2}}{2}\mathbf{i} + \frac{\sqrt{2}}{2}\mathbf{j}$

$\|\mathbf{u}\| = \|\mathbf{v}\| = 1$

$$\cos \theta = \frac{\mathbf{u} \cdot \mathbf{v}}{\|\mathbf{u}\| \|\mathbf{v}\|} = \mathbf{u} \cdot \mathbf{v} = \left(\frac{1}{2}\right)\left(-\frac{\sqrt{2}}{2}\right) + \left(\frac{\sqrt{3}}{2}\right)\left(\frac{\sqrt{2}}{2}\right) = \frac{-\sqrt{2} + \sqrt{6}}{4}$$

$$\theta = \arccos\left(\frac{-\sqrt{2} + \sqrt{6}}{4}\right) = 75° = \frac{5\pi}{12}$$

25. $P = (1, 2)$, $Q = (3, 4)$, $R = (2, 5)$

$\overrightarrow{PQ} = \langle 2, 2 \rangle$, $\overrightarrow{PR} = \langle 1, 3 \rangle$, $\overrightarrow{QR} = \langle -1, 1 \rangle$

$$\cos \alpha = \frac{\overrightarrow{PQ} \cdot \overrightarrow{PR}}{\|\overrightarrow{PQ}\| \|\overrightarrow{PR}\|} = \frac{8}{(2\sqrt{2})(\sqrt{10})} \implies \alpha = \arccos \frac{2}{\sqrt{5}} \approx 26.6°$$

$$\cos \beta = \frac{\overrightarrow{PQ} \cdot \overrightarrow{QR}}{\|\overrightarrow{PQ}\| \|\overrightarrow{QR}\|} = 0 \implies \beta = 90°. \text{ Thus, } \gamma = 180° - 26.6° - 90° = 63.4°.$$

27. $P = (-3, 0)$, $Q = (2, 2)$, $R = (0, 6)$

$\overrightarrow{QP} = \langle -5, -2 \rangle$, $\overrightarrow{PR} = \langle 3, 6 \rangle$, $\overrightarrow{QR} = \langle -2, 4 \rangle$

$$\cos \alpha = \frac{\overrightarrow{PQ} \cdot \overrightarrow{PR}}{\|\overrightarrow{PQ}\| \|\overrightarrow{PR}\|} = \frac{27}{\sqrt{29}\sqrt{45}} \implies \alpha \approx 41.6°$$

$$\cos \beta = \frac{\overrightarrow{QP} \cdot \overrightarrow{QR}}{\|\overrightarrow{QP}\| \|\overrightarrow{GR}\|} = \frac{2}{\sqrt{29}\sqrt{20}} \implies \beta \approx 85.2°$$

$$\delta = 180° - 41.6° - 85.2° = 53.2°$$

29. $\mathbf{u} \cdot \mathbf{v} = \|\mathbf{u}\| \|\mathbf{v}\| \cos \theta$

$$= (4)(10) \cos \frac{2\pi}{3}$$

$$= 40\left(-\frac{1}{2}\right)$$

$$= -20$$

31. $\mathbf{u} \cdot \mathbf{v} = \|\mathbf{u}\| \|\mathbf{v}\| \cos \theta$

$$= (81)(64)\cos \frac{\pi}{4}$$

$$= 5184\left(\frac{\sqrt{2}}{2}\right)$$

$$= 2592\sqrt{2}$$

33. $\mathbf{u} = \langle -12, 30 \rangle$, $\mathbf{v} = \langle \frac{1}{2}, -\frac{5}{4} \rangle$

$\mathbf{u} = -24\mathbf{v} \implies \mathbf{u}$ and $\mathbf{v}$ are parallel.

35. $\mathbf{u} = \frac{1}{4}(3\mathbf{i} - \mathbf{j})$, $\mathbf{v} = 5\mathbf{i} + 6\mathbf{j}$

$\mathbf{u} \neq k\mathbf{v} \implies$ Not parallel

$\mathbf{u} \cdot \mathbf{v} \neq 0 \implies$ Not orthogonal

Neither

37. $\mathbf{u} = 2\mathbf{i} - 2\mathbf{j}$, $\mathbf{v} = -\mathbf{i} - \mathbf{j}$

$\mathbf{u} \cdot \mathbf{v} = 0 \Rightarrow \mathbf{u}$ and $\mathbf{v}$ are orthogonal.

39. $\mathbf{u} = \langle 2, 2 \rangle$, $\mathbf{v} = \langle 6, 1 \rangle$

$\mathbf{w}_1 = \text{proj}_{\mathbf{v}}\mathbf{u} = \left(\dfrac{\mathbf{u} \cdot \mathbf{v}}{\|\mathbf{v}\|^2}\right)\mathbf{v} = \dfrac{14}{37}\mathbf{v} = \dfrac{14}{37}\langle 6, 1 \rangle$

$\mathbf{w}_2 = \mathbf{u} - \mathbf{w}_1 = \langle 2, 2 \rangle - \dfrac{14}{37}\langle 6, 1 \rangle = \left\langle -\dfrac{10}{37}, \dfrac{60}{37} \right\rangle = \dfrac{10}{37}\langle -1, 6 \rangle$

41. $\mathbf{u} = \langle 0, 3 \rangle$, $\mathbf{v} = \langle 2, 15 \rangle$

$\mathbf{w}_1 = \text{proj}_{\mathbf{v}}\mathbf{u} = \left(\dfrac{\mathbf{u} \cdot \mathbf{v}}{\|\mathbf{v}\|^2}\right)\mathbf{v} = \dfrac{45}{229}\langle 2, 15 \rangle$

$\mathbf{w}_2 = \mathbf{u} - \mathbf{w}_1 = \langle 0, 3 \rangle - \dfrac{45}{229}\langle 2, 15 \rangle = \left\langle -\dfrac{90}{229}, \dfrac{12}{229} \right\rangle = \dfrac{6}{229}\langle -15, 2 \rangle$

43. $\mathbf{u} = \langle 3, 5 \rangle$

For $\mathbf{v}$ to be orthogonal to $\mathbf{u}$, $\mathbf{u} \cdot \mathbf{v}$ must equal 0.

Two possibilities: $\langle -5, 3 \rangle$ and $\langle 5, -3 \rangle$

45. $\mathbf{u} = \dfrac{2}{3}\mathbf{i} - \dfrac{1}{2}\mathbf{j}$

For $\mathbf{u}$ and $\mathbf{v}$ to be orthogonal, $\mathbf{u} \cdot \mathbf{v}$ must equal 0.

Two possibilities: $\dfrac{1}{2}\mathbf{i} + \dfrac{2}{3}\mathbf{j}$ and $-\dfrac{1}{2}\mathbf{i} - \dfrac{2}{3}\mathbf{j}$

47. $w = \|\text{proj}_{\overrightarrow{PQ}}\mathbf{v}\|\|\overrightarrow{PQ}\|$ where $\overrightarrow{PQ} = \langle 4, 7 \rangle$ and $\mathbf{v} = \langle 1, 4 \rangle$.

$\text{proj}_{\overrightarrow{PQ}}\mathbf{v} = \left(\dfrac{\mathbf{v} \cdot \overrightarrow{PQ}}{\|\overrightarrow{PQ}\|^2}\right)\overrightarrow{PQ} = \left(\dfrac{32}{65}\right)\langle 4, 7 \rangle$

$w = \|\text{proj}_{\overrightarrow{PQ}}\mathbf{v}\|\|\overrightarrow{PQ}\| = \left(\dfrac{32\sqrt{65}}{65}\right)\left(\sqrt{65}\right) = 32$

49. Since $\dfrac{\mathbf{u} \cdot \mathbf{v}}{\|\mathbf{u}\|\|\mathbf{v}\|} = \cos\theta$, we have $\mathbf{u} \cdot \mathbf{v} = \|\mathbf{u}\|\|\mathbf{v}\|\cos\theta$.

The dot product of $\mathbf{u}$ and $\mathbf{v}$ equals the product of their lengths (magnitudes) when $\cos\theta = 1$. This happens when the angle between them is 0.

51. $\mathbf{u} = \langle 1650, 3200 \rangle$, $\mathbf{v} = \langle 15.25, 10.50 \rangle$

$\mathbf{u} \cdot \mathbf{v} = 1650(15.25) + 3200(10.50) = \$58{,}762.50$

This gives the total revenue that can be earned by selling all of the units.

53. (a) $\mathbf{F} = -30{,}000\mathbf{j}$

$\mathbf{v} = (\cos d°)\mathbf{i} + (\sin d°)\mathbf{j}$

$\mathbf{w}_1 = \text{proj}_{\mathbf{v}}\mathbf{F} = \left(\dfrac{\mathbf{F} \cdot \mathbf{v}}{\|\mathbf{v}\|^2}\right)\mathbf{v} = (\mathbf{F} \cdot \mathbf{v})\mathbf{v} = -30{,}000(\sin d°)\mathbf{v}$

The magnitude of this force is $30{,}000(\sin d°)$ and this force, in pounds, is needed to keep the truck from rolling down the hill.

—CONTINUED—

53. —CONTINUED—

(b)

d(degrees)	Force(pounds)
0°	0
1°	523.6
2°	1047.0
3°	1570.1
4°	2092.7
5°	2614.7
6°	3135.9
7°	3656.1
8°	4175.2
9°	4693.0
10°	5209.4

(c) $\mathbf{w}_2 = \mathbf{F} - \mathbf{w}_1 = -30{,}000\mathbf{j} + 30{,}000\sin 5°(\cos 5°\mathbf{i} + \sin 5°\mathbf{j})$

$\approx 2604.72\mathbf{i} - 29{,}772.12\mathbf{j}$

$\|\mathbf{w}_2\| \approx 29{,}885.8$ pounds

55. $\mathbf{w} = (245)(3) = 735$ Newton-meters

57. $\mathbf{w} = (\cos 30°)(45)(20) \approx 779.4$ foot-pounds

59. False. The dot product of two vectors is a scalar which can be positive, zero, or negative.

61. False. Work is represented by a scalar.

63. Let $\mathbf{u} = \langle u_1, u_2\rangle$, $\mathbf{v} = \langle v_1, v_2\rangle$, $\mathbf{w} = \langle w_1, w_2\rangle$ and $\mathbf{0} = \langle 0, 0\rangle$.

(a) $\mathbf{0} \cdot \mathbf{v} = \langle 0, 0\rangle \cdot \langle v_1, v_2\rangle$
$= 0(v_1) + 0(v_2)$
$= 0 + 0$
$= 0$

(b) $\mathbf{u} \cdot (\mathbf{v} + \mathbf{w}) = \langle u_1, u_2\rangle \cdot (\langle v_1, v_2\rangle + \langle w_1, w_2\rangle)$
$= \langle u_1, u_2\rangle \cdot \langle v_1 + w_1, v_2 + w_2\rangle$
$= u_1(v_1 + w_1) + u_2(v_2 + w_2)$
$= u_1 v_1 + u_1 w_1 + u_2 v_2 + u_2 w_2$
$= (u_1 v_1 + u_2 v_2) + (u_1 w_1 + u_2 w_2)$
$= \mathbf{u} \cdot \mathbf{v} + \mathbf{u} \cdot \mathbf{w}$

(c) $c(\mathbf{u} \cdot \mathbf{v}) = c(\langle u_1, u_2\rangle \cdot \langle v_1, v_2\rangle)$
$= c(u_1 v_1 + u_2 v_2)$
$= u_1(cv_1) + u_2(cv_2)$
$= \langle u_1, u_2\rangle \cdot \langle cv_1, cv_2\rangle$
$= \mathbf{u} \cdot (c\mathbf{v})$

Section 13.5   Trigonometric Form of a Complex Number

1. $|-7i| = \sqrt{0^2 + (-7)^2}$
$= \sqrt{49} = 7$

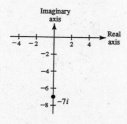

3. $|-4 + 4i| = \sqrt{(-4)^2 + (4)^2}$
$= \sqrt{32} = 4\sqrt{2}$

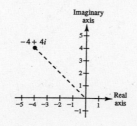

5. $|6 - 7i| = \sqrt{6^2 + (-7)^2}$
$= \sqrt{85}$

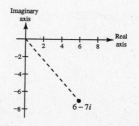

7. $z = 3i$

$r = \sqrt{0^2 + 3^2} = \sqrt{9} = 3$

$\tan \theta = \dfrac{3}{0}$, undefined $\implies \theta = \dfrac{\pi}{2}$

$z = 3\left(\cos\dfrac{\pi}{2} + i\sin\dfrac{\pi}{2}\right)$

9. $z = 3 - i$

$r = \sqrt{(3)^2 + (-1)^2} = \sqrt{10}$

$\tan \theta = -\dfrac{1}{3}$, θ is in Quadrant IV.

$\theta \approx 5.96$ radians

$z \approx \sqrt{10}(\cos 5.96 + i\sin 5.96)$

11. $z = 3 - 3i$

$r = \sqrt{3^2 + (-3)^2} = \sqrt{18} = 3\sqrt{2}$

$\tan \theta = \dfrac{-3}{3} = -1$, θ is in Quadrant IV $\implies \theta = \dfrac{7\pi}{4}$.

$z = 3\sqrt{2}\left(\cos\dfrac{7\pi}{4} + i\sin\dfrac{7\pi}{4}\right)$

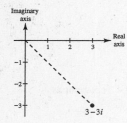

13. $z = \sqrt{3} + i$

$r = \sqrt{(\sqrt{3})^2 + 1^2} = \sqrt{4} = 2$

$\tan \theta = \dfrac{1}{\sqrt{3}} = \dfrac{\sqrt{3}}{3} \implies \theta = \dfrac{\pi}{6}$

$z = 2\left(\cos\dfrac{\pi}{6} + i\sin\dfrac{\pi}{6}\right)$

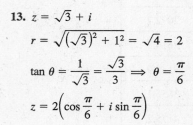

15. $z = -2(1 + \sqrt{3}i)$

$r = \sqrt{(-2)^2 + (-2\sqrt{3})^2} = \sqrt{16} = 4$

$\tan \theta = \dfrac{\sqrt{3}}{1} = \sqrt{3}$, θ is in Quadrant III $\implies \theta = \dfrac{4\pi}{3}$.

$z = 4\left(\cos\dfrac{4\pi}{3} + i\sin\dfrac{4\pi}{3}\right)$

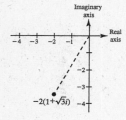

17. $z = -5i$

$r = \sqrt{0^2 + (-5)^2} = \sqrt{25} = 5$

$\tan \theta = \dfrac{-5}{0}$, undefined $\implies \theta = \dfrac{3\pi}{2}$

$z = 5\left(\cos\dfrac{3\pi}{2} + i\sin\dfrac{3\pi}{2}\right)$

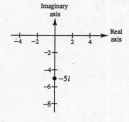

19. $z = -7 + 4i$

$r = \sqrt{(-7)^2 + (4)^2} = \sqrt{65}$

$\tan \theta = \dfrac{4}{-7}$, θ is in Quadrant II $\implies \theta \approx 2.62$.

$z \approx \sqrt{65}(\cos 2.62 + i\sin 2.62)$

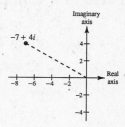

21. $z = 7 + 0i$

$r = \sqrt{(7)^2 + (0)^2} = \sqrt{49} = 7$

$\tan \theta = \dfrac{0}{7} = 0 \implies \theta = 0$

$z = 7(\cos 0 + i \sin 0)$

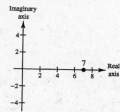

23. $z = 3 + \sqrt{3}i$

$r = \sqrt{(3)^2 + \left(\sqrt{3}\right)^2} = \sqrt{12}$

$\qquad = 2\sqrt{3}$

$\tan \theta = \dfrac{\sqrt{3}}{3} \implies \theta = \dfrac{\pi}{6}$

$z = 2\sqrt{3}\left(\cos \dfrac{\pi}{6} + i \sin \dfrac{\pi}{6}\right)$

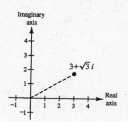

25. $z = -3 - i$

$r = \sqrt{(-3)^2 + (-1)^2} = \sqrt{10}$

$\tan \theta = \dfrac{-1}{-3} = \dfrac{1}{3},\ \theta$ is in Quadrant III $\implies \theta \approx 3.46.$

$z \approx \sqrt{10}\,(\cos 3.46 + i \sin 3.46)$

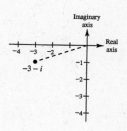

27. $z = 5 + 2i$

$r \approx 5.39$

$\theta \approx 0.38$

$z \approx 5.39(\cos 0.38 + i \sin 0.38)$

29. $z = -3 + i$

$r \approx 3.16$

$\theta \approx 2.82$

$z \approx 3.16(\cos 2.82 + i \sin 2.82)$

31. $z = 3\sqrt{2} - 7i$

$r \approx 8.19$

$\theta \approx 5.26$

$z \approx 8.19(\cos 5.26 + i \sin 5.26)$

33. $z = -8 - 5\sqrt{3}i$

$r \approx 11.79$

$\theta \approx 3.97$

$z \approx 11.79(\cos 3.97 + i \sin 3.97)$

35. $3(\cos 120° + i \sin 120°) = 3\left(-\dfrac{1}{2} + \dfrac{\sqrt{3}}{2}i\right)$

$\qquad\qquad\qquad\qquad\qquad = -\dfrac{3}{2} + \dfrac{3\sqrt{3}}{2}i$

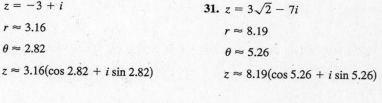

37. $\dfrac{3}{2}(\cos 300° + i \sin 300°) = \dfrac{3}{2}\left[\dfrac{1}{2} + i\left(-\dfrac{\sqrt{3}}{2}\right)\right]$

$\qquad\qquad\qquad\qquad\qquad = \dfrac{3}{4} - \dfrac{3\sqrt{3}}{4}i$

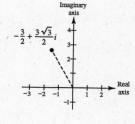

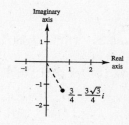

39. $3.75\left(\cos\dfrac{3\pi}{4} + i\sin\dfrac{3\pi}{4}\right) = -\dfrac{15\sqrt{2}}{8} + \dfrac{15\sqrt{2}}{8}i$

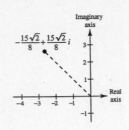

41. $8\left(\cos\dfrac{\pi}{2} + i\sin\dfrac{\pi}{2}\right) = 8(0 + i) = 8i$

43. $3[\cos(18°45') + i\sin(18°45')] \approx 2.8408 + 0.9643i$

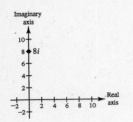

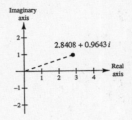

45. $5\left(\cos\dfrac{\pi}{9} + i\sin\dfrac{\pi}{9}\right) \approx 4.70 + 1.71i$

47. $3(\cos 165.5° + i\sin 165.5°) \approx -2.90 + 0.75i$

49. $\left[2\left(\cos\dfrac{\pi}{4} + i\sin\dfrac{\pi}{4}\right)\right]\left[6\left(\cos\dfrac{\pi}{12} + i\sin\dfrac{\pi}{12}\right)\right] = (2)(6)\left[\cos\left(\dfrac{\pi}{4} + \dfrac{\pi}{12}\right) + i\sin\left(\dfrac{\pi}{4} + \dfrac{\pi}{12}\right)\right]$

$$= 12\left(\cos\dfrac{\pi}{3} + i\sin\dfrac{\pi}{3}\right)$$

51. $\left[\dfrac{5}{3}(\cos 140° + i\sin 140°)\right]\left[\dfrac{2}{3}(\cos 60° + i\sin 60°)\right] = \left(\dfrac{5}{3}\right)\left(\dfrac{2}{3}\right)[\cos(140° + 60°) + i\sin(140° + 60°)]$

$$= \dfrac{10}{9}(\cos 200° + i\sin 200°)$$

53. $[0.45(\cos 310° + i\sin 310°)][0.60(\cos 200° + i\sin 200°)] = (0.45)(0.60)[\cos(310° + 200°) + i\sin(310° + 200°)]$

$$= 0.27(\cos 510° + i\sin 510°)$$

$$= 0.27(\cos 150° + i\sin 150°)$$

55. $\dfrac{\cos 50° + i\sin 50°}{\cos 20° + i\sin 20°} = \cos(50° - 20°) + i\sin(50° - 20°)$

$$= \cos 30° + i\sin 30°$$

57. $\dfrac{\cos\dfrac{5\pi}{3} + i\sin\dfrac{5\pi}{3}}{\cos\pi + i\sin\pi} = \cos\left(\dfrac{5\pi}{3} - \pi\right) + i\sin\left(\dfrac{5\pi}{3} - \pi\right) = \cos\left(\dfrac{2\pi}{3}\right) + i\sin\left(\dfrac{2\pi}{3}\right)$

59. $\dfrac{12(\cos 52° + i\sin 52°)}{3(\cos 110° + i\sin 110°)} = 4[\cos(52° - 110°) + i\sin(52° - 110°)]$

$$= 4[\cos(-58°) + i\sin(-58°)]$$

61. (a) $2 + 2i = 2\sqrt{2}\left(\cos\frac{\pi}{4} + i\sin\frac{\pi}{4}\right)$

   $1 - i = \sqrt{2}\left[\cos\left(-\frac{\pi}{4}\right) + i\sin\left(-\frac{\pi}{4}\right)\right]$

   (b) $(2 + 2i)(1 - i) = \left[2\sqrt{2}\left(\cos\frac{\pi}{4} + i\sin\frac{\pi}{4}\right)\right]\left[\sqrt{2}\left(\cos\left(-\frac{\pi}{4}\right) + i\sin\left(-\frac{\pi}{4}\right)\right)\right] = 4(\cos 0 + i\sin 0) = 4$

   (c) $(2 + 2i)(1 - i) = 2 - 2i + 2i - 2i^2 = 2 + 2 = 4$

63. (a) $-2i = 2\left[\cos\left(-\frac{\pi}{2}\right) + i\sin\left(-\frac{\pi}{2}\right)\right]$

   $1 + i = \sqrt{2}\left(\cos\frac{\pi}{4} + i\sin\frac{\pi}{4}\right)$

   (b) $-2i(1 + i) = 2\left[\cos\left(-\frac{\pi}{2}\right) + i\sin\left(-\frac{\pi}{2}\right)\right]\left[\sqrt{2}\left(\cos\frac{\pi}{4} + i\sin\frac{\pi}{4}\right)\right]$

   $= 2\sqrt{2}\left[\cos\left(-\frac{\pi}{4}\right) + i\sin\left(-\frac{\pi}{4}\right)\right]$

   $= 2\sqrt{2}\left[\frac{1}{\sqrt{2}} - \frac{1}{\sqrt{2}}i\right] = 2 - 2i$

   (c) $-2i(1 + i) = -2i - 2i^2 = -2i + 2 = 2 - 2i$

65. (a) $3 + 4i \approx 5(\cos 0.9273 + i\sin 0.9273)$

   $1 - \sqrt{3}i = 2\left(\cos\frac{5\pi}{3} + i\sin\frac{5\pi}{3}\right)$

   (b) $\dfrac{3 + 4i}{1 - \sqrt{3}i} \approx \dfrac{5(\cos 0.9273 + i\sin 0.9273)}{2\left(\cos\dfrac{5\pi}{3} + i\sin\dfrac{5\pi}{3}\right)}$

   $\approx 2.5[\cos(-4.3087) + i\sin(-4.3087)]$

   $\approx -0.982 + 2.299i$

   (c) $\dfrac{3 + 4i}{1 - \sqrt{3}i} = \dfrac{3 + 4i}{1 - \sqrt{3}i} \cdot \dfrac{1 + \sqrt{3}i}{1 + \sqrt{3}i}$

   $= \dfrac{3 + \left(4 + 3\sqrt{3}\right)i + 4\sqrt{3}i^2}{1 + 3}$

   $= \dfrac{3 - 4\sqrt{3}}{4} + \dfrac{4 + 3\sqrt{3}}{4}i$

   $\approx -0.982 + 2.299i$

67. (a) $5 = 5(\cos 0 + i\sin 0)$

   $2 + 3i \approx \sqrt{13}(\cos 0.9828 + i\sin 0.9828)$

   (b) $\dfrac{5}{2 + 3i} \approx \dfrac{5(\cos 0 + i\sin 0)}{\sqrt{13}(\cos 0.9828 + i\sin 0.9828)} = \dfrac{5\sqrt{13}}{13}[\cos(-0.9828) + i\sin(-0.9828)] \approx 0.769 - 1.154i$

   (c) $\dfrac{5}{2 + 3i} = \dfrac{5}{2 + 3i} \cdot \dfrac{2 - 3i}{2 - 3i} = \dfrac{10 - 15i}{13} = \dfrac{10}{13} - \dfrac{15}{13}i \approx 0.769 - 1.154i$

69. Let $z = x + iy$ such that:

$|z| = 2 \implies 2 = \sqrt{x^2 + y^2}$

$\implies 4 = x^2 + y^2$:

circle with radius of 2

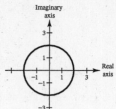

71. Let $\theta = \dfrac{\pi}{6}$.

Let $z = x + iy$ such that:

$\tan \dfrac{\pi}{6} = \dfrac{y}{x}$; line $y = \dfrac{\sqrt{3}}{3}x$

73. $(1 + i)^5 = \left[\sqrt{2}\left(\cos \dfrac{\pi}{4} + i \sin \dfrac{\pi}{4} \right) \right]^5$

$= \left(\sqrt{2} \right)^5 \left(\cos \dfrac{5\pi}{4} + i \sin \dfrac{5\pi}{4} \right)$

$= 4\sqrt{2}\left(-\dfrac{\sqrt{2}}{2} - \dfrac{\sqrt{2}}{2}i \right)$

$= -4 - 4i$

75. $(-1 + i)^{10} = \left[\sqrt{2}\left(\cos \dfrac{3\pi}{4} + i \sin \dfrac{3\pi}{4} \right) \right]^{10}$

$= \left(\sqrt{2} \right)^{10} \left(\cos \dfrac{30\pi}{4} + i \sin \dfrac{30\pi}{4} \right)$

$= 32\left[\cos\left(\dfrac{3\pi}{2} + 6\pi \right) + i \sin\left(\dfrac{3\pi}{2} + 6\pi \right) \right]$

$= 32\left(\cos \dfrac{3\pi}{2} + i \sin \dfrac{3\pi}{2} \right)$

$= 32[0 + i(-1)]$

$= -32i$

77. $2\left(\sqrt{3} + i \right)^7 = 2\left[2\left(\cos \dfrac{\pi}{6} + i \sin \dfrac{\pi}{6} \right) \right]^7$

$= 2\left[2^7\left(\cos \dfrac{7\pi}{6} + i \sin \dfrac{7\pi}{6} \right) \right]$

$= 256\left(-\dfrac{\sqrt{3}}{2} - \dfrac{1}{2}i \right)$

$= -128\sqrt{3} - 128i$

79. $[5(\cos 20° + i \sin 20°)]^3 = 5^3(\cos 60° + i \sin 60°) = \dfrac{125}{2} + \dfrac{125\sqrt{3}}{2}i$

81. $\left(\cos \dfrac{\pi}{4} + i \sin \dfrac{\pi}{4} \right)^{12} = \cos \dfrac{12\pi}{4} + i \sin \dfrac{12\pi}{4}$

$= \cos 3\pi + i \sin 3\pi$

$= -1$

83. $[5(\cos 3.2 + i \sin 3.2)]^4 = 5^4(\cos 12.8 + i \sin 12.8)$

$\approx 608.02 + 144.69i$

85. $(3 - 2i)^5 \approx [3.6056[\cos(-0.588) + i \sin(-0.588)]]^5$

$\approx (3.6056)^5[\cos(-2.94) + i \sin(-2.94)]$

$\approx -597 - 122i$

87. $[3(\cos 15° + i \sin 15°)]^4 = 81(\cos 60° + i \sin 60°)$

$= \dfrac{81}{2} + \dfrac{81\sqrt{3}}{2}i$

89. $\left[2\left(\cos \dfrac{\pi}{10} + i \sin \dfrac{\pi}{10} \right) \right]^5 = 2^5\left(\cos \dfrac{\pi}{2} + i \sin \dfrac{\pi}{2} \right)$

$= 32i$

91. (a) Square roots of $5(\cos 120° + i \sin 120°)$:

$$\sqrt{5}\left[\cos\left(\frac{120° + 360°k}{2}\right) + i \sin\left(\frac{120° + 360°k}{2}\right)\right], \ k = 0, \ 1$$

$k = 0: \ \sqrt{5}(\cos 60° + i \sin 60°)$

$k = 1: \ \sqrt{5}(\cos 240° + i \sin 240°)$

(b)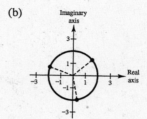

(c) $\dfrac{\sqrt{5}}{2} + \dfrac{\sqrt{15}}{2}i, \ -\dfrac{\sqrt{5}}{2} - \dfrac{\sqrt{15}}{2}i$

93. (a) Cube roots of $8\left(\cos\dfrac{2\pi}{3} + i \sin\dfrac{2\pi}{3}\right)$:

$$\sqrt[3]{8}\left[\cos\left(\frac{\frac{2\pi}{3} + 2\pi k}{3}\right) + i \sin\left(\frac{\frac{2\pi}{3} + 2\pi k}{3}\right)\right], k = 0, 1, 2$$

$k = 0: \ 2\left(\cos\dfrac{2\pi}{9} + i \sin\dfrac{2\pi}{9}\right)$

$k = 1: \ 2\left(\cos\dfrac{8\pi}{9} + i \sin\dfrac{8\pi}{9}\right)$

$k = 2: \ 2\left(\cos\dfrac{14\pi}{9} + i \sin\dfrac{14\pi}{9}\right)$

(b)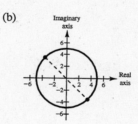

(c) $1.5321 + 1.2856i$

    $-1.8794 + 0.6840i$

    $0.3473 - 1.9696i$

95. (a) Square roots of $-25i = 25\left(\cos\dfrac{3\pi}{2} + i \sin\dfrac{3\pi}{2}\right)$:

$$\sqrt{25}\left[\cos\left(\frac{\frac{3\pi}{2} + 2k\pi}{2}\right) + i \sin\left(\frac{\frac{3\pi}{2} + 2k\pi}{2}\right)\right], k = 0, 1$$

$k = 0: \ 5\left(\cos\dfrac{3\pi}{4} + i \sin\dfrac{3\pi}{4}\right)$

$k = 1: \ 5\left(\cos\dfrac{7\pi}{4} + i \sin\dfrac{7\pi}{4}\right)$

(b)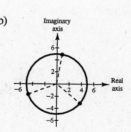

(c) $-\dfrac{5\sqrt{2}}{2} + \dfrac{5\sqrt{2}}{2}i, \ \dfrac{5\sqrt{2}}{2} - \dfrac{5\sqrt{2}}{2}i$

97. (a) Cube roots of $-\dfrac{125}{2}(1 + \sqrt{3}i) = 125\left(\cos\dfrac{4\pi}{3} + i \sin\dfrac{4\pi}{3}\right)$:

$$\sqrt[3]{125}\left[\cos\left(\frac{\frac{4\pi}{3} + 2k\pi}{3}\right) + i \sin\left(\frac{\frac{4\pi}{3} + 2k\pi}{3}\right)\right], \ k = 0, 1, 2$$

$k = 0: \ 5\left(\cos\dfrac{4\pi}{9} + i \sin\dfrac{4\pi}{9}\right)$

$k = 1: \ 5\left(\cos\dfrac{10\pi}{9} + i \sin\dfrac{10\pi}{9}\right)$

$k = 2: \ 5\left(\cos\dfrac{16\pi}{9} + i \sin\dfrac{16\pi}{9}\right)$

(b)

(c) $0.8682 + 4.924i, \ -4.6985 - 1.710i, \ 3.8302 - 3.214i$

99. (a) Fourth roots of $16 = 16(\cos 0 + i \sin 0)$:

$$\sqrt[4]{16}\left[\cos\frac{0 + 2\pi k}{4} + i \sin\frac{0 + 2\pi k}{4}\right], k = 0, 1, 2, 3$$

$k = 0$: $2(\cos 0 + i \sin 0)$

$k = 1$: $2\left(\cos\frac{\pi}{2} + i \sin\frac{\pi}{2}\right)$

$k = 2$: $2(\cos \pi + i \sin \pi)$

$k = 3$: $2\left(\cos\frac{3\pi}{2} + i \sin\frac{3\pi}{2}\right)$

(c) $2, 2i, -2, -2i$

(b)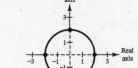

101. (a) Fifth roots of $1 = \cos 0 + i \sin 0$:

$$\cos\left(\frac{2k\pi}{5}\right) + i \sin\left(\frac{2k\pi}{5}\right), k = 0, 1, 2, 3, 4$$

$k = 0$: $\cos 0 + i \sin 0$

$k = 1$: $\cos\frac{2\pi}{5} + i \sin\frac{2\pi}{5}$

$k = 2$: $\cos\frac{4\pi}{5} + i \sin\frac{4\pi}{5}$

$k = 3$: $\cos\frac{6\pi}{5} + i \sin\frac{6\pi}{5}$

$k = 4$: $\cos\frac{8\pi}{5} + i \sin\frac{8\pi}{5}$

(c) $1, 0.3090 + 0.9511i, -0.8090 + 0.5878i, -0.8090 - 0.5878i, 0.3090 - 0.9511i$

(b)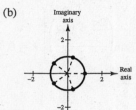

103. (a) Cube roots of $-125 = 125(\cos \pi + i \sin \pi)$:

$$\sqrt[3]{125}\left[\cos\left(\frac{\pi + 2\pi k}{3}\right) + i \sin\left(\frac{\pi + 2\pi k}{3}\right)\right], k = 0, 1, 2$$

$k = 0$: $5\left(\cos\frac{\pi}{3} + i \sin\frac{\pi}{3}\right)$

$k = 1$: $5(\cos \pi + i \sin \pi)$

$k = 2$: $5\left(\cos\frac{5\pi}{3} + i \sin\frac{5\pi}{3}\right)$

(b)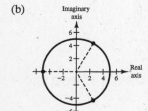

(c) $\dfrac{5}{2} + \dfrac{5\sqrt{3}}{2}i, -5, \dfrac{5}{2} - \dfrac{5\sqrt{3}}{2}i$

105. (a) Fifth roots of $128(-1 + i) = 128\sqrt{2}\left(\cos\dfrac{3\pi}{4} + i\sin\dfrac{3\pi}{4}\right)$:

$$\sqrt[5]{128\sqrt{2}}\left[\cos\left(\dfrac{\dfrac{3\pi}{4} + 2\pi k}{5}\right) + i\sin\left(\dfrac{\dfrac{3\pi}{4} + 2\pi k}{5}\right)\right], \ k = 0, 1, 2, 3, 4\,; \ \sqrt[5]{128\sqrt{2}} = 2\sqrt[5]{4\sqrt{2}} = 2(2^{5/2})^{1/5}$$

$$= 2(2^{1/2}) = 2\sqrt{2}$$

$$k = 0: \ 2\sqrt{2}\left(\cos\dfrac{3\pi}{20} + i\sin\dfrac{3\pi}{20}\right)$$

$$k = 1: \ 2\sqrt{2}\left(\cos\dfrac{11\pi}{20} + i\sin\dfrac{11\pi}{20}\right)$$

$$k = 2: \ 2\sqrt{2}\left(\cos\dfrac{19\pi}{20} + i\sin\dfrac{19\pi}{20}\right)$$

$$k = 3: \ 2\sqrt{2}\left(\cos\dfrac{27\pi}{20} + i\sin\dfrac{27\pi}{20}\right)$$

$$k = 4: \ 2\sqrt{2}\left(\cos\dfrac{7\pi}{4} + i\sin\dfrac{7\pi}{4}\right)$$

(b)

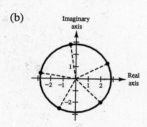

(c) $2.5201 + 1.2841i, -0.4425 + 2.7936i, -2.7936 + 0.4425i, -1.2841 - 2.5201i, 2 - 2i$

107. A complex number z has exactly n distinct n^{th} roots. Since one fourth root is shown, that means that there are **three** more fourth roots which are not shown.

109. $x^4 - i = 0$

$\qquad x^4 = i$

The solutions are the fourth roots of

$i = \cos\dfrac{\pi}{2} + i\sin\dfrac{\pi}{2}$:

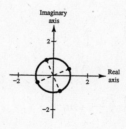

$$\sqrt[4]{1}\left[\cos\left(\dfrac{\dfrac{\pi}{2} + 2k\pi}{4}\right) + i\sin\left(\dfrac{\dfrac{\pi}{2} + 2k\pi}{4}\right)\right], k = 0, 1, 2, 3$$

$$k = 0: \cos\dfrac{\pi}{8} + i\sin\dfrac{\pi}{8}$$

$$k = 1: \cos\dfrac{5\pi}{8} + i\sin\dfrac{5\pi}{8}$$

$$k = 2: \cos\dfrac{9\pi}{8} + i\sin\dfrac{9\pi}{8}$$

$$k = 3: \cos\dfrac{13\pi}{8} + i\sin\dfrac{13\pi}{8}$$

111. $x^5 + 243 = 0$

$$x^5 = -243$$

The solutions are the fifth roots of

$$-243 = 243(\cos \pi + i \sin \pi):$$

$$\sqrt[5]{243}\left[\cos\left(\frac{\pi + 2k\pi}{5}\right) + i \sin\left(\frac{\pi + 2k\pi}{5}\right)\right], k = 0, 1, 2, 3, 4$$

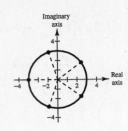

$k = 0: 3\left(\cos \dfrac{\pi}{5} + i \sin \dfrac{\pi}{5}\right)$

$k = 1: 3\left(\cos \dfrac{3\pi}{5} + i \sin \dfrac{3\pi}{5}\right)$

$k = 2: 3(\cos \pi + i \sin \pi) = -3$

$k = 3: 3\left(\cos \dfrac{7\pi}{5} + i \sin \dfrac{7\pi}{5}\right)$

$k = 4: 3\left(\cos \dfrac{9\pi}{5} + i \sin \dfrac{9\pi}{5}\right)$

113. $x^4 + 16i = 0$

$$x^4 = -16i$$

The solutions are the fourth roots of $-16i = 16\left(\cos \dfrac{3\pi}{2} + i \sin \dfrac{3\pi}{2}\right):$

$$\sqrt[4]{16}\left[\cos \frac{\frac{3\pi}{2} + 2\pi k}{4} + i \sin \frac{\frac{3\pi}{2} + 2\pi k}{4}\right], k = 0, 1, 2, 3$$

$k = 0: 2\left(\cos \dfrac{3\pi}{8} + i \sin \dfrac{3\pi}{8}\right)$

$k = 1: 2\left(\cos \dfrac{7\pi}{8} + i \sin \dfrac{7\pi}{8}\right)$

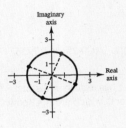

$k = 2: 2\left(\cos \dfrac{11\pi}{8} + i \sin \dfrac{11\pi}{8}\right)$

$k = 3: 2\left(\cos \dfrac{15\pi}{8} + i \sin \dfrac{15\pi}{8}\right)$

115. $x^3 - (1 - i) = 0$

$$x^3 = 1 - i = \sqrt{2}\left(\cos \frac{7\pi}{4} + i \sin \frac{7\pi}{4}\right)$$

The solutions are the cube roots of $1 - i$:

$$\sqrt[3]{\sqrt{2}}\left[\cos\left(\frac{\frac{7\pi}{4} + 2\pi k}{3}\right) + i \sin\left(\frac{\frac{7\pi}{4} + 2\pi k}{3}\right)\right], k = 0, 1, 2$$

$k = 0: \sqrt[6]{2}\left(\cos \dfrac{7\pi}{12} + i \sin \dfrac{7\pi}{12}\right)$

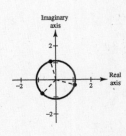

$k = 1: \sqrt[6]{2}\left(\cos \dfrac{5\pi}{4} + i \sin \dfrac{5\pi}{4}\right)$

$k = 2: \sqrt[6]{2}\left(\cos \dfrac{23\pi}{12} + i \sin \dfrac{23\pi}{12}\right)$

117. True, by the definition of the absolute value of a complex number.

119. True. $z_1z_2 = r_1r_2[\cos(\theta_1 + \theta_2) + i\sin(\theta_1 + \theta_2)]$ and $z_1z_2 = 0$ if and only if $r_1 = 0$ and/or $r_2 = 0$.

121. True.

$$2 - 2\sqrt{3}i = 4\left(\cos\frac{5\pi}{3} + i\sin\frac{5\pi}{3}\right)$$

$$\left(2 - 2\sqrt{3}i\right)^3 = 64(\cos 5\pi + i\sin 5\pi)$$

$$= 64(\cos \pi + i\sin \pi)$$

123.

$$2^{-1/4}(1 - i) = 2^{-1/4}\left[\sqrt{2}\left(\cos\frac{7\pi}{4} + i\sin\frac{7\pi}{4}\right)\right]$$

$$= 2^{1/4}\left(\cos\frac{7\pi}{4} + i\sin\frac{7\pi}{4}\right)$$

$$\left[2^{1/4}\left(\cos\frac{7\pi}{4} + i\sin\frac{7\pi}{4}\right)\right]^4 = (2^{1/4})^4(\cos 7\pi + i\sin 7\pi)$$

$$= 2(\cos \pi + i\sin \pi)$$

$$= -2$$

125. (a) $2(\cos 30° + i\sin 30°)$

$2(\cos 150° + i\sin 150°)$

$2(\cos 270° + i\sin 270°)$

(b) These are the cube roots of $8i$.

Review Exercises for Chapter 13

1. Given: $A = 35°, B = 71°, a = 8$

$C = 180° - 35° - 71° = 74°$

$b = \dfrac{a\sin B}{\sin A} = \dfrac{8\sin 71°}{\sin 35°} \approx 13.19$

$c = \dfrac{a\sin C}{\sin A} = \dfrac{8\sin 74°}{\sin 35°} \approx 13.41$

3. Given: $B = 72°, C = 82°, b = 54$

$A = 180° - 72° - 82° = 26°$

$a = \dfrac{b\sin A}{\sin B} = \dfrac{54\sin 26°}{\sin 72°} \approx 24.89$

$c = \dfrac{b\sin C}{\sin B} = \dfrac{54\sin 82°}{\sin 72°} \approx 56.23$

5. Given: $A = 16°, B = 98°, c = 8.4$

$C = 180° - 16° - 98° = 66°$

$a = \dfrac{c\sin A}{\sin C} = \dfrac{8.4\sin 16°}{\sin 66°} \approx 2.53$

$b = \dfrac{c\sin B}{\sin C} = \dfrac{8.4\sin 98°}{\sin 66°} \approx 9.11$

7. Given: $A = 24°, C = 48°, b = 27.5$

$B = 180° - 24° - 48° = 108°$

$a = \dfrac{b\sin A}{\sin B} = \dfrac{27.5\sin 24°}{\sin 108°} \approx 11.76$

$c = \dfrac{b\sin C}{\sin B} = \dfrac{27.5\sin 48°}{\sin 108°} \approx 21.49$

9. Given: $B = 150°, b = 30, c = 10$

$\sin C = \dfrac{c\sin B}{b} = \dfrac{10\sin 150°}{30} \approx 0.1667 \Rightarrow C \approx 9.59°$

$A \approx 180° - 150° - 9.59° = 20.41°$

$a = \dfrac{b\sin A}{\sin B} = \dfrac{30\sin 20.41°}{\sin 150°} \approx 20.92$

11. $A = 75°, a = 51.2, b = 33.7$

$\sin B = \dfrac{b\sin A}{a} = \dfrac{33.7\sin 75°}{51.2} \approx 0.6358 \Rightarrow B \approx 39.48°$

$C \approx 180° - 75° - 39.48° = 65.52°$

$c = \dfrac{a\sin C}{\sin A} = \dfrac{51.2\sin 65.52°}{\sin 75°} \approx 48.24$

13. Area $= \frac{1}{2}bc \sin A = \frac{1}{2}(5)(7)\sin 27° \approx 7.945$

15. Area $= \frac{1}{2}ab \sin C = \frac{1}{2}(16)(5)\sin 123° \approx 33.547$

17. $\tan 17° = \dfrac{h}{x + 50} \implies h = (x + 50)\tan 17°$

$\qquad\qquad\qquad\qquad h = x\tan 17° + 50\tan 17°$

$\qquad \tan 31° = \dfrac{h}{x} \implies h = x\tan 31°$

$\qquad\qquad\qquad x\tan 17° + 50\tan 17° = x\tan 31°$

$\qquad\qquad\qquad 50\tan 17° = x(\tan 31° - \tan 17°)$

$\qquad\qquad\qquad \dfrac{50\tan 17°}{\tan 31° - \tan 17°} = x$

$\qquad\qquad\qquad x \approx 51.7959$

$\qquad\qquad\qquad h = x\tan 31°$

$\qquad\qquad\qquad\quad \approx 51.7959\tan 31°$

$\qquad\qquad\qquad\quad \approx 31.1 \text{ meters}$

19.

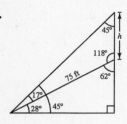

$\dfrac{h}{\sin 17°} = \dfrac{75}{\sin 45°}$

$h = \dfrac{75\sin 17°}{\sin 45°}$

$h \approx 31.01 \text{ feet}$

21. Given: $a = 80, b = 60, c = 100$

$\cos C = \dfrac{a^2 + b^2 - c^2}{2ab} = \dfrac{6400 + 3600 - 10,000}{2(80)(60)} = 0 \implies C = 90°$

$\sin A = \dfrac{80}{100} = 0.8 \implies A \approx 53.13°$

$\sin B = \dfrac{60}{100} = 0.6 \implies B \approx 36.87°$

23. Given: $a = 16.4, b = 8.8, c = 12.2$

$\cos A = \dfrac{b^2 + c^2 - a^2}{2bc} = \dfrac{8.8^2 + 12.2^2 - 16.4^2}{2(8.8)(12.2)} \approx -0.1988 \implies A \approx 101.47°$

$\sin B = \dfrac{b\sin A}{a} \approx \dfrac{8.8\sin 101.47°}{16.4} \approx 0.5259 \implies B \approx 31.73°$

$C \approx 180° - 101.47° - 31.73° = 46.80°$

25. Given: $B = 150°, a = 10, c = 20$

$b^2 = 10^2 + 20^2 - 2(10)(20)\cos 150° \implies b \approx 29.09$

$\sin A = \dfrac{a\sin B}{b} \approx \dfrac{10\sin 150°}{29.09} \implies A \approx 9.90°$

$C \approx 180° - 150° - 9.90° = 20.10°$

27. Given: $A = 62°$, $b = 11.34$, $c = 19.52$

$a^2 = 11.34^2 + 19.52^2 - 2(11.34)(19.52) \cos 62° \implies a \approx 17.37$

$\sin B = \dfrac{b \sin A}{a} \approx \dfrac{11.34 \sin 62°}{17.37} \implies B \approx 35.20°$

$C \approx 180° - 62° - 35.20° = 82.80°$

29. (a)

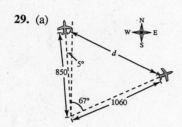

(b) $d^2 = 850^2 + 1060^2 - 2(850)(1060) \cos 72°$

$\approx 1{,}289{,}251$

$d \approx 1135$ miles

31. $a = 15$, $b = 8$, $c = 10$

$s = \dfrac{15 + 8 + 10}{2} = 16.5$

Area $= \sqrt{16.5(1.5)(8.5)(6.5)} \approx 36.979$

33. $a = 38.1$, $b = 26.7$, $c = 19.4$

$s = \dfrac{38.1 + 26.7 + 19.4}{2} = 42.1$

Area $= \sqrt{42.1(4)(15.4)(22.7)} \approx 242.630$

35. Initial point: $(3, 4)$

Terminal point: $(-5, -7)$

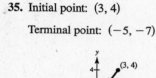

37. Initial point: $(-6, -8)$

Terminal point: $(8, 3)$

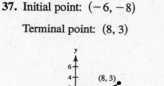

39. Initial point: $(0, 1)$

Terminal point: $\left(6, \frac{7}{2}\right)$

$\mathbf{v} = \left\langle 6 - 0, \frac{7}{2} - 1 \right\rangle = \left\langle 6, \frac{5}{2} \right\rangle$

41. Initial point: $(1, 5)$

Terminal point: $(15, 9)$

$\mathbf{v} = \langle 15 - 1, 9 - 5 \rangle = \langle 14, 4 \rangle$

43. $\|\mathbf{v}\| = \dfrac{1}{3}$, $\theta = 210°$

$\left\langle \dfrac{1}{3} \cos 210°, \dfrac{1}{3} \sin 210° \right\rangle = \left\langle -\dfrac{\sqrt{3}}{6}, -\dfrac{1}{6} \right\rangle$

45. $\mathbf{u} = 6\mathbf{i} - 5\mathbf{j}$, $\mathbf{v} = 10\mathbf{i} + 3\mathbf{j}$

$4\mathbf{u} - 5\mathbf{v} = (24\mathbf{i} - 20\mathbf{j}) - (50\mathbf{i} + 15\mathbf{j}) = -26\mathbf{i} - 35\mathbf{j}$

$= \langle -26, -35 \rangle$

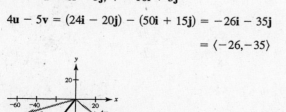

47. $\mathbf{v} = 10\mathbf{i} + 3\mathbf{j}$

$\frac{1}{2}\mathbf{v} = 5\mathbf{i} + \frac{3}{2}\mathbf{j} = \left\langle 5, \frac{3}{2} \right\rangle$

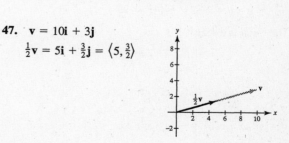

49. $\mathbf{u} = \langle -6, -8 \rangle = -6\mathbf{i} - 8\mathbf{j}$

51. Initial point: $(-2, 7)$

Terminal point: $(5, -9)$

$\mathbf{u} = \langle 5 - (-2), -9 - 7 \rangle = \langle 7, -16 \rangle = 7\mathbf{i} - 16\mathbf{j}$

53. $\mathbf{v} = 4\mathbf{i} - \mathbf{j}$

$\|\mathbf{v}\| = \sqrt{4^2 + (-1)^2} = \sqrt{17}$

$\tan \theta = \dfrac{-1}{4}, \theta$ in Quadrant IV $\implies \theta \approx 346°$

$\mathbf{v} \approx \sqrt{17}(\cos 346° \, \mathbf{i} + \sin 346° \, \mathbf{j})$

55. $\mathbf{v} = 3(\cos 150° \mathbf{i} + \sin 150° \, \mathbf{j})$

$\|\mathbf{v}\| = 3, \theta = 150°$

57. $\mathbf{v} = -4\mathbf{i} + 7\mathbf{j}$

$\|\mathbf{v}\| = \sqrt{(-4)^2 + 7^2} = \sqrt{65}$

$\tan \theta = \dfrac{7}{-4}, \theta$ in Quadrant II $\implies \theta \approx 119.7°$

59. $\mathbf{v} = 8\mathbf{i} - \mathbf{j}$

$\|\mathbf{v}\| = \sqrt{8^2 + (-1)^2} = \sqrt{65}$

$\tan \theta = \dfrac{-1}{8}, \theta$ in Quadrant IV $\implies \theta \approx 352.9°$

61. Rope One: $\mathbf{u} = \|\mathbf{u}\|(\cos 30°\mathbf{i} - \sin 30°\mathbf{j}) = \|\mathbf{u}\|\left(\dfrac{\sqrt{3}}{2}\mathbf{i} - \dfrac{1}{2}\mathbf{j} \right)$

Rope Two: $\mathbf{v} = \|\mathbf{u}\|(-\cos 30°\mathbf{i} - \sin 30°\mathbf{j}) = \|\mathbf{u}\|\left(-\dfrac{\sqrt{3}}{2}\mathbf{i} - \dfrac{1}{2}\mathbf{j} \right)$

Resultant: $\mathbf{u} + \mathbf{v} = -\|\mathbf{u}\|\mathbf{j} = -180\mathbf{j}$

$\|\mathbf{u}\| = 180$

Therefore, the tension on each rope is $\|\mathbf{u}\| = 180$ lbs.

63. $\mathbf{u} = \langle 6, 7 \rangle$

$\mathbf{v} = \langle -3, 9 \rangle$

$\mathbf{u} \cdot \mathbf{v} = 6(-3) + 7(9) = 45$

65. $\mathbf{u} = 3\mathbf{i} + 7\mathbf{j}$

$\mathbf{v} = 11\mathbf{i} - 5\mathbf{j}$

$\mathbf{u} \cdot \mathbf{v} = 3(11) + 7(-5) = -2$

67. $\mathbf{u} = \langle -3, 4 \rangle$

$2\mathbf{u} = \langle -6, 8 \rangle$

$2\mathbf{u} \cdot \mathbf{u} = (-6)(-3) + 8(4) = 50$

The result is a scalar.

69. $\mathbf{u} = \langle -3, 4 \rangle, \mathbf{v} = \langle 2, 1 \rangle$

$\mathbf{u} \cdot \mathbf{v} = (-3)(2) + 4(1) = -2$

$\mathbf{u}(\mathbf{u} \cdot \mathbf{v}) = \mathbf{u}(-2) = -2\mathbf{u} = \langle 6, -8 \rangle$

The result is a vector.

71. $\mathbf{u} = \langle -3, 4 \rangle$

$\|\mathbf{u}\|^2 = \mathbf{u} \cdot \mathbf{u} = (-3)^2 + (4)^2 = 25$

The result is a scalar.

73. $\mathbf{u} = \cos \dfrac{7\pi}{4}\mathbf{i} + \sin \dfrac{7\pi}{4}\mathbf{j} = \left\langle \dfrac{1}{\sqrt{2}}, -\dfrac{1}{\sqrt{2}} \right\rangle$

$\mathbf{v} = \cos \dfrac{5\pi}{6}\mathbf{i} + \sin \dfrac{5\pi}{6}\mathbf{j} = \left\langle -\dfrac{\sqrt{3}}{2}, \dfrac{1}{2} \right\rangle$

$\cos \theta = \dfrac{\mathbf{u} \cdot \mathbf{v}}{\|\mathbf{u}\|\|\mathbf{v}\|} = \dfrac{-\sqrt{3} - 1}{2\sqrt{2}} \implies \theta = \dfrac{11\pi}{12}$

75. $\mathbf{u} = \langle 2\sqrt{2}, -4 \rangle, \mathbf{v} = \langle -\sqrt{2}, 1 \rangle$

$\cos \theta = \dfrac{\mathbf{u} \cdot \mathbf{v}}{\|\mathbf{u}\| \|\mathbf{v}\|} = \dfrac{-8}{(\sqrt{24})(\sqrt{3})} \implies \theta \approx 160.5°$

77. $\mathbf{u} = \langle -3, 8 \rangle$

$\mathbf{v} = \langle 8, 3 \rangle$

$\mathbf{u} \cdot \mathbf{v} = -3(8) + 8(3) = 0$

$\mathbf{u}$ and $\mathbf{v}$ are orthogonal.

79. $\mathbf{u} = -\mathbf{i}$

$\mathbf{v} = \mathbf{i} + 2\mathbf{j}$

$\mathbf{u} \cdot \mathbf{v} \neq 0 \implies$ Not orthogonal

$\mathbf{v} \neq k\mathbf{u} \implies$ Not parallel

Neither

81. $\mathbf{u} = \langle -4, 3 \rangle, \; \mathbf{v} = \langle -8, -2 \rangle$

$\mathbf{w}_1 = \text{proj}_\mathbf{v}\mathbf{u} = \left(\dfrac{\mathbf{u} \cdot \mathbf{v}}{\|\mathbf{v}\|^2} \right)\mathbf{v} = \left(\dfrac{26}{68} \right)\langle -8, -2 \rangle$

$= -\dfrac{13}{17}\langle 4, 1 \rangle$

$\mathbf{w}_2 = \mathbf{u} - \mathbf{w}_1 = \langle -4, 3 \rangle - \left(-\dfrac{13}{17} \right)\langle 4, 1 \rangle$

$= \dfrac{16}{17}\langle -1, 4 \rangle$

83. $\mathbf{u} = \langle 2, 7 \rangle, \; \mathbf{v} = \langle 1, -1 \rangle$

$\mathbf{w}_1 = \text{proj}_\mathbf{v}\mathbf{u} = \left(\dfrac{\mathbf{u} \cdot \mathbf{v}}{\|\mathbf{v}\|^2} \right)\mathbf{v} = -\dfrac{5}{2}\langle 1, -1 \rangle$

$= \dfrac{5}{2}\langle -1, 1 \rangle$

$\mathbf{w}_2 = \mathbf{u} - \mathbf{w}_1 = \langle 2, 7 \rangle - \left(\dfrac{5}{2} \right)\langle -1, 1 \rangle$

$= \dfrac{9}{2}\langle 1, 1 \rangle$

85. $P = (5, 3), Q = (8, 9) \implies \overrightarrow{PQ} = \langle 3, 6 \rangle$

$W = \mathbf{F} \cdot \overrightarrow{PQ} = \langle 2, 7 \rangle \cdot \langle 3, 6 \rangle = 48$

87. $|7i| = \sqrt{0^2 + 7^2} = 7$

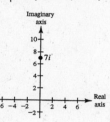

89. $|5 + 3i| = \sqrt{5^2 + 3^2} = \sqrt{34}$

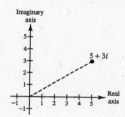

91. $5 - 5i$

$r = \sqrt{5^2 + (-5)^2} = \sqrt{50} = 5\sqrt{2}$

$\tan \theta = \dfrac{-5}{5} = -1 \implies \theta = \dfrac{7\pi}{4}$ since the

complex number is in Quadrant IV.

$5 - 5i = 5\sqrt{2}\left(\cos \dfrac{7\pi}{4} + i \sin \dfrac{7\pi}{4} \right)$

93. $-3\sqrt{3} + 3i$

$r = \sqrt{(-3\sqrt{3})^2 + 3^2} = \sqrt{36} = 6$

$\tan \theta = \dfrac{3}{-3\sqrt{3}} = -\dfrac{1}{\sqrt{3}} \implies \theta = \dfrac{5\pi}{6}$ since the

complex number is in Quadrant II.

$-3\sqrt{3} + 3i = 6\left(\cos \dfrac{5\pi}{6} + i \sin \dfrac{5\pi}{6} \right)$

95. (a) $z_1 = 2\sqrt{3} - 2i = 4\left(\cos \dfrac{11\pi}{6} + i \sin \dfrac{11\pi}{6} \right)$

$z_2 = -10i = 10\left(\cos \dfrac{3\pi}{2} + i \sin \dfrac{3\pi}{2} \right)$

(b) $z_1 z_2 = \left[4\left(\cos \dfrac{11\pi}{6} + i \sin \dfrac{11\pi}{6} \right) \right]\left[10\left(\cos \dfrac{3\pi}{2} + i \sin \dfrac{3\pi}{2} \right) \right] = 40\left(\cos \dfrac{10\pi}{3} + i \sin \dfrac{10\pi}{3} \right)$

$\dfrac{z_1}{z_2} = \dfrac{4\left(\cos \dfrac{11\pi}{6} + i \sin \dfrac{11\pi}{6} \right)}{10\left(\cos \dfrac{3\pi}{2} + i \sin \dfrac{3\pi}{2} \right)} = \dfrac{2}{5}\left(\cos \dfrac{\pi}{3} + i \sin \dfrac{\pi}{3} \right)$

97. $\left[5\left(\cos\dfrac{\pi}{12} + i\sin\dfrac{\pi}{12}\right)\right]^4 = 5^4\left(\cos\dfrac{4\pi}{12} + i\sin\dfrac{4\pi}{12}\right)$

$$= 625\left(\cos\dfrac{\pi}{3} + i\sin\dfrac{\pi}{3}\right)$$

$$= 625\left(\dfrac{1}{2} + \dfrac{\sqrt{3}}{2}i\right)$$

$$= \dfrac{625}{2} + \dfrac{625\sqrt{3}}{2}i$$

99. $(2 + 3i)^6 \approx [\sqrt{13}(\cos 56.3° + i\sin 56.3°)]^6$

$$= 13^3(\cos 337.9° + i\sin 337.9°)$$

$$\approx 13^3(0.9263 - 0.3769i)$$

$$\approx 2035 - 828i$$

101. (a) The trigonometric form of the three roots shown is:

$4(\cos 60° + i\sin 60°)$

$4(\cos 180° + i\sin 180°)$

$4(\cos 300° + i\sin 300°)$

(b) Since there are three evenly spaced roots on the circle of radius 4, they are cube roots of a complex number of modulus $4^3 = 64$. Cubing them yields -64.

$[4(\cos 60° + i\sin 60°)]^3 = -64$

$[4(\cos 180° + i\sin 180°)]^3 = -64$

$[4(\cos 300° + i\sin 300°)]^3 = -64$

103. Sixth roots of $-729i = 729\left(\cos\dfrac{3\pi}{2} + i\sin\dfrac{3\pi}{2}\right)$:

$$\sqrt[6]{729}\left[\cos\left(\dfrac{\dfrac{3\pi}{2} + 2k\pi}{6}\right) + i\sin\left(\dfrac{\dfrac{3\pi}{2} + 2k\pi}{6}\right)\right], k = 0, 1, 2, 3, 4, 5$$

$k = 0$: $3\left(\cos\dfrac{\pi}{4} + i\sin\dfrac{\pi}{4}\right)$

$k = 1$: $3\left(\cos\dfrac{7\pi}{12} + i\sin\dfrac{7\pi}{12}\right)$

$k = 2$: $3\left(\cos\dfrac{11\pi}{12} + i\sin\dfrac{11\pi}{12}\right)$

$k = 3$: $3\left(\cos\dfrac{5\pi}{4} + i\sin\dfrac{5\pi}{4}\right)$

$k = 4$: $3\left(\cos\dfrac{19\pi}{12} + i\sin\dfrac{19\pi}{12}\right)$

$k = 5$: $3\left(\cos\dfrac{23\pi}{12} + i\sin\dfrac{23\pi}{12}\right)$

105. Cube roots of $64 = 64(\cos 0 + i\sin 0)$:

$$\sqrt[3]{64}\left[\cos\left(\dfrac{0 + 2\pi k}{3}\right) + i\sin\left(\dfrac{0 + 2\pi k}{3}\right)\right],$$

$k = 0, 1, 2$

$k = 0$: $4(\cos 0 + i\sin 0) = 4$

$k = 1$: $4\left(\cos\dfrac{2\pi}{3} + i\sin\dfrac{2\pi}{3}\right) = -2 + 2\sqrt{3}i$

$k = 2$: $4\left(\cos\dfrac{4\pi}{3} + i\sin\dfrac{4\pi}{3}\right) = -2 - 2\sqrt{3}i$

107. $x^4 + 81 = 0$

$x^4 = -81$ Solve by finding the fourth roots of -81.

$-81 = 81(\cos \pi + i \sin \pi)$

$\sqrt[4]{-81} = \sqrt[4]{81}\left[\cos\left(\dfrac{\pi + 2\pi k}{4}\right) + i \sin\left(\dfrac{\pi + 2\pi k}{4}\right)\right]$, $k = 0, 1, 2, 3$

$k = 0: 3\left(\cos\dfrac{\pi}{4} + i \sin\dfrac{\pi}{4}\right) = \dfrac{3\sqrt{2}}{2} + \dfrac{3\sqrt{2}}{2}i$

$k = 1: 3\left(\cos\dfrac{3\pi}{4} + i \sin\dfrac{3\pi}{4}\right) = -\dfrac{3\sqrt{2}}{2} + \dfrac{3\sqrt{2}}{2}i$

$k = 2: 3\left(\cos\dfrac{5\pi}{4} + i \sin\dfrac{5\pi}{4}\right) = -\dfrac{3\sqrt{2}}{2} - \dfrac{3\sqrt{2}}{2}i$

$k = 3: 3\left(\cos\dfrac{7\pi}{4} + i \sin\dfrac{7\pi}{4}\right) = \dfrac{3\sqrt{2}}{2} - \dfrac{3\sqrt{2}}{2}i$

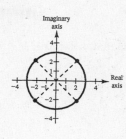

109. $x^3 + 8i = 0$

$x^3 = -8i$ Solve by finding the cube roots of $-8i$.

$-8i = 8\left(\cos\dfrac{3\pi}{2} + i \sin\dfrac{3\pi}{2}\right)$

$\sqrt[3]{-8i} = \sqrt[3]{8}\left[\cos\left(\dfrac{\dfrac{3\pi}{2} + 2\pi k}{3}\right) + i \sin\left(\dfrac{\dfrac{3\pi}{2} + 2\pi k}{3}\right)\right]$, $k = 0, 1, 2$

$k = 0: 2\left(\cos\dfrac{\pi}{2} + i \sin\dfrac{\pi}{2}\right) = 2i$

$k = 1: 2\left(\cos\dfrac{7\pi}{6} + i \sin\dfrac{7\pi}{6}\right) = -\sqrt{3} - i$

$k = 2: 2\left(\cos\dfrac{11\pi}{6} + i \sin\dfrac{11\pi}{6}\right) = \sqrt{3} - i$

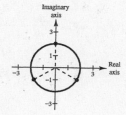

Problem Solving for Chapter 13

1. (a) $\dfrac{\sin \alpha}{9} = \dfrac{\sin \beta}{18}$

    $\sin \alpha = 0.5 \sin \beta$

      $\alpha = \arcsin(0.5 \sin \beta)$; Domain: $0 < \beta < \pi$

(b) $\dfrac{d\alpha}{d\beta} = \dfrac{0.5 \cos \beta}{\sqrt{1 - (0.5 \sin \beta)^2}}$

      $= \dfrac{0.5 \cos \beta}{\sqrt{1 - \dfrac{\sin^2\beta}{4}}} \cdot \dfrac{2}{2}$

      $= \dfrac{\cos \beta}{\sqrt{4 - \sin^2\beta}}$

Maximum: $\left(\dfrac{\pi}{2}, \dfrac{\pi}{6}\right)$

Range: $0 < \alpha \le \dfrac{\pi}{6}$

(c)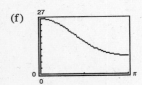

(d) $\dfrac{d\alpha}{dt} = \dfrac{\cos \beta}{\sqrt{4 - \sin^2\beta}}\left(\dfrac{d\beta}{dt}\right)$

    When $\dfrac{d\beta}{dt} = 0.2$ and $\beta = \dfrac{\pi}{4}$ we have $\dfrac{d\alpha}{dt} = \dfrac{\dfrac{\sqrt{2}}{2}(0.2)}{\sqrt{4 - \dfrac{1}{2}}} = \dfrac{\sqrt{7}}{35} \approx 0.0756$ radians per second.

(e) $\gamma = \pi - \alpha - \beta = \pi - \beta - \arcsin(0.5 \sin \beta)$

    $\dfrac{c}{\sin \gamma} = \dfrac{18}{\sin \beta}$

      $c = \dfrac{18 \sin \gamma}{\sin \beta} = \dfrac{18 \sin[\pi - \beta - \arcsin(0.5 \sin \beta)]}{\sin \beta}$; Domain: $0 < \beta < \pi$

(f)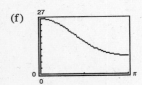

    Range: $9 < c < 27$

(g) $\dfrac{dc}{dt} = \dfrac{18\left[\sin \beta \cos [\pi - \beta - \arcsin(0.5 \sin \beta)]\left(-1 - \dfrac{\cos \beta}{\sqrt{4 - \sin^2\beta}}\right) - \sin [\pi - \beta - \arcsin(0.5 \sin \beta)] \cos \beta\right]}{\sin^2\beta} \cdot \dfrac{d\beta}{dt}$

    When $\dfrac{d\beta}{dt} = 0.2$ and $\beta = \dfrac{\pi}{4}$, $\dfrac{dc}{dt} \approx -1.7539$ radians per second.

(h)

β	0	0.4	0.8	1.2	1.6	2.0	2.4	2.8
α	0	0.1960	0.3669	0.4848	0.5234	0.4720	0.3445	0.1683
c	Undef.	25.95	23.07	19.19	15.33	12.29	10.31	9.27

(i) When $\beta \to 0$, $c \to 27$. On the graph it is undefined. The function obtained by using the Law of Sines is not valid when $\beta = 0$ since we would no longer have a triangle.

3. (a) Initial point: $(0, 0)$

Terminal point: $\left(\dfrac{u_1 + v_1}{2}, \dfrac{u_2 + v_2}{2}\right)$

$W = \left\langle \dfrac{u_1 + v_1}{2}, \dfrac{u_2 + v_2}{2} \right\rangle = \dfrac{1}{2}(\mathbf{u} + \mathbf{v})$

(b) Initial point: (u_1, u_2)

Terminal point: $\dfrac{1}{2}(u_1 + v_1, u_2 + v_2)$

$W = \left\langle \dfrac{u_1 + v_1}{2} - u_1, \dfrac{u_2 + v_2}{2} - u_2 \right\rangle$

$\quad = \left\langle \dfrac{v_1 - u_1}{2}, \dfrac{v_2 - u_2}{2} \right\rangle = \dfrac{1}{2}(\mathbf{v} - \mathbf{u})$

5. $\dfrac{1}{2}bc(1 - \cos A) = \dfrac{1}{2}bc\left[1 + \dfrac{a^2 - (b^2 + c^2)}{2bc}\right]$

$\quad = \dfrac{1}{2}bc\left[\dfrac{2bc + a^2 - b^2 - c^2}{2bc}\right]$

$\quad = \dfrac{a^2 - (b^2 - 2bc + c^2)}{4}$

$\quad = \dfrac{a^2 - (b - c)^2}{4}$

$\quad = \left(\dfrac{a - (b - c)}{2}\right)\left(\dfrac{a + (b - c)}{2}\right)$

$\quad = \dfrac{a - b + c}{2} \cdot \dfrac{a + b - c}{2}$

7. (a) $A = \dfrac{1}{2}(30)(20)\sin\left(\theta + \dfrac{\theta}{2}\right) - \dfrac{1}{2}(8)(20)\sin\dfrac{\theta}{2} - \dfrac{1}{2}(8)(30)\sin\theta$

$\quad = 300 \sin\dfrac{3\theta}{2} - 80\sin\dfrac{\theta}{2} - 120\sin\theta$

$\quad = 20\left[15\sin\dfrac{3\theta}{2} - 4\sin\dfrac{\theta}{2} - 6\sin\theta\right]$

(b)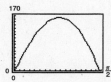

(c) Domain: $0 \leq \theta \leq 1.6690$

The area would increase and the domain would increase at the right endpoint of its interval.

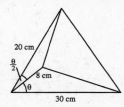

(d) $\dfrac{dA}{d\theta} = 20\left[15\left(\dfrac{3}{2}\right)\cos\dfrac{3\theta}{2} - 4\left(\dfrac{1}{2}\right)\cos\dfrac{\theta}{2} - 6\cos\theta\right]$

$\quad = 20\left[\dfrac{45}{2}\cos\dfrac{3\theta}{2} - 2\cos\dfrac{\theta}{2} - 6\cos\theta\right]$

$\quad = 10\left[45\cos\dfrac{3\theta}{2} - 4\cos\dfrac{\theta}{2} - 12\cos\theta\right]$

Critical number: $\theta \approx 0.8782$

9. (a) Let $\mathbf{u} = \langle u_1, u_2 \rangle$ and $\mathbf{v} = \langle v_1, v_2 \rangle$, then

$\|\mathbf{u} + \mathbf{v}\|^2 + \|\mathbf{u} - \mathbf{v}\|^2 = (u_1 + v_1)^2 + (u_2 + v_2)^2 + (u_1 - v_1)^2 + (u_2 - v_2)^2$

$\quad = u_1{}^2 + 2u_1 v_1 + v_1{}^2 + u_2{}^2 + 2u_2v_2 + v_2{}^2 + u_1{}^2 - 2u_1v_1 + v_1{}^2 + u_2{}^2 - 2u_2v_2 + v_2{}^2$

$\quad = 2(u_1{}^2 + u_2{}^2) + 2(v_1{}^2 + v_2{}^2)$

$\quad = 2\|\mathbf{u}\|^2 + 2\|\mathbf{v}\|^2.$

(b) The sum of the squares of the lengths of the diagonals of a parallelagram is equal to the sum of the squares of the lengths of all four sides.

11. (a) $z\bar{z} = [r(\cos\theta + i\sin\theta)][r(\cos(-\theta) + i\sin(-\theta))]$

$\quad = r^2[\cos(\theta - \theta) + i\sin(\theta - \theta)]$

$\quad = r^2[\cos 0 + i\sin 0]$

$\quad = r^2$

(b) $\dfrac{z}{\bar{z}} = \dfrac{r(\cos\theta + i\sin\theta)}{r[\cos(-\theta) + i\sin(-\theta)]}, z \neq 0$

$\quad = \dfrac{r}{r}[\cos(\theta - (-\theta)) + i\sin(\theta - (-\theta))]$

$\quad = \cos 2\theta + i\sin 2\theta$

13. Euler's Formula: $e^{a+bi} = e^a(\cos b + i \sin b)$

$$e^{\pi i} + 1 = e^0(\cos \pi + i \sin \pi) + 1 = -1 + i(0) + 1 = 0$$

15. $z_1 = 2(\cos \theta + i \sin \theta)$

$\quad z_2 = 2(\cos(\pi - \theta) + i \sin(\pi - \theta))$

$\quad z_1 z_2 = (2)(2)[\cos(\theta + (\pi - \theta)) + i \sin(\theta + (\pi - \theta))]$

$\qquad = 4(\cos \pi + i \sin \pi)$

$\qquad = -4$

$\dfrac{z_1}{z_2} = \dfrac{2(\cos \theta + i \sin \theta)}{2(\cos(\pi - \theta) + i \sin(\pi - \theta))}$

$\qquad = 1[\cos(\theta - (\pi - \theta)) + i \sin(\theta - (\pi - \theta))]$

$\qquad = \cos(2\theta - \pi) + i \sin(2\theta - \pi)$

$\qquad = \cos 2\theta \cos \pi + \sin 2\theta \sin \pi + i(\sin 2\theta \cos \pi - \cos 2\theta \sin \pi)$

$\qquad = -\cos 2\theta - i \sin 2\theta$

17. The base is an equilateral triangle

$$\Rightarrow B = \frac{\sqrt{3}}{4}x^2 \tan 35° = \frac{h}{x} \Rightarrow h = x \tan 35°$$

Thus, $V = \dfrac{1}{3}Bh$

$\qquad = \dfrac{1}{3}\left(\dfrac{\sqrt{3}}{4}x^2\right)(x \tan 35°)$

$\qquad = \dfrac{\sqrt{3}}{12}x^3 \tan 35°$

CHAPTER 14
Systems of Equations and Matrices

CHAPTER 14
Systems of Equations and Matrices

Section 14.1   Systems of Linear Equations in Two Variables

1. $\begin{cases} 4x - y = 1 \\ 6x + y = -6 \end{cases}$

   (a) $4(0) - (-3) \neq 1$

      $(0, -3)$ is **not** a solution.

   (b) $4(-1) - (-4) \neq 1$

      $(-1, -4)$ is **not** a solution.

   (c) $4\left(-\frac{3}{2}\right) - (-2) \neq 1$

      $\left(-\frac{3}{2}, -2\right)$ is **not** a solution.

   (d) $4\left(-\frac{1}{2}\right) - (-3) = 1$

      $6\left(-\frac{1}{2}\right) + (-3) = -6$

      $\left(-\frac{1}{2}, -3\right)$ **is** a solution.

3. $\begin{cases} y = -2e^x \\ 3x - y = 2 \end{cases}$

   (a) $0 \neq -2e^{-2}$

      $(-2, 0)$ is **not** a solution.

   (b) $\qquad -2 = -2e^0$

      $3(0) - (-2) = 2$

      $(0, -2)$ **is** a solution.

   (c) $-3 \neq -2e^0$

      $(0, -3)$ is **not** a solution.

   (d) $2 \neq -2e^{-1}$

      $(-1, 2)$ is **not** a solution.

5. $\begin{cases} x - y = 0 & \text{Equation 1} \\ 5x - 3y = 10 & \text{Equation 2} \end{cases}$

Solve for y in Equation 1: $y = x$

Substitute for y in Equation 2: $5x - 3x = 10$

Solve for x: $2x = 10 \implies x = 5$

Back-substitute in Equation 1: $y = x = 5$

Solution: $(5, 5)$

7. $\begin{cases} 2x - y + 2 = 0 & \text{Equation 1} \\ 4x + y - 5 = 0 & \text{Equation 2} \end{cases}$

Solve for y in Equation 1: $y = 2x + 2$

Substitute for y in Equation 2: $4x + (2x + 2) - 5 = 0$

Solve for x: $6x - 3 = 0 \implies x = \frac{1}{2}$

Back-substitute $x = \frac{1}{2}$: $y = 2x + 2 = 2\left(\frac{1}{2}\right) + 2 = 3$

Solution: $\left(\frac{1}{2}, 3\right)$

9. $\begin{cases} 1.5x + 0.8y = 2.3 & \text{Equation 1} \\ 0.3x - 0.2y = 0.1 & \text{Equation 2} \end{cases}$

Multiply the equations by 10.

     $15x + 8y = 23 \qquad$ Revised Equation 1

     $3x - 2y = 1 \qquad$ Revised Equation 2

Solve for y in revised Equation 2: $y = \frac{3}{2}x - \frac{1}{2}$

Substitute for y in revised Equation 1: $15x + 8\left(\frac{3}{2}x - \frac{1}{2}\right) = 23$

Solve for x: $15x + 12x - 4 = 23 \implies 27x = 27 \implies x = 1$

Back-substitute $x = 1$: $y = \frac{3}{2}(1) - \frac{1}{2} = 1$

Solution: $(1, 1)$

11. $\begin{cases} \frac{1}{5}x + \frac{1}{2}y = 8 & \text{Equation 1} \\ x + y = 20 & \text{Equation 2} \end{cases}$

Solve for x in Equation 2: $x = 20 - y$

Substitute for x in Equation 1: $\frac{1}{5}(20 - y) + \frac{1}{2}y = 8$

Solve for y: $4 + \frac{3}{10}y = 8 \implies y = \frac{40}{3}$

Back-substitute $y = \frac{40}{3}$: $x = 20 - y = 20 - \frac{40}{3} = \frac{20}{3}$

Solution: $\left(\frac{20}{3}, \frac{40}{3}\right)$

13. $\begin{cases} 6x + 5y = -3 & \text{Equation 1} \\ -x - \frac{5}{6}y = -7 & \text{Equation 2} \end{cases}$

Solve for x in Equation 2: $x = 7 - \frac{5}{6}y$

Substitute for x in Equation 1: $6\left(7 - \frac{5}{6}y\right) + 5y = -3$

Solve for y: $42 - 5y + 5y = -3 \implies 42 = -3$ (False)

No solution

15. $\begin{cases} x^2 - y = 0 & \text{Equation 1} \\ 2x + y = 0 & \text{Equation 2} \end{cases}$

Solve for y in Equation 2: $y = -2x$

Substitute for y in Equation 1: $x^2 - (-2x) = 0$

Solve for x:
$x^2 + 2x = 0 \implies x(x + 2) = 0 \implies x = 0, -2$

Back-substitute $x = 0$: $y = -2(0) = 0$

Back-substitute $x = -2$: $y = -2(-2) = 4$

Solutions: $(0, 0), (-2, 4)$

17. $\begin{cases} x^3 - y = 0 & \text{Equation 1} \\ x - y = 0 & \text{Equation 2} \end{cases}$

Solve for y in Equation 2: $y = x$

Substitute for y in Equation 1: $x^3 - x = 0$

Solve for x: $x(x + 1)(x - 1) = 0 \implies x = 0, \pm 1$

Back-substitute $x = 0$: $y = 0$

Back-substitute $x = 1$: $y = 1$

Back-substitute $x = -1$: $y = -1$

Solutions: $(0, 0), (1, 1), (-1, -1)$

19. $\begin{cases} y = x^3 - 2x + 1 & \text{Equation 1} \\ y = x^2 - 1 & \text{Equation 2} \end{cases}$

Substitute for y in Equation 2: $x^3 - 2x + 1 = x^2 - 1$

Solve for x: $x^3 - x^2 - 2x + 2 = 0 \implies x^2(x - 1) - 2(x - 1) = 0$

$\implies (x - 1)(x^2 - 2) = 0$

$\implies x = 1, \pm\sqrt{2}$

Back-substitute $x = 1 \implies y = 0$

$x = \pm\sqrt{2} \implies y = 1$

Solutions: $(1, 0), \left(\pm\sqrt{2}, 1\right)$

21. $\begin{cases} 2x + y = 5 & \text{Equation 1} \\ x - y = 1 & \text{Equation 2} \end{cases}$

Add Equation 1 to Equation 2: $3x = 6 \implies x = 2$

Back-substitute $x = 2 \implies y = 1$

Solution: $(2, 1)$

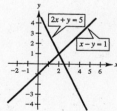

23. $\begin{cases} x + y = 0 & \text{Equation 1} \\ 3x + 2y = 1 & \text{Equation 2} \end{cases}$

Multiply Equation 1 by -2: $-2x - 2y = 0$

Add this to Equation 2 to eliminate y: $x = 1$

Substitute $x = 1$ in Equation 1: $1 + y = 0 \implies y = -1$

Solution: $(1, -1)$

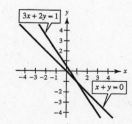

25. $\begin{cases} x - y = 2 & \text{Equation 1} \\ -2x + 2y = 5 & \text{Equation 2} \end{cases}$

Multiply Equation 1 by 2: $2x - 2y = 4$

Add this to Equation 2: $0 = 9$

There are no solutions.

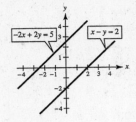

27. $\begin{cases} 3x - 2y = 5 & \text{Equation 1} \\ -6x + 4y = -10 & \text{Equation 2} \end{cases}$

Multiply Equation 1 by 2 and add to Equation 2: $0 = 0$

The equations are dependent. There are infinitely many solutions. Let $x = a$.

$3a - 2y = 5 \implies y = \frac{1}{2}(3a - 5)$

Solution: $\left(a, \frac{1}{2}(3a - 5)\right)$ where a is any real number.

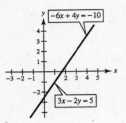

29. $\begin{cases} 9x + 3y = 1 & \text{Equation 1} \\ 3x - 6y = 5 & \text{Equation 2} \end{cases}$

Multiply Equation 2 by (-3): $\quad\begin{aligned} 9x + 3y &= 1 \\ -9x + 18y &= -15 \end{aligned}$

Add to eliminate x: $21y = -14 \implies y = -\frac{2}{3}$

Substitute $y = -\frac{2}{3}$ in Equation 1: $9x + 3\left(-\frac{2}{3}\right) = 1$

$$x = \frac{1}{3}$$

Solution: $\left(\frac{1}{3}, -\frac{2}{3}\right)$

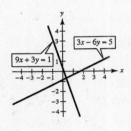

31. $\begin{cases} x + 2y = 4 & \text{Equation 1} \\ x - 2y = 1 & \text{Equation 2} \end{cases}$

Add to eliminate y:

$$2x = 5$$
$$x = \frac{5}{2}$$

Substitute $x = \frac{5}{2}$ in Equation 1:

$$\frac{5}{2} + 2y = 4 \implies y = \frac{3}{4}$$

Solution: $\left(\frac{5}{2}, \frac{3}{4}\right)$

33. $\begin{cases} 2x + 3y = 18 & \text{Equation 1} \\ 5x - y = 11 & \text{Equation 2} \end{cases}$

Multiply Equation 2 by 3: $15x - 3y = 33$

Add this to Equation 1 to eliminate y:

$$17x = 51 \implies x = 3$$

Substitute $x = 3$ in Equation 1:

$$6 + 3y = 18 \implies y = 4$$

Solution: $(3, 4)$

35. $\begin{cases} 3x + 2y = 10 & \text{Equation 1} \\ 2x + 5y = 3 & \text{Equation 2} \end{cases}$

Multiply Equation 1 by 2 and Equation 2 by (-3):

$$\begin{cases} 6x + 4y = 20 \\ -6x - 15y = -9 \end{cases}$$

Add to eliminate x: $-11y = 11 \implies y = -1$

Substitute $y = -1$ in Equation 1:

$$3x - 2 = 10 \implies x = 4$$

Solution: $(4, -1)$

37. $\begin{cases} 5u + 6v = 24 & \text{Equation 1} \\ 3u + 5v = 18 & \text{Equation 2} \end{cases}$

Multiply Equation 1 by 5 and Equation 2 by -6:

$$\begin{cases} 25u + 30v = 120 \\ -18u - 30v = -108 \end{cases}$$

Add to eliminate v: $7u = 12 \implies u = \frac{12}{7}$

Substitute $u = \frac{12}{7}$ in Equation 1:

$$5\left(\frac{12}{7}\right) + 6v = 24 \implies 6v = \frac{108}{7} \implies v = \frac{18}{7}$$

Solution: $\left(\frac{12}{7}, \frac{18}{7}\right)$

39. $\begin{cases} 1.8x + 1.2y = 4 & \text{Equation 1} \\ 9x + 6y = 3 & \text{Equation 2} \end{cases}$

Multiply Equation 1 by 10 and Equation 2 by -2:

$\begin{cases} 18x + 12y = 40 \\ -18x - 12y = -6 \end{cases}$

Add to eliminate x and y: $0 = 34$

Inconsistent

No solution

41. $\begin{cases} \dfrac{x}{4} + \dfrac{y}{6} = 1 & \text{Equation 1} \\ x - y = 3 & \text{Equation 2} \end{cases}$

Multiply Equation 1 by 6: $\dfrac{3}{2}x + y = 6$

Add this to Equation 2 to eliminate y: $\dfrac{5}{2}x = 9 \implies x = \dfrac{18}{5}$

Substitute $x = \dfrac{18}{5}$ in Equation 2:

$$\dfrac{18}{5} - y = 3$$

$$y = \dfrac{3}{5}$$

Solution: $\left(\dfrac{18}{5}, \dfrac{3}{5}\right)$

43. $\begin{cases} 2.5x - 3y = 1.5 & \text{Equation 1} \\ 2x - 2.4y = 1.2 & \text{Equation 2} \end{cases}$

Multiply Equation 1 by 20 and Equation 2 by -25:

$50x - 60y = 30$
$-50x + 60y = -30$

Add to eliminate x and y: $0 = 0$

The equations are dependent.

There are infinitely many solutions.

Let $x = a$, then $2.5a - 3y = 1.5 \implies y = \dfrac{2.5a - 1.5}{3} = \dfrac{5}{6}a - \dfrac{1}{2}$.

Solution: $\left(a, \dfrac{5}{6}a - \dfrac{1}{2}\right)$ where a is any real number.

45. $\begin{cases} 0.05x - 0.03y = 0.21 & \text{Equation 1} \\ 0.07x + 0.02y = 0.16 & \text{Equation 2} \end{cases}$

Multiply Equation 1 by 200 and Equation 2 by 300.

$\begin{cases} 10x - 6y = 42 \\ 21x + 6y = 48 \end{cases}$

Add to eliminate y: $31x = 90 \implies x = \dfrac{90}{31}$

Back-substitute $x = \dfrac{90}{31}$ into the revised Equation 1

$10\left(\dfrac{90}{31}\right) - 6y = 42 \implies y = -\dfrac{1}{6}\left(42 - \dfrac{900}{31}\right) = -\dfrac{67}{31}$

Solution: $\left(\dfrac{90}{31}, -\dfrac{67}{31}\right)$

47. $\begin{cases} 4b + 3m = 3 & \text{Equation 1} \\ 3b + 11m = 13 & \text{Equation 2} \end{cases}$

Multiply Equation 1 by 11 and Equation 2 by -3.

$\begin{cases} 44b + 33m = 33 \\ -9b - 33m = -39 \end{cases}$

Add to eliminate y: $35b = -6 \implies b = -\dfrac{6}{35}$

Back-substitute $b = -\dfrac{6}{35}$ into Equation 1.

$4\left(-\dfrac{6}{35}\right) + 3m = 3 \implies m = \dfrac{1}{3}\left(3 + \dfrac{24}{35}\right) = \dfrac{43}{35}$

Solution: $\left(-\dfrac{6}{35}, \dfrac{43}{35}\right)$

49. $\begin{cases} \dfrac{x+3}{4} + \dfrac{y-1}{3} = 1 & \text{Equation 1} \\ 2x - y = 12 & \text{Equation 2} \end{cases}$

Multiply Equation 1 by 12 to eliminate the fractions.

$3(x+3) + 4(y-1) = 12 \implies 3x + 4y = 12 - 9 + 4 = 7$

$\begin{cases} 3x + 4y = 7 \\ 2x - y = 12 \end{cases}$

Multiply Equation 2 by 4.

$\begin{cases} 3x + 4y = 7 \\ 8x - 4y = 48 \end{cases}$

Add to eliminate y: $11x = 55 \implies x = 5$

Back-substitute $x = 5$ into Equation 2.

$2(5) - y = 12 \implies y = -(12 - 10) = -2$

Solution: $(5, -2)$

51. $\begin{cases} 2x - 5y = 0 \implies y = \frac{2}{5}x \\ x - y = 3 \implies y = x - 3 \end{cases}$

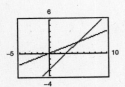

The system is consistent. There is one solution.

Solution: $(5, 2)$

53. $\begin{cases} \frac{3}{5}x - y = 3 \implies y = \frac{3}{5}x - 3 \\ -3x + 5y = 9 \implies y = \frac{3}{5}x + \frac{9}{5} \end{cases}$

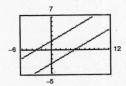

The lines are parallel. The system is inconsistent.

There are no solutions.

55. $\begin{cases} x + 7y = 2 \implies y = -\frac{1}{7}x + \frac{2}{7} \\ 4x - y = 9 \implies y = 4x - 9 \end{cases}$

The system is consistent.

There is one solution.

Solution: $\left(\frac{65}{29}, -\frac{1}{29}\right)$

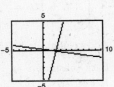

57. $\begin{cases} -x + 7y = 3 \implies y = \frac{1}{7}x + \frac{3}{7} \\ -\frac{1}{7}x + y = 5 \implies y = \frac{1}{7}x + 5 \end{cases}$

The lines are parallel.

The system is inconsistent.

There are no solutions.

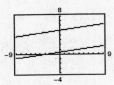

59. $\begin{cases} 8x + 9y = 42 \implies y = -\frac{8}{9}x + \frac{14}{3} \\ 6x - y = 16 \implies y = 6x - 16 \end{cases}$

The system is consistent.

There is one solution.

Solution: $(3, 2)$

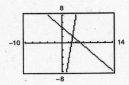

61. $\begin{cases} \frac{3}{2}x - \frac{1}{5}y = 8 \implies y = \frac{15}{2}x - 40 \\ -2x + 3y = 3 \implies y = \frac{2}{3}x + 1 \end{cases}$

The system is consistent.

There is one solution.

Solution: $(6, 5)$

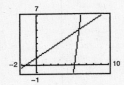

63. $\begin{cases} 0.5x + 2.2y = 9 \implies y = -\frac{5}{22}x + \frac{45}{11} \\ 6x + 0.4y = -22 \implies y = -15x - 55 \end{cases}$

The system is consistent.

There is one solution.

Solution: $(-4, 5)$

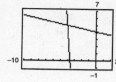

65. $\begin{cases} 3x - 5y = 7 & \text{Equation 1} \\ 2x + y = 9 & \text{Equation 2} \end{cases}$

Multiply Equation 2 by 5:

$10x + 5y = 45$

Add this to Equation 1:

$13x = 52 \implies x = 4$

Back-substitute $x = 4$ into Equation 2:

$2(4) + y = 9 \implies y = 1$

Solution: $(4, 1)$

67. $\begin{cases} y = 2x - 5 & \text{Equation 1} \\ y = 5x - 11 & \text{Equation 2} \end{cases}$

Since both equations are solved for y, set them equal to one another and solve for x.

$2x - 5 = 5x - 11$

$6 = 3x$

$2 = x$

Back-substitute $x = 2$ into Equation 1:

$y = 2(2) - 5 = -1$

Solution: $(2, -1)$

69. $\begin{cases} x - 5y = 21 & \text{Equation 1} \\ 6x + 5y = 21 & \text{Equation 2} \end{cases}$

Add the equations: $7x = 42 \implies x = 6$

Back-substitute $x = 6$ into Equation 1:

$6 - 5y = 21 \implies -5y = 15 \implies y = -3$

Solution: $(6, -3)$

71. $\begin{cases} -2x + 8y = 19 & \text{Equation 1} \\ y = x - 3 & \text{Equation 2} \end{cases}$

Substitute the expression for y from Equation 2 into Equation 1.

$-2x + 8(x - 3) = 19 \implies -2x + 8x - 24 = 19 \implies 6x = 43 \implies x = \frac{43}{6}$

Back-substitute $x = \frac{43}{6}$ into Equation 2:

$y = \frac{43}{6} - 3 \implies y = \frac{25}{6}$

Solution: $\left(\frac{43}{6}, \frac{25}{6}\right)$

73.   Demand = Supply

$50 - 0.5x = 0.125x$

$50 = 0.625x$

$x = 80$ units

$p = \$10$

Point of Equilibrium: $(80, 10)$

75.   Demand = Supply

$140 - 0.00002x = 80 + 0.00001x$

$60 = 0.00003x$

$x = 2{,}000{,}000$ units

$p = \$100$

Point of Equilibrium: $(2{,}000{,}000, 100)$

77. A solution of a system is an ordered pair that satisfies each equation in the system.

79. The advantage of the method of substitution over the graphical method is that substitution gives exact solutions but graphical solutions, may only be approximate.

81. $\begin{cases} 100y - x = 200 & \text{Equation 1} \\ 99y - x = -198 & \text{Equation 2} \end{cases}$

Subtract Equation 2 from Equation 1 to eliminate x:
$$\begin{array}{r} 100y - x = 200 \\ -99y + x = 198 \\ \hline y \qquad = 398 \end{array}$$

Substitute $y = 398$ into Equation 1: $100(398) - x = 200 \implies x = 39{,}600$

Solution: $(39{,}600, 398)$

The lines are not parallel. The scale on the axes must be changed to see the point of intersection.

83. No, it is not possible for a consistent system of linear equations to have exactly two solutions. Either the lines will intersect once or they will coincide and then the system would have infinite solutions.

85. (a) $\begin{cases} x + y = 25{,}000 \\ 0.06x + 0.085y = 2000 \end{cases}$

As the amount at 6% increases, the amount at 8.5% decreases. The amount of interest is fixed at $2000.

(b) The point of intersection occurs when $x = 5000$, so the most that can be invested at 6% and still earn $2000 per year in interest is $5000.

87. $0.06x = 0.03x + 350$

$0.03x = 350$

$x \approx \$11{,}666.67$

To make the straight commission offer the better offer, you would have to sell more than $11,666.67 per week.

89. $2l + 2w = 40 \implies l + w = 20 \implies w = 20 - l$

$lw = 96 \implies l(20 - l) = 96$

$20l - l^2 = 96$

$0 = l^2 - 20l + 96$

$0 = (l - 8)(l - 12)$

$l = 8 \text{ or } l = 12$

If $l = 8$, then $w = 12$.

If $l = 12$, then $w = 8$.

Since the length is supposed to be greater than the width, we have $l = 12$ kilometers and $w = 8$ kilometers. Dimensions: 8×12 kilometers

91. Let $r_1 = $ the air speed of the plane and $r_2 = $ the wind air speed.

$3.6(r_1 - r_2) = 1800$ Equation 1

$3(r_1 + r_2) = 1800$ Equation 2

$$\begin{array}{r} r_1 - r_2 = 500 \\ r_1 + r_2 = 600 \\ \hline 2r_1 \qquad = 1100 \\ r_1 \qquad = 550 \\ 550 + r_2 = 600 \\ r_2 = 50 \end{array}$$

The air speed of the plane is 550 mph and the speed of the wind is 50 mph.

93. Let $x = $ number of deodorant containers produced by Machine 1

$y = $ number of deodorant containers produced by Machine 2

$x + y = 1764$ Equation 1

$x = 1.8y$ Equation 2

Substitute $1.8y$ for x in Equation 1.

$1.8y + y = 1764 \implies 2.8y = 1764 \implies y = 630$

Back-substitute $y = 630$ into Equation 2.

$x = 1.8(630) \implies x = 1134$

Solution: Machine 1 produces 1134 deodorant containers and Machine 2 produces 630 deodorant containers.

95. Let x = the number of liters at 20%

Let y = the number of liters at 50%.

(a) $\begin{cases} x + y = 10 \\ 0.2x + 0.5y = 0.3(10) \end{cases}$

(c) $-2 \cdot$ Equation 1 $-2x - 2y = -20$

     $10 \cdot$ Equation 2 $\underline{2x + 5y = \;\;\; 30}$

$$3y = 10$$
$$y = \tfrac{10}{3}$$
$$x + \tfrac{10}{3} = 10$$
$$x = \tfrac{20}{3}$$

In order to obtain the specified concentration of the final mixture, $6\tfrac{2}{3}$ liters of the 20% solution and $3\tfrac{1}{3}$ liters of the 50% solution are required.

(b)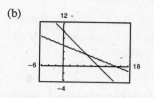

As x increases, y decreases.

97. Let x = the number of student tickets;
y = the number of adult tickets

$\begin{cases} x + y = 1435 & \text{Equation 1} \\ 1.50x + 5.00y = 3552.50 & \text{Equation 2} \end{cases}$

Multiply Equation 1 by -1.50.

$\begin{cases} -1.50x - 1.50y = -2152.50 \\ \underline{1.50x + 5.00y = \;\;\; 3552.50} \end{cases}$ Add the equations

$$3.50y = 1400.00$$
$$y = 400$$
$$x + 400 = 1435 \Longrightarrow x = 1035$$

Solution: 1035 student tickets and 400 adult tickets were sold.

99. $(1.00, 450), (1.25, 375), (1.50, 330)$

(a) $\begin{cases} 3b + 3.75a = 1155 \\ 3.75b + 4.8125a = 1413.75 \end{cases}$

By elimination we have $a = -240$ and $b = 685$.

(b) Least squares regression line:
$$y = -240x + 685$$

(c)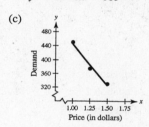

(d) $y = -240(1.40) + 685$

     $= 349$ units

101. There are infinitely many systems that have the solution $(6, 3)$. One possible system is:

$$\begin{cases} x + y = 9 \\ 3x - 2y = 12 \end{cases}$$

103. There are infinitely many systems that have the solution $\left(3, \tfrac{5}{2}\right)$. One possible system is:

$$\begin{cases} 2x + 2y = 11 \\ x - 4y = -7 \end{cases}$$

105. $\begin{cases} 4x - 8y = -3 & \text{Equation 1} \\ 2x + ky = 16 & \text{Equation 2} \end{cases}$

Multiply Equation 2 by -2: $-4x - 2ky = -32$

Add this to Equation 1: $4x - 8y = -3$
$$\underline{-4x - 2ky = -32}$$
$$-8y - 2ky = -35$$

The system is inconsistent if $-8y - 2ky = 0$. This occurs when $k = -4$.

107. False. To solve a system of equations by substitution, you can solve for either variable in one of the two equations and then back-substitute.

109. False. Two lines that coincide have infinitely many points of intersection.

111. True. If there are no points of intersection (solutions) then the lines must be parallel.

Section 14.2   Multivariable Linear Systems

1. $\begin{cases} 3x - y + z = 1 \\ 2x \quad\ - 3z = -14 \\ \quad 5y + 2z = 8 \end{cases}$

(a) $3(2) - (0) + (-3) \neq 1$

$(2, 0, -3)$ is **not** a solution.

(b) $3(-2) - (0) + 8 \neq 1$

$(-2, 0, 8)$ is **not** a solution.

(c) $3(0) - (-1) + 3 \neq 1$

$(0, -1, 3)$ is **not** a solution.

(d) $3(-1) - (0) + 4 = 1$
$2(-1) \quad\quad - 3(4) = -14$
$\quad\quad 5(0) + 2(4) = 8$

$(-1, 0, 4)$ **is** a solution.

3. $\begin{cases} 4x + y - z = 0 \\ -8x - 6y + z = -\frac{7}{4} \\ 3x - y = -\frac{9}{4} \end{cases}$

(a) $4\left(\frac{1}{2}\right) + \left(-\frac{3}{4}\right) - \left(-\frac{7}{4}\right) \neq 0$

$\left(\frac{1}{2}, -\frac{3}{4}, -\frac{7}{4}\right)$ is **not** a solution.

(b) $4\left(-\frac{3}{2}\right) + \left(\frac{5}{4}\right) - \left(-\frac{5}{4}\right) \neq 0$

$\left(-\frac{3}{2}, \frac{5}{4}, -\frac{5}{4}\right)$ is **not** a solution.

(c) $4\left(-\frac{1}{2}\right) + \left(\frac{3}{4}\right) - \left(-\frac{5}{4}\right) = 0$
$-8\left(-\frac{1}{2}\right) - 6\left(\frac{3}{4}\right) + \left(-\frac{5}{4}\right) = -\frac{7}{4}$
$3\left(-\frac{1}{2}\right) - \left(\frac{3}{4}\right) = -\frac{9}{4}$

$\left(-\frac{1}{2}, \frac{3}{4}, -\frac{5}{4}\right)$ **is** a solution.

(d) $4\left(-\frac{1}{2}\right) + \left(\frac{1}{6}\right) - \left(-\frac{3}{4}\right) \neq 0$

$\left(-\frac{1}{2}, \frac{1}{6}, -\frac{3}{4}\right)$ is **not** a solution.

5. $\begin{cases} 2x - y + 5z = 24 & \text{Equation 1} \\ y + 2z = 6 & \text{Equation 2} \\ z = 4 & \text{Equation 3} \end{cases}$

Back-substitute $z = 4$ into Equation 2.

$y + 2(4) = 6$

$y = -2$

Back-substitute $y = -2$ and $z = 4$ into Equation 1.

$2x - (-2) + 5(4) = 24$

$2x + 22 = 24$

$x = 1$

Solution: $(1, -2, 4)$

7. $\begin{cases} 2x + y - 3z = 10 & \text{Equation 1} \\ y = 2 & \text{Equation 2} \\ y - z = 4 & \text{Equation 3} \end{cases}$

Back-substitute $y = 2$ into Equation 3.

$2 - z = 4$

$z = -2$

Back-substitute $y = 2$ and $z = -2$ into Equation 1.

$2x + 2 - 3(-2) = 10$

$2x + 8 = 10$

$x = 1$

Solution: $(1, 2, -2)$

9. $\begin{cases} 4x - 2y + z = 8 & \text{Equation 1} \\ 2z = 4 & \text{Equation 2} \\ -y + z = 4 & \text{Equation 3} \end{cases}$

From Equation 2 we have $z = 2$. Back-substitute $z = 2$ into Equation 3.

$-y + 2 = 4$

$y = -2$

Back-substitute $y = -2$ and $z = 2$ into Equation 1.

$4x - 2(-2) + 2 = 8$

$4x + 6 = 8$

$x = \frac{1}{2}$

Solution: $\left(\frac{1}{2}, -2, 2\right)$

11. $\begin{cases} x + y + z = 6 \\ 2x - y + z = 3 \\ 3x - z = 0 \end{cases}$ Equation 1
Equation 2
Equation 3

$\begin{cases} x + y + z = 6 \\ -3y - z = -9 \\ -3y - 4z = -18 \end{cases}$ -2Eq.1 $+$ Eq.2
-3Eq.1 $+$ Eq.3

$\begin{cases} x + y + z = 6 \\ -3y - z = -9 \\ -3z = -9 \end{cases}$ $-$Eq.2 $+$ Eq.3

$\begin{cases} x + y + z = 6 \\ -3y - z = -9 \\ z = 3 \end{cases}$ $-\frac{1}{3}$Eq.3

$-3y - 3 = -9 \Longrightarrow y = 2$

$x + 2 + 3 = 6 \Longrightarrow x = 1$

Solution: $(1, 2, 3)$

13. $\begin{cases} 2x + 2z = 2 \\ 5x + 3y = 4 \\ 3y - 4z = 4 \end{cases}$

$\begin{cases} x + z = 1 \\ 5x + 3y = 4 \\ 3y - 4z = 4 \end{cases}$ $\frac{1}{2}$Eq.1

$\begin{cases} x + z = 1 \\ 3y - 5z = -1 \\ 3y - 4z = 4 \end{cases}$ -5Eq.1 $+$ Eq.2

$\begin{cases} x + z = 1 \\ 3y - 5z = -1 \\ z = 5 \end{cases}$ $-$Eq.2 $+$ Eq.3

$3y - 5(5) = -1 \Longrightarrow y = 8$

$x + 5 = 1 \Longrightarrow x = -4$

Solution: $(-4, 8, 5)$

15. $\begin{cases} 3x + 3y = 9 \\ 2x - 3z = 10 \\ 6y + 4z = -12 \end{cases}$ Interchange Equations.

$\begin{cases} x + y = 3 \\ 2x - 3z = 10 \\ 6y + 4z = -12 \end{cases}$ $\frac{1}{3}$Eq.1

$\begin{cases} x + y = 3 \\ -2y - 3z = 4 \\ 6y + 4z = -12 \end{cases}$ -2Eq.1 $+$ Eq.2

$\begin{cases} x + y = 3 \\ -2y - 3z = 4 \\ -5z = 0 \end{cases}$ 3Eq.2 $+$ Eq.3

$\begin{cases} x + y = 3 \\ -2y - 3z = 4 \\ z = 0 \end{cases}$ $-\frac{1}{5}$Eq.3

$-2y - 3(0) = 4 \Longrightarrow y = -2$

$x - 2 = 3 \Longrightarrow x = 5$

Solution: $(5, -2, 0)$

17. $\begin{cases} x - 2y + 2z = -9 \\ 2x + y - z = 7 \\ 3x - y + z = 5 \end{cases}$ Interchange Equations.

$\begin{cases} x - 2y + 2z = -9 \\ 5y - 5z = 25 \\ 5y - 5z = 32 \end{cases}$ -2Eq.1 $+$ Eq.2
-3Eq.1 $+$ Eq.3

$\begin{cases} x - 2y + 2z = -9 \\ 5y - 5z = 25 \\ 0 = 7 \end{cases}$ $-$Eq.2 $+$ Eq.3

Inconsistent, no solution.

19. $\begin{cases} 3x - 5y + 5z = 1 \\ 5x - 2y + 3z = 0 \\ 7x - y + 3z = 0 \end{cases}$

$\begin{cases} 6x - 10y + 10z = 2 \\ 5x - 2y + 3z = 0 \\ 7x - y + 3z = 0 \end{cases}$ 2Eq.1

$\begin{cases} x - 8y + 7z = 2 \\ 5x - 2y + 3z = 0 \\ 7x - y + 3z = 0 \end{cases}$ $-$Eq.2 $+$ Eq.1

$\begin{cases} x - 8y + 7z = 2 \\ 38y - 32z = -10 \\ 55y - 46z = -14 \end{cases}$ -5Eq.1 $+$ Eq.2
-7Eq.1 $+$ Eq.3

$\begin{cases} x - 8y + 7z = 2 \\ 2090y - 1760z = -550 \\ -2090y + 1748z = 532 \end{cases}$ 55Eq.2
-38Eq.3

$\begin{cases} x - 8y + 7z = 2 \\ 2090y - 1760z = -550 \\ -12z = -18 \end{cases}$ Eq.2 $+$ Eq.3

$-12z = -18 \Longrightarrow z = \frac{3}{2}$

$38y - 32\left(\frac{3}{2}\right) = -10 \Longrightarrow y = 1$

$x - 8(1) + 7\left(\frac{3}{2}\right) = 2 \Longrightarrow x = -\frac{1}{2}$

Solution: $\left(-\frac{1}{2}, 1, \frac{3}{2}\right)$

21. $\begin{cases} x + 2y - 7z = -4 \\ 2x + y + z = 13 \\ 3x + 9y - 36z = -33 \end{cases}$

$\begin{cases} x + 2y - 7z = -4 \\ -3y + 15z = 21 \\ 3y - 15z = -21 \end{cases}$ $\begin{matrix} \\ -2\text{Eq.1} + \text{Eq.2} \\ -3\text{Eq.1} + \text{Eq.3} \end{matrix}$

$\begin{cases} x + 2y - 7z = -4 \\ -3y + 15z = 21 \\ 0 = 0 \end{cases}$ $\begin{matrix} \\ \\ \text{Eq.2} + \text{Eq.3} \end{matrix}$

$\begin{cases} x + 2y - 7z = -4 \\ y - 5z = -7 \end{cases}$ $\begin{matrix} \\ -\frac{1}{3}\text{Eq.2} \end{matrix}$

$\begin{cases} x + 3z = 10 \\ y - 5z = -7 \end{cases}$ $\begin{matrix} -2\text{Eq.2} + \text{Eq.1} \\ \end{matrix}$

Let $z = a$, then:

$y = 5a - 7$

$x = -3a + 10$

Solution: $(-3a + 10, 5a - 7, a)$

23. $\begin{cases} 3x - 3y + 6z = 6 \\ x + 2y - z = 5 \\ 5x - 8y + 13z = 7 \end{cases}$

$\begin{cases} x - y + 2z = 2 \\ x + 2y - z = 5 \\ 5x - 8y + 13z = 7 \end{cases}$ $\frac{1}{3}\text{Eq.1}$

$\begin{cases} x - y + 2z = 2 \\ 3y - 3z = 3 \\ -3y + 3z = -3 \end{cases}$ $\begin{matrix} \\ -\text{Eq.1} + \text{Eq.2} \\ -5\text{Eq.1} + \text{Eq.3} \end{matrix}$

$\begin{cases} x - y + 2z = 2 \\ y - z = 1 \\ 0 = 0 \end{cases}$ $\begin{matrix} \\ \frac{1}{3}\text{Eq.2} \\ \text{Eq.2} + \text{Eq.3} \end{matrix}$

$\begin{cases} x + z = 3 \\ y - z = 1 \end{cases}$ $\begin{matrix} \text{Eq.2} + \text{Eq.1} \\ \end{matrix}$

Let $z = a$, then:

$y = a + 1$

$x = -a + 3$

Solution: $(-a + 3, a + 1, a)$

25. $\begin{cases} x - 2y + 5z = 2 \\ 4x - z = 0 \end{cases}$

Let $z = a$, then $x = \frac{1}{4}a$.

$\frac{1}{4}a - 2y + 5a = 2$

$a - 8y + 20a = 8$

$-8y = -21a + 8$

$y = \frac{21}{8}a - 1$

Answer: $\left(\frac{1}{4}a, \frac{21}{8}a - 1, a\right)$

To avoid fractions, we could go back and let $z = 8a$, then $4x - 8a = 0 \implies x = 2a$.

$2a - 2y + 5(8a) = 2$

$-2y + 42a = 2$

$y = 21a - 1$

Solution: $(2a, 21a - 1, 8a)$

27. $\begin{cases} 2x - 3y + z = -2 \\ -4x + 9y = 7 \end{cases}$

$\begin{cases} 2x - 3y + z = -2 \\ 3y + 2z = 3 \end{cases}$ $\begin{matrix} \\ 2\text{Eq.1} + \text{Eq.2} \end{matrix}$

$\begin{cases} 2x + 3z = 1 \\ 3y + 2z = 3 \end{cases}$ $\begin{matrix} \text{Eq.2} + \text{Eq.1} \\ \end{matrix}$

Let $x = a$, then:

$y = -\frac{2}{3}a + 1$

$x = -\frac{3}{2}a + \frac{1}{2}$

Solution: $\left(-\frac{3}{2}a + \frac{1}{2}, -\frac{2}{3}a + 1, a\right)$

29. $\begin{cases} x \qquad\quad + 3w = 4 \\ \quad 2y - z - w = 0 \\ \quad 3y \quad\;\; - 2w = 1 \\ 2x - y + 4z \quad\;\;\; = 5 \end{cases}$

$\begin{cases} x \qquad\quad + 3w = 4 \\ \quad 2y - z - w = 0 \\ \quad 3y \quad\;\; - 2w = 1 \\ \quad -y + 4z - 6w = -3 \end{cases}$ $\;\;-2\text{Eq.1} + \text{Eq.4}$

$\begin{cases} x \qquad\quad + 3w = 4 \\ \quad y - 4z + 6w = 3 \\ \quad 2y - z - w = 0 \\ \quad 3y \quad\;\; - 2w = 1 \end{cases}$ $\begin{matrix} -\text{Eq.4 and} \\ \text{interchange} \\ \text{the equations.} \end{matrix}$

$\begin{cases} x \qquad\quad + 3w = 4 \\ \quad y - 4z + 6w = 3 \\ \qquad 7z - 13w = -6 \\ \qquad 12z - 20w = -8 \end{cases}$ $\begin{matrix} -2\text{Eq.2} + \text{Eq.3} \\ -3\text{Eq.2} + \text{Eq.4} \end{matrix}$

$\begin{cases} x \qquad\quad + 3w = 4 \\ \quad y - 4z + 6w = 3 \\ \qquad z - 3w = -2 \\ \qquad 12z - 20w = -8 \end{cases}$ $\;\;-\tfrac{1}{2}\text{Eq.4} + \text{Eq.3}$

$\begin{cases} x \qquad\quad + 3w = 4 \\ \quad y - 4z + 6w = 3 \\ \qquad z - 3w = -2 \\ \qquad\quad 16w = 16 \end{cases}$ $\;\;-12\text{Eq.3} + \text{Eq.4}$

$16w = 16 \implies w = 1$

$z - 3(1) = -2 \implies z = 1$

$y - 4(1) + 6(1) = 3 \implies x = 1$

$x + 3(1) = 4 \implies x = 1$

Solution: $(1, 1, 1, 1)$

31. $\begin{cases} x \quad\;\; + 4z = 1 \\ x + y + 10z = 10 \\ 2x - y + 2z = -5 \end{cases}$

$\begin{cases} x \quad\;\; + 4z = 1 \\ \quad y + 6z = 9 \\ \quad -y - 6z = -7 \end{cases}$ $\begin{matrix} -\text{Eq.1} + \text{Eq.2} \\ -2\text{Eq.1} + \text{Eq.3} \end{matrix}$

$\begin{cases} x \quad\;\; + 4z = 1 \\ \quad y + 6z = 9 \\ \qquad 0 = 2 \end{cases}$ $\;\;\text{Eq.2} + \text{Eq.3}$

No solution, inconsistent

33. $\begin{cases} 2x + 3y \qquad = 0 \\ 4x + 3y - z = 0 \\ 8x + 3y + 3z = 0 \end{cases}$

$\begin{cases} 2x + 3y \qquad = 0 \\ \quad -3y - z = 0 \\ \quad -9y + 3z = 0 \end{cases}$ $\begin{matrix} -2\text{Eq.1} + \text{Eq.2} \\ -4\text{Eq.1} + \text{Eq.3} \end{matrix}$

$\begin{cases} 2x + 3y \qquad = 0 \\ \quad -3y - z = 0 \\ \qquad 6z = 0 \end{cases}$ $\;\;-3\text{Eq.2} + \text{Eq.3}$

$6z = 0 \implies z = 0$

$-3y - 0 = 0 \implies y = 0$

$2x + 3(0) = 0 \implies x = 0$

Solution: $(0, 0, 0)$

35. $\begin{cases} 12x + 5y + z = 0 \\ 23x + 4y - z = 0 \end{cases}$

$\begin{cases} 24x + 10y + 2z = 0 \\ 23x + 4y - z = 0 \end{cases}$ $\;\;2\text{Eq.1}$

$\begin{cases} x + 6y + 3z = 0 \\ 23x + 4y - z = 0 \end{cases}$ $\;\;-\text{Eq.2} + \text{Eq.1}$

$\begin{cases} x + 6y + 3z = 0 \\ \quad -134y - 70z = 0 \end{cases}$ $\;\;-23\text{Eq.1} + \text{Eq.2}$

$\begin{cases} x + 6y + 3z = 0 \\ \quad -67y - 35z = 0 \end{cases}$ $\;\;\tfrac{1}{2}\text{Eq.2}$

To avoid fractions, let $z = 67a$, then:

$-67y - 35(67a) = 0$

$y = -35a$

$x + 6(-35a) + 3(67a) = 0$

$x = 9a$

Solution: $(9a, -35a, 67a)$

37. $\begin{cases} x - 2y + 3z = 5 \qquad \text{Equation 1} \\ -x + 3y - 5z = 4 \qquad \text{Equation 2} \\ 2x \qquad - 3z = 0 \qquad \text{Equation 3} \end{cases}$

Add Equation 1 to Equation 2.

$\begin{cases} x - 2y + 3z = 5 \\ \quad y - 2z = 9 \\ 2x \quad\;\; - 3z = 0 \end{cases}$

This is a first step in putting the system in row-echelon form.

39. No, they are not equivalent. For the second system to be equivalent to the first, the constant in the second equation should be -11 and the coefficient of z in the third equation should be 2.

41. When using Gaussian elimination to solve a system of linear equations, a system has no solution when there is a row representing a contradictory equation such as $0 = N$, where N is a nonzero real number.

For instance:
$$x + y = 3 \qquad \text{Equation 1}$$
$$-x - y = 3 \qquad \text{Equation 2}$$

$$x + y = 0$$
$$ 0 = 6 \qquad \text{Eq.1 + Eq.2}$$

No solution

43. $y = ax^2 + bx + c$ passing through $(0, 0), (2, -2), (4, 0)$

$(0, \ 0)$: $0 = c$

$(2, -2)$: $-2 = \ 4a + 2b + c \implies -1 = 2a + b$

$(4, \ 0)$: $0 = 16a + 4b + c \implies 0 = 4a + b$

Solution: $a = \frac{1}{2}, b = -2, c = 0$

The equation of the parabola is $y = \frac{1}{2}x^2 - 2x$.

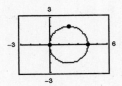

45. $y = ax^2 + bx + c$ passing through $(2, 0), (3, -1), (4, 0)$

$(2, \ 0)$: $0 = \ 4a + 2b + c$

$(3, -1)$: $-1 = \ 9a + 3b + c$

$(4, \ 0)$: $0 = 16a + 4b + c$

$$\begin{cases} 0 = \ 4a + 2b + c \\ -1 = \ 5a + b \qquad\qquad -\text{Eq.1} + \text{Eq.2} \\ 0 = 12a + 2b \qquad\qquad -\text{Eq.1} + \text{Eq.3} \end{cases}$$

$0 = \ 4a + 2b + c$

$-1 = \ 5a + b$

$2 = \ 2a \qquad\qquad\quad -2\text{Eq.2} - 3\text{Eq.}$

Solution: $a = 1, b = -6, c = 8$

The equation of the parabola is $y = x^2 - 6x + 8$.

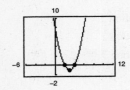

47. $x^2 + y^2 + Dx + Ey + F = 0$ passing through $(0, 0), (2, 2), (4, 0)$

$(0, 0)$: $ F = 0$

$(2, 2)$: $8 + 2D + 2E + F = 0 \implies D + E = -4$

$(4, 0)$: $16 + 4D + F = 0 \implies D = -4$ and $E = 0$

The equation of the circle is $x^2 + y^2 - 4x = 0$.

To graph, let $y_1 = \sqrt{4x - x^2}$ and $y_2 = -\sqrt{4x - x^2}$.

49. $x^2 + y^2 + Dx + Ey + F = 0$ passing through $(-3, -1), (2, 4), (-6, 8)$

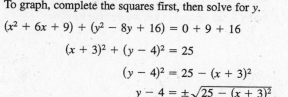

$(-3, -1)$: $10 - 3D - E + F = 0 \implies 10 = 3D + E - F$

$(\ \ 2, \ \ 4)$: $20 + 2D + 4E + F = 0 \implies 20 = -2D - 4E - F$

$(-6, \ \ 8)$: $100 - 6D + 8E + F = 0 \implies 100 = 6D - 8E - F$

Solution: $D = 6, E = -8, F = 0$

The equation of the circle is $x^2 + y^2 + 6x - 8y = 0$.

To graph, complete the squares first, then solve for y.

$$(x^2 + 6x + 9) + (y^2 - 8y + 16) = 0 + 9 + 16$$
$$(x + 3)^2 + (y - 4)^2 = 25$$
$$(y - 4)^2 = 25 - (x + 3)^2$$
$$y - 4 = \pm\sqrt{25 - (x + 3)^2}$$
$$y = 4 \pm \sqrt{25 - (x + 3)^2}$$

Let $y_1 = 4 + \sqrt{25 - (x + 3)^2}$ and $y_2 = 4 - \sqrt{25 - (x + 3)^2}$.

51. $s = \frac{1}{2}at^2 + v_0 t + s_0$

$(1, 128), (2, 80), (3, 0)$

$128 = \frac{1}{2}a + v_0 + s_0 \implies a + 2v_0 + 2s_0 = 256$

$80 = 2a + 2v_0 + s_0 \implies 2a + 2v_0 + s_0 = 80$

$0 = \frac{9}{2}a + 3v_0 + s_0 \implies 9a + 6v_0 + 2s_0 = 0$

Solving this system yields $a = -32, v_0 = 0, s_0 = 144$.

Thus, $s = \frac{1}{2}(-32)t^2 + (0)t + 144$

$= -16t^2 + 144$.

53. $s = \frac{1}{2}at^2 + v_0 t + s_0$

$(1, 452), (2, 372), (3, 260)$

$452 = \frac{1}{2}a + v_0 + s_0 \implies a + 2v_0 + 2s_0 = 904$

$372 = 2a + 2v_0 + s_0 \implies 2a + 2v_0 + s_0 = 372$

$260 = \frac{9}{2}a + 3v_0 + s_0 \implies 9a + 6v_0 + 2s_0 = 520$

Solving this system yields $a = -32, v_0 = -32, s_0 = 500$.

Thus, $s = \frac{1}{2}(-32)t^2 + (-32)t + 500$

$= -16t^2 - 32t + 500$.

55. Let $x =$ number of touchdowns.
Let $y =$ number of extra-point kicks.
Let $z =$ number of field goals.

$$\begin{cases} x + y + z = 9 \\ 6x + y + 3z = 30 \\ x - y = 0 \end{cases}$$

$$\begin{cases} x + y + z = 9 \\ -5y - 3z = -24 \\ -2y - z = -9 \end{cases} \quad \begin{matrix} -6\text{Eq.1} + \text{Eq.2} \\ -\text{Eq.1} + \text{Eq.3} \end{matrix}$$

$$\begin{cases} x + y + z = 9 \\ y + \frac{3}{5}z = \frac{24}{5} \\ -2y - z = -9 \end{cases} \quad -\frac{1}{5}\text{Eq.2}$$

$$\begin{cases} x + y + z = 9 \\ y + \frac{3}{5}z = \frac{24}{5} \\ \frac{1}{5}z = \frac{3}{5} \end{cases} \quad 2\text{Eq.2} + \text{Eq.3}$$

$\frac{1}{5}z = \frac{3}{5} \implies z = 3$

$y + \frac{3}{5}(3) = \frac{24}{5} \implies y = 3$

$x + 3 + 3 = 9 \implies x = 3$

Thus, 3 touchdowns, 3 extra-kick points, and 3 field goals were scored.

57. Let $x =$ amount at 8%.
Let $y =$ amount at 9%.
Let $z =$ amount at 10%.

$$\begin{cases} x + y + z = 775{,}000 \\ 0.08x + 0.09y + 0.10z = 67{,}500 \\ x = 4z \end{cases}$$

$y + 5z = 775{,}000$ Substitute $4z$ for x in

$0.09y + 0.42z = 67{,}500$ Eq.1 and Eq.2.

$z = 75{,}000$

$y = 775{,}000 - 5z = 400{,}000$

$x = 4z = 300{,}000$

$300{,}000 was borrowed at 8%.

$400{,}000 was borrowed at 9%.

$75{,}000 was borrowed at 10%.

59. Let C = amount in certificates of deposit.
Let M = amount in municipal bonds.
Let B = amount in blue-chip stocks.
Let G = amount in growth or speculative stocks.

$$\begin{cases} C + M + B + G = 500{,}000 \\ 0.10C + 0.08M + 0.12B + 0.13G = 0.10(500{,}000) \\ B + G = \tfrac{1}{4}(500{,}000) \end{cases}$$

This system has infinitely many solutions.

Let $G = s$, then $B = 125{,}000 - s$

$$M = 125{,}000 + \tfrac{1}{2}s$$
$$C = 250{,}000 - \tfrac{1}{2}s.$$

One possible solution is to let $s = 50{,}000$.

Certificates of deposit: $225,000

Municipal bonds: $150,000

Blue-chip stocks: $75,000

Growth or speculative stocks: $50,000

61. Let x = liters of spray X.

Let y = liters of spray Y.

Let z = liters of spray Z.

Chemical A: $\tfrac{1}{5}x + \tfrac{1}{2}z = 12$
Chemical B: $\tfrac{2}{5}x + \tfrac{1}{2}z = 16$ $\Big\}$ $\Rightarrow x = 20, z = 16$

Chemical C: $\tfrac{2}{5}x + \ y = 26$ $\Rightarrow y = 18$

20 liters of spray X, 18 liters of spray Y, and 16 liters of spray Z are needed to get the desired mixture.

63.

	Product	
Truck	A	B
Large	6	3
Medium	4	4
Small	0	3

Let x = number of large trucks.

Let y = number of medium trucks.

Let z = number of small trucks.

$$\begin{cases} 6x + 4y \geq 15 \\ 3x + 4y + 3z \geq 16 \end{cases}$$

Possible solutions:

(1) 4 medium trucks

(2) 2 large trucks, 1 medium truck, 2 small trucks

(3) 3 large trucks, 1 medium truck, 1 small truck

(4) 3 large trucks, 3 small trucks

65. (a) To use as little of the 50% solution as possible, the chemist should use no 10% solution.

$$x(0.20) + (10 - x)(0.50) = 10(0.25)$$
$$x(0.20) + 5 - 0.50x = 2.5$$
$$-0.30x = -2.5$$
$$x = 8\tfrac{1}{3} \text{ liters of 20\% solution}$$
$$10 - x = 1\tfrac{2}{3} \text{ liters of 50\% solution}$$

(b) To use as much 50% solution as possible, the chemist should use no 20% solution.

$$x(0.10) + (10 - x)0.50 = 10(0.25)$$
$$0.10x + 5 - 0.50x = 2.5$$
$$-0.40x = -2.5$$
$$x = 6\tfrac{1}{4} \text{ liters of 10\% solution}$$
$$10 - x = 3\tfrac{3}{4} \text{ liters of 50\% solution}$$

(c) To use 2 liters of 50% solution we let x = the number of liters at 10% and y = the number of liters at 20%.

$$0.10x + 0.20y + 2(0.50) = 10(0.25) \qquad \text{Equation 1}$$
$$x + y = 8 \qquad \text{Equation 2}$$

Solution: $y = 7$ liters of 20% solution; $x = 1$ liter of 10% solution

67.
$$\begin{cases} t_1 - 2t_2 & = 0 \\ t_1 & - 2a = 128 \\ & t_2 + 2a = 64 \end{cases}$$
Equation 1
Equation 2
Equation 3

$$\begin{cases} t_1 - 2t_2 & = 0 \\ 2t_2 - 2a = 128 \\ t_2 + 2a = 64 \end{cases} \quad (-1)\text{Eq.1} + \text{Eq.2}$$

$$\begin{cases} t_1 - 2t_2 & = 0 \\ 2t_2 - 2a = 128 \\ 3a = 0 \end{cases} \quad \left(-\tfrac{1}{2}\right)\text{Eq.2} + \text{Eq.3}$$

$3a = 0 \implies a = 0$
$2t_2 - 2(0) = 128 \implies t_2 = 64$
$t_1 - 2(64) = 0 \implies t_1 = 128$

Solution: $a = 0$ ft/sec^2

$\qquad t_1 = 128$ lb

$\qquad t_2 = 64$ lb

69. (a) $(100, 75), (120, 68), (140, 55)$

$$\begin{cases} c + 100b + 10{,}000a = 75 \\ c + 120b + 14{,}400a = 68 \\ c + 140b + 19{,}600a = 55 \end{cases}$$
Equation 1
Equation 2
Equation 3

$$\begin{cases} c + 100b + 10{,}000a = 75 \\ 20b + 4400a = -7 \\ 40b + 9600a = -20 \end{cases} \begin{array}{l} \\ (-1)\text{Eq.1} + \text{Eq.2} \\ (-1)\text{Eq.1} + \text{Eq.3} \end{array}$$

$$\begin{cases} c + 100b + 10{,}000a = 75 \\ 20b + 4400a = -7 \\ 800a = -6 \end{cases} \begin{array}{l} \\ \\ (-2)\text{Eq.2} + \text{Eq.3} \end{array}$$

$$\begin{cases} c + 100b + 10{,}000a = 75 \\ b + 220a = -\tfrac{7}{20} \\ a = -\tfrac{6}{800} \end{cases} \begin{array}{l} \\ \left(\tfrac{1}{20}\right)\text{Eq.2} \\ \left(\tfrac{1}{800}\right)\text{Eq.2} \end{array}$$

$a = -\tfrac{6}{800} = -0.0075$

$b + 220(-0.0075) = -\tfrac{7}{20} \implies b = \tfrac{26}{20} = 1.3$

$c + 100(1.3) + 10{,}000(-0.0075) = 75 \implies c = 20$

So, $y = -0.0075x^2 + 1.3x + 20$

(b)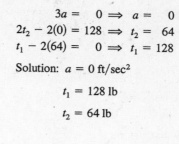

(c) For $x = 170$:

$\quad y = -0.0075(170)^2 + 1.3(170) + 20$

$\quad\ \ = 24.25\%$

71. There are an infinite number of linear systems that have $(4, -1, 2)$ as their solution. One such system is as follows:

$$\begin{cases} 3x + y - z = 9 \\ x + 2y - z = 0 \\ -x + y + 3z = 1 \end{cases}$$

73. There are an infinite number of linear systems that have $\left(3, -\tfrac{1}{2}, \tfrac{7}{4}\right)$ as their solution. One such system is as follows:

$$\begin{cases} x + 2y - 4z = -5 \\ -x - 4y + 8z = 13 \\ x + 6y + 4z = 7 \end{cases}$$

75.
$$\begin{cases} y + \lambda = 0 \\ x + \lambda = 0 \\ x + y - 10 = 0 \end{cases} \begin{array}{l} \Big\} \implies x = y = -\lambda \\ \implies 2x - 10 = 0 \\ \qquad\quad x = 5 \\ \qquad\quad y = 5 \\ \qquad\quad \lambda = -5 \end{array}$$

77. $\begin{cases} 2x - 2x\lambda = 0 \implies 2x(1 - \lambda) = 0 \implies \lambda = 1 \text{ or } x = 0 \\ -2y + \lambda = 0 \\ y - x^2 = 0 \end{cases}$

If $\lambda = 1$:

$$2y = \lambda \implies y = \frac{1}{2}$$

$$x^2 = y \implies x = \pm\sqrt{\frac{1}{2}} = \pm\frac{\sqrt{2}}{2}$$

If $x = 0$:

$$x^2 = y \implies y = 0$$

$$2y = \lambda \implies \lambda = 0$$

Solution: $x = \pm\dfrac{\sqrt{2}}{2}$ or $x = 0$

$y = \dfrac{1}{2}$ $y = 0$

$\lambda = 1$ $\lambda = 0$

79. False. Equation 2 does not have a leading coefficient of 1.

81. False. A linear system of three equations with three unknowns can have exactly one solution, infinitely many solutions, or no solution. In the case of no solution, the system would be inconsistant.

Section 14.3   Systems of Inequalities

1. $x \geq 2$

Using a solid line, graph the vertical line $x = 2$ and shade to the right of this line.

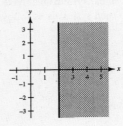

3. $y \geq -1$

Using a solid line, graph the horizontal line $y = -1$ and shade above this line.

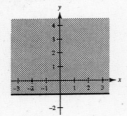

5. $y < 2 - x$

Using a dashed line, graph $y = 2 - x$, and then shade below the line. (Use $(0, 0)$ as a test point.)

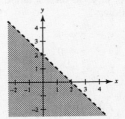

7. $2y - x \geq 4$

Using a solid line, graph $2y - x = 4$, and then shade above the line. (Use $(0, 0)$ as a test point.)

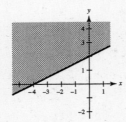

9. $(x + 1)^2 + (y - 2)^2 < 9$

Using a dashed line, sketch the circle $(x + 1)^2 + (y - 2)^2 = 9$.

Center: $(-1, 2)$

Radius: 3

Test point: $(0, 0)$. Shade the inside of the circle.

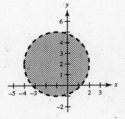

11. $y \leq \dfrac{1}{1 + x^2}$

Using a solid line,

graph $y = \dfrac{1}{1 + x^2}$, and then

shade below the curve.
(Use $(0, 0)$ as a test point.)

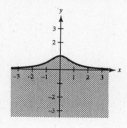

13. $y < \ln x$

Using a dashed line, sketch
$y = \ln x$, and then shade
below the curve.
(Use $(e, 0)$ as a test point.)

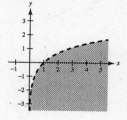

15. $y < 3^{-x-4}$

Using a dashed line, sketch
$y = 3^{-x-4}$, and then shade
below the curve.
(Use $(0, 0)$ as a test point.)

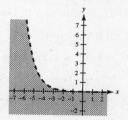

17. $y \geq \dfrac{2}{3}x - 1$

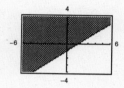

19. $y < -3.8x + 1.1$

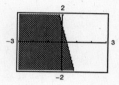

21. $x^2 + 5y - 10 \leq 0$

$$y \leq 2 - \dfrac{x^2}{5}$$

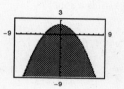

23. $\dfrac{5}{2}y - 3x^2 - 6 \geq 0$

$$y \geq \dfrac{2}{5}(3x^2 + 6)$$

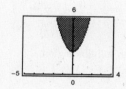

25. The line through $(-4, 0)$ and
$(0, 2)$ is $y = \frac{1}{2}x + 2$. For the
shaded region below the line,
we have $y \leq \frac{1}{2}x + 2$.

27. The line through $(0, 2)$ and $(3, 0)$
is $y = -\frac{2}{3}x + 2$. For the shaded
region above the line, we have
$y \geq -\frac{2}{3}x + 2$.

29. $\begin{cases} x \geq -4 \\ y > -3 \\ y \leq -8x - 3 \end{cases}$

(a) $0 \leq -8(0) - 3$, False

$(0, 0)$ is **not** a solution.

(c) $-4 \geq -4$, True

$0 > -3$, True

$0 \leq -8(-4) - 3$, True

$(-4, 0)$ **is** a solution.

(b) $-3 > -3$, False

$(-1, -3)$ is **not** a solution.

(d) $-3 \geq -4$, True

$11 > -3$, True

$11 \leq -8(-3) - 3$, True

$(-3, 11)$ **is** a solution.

31. $\begin{cases} 3x + y > 1 \\ -y - \frac{1}{2}x^2 \leq -4 \\ -15x + 4y > 0 \end{cases}$

(a) $3(0) + (10) > 1$, True

$-10 - \frac{1}{2}(0)^2 \leq -4$, True

$-15(0) + 4(10) > 0$, True

$(0, 10)$ **is** a solution.

(b) $3(0) + (-1) > 1$, False $\Rightarrow$ $(0, -1)$ is **not** a solution.

(c) $3(2) + (9) > 1$, True

$-9 - \frac{1}{2}(2)^2 \leq -4$, True

$-15(2) + 4(9) > 0$, True

$(2, 9)$ **is** a solution.

(d) $3(-1) + 6 > 1$, True

$-6 - \frac{1}{2}(-1)^2 \leq -4$, True

$-15(-1) + 4(6) > 0$, True

$(-1, 6)$ **is** a solution.

33. $\begin{cases} x + y \leq 1 \\ -x + y \leq 1 \\ y \geq 0 \end{cases}$

First, find the points of intersection of each pair of equations.

Vertex A	Vertex B	Vertex C
$x + y = 1$	$x + y = 1$	$-x + y = 1$
$-x + y = 1$	$y = 0$	$y = 0$
$(0, 1)$	$(1, 0)$	$(-1, 0)$

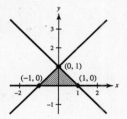

35. $\begin{cases} x^2 + y \leq 5 \\ x \geq -1 \\ y \geq 0 \end{cases}$

First, find the points of intersection of each pair of equations.

Vertex A	Vertex B	Vertex C
$x^2 + y = 5$	$x^2 + y = 5$	$x = -1$
$x = -1$	$y = 0$	$y = 0$
$(-1, 4)$	$(\pm\sqrt{5}, 0)$	$(-1, 0)$

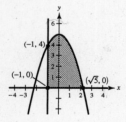

37. $\begin{cases} -3x + 2y < 6 \\ x - 4y > -2 \\ 2x + y < 3 \end{cases}$

First, find the points of intersection of each pair of equations.

Vertex A	Point B	Vertex C
$-3x + 2y = 6$	$-3x + 2y = 6$	$x - 4y = -2$
$x - 4y = -2$	$2x + y = 3$	$2x + y = 3$
$(-2, 0)$	$(0, 3)$	$\left(\frac{10}{9}, \frac{7}{9}\right)$

Note that B is not a vertex of the solution region.

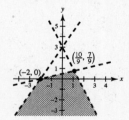

39. $\begin{cases} 2x + y > 2 \\ 6x + 3y < 2 \end{cases}$

The lines are parallel. There are no points of intersection. There is no region common to both inequalities.

The system has no solution.

41. $\begin{cases} x > y^2 \\ x < y + 2 \end{cases}$

Points of intersection:

$$y^2 = y + 2$$
$$y^2 - y - 2 = 0$$
$$(y + 1)(y - 2) = 0$$
$$y = -1, 2$$
$$(1, -1), (4, 2)$$

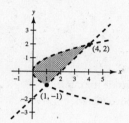

43. $\begin{cases} x^2 + y^2 \le 9 \\ x^2 + y^2 \ge 1 \end{cases}$

There are no points of intersection. The region common to both inequalities is the region between the circles.

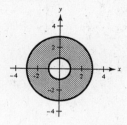

45. $3x + 4 \ge y^2$

$x - y < 0$

Points of intersection:

$$x - y = 0 \implies y = x$$
$$3y + 4 = y^2$$
$$0 = y^2 - 3y - 4$$
$$0 = (y - 4)(y + 1)$$
$$y = 4 \text{ or } y = -1$$
$$x = 4 \qquad x = -1$$
$$(4, 4) \text{ and } (-1, -1)$$

47. $\begin{cases} y \le \sqrt{3x} + 1 \\ y \ge x^2 + 1 \end{cases}$

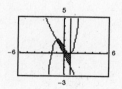

49. $\begin{cases} y < x^3 - 2x + 1 \\ y > -2x \\ x \le 1 \end{cases}$

51. $\begin{cases} x^2 y \ge 1 \implies y \ge \dfrac{1}{x^2} \\ 0 < x \le 4 \\ y \le 4 \end{cases}$

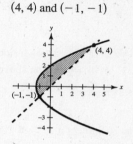

53. $\begin{cases} y \le \sqrt{x} \\ y \ge x^2 \end{cases}$

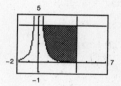

55. $\begin{cases} y \le 4 - x \\ x \ge 0 \\ y \ge 0 \end{cases}$

57. Line through points $(0, 4)$ and $(4, 0)$: $y = 4 - x$

Line through points $(0, 2)$ and $(8, 0)$: $y = 2 - \frac{1}{4}x$

$$\begin{cases} y \ge 4 - x \\ y \ge 2 - \frac{1}{4}x \\ x \ge 0 \\ y \ge 0 \end{cases}$$

59. $\begin{cases} x^2 + y^2 \le 16 \\ x \ge 0 \\ y \ge 0 \end{cases}$

61. Rectangular region with vertices at
$(2, 1), (5, 1), (5, 7),$ and $(2, 7)$

$$\begin{cases} x \geq 2 \\ x \leq 5 \\ y \geq 1 \\ y \leq 7 \end{cases}$$

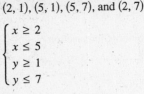

This system may be written as:

$$\begin{cases} 2 \leq x \leq 5 \\ 1 \leq y \leq 7 \end{cases}$$

63. Triangle with vertices at $(0, 0), (5, 0), (2, 3)$

$(0, 0), (5, 0)$ Line: $y = 0$

$(0, 0), (2, 3)$ Line: $y = \frac{3}{2}x$

$(2, 3), (5, 0)$ Line: $y = -x + 5$

$$\begin{cases} y \leq \frac{3}{2}x \\ y \leq -x + 5 \\ y \geq 0 \end{cases}$$

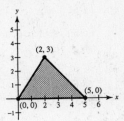

65. $\begin{cases} x^2 + y^2 \leq 16 \implies \text{region inside the circle} \\ x + y \geq 4 \implies \text{region above the line} \end{cases}$

Matches graph (d).

67. $\begin{cases} x^2 + y^2 \geq 16 \implies \text{region outside the circle} \\ x + y \geq 4 \implies \text{region above the line} \end{cases}$

Matches graph (c).

69. (a) The solution of $x + 2y \leq 6$ would include the points
on the line $x + 2y = 6$. The line would be drawn solid
in the graph and shading remains the same.

(b) Shading would be above the dashed line in the graph
of $x + 2y > 6$.

71. Demand = Supply

$$50 - 0.5x = 0.125x$$

$$50 = 0.625x$$

$$80 = x$$

$$10 = p$$

Point of equilibrium:
$(80, 10)$

The consumer surplus is the area of the triangular region
defined by

$$\begin{cases} p \leq 50 - 0.5x \\ p \geq 10 \\ x \geq 0. \end{cases}$$

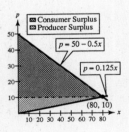

Consumer surplus $= \frac{1}{2}(\text{base})(\text{height}) = \frac{1}{2}(80)(40) = \$1600$

The producer surplus is the area of the triangular region
defined by

$$\begin{cases} p \geq 0.125x \\ p \leq 10 \\ x \geq 0. \end{cases}$$

Producer surplus $= \frac{1}{2}(\text{base})(\text{height}) = \frac{1}{2}(80)(10) = \$400$

73.   Demand = Supply

$$140 - 0.00002x = 80 + 0.00001x$$

$$60 = 0.00003x$$

$$2{,}000{,}000 = x$$

$$100 = p$$

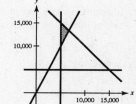

Point of equilibrium: (2,000,000, 100)

The consumer surplus is the area of the triangular region defined by

$$\begin{cases} p \le 140 - 0.00002x \\ p \ge 100 \\ x \ge \quad 0. \end{cases}$$

Consumer surplus = $\frac{1}{2}$(base)(height) = $\frac{1}{2}$(2,000,000)(40) = \$40,000,000 or \$40 million

The producer surplus is the area of the triangular region defined by

$$\begin{cases} p \ge \quad 80 + 0.00001x \\ p \le 100 \\ x \ge \quad 0. \end{cases}$$

Producer surplus = $\frac{1}{2}$(base)(height) = $\frac{1}{2}$(2,000,000)(20) = \$20,000,000 or \$20 million

75. x = number of tables
y = number of chairs

$$\begin{cases} x + \frac{3}{2}y \le 12 & \text{Assembly center} \\ \frac{4}{3}x + \frac{3}{2}y \le 15 & \text{Finishing center} \\ \quad x \ge \quad 0 \\ \quad y \ge \quad 0 \end{cases}$$

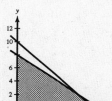

77. x = amount in smaller account
y = amount in larger account

Account constraints:

$$\begin{cases} x + y \le 20{,}000 \\ \quad y \ge \quad 2x \\ \quad x \ge \quad 5{,}000 \\ \quad y \ge \quad 5{,}000 \end{cases}$$

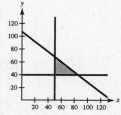

79. x = number of packages of gravel
y = number of bags of stone

$$\begin{cases} 55x + 70y \le 7500 & \text{Weight} \\ \quad x \ge \quad 50 \\ \quad y \ge \quad 40 \end{cases}$$

81. $\begin{cases} xy \geq 500 \\ 2x + \pi y \geq 125 \\ x \geq 0 \\ y \geq 0 \end{cases}$ Body = building space

Track (Two semi–circles and two lengths)

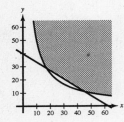

83. True. The figure is a rectangle with length of 9 units and width of 11 units.

85. False.

The graph shows the solution to the following system:

$\begin{cases} y \geq x^2 \\ x \geq 0 \\ y \leq 2 \end{cases}$

Section 14.4   Matrices and Systems of Equations

1. Since the matrix has one row and two columns, its order is 1×2.

3. Since the matrix has three rows and one column, its order is 3×1.

5. Since the matrix has two rows and two columns, its order is 2×2.

7. $\begin{cases} 4x - 3y = -5 \\ -x + 3y = 12 \end{cases}$

$\begin{bmatrix} 4 & -3 & \vdots & -5 \\ -1 & 3 & \vdots & 12 \end{bmatrix}$

9. $\begin{cases} x + 10y - 2z = 2 \\ 5x - 3y + 4z = 0 \\ 2x + y = 6 \end{cases}$

$\begin{bmatrix} 1 & 10 & -2 & \vdots & 2 \\ 5 & -3 & 4 & \vdots & 0 \\ 2 & 1 & 0 & \vdots & 6 \end{bmatrix}$

11. $\begin{cases} 7x - 5y + z = 13 \\ 19x - 8z = 10 \end{cases}$

$\begin{bmatrix} 7 & -5 & 1 & \vdots & 13 \\ 19 & 0 & -8 & \vdots & 10 \end{bmatrix}$

13. $\begin{bmatrix} 1 & 2 & \vdots & 7 \\ 2 & -3 & \vdots & 4 \end{bmatrix}$

$\begin{cases} x + 2y = 7 \\ 2x - 3y = 4 \end{cases}$

15. $\begin{bmatrix} 2 & 0 & 5 & \vdots & -12 \\ 0 & 1 & -2 & \vdots & 7 \\ 6 & 3 & 0 & \vdots & 2 \end{bmatrix}$

$\begin{cases} 2x + 5z = -12 \\ y - 2z = 7 \\ 6x + 3y = 2 \end{cases}$

17. $\begin{bmatrix} 9 & 12 & 3 & 0 & \vdots & 0 \\ -2 & 18 & 5 & 2 & \vdots & 10 \\ 1 & 7 & -8 & 0 & \vdots & -4 \\ 3 & 0 & 2 & 0 & \vdots & -10 \end{bmatrix}$

$\begin{cases} 9x + 12y + 3z = 0 \\ -2x + 18y + 5z + 2w = 10 \\ x + 7y - 8z = -4 \\ 3x + 2z = -10 \end{cases}$

19. $\begin{bmatrix} 1 & 0 & 0 & 0 \\ 0 & 1 & 1 & 5 \\ 0 & 0 & 0 & 0 \end{bmatrix}$

This matrix is in reduced row–echelon form.

21. $\begin{bmatrix} 2 & 0 & 4 & 0 \\ 0 & -1 & 3 & 6 \\ 0 & 0 & 1 & 5 \end{bmatrix}$

The first nonzero entries in rows one and two are not one. The matrix is not in row–echelon form.

23. $\begin{bmatrix} 1 & 4 & 3 \\ 2 & 10 & 5 \end{bmatrix}$

$-2R_1 + R_2 \rightarrow \begin{bmatrix} 1 & 4 & 3 \\ 0 & \boxed{2} & -1 \end{bmatrix}$

25.
$$\begin{bmatrix} 1 & 1 & 4 & -1 \\ 3 & 8 & 10 & 3 \\ -2 & 1 & 12 & 6 \end{bmatrix}$$

$$\begin{matrix} -3R_1 + R_2 \rightarrow \\ 2R_1 + R_3 \rightarrow \end{matrix} \begin{bmatrix} 1 & 1 & 4 & -1 \\ 0 & 5 & \boxed{-2} & \boxed{6} \\ 0 & 3 & \boxed{20} & \boxed{4} \end{bmatrix}$$

$$\tfrac{1}{5}R_2 \rightarrow \begin{bmatrix} 1 & 1 & 4 & -1 \\ 0 & 1 & -\tfrac{2}{5} & \tfrac{6}{5} \\ 0 & 3 & \boxed{20} & \boxed{4} \end{bmatrix}$$

27. $\begin{bmatrix} -2 & 5 & 1 \\ 3 & -1 & -8 \end{bmatrix} \rightarrow \begin{bmatrix} 13 & 0 & -39 \\ 3 & -1 & -8 \end{bmatrix}$

Add five times Row 2 to Row 1.

29. $\begin{bmatrix} 0 & -1 & -5 & 5 \\ -1 & 3 & -7 & 6 \\ 4 & -5 & 1 & 3 \end{bmatrix} \rightarrow \begin{bmatrix} -1 & 3 & -7 & 6 \\ 0 & -1 & -5 & 5 \\ 0 & 7 & -27 & 27 \end{bmatrix}$

Interchange Row 1 and Row 2. Then add four times the new Row 1 to Row 3.

31. $\begin{bmatrix} 1 & 2 & 3 \\ 2 & -1 & -4 \\ 3 & 1 & -1 \end{bmatrix}$

(a) $\begin{bmatrix} 1 & 2 & 3 \\ 0 & -5 & -10 \\ 3 & 1 & -1 \end{bmatrix}$

(b) $\begin{bmatrix} 1 & 2 & 3 \\ 0 & -5 & -10 \\ 0 & -5 & -10 \end{bmatrix}$

(c) $\begin{bmatrix} 1 & 2 & 3 \\ 0 & -5 & -10 \\ 0 & 0 & 0 \end{bmatrix}$

(d) $\begin{bmatrix} 1 & 2 & 3 \\ 0 & 1 & 2 \\ 0 & 0 & 0 \end{bmatrix}$

(e) $\begin{bmatrix} 1 & 0 & -1 \\ 0 & 1 & 2 \\ 0 & 0 & 0 \end{bmatrix}$

This system is in reduced row-echelon form.

33.
$$\begin{bmatrix} 1 & 1 & 0 & 5 \\ -2 & -1 & 2 & -10 \\ 3 & 6 & 7 & 14 \end{bmatrix}$$

$$\begin{matrix} 2R_1 + R_2 \rightarrow \\ -3R_1 + R_3 \rightarrow \end{matrix} \begin{bmatrix} 1 & 1 & 0 & 5 \\ 0 & 1 & 2 & 0 \\ 0 & 3 & 7 & -1 \end{bmatrix}$$

$$-3R_2 + R_3 \rightarrow \begin{bmatrix} 1 & 1 & 0 & 5 \\ 0 & 1 & 2 & 0 \\ 0 & 0 & 1 & -1 \end{bmatrix}$$

35.
$$\begin{bmatrix} 1 & -1 & -1 & 1 \\ 5 & -4 & 1 & 8 \\ -6 & 8 & 18 & 0 \end{bmatrix}$$

$$\begin{matrix} -5R_1 + R_2 \rightarrow \\ 6R_1 + R_3 \rightarrow \end{matrix} \begin{bmatrix} 1 & -1 & -1 & 1 \\ 0 & 1 & 6 & 3 \\ 0 & 2 & 12 & 6 \end{bmatrix}$$

$$-2R_2 + R_3 \rightarrow \begin{bmatrix} 1 & -1 & -1 & 1 \\ 0 & 1 & 6 & 3 \\ 0 & 0 & 0 & 0 \end{bmatrix}$$

37. Use the reduced row–echelon form feature of a graphing utility.

$$\begin{bmatrix} 3 & 3 & 3 \\ -1 & 0 & -4 \\ 2 & 4 & -2 \end{bmatrix} \Rightarrow \begin{bmatrix} 1 & 0 & 0 \\ 0 & 1 & 0 \\ 0 & 0 & 1 \end{bmatrix}$$

39. Use the reduced row–echelon form feature of a graphing utility.

$$\begin{bmatrix} 1 & 2 & 3 & -5 \\ 1 & 2 & 4 & -9 \\ -2 & -4 & -4 & 3 \\ 4 & 8 & 11 & -14 \end{bmatrix} \Rightarrow \begin{bmatrix} 1 & 2 & 0 & 0 \\ 0 & 0 & 1 & 0 \\ 0 & 0 & 0 & 1 \\ 0 & 0 & 0 & 0 \end{bmatrix}$$

41. Use the reduced row–echelon form feature of a graphing utility.

$$\begin{bmatrix} -3 & 5 & 1 & 12 \\ 1 & -1 & 1 & 4 \end{bmatrix} \Rightarrow \begin{bmatrix} 1 & 0 & 3 & 16 \\ 0 & 1 & 2 & 12 \end{bmatrix}$$

43. $\begin{cases} x - 2y = 4 \\ y = -3 \end{cases}$

$x - 2(-3) = 4$

$x = -2$

Solution: $(-2, -3)$

45. $\begin{cases} x - y + 2z = 4 \\ y - z = 2 \\ z = -2 \end{cases}$

$y - (-2) = 2$

$y = 0$

$x - 0 + 2(-2) = 4$

$x = 8$

Solution: $(8, 0, -2)$

47. $\begin{bmatrix} 1 & 0 & \vdots & 3 \\ 0 & 1 & \vdots & -4 \end{bmatrix}$

$x = 3$

$y = -4$

Solution: $(3, -4)$

49. $\begin{bmatrix} 1 & 0 & 0 & \vdots & -4 \\ 0 & 1 & 0 & \vdots & -10 \\ 0 & 0 & 1 & \vdots & 4 \end{bmatrix}$

$x = -4$

$y = -10$

$z = 4$

Solution: $(-4, -10, 4)$

51. $\begin{cases} x + 2y = 7 \\ 2x + y = 8 \end{cases}$

$\begin{bmatrix} 1 & 2 & \vdots & 7 \\ 2 & 1 & \vdots & 8 \end{bmatrix}$

$-2R_1 + R_2 \rightarrow \begin{bmatrix} 1 & 2 & \vdots & 7 \\ 0 & -3 & \vdots & -6 \end{bmatrix}$

$-\frac{1}{3}R_2 \rightarrow \begin{bmatrix} 1 & 2 & \vdots & 7 \\ 0 & 1 & \vdots & 2 \end{bmatrix}$

$\begin{cases} x + 2y = 7 \\ y = 2 \end{cases}$

$y = 2$

$x + 2(2) = 7 \implies x = 3$

Solution: $(3, 2)$

53. $\begin{cases} 3x - 2y = -27 \\ x + 3y = 13 \end{cases}$

$\begin{bmatrix} 3 & -2 & \vdots & -27 \\ 1 & 3 & \vdots & 13 \end{bmatrix}$

$\begin{matrix} R_1 \\ R_2 \end{matrix} \begin{bmatrix} 1 & 3 & \vdots & 13 \\ 3 & -2 & \vdots & -27 \end{bmatrix}$

$-3R_1 + R_2 \rightarrow \begin{bmatrix} 1 & 3 & & 13 \\ 0 & -11 & & -66 \end{bmatrix}$

$-\frac{1}{11}R_2 \rightarrow \begin{bmatrix} 1 & 3 & \vdots & 13 \\ 0 & 1 & \vdots & 6 \end{bmatrix}$

$\begin{cases} x + 3y = 13 \\ y = 6 \end{cases}$

$y = 6$

$x + 3(6) = 13 \implies x = -5$

Solution: $(-5, 6)$

55. $\begin{cases} -2x + 6y = -22 \\ x + 2y = -9 \end{cases}$

$\begin{bmatrix} -2 & 6 & \vdots & -22 \\ 1 & 2 & \vdots & -9 \end{bmatrix}$

$\begin{matrix} R_1 \\ R_2 \end{matrix} \begin{bmatrix} 1 & 2 & \vdots & -9 \\ -2 & 6 & \vdots & -22 \end{bmatrix}$

$2R_1 + R_2 \rightarrow \begin{bmatrix} 1 & 2 & \vdots & -9 \\ 0 & 10 & \vdots & -40 \end{bmatrix}$

$\frac{1}{10}R_2 \rightarrow \begin{bmatrix} 1 & 2 & \vdots & -9 \\ 0 & 1 & \vdots & -4 \end{bmatrix}$

$\begin{cases} x + 2y = -9 \\ y = -4 \end{cases}$

$y = -4$

$x + 2(-4) = -9 \implies x = -1$

Solution: $(-1, -4)$

57. $\begin{cases} -x + 2y = 1.5 \\ 2x - 4y = 3.0 \end{cases}$

$$\begin{bmatrix} -1 & 2 & \vdots & 1.5 \\ 2 & -4 & \vdots & 3.0 \end{bmatrix}$$

$$2R_1 + R_2 \rightarrow \begin{bmatrix} -1 & 2 & \vdots & 1.5 \\ 0 & 0 & \vdots & 6.0 \end{bmatrix}$$

The system is inconsistent and there is no solution.

59. $\begin{cases} x & - 3z = -2 \\ 3x + y - 2z = 5 \\ 2x + 2y + z = 4 \end{cases}$

$$\begin{bmatrix} 1 & 0 & -3 & \vdots & -2 \\ 3 & 1 & -2 & \vdots & 5 \\ 2 & 2 & 1 & \vdots & 4 \end{bmatrix}$$

$$\begin{matrix} -3R_1 + R_2 \rightarrow \\ -2R_1 + R_3 \rightarrow \end{matrix} \begin{bmatrix} 1 & 0 & -3 & \vdots & -2 \\ 0 & 1 & 7 & \vdots & 11 \\ 0 & 2 & 7 & \vdots & 8 \end{bmatrix}$$

$$-2R_2 + R_3 \rightarrow \begin{bmatrix} 1 & 0 & -3 & \vdots & -2 \\ 0 & 1 & 7 & \vdots & 11 \\ 0 & 0 & -7 & \vdots & -14 \end{bmatrix}$$

$$-\tfrac{1}{7}R_3 \rightarrow \begin{bmatrix} 1 & 0 & -3 & \vdots & -2 \\ 0 & 1 & 7 & \vdots & 11 \\ 0 & 0 & 1 & \vdots & 2 \end{bmatrix}$$

$$\begin{cases} x & - 3z = -2 \\ y + 7z = 11 \\ z = 2 \end{cases}$$

$z = 2$

$y + 7(2) = 11 \implies y = -3$

$x - 3(2) = -2 \implies x = 4$

Solution: $(4, -3, 2)$

61. $\begin{cases} -x + y - z = -14 \\ 2x - y + z = 21 \\ 3x + 2y + z = 19 \end{cases}$

$$\begin{bmatrix} -1 & 1 & -1 & \vdots & -14 \\ 2 & -1 & 1 & \vdots & 21 \\ 3 & 2 & 1 & \vdots & 19 \end{bmatrix}$$

$$-R_1 \rightarrow \begin{bmatrix} 1 & -1 & 1 & \vdots & 14 \\ 2 & -1 & 1 & \vdots & 21 \\ 3 & 2 & 1 & \vdots & 19 \end{bmatrix}$$

$$\begin{matrix} -2R_1 + R_2 \rightarrow \\ -3R_1 + R_3 \rightarrow \end{matrix} \begin{bmatrix} 1 & -1 & 1 & \vdots & 14 \\ 0 & 1 & -1 & \vdots & -7 \\ 0 & 5 & -2 & \vdots & -23 \end{bmatrix}$$

$$-5R_2 + R_3 \rightarrow \begin{bmatrix} 1 & -1 & 1 & \vdots & 14 \\ 0 & 1 & -1 & \vdots & -7 \\ 0 & 0 & 3 & \vdots & 12 \end{bmatrix}$$

$$\tfrac{1}{3}R_3 \rightarrow \begin{bmatrix} 1 & -1 & 1 & \vdots & 14 \\ 0 & 1 & -1 & \vdots & -7 \\ 0 & 0 & 1 & \vdots & 4 \end{bmatrix}$$

$$\begin{cases} x - y + z = 14 \\ y - z = -7 \\ z = 4 \end{cases}$$

$z = 4$

$y - 4 = -7 \implies y = -3$

$x - (-3) + 4 = 14 \implies x = 7$

Solution: $(7, -3, 4)$

63. $\begin{cases} x + 2y - 3z = -28 \\ 4y + 2z = 0 \\ -x + y - z = -5 \end{cases}$

$$\begin{bmatrix} 1 & 2 & -3 & : & -28 \\ 0 & 4 & 2 & : & 0 \\ -1 & 1 & -1 & : & -5 \end{bmatrix}$$

$$\begin{matrix} \frac{1}{4}R_2 \to \\ R_1 + R_3 \to \end{matrix} \begin{bmatrix} 1 & 2 & -3 & : & -28 \\ 0 & 1 & \frac{1}{2} & : & 0 \\ 0 & 3 & -4 & : & -33 \end{bmatrix}$$

$$-3R_2 + R_3 \to \begin{bmatrix} 1 & 2 & -3 & & -28 \\ 0 & 1 & \frac{1}{2} & & 0 \\ 0 & 0 & -\frac{11}{2} & & -33 \end{bmatrix}$$

$$-\frac{2}{11}R_3 \to \begin{bmatrix} 1 & 2 & -3 & : & -28 \\ 0 & 1 & \frac{1}{2} & : & 0 \\ 0 & 0 & 1 & : & 6 \end{bmatrix}$$

$\begin{cases} x + 2y - 3z = -28 \\ y + \frac{1}{2}z = 0 \\ \phantom{x + 2y + \frac{1}{2}}z = 6 \end{cases}$

$z = 6$

$y + \frac{1}{2}(6) = 0 \implies y = -3$

$x + 2(-3) - 3(6) = -28 \implies x = -4$

Solution: $(-4, -3, 6)$

65. $\begin{cases} x + y - 5z = 3 \\ x - 2z = 1 \\ 2x - y - z = 0 \end{cases}$

$$\begin{bmatrix} 1 & 1 & -5 & : & 3 \\ 1 & 0 & -2 & : & 1 \\ 2 & -1 & -1 & : & 0 \end{bmatrix}$$

$$\begin{matrix} -R_1 + R_2 \to \\ -2R_1 + R_3 \to \end{matrix} \begin{bmatrix} 1 & 1 & -5 & : & 3 \\ 0 & -1 & 3 & : & -2 \\ 0 & -3 & 9 & : & -6 \end{bmatrix}$$

$$-R_2 \to \begin{bmatrix} 1 & 1 & -5 & : & 3 \\ 0 & 1 & -3 & : & 2 \\ 0 & -3 & 9 & : & -6 \end{bmatrix}$$

$$\begin{matrix} -R_2 + R_1 \to \\ 3R_2 + R_3 \to \end{matrix} \begin{bmatrix} 1 & 0 & -2 & : & 1 \\ 0 & 1 & -3 & : & 2 \\ 0 & 0 & 0 & : & 0 \end{bmatrix}$$

$\begin{cases} x - 2z = 1 \\ y - 3z = 2 \end{cases}$

$z = a$

$y - 3a = 2 \implies y = 3a + 2$

$x - 2a = 1 \implies x = 2a + 1$

Solution: $(2a + 1, 3a + 2, a)$

67. $\begin{cases} x + 2y + z + 2w = 8 \\ 3x + 7y + 6z + 9w = 26 \end{cases}$

$$\begin{bmatrix} 1 & 2 & 1 & 2 & : & 8 \\ 3 & 7 & 6 & 9 & : & 26 \end{bmatrix}$$

$$-3R_1 + R_2 \to \begin{bmatrix} 1 & 2 & 1 & 2 & : & 8 \\ 0 & 1 & 3 & 3 & : & 2 \end{bmatrix}$$

$$-2R_2 + R_1 \to \begin{bmatrix} 1 & 0 & -5 & -4 & : & 4 \\ 0 & 1 & 3 & 3 & : & 2 \end{bmatrix}$$

$\begin{cases} x - 5z - 4w = 4 \\ y + 3z + 3w = 2 \end{cases}$

$w = a, z = b$

$y + 3b + 3a = 2 \implies y = 2 - 3b - 3a$

$x - 5b - 4a = 4 \implies x = 4 + 5b + 4a$

Solution: $(4 + 5b + 4a, 2 - 3b - 3a, b, a)$,
where a and b are real numbers.

69. $\begin{cases} -x + y = -22 \\ 3x + 4y = 4 \\ 4x - 8y = 32 \end{cases}$

$$\begin{bmatrix} -1 & 1 & \vdots & -22 \\ 3 & 4 & \vdots & 4 \\ 4 & -8 & \vdots & 32 \end{bmatrix}$$

$$-R_1 \rightarrow \begin{bmatrix} 1 & -1 & \vdots & 22 \\ 3 & 4 & \vdots & 4 \\ 4 & -8 & \vdots & 32 \end{bmatrix}$$

$$\begin{matrix} \\ -3R_1 + R_2 \rightarrow \\ -4R_1 + R_3 \rightarrow \end{matrix} \begin{bmatrix} 1 & -1 & \vdots & 22 \\ 0 & 7 & \vdots & -62 \\ 0 & -4 & \vdots & -56 \end{bmatrix}$$

$$\begin{matrix} \\ \frac{1}{7}R_2 \rightarrow \\ -\frac{1}{4}R_3 \rightarrow \end{matrix} \begin{bmatrix} 1 & -1 & \vdots & 22 \\ 0 & 1 & \vdots & -\frac{62}{7} \\ 0 & 1 & \vdots & 14 \end{bmatrix}$$

$$\begin{matrix} \\ \\ -R_2 + R_3 \rightarrow \end{matrix} \begin{bmatrix} 1 & -1 & & 22 \\ 0 & 1 & & -\frac{62}{7} \\ 0 & 0 & & \frac{160}{7} \end{bmatrix}$$

The system is inconsistent and there is no solution.

71. Use the reduced row–echelon form feature of a graphing utility.

$$\begin{cases} 3x + 3y + 12z = 6 \\ x + y + 4z = 2 \\ 2x + 5y + 20z = 10 \\ -x + 2y + 8z = 4 \end{cases} \quad \begin{bmatrix} 3 & 3 & 12 & \vdots & 6 \\ 1 & 1 & 4 & \vdots & 2 \\ 2 & 5 & 20 & \vdots & 10 \\ -1 & 2 & 8 & \vdots & 4 \end{bmatrix} \Rightarrow \begin{bmatrix} 1 & 0 & 0 & \vdots & 0 \\ 0 & 1 & 4 & \vdots & 2 \\ 0 & 0 & 0 & \vdots & 0 \\ 0 & 0 & 0 & \vdots & 0 \end{bmatrix}$$

$z = a$
$y = 2 - 4a$
$x = 0$

$\begin{cases} x = 0 \\ y + 4z = 2 \end{cases}$

Solution: $(0, 2 - 4a, a)$

73. Use the reduced row–echelon form feature of a graphing utility.

$$\begin{cases} 2x + y - z + 2w = -6 \\ 3x + 4y + w = 1 \\ x + 5y + 2z + 6w = -3 \\ 5x + 2y - z - w = 3 \end{cases} \quad \begin{bmatrix} 2 & 1 & -1 & 2 & \vdots & -6 \\ 3 & 4 & 0 & 1 & \vdots & 1 \\ 1 & 5 & 2 & 6 & \vdots & -3 \\ 5 & 2 & -1 & -1 & \vdots & 3 \end{bmatrix} \Rightarrow \begin{bmatrix} 1 & 0 & 0 & 0 & \vdots & 1 \\ 0 & 1 & 0 & 0 & \vdots & 0 \\ 0 & 0 & 1 & 0 & \vdots & 4 \\ 0 & 0 & 0 & 1 & \vdots & -2 \end{bmatrix}$$

$x = 1$

$y = 0$

$z = 4$

$w = -2$

Solution: $(1, 0, 4, -2)$

75. Use the reduced row–echelon form feature of a graphing utility.

$$\begin{cases} x + y + z + w = 0 \\ 2x + 3y + z - 2w = 0 \\ 3x + 5y + z = 0 \end{cases}$$

$$\begin{bmatrix} 1 & 1 & 1 & 1 & \vdots & 0 \\ 2 & 3 & 1 & -2 & \vdots & 0 \\ 3 & 5 & 1 & 0 & \vdots & 0 \end{bmatrix} \Rightarrow \begin{bmatrix} 1 & 0 & 2 & 0 & \vdots & 0 \\ 0 & 1 & -1 & 0 & \vdots & 0 \\ 0 & 0 & 0 & 1 & \vdots & 0 \end{bmatrix}$$

$$\begin{cases} x + 2z = 0 \\ y - z = 0 \\ w = 0 \end{cases}$$

Let $z = a$. Then $x = -2a$ and $y = a$.

Solution: $(-2a, a, a, 0)$, where a is a real number.

77. (a) $\begin{cases} x - 2y + z = -6 \\ y - 5z = 16 \\ z = -3 \end{cases}$

$$y - 5(-3) = 16$$
$$y = 1$$
$$x - 2(1) + (-3) = -6$$
$$x = -1$$

Solution: $(-1, 1, -3)$

(b) $\begin{cases} x + y - 2z = 6 \\ y + 3z = -8 \\ z = -3 \end{cases}$

$$y + 3(-3) = -8$$
$$y = 1$$
$$x + (1) - 2(-3) = 6$$
$$x = -1$$

Solution: $(-1, 1, -3)$

Both systems yield the same solution, namely $(-1, 1, -3)$.

79. (a) $\begin{cases} x - 4y + 5z = 27 \\ y - 7z = -54 \\ z = 8 \end{cases}$

$$y - 7(8) = -54$$
$$y = 2$$
$$x - 4(2) + 5(8) = 27$$
$$x = -5$$

Solution: $(-5, 2, 8)$

(b) $\begin{cases} x - 6y + z = 15 \\ y + 5z = 42 \\ z = 8 \end{cases}$

$$y + 5(8) = 42$$
$$y = 2$$
$$x - 6(2) + (8) = 15$$
$$x = 19$$

Solution: $(19, 2, 8)$

The systems do **not** yield the same solution.

81. (a) In the row-echelon form of an augmented matrix that corresponds to an inconsistent sytem of linear equations, there exists a row consisting of all zeros except for the entry in the last column.

(b) In the row-echelon form of an augmented matrix that corresponds to a system with an infinte number of solutions, there are fewer rows with nonzero entries than there are variables.

83. They are the same.

85.
$$\begin{cases} I_1 - I_2 + I_3 = 0 \\ 3I_1 + 4I_2 \quad\;\; = 18 \\ \quad\;\; I_2 + 3I_3 = 6 \end{cases}$$

$$\begin{bmatrix} 1 & -1 & 1 & \vdots & 0 \\ 3 & 4 & 0 & \vdots & 18 \\ 0 & 1 & 3 & \vdots & 6 \end{bmatrix}$$

$$-3R_1 + R_2 \to \begin{bmatrix} 1 & -1 & 1 & \vdots & 0 \\ 0 & 7 & -3 & \vdots & 18 \\ 0 & 1 & 3 & \vdots & 6 \end{bmatrix}$$

$$\begin{matrix} R_3 \\ R_2 \end{matrix} \begin{bmatrix} 1 & -1 & 1 & \vdots & 0 \\ 0 & 1 & 3 & \vdots & 6 \\ 0 & 7 & -3 & \vdots & 18 \end{bmatrix}$$

$$-7R_2 + R_3 \to \begin{bmatrix} 1 & -1 & 1 & \vdots & 0 \\ 0 & 1 & 3 & \vdots & 6 \\ 0 & 0 & -24 & \vdots & -24 \end{bmatrix}$$

$$-\tfrac{1}{24}R_3 \to \begin{bmatrix} 1 & -1 & 1 & \vdots & 0 \\ 0 & 1 & 3 & \vdots & 6 \\ 0 & 0 & 1 & \vdots & 1 \end{bmatrix}$$

$$\begin{cases} I_1 - I_2 + I_3 = 0 \\ \quad\;\; I_2 + 3I_3 = 6 \\ \quad\quad\quad I_3 = 1 \end{cases}$$

$I_3 = 1$

$I_2 + 3(1) = 6 \Rightarrow I_2 = 3$

$I_1 - 3 + 1 = 0 \Rightarrow I_1 = 2$

89. $f(x) = ax^2 + bx + c$

$f(1) = a + b + c = 9$

$f(2) = 4a + 2b + c = 8$

$f(3) = 9a + 3b + c = 5$

$$\begin{bmatrix} 1 & 1 & 1 & \vdots & 9 \\ 4 & 2 & 1 & \vdots & 8 \\ 9 & 3 & 1 & \vdots & 5 \end{bmatrix}$$

$$\begin{matrix} -4R_1 + R_2 \to \\ -9R_1 + R_3 \to \end{matrix} \begin{bmatrix} 1 & 1 & 1 & \vdots & 9 \\ 0 & -2 & -3 & \vdots & -28 \\ 0 & -6 & -8 & \vdots & -76 \end{bmatrix}$$

$$-\tfrac{1}{2}R_2 \to \begin{bmatrix} 1 & 1 & 1 & \vdots & 9 \\ 0 & 1 & \tfrac{3}{2} & \vdots & 14 \\ 0 & -6 & -8 & \vdots & -76 \end{bmatrix}$$

$$6R_2 + R_3 \to \begin{bmatrix} 1 & 1 & 1 & \vdots & 9 \\ 0 & 1 & \tfrac{3}{2} & \vdots & 14 \\ 0 & 0 & 1 & \vdots & 8 \end{bmatrix}$$

87. x = amount at 9%, y = amount at 10%, z = amount at 12%

$x + y + z = 500{,}000$

$0.09x + 0.10y + 0.12z = 52{,}000$

$2.5x - y = 0$

$$\begin{bmatrix} 1 & 1 & 1 & \vdots & 500{,}000 \\ 0.09 & 0.10 & 0.12 & \vdots & 52{,}000 \\ 2.5 & -1 & 0 & \vdots & 0 \end{bmatrix}$$

$$\begin{matrix} -0.09R_1 + R_2 \to \\ -2.5R_1 + R_3 \to \end{matrix} \begin{bmatrix} 1 & 1 & 1 & \vdots & 500{,}000 \\ 0 & 0.01 & 0.03 & \vdots & 7{,}000 \\ 0 & -3.5 & -2.5 & \vdots & -1{,}250{,}000 \end{bmatrix}$$

$$\begin{matrix} 100R_2 \to \\ 2R_3 \to \end{matrix} \begin{bmatrix} 1 & 1 & 1 & \vdots & 500{,}000 \\ 0 & 1 & 3 & \vdots & 700{,}000 \\ 0 & -7 & -5 & \vdots & -2{,}500{,}000 \end{bmatrix}$$

$$\begin{matrix} -R_2 + R_1 \to \\ 7R_2 + R_3 \to \end{matrix} \begin{bmatrix} 1 & 0 & -2 & \vdots & -200{,}000 \\ 0 & 1 & 3 & \vdots & 700{,}000 \\ 0 & 0 & 16 & \vdots & 2{,}400{,}000 \end{bmatrix}$$

$$\tfrac{1}{16}R_3 \to \begin{bmatrix} 1 & 0 & -2 & \vdots & -200{,}000 \\ 0 & 1 & 3 & \vdots & 700{,}000 \\ 0 & 0 & 1 & \vdots & 150{,}000 \end{bmatrix}$$

$$\begin{cases} x - 2z = -200{,}000 \\ y + 3z = 700{,}000 \\ \quad\;\; z = 150{,}000 \end{cases}$$

$y + 3(150{,}000) = 700{,}000 \Rightarrow y = 250{,}000$

$x - 2(150{,}000) = -200{,}000 \Rightarrow x = 100{,}000$

Solution: $(100{,}000, 250{,}000, 150{,}000)$

Answer: \$100,000 at 9%, \$250,000 at 10%, \$150,000 at 12%

$$\begin{cases} a + b + c = 9 \\ \quad\; b + \tfrac{3}{2}c = 14 \\ \quad\quad\quad c = 8 \end{cases}$$

$c = 8$

$b + \tfrac{3}{2}(8) = 14 \Rightarrow b = 2$

$a + (2) + (8) = 9 \Rightarrow a = -1$

Equation of parabola: $y = -x^2 + 2x + 8$

91. (a) $(9, 121.1)$, $(10, 130.9)$, $(11, 141.5)$

$$f(t) = at^2 + bt + c$$

$$f(9) = 81a + 9b + c = 121.1$$

$$f(10) = 100a + 10b + c = 130.9$$

$$f(11) = 121a + 11b + c = 141.5$$

$$\begin{bmatrix} 81 & 9 & 1 & \vdots & 121.1 \\ 100 & 10 & 1 & \vdots & 130.9 \\ 121 & 11 & 1 & \vdots & 141.5 \end{bmatrix} \Rightarrow \begin{bmatrix} 1 & 0 & 0 & \vdots & 0.4 \\ 0 & 1 & 0 & \vdots & 2.2 \\ 0 & 0 & 1 & \vdots & 68.9 \end{bmatrix}$$

$$f(t) = 0.4t^2 + 2.2t + 68.9$$

(b)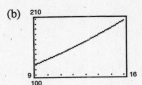

(c) $f(12) = \$152.9$ billion

This is close to the actual value of \$153.0 billion for 2002 sales.

(d) $f(16) = \$206.5$ billion

The parabola grows without bound as t increases. Even though drug store sales are also growing, this figure represents an increase in sales of nearly 35% in four years, which seems excessive.

93. False. The rows are in the wrong order. To change this matrix to reduced row–echelon form, interchange Row 1 and Row 4, and interchange Row 2 and Row 3.

95. False. See Example 8 in this section.

Section 14.5   Operations with Matrices

1. $x = -4$, $y = 22$

3. $x + 1 = 2x - 1 \Rightarrow x = 2$

$-y = y + 1 \Rightarrow y = -\frac{1}{2}$

5. $2x + 1 = 5$, $3x = 6$, $3y - 5 = 4$

$x = 2$, $y = 3$

7. (a) $A + B = \begin{bmatrix} 1 & -1 \\ 2 & -1 \end{bmatrix} + \begin{bmatrix} 2 & -1 \\ -1 & 8 \end{bmatrix} = \begin{bmatrix} 1+2 & -1-1 \\ 2-1 & -1+8 \end{bmatrix} = \begin{bmatrix} 3 & -2 \\ 1 & 7 \end{bmatrix}$

(b) $A - B = \begin{bmatrix} 1 & -1 \\ 2 & -1 \end{bmatrix} - \begin{bmatrix} 2 & -1 \\ -1 & 8 \end{bmatrix} = \begin{bmatrix} 1-2 & -1+1 \\ 2+1 & -1-8 \end{bmatrix} = \begin{bmatrix} -1 & 0 \\ 3 & -9 \end{bmatrix}$

(c) $3A = 3\begin{bmatrix} 1 & -1 \\ 2 & -1 \end{bmatrix} = \begin{bmatrix} 3(1) & 3(-1) \\ 3(2) & 3(-1) \end{bmatrix} = \begin{bmatrix} 3 & -3 \\ 6 & -3 \end{bmatrix}$

(d) $3A - 2B = \begin{bmatrix} 3 & -3 \\ 6 & -3 \end{bmatrix} - 2\begin{bmatrix} 2 & -1 \\ -1 & 8 \end{bmatrix} = \begin{bmatrix} 3 & -3 \\ 6 & -3 \end{bmatrix} + \begin{bmatrix} -4 & 2 \\ 2 & -16 \end{bmatrix} = \begin{bmatrix} -1 & -1 \\ 8 & -19 \end{bmatrix}$

9. $A = \begin{bmatrix} 6 & -1 \\ 2 & 4 \\ -3 & 5 \end{bmatrix}$, $B = \begin{bmatrix} 1 & 4 \\ -1 & 5 \\ 1 & 10 \end{bmatrix}$

(a) $A + B = \begin{bmatrix} 7 & 3 \\ 1 & 9 \\ -2 & 15 \end{bmatrix}$

(b) $A - B = \begin{bmatrix} 5 & -5 \\ 3 & -1 \\ -4 & -5 \end{bmatrix}$

(c) $3A = \begin{bmatrix} 18 & -3 \\ 6 & 12 \\ -9 & 15 \end{bmatrix}$

(d) $3A - 2B = \begin{bmatrix} 18 & -3 \\ 6 & 12 \\ -9 & 15 \end{bmatrix} - \begin{bmatrix} 2 & 8 \\ -2 & 10 \\ 2 & 20 \end{bmatrix} = \begin{bmatrix} 16 & -11 \\ 8 & 2 \\ -11 & -5 \end{bmatrix}$

11. $A = \begin{bmatrix} 2 & 2 & -1 & 0 & 1 \\ 1 & 1 & -2 & 0 & -1 \end{bmatrix}$, $B = \begin{bmatrix} 1 & 1 & -1 & 1 & 0 \\ -3 & 4 & 9 & -6 & -7 \end{bmatrix}$

   (a) $A + B = \begin{bmatrix} 3 & 3 & -2 & 1 & 1 \\ -2 & 5 & 7 & -6 & -8 \end{bmatrix}$

   (b) $A - B = \begin{bmatrix} 1 & 1 & 0 & -1 & 1 \\ 4 & -3 & -11 & 6 & 6 \end{bmatrix}$

   (c) $3A = \begin{bmatrix} 6 & 6 & -3 & 0 & 3 \\ 3 & 3 & -6 & 0 & -3 \end{bmatrix}$

   (d) $3A - 2B = \begin{bmatrix} 6 & 6 & -3 & 0 & 3 \\ 3 & 3 & -6 & 0 & -3 \end{bmatrix} - \begin{bmatrix} 2 & 2 & -2 & 2 & 0 \\ -6 & 8 & 18 & -12 & -14 \end{bmatrix} = \begin{bmatrix} 4 & 4 & -1 & -2 & 3 \\ 9 & -5 & -24 & 12 & 11 \end{bmatrix}$

13. $A = \begin{bmatrix} 6 & 0 & 3 \\ -1 & -4 & 0 \end{bmatrix}$, $B = \begin{bmatrix} 8 & -1 \\ 4 & -3 \end{bmatrix}$

   (a) $A + B$ is not possible. A and B do not have the same order.

   (b) $A - B$ is not possible. A and B do not have the same order.

   (c) $3A = \begin{bmatrix} 18 & 0 & 9 \\ -3 & -12 & 0 \end{bmatrix}$

   (d) $3A - 2B$ is not possible. A and B do not have the same order.

15. $\begin{bmatrix} -5 & 0 \\ 3 & -6 \end{bmatrix} + \begin{bmatrix} 7 & 1 \\ -2 & -1 \end{bmatrix} + \begin{bmatrix} -10 & -8 \\ 14 & 6 \end{bmatrix} = \begin{bmatrix} -5 + 7 + (-10) & 0 + 1 + (-8) \\ 3 + (-2) + 14 & -6 + (-1) + 6 \end{bmatrix} = \begin{bmatrix} -8 & -7 \\ 15 & -1 \end{bmatrix}$

17. $4\left(\begin{bmatrix} -4 & 0 & 1 \\ 0 & 2 & 3 \end{bmatrix} - \begin{bmatrix} 2 & 1 & -2 \\ 3 & -6 & 0 \end{bmatrix} \right) = 4\begin{bmatrix} -6 & -1 & 3 \\ -3 & 8 & 3 \end{bmatrix} = \begin{bmatrix} -24 & -4 & 12 \\ -12 & 32 & 12 \end{bmatrix}$

19. $-3\left(\begin{bmatrix} 0 & -3 \\ 7 & 2 \end{bmatrix} + \begin{bmatrix} -6 & 3 \\ 8 & 1 \end{bmatrix} \right) - 2\begin{bmatrix} 4 & -4 \\ 7 & -9 \end{bmatrix} = -3\begin{bmatrix} -6 & 0 \\ 15 & 3 \end{bmatrix} - \begin{bmatrix} 8 & -8 \\ 14 & -18 \end{bmatrix} = \begin{bmatrix} 18 & 0 \\ -45 & -9 \end{bmatrix} - \begin{bmatrix} 8 & -8 \\ 14 & -18 \end{bmatrix} = \begin{bmatrix} 10 & 8 \\ -59 & 9 \end{bmatrix}$

21. $\frac{3}{7}\begin{bmatrix} 2 & 5 \\ -1 & -4 \end{bmatrix} + 6\begin{bmatrix} -3 & 0 \\ 2 & 2 \end{bmatrix} \approx \begin{bmatrix} -17.143 & 2.143 \\ 11.571 & 10.286 \end{bmatrix}$

23. $-\begin{bmatrix} 3.211 & 6.829 \\ -1.004 & 4.914 \\ 0.055 & -3.889 \end{bmatrix} - \begin{bmatrix} -1.630 & -3.090 \\ 5.256 & 8.335 \\ -9.768 & 4.251 \end{bmatrix} = \begin{bmatrix} -1.581 & -3.739 \\ -4.252 & -13.249 \\ 9.713 & -0.362 \end{bmatrix}$

25. $X = 3\begin{bmatrix} -2 & -1 \\ 1 & 0 \\ 3 & -4 \end{bmatrix} - 2\begin{bmatrix} 0 & 3 \\ 2 & 0 \\ -4 & -1 \end{bmatrix} = \begin{bmatrix} -6 & -3 \\ 3 & 0 \\ 9 & -12 \end{bmatrix} - \begin{bmatrix} 0 & 6 \\ 4 & 0 \\ -8 & -2 \end{bmatrix} = \begin{bmatrix} -6 & -9 \\ -1 & 0 \\ 17 & -10 \end{bmatrix}$

27. $X = -\frac{3}{2}A + \frac{1}{2}B = -\frac{3}{2}\begin{bmatrix} -2 & -1 \\ 1 & 0 \\ 3 & -4 \end{bmatrix} + \frac{1}{2}\begin{bmatrix} 0 & 3 \\ 2 & 0 \\ -4 & -1 \end{bmatrix} = \begin{bmatrix} 3 & \frac{3}{2} \\ -\frac{3}{2} & 0 \\ -\frac{9}{2} & 6 \end{bmatrix} + \begin{bmatrix} 0 & \frac{3}{2} \\ 1 & 0 \\ -2 & -\frac{1}{2} \end{bmatrix} = \begin{bmatrix} 3 & 3 \\ -\frac{1}{2} & 0 \\ -\frac{13}{2} & \frac{11}{2} \end{bmatrix}$

29. (a) $AB = \begin{bmatrix} 1 & 2 \\ 4 & 2 \end{bmatrix}\begin{bmatrix} 2 & -1 \\ -1 & 8 \end{bmatrix} = \begin{bmatrix} (1)(2) + (2)(-1) & (1)(-1) + (2)(8) \\ (4)(2) + (2)(-1) & (4)(-1) + (2)(8) \end{bmatrix} = \begin{bmatrix} 0 & 15 \\ 6 & 12 \end{bmatrix}$

(b) $BA = \begin{bmatrix} 2 & -1 \\ -1 & 8 \end{bmatrix}\begin{bmatrix} 1 & 2 \\ 4 & 2 \end{bmatrix} = \begin{bmatrix} (2)(1) + (-1)(4) & (2)(2) + (-1)(2) \\ (-1)(1) + (8)(4) & (-1)(2) + (8)(2) \end{bmatrix} = \begin{bmatrix} -2 & 2 \\ 31 & 14 \end{bmatrix}$

(c) $A^2 = \begin{bmatrix} 1 & 2 \\ 4 & 2 \end{bmatrix}\begin{bmatrix} 1 & 2 \\ 4 & 2 \end{bmatrix} = \begin{bmatrix} (1)(1) + (2)(4) & (1)(2) + (2)(2) \\ (4)(1) + (2)(4) & (4)(2) + (2)(2) \end{bmatrix} = \begin{bmatrix} 9 & 6 \\ 12 & 12 \end{bmatrix}$

31. (a) $AB = \begin{bmatrix} 3 & -1 \\ 1 & 3 \end{bmatrix}\begin{bmatrix} 1 & -3 \\ 3 & 1 \end{bmatrix} = \begin{bmatrix} (3)(1) + (-1)(3) & (3)(-3) + (-1)(1) \\ (1)(1) + (3)(3) & (1)(-3) + (3)(1) \end{bmatrix} = \begin{bmatrix} 0 & -10 \\ 10 & 0 \end{bmatrix}$

(b) $BA = \begin{bmatrix} 1 & -3 \\ 3 & 1 \end{bmatrix}\begin{bmatrix} 3 & -1 \\ 1 & 3 \end{bmatrix} = \begin{bmatrix} (1)(3) + (-3)(1) & (1)(-1) + (-3)(3) \\ (3)(3) + (1)(1) & (3)(-1) + (1)(3) \end{bmatrix} = \begin{bmatrix} 0 & -10 \\ 10 & 0 \end{bmatrix}$

(c) $A^2 = \begin{bmatrix} 3 & -1 \\ 1 & 3 \end{bmatrix}\begin{bmatrix} 3 & -1 \\ 1 & 3 \end{bmatrix} = \begin{bmatrix} (3)(3) + (-1)(1) & (3)(-1) + (-1)(3) \\ (1)(3) + (3)(1) & (1)(-1) + (3)(3) \end{bmatrix} = \begin{bmatrix} 8 & -6 \\ 6 & 8 \end{bmatrix}$

33. (a) $AB = \begin{bmatrix} 7 \\ 8 \\ -1 \end{bmatrix}\begin{bmatrix} 1 & 1 & 2 \end{bmatrix} = \begin{bmatrix} 7(1) & 7(1) & 7(2) \\ 8(1) & 8(1) & 8(2) \\ -1(1) & -1(1) & -1(2) \end{bmatrix} = \begin{bmatrix} 7 & 7 & 14 \\ 8 & 8 & 16 \\ -1 & -1 & -2 \end{bmatrix}$

(b) $BA = \begin{bmatrix} 1 & 1 & 2 \end{bmatrix}\begin{bmatrix} 7 \\ 8 \\ -1 \end{bmatrix} = [(1)(7) + (1)(8) + (2)(-1)] = [13]$

(c) A^2 is not possible.

35. A is 3×2 and B is 3×3. AB is not possible.

37. A is 3×3, B is $3 \times 2 \implies AB$ is 3×2.

$\begin{bmatrix} 0 & -1 & 0 \\ 4 & 0 & 2 \\ 8 & -1 & 7 \end{bmatrix}\begin{bmatrix} 2 & 1 \\ -3 & 4 \\ 1 & 6 \end{bmatrix} = \begin{bmatrix} (0)(2) + (-1)(-3) + (0)(1) & (0)(1) + (-1)(4) + (0)(6) \\ (4)(2) + (0)(-3) + (2)(1) & (4)(1) + (0)(4) + (2)(6) \\ (8)(2) + (-1)(-3) + (7)(1) & (8)(1) + (-1)(4) + (7)(6) \end{bmatrix} = \begin{bmatrix} 3 & -4 \\ 10 & 16 \\ 26 & 46 \end{bmatrix}$

39. A is 3×3, B is $3 \times 3 \implies AB$ is 3×3.

$\begin{bmatrix} 1 & 0 & 0 \\ 0 & 4 & 0 \\ 0 & 0 & -2 \end{bmatrix}\begin{bmatrix} 3 & 0 & 0 \\ 0 & -1 & 0 \\ 0 & 0 & 5 \end{bmatrix} = \begin{bmatrix} (1)(3) + (0)(0) + (0)(0) & (1)(0) + (0)(-1) + (0)(0) & (1)(0) + (0)(0) + (0)(5) \\ (0)(3) + (4)(0) + (0)(0) & (0)(0) + (4)(-1) + (0)(0) & (0)(0) + (4)(0) + (0)(5) \\ (0)(3) + (0)(0) + (-2)(0) & (0)(0) + (0)(-1) + (-2)(0) & (0)(0) + (0)(0) + (-2)(5) \end{bmatrix}$

$= \begin{bmatrix} 3 & 0 & 0 \\ 0 & -4 & 0 \\ 0 & 0 & -10 \end{bmatrix}$

41. A is 3×3, B is $3 \times 3 \implies AB$ is 3×3.

$\begin{bmatrix} 0 & 0 & 5 \\ 0 & 0 & -3 \\ 0 & 0 & 4 \end{bmatrix}\begin{bmatrix} 6 & -11 & 4 \\ 8 & 16 & 4 \\ 0 & 0 & 0 \end{bmatrix} =$

$\begin{bmatrix} (0)(6) + (0)(8) + (5)(0) & (0)(-11) + (0)(16) + (5)(0) & (0)(4) + (0)(4) + (5)(0) \\ (0)(6) + (0)(8) + (-3)(0) & (0)(-11) + (0)(16) + (-3)(0) & (0)(4) + (0)(4) + (-3)(0) \\ (0)(6) + (0)(8) + (4)(0) & (0)(-11) + (0)(16) + (4)(0) & (0)(4) + (0)(4) + (4)(0) \end{bmatrix} = \begin{bmatrix} 0 & 0 & 0 \\ 0 & 0 & 0 \\ 0 & 0 & 0 \end{bmatrix}$

43. $\begin{bmatrix} 5 & 6 & -3 \\ -2 & 5 & 1 \\ 10 & -5 & 5 \end{bmatrix}\begin{bmatrix} 1 & -1 & 2 \\ 8 & 1 & 4 \\ 4 & -2 & 9 \end{bmatrix} = \begin{bmatrix} 41 & 7 & 7 \\ 42 & 5 & 25 \\ -10 & -25 & 45 \end{bmatrix}$

45. $\begin{bmatrix} -3 & 8 & -6 & 8 \\ -12 & 15 & 9 & 6 \\ 5 & -1 & 1 & 5 \end{bmatrix}\begin{bmatrix} 3 & 1 & 6 \\ 24 & 15 & 14 \\ 16 & 10 & 21 \\ 8 & -4 & 10 \end{bmatrix} = \begin{bmatrix} 151 & 25 & 48 \\ 516 & 279 & 387 \\ 47 & -20 & 87 \end{bmatrix}$

47. A is 2×4 and B is $2 \times 4 \Rightarrow AB$ is not possible.

49. $\begin{bmatrix} 3 & 1 \\ 0 & -2 \end{bmatrix}\begin{bmatrix} 1 & 0 \\ -2 & 2 \end{bmatrix}\begin{bmatrix} 1 & 0 \\ 2 & 4 \end{bmatrix} = \begin{bmatrix} 5 & 8 \\ -4 & -16 \end{bmatrix}$

51. $\begin{bmatrix} 0 & 2 & -2 \\ 4 & 1 & 2 \end{bmatrix}\left(\begin{bmatrix} 4 & 0 \\ 0 & -1 \\ -1 & 2 \end{bmatrix} + \begin{bmatrix} -2 & 3 \\ -3 & 5 \\ 0 & -3 \end{bmatrix} \right) = \begin{bmatrix} -4 & 10 \\ 3 & 14 \end{bmatrix}$

53. (a) $\begin{bmatrix} -1 & 1 \\ -2 & 1 \end{bmatrix}\begin{bmatrix} x_1 \\ x_2 \end{bmatrix} = \begin{bmatrix} 4 \\ 0 \end{bmatrix}$

(b) $\begin{bmatrix} -1 & 1 & \vdots & 4 \\ -2 & 1 & \vdots & 0 \end{bmatrix}$

$-R_2 + R_1 \rightarrow \begin{bmatrix} 1 & 0 & \vdots & 4 \\ -2 & 1 & \vdots & 0 \end{bmatrix}$

$2R_1 + R_2 \rightarrow \begin{bmatrix} 1 & 0 & \vdots & 4 \\ 0 & 1 & \vdots & 8 \end{bmatrix}$

$X = \begin{bmatrix} 4 \\ 8 \end{bmatrix}$

55. (a) $\begin{bmatrix} -2 & -3 \\ 6 & 1 \end{bmatrix}\begin{bmatrix} x_1 \\ x_2 \end{bmatrix} = \begin{bmatrix} -4 \\ -36 \end{bmatrix}$

(b) $\begin{bmatrix} -2 & -3 & \vdots & -4 \\ 6 & 1 & \vdots & -36 \end{bmatrix}$

$3R_1 + R_2 \rightarrow \begin{bmatrix} -2 & -3 & \vdots & -4 \\ 0 & -8 & \vdots & -48 \end{bmatrix}$

$\begin{matrix} -\frac{1}{2}R_1 \rightarrow \\ -\frac{1}{8}R_2 \rightarrow \end{matrix} \begin{bmatrix} 1 & \frac{3}{2} & \vdots & 2 \\ 0 & 1 & \vdots & 6 \end{bmatrix}$

$-\frac{3}{2}R_2 + R_1 \rightarrow \begin{bmatrix} 1 & 0 & \vdots & -7 \\ 0 & 1 & \vdots & 6 \end{bmatrix}$

$X = \begin{bmatrix} -7 \\ 6 \end{bmatrix}$

57. (a) $\begin{bmatrix} 1 & -2 & 3 \\ -1 & 3 & -1 \\ 2 & -5 & 5 \end{bmatrix}\begin{bmatrix} x_1 \\ x_2 \\ x_3 \end{bmatrix} = \begin{bmatrix} 9 \\ -6 \\ 17 \end{bmatrix}$

(b) $\begin{bmatrix} 1 & -2 & 3 & \vdots & 9 \\ -1 & 3 & -1 & \vdots & -6 \\ 2 & -5 & 5 & \vdots & 17 \end{bmatrix}$

$\begin{matrix} R_1 + R_2 \rightarrow \\ -2R_2 + R_3 \rightarrow \end{matrix} \begin{bmatrix} 1 & -2 & 3 & \vdots & 9 \\ 0 & 1 & 2 & \vdots & 3 \\ 0 & -1 & -1 & \vdots & -1 \end{bmatrix}$

$\begin{matrix} 2R_2 + R_1 \rightarrow \\ \\ R_2 + R_3 \rightarrow \end{matrix} \begin{bmatrix} 1 & 0 & 7 & \vdots & 15 \\ 0 & 1 & 2 & \vdots & 3 \\ 0 & 0 & 1 & \vdots & 2 \end{bmatrix}$

$\begin{matrix} -7R_3 + R_1 \rightarrow \\ -2R_3 + R_2 \rightarrow \end{matrix} \begin{bmatrix} 1 & 0 & 0 & \vdots & 1 \\ 0 & 1 & 0 & \vdots & -1 \\ 0 & 0 & 1 & \vdots & 2 \end{bmatrix}$

$X = \begin{bmatrix} 1 \\ -1 \\ 2 \end{bmatrix}$

59. (a) $\begin{bmatrix} 1 & -5 & 2 \\ -3 & 1 & -1 \\ 0 & -2 & 5 \end{bmatrix} \begin{bmatrix} x_1 \\ x_2 \\ x_3 \end{bmatrix} = \begin{bmatrix} -20 \\ 8 \\ -16 \end{bmatrix}$

(b) $\begin{bmatrix} 1 & -5 & 2 & \vdots & -20 \\ -3 & 1 & -1 & \vdots & 8 \\ 0 & -2 & 5 & \vdots & -16 \end{bmatrix}$

$3R_1 + R_2 \rightarrow \begin{bmatrix} 1 & -5 & 2 & \vdots & -20 \\ 0 & -14 & 5 & \vdots & -52 \\ 0 & -2 & 5 & \vdots & -16 \end{bmatrix}$

$-R_3 + R_2 \rightarrow \begin{bmatrix} 1 & -5 & 2 & \vdots & -20 \\ 0 & -12 & 0 & \vdots & -36 \\ 0 & -2 & 5 & \vdots & -16 \end{bmatrix}$

$-\frac{1}{12}R_2 \rightarrow \begin{bmatrix} 1 & -5 & 2 & \vdots & -20 \\ 0 & 1 & 0 & \vdots & 3 \\ 0 & -2 & 5 & \vdots & -16 \end{bmatrix}$

$\begin{matrix} 5R_2 + R_1 \rightarrow \\ \\ 2R_2 + R_3 \rightarrow \end{matrix} \begin{bmatrix} 1 & 0 & 2 & \vdots & -5 \\ 0 & 1 & 0 & \vdots & 3 \\ 0 & 0 & 5 & \vdots & -10 \end{bmatrix}$

$\frac{1}{5}R_3 \rightarrow \begin{bmatrix} 1 & 0 & 2 & \vdots & -5 \\ 0 & 1 & 0 & \vdots & 3 \\ 0 & 0 & 1 & \vdots & -2 \end{bmatrix}$

$-2R_3 + R_1 \rightarrow \begin{bmatrix} 1 & 0 & 0 & \vdots & -1 \\ 0 & 1 & 0 & \vdots & 3 \\ 0 & 0 & 1 & \vdots & -2 \end{bmatrix}$

$X = \begin{bmatrix} -1 \\ 3 \\ -2 \end{bmatrix}$

For 61–69, A is of order 2×3, B is of order 2×3, C is of order 3×2, and D is of order 2×2.

61. $A + 2C$ is not possible. A and C are not of the same order.

63. AB is not possible. The number of columns of A does not equal the number of rows of B.

65. $BC - D$ is possible. The resulting order is 2×2.

67. (CA) is 3×3 so $(CA)D$ is not possible.

69. $D(A - 3B)$ is possible. The resulting order is 2×3.

71. The product of two diagonal matrices of the same order is a diagonal matrix whose entries are the products of the corresponding diagonal entries of A and B.

73. $1.2 \begin{bmatrix} 70 & 50 & 25 \\ 35 & 100 & 70 \end{bmatrix} = \begin{bmatrix} 84 & 60 & 30 \\ 42 & 120 & 84 \end{bmatrix}$

75. $BA = \begin{bmatrix} 3.50 & 6.00 \end{bmatrix} \begin{bmatrix} 125 & 100 & 75 \\ 100 & 175 & 125 \end{bmatrix} = \begin{bmatrix} \$1037.50 & \$1400.00 & \$1012.50 \end{bmatrix}$

The entries in the matrix represent the profits for both crops at the three outlets.

77. $ST = \begin{bmatrix} 3 & 2 & 2 & 3 & 0 \\ 0 & 2 & 3 & 4 & 3 \\ 4 & 2 & 1 & 3 & 2 \end{bmatrix} \begin{bmatrix} 840 & 1100 \\ 1200 & 1350 \\ 1450 & 1650 \\ 2650 & 3000 \\ 3050 & 3200 \end{bmatrix} = \begin{bmatrix} \$15,770 & \$18,300 \\ \$26,500 & \$29,250 \\ \$21,260 & \$24,150 \end{bmatrix}$

The entries represent the wholesale and retail inventory values of the inventories at the three outlets.

79. True. The sum of two matrices of different orders is undefined.

81. False. $\begin{bmatrix} -2 & 4 \\ -3 & 0 \\ 6 & 1 \end{bmatrix} \begin{bmatrix} 1 & 1 \\ 1 & 1 \end{bmatrix} = \begin{bmatrix} 2 & 2 \\ -3 & -3 \\ 7 & 7 \end{bmatrix}$

Section 14.6   The Inverse of a Square Matrix

1. $AB = \begin{bmatrix} 2 & 1 \\ 5 & 3 \end{bmatrix}\begin{bmatrix} 3 & -1 \\ -5 & 2 \end{bmatrix} = \begin{bmatrix} 6-5 & -2+2 \\ 15-15 & -5+6 \end{bmatrix} = \begin{bmatrix} 1 & 0 \\ 0 & 1 \end{bmatrix}$

$BA = \begin{bmatrix} 3 & -1 \\ -5 & 2 \end{bmatrix}\begin{bmatrix} 2 & 1 \\ 5 & 3 \end{bmatrix} = \begin{bmatrix} 6-5 & 3-3 \\ -10+10 & -5+6 \end{bmatrix} = \begin{bmatrix} 1 & 0 \\ 0 & 1 \end{bmatrix}$

3. $AB = \begin{bmatrix} 1 & 2 \\ 3 & 4 \end{bmatrix}\begin{bmatrix} -2 & 1 \\ \frac{3}{2} & -\frac{1}{2} \end{bmatrix} = \begin{bmatrix} -2+3 & 1-1 \\ -6+6 & 3-2 \end{bmatrix} = \begin{bmatrix} 1 & 0 \\ 0 & 1 \end{bmatrix}$

$BA = \begin{bmatrix} -2 & 1 \\ \frac{3}{2} & -\frac{1}{2} \end{bmatrix}\begin{bmatrix} 1 & 2 \\ 3 & 4 \end{bmatrix} = \begin{bmatrix} -2+3 & -4+4 \\ \frac{3}{2}-\frac{3}{2} & 3-2 \end{bmatrix} = \begin{bmatrix} 1 & 0 \\ 0 & 1 \end{bmatrix}$

5. $AB = \begin{bmatrix} 2 & -17 & 11 \\ -1 & 11 & -7 \\ 0 & 3 & -2 \end{bmatrix}\begin{bmatrix} 1 & 1 & 2 \\ 2 & 4 & -3 \\ 3 & 6 & -5 \end{bmatrix} = \begin{bmatrix} 2-17+33 & 2-68+66 & 4+51-55 \\ -1+22-21 & -1+44-42 & -2-33+35 \\ 6-6 & 12-12 & -9+10 \end{bmatrix} = \begin{bmatrix} 1 & 0 & 0 \\ 0 & 1 & 0 \\ 0 & 0 & 1 \end{bmatrix}$

$BA = \begin{bmatrix} 1 & 1 & 2 \\ 2 & 4 & -3 \\ 3 & 6 & -5 \end{bmatrix}\begin{bmatrix} 2 & -17 & 11 \\ -1 & 11 & -7 \\ 0 & 3 & -2 \end{bmatrix} = \begin{bmatrix} 2-1 & -17+11+6 & 11-7-4 \\ 4-4 & -34+44-9 & 22-28+6 \\ 6-6 & -51+66-15 & 33-42+10 \end{bmatrix} = \begin{bmatrix} 1 & 0 & 0 \\ 0 & 1 & 0 \\ 0 & 0 & 1 \end{bmatrix}$

7. $AB = \begin{bmatrix} 2 & 0 & 1 & 1 \\ 3 & 0 & 0 & 1 \\ -1 & 1 & -2 & 1 \\ 4 & -1 & 1 & 0 \end{bmatrix}\begin{bmatrix} -1 & 2 & -1 & -1 \\ -4 & 9 & -5 & -6 \\ 0 & 1 & -1 & -1 \\ 3 & -5 & 3 & 3 \end{bmatrix}$

$= \begin{bmatrix} -2+3 & 4+1-5 & -2-1+3 & -2-1+3 \\ 0 & 6-5 & 0 & 0 \\ 1-4+3 & -2+9-2-5 & 1-5+2+3 & 1-6+2+3 \\ 0 & 8-9+1 & -4+5-1 & -4+6-1 \end{bmatrix} = \begin{bmatrix} 1 & 0 & 0 & 0 \\ 0 & 1 & 0 & 0 \\ 0 & 0 & 1 & 0 \\ 0 & 0 & 0 & 1 \end{bmatrix}$

$BA = \begin{bmatrix} -1 & 2 & -1 & -1 \\ -4 & 9 & -5 & -6 \\ 0 & 1 & -1 & -1 \\ 3 & -5 & 3 & 3 \end{bmatrix}\begin{bmatrix} 2 & 0 & 1 & 1 \\ 3 & 0 & 0 & 1 \\ -1 & 1 & -2 & 1 \\ 4 & -1 & 1 & 0 \end{bmatrix}$

$= \begin{bmatrix} -2+6+1-4 & 0 & -1+2-1 & -1+2-1 \\ -8+27+5-24 & -5+6 & -4+10-6 & -4+9-5 \\ 3+1-4 & 0 & 2-1 & 0 \\ 6-15-3+12 & 0 & 3-6+3 & 3-5+3 \end{bmatrix} = \begin{bmatrix} 1 & 0 & 0 & 0 \\ 0 & 1 & 0 & 0 \\ 0 & 0 & 1 & 0 \\ 0 & 0 & 0 & 1 \end{bmatrix}$

9. $AB = \frac{1}{3}\begin{bmatrix} -2 & 2 & 3 \\ 1 & -1 & 0 \\ 0 & 1 & 4 \end{bmatrix}\begin{bmatrix} -4 & -5 & 3 \\ -4 & -8 & 3 \\ 1 & 2 & 0 \end{bmatrix} = \frac{1}{3}\begin{bmatrix} -8+8+3 & 10-16+6 & -6+6 \\ -4+4 & -5+8 & 3-3 \\ -4+4 & -8+8 & 3 \end{bmatrix}$

$= \frac{1}{3}\begin{bmatrix} 3 & 0 & 0 \\ 0 & 3 & 0 \\ 0 & 0 & 3 \end{bmatrix} = \begin{bmatrix} 1 & 0 & 0 \\ 0 & 1 & 0 \\ 0 & 0 & 1 \end{bmatrix}$

$BA = \frac{1}{3}\begin{bmatrix} -4 & -5 & 3 \\ -4 & -8 & 3 \\ 1 & 2 & 0 \end{bmatrix}\begin{bmatrix} -2 & 2 & 3 \\ 1 & -1 & 0 \\ 0 & 1 & 4 \end{bmatrix} = \frac{1}{3}\begin{bmatrix} 8-5 & -8+5+3 & -12+12 \\ 8-8 & -8+8+3 & -12+12 \\ -2+2 & 2-2 & 3 \end{bmatrix} = \begin{bmatrix} 1 & 0 & 0 \\ 0 & 1 & 0 \\ 0 & 0 & 1 \end{bmatrix}$

11. $[A \;\vdots\; I] = \begin{bmatrix} 2 & 0 & \vdots & 1 & 0 \\ 0 & 3 & \vdots & 0 & 1 \end{bmatrix}$

$\begin{matrix} \frac{1}{2}R_1 \to \\ \frac{1}{3}R_2 \to \end{matrix} \begin{bmatrix} 1 & 0 & \vdots & \frac{1}{2} & 0 \\ 0 & 1 & \vdots & 0 & \frac{1}{3} \end{bmatrix} = [I \;\vdots\; A^{-1}]$

$A^{-1} = \begin{bmatrix} \frac{1}{2} & 0 \\ 0 & \frac{1}{3} \end{bmatrix}$

13. $[A \;\vdots\; I] = \begin{bmatrix} 1 & -2 & \vdots & 1 & 0 \\ 2 & -3 & \vdots & 0 & 1 \end{bmatrix}$

$-2R_1 + R_2 \to \begin{bmatrix} 1 & -2 & \vdots & 1 & 0 \\ 0 & 1 & \vdots & -2 & 1 \end{bmatrix}$

$2R_2 + R_1 \to \begin{bmatrix} 1 & 0 & \vdots & -3 & 2 \\ 0 & 1 & \vdots & -2 & 1 \end{bmatrix} = [I \;\vdots\; A^{-1}]$

$A^{-1} = \begin{bmatrix} -3 & 2 \\ -2 & 1 \end{bmatrix}$

15. $[A \;\vdots\; I] = \begin{bmatrix} -1 & 1 & \vdots & 1 & 0 \\ -2 & 1 & \vdots & 0 & 1 \end{bmatrix}$

$-R_2 + R_1 \to \begin{bmatrix} 1 & 0 & \vdots & 1 & -1 \\ -2 & 1 & \vdots & 0 & 1 \end{bmatrix}$

$2R_1 + R_2 \to \begin{bmatrix} 1 & 0 & \vdots & 1 & -1 \\ 0 & 1 & \vdots & 2 & -1 \end{bmatrix} = [I \;\vdots\; A^{-1}]$

$A^{-1} = \begin{bmatrix} 1 & -1 \\ 2 & -1 \end{bmatrix}$

17. $[A \;\vdots\; I] = \begin{bmatrix} 2 & 4 & \vdots & 1 & 0 \\ 4 & 8 & \vdots & 0 & 1 \end{bmatrix}$

$-2R_1 + R_2 \to \begin{bmatrix} 2 & 4 & \vdots & 1 & 0 \\ 0 & 0 & \vdots & -2 & 1 \end{bmatrix}$

The two zeros in the second row imply that the inverse does not exist.

19. $A = \begin{bmatrix} 2 & 7 & 1 \\ -3 & -9 & 2 \end{bmatrix}$ *A* has no inverse because it is not square.

21. $[A \;\vdots\; I] = \begin{bmatrix} 1 & 1 & 1 & \vdots & 1 & 0 & 0 \\ 3 & 5 & 4 & \vdots & 0 & 1 & 0 \\ 3 & 6 & 5 & \vdots & 0 & 0 & 1 \end{bmatrix}$

$\begin{matrix} -3R_1 + R_2 \to \\ -3R_1 + R_3 \to \end{matrix} \begin{bmatrix} 1 & 1 & 1 & \vdots & 1 & 0 & 0 \\ 0 & 2 & 1 & \vdots & -3 & 1 & 0 \\ 0 & 3 & 2 & \vdots & -3 & 0 & 1 \end{bmatrix}$

$\frac{1}{2}R_2 \to \begin{bmatrix} 1 & 1 & 1 & \vdots & 1 & 0 & 0 \\ 0 & 1 & \frac{1}{2} & \vdots & -\frac{3}{2} & \frac{1}{2} & 0 \\ 0 & 3 & 2 & \vdots & -3 & 0 & 1 \end{bmatrix}$

$\begin{matrix} -R_2 + R_1 \to \\ \\ -3R_2 + R_3 \to \end{matrix} \begin{bmatrix} 1 & 0 & \frac{1}{2} & \vdots & \frac{5}{2} & -\frac{1}{2} & 0 \\ 0 & 1 & \frac{1}{2} & \vdots & -\frac{3}{2} & \frac{1}{2} & 0 \\ 0 & 0 & \frac{1}{2} & \vdots & \frac{3}{2} & -\frac{3}{2} & 1 \end{bmatrix}$

$\begin{matrix} -R_3 + R_1 \to \\ -R_3 + R_2 \to \end{matrix} \begin{bmatrix} 1 & 0 & 0 & \vdots & 1 & 1 & -1 \\ 0 & 1 & 0 & \vdots & -3 & 2 & -1 \\ 0 & 0 & \frac{1}{2} & \vdots & \frac{3}{2} & -\frac{3}{2} & 1 \end{bmatrix}$

$2R_3 \to \begin{bmatrix} 1 & 0 & 0 & \vdots & 1 & 1 & -1 \\ 0 & 1 & 0 & \vdots & -3 & 2 & -1 \\ 0 & 0 & 1 & \vdots & 3 & -3 & 2 \end{bmatrix} = [I \;\vdots\; A^{-1}]$

$A^{-1} = \begin{bmatrix} 1 & 1 & -1 \\ -3 & 2 & -1 \\ 3 & -3 & 2 \end{bmatrix}$

23. $[A \; : \; I] = \begin{bmatrix} 1 & 0 & 0 & : & 1 & 0 & 0 \\ 3 & 4 & 0 & : & 0 & 1 & 0 \\ 2 & 5 & 5 & : & 0 & 0 & 1 \end{bmatrix}$

$\begin{matrix} -3R_1 + R_2 \rightarrow \\ -2R_1 + R_3 \rightarrow \end{matrix} \begin{bmatrix} 1 & 0 & 0 & : & 1 & 0 & 0 \\ 0 & 4 & 0 & : & -3 & 1 & 0 \\ 0 & 5 & 5 & : & -2 & 0 & 1 \end{bmatrix}$

$-\tfrac{5}{4}R_2 + R_3 \rightarrow \begin{bmatrix} 1 & 0 & 0 & : & 1 & 0 & 0 \\ 0 & 4 & 0 & : & -3 & 1 & 0 \\ 0 & 0 & 5 & : & \tfrac{7}{4} & -\tfrac{5}{4} & 1 \end{bmatrix}$

$\begin{matrix} \tfrac{1}{4}R_2 \rightarrow \\ \tfrac{1}{5}R_3 \rightarrow \end{matrix} \begin{bmatrix} 1 & 0 & 0 & : & 1 & 0 & 0 \\ 0 & 1 & 0 & : & -\tfrac{3}{4} & \tfrac{1}{4} & 0 \\ 0 & 0 & 1 & : & \tfrac{7}{20} & -\tfrac{1}{4} & \tfrac{1}{5} \end{bmatrix} = [I \; : \; A^{-1}]$

$A^{-1} = \begin{bmatrix} 1 & 0 & 0 \\ -\tfrac{3}{4} & \tfrac{1}{4} & 0 \\ \tfrac{7}{20} & -\tfrac{1}{4} & \tfrac{1}{5} \end{bmatrix}$

25. $[A \; : \; I] = \begin{bmatrix} -8 & 0 & 0 & 0 & : & 1 & 0 & 0 & 0 \\ 0 & 1 & 0 & 0 & : & 0 & 1 & 0 & 0 \\ 0 & 0 & 4 & 0 & : & 0 & 0 & 1 & 0 \\ 0 & 0 & 0 & -5 & : & 0 & 0 & 0 & 1 \end{bmatrix}$

$\begin{matrix} -\tfrac{1}{8}R_1 \rightarrow \\ \\ \tfrac{1}{4}R_3 \rightarrow \\ -\tfrac{1}{5}R_4 \rightarrow \end{matrix} \begin{bmatrix} 1 & 0 & 0 & 0 & : & -\tfrac{1}{8} & 0 & 0 & 0 \\ 0 & 1 & 0 & 0 & : & 0 & 1 & 0 & 0 \\ 0 & 0 & 1 & 0 & : & 0 & 0 & \tfrac{1}{4} & 0 \\ 0 & 0 & 0 & 1 & : & 0 & 0 & 0 & -\tfrac{1}{5} \end{bmatrix} = [I \; : \; A^{-1}]$

$A^{-1} = \begin{bmatrix} -\tfrac{1}{8} & 0 & 0 & 0 \\ 0 & 1 & 0 & 0 \\ 0 & 0 & \tfrac{1}{4} & 0 \\ 0 & 0 & 0 & -\tfrac{1}{5} \end{bmatrix}$

27. $A = \begin{bmatrix} 1 & 2 & -1 \\ 3 & 7 & -10 \\ -5 & -7 & -15 \end{bmatrix}$

$A^{-1} = \begin{bmatrix} -175 & 37 & -13 \\ 95 & -20 & 7 \\ 14 & -3 & 1 \end{bmatrix}$

29. $A = \begin{bmatrix} 1 & 1 & 2 \\ 3 & 1 & 0 \\ -2 & 0 & 3 \end{bmatrix}$

$A^{-1} = \tfrac{1}{2} \begin{bmatrix} -3 & 3 & 2 \\ 9 & -7 & -6 \\ -2 & 2 & 2 \end{bmatrix} = \begin{bmatrix} -1.5 & 1.5 & 1 \\ 4.5 & -3.5 & -3 \\ -1 & 1 & 1 \end{bmatrix}$

31. $A = \begin{bmatrix} -\tfrac{1}{2} & \tfrac{3}{4} & \tfrac{1}{4} \\ 1 & 0 & -\tfrac{3}{2} \\ 0 & -1 & \tfrac{1}{2} \end{bmatrix}$

$A^{-1} = \begin{bmatrix} -12 & -5 & -9 \\ -4 & -2 & -4 \\ -8 & -4 & -6 \end{bmatrix}$

33. $A = \begin{bmatrix} 0.1 & 0.2 & 0.3 \\ -0.3 & 0.2 & 0.2 \\ 0.5 & 0.4 & 0.4 \end{bmatrix}$

$A^{-1} = \tfrac{5}{11} \begin{bmatrix} 0 & -4 & 2 \\ -22 & 11 & 11 \\ 22 & -6 & -8 \end{bmatrix} = \begin{bmatrix} 0 & -1.\overline{81} & 0.\overline{90} \\ -10 & 5 & 5 \\ 10 & -2.\overline{72} & -3.\overline{63} \end{bmatrix}$

35. $A = \begin{bmatrix} 1 & 0 & 3 & 0 \\ 0 & 2 & 0 & 4 \\ 1 & 0 & 3 & 0 \\ 0 & 2 & 0 & 4 \end{bmatrix}$

A^{-1} does not exist.

37. $A = \begin{bmatrix} -1 & 0 & 1 & 0 \\ 0 & 2 & 0 & -1 \\ 2 & 0 & -1 & 0 \\ 0 & -1 & 0 & 1 \end{bmatrix}$

$A^{-1} = \begin{bmatrix} 1 & 0 & 1 & 0 \\ 0 & 1 & 0 & 1 \\ 2 & 0 & 1 & 0 \\ 0 & 1 & 0 & 2 \end{bmatrix}$

39. $A = \begin{bmatrix} a & b \\ c & d \end{bmatrix}, A^{-1} = \dfrac{1}{ad - bc} \begin{bmatrix} d & -b \\ -c & a \end{bmatrix}$

$A = \begin{bmatrix} 5 & -2 \\ 2 & 3 \end{bmatrix}$

$ad - bc = (5)(3) - (-2)(2) = 19$

$A^{-1} = \dfrac{1}{19} \begin{bmatrix} 3 & 2 \\ -2 & 5 \end{bmatrix} = \begin{bmatrix} \frac{3}{19} & \frac{2}{19} \\ -\frac{2}{19} & \frac{5}{19} \end{bmatrix}$

41. $A = \begin{bmatrix} -4 & -6 \\ 2 & 3 \end{bmatrix}$

$ad - bc = (-4)(3) - (2)(-6) = 0$

Since $ad - bc = 0$, A^{-1} does not exist.

43. $A = \begin{bmatrix} \frac{7}{2} & -\frac{3}{4} \\ \frac{1}{5} & \frac{4}{5} \end{bmatrix}$

$ad - bc = \left(\frac{7}{2}\right)\left(\frac{4}{5}\right) - \left(-\frac{3}{4}\right)\left(\frac{1}{5}\right) = \frac{28}{10} + \frac{3}{20} = \frac{59}{20}$

$A^{-1} = \dfrac{1}{\frac{59}{20}} \begin{bmatrix} \frac{4}{5} & \frac{3}{4} \\ -\frac{1}{5} & \frac{7}{2} \end{bmatrix} = \dfrac{20}{59} \begin{bmatrix} \frac{4}{5} & \frac{3}{4} \\ -\frac{1}{5} & \frac{7}{2} \end{bmatrix} = \begin{bmatrix} \frac{16}{59} & \frac{15}{59} \\ -\frac{4}{59} & \frac{70}{59} \end{bmatrix}$

45. $\begin{bmatrix} x \\ y \end{bmatrix} = \begin{bmatrix} -3 & 2 \\ -2 & 1 \end{bmatrix} \begin{bmatrix} 5 \\ 10 \end{bmatrix} = \begin{bmatrix} 5 \\ 0 \end{bmatrix}$

Solution: $(5, 0)$

47. $\begin{bmatrix} x \\ y \end{bmatrix} = \begin{bmatrix} -3 & 2 \\ -2 & 1 \end{bmatrix} \begin{bmatrix} 4 \\ 2 \end{bmatrix} = \begin{bmatrix} -8 \\ -6 \end{bmatrix}$

Solution: $(-8, -6)$

49. $\begin{bmatrix} x \\ y \\ z \end{bmatrix} = \begin{bmatrix} 1 & 1 & -1 \\ -3 & 2 & -1 \\ 3 & -3 & 2 \end{bmatrix} \begin{bmatrix} 0 \\ 5 \\ 2 \end{bmatrix} = \begin{bmatrix} 3 \\ 8 \\ -11 \end{bmatrix}$

Solution: $(3, 8, -11)$

51. $\begin{bmatrix} x_1 \\ x_2 \\ x_3 \\ x_4 \end{bmatrix} = \begin{bmatrix} -24 & 7 & 1 & -2 \\ -10 & 3 & 0 & -1 \\ -29 & 7 & 3 & -2 \\ 12 & -3 & -1 & 1 \end{bmatrix} \begin{bmatrix} 0 \\ 1 \\ -1 \\ 2 \end{bmatrix} = \begin{bmatrix} 2 \\ 1 \\ 0 \\ 0 \end{bmatrix}$

Solution: $(2, 1, 0, 0)$

53. $A = \begin{bmatrix} 3 & 4 \\ 5 & 3 \end{bmatrix}$

$A^{-1} = \dfrac{1}{9 - 20} \begin{bmatrix} 3 & -4 \\ -5 & 3 \end{bmatrix}$

$\begin{bmatrix} x \\ y \end{bmatrix} = -\dfrac{1}{11} \begin{bmatrix} 3 & -4 \\ -5 & 3 \end{bmatrix} \begin{bmatrix} -2 \\ 4 \end{bmatrix} = -\dfrac{1}{11} \begin{bmatrix} -22 \\ 22 \end{bmatrix} = \begin{bmatrix} 2 \\ -2 \end{bmatrix}$

Solution: $(2, -2)$

55. $A = \begin{bmatrix} -0.4 & 0.8 \\ 2 & -4 \end{bmatrix}$

$A^{-1} = \dfrac{1}{1.6 - 1.6} \begin{bmatrix} -4 & -0.8 \\ -2 & -0.4 \end{bmatrix}$

A^{-1} does not exist.

This implies that there is no unique solution; that is, either the system is inconsistent *or* there are infinitely many solutions.

Find the reduced row–echelon form of the matrix corresponding to the system.

$\begin{bmatrix} -0.4 & 0.8 & \vdots & 1.6 \\ 2 & -4 & \vdots & 5 \end{bmatrix}$

$-2.5R_1 \rightarrow \begin{bmatrix} 1 & -2 & \vdots & -4 \\ 2 & -4 & \vdots & 5 \end{bmatrix}$

$-2R_1 + R_2 \rightarrow \begin{bmatrix} 1 & -2 & \vdots & -4 \\ 0 & 0 & \vdots & 13 \end{bmatrix}$

The given system is inconsistent and there is no solution.

57. $\quad A = \begin{bmatrix} 3 & 6 \\ 6 & 14 \end{bmatrix}$

$$A^{-1} = \frac{1}{42-36}\begin{bmatrix} 14 & -6 \\ -6 & 3 \end{bmatrix}$$

$$\begin{bmatrix} x \\ y \end{bmatrix} = \frac{1}{6}\begin{bmatrix} 14 & -6 \\ -6 & 3 \end{bmatrix}\begin{bmatrix} 6 \\ 11 \end{bmatrix} = \frac{1}{6}\begin{bmatrix} 18 \\ -3 \end{bmatrix} = \begin{bmatrix} 3 \\ -\frac{1}{2} \end{bmatrix}$$

Solution: $\left(3, -\frac{1}{2}\right)$

59. $\quad A = \begin{bmatrix} -\frac{1}{4} & \frac{3}{8} \\ \frac{3}{2} & \frac{3}{4} \end{bmatrix}$

$$A^{-1} = \frac{1}{-\frac{3}{16}-\frac{9}{16}}\begin{bmatrix} \frac{3}{4} & -\frac{3}{8} \\ -\frac{3}{2} & -\frac{1}{4} \end{bmatrix} = -\frac{4}{3}\begin{bmatrix} \frac{3}{4} & -\frac{3}{8} \\ -\frac{3}{2} & -\frac{1}{4} \end{bmatrix} = \begin{bmatrix} -1 & \frac{1}{2} \\ 2 & \frac{1}{3} \end{bmatrix}$$

$$\begin{bmatrix} x \\ y \end{bmatrix} = \begin{bmatrix} -1 & \frac{1}{2} \\ 2 & \frac{1}{3} \end{bmatrix}\begin{bmatrix} -2 \\ -12 \end{bmatrix} = \begin{bmatrix} -4 \\ -8 \end{bmatrix}$$

Solution: $(-4, -8)$

61. $\quad A = \begin{bmatrix} 4 & -1 & 1 \\ 2 & 2 & 3 \\ 5 & -2 & 6 \end{bmatrix}$

Find A^{-1}.

$$[A \; \vdots \; I] = \begin{bmatrix} 4 & -1 & 1 & \vdots & 1 & 0 & 0 \\ 2 & 2 & 3 & \vdots & 0 & 1 & 0 \\ 5 & -2 & 6 & \vdots & 0 & 0 & 1 \end{bmatrix}$$

$$\begin{matrix} R_1 \\ \\ R_3 \end{matrix} \begin{bmatrix} 5 & -2 & 6 & \vdots & 0 & 0 & 1 \\ 2 & 2 & 3 & \vdots & 0 & 1 & 0 \\ 4 & -1 & 1 & \vdots & 1 & 0 & 0 \end{bmatrix}$$

$$-R_3 + R_1 \rightarrow \begin{bmatrix} 1 & -1 & 5 & \vdots & -1 & 0 & 1 \\ 2 & 2 & 3 & \vdots & 0 & 1 & 0 \\ 4 & -1 & 1 & \vdots & 1 & 0 & 0 \end{bmatrix}$$

$$\begin{matrix} -2R_1 + R_2 \rightarrow \\ -4R_1 + R_3 \rightarrow \end{matrix} \begin{bmatrix} 1 & -1 & 5 & \vdots & -1 & 0 & 1 \\ 0 & 4 & -7 & \vdots & 2 & 1 & -2 \\ 0 & 3 & -19 & \vdots & 5 & 0 & -4 \end{bmatrix}$$

$$-R_3 + R_2 \rightarrow \begin{bmatrix} 1 & -1 & 5 & \vdots & -1 & 0 & 1 \\ 0 & 1 & 12 & \vdots & -3 & 1 & 2 \\ 0 & 3 & -19 & \vdots & 5 & 0 & -4 \end{bmatrix}$$

$$\begin{matrix} R_2 + R_1 \rightarrow \\ \\ -3R_2 + R_3 \rightarrow \end{matrix} \begin{bmatrix} 1 & 0 & 17 & \vdots & -4 & 1 & 3 \\ 0 & 1 & 12 & \vdots & -3 & 1 & 2 \\ 0 & 0 & -55 & \vdots & 14 & -3 & -10 \end{bmatrix}$$

$$-\frac{1}{55}R_3 \rightarrow \begin{bmatrix} 1 & 0 & 17 & \vdots & -4 & 1 & 3 \\ 0 & 1 & 12 & \vdots & -3 & 1 & 2 \\ 0 & 0 & 1 & \vdots & -\frac{14}{55} & \frac{3}{55} & \frac{2}{11} \end{bmatrix}$$

$$\begin{matrix} -17R_3 + R_1 \rightarrow \\ -12R_3 + R_2 \rightarrow \\ \\ \end{matrix} \begin{bmatrix} 1 & 0 & 0 & \vdots & \frac{18}{55} & \frac{4}{55} & -\frac{1}{11} \\ 0 & 1 & 0 & \vdots & \frac{3}{55} & \frac{19}{55} & -\frac{2}{11} \\ 0 & 0 & 1 & \vdots & -\frac{14}{55} & \frac{3}{55} & \frac{2}{11} \end{bmatrix} = [I \; \vdots \; A^{-1}]$$

$$A^{-1} = \frac{1}{55}\begin{bmatrix} 18 & 4 & -5 \\ 3 & 19 & -10 \\ -14 & 3 & 10 \end{bmatrix}$$

$$\begin{bmatrix} x \\ y \\ z \end{bmatrix} = \frac{1}{55}\begin{bmatrix} 18 & 4 & -5 \\ 3 & 19 & -10 \\ -14 & 3 & 10 \end{bmatrix}\begin{bmatrix} -5 \\ 10 \\ 1 \end{bmatrix} = \frac{1}{55}\begin{bmatrix} -55 \\ 165 \\ 110 \end{bmatrix} = \begin{bmatrix} -1 \\ 3 \\ 2 \end{bmatrix} \qquad \text{Solution: } (-1, 3, 2)$$

63. $A = \begin{bmatrix} 5 & -3 & 2 \\ 2 & 2 & -3 \\ 1 & -7 & 8 \end{bmatrix}$

A^{-1} does not exist. This implies that there is no unique solution; that is, either the system is inconsistent *or* the system has infinitely many solution. Use a graphing utility to find the reduced row–echelon form of the matrix corresponding to the system.

$\begin{bmatrix} 5 & -3 & 2 & : & 2 \\ 2 & 2 & -3 & : & 3 \\ 1 & -7 & 8 & : & -4 \end{bmatrix}$

$\begin{bmatrix} 1 & 0 & -\frac{5}{16} & : & \frac{13}{16} \\ 0 & 1 & -\frac{19}{16} & : & \frac{11}{16} \\ 0 & 0 & 0 & : & 0 \end{bmatrix}$

$\begin{cases} x - \frac{5}{16}z = \frac{13}{16} \\ y - \frac{19}{16}z = \frac{11}{16} \end{cases}$

Let $z = a$. Then $x = \frac{5}{16}a + \frac{13}{16}$ and $y = \frac{19}{16}a + \frac{11}{16}$.

Solution: $\left(\frac{5}{16}a + \frac{13}{16}, \frac{19}{16}a + \frac{11}{16}, 16a \right)$, where a is a real number.

67. $A = \begin{bmatrix} 7 & -3 & 0 & 2 \\ -2 & 1 & 0 & -1 \\ 4 & 0 & 1 & -2 \\ -1 & 1 & 0 & -1 \end{bmatrix}$

$A^{-1} = \begin{bmatrix} 0 & -1 & 0 & 1 \\ -1 & -5 & 0 & 3 \\ -2 & -4 & 1 & -2 \\ -1 & -4 & 0 & 1 \end{bmatrix}$

$\begin{bmatrix} x \\ y \\ z \\ w \end{bmatrix} = \begin{bmatrix} 0 & -1 & 0 & 1 \\ -1 & -5 & 0 & 3 \\ -2 & -4 & 1 & -2 \\ -1 & -4 & 0 & 1 \end{bmatrix} \begin{bmatrix} 41 \\ -13 \\ 12 \\ -8 \end{bmatrix} = \begin{bmatrix} 5 \\ 0 \\ -2 \\ 3 \end{bmatrix}$

Solution: $(5, 0, -2, 3)$

69. The inverse matrix can be calculated once and used for more than one exercise.

65. $A = \begin{bmatrix} 2 & 3 & 5 \\ 3 & 5 & 9 \\ 5 & 9 & 17 \end{bmatrix}$

A^{-1} does not exist. This implies that there is no unique solution; that is, either the system is inconsistent *or* the system has infinitely many solution. Use a graphing utility to find the reduced row–echelon form of the matrix corresponding to the system.

$\begin{bmatrix} 2 & 3 & 5 & : & 4 \\ 3 & 5 & 9 & : & 7 \\ 5 & 9 & 17 & : & 13 \end{bmatrix}$

$\begin{bmatrix} 1 & 0 & -2 & : & -1 \\ 0 & 1 & 3 & : & 2 \\ 0 & 0 & 0 & : & 0 \end{bmatrix}$

$\begin{cases} x - 2z = -1 \\ y + 3z = 2 \end{cases}$

Let $z = a$. Then $x = 2a - 1$ and $y = -3a + 2$.

Solution: $(2a - 1, -3a + 2, a)$, where a is a real number.

71. A matrix is singular if it does not have an inverse.

73. $A = \begin{bmatrix} 1 & 1 & 1 \\ 0.065 & 0.07 & 0.09 \\ 0 & 2 & -1 \end{bmatrix}$

$[A \vdots I] = \begin{bmatrix} 1 & 1 & 1 & \vdots & 1 & 0 & 0 \\ 0.065 & 0.07 & 0.09 & \vdots & 0 & 1 & 0 \\ 0 & 2 & -1 & \vdots & 0 & 0 & 1 \end{bmatrix}$

$200R_2 \rightarrow \begin{bmatrix} 1 & 1 & 1 & \vdots & 1 & 0 & 0 \\ 13 & 14 & 18 & \vdots & 0 & 200 & 0 \\ 0 & 2 & -1 & \vdots & 0 & 0 & 1 \end{bmatrix}$

$-13R_1 + R_2 \rightarrow \begin{bmatrix} 1 & 1 & 1 & \vdots & 1 & 0 & 0 \\ 0 & 1 & 5 & \vdots & -13 & 200 & 0 \\ 0 & 2 & -1 & \vdots & 0 & 0 & 1 \end{bmatrix}$

$\begin{matrix} -R_2 + R_1 \rightarrow \\ \\ -2R_2 + R_3 \rightarrow \end{matrix} \begin{bmatrix} 1 & 0 & -4 & \vdots & 14 & -200 & 0 \\ 0 & 1 & 5 & \vdots & -13 & 200 & 0 \\ 0 & 0 & -11 & \vdots & 26 & -400 & 1 \end{bmatrix}$

$-\frac{1}{11}R_3 \rightarrow \begin{bmatrix} 1 & 0 & -4 & \vdots & 14 & -200 & 0 \\ 0 & 1 & 5 & \vdots & -13 & 200 & 0 \\ 0 & 0 & 1 & \vdots & -\frac{26}{11} & \frac{400}{11} & -\frac{1}{11} \end{bmatrix}$

$\begin{matrix} 4R_3 + R_1 \rightarrow \\ -5R_3 + R_2 \rightarrow \\ \\ \end{matrix} \begin{bmatrix} 1 & 0 & 0 & \vdots & \frac{50}{11} & -\frac{600}{11} & -\frac{4}{11} \\ 0 & 1 & 0 & \vdots & -\frac{13}{11} & \frac{200}{11} & \frac{5}{11} \\ 0 & 0 & 1 & \vdots & -\frac{26}{11} & \frac{400}{11} & -\frac{1}{11} \end{bmatrix} = [I \vdots A^{-1}]$

$X = A^{-1}B = \frac{1}{11}\begin{bmatrix} 50 & -600 & -4 \\ -13 & 200 & 5 \\ -26 & 400 & -1 \end{bmatrix}\begin{bmatrix} 10{,}000 \\ 705 \\ 0 \end{bmatrix} = \begin{bmatrix} 7000 \\ 1000 \\ 2000 \end{bmatrix}$

Answer: $7000 in AAA–rated bonds, $1000 in A–rated bonds, $2000 in B–rated bonds

75. Use the inverse matrix A^{-1} from Exercise 73.

$X = A^{-1}B = \frac{1}{11}\begin{bmatrix} 50 & -600 & -4 \\ -13 & 200 & 5 \\ -26 & 400 & -1 \end{bmatrix}\begin{bmatrix} 12{,}000 \\ 835 \\ 0 \end{bmatrix} = \begin{bmatrix} 9000 \\ 1000 \\ 2000 \end{bmatrix}$

Answer: $9000 in AAA–rated bonds, $1000 in A–rated bonds, $2000 in B–rated bonds

77. $A = \begin{bmatrix} 2 & 0 & 4 \\ 0 & 1 & 4 \\ 1 & 1 & -1 \end{bmatrix}$

$[A \vdots I] = \begin{bmatrix} 2 & 0 & 4 & \vdots & 1 & 0 & 0 \\ 0 & 1 & 4 & \vdots & 0 & 1 & 0 \\ 1 & 1 & -1 & \vdots & 0 & 0 & 1 \end{bmatrix}$

$\begin{matrix} R_1 \\ \\ R_3 \end{matrix} \begin{bmatrix} 1 & 1 & -1 & \vdots & 0 & 0 & 1 \\ 0 & 1 & 4 & \vdots & 0 & 1 & 0 \\ 2 & 0 & 4 & \vdots & 1 & 0 & 0 \end{bmatrix}$

$-2R_1 + R_3 \rightarrow \begin{bmatrix} 1 & 1 & -1 & \vdots & 0 & 0 & 1 \\ 0 & 1 & 4 & \vdots & 0 & 1 & 0 \\ 0 & -2 & 6 & \vdots & 1 & 0 & -2 \end{bmatrix}$

$\begin{matrix} -R_2 + R_1 \rightarrow \\ \\ 2R_2 + R_3 \rightarrow \end{matrix} \begin{bmatrix} 1 & 0 & -5 & \vdots & 0 & -1 & 1 \\ 0 & 1 & 4 & \vdots & 0 & 1 & 0 \\ 0 & 0 & 14 & \vdots & 1 & 2 & -2 \end{bmatrix}$

$\frac{1}{14}R_3 \rightarrow \begin{bmatrix} 1 & 0 & -5 & \vdots & 0 & -1 & 1 \\ 0 & 1 & 4 & \vdots & 0 & 1 & 0 \\ 0 & 0 & 1 & \vdots & \frac{1}{14} & \frac{1}{7} & -\frac{1}{7} \end{bmatrix}$

—CONTINUED—

77. **—CONTINUED—**

$$\begin{array}{c} 5R_3 + R_1 \rightarrow \\ -4R_3 + R_2 \rightarrow \\ \end{array} \begin{bmatrix} 1 & 0 & 0 & \vdots & \frac{5}{14} & -\frac{2}{7} & \frac{2}{7} \\ 0 & 1 & 0 & \vdots & -\frac{2}{7} & \frac{3}{7} & \frac{4}{7} \\ 0 & 0 & 1 & \vdots & \frac{1}{14} & \frac{1}{7} & -\frac{1}{7} \end{bmatrix} = \begin{bmatrix} I & \vdots & A^{-1} \end{bmatrix}$$

$$A^{-1} = \frac{1}{14} \begin{bmatrix} 5 & -4 & 4 \\ -4 & 6 & 8 \\ 1 & 2 & -2 \end{bmatrix}$$

$$\begin{bmatrix} I_1 \\ I_2 \\ I_3 \end{bmatrix} = \frac{1}{14} \begin{bmatrix} 5 & -4 & 4 \\ -4 & 6 & 8 \\ 1 & 2 & -2 \end{bmatrix} \begin{bmatrix} 14 \\ 28 \\ 0 \end{bmatrix} = \begin{bmatrix} -3 \\ 8 \\ 5 \end{bmatrix}$$

Answer: $I_1 = -3$ amperes, $I_2 = 8$ amperes, $I_3 = 5$ amperes

79. The sum of two invertible matrices is not necessarily invertible. For example, let

$$A = \begin{bmatrix} 1 & 0 \\ 0 & 1 \end{bmatrix} \text{ and } B = \begin{bmatrix} -1 & 0 \\ 0 & -1 \end{bmatrix}.$$

Then $A + B = \begin{bmatrix} 0 & 0 \\ 0 & 0 \end{bmatrix}$ which is clearly singular.

81. If $A = \begin{bmatrix} 4 & x \\ -2 & -3 \end{bmatrix}$ is singular then $ad - bc = -12 + 2x = 0$. Thus, $x = 6$.

83. True. If B is the inverse of A, then $AB = I = BA$.

85. True. If A is of order $m \times n$ and B is of order $n \times m$ (where $m \neq n$), the products AB and BA are of different orders and so cannot be equal to each other.

87. True. If A can be row reduced to the identity matrix, then A has an inverse and is nonsingular.

Section 14.7   The Determinant of a Square Matrix

1. 5

3. $\begin{vmatrix} 2 & 1 \\ 3 & 4 \end{vmatrix} = 2(4) - 1(3) = 8 - 3 = 5$

5. $\begin{vmatrix} 5 & 2 \\ -6 & 3 \end{vmatrix} = 5(3) - 2(-6) = 15 + 12 = 27$

7. $\begin{vmatrix} -7 & 0 \\ 3 & 0 \end{vmatrix} = -7(0) - 0(3) = 0$

9. $\begin{vmatrix} 2 & -3 \\ -6 & 9 \end{vmatrix} = (2)(9) - (-6)(-3) = 0$

11. $\begin{vmatrix} 4 & 7 \\ -2 & 5 \end{vmatrix} = (4)(5) - (-2)(7) = 34$

13. $\begin{vmatrix} -\frac{1}{2} & \frac{1}{3} \\ -6 & \frac{1}{3} \end{vmatrix} = \left(-\frac{1}{2}\right)\left(\frac{1}{3}\right) - \left(\frac{1}{3}\right)(-6) = \frac{11}{6}$

15. $\begin{vmatrix} 0.3 & 0.2 & 0.2 \\ 0.2 & 0.2 & 0.2 \\ -0.4 & 0.4 & 0.3 \end{vmatrix} = -0.002$

17. $\begin{vmatrix} 0.9 & 0.7 & 0 \\ -0.1 & 0.3 & 1.3 \\ -2.2 & 4.2 & 6.1 \end{vmatrix} = -4.842$

19. $\begin{bmatrix} 3 & 4 \\ 2 & -5 \end{bmatrix}$

(a) $M_{11} = -5$
$M_{12} = 2$
$M_{21} = 4$
$M_{22} = 3$

(b) $C_{11} = M_{11} = -5$
$C_{12} = -M_{12} = -2$
$C_{21} = -M_{21} = -4$
$C_{22} = M_{22} = 3$

21. $\begin{bmatrix} 3 & 1 \\ -2 & -4 \end{bmatrix}$

(a) $M_{11} = -4$
$M_{12} = -2$
$M_{21} = 1$
$M_{22} = 3$

(b) $C_{11} = M_{11} = -4$
$C_{12} = -M_{12} = 2$
$C_{21} = -M_{21} = -1$
$C_{22} = M_{22} = 3$

23. $\begin{bmatrix} 4 & 0 & 2 \\ -3 & 2 & 1 \\ 1 & -1 & 1 \end{bmatrix}$

(a) $M_{11} = \begin{vmatrix} 2 & 1 \\ -1 & 1 \end{vmatrix} = 2 - (-1) = 3$

$M_{12} = \begin{vmatrix} -3 & 1 \\ 1 & 1 \end{vmatrix} = -3 - 1 = -4$

$M_{13} = \begin{vmatrix} -3 & 2 \\ 1 & -1 \end{vmatrix} = 3 - 2 = 1$

$M_{21} = \begin{vmatrix} 0 & 2 \\ -1 & 1 \end{vmatrix} = 0 - (-2) = 2$

$M_{22} = \begin{vmatrix} 4 & 2 \\ 1 & 1 \end{vmatrix} = 4 - 2 = 2$

$M_{23} = \begin{vmatrix} 4 & 0 \\ 1 & -1 \end{vmatrix} = -4 - 0 = -4$

$M_{31} = \begin{vmatrix} 0 & 2 \\ 2 & 1 \end{vmatrix} = 0 - 4 = -4$

$M_{32} = \begin{vmatrix} 4 & 2 \\ -3 & 1 \end{vmatrix} = 4 - (-6) = 10$

$M_{33} = \begin{vmatrix} 4 & 0 \\ -3 & 2 \end{vmatrix} = 8 - 0 = 8$

(b) $C_{11} = (-1)^2 M_{11} = 3$
$C_{12} = (-1)^3 M_{12} = 4$
$C_{13} = (-1)^4 M_{13} = 1$
$C_{21} = (-1)^3 M_{21} = -2$
$C_{22} = (-1)^4 M_{22} = 2$
$C_{23} = (-1)^5 M_{23} = 4$
$C_{31} = (-1)^4 M_{31} = -4$
$C_{32} = (-1)^5 M_{32} = -10$
$C_{33} = (-1)^6 M_{33} = 8$

25. $\begin{bmatrix} 3 & -2 & 8 \\ 3 & 2 & -6 \\ -1 & 3 & 6 \end{bmatrix}$

(a) $M_{11} = \begin{vmatrix} 2 & -6 \\ 3 & 6 \end{vmatrix} = 12 + 18 = 30$

$M_{12} = \begin{vmatrix} 3 & -6 \\ -1 & 6 \end{vmatrix} = 18 - 6 = 12$

$M_{13} = \begin{vmatrix} 3 & 2 \\ -1 & 3 \end{vmatrix} = 9 + 2 = 11$

$M_{21} = \begin{vmatrix} -2 & 8 \\ 3 & 6 \end{vmatrix} = -12 - 24 = -36$

$M_{22} = \begin{vmatrix} 3 & 8 \\ -1 & 6 \end{vmatrix} = 18 + 8 = 26$

$M_{23} = \begin{vmatrix} 3 & -2 \\ -1 & 3 \end{vmatrix} = 9 - 2 = 7$

$M_{31} = \begin{vmatrix} -2 & 8 \\ 2 & -6 \end{vmatrix} = 12 - 16 = -4$

$M_{32} = \begin{vmatrix} 3 & 8 \\ 3 & -6 \end{vmatrix} = -18 - 24 = -42$

$M_{33} = \begin{vmatrix} 3 & -2 \\ 3 & 2 \end{vmatrix} = 6 + 6 = 12$

(b) $C_{11} = (-1)^2 M_{11} = 30$
$C_{12} = (-1)^3 M_{12} = -12$
$C_{13} = (-1)^4 M_{13} = 11$
$C_{21} = (-1)^3 M_{21} = 36$
$C_{22} = (-1)^4 M_{22} = 26$
$C_{23} = (-1)^5 M_{23} = -7$
$C_{31} = (-1)^4 M_{31} = -4$
$C_{32} = (-1)^5 M_{32} = 42$
$C_{33} = (-1)^6 M_{33} = 12$

27. (a) $\begin{vmatrix} -3 & 2 & 1 \\ 4 & 5 & 6 \\ 2 & -3 & 1 \end{vmatrix} = -3 \begin{vmatrix} 5 & 6 \\ -3 & 1 \end{vmatrix} - 2 \begin{vmatrix} 4 & 6 \\ 2 & 1 \end{vmatrix} + \begin{vmatrix} 4 & 5 \\ 2 & -3 \end{vmatrix} = -3(23) - 2(-8) - 22 = -75$

(b) $\begin{vmatrix} -3 & 2 & 1 \\ 4 & 5 & 6 \\ 2 & -3 & 1 \end{vmatrix} = -2 \begin{vmatrix} 4 & 6 \\ 2 & 1 \end{vmatrix} + 5 \begin{vmatrix} -3 & 1 \\ 2 & 1 \end{vmatrix} + 3 \begin{vmatrix} -3 & 1 \\ 4 & 6 \end{vmatrix} = -2(-8) + 5(-5) + 3(-22) = -75$

29. (a) $\begin{vmatrix} 5 & 0 & -3 \\ 0 & 12 & 4 \\ 1 & 6 & 3 \end{vmatrix} = 0 \begin{vmatrix} 0 & -3 \\ 6 & 3 \end{vmatrix} + 12 \begin{vmatrix} 5 & -3 \\ 1 & 3 \end{vmatrix} - 4 \begin{vmatrix} 5 & 0 \\ 1 & 6 \end{vmatrix} = 0(18) + 12(18) - 4(30) = 96$

(b) $\begin{vmatrix} 5 & 0 & -3 \\ 0 & 12 & 4 \\ 1 & 6 & 3 \end{vmatrix} = 0 \begin{vmatrix} 0 & 4 \\ 1 & 3 \end{vmatrix} + 12 \begin{vmatrix} 5 & -3 \\ 1 & 3 \end{vmatrix} - 6 \begin{vmatrix} 5 & -3 \\ 0 & 4 \end{vmatrix} = 0(-4) + 12(18) - 6(20) = 96$

31. (a) $\begin{vmatrix} 6 & 0 & -3 & 5 \\ 4 & 13 & 6 & -8 \\ -1 & 0 & 7 & 4 \\ 8 & 6 & 0 & 2 \end{vmatrix} = -4 \begin{vmatrix} 0 & -3 & 5 \\ 0 & 7 & 4 \\ 6 & 0 & 2 \end{vmatrix} + 13 \begin{vmatrix} 6 & -3 & 5 \\ -1 & 7 & 4 \\ 8 & 0 & 2 \end{vmatrix} - 6 \begin{vmatrix} 6 & 0 & 5 \\ -1 & 0 & 4 \\ 8 & 6 & 2 \end{vmatrix} - 8 \begin{vmatrix} 6 & 0 & -3 \\ -1 & 0 & 7 \\ 8 & 6 & 0 \end{vmatrix}$

$= -4(-282) + 13(-298) - 6(-174) - 8(-234) = 170$

(b) $\begin{vmatrix} 6 & 0 & -3 & 5 \\ 4 & 13 & 6 & -8 \\ -1 & 0 & 7 & 4 \\ 8 & 6 & 0 & 2 \end{vmatrix} = 0 \begin{vmatrix} 4 & 6 & -8 \\ -1 & 7 & 4 \\ 8 & 0 & 2 \end{vmatrix} + 13 \begin{vmatrix} 6 & -3 & 5 \\ -1 & 7 & 4 \\ 8 & 0 & 2 \end{vmatrix} + 0 \begin{vmatrix} 6 & -3 & 5 \\ 4 & 6 & -8 \\ 8 & 0 & 2 \end{vmatrix} + 6 \begin{vmatrix} 6 & -3 & 5 \\ 4 & 6 & -8 \\ -1 & 7 & 4 \end{vmatrix}$

$= 0 + 13(-298) + 0 + 6(674) = 170$

33. Expand along Column 1.

$\begin{vmatrix} 2 & -1 & 0 \\ 4 & 2 & 1 \\ 4 & 2 & 1 \end{vmatrix} = 2 \begin{vmatrix} 2 & 1 \\ 2 & 1 \end{vmatrix} - 4 \begin{vmatrix} -1 & 0 \\ 2 & 1 \end{vmatrix} + 4 \begin{vmatrix} -1 & 0 \\ 2 & 1 \end{vmatrix} = 2(0) - 4(-1) + 4(-1) = 0$

35. Expand along Row 2.

$\begin{vmatrix} 6 & 3 & -7 \\ 0 & 0 & 0 \\ 4 & -6 & 3 \end{vmatrix} = 0 \begin{vmatrix} 3 & -7 \\ -6 & 3 \end{vmatrix} - 0 \begin{vmatrix} 6 & -7 \\ 4 & 3 \end{vmatrix} + 0 \begin{vmatrix} 6 & 3 \\ 4 & -6 \end{vmatrix} = 0$

37. $\begin{vmatrix} -1 & 2 & 5 \\ 0 & 3 & 4 \\ 0 & 0 & 3 \end{vmatrix} = -1 \begin{vmatrix} 3 & 4 \\ 0 & 3 \end{vmatrix} = -1(9) = -9$

39. Expand along Column 3.

$\begin{vmatrix} 1 & 4 & -2 \\ 3 & 2 & 0 \\ -1 & 4 & 3 \end{vmatrix} = -2 \begin{vmatrix} 3 & 2 \\ -1 & 4 \end{vmatrix} + 3 \begin{vmatrix} 1 & 4 \\ 3 & 2 \end{vmatrix} = -2(14) + 3(-10) = -58$

41. Expand along Column 3.

$\begin{vmatrix} 2 & 6 & 6 & 2 \\ 2 & 7 & 3 & 6 \\ 1 & 5 & 0 & 1 \\ 3 & 7 & 0 & 7 \end{vmatrix} = 6 \begin{vmatrix} 2 & 7 & 6 \\ 1 & 5 & 1 \\ 3 & 7 & 7 \end{vmatrix} - 3 \begin{vmatrix} 2 & 6 & 2 \\ 1 & 5 & 1 \\ 3 & 7 & 7 \end{vmatrix} = 6(-20) - 3(16) = -168$

43. Expand along Column 1.

$$\begin{vmatrix} 5 & 3 & 0 & 6 \\ 4 & 6 & 4 & 12 \\ 0 & 2 & -3 & 4 \\ 0 & 1 & -2 & 2 \end{vmatrix} = 5\begin{vmatrix} 6 & 4 & 12 \\ 2 & -3 & 4 \\ 1 & -2 & 2 \end{vmatrix} - 4\begin{vmatrix} 3 & 0 & 6 \\ 2 & -3 & 4 \\ 1 & -2 & 2 \end{vmatrix} = 5(0) - 4(0) = 0$$

45. Expand along Column 2, then along Column 4.

$$\begin{vmatrix} 3 & 2 & 4 & -1 & 5 \\ -2 & 0 & 1 & 3 & 2 \\ 1 & 0 & 0 & 4 & 0 \\ 6 & 0 & 2 & -1 & 0 \\ 3 & 0 & 5 & 1 & 0 \end{vmatrix} = -2\begin{vmatrix} -2 & 1 & 3 & 2 \\ 1 & 0 & 4 & 0 \\ 6 & 2 & -1 & 0 \\ 3 & 5 & 1 & 0 \end{vmatrix} = (-2)(-2)\begin{vmatrix} 1 & 0 & 4 \\ 6 & 2 & -1 \\ 3 & 5 & 1 \end{vmatrix} = 4(103) = 412$$

47. $\begin{vmatrix} 3 & 8 & -7 \\ 0 & -5 & 4 \\ 8 & 1 & 6 \end{vmatrix} = -126$ **49.** $\begin{vmatrix} 7 & 0 & -14 \\ -2 & 5 & 4 \\ -6 & 2 & 12 \end{vmatrix} = 0$ **51.** $\begin{vmatrix} 1 & -1 & 8 & 4 \\ 2 & 6 & 0 & -4 \\ 2 & 0 & 2 & 6 \\ 0 & 2 & 8 & 0 \end{vmatrix} = -336$

53. (a) $\begin{vmatrix} -1 & 0 \\ 0 & 3 \end{vmatrix} = -3$ (b) $\begin{vmatrix} 2 & 0 \\ 0 & -1 \end{vmatrix} = -2$

(c) $\begin{bmatrix} -1 & 0 \\ 0 & 3 \end{bmatrix}\begin{bmatrix} 2 & 0 \\ 0 & -1 \end{bmatrix} = \begin{bmatrix} -2 & 0 \\ 0 & -3 \end{bmatrix}$ (d) $\begin{vmatrix} -2 & 0 \\ 0 & -3 \end{vmatrix} = 6$

55. (a) $\begin{vmatrix} 0 & 1 & 2 \\ -3 & -2 & 1 \\ 0 & 4 & 1 \end{vmatrix} = -21$ **57.** (a) $\begin{vmatrix} -1 & 2 & 1 \\ 1 & 0 & 1 \\ 0 & 1 & 0 \end{vmatrix} = 2$

(b) $\begin{vmatrix} 3 & -2 & 0 \\ 1 & -1 & 2 \\ 3 & 1 & 1 \end{vmatrix} = -19$ (b) $\begin{vmatrix} -1 & 0 & 0 \\ 0 & 2 & 0 \\ 0 & 0 & 3 \end{vmatrix} = -6$

(c) $\begin{bmatrix} 0 & 1 & 2 \\ -3 & -2 & 1 \\ 0 & 4 & 1 \end{bmatrix}\begin{bmatrix} 3 & -2 & 0 \\ 1 & -1 & 2 \\ 3 & 1 & 1 \end{bmatrix} = \begin{bmatrix} 7 & 1 & 4 \\ -8 & 9 & -3 \\ 7 & -3 & 9 \end{bmatrix}$ (c) $\begin{bmatrix} -1 & 2 & 1 \\ 1 & 0 & 1 \\ 0 & 1 & 0 \end{bmatrix}\begin{bmatrix} -1 & 0 & 0 \\ 0 & 2 & 0 \\ 0 & 0 & 3 \end{bmatrix} = \begin{bmatrix} 1 & 4 & 3 \\ -1 & 0 & 3 \\ 0 & 2 & 0 \end{bmatrix}$

(d) $\begin{vmatrix} 7 & 1 & 4 \\ -8 & 9 & -3 \\ 7 & -3 & 9 \end{vmatrix} = 399$ (d) $\begin{vmatrix} 1 & 4 & 3 \\ -1 & 0 & 3 \\ 0 & 2 & 0 \end{vmatrix} = -12$

59. $\begin{vmatrix} w & x \\ y & z \end{vmatrix} = wz - xy$

$-\begin{vmatrix} y & z \\ w & x \end{vmatrix} = -(xy - wz) = wz - xy$

Thus, $\begin{vmatrix} w & x \\ y & z \end{vmatrix} = -\begin{vmatrix} y & z \\ w & x \end{vmatrix}$.

61. $\begin{vmatrix} 1 & x & x^2 \\ 1 & y & y^2 \\ 1 & z & z^2 \end{vmatrix} = \begin{vmatrix} y & y^2 \\ z & z^2 \end{vmatrix} - \begin{vmatrix} x & x^2 \\ z & z^2 \end{vmatrix} + \begin{vmatrix} x & x^2 \\ y & y^2 \end{vmatrix}$

$= (yz^2 - y^2z) - (xz^2 - x^2z) + (xy^2 - x^2y)$

$= yz^2 - xz^2 - y^2z + x^2z + xy(y - x)$

$= z^2(y - x) - z(y^2 - x^2) + xy(y - x)$

$= z^2(y - x) - z(y - x)(y + x) + xy(y - x)$

$= (y - x)[z^2 - z(y + x) + xy]$

$= (y - x)[z^2 - zy - zx + xy]$

$= (y - x)[z^2 - zx - zy + xy]$

$= (y - x)[z(z - x) - y(z - x)]$

$= (y - x)(z - x)(z - y)$

63. $\begin{vmatrix} x - 1 & 2 \\ 3 & x - 2 \end{vmatrix} = 0$

$(x - 1)(x - 2) - 6 = 0$

$x^2 - 3x - 4 = 0$

$(x + 1)(x - 4) = 0$

$x = -1 \text{ or } x = 4$

65. $\begin{vmatrix} x + 3 & 2 \\ 1 & x + 2 \end{vmatrix} = 0$

$(x + 3)(x + 2) - 2 = 0$

$x^2 + 5x + 4 = 0$

$(x + 1)(x + 4) = 0$

$x = -1 \text{ or } x = -4$

67. A square matrix is a square array of numbers. The determinant of a square matrix is a real number.

69. Points: $(-4, 0)$ and $(4, 4)$

Equation: $\begin{vmatrix} x & y & 1 \\ -4 & 0 & 1 \\ 4 & 4 & 1 \end{vmatrix} = 0$

$x\begin{vmatrix} 0 & 1 \\ 4 & 1 \end{vmatrix} - y\begin{vmatrix} -4 & 1 \\ 4 & 1 \end{vmatrix} + 1\begin{vmatrix} -4 & 0 \\ 4 & 4 \end{vmatrix} = 0$

$-4x + 8y - 16 = 0 \implies x - 2y + 4 = 0$

71. Points: $\left(-\frac{5}{2}, 3\right)$ and $\left(\frac{7}{2}, 1\right)$

Equation: $\begin{vmatrix} x & y & 1 \\ -\frac{5}{2} & 3 & 1 \\ \frac{7}{2} & 1 & 1 \end{vmatrix}$

$x\begin{vmatrix} 3 & 1 \\ 1 & 1 \end{vmatrix} - y\begin{vmatrix} -\frac{5}{2} & 1 \\ \frac{7}{2} & 1 \end{vmatrix} + 1\begin{vmatrix} -\frac{5}{2} & 3 \\ \frac{7}{2} & 1 \end{vmatrix} = 0$

$2x + 6y - 13 = 0$

73. $\begin{vmatrix} 4u & -1 \\ -1 & 2v \end{vmatrix} = 8uv - 1$

75. $\begin{vmatrix} e^{2x} & e^{3x} \\ 2e^{2x} & 3e^{3x} \end{vmatrix} = 3e^{5x} - 2e^{5x} = e^{5x}$

77. True. If an entire row is zero, then each cofactor in the expansion is multiplied by zero.

79. Let $A = \begin{bmatrix} x_{11} & x_{12} & x_{13} \\ x_{21} & x_{22} & x_{23} \\ x_{31} & x_{32} & x_{33} \end{bmatrix}$ and $|A| = 5$.

$2A = \begin{bmatrix} 2x_{11} & 2x_{12} & 2x_{13} \\ 2x_{21} & 2x_{22} & 2x_{23} \\ 2x_{31} & 2x_{32} & 2x_{33} \end{bmatrix}$

$|2A| = 2x_{11}\begin{vmatrix} 2x_{22} & 2x_{23} \\ 2x_{32} & 2x_{33} \end{vmatrix} - 2x_{12}\begin{vmatrix} 2x_{21} & 2x_{23} \\ 2x_{31} & 2x_{33} \end{vmatrix} + 2x_{13}\begin{vmatrix} 2x_{21} & 2x_{22} \\ 2x_{31} & 2x_{32} \end{vmatrix}$

$= 2[x_{11}(4x_{22}x_{33} - 4x_{32}x_{23}) - x_{12}(4x_{21}x_{33} - 4x_{31}x_{23}) + x_{13}(4x_{21}x_{32} - 4x_{31}x_{22})]$

$= 8[x_{11}(x_{22}x_{33} - x_{32}x_{23}) - x_{12}(x_{21}x_{33} - x_{31}x_{23}) + x_{13}(x_{21}x_{32} - x_{31}x_{22})]$

$= 8|A|$

Thus, $|2A| = 8|A| = 8(5) = 40.$

Review Exercises for Chapter 14

1. $\begin{cases} x^2 - y^2 = 9 \\ x - y = 1 \implies x = y + 1 \end{cases}$

$$(y + 1)^2 - y^2 = 9$$
$$2y + 1 = 9$$
$$y = 4$$
$$x = 5$$

Solution: $(5, 4)$

3. $\begin{cases} y = 2x^2 \\ y = x^4 - 2x^2 \implies 2x^2 = x^4 - 2x^2 \end{cases}$

$$0 = x^4 - 4x^2$$
$$0 = x^2(x^2 - 4)$$
$$0 = x^2(x + 2)(x - 2)$$
$$x = 0, x = -2, x = 2$$
$$y = 0, y = 8, y = 8$$

Solutions: $(0, 0), (-2, 8), (2, 8)$

5. $\begin{cases} 2x - y = 10 \\ x + 5y = -6 \end{cases}$

Point of intersection: $(4, -2)$

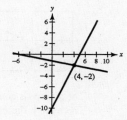

7. $\begin{cases} y = -2e^{-x} \\ 2e^x + y = 0 \implies y = -2e^x \end{cases}$

Point of intersection: $(0, -2)$

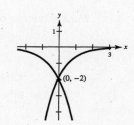

9. $\begin{cases} y = 2x^2 - 4x + 1 \\ y = x^2 - 4x + 3 \end{cases}$

Point of intersection:
$(1.41, -0.66), (-1.41, 10.66)$

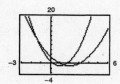

11. $\begin{cases} 2x - y = 2 \implies 16x - 8y = 16 \\ 6x + 8y = 39 \implies \underline{\quad 6x + 8y = 39} \end{cases}$

$$ 22x = 55$$
$$x = \tfrac{55}{22} = \tfrac{5}{2}$$

Back-substitute $x = \tfrac{5}{2}$ into Equation 1

$$2\left(\tfrac{5}{2}\right) - y = 2$$
$$y = 3$$

Solution: $\left(\tfrac{5}{2}, 3\right)$

13. $\begin{cases} 0.2x + 0.3y = 0.14 \implies 20x + 30y = 14 \implies 20x + 30y = 14 \\ 0.4x + 0.5y = 0.20 \implies 4x + 5y = 2 \implies \underline{-20x - 25y = -10} \end{cases}$

$$5y = 4$$
$$y = \tfrac{4}{5}$$

Back-substitute $y = \tfrac{4}{5}$ into Equation 2.

$$4x + 5\left(\tfrac{4}{5}\right) = 2$$
$$4x = -2$$
$$x = -\tfrac{1}{2}$$

Solution: $\left(-\tfrac{1}{2}, \tfrac{4}{5}\right) = (-0.5, 0.8)$

15. $\begin{cases} 3x - 2y = 0 \Rightarrow 3x - 2y = 0 \\ 3x + 2(y + 5) = 10 \Rightarrow \underline{3x + 2y = 0} \\ 6x = 0 \\ x = 0 \end{cases}$

Back-substitute $x = 0$ into Equation 1.
$$3(0) - 2y = 0$$
$$2y = 0$$
$$y = 0$$

Solution: $(0, 0)$

19. $\begin{cases} -3x - 5y = -1 \\ 6x + y = 4 \end{cases}$

The system is consistent.
The lines intersect at one
point so there is one solution.

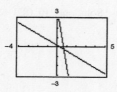

23. $C = 2.15x + 50,000$

$R = 6.95x$

Break-Even: $R = C$

$$6.95x = 2.15x + 50,000$$
$$4.80x = 50,000$$
$$x \approx 10,417 \text{ units}$$

27. $37 - 0.0002x = 22 + 0.00001x$

$$15 = 0.00021x$$

$$x = \frac{500,000}{7}, p = \frac{159}{7}$$

Point of equilibrium: $\left(\dfrac{500,000}{7}, \dfrac{159}{7}\right)$

31. $\begin{cases} x + 2y + 6z = 4 \\ -3x + 2y - z = -4 \\ 4x + 2z = 16 \end{cases}$

$\begin{cases} x + 2y + 6z = 4 \\ 8y + 17z = 8 \\ -8y - 22z = 0 \end{cases}$ $\begin{aligned} &3\text{Eq.1} + \text{Eq.2} \\ &-4\text{Eq.1} + \text{Eq.3} \end{aligned}$

$\begin{cases} x + 2y + 6z = 4 \\ 8y + 17z = 8 \\ -5z = 8 \end{cases}$ $\text{Eq.2} + \text{Eq.3}$

17. $\begin{cases} 1.25x - 2y = 3.5 \Rightarrow 5x - 8y = 14 \\ 5x - 8y = 14 \Rightarrow \underline{-5x + 8y = -14} \\ 0 = 0 \end{cases}$

There are infinitely many solutions.

Let $y = a$, then $5x - 8a = 14 \Rightarrow x = \frac{8}{5}a + \frac{14}{5}a$.

Solution: $\left(\frac{8}{5}a + \frac{14}{5}, a\right)$

21. $\begin{cases} 6x - 14.4y = 1.8 \\ 1.2x - 2.88y = 0.36 \end{cases}$

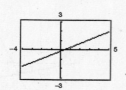

The system is consistent. The lines are identical.
There are infinitely many solutions.

25. $x =$ number of \$9.95 compact discs

$y =$ number of \$14.95 compact discs

$$x + y = 650 \Rightarrow y = 650 - x$$
$$9.95x + 14.95y = 7717.50$$
$$9.95x + 14.95(650 - x) = 7717.50$$
$$-5x = -2000$$
$$x = 400$$
$$y = 250$$

Solution: 400 at \$9.95 and 250 at \$14.95

29. $\begin{cases} x - 4y + 3z = 3 \\ -y + z = -1 \\ z = -5 \end{cases}$

$$-y + (-5) = -1 \Rightarrow y = -4$$
$$x - 4(-4) + 3(-5) = 3 \Rightarrow x = 2$$

Solution: $(2, -4, -5)$

$\begin{cases} x + 2y + 6z = 4 \\ 8y + 17z = 8 \\ z = -\frac{8}{5} \end{cases}$ $-\frac{1}{5}\text{Eq.3}$

$$8y + 17\left(-\frac{8}{5}\right) = 8 \Rightarrow y = \frac{22}{5}$$
$$x + 2\left(\frac{22}{5}\right) + 6\left(-\frac{8}{5}\right) = 4 \Rightarrow x = \frac{24}{5}$$

Solution: $\left(\frac{24}{5}, \frac{22}{5}, -\frac{8}{5}\right)$

33. $\begin{cases} x - 2y + z = -6 \\ 2x - 3y = -7 \\ -x + 3y - 3z = 11 \end{cases}$

$\begin{cases} x - 2y + z = -6 \\ \quad\;\; y - 2z = 5 \quad -2\text{Eq.1} + \text{Eq.2} \\ \quad\;\; y - 2z = 5 \quad \text{Eq.1} + \text{Eq.3} \end{cases}$

$\begin{cases} x - 2y + z = -6 \\ \quad\;\; y - 2z = 5 \\ \quad\quad\quad 0 = 0 \quad -\text{Eq.2} + \text{Eq.3} \end{cases}$

Let $z = a$, then:

$$y = 2a + 5$$
$$x - 2(2a + 5) + a = -6$$
$$x - 3a - 10 = -6$$
$$x = 3a + 4$$

Solution: $(3a + 4, 2a + 5, a)$ where a is any real number.

37. $y = ax^2 + bx + c$ through $(0, -5)$, $(1, -2)$, and $(2, 5)$.

$(0, -5): -5 = \qquad\quad + c \Rightarrow \qquad c = -5$
$(1, -2): -2 = a + b + c \Rightarrow \begin{cases} a + b = 3 \\ 2a + b = 5 \end{cases}$
$(2, \;\; 5): \;\; 5 = 4a + 2b + c \Rightarrow$

$\begin{cases} 2a + b = 5 \\ -a - b = -3 \end{cases}$
$\qquad\qquad a = 2$
$\qquad\qquad\quad b = 1$

The equation of the parabola is $y = 2x^2 + x - 5$.

39. $x^2 + y^2 + Dx + Ey + F = 0$ through $(-1, -2)$, $(5, -2)$ and $(2, 1)$.

$(-1,-2): \;\; 5 - D - 2E + F = 0 \Rightarrow \begin{cases} D + 2E - F = 5 \\ 5D - 2E + F = -29 \\ 2D + E + F = -5 \end{cases}$
$(\;\; 5,-2): 29 + 5D - 2E + F = 0 \Rightarrow$
$(\;\; 2, \;\; 1): \;\; 5 + 2D + 2E + F = 0 \Rightarrow$

From the first two equations we have

$$6D = -24$$
$$D = -4.$$

Substituting $D = -4$ into the second and third equations yields:

$-20 - 2E + F = -29 \Rightarrow \begin{cases} -2E + F = -9 \\ -E - F = -3 \end{cases}$
$-8 + E + F = -5 \Rightarrow$
$\qquad\qquad\qquad -3E = -12$
$\qquad\qquad\qquad\quad E = 4$
$\qquad\qquad\qquad\quad F = -1$

The equation of the circle is $x^2 + y^2 - 4x + 4y - 1 = 0$.

35. $\begin{cases} 5x - 12y + 7z = 16 \Rightarrow \\ 3x - 7y + 4z = 9 \Rightarrow \end{cases} \begin{cases} 15x - 36y + 21z = 48 \\ \underline{-15x + 35y - 20z = -45} \\ \qquad\quad -y + z = 3 \end{cases}$

Let $y = a$.

Then $z = a + 3$
and $5x - 12a + 7(a + 3) = 16 \Rightarrow x = a - 1.$

Solution: $(a - 1, a, a + 3)$ where a is any real number.

41. From the following chart we obtain our system of equations.

	A	B	C
Mixture X	$\frac{1}{5}$	$\frac{2}{5}$	$\frac{2}{5}$
Mixture Y	0	0	1
Mixture Z	$\frac{1}{3}$	$\frac{1}{3}$	$\frac{1}{3}$
Desired Mixture	$\frac{6}{27}$	$\frac{8}{27}$	$\frac{13}{27}$

$$\left.\begin{array}{l} \frac{1}{5}x + \frac{1}{3}z = \frac{6}{27} \\ \frac{2}{5}x + \frac{1}{3}z = \frac{8}{27} \end{array}\right\} x = \frac{10}{27},\ z = \frac{12}{27}$$

$$\frac{2}{5}x + y + \frac{1}{3}z = \frac{13}{27} \implies y = \frac{5}{27}$$

To obtain the desired mixture, use 10 gallons of spray X, 5 gallons of spray Y, and 12 gallons of spray Z.

43. $s(t) = \frac{1}{2}at^2 + v_0 t + s_0$

$(1, 52),\ (2, 72),\ (3, 60)$

$52 = \frac{1}{2}a + v_0 + s_0 \implies a + 2v_0 + 2s_0 = 104$

$72 = 2a + 2v_0 + s_0 \implies 2a + 2v_0 + s_0 = 72$

$60 = \frac{9}{2}a + 3v_0 + s_0 \implies 9a + 6v_0 + 2s_0 = 120$

Solving this system yields $a = -32$, $v_0 = 68$, $s_0 = 0$.

Thus, $s(t) = \frac{1}{2}(-32)t^2 + 68t + 0 = -16t^2 + 68t$.

45. $y \le 5 - \frac{1}{2}x$

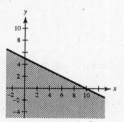

47. $y - 4x^2 > -1$

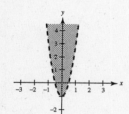

49. $\begin{cases} x + 2y \le 160 \\ 3x + y \le 180 \\ \quad\ x \ge 0 \\ \quad\ y \ge 0 \end{cases}$

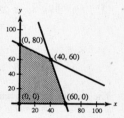

Vertex A	Vertex B	Vertex C	Vertex D
$x + 2y = 160$	$x + 2y = 160$	$3x + y = 180$	$x = 0$
$3x + y = 180$	$x = 0$	$y = 0$	$y = 0$
$(40, 60)$	$(0, 80)$	$(60, 0)$	$(0, 0)$

51. $\begin{cases} 3x + 2y \ge 24 \\ x + 2y \ge 12 \\ 2 \le x \le 15 \\ \quad\ y \le 15 \end{cases}$

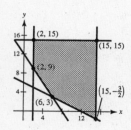

Vertex A	Vertex B
$3x + 2y = 24$	$3x + 2y = 24$
$x + 2y = 12$	$x = 2$
$(6, 3)$	$(2, 9)$

Vertex C	Vertex D	Vertex E
$x = 2$	$x = 15$	$x + 2y = 12$
$y = 15$	$y = 15$	$x = 15$
$(2, 15)$	$(15, 15)$	$\left(15, -\frac{3}{2}\right)$

53. $\begin{cases} y < x + 1 \\ y > x^2 - 1 \end{cases}$

Vertices:

$x + 1 = x^2 - 1$

$0 = x^2 - x - 2 = (x + 1)(x - 2)$

$x = -1$ or $x = 2$

$y = 0 \qquad y = 3$

$(-1, 0) \qquad (2, 3)$

55. $\begin{cases} 2x - 3y \geq 0 \\ 2x - y \leq 8 \\ y \geq 0 \end{cases}$

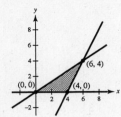

Vertex A	Vertex B	Vertex C
$2x - 3y = 0$	$2x - 3y = 0$	$2x - y = 8$
$2x - y = 8$	$y = 0$	$y = 0$
$(6, 4)$	$(0, 0)$	$(4, 0)$

57. Let x = the number of bushels for Harrisburg, and
y = the number of bushels for Philadelphia.

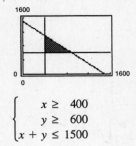

$\begin{cases} x \geq 400 \\ y \geq 600 \\ x + y \leq 1500 \end{cases}$

59. $160 - 0.0001x = 70 + 0.0002x$

$90 = 0.0003x$

$x = 300,000 \text{ units}$

$p = \$130$

Point of equilibrium: $(300,000, 130)$

Consumer surplus: $\frac{1}{2}(300,000)(30) = \$4,500,000$

Producer surplus: $\frac{1}{2}(300,000)(60) = \$9,000,000$

61. $\begin{bmatrix} -4 \\ 0 \\ 5 \end{bmatrix}$

Order: 3×1

63. $\begin{bmatrix} 3 \end{bmatrix}$

Order: 1×1

65. $\begin{bmatrix} 2 & 3 & -2 \\ 1 & 0 & 1 \end{bmatrix}$

The matrix has two rows and three columns.

Order: 2×3

67. $\begin{cases} 3x - 10y = 15 \\ 5x + 4y = 22 \end{cases}$

$\begin{bmatrix} 3 & -10 & \vdots & 15 \\ 5 & 4 & \vdots & 22 \end{bmatrix}$

69. $\begin{bmatrix} 5 & 1 & 7 & \vdots & -9 \\ 4 & 2 & 0 & \vdots & 10 \\ 9 & 4 & 2 & \vdots & 3 \end{bmatrix}$

$\begin{cases} 5x + y + 7z = -9 \\ 4x + 2y = 10 \\ 9x + 4y + 2z = 3 \end{cases}$

$\begin{matrix} -2R_2 + R_1 \rightarrow \\ \\ 2R_2 + R_3 \rightarrow \end{matrix} \begin{bmatrix} 1 & 0 & 1 \\ 0 & 1 & 1 \\ 0 & 0 & -2 \end{bmatrix}$

$\begin{matrix} \\ \\ -\frac{1}{2}R_3 \rightarrow \end{matrix} \begin{bmatrix} 1 & 0 & 1 \\ 0 & 1 & 1 \\ 0 & 0 & 1 \end{bmatrix}$

$\begin{matrix} -R_3 + R_1 \rightarrow \\ -R_3 + R_2 \rightarrow \\ \end{matrix} \begin{bmatrix} 1 & 0 & 0 \\ 0 & 1 & 0 \\ 0 & 0 & 1 \end{bmatrix}$

71. $\begin{bmatrix} 0 & 1 & 1 \\ 1 & 2 & 3 \\ 2 & 2 & 2 \end{bmatrix}$

$\begin{matrix} R_1 \\ R_2 \end{matrix} \begin{bmatrix} 1 & 2 & 3 \\ 0 & 1 & 1 \\ 2 & 2 & 2 \end{bmatrix}$

$-2R_1 + R_3 \rightarrow \begin{bmatrix} 1 & 2 & 3 \\ 0 & 1 & 1 \\ 0 & -2 & -4 \end{bmatrix}$

73. $\begin{bmatrix} 1 & 2 & 3 & \vdots & 9 \\ 0 & 1 & -2 & \vdots & 2 \\ 0 & 0 & 0 & \vdots & 0 \end{bmatrix}$

Consistent

Infinitely many solutions

75. $\begin{bmatrix} 1 & 2 & 3 & \vdots & 9 \\ 0 & 1 & -2 & \vdots & 2 \\ 0 & 0 & 1 & \vdots & -3 \end{bmatrix}$

Consistent

One solution

77. $\begin{bmatrix} 5 & 4 & \vdots & 2 \\ -1 & 1 & \vdots & -22 \end{bmatrix}$

$4R_2 + R_1 \to \begin{bmatrix} 1 & 8 & \vdots & -86 \\ -1 & 1 & \vdots & -22 \end{bmatrix}$

$R_1 + R_2 \to \begin{bmatrix} 1 & 8 & \vdots & -86 \\ 0 & 9 & \vdots & -108 \end{bmatrix}$

$\frac{1}{9}R_2 \to \begin{bmatrix} 1 & 8 & \vdots & -86 \\ 0 & 1 & \vdots & -12 \end{bmatrix}$

$\begin{cases} x + 8y = -86 \\ \quad\ y = -12 \end{cases}$

$y = -12$

$x + 8(-12) = -86 \implies x = 10$

Solution: $(10, -12)$

79. $\begin{bmatrix} 0.3 & -0.1 & \vdots & -0.13 \\ 0.2 & -0.3 & \vdots & -0.25 \end{bmatrix}$

$\begin{matrix} 10R_1 \to \\ 10R_2 \to \end{matrix} \begin{bmatrix} 3 & -1 & \vdots & -1.3 \\ 2 & -3 & \vdots & -2.5 \end{bmatrix}$

$-R_2 + R_1 \to \begin{bmatrix} 1 & 2 & \vdots & 1.2 \\ 2 & -3 & \vdots & -2.5 \end{bmatrix}$

$-2R_1 + R_2 \to \begin{bmatrix} 1 & 2 & \vdots & 1.2 \\ 0 & -7 & \vdots & -4.9 \end{bmatrix}$

$-\frac{1}{7}R_2 \to \begin{bmatrix} 1 & 2 & \vdots & 1.2 \\ 0 & 1 & \vdots & 0.7 \end{bmatrix}$

$\begin{cases} x + 2y = 1.2 \\ \quad\ y = 0.7 \end{cases}$

$y = 0.7$

$x + 2(0.7) = 1.2 \implies x = -0.2$

Solution: $(-0.2, 0.7)$

81. $\begin{bmatrix} 2 & 3 & 1 & \vdots & 10 \\ 2 & -3 & -3 & \vdots & 22 \\ 4 & -2 & 3 & \vdots & -2 \end{bmatrix}$

$\begin{matrix} -R_1 + R_2 \to \\ -2R_1 + R_3 \to \end{matrix} \begin{bmatrix} 2 & 3 & 1 & \vdots & 10 \\ 0 & -6 & -4 & \vdots & 12 \\ 0 & -8 & 1 & \vdots & -22 \end{bmatrix}$

$\begin{matrix} \frac{1}{2}R_1 \to \\ -\frac{1}{6}R_2 \to \end{matrix} \begin{bmatrix} 1 & \frac{3}{2} & \frac{1}{2} & \vdots & 5 \\ 0 & 1 & \frac{2}{3} & \vdots & -2 \\ 0 & -8 & 1 & \vdots & -22 \end{bmatrix}$

$8R_2 + R_3 \to \begin{bmatrix} 1 & \frac{3}{2} & \frac{1}{2} & & 5 \\ 0 & 1 & \frac{2}{3} & & -2 \\ 0 & 0 & \frac{19}{3} & & -38 \end{bmatrix}$

$\frac{3}{19}R_3 \to \begin{bmatrix} 1 & \frac{3}{2} & \frac{1}{2} & \vdots & 5 \\ 0 & 1 & \frac{2}{3} & \vdots & -2 \\ 0 & 0 & 1 & \vdots & -6 \end{bmatrix}$

$z = -6$

$y + \frac{2}{3}(-6) = -2 \implies y = 2$

$x + \frac{3}{2}(2) + \frac{1}{2}(-6) = 5 \implies x = 5$

Solution: $(5, 2, -6)$

83. $\begin{bmatrix} 2 & 1 & 2 & \vdots & 4 \\ 2 & 2 & 0 & \vdots & 5 \\ 2 & -1 & 6 & \vdots & 2 \end{bmatrix}$

$\begin{matrix} -R_1 + R_2 \to \\ -R_1 + R_3 \to \end{matrix} \begin{bmatrix} 2 & 1 & 2 & \vdots & 4 \\ 0 & 1 & -2 & \vdots & 1 \\ 0 & -2 & 4 & \vdots & -2 \end{bmatrix}$

$\begin{matrix} -R_2 + R_1 \to \\ \\ 2R_2 + R_3 \to \end{matrix} \begin{bmatrix} 2 & 0 & 4 & \vdots & 3 \\ 0 & 1 & -2 & \vdots & 1 \\ 0 & 0 & 0 & \vdots & 0 \end{bmatrix}$

$\frac{1}{2}R_1 \to \begin{bmatrix} 1 & 0 & 2 & \vdots & \frac{3}{2} \\ 0 & 1 & -2 & \vdots & 1 \\ 0 & 0 & 0 & \vdots & 0 \end{bmatrix}$

Let $z = a$, then:

$y - 2a = 1 \implies y = 2a + 1$

$x + 2a = \frac{3}{2} \implies x = -2a + \frac{3}{2}$

Solution: $\left(-2a + \frac{3}{2}, 2a + 1, a\right)$

85.
$$\begin{bmatrix} 2 & 1 & 1 & 0 & \vdots & 6 \\ 0 & -2 & 3 & -1 & \vdots & 9 \\ 3 & 3 & -2 & -2 & \vdots & -11 \\ 1 & 0 & 1 & 3 & \vdots & 14 \end{bmatrix}$$

$$-R_4 + R_1 \begin{bmatrix} 1 & 1 & 0 & -3 & \vdots & -8 \\ 0 & -2 & 3 & -1 & \vdots & 9 \\ 3 & 3 & -2 & -2 & \vdots & -11 \\ 1 & 0 & 1 & 3 & \vdots & 14 \end{bmatrix}$$

$$\begin{matrix} \\ \\ -3R_1 + R_3 \to \\ -R_1 + R_4 \to \end{matrix} \begin{bmatrix} 1 & 1 & 0 & -3 & \vdots & -8 \\ 0 & -2 & 3 & -1 & \vdots & 9 \\ 0 & 0 & -2 & 7 & \vdots & 13 \\ 0 & -1 & 1 & 6 & \vdots & 22 \end{bmatrix}$$

$$-3R_4 + R_2 \to \begin{bmatrix} 1 & 1 & 0 & -3 & \vdots & -8 \\ 0 & 1 & 0 & -19 & \vdots & -57 \\ 0 & 0 & -2 & 7 & \vdots & 13 \\ 0 & -1 & 1 & 6 & \vdots & 22 \end{bmatrix}$$

$$R_2 + R_4 \to \begin{bmatrix} 1 & 1 & 0 & -3 & \vdots & -8 \\ 0 & 1 & 0 & -19 & \vdots & -57 \\ 0 & 0 & -2 & 7 & \vdots & 13 \\ 0 & 0 & 1 & -13 & \vdots & -35 \end{bmatrix}$$

$$\begin{matrix} \\ \\ R_4 \\ R_3 \end{matrix} \begin{bmatrix} 1 & 1 & 0 & -3 & \vdots & -8 \\ 0 & 1 & 0 & -19 & \vdots & -57 \\ 0 & 0 & 1 & -13 & \vdots & -35 \\ 0 & 0 & -2 & 7 & \vdots & 13 \end{bmatrix}$$

$$2R_3 + R_4 \to \begin{bmatrix} 1 & 1 & 0 & -3 & \vdots & -8 \\ 0 & 1 & 0 & -19 & \vdots & -57 \\ 0 & 0 & 1 & -13 & \vdots & -35 \\ 0 & 0 & 0 & -19 & \vdots & -57 \end{bmatrix}$$

$$\tfrac{1}{19}R_4 \to \begin{bmatrix} 1 & 1 & 0 & -3 & \vdots & -8 \\ 0 & 1 & 0 & -19 & \vdots & -57 \\ 0 & 0 & 1 & -13 & \vdots & -35 \\ 0 & 0 & 0 & 1 & \vdots & 3 \end{bmatrix}$$

$w = 3$

$z - 13(3) = -35 \implies z = 4$

$y - 19(3) = -57 \implies y = 0$

$x + 0 - 3(3) = -8 \implies x = 1$

Solution: $(1, 0, 4, 3)$

87.
$$\begin{bmatrix} -1 & 1 & 2 & \vdots & 1 \\ 2 & 3 & 1 & \vdots & -2 \\ 5 & 4 & 2 & \vdots & 4 \end{bmatrix}$$

$$-R_1 \to \begin{bmatrix} 1 & -1 & -2 & \vdots & -1 \\ 2 & 3 & 1 & \vdots & -2 \\ 5 & 4 & 2 & \vdots & 4 \end{bmatrix}$$

$$\begin{matrix} -2R_1 + R_2 \to \\ -5R_1 + R_3 \to \end{matrix} \begin{bmatrix} 1 & -1 & -2 & \vdots & -1 \\ 0 & 5 & 5 & \vdots & 0 \\ 0 & 9 & 12 & \vdots & 9 \end{bmatrix}$$

$$\tfrac{1}{5}R_2 \to \begin{bmatrix} 1 & -1 & -2 & \vdots & -1 \\ 0 & 1 & 1 & \vdots & 0 \\ 0 & 9 & 12 & \vdots & 9 \end{bmatrix}$$

$$\begin{matrix} R_2 + R_1 \to \\ \\ -9R_2 + R_3 \to \end{matrix} \begin{bmatrix} 1 & 0 & -1 & \vdots & -1 \\ 0 & 1 & 1 & \vdots & 0 \\ 0 & 0 & 3 & \vdots & 9 \end{bmatrix}$$

$$\tfrac{1}{3}R_3 \to \begin{bmatrix} 1 & 0 & -1 & \vdots & -1 \\ 0 & 1 & 1 & \vdots & 0 \\ 0 & 0 & 1 & \vdots & 3 \end{bmatrix}$$

$$\begin{matrix} R_3 + R_1 \to \\ -R_3 + R_2 \to \end{matrix} \begin{bmatrix} 1 & 0 & 0 & \vdots & 2 \\ 0 & 1 & 0 & \vdots & -3 \\ 0 & 0 & 1 & \vdots & 3 \end{bmatrix}$$

$x = 2, y = -3, z = 3$

Solution: $(2, -3, 3)$

89.
$$\begin{bmatrix} 2 & -1 & 9 & \vdots & -8 \\ -1 & -3 & 4 & \vdots & -15 \\ 5 & 2 & -1 & \vdots & 17 \end{bmatrix}$$

$$R_2 + R_1 \rightarrow \begin{bmatrix} 1 & -4 & 13 & \vdots & -23 \\ -1 & -3 & 4 & \vdots & -15 \\ 5 & 2 & -1 & \vdots & 17 \end{bmatrix}$$

$$\begin{matrix} R_1 + R_2 \rightarrow \\ -5R_1 + R_3 \rightarrow \end{matrix} \begin{bmatrix} 1 & -4 & 13 & \vdots & -23 \\ 0 & -7 & 17 & \vdots & -38 \\ 0 & 22 & -66 & \vdots & 132 \end{bmatrix}$$

$$\begin{matrix} R_3 \\ R_2 \end{matrix} \begin{bmatrix} 1 & -4 & 13 & \vdots & -23 \\ 0 & 22 & -66 & \vdots & 132 \\ 0 & -7 & 17 & \vdots & 38 \end{bmatrix}$$

$$\tfrac{1}{22}R_2 \rightarrow \begin{bmatrix} 1 & -4 & 13 & \vdots & -23 \\ 0 & 1 & -3 & \vdots & 6 \\ 0 & -7 & 17 & \vdots & -38 \end{bmatrix}$$

$$7R_2 + R_3 \rightarrow \begin{bmatrix} 1 & -4 & 13 & \vdots & -23 \\ 0 & 1 & -3 & \vdots & 6 \\ 0 & 0 & -4 & \vdots & 4 \end{bmatrix}$$

$$-\tfrac{1}{4}R_3 \begin{bmatrix} 1 & -4 & 13 & \vdots & -23 \\ 0 & 1 & -3 & \vdots & 6 \\ 0 & 0 & 1 & \vdots & -1 \end{bmatrix}$$

$$4R_2 + R_1 \rightarrow \begin{bmatrix} 1 & 0 & 1 & \vdots & 1 \\ 0 & 1 & -3 & \vdots & 6 \\ 0 & 0 & 1 & \vdots & -1 \end{bmatrix}$$

$$\begin{matrix} -R_3 + R_1 \rightarrow \\ 3R_3 + R_2 \rightarrow \end{matrix} \begin{bmatrix} 1 & 0 & 0 & \vdots & 2 \\ 0 & 1 & 0 & \vdots & 3 \\ 0 & 0 & 1 & \vdots & -1 \end{bmatrix}$$

$x = 2, y = 3, z = -1$

Solution: $(2, 3, -1)$

91. Use the reduced row–echelon form feature of a graphing utility.

$$\begin{bmatrix} 3 & -1 & 5 & -2 & \vdots & -44 \\ 1 & 6 & 4 & -1 & \vdots & 1 \\ 5 & -1 & 1 & 3 & \vdots & -15 \\ 0 & 4 & -1 & -8 & \vdots & 58 \end{bmatrix} \Rightarrow \begin{bmatrix} 1 & 0 & 0 & 0 & \vdots & 2 \\ 0 & 1 & 0 & 0 & \vdots & 6 \\ 0 & 0 & 1 & 0 & \vdots & -10 \\ 0 & 0 & 0 & 1 & \vdots & -3 \end{bmatrix}$$

$x = 2, y = 6, z = -10, w = -3$

Solution: $(2, 6, -10, -3)$

93. $\begin{bmatrix} -1 & x \\ y & 9 \end{bmatrix} = \begin{bmatrix} -1 & 12 \\ -7 & 9 \end{bmatrix} \Rightarrow x = 12$ and $y = -7$

95. Since A and B are both of order 2×2, $A + 3B$ can be performed.

97. $\begin{bmatrix} 7 & 3 \\ -1 & 5 \end{bmatrix} + \begin{bmatrix} 10 & -20 \\ 14 & -3 \end{bmatrix} = \begin{bmatrix} 7+10 & 3-20 \\ -1+14 & 5-3 \end{bmatrix} = \begin{bmatrix} 17 & -17 \\ 13 & 2 \end{bmatrix}$

99. $-2\begin{bmatrix} 1 & 2 \\ 5 & -4 \\ 6 & 0 \end{bmatrix} + 8\begin{bmatrix} 7 & 1 \\ 1 & 2 \\ 1 & 4 \end{bmatrix} = \begin{bmatrix} -2 & -4 \\ -10 & 8 \\ -12 & 0 \end{bmatrix} + \begin{bmatrix} 56 & 8 \\ 8 & 16 \\ 8 & 32 \end{bmatrix} = \begin{bmatrix} 54 & 4 \\ -2 & 24 \\ -4 & 32 \end{bmatrix}$

101. $X = 3A - 2B = 3\begin{bmatrix} -4 & 0 \\ 1 & -5 \\ -3 & 2 \end{bmatrix} - 2\begin{bmatrix} 1 & 2 \\ -2 & 1 \\ 4 & 4 \end{bmatrix}$

$$= \begin{bmatrix} -14 & -4 \\ 7 & -17 \\ -17 & -2 \end{bmatrix}$$

103. $X = \dfrac{1}{3}[B - 2A] = \dfrac{1}{3}\left(\begin{bmatrix} 1 & 2 \\ -2 & 1 \\ 4 & 4 \end{bmatrix} - 2\begin{bmatrix} -4 & 0 \\ 1 & -5 \\ -3 & 2 \end{bmatrix} \right)$

$$= \dfrac{1}{3}\begin{bmatrix} 9 & 2 \\ -4 & 11 \\ 10 & 0 \end{bmatrix}$$

105. Not possible because the number of columns of A does not equal the number of rows of B.

107.
$$\begin{bmatrix} 1 & 2 \\ 5 & -4 \\ 6 & 0 \end{bmatrix} \begin{bmatrix} 6 & -2 & 8 \\ 4 & 0 & 0 \end{bmatrix} = \begin{bmatrix} 1(6) + 2(4) & 1(-2) + 2(0) & 1(8) + 2(0) \\ 5(6) + (-4)(4) & 5(-2) + (-4)(0) & 5(8) + (-4)(0) \\ 6(6) + (0)(4) & 6(-2) + (0)(0) & 6(8) + (0)(0) \end{bmatrix}$$

$$= \begin{bmatrix} 14 & -2 & 8 \\ 14 & -10 & 40 \\ 36 & -12 & 48 \end{bmatrix}$$

109.
$$\begin{bmatrix} 1 & 5 & 6 \\ 2 & -4 & 0 \end{bmatrix} \begin{bmatrix} 6 & 4 \\ -2 & 0 \\ 8 & 0 \end{bmatrix} = \begin{bmatrix} 1(6) + 5(-2) + 6(8) & 1(4) + 5(0) + 6(0) \\ 2(6) - 4(-2) + 0(8) & 2(4) - 4(0) + 0(0) \end{bmatrix}$$

$$= \begin{bmatrix} 44 & 4 \\ 20 & 8 \end{bmatrix}$$

111.
$$\begin{bmatrix} 4 \\ 6 \end{bmatrix} \begin{bmatrix} 6 & -2 \end{bmatrix} = \begin{bmatrix} 4(6) & 4(-2) \\ 6(6) & 6(-2) \end{bmatrix} = \begin{bmatrix} 24 & -8 \\ 36 & -12 \end{bmatrix}$$

113.
$$\begin{bmatrix} 2 & 1 \\ 6 & 0 \end{bmatrix} \left(\begin{bmatrix} 4 & 2 \\ -3 & 1 \end{bmatrix} + \begin{bmatrix} -2 & 4 \\ 0 & 4 \end{bmatrix} \right) = \begin{bmatrix} 2 & 1 \\ 6 & 0 \end{bmatrix} \begin{bmatrix} 2 & 6 \\ -3 & 5 \end{bmatrix}$$

$$= \begin{bmatrix} 2(2) + 1(-3) & 2(6) + 1(5) \\ 6(2) + 0 & 6(6) + 0 \end{bmatrix}$$

$$= \begin{bmatrix} 1 & 17 \\ 12 & 36 \end{bmatrix}$$

115.
$$\begin{bmatrix} 4 & 1 \\ 11 & -7 \\ 12 & 3 \end{bmatrix} \begin{bmatrix} 3 & -5 & 6 \\ 2 & -2 & -2 \end{bmatrix} = \begin{bmatrix} 14 & -22 & 22 \\ 19 & -41 & 80 \\ 42 & -66 & 66 \end{bmatrix}$$

117.
$$\begin{cases} 5x + 4y = 2 \\ -x + y = -22 \end{cases}$$

(a)
$$\begin{bmatrix} 5 & 4 \\ -1 & 1 \end{bmatrix} \begin{bmatrix} x \\ y \end{bmatrix} = \begin{bmatrix} 2 \\ -22 \end{bmatrix}$$

(b)
$$\begin{bmatrix} 5 & 4 & \vdots & 2 \\ -1 & 1 & \vdots & -22 \end{bmatrix} \; 4R_2 + R_1 \to \begin{bmatrix} 1 & 8 & \vdots & -86 \\ -1 & 1 & \vdots & -22 \end{bmatrix} \; R_1 + R_2 \to \begin{bmatrix} 1 & 8 & \vdots & -86 \\ 0 & 9 & \vdots & -108 \end{bmatrix}$$

$$\tfrac{1}{9}R_2 \to \begin{bmatrix} 1 & 8 & \vdots & -86 \\ 0 & 1 & \vdots & -12 \end{bmatrix} \; -8R_2 + R_1 \to \begin{bmatrix} 1 & 0 & \vdots & 10 \\ 0 & 1 & \vdots & -12 \end{bmatrix}$$

Solution: $x = 10$, $y = -12$

119. $BA = \begin{bmatrix} 10.25 & 14.50 & 17.75 \end{bmatrix} \begin{bmatrix} 8200 & 7400 \\ 6500 & 9800 \\ 5400 & 4800 \end{bmatrix} = \begin{bmatrix} \$274{,}150 & \$303{,}150 \end{bmatrix}$

The merchandise shipped to warehouse 1 is worth \$274,150, and the merchandise shipped to warehouse 2 is worth \$303,150.

121. $AB = \begin{bmatrix} -4 & -1 \\ 7 & 2 \end{bmatrix}\begin{bmatrix} -2 & -1 \\ 7 & 4 \end{bmatrix} = \begin{bmatrix} -4(-2) + (-1)(7) & -4(-1) + (-1)(4) \\ 7(-2) + 2(7) & 7(-1) + 2(4) \end{bmatrix}$

$ = \begin{bmatrix} 1 & 0 \\ 0 & 1 \end{bmatrix} = I$

$BA = \begin{bmatrix} -2 & -1 \\ 7 & 4 \end{bmatrix}\begin{bmatrix} -4 & -1 \\ 7 & 2 \end{bmatrix} = \begin{bmatrix} -2(-4) + (-1)(7) & -2(-1) + (-1)(2) \\ 7(-4) + 4(7) & 7(-1) + 4(2) \end{bmatrix}$

$ = \begin{bmatrix} 1 & 0 \\ 0 & 1 \end{bmatrix} = I$

123. $[A \; : \; I] = \begin{bmatrix} -6 & 5 & : & 1 & 0 \\ -5 & 4 & : & 0 & 1 \end{bmatrix}$

$-\tfrac{1}{6}R_1 \rightarrow \begin{bmatrix} 1 & -\tfrac{5}{6} & : & -\tfrac{1}{6} & 0 \\ -5 & 4 & : & 0 & 1 \end{bmatrix}$

$5R_1 + R_2 \rightarrow \begin{bmatrix} 1 & -\tfrac{5}{6} & : & -\tfrac{1}{6} & 0 \\ 0 & -\tfrac{1}{6} & : & -\tfrac{5}{6} & 1 \end{bmatrix}$

$-6R_2 \rightarrow \begin{bmatrix} 1 & -\tfrac{5}{6} & & -\tfrac{1}{6} & 0 \\ 0 & 1 & & 5 & -6 \end{bmatrix}$

$\tfrac{5}{6}R_2 + R_1 \rightarrow \begin{bmatrix} 1 & 0 & : & 4 & -5 \\ 0 & 1 & : & 5 & -6 \end{bmatrix} = [I \; : \; A^{-1}]$

$A^{-1} = \begin{bmatrix} 4 & -5 \\ 5 & -6 \end{bmatrix}$

125. $\begin{bmatrix} 2 & 0 & 3 \\ -1 & 1 & 1 \\ 2 & -2 & 1 \end{bmatrix}^{-1} = \begin{bmatrix} \tfrac{1}{2} & -1 & -\tfrac{1}{2} \\ \tfrac{1}{2} & -\tfrac{2}{3} & -\tfrac{5}{6} \\ 0 & \tfrac{2}{3} & \tfrac{1}{3} \end{bmatrix}$

127. $A = \begin{bmatrix} -7 & 2 \\ -8 & 2 \end{bmatrix}$

$A^{-1} = \dfrac{1}{-7(2) - 2(-8)}\begin{bmatrix} 2 & -2 \\ 8 & -7 \end{bmatrix} = \dfrac{1}{2}\begin{bmatrix} 2 & -2 \\ 8 & -7 \end{bmatrix} = \begin{bmatrix} 1 & -1 \\ 4 & -\tfrac{7}{2} \end{bmatrix}$

129. $A = \begin{bmatrix} -\tfrac{1}{2} & 20 \\ \tfrac{3}{10} & -6 \end{bmatrix}$

$A^{-1} = \dfrac{1}{-\tfrac{1}{2}(-6) - 20\left(\tfrac{3}{10}\right)}\begin{bmatrix} -6 & -20 \\ -\tfrac{3}{10} & -\tfrac{1}{2} \end{bmatrix} = -\dfrac{1}{3}\begin{bmatrix} -6 & -20 \\ -\tfrac{3}{10} & -\tfrac{1}{2} \end{bmatrix}$

$ = \begin{bmatrix} 2 & \tfrac{20}{3} \\ \tfrac{1}{10} & \tfrac{1}{6} \end{bmatrix}$

131. $\begin{cases} -x + 4y = 8 \\ 2x - 7y = -5 \end{cases}$

$\begin{bmatrix} x \\ y \end{bmatrix} = \begin{bmatrix} -1 & 4 \\ 2 & -7 \end{bmatrix}^{-1}\begin{bmatrix} 8 \\ -5 \end{bmatrix} = \begin{bmatrix} 7 & 4 \\ 2 & 1 \end{bmatrix}\begin{bmatrix} 8 \\ -5 \end{bmatrix}$

$ = \begin{bmatrix} 7(8) + 4(-5) \\ 2(8) + 1(-5) \end{bmatrix} = \begin{bmatrix} 36 \\ 11 \end{bmatrix}$

Solution: $(36, 11)$

133. $\begin{cases} -3x + 10y = 8 \\ 5x - 17y = -13 \end{cases}$

$\begin{bmatrix} x \\ y \end{bmatrix} = \begin{bmatrix} -3 & 10 \\ 5 & -17 \end{bmatrix}^{-1}\begin{bmatrix} 8 \\ -13 \end{bmatrix} = \begin{bmatrix} -17 & -10 \\ -5 & -3 \end{bmatrix}\begin{bmatrix} 8 \\ -13 \end{bmatrix}$

$ = \begin{bmatrix} -17(8) + (-10)(-13) \\ -5(8) + (-3)(-13) \end{bmatrix} = \begin{bmatrix} -6 \\ -1 \end{bmatrix}$

Solution: $(-6, -1)$

135. $\begin{cases} 3x + 2y - z = 6 \\ x - y + 2z = -1 \\ 5x + y + z = 7 \end{cases}$

$\begin{bmatrix} x \\ y \\ z \end{bmatrix} = \begin{bmatrix} 3 & 2 & -1 \\ 1 & -1 & 2 \\ 5 & 1 & 1 \end{bmatrix}^{-1} \begin{bmatrix} 6 \\ -1 \\ 7 \end{bmatrix} = \begin{bmatrix} -1 & -1 & 1 \\ 3 & \frac{8}{3} & -\frac{7}{3} \\ 2 & \frac{7}{3} & -\frac{5}{3} \end{bmatrix} \begin{bmatrix} 6 \\ -1 \\ 7 \end{bmatrix}$

$= \begin{bmatrix} -1(6) - 1(-1) + 1(7) \\ 3(6) + \frac{8}{3}(-1) - \frac{7}{3}(7) \\ 2(6) + \frac{7}{3}(-1) - \frac{5}{3}(7) \end{bmatrix} = \begin{bmatrix} 2 \\ -1 \\ -2 \end{bmatrix}$

Solution: $(2, -1, -2)$

137. $\begin{cases} -2x + y + 2z = -13 \\ -x - 4y + z = -11 \\ -y - z = 0 \end{cases}$

$\begin{bmatrix} x \\ y \\ z \end{bmatrix} = \begin{bmatrix} -2 & 1 & 2 \\ -1 & -4 & 1 \\ 0 & -1 & -1 \end{bmatrix}^{-1} \begin{bmatrix} -13 \\ -11 \\ 0 \end{bmatrix} = \begin{bmatrix} -\frac{5}{9} & \frac{1}{9} & -1 \\ \frac{1}{9} & -\frac{2}{9} & 0 \\ -\frac{1}{9} & \frac{2}{9} & -1 \end{bmatrix} \begin{bmatrix} -13 \\ -11 \\ 0 \end{bmatrix}$

$= \begin{bmatrix} -\frac{5}{9}(-13) + \frac{1}{9}(-11) - 1(0) \\ \frac{1}{9}(-13) - \frac{2}{9}(-11) + 0(0) \\ -\frac{1}{9}(-13) + \frac{2}{9}(-11) - 1(0) \end{bmatrix} = \begin{bmatrix} 6 \\ 1 \\ -1 \end{bmatrix}$

Solution: $(6, 1, -1)$

139. $\begin{cases} x + 2y = -1 \\ 3x + 4y = -5 \end{cases}$

$\begin{bmatrix} x \\ y \end{bmatrix} = \begin{bmatrix} 1 & 2 \\ 3 & 4 \end{bmatrix}^{-1} \begin{bmatrix} -1 \\ -5 \end{bmatrix} = \begin{bmatrix} -2 & 1 \\ \frac{3}{2} & -\frac{1}{2} \end{bmatrix} \begin{bmatrix} -1 \\ -5 \end{bmatrix} = \begin{bmatrix} -3 \\ 1 \end{bmatrix}$

Solution: $(-3, 1)$

141. $\begin{cases} -3x - 3y - 4z = 2 \\ y + z = -1 \\ 4x + 3y + 4z = -1 \end{cases}$

$\begin{bmatrix} x \\ y \\ z \end{bmatrix} = \begin{bmatrix} -3 & -3 & -4 \\ 0 & 1 & 1 \\ 4 & 3 & 4 \end{bmatrix}^{-1} \begin{bmatrix} 2 \\ -1 \\ -1 \end{bmatrix} = \begin{bmatrix} 1 & 0 & 1 \\ 4 & 4 & 3 \\ -4 & -3 & -3 \end{bmatrix} \begin{bmatrix} 2 \\ -1 \\ -1 \end{bmatrix} = \begin{bmatrix} 1 \\ 1 \\ -2 \end{bmatrix}$

Solution: $(1, 1, -2)$

143. $\begin{vmatrix} 8 & 5 \\ 2 & -4 \end{vmatrix} = 8(-4) - 5(2) = -42$

145. $\begin{vmatrix} 50 & -30 \\ 10 & 5 \end{vmatrix} = 50(5) - (-30)(10) = 550$

147. $\begin{bmatrix} 2 & -1 \\ 7 & 4 \end{bmatrix}$

(a) $M_{11} = 4$
$M_{12} = 7$
$M_{21} = -1$
$M_{22} = 2$

(b) $C_{11} = M_{11} = 4$
$C_{12} = -M_{12} = -7$
$C_{21} = -M_{21} = 1$
$C_{22} = M_{22} = 2$

149. $\begin{bmatrix} 3 & 2 & -1 \\ -2 & 5 & 0 \\ 1 & 8 & 6 \end{bmatrix}$

(a) $M_{11} = \begin{vmatrix} 5 & 0 \\ 8 & 6 \end{vmatrix} = 30$

     $M_{12} = \begin{vmatrix} -2 & 0 \\ 1 & 6 \end{vmatrix} = -12$

     $M_{13} = \begin{vmatrix} -2 & 5 \\ 1 & 8 \end{vmatrix} = -21$

     $M_{21} = \begin{vmatrix} 2 & -1 \\ 8 & 6 \end{vmatrix} = 20$

     $M_{22} = \begin{vmatrix} 3 & -1 \\ 1 & 6 \end{vmatrix} = 19$

     $M_{23} = \begin{vmatrix} 3 & 2 \\ 1 & 8 \end{vmatrix} = 22$

     $M_{31} = \begin{vmatrix} 2 & -1 \\ 5 & 0 \end{vmatrix} = 5$

     $M_{32} = \begin{vmatrix} 3 & -1 \\ -2 & 0 \end{vmatrix} = -2$

     $M_{33} = \begin{vmatrix} 3 & 2 \\ -2 & 5 \end{vmatrix} = 19$

(b) $C_{11} = M_{11} = 30$

     $C_{12} = -M_{12} = 12$

     $C_{13} = M_{13} = -21$

     $C_{21} = -M_{21} = -20$

     $C_{22} = M_{22} = 19$

     $C_{23} = -M_{23} = -22$

     $C_{31} = M_{31} = 5$

     $C_{32} = -M_{32} = 2$

     $C_{33} = M_{33} = 19$

151. Expand using Column 2.

$$\begin{vmatrix} -2 & 4 & 1 \\ -6 & 0 & 2 \\ 5 & 3 & 4 \end{vmatrix} = -4\begin{vmatrix} -6 & 2 \\ 5 & 4 \end{vmatrix} - 3\begin{vmatrix} -2 & 1 \\ -6 & 2 \end{vmatrix}$$

$$= -4(-34) - 3(2) = 130$$

153. Expand along Row 1.

$$\begin{vmatrix} 3 & 0 & -4 & 0 \\ 0 & 8 & 1 & 2 \\ 6 & 1 & 8 & 2 \\ 0 & 3 & -4 & 1 \end{vmatrix} = 3\begin{vmatrix} 8 & 1 & 2 \\ 1 & 8 & 2 \\ 3 & -4 & 1 \end{vmatrix} + (-4)\begin{vmatrix} 0 & 8 & 2 \\ 6 & 1 & 2 \\ 0 & 3 & 1 \end{vmatrix}$$

$$= 3[8(8 - (-8)) - 1(1 - 6) + 2(-4 - 24)] - 4[0 - 6(8 - 6) + 0]$$

$$= 3[128 + 5 - 56] - 4[-12]$$

$$= 279$$

155. Points: $(0, 0), (-2, 2)$

Equation: $\begin{vmatrix} x & y & 1 \\ 0 & 0 & 1 \\ -2 & 2 & 1 \end{vmatrix} - \begin{vmatrix} x & y \\ -2 & 2 \end{vmatrix} = -(2x + 2y) = 0$ or $x + y = 0$

157. Points: $(10, 7), (-2, -7)$

Equation:

$$\begin{vmatrix} x & y & 1 \\ 10 & 7 & 1 \\ -2 & -7 & 1 \end{vmatrix} = \begin{vmatrix} 10 & 7 \\ -2 & -7 \end{vmatrix} - \begin{vmatrix} x & y \\ -2 & -7 \end{vmatrix} + \begin{vmatrix} x & y \\ 10 & 7 \end{vmatrix} = -70 + 14 - (-7x + 2y) + 7x - 10y = 0 \text{ or } 7x - 6y - 28 = 0$$

159. Points: $(-3, 10)$, $(1, -2)$

Equation: $\begin{vmatrix} x & y & 1 \\ -3 & 10 & 1 \\ 1 & -2 & 1 \end{vmatrix} = x \begin{vmatrix} 10 & 1 \\ -2 & 1 \end{vmatrix} - y \begin{vmatrix} -3 & 1 \\ 1 & 1 \end{vmatrix} + \begin{vmatrix} -3 & 10 \\ 1 & -2 \end{vmatrix} = 12x + 4y - 4 = 0 \Longrightarrow 3x + y - 1 = 0$

Problem Solving for Chapter 14

1. (a)

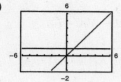

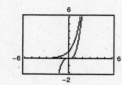

(b) If b is even, there are three points of intersection.
If b is odd, there are two points of intersection.

3.

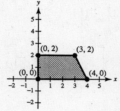

System of Inequalities

$\begin{cases} 0 \le y \le 2 \\ x \ge 0 \\ y \le -2x + 8 \end{cases}$

Find the equations of the lines through each pair of adjacent vertices.

$(0, 0)$, $(0, 2)$: $x = 0$

$(0, 0)$, $(4, 0)$: $y = 0$

$(0, 2)$, $(3, 2)$: $y = 2$

$(3, 2)$, $(4, 0)$: $y = -2x + 8$

5. $z = a$
$y = -4a + 1$
$x = -3a - 2$

One possible system is:

$\begin{cases} x + y + 7z = (-3a - 2) + (-4a + 1) + 7a = -1 \\ x + 2y + 11z = (-3a - 2) + 2(-4a + 1) + 11a = 0 \\ 2x + y + 10z = 2(-3a - 2) + (-4a + 1) + 10a = -3 \end{cases}$

or

$\begin{cases} x + y + 7z = -1 \\ x + 2y + 11z = 0 \\ 2x + y + 10z = -3 \end{cases}$

(Note that the coefficients of x, y, and z have been chosen so that the a–terms cancel.)

7. $AC = \begin{bmatrix} 0 & 1 \\ 0 & 1 \end{bmatrix}\begin{bmatrix} 2 & 3 \\ 2 & 3 \end{bmatrix} = \begin{bmatrix} 2 & 3 \\ 2 & 3 \end{bmatrix}$

$BC = \begin{bmatrix} 1 & 0 \\ 1 & 0 \end{bmatrix}\begin{bmatrix} 2 & 3 \\ 2 & 3 \end{bmatrix} = \begin{bmatrix} 2 & 3 \\ 2 & 3 \end{bmatrix}$

Thus, $AC = BC$ even though $A \ne B$.

9. $A = \begin{bmatrix} 1 & 2 \\ -2 & 1 \end{bmatrix}$

(a) $A^2 - 2A + 5I = \begin{bmatrix} 1 & 2 \\ -2 & 1 \end{bmatrix}\begin{bmatrix} 1 & 2 \\ -2 & 1 \end{bmatrix} - 2\begin{bmatrix} 1 & 2 \\ -2 & 1 \end{bmatrix} + 5\begin{bmatrix} 1 & 0 \\ 0 & 1 \end{bmatrix}$

$= \begin{bmatrix} -3 & 4 \\ -4 & -3 \end{bmatrix} + \begin{bmatrix} -2 & -4 \\ 4 & -2 \end{bmatrix} + \begin{bmatrix} 5 & 0 \\ 0 & 5 \end{bmatrix}$

$= \begin{bmatrix} 0 & 0 \\ 0 & 0 \end{bmatrix} = 0$

(b) $A^{-1} = \frac{1}{(1) - (-4)}\begin{bmatrix} 1 & -2 \\ 2 & 1 \end{bmatrix} = \frac{1}{5}\begin{bmatrix} 1 & -2 \\ 2 & 1 \end{bmatrix}$

$\frac{1}{5}(2I - A) = \frac{1}{5}\left[\begin{bmatrix} 2 & 0 \\ 0 & 2 \end{bmatrix} - \begin{bmatrix} 1 & 2 \\ -2 & 1 \end{bmatrix}\right] = \frac{1}{5}\begin{bmatrix} 1 & -2 \\ 2 & 1 \end{bmatrix}$

Thus, $A^{-1} = \frac{1}{5}(2I - A)$.

(c) $A^2 - 2A + 5I = 0$

$A^2 - 2A = -5I$

$(A - 2I)A = -5I$

$-\frac{1}{5}(A - 2I)A = -5I$

$\frac{1}{5}(2I - A)A = I$

Thus, $A^{-1} = \frac{1}{5}(2I - A)$.

11. $AA^{-1} = \begin{bmatrix} a & b \\ c & d \end{bmatrix}\left(\frac{1}{ad - bc}\right)\begin{bmatrix} d & -b \\ -c & a \end{bmatrix} = \frac{1}{ad - bc}\begin{bmatrix} a & b \\ c & d \end{bmatrix}\begin{bmatrix} d & -b \\ -c & a \end{bmatrix}$

$= \frac{1}{ad - bc}\begin{bmatrix} ad - bc & 0 \\ 0 & ad - bc \end{bmatrix} = \begin{bmatrix} 1 & 0 \\ 0 & 1 \end{bmatrix}$

$A^{-1}A = \frac{1}{ad - bc}\begin{bmatrix} d & -b \\ -c & a \end{bmatrix}\begin{bmatrix} a & b \\ c & d \end{bmatrix} = \frac{1}{ad - bc}\begin{bmatrix} ad - bc & 0 \\ 0 & ad - bc \end{bmatrix} = \begin{bmatrix} 1 & 0 \\ 0 & 1 \end{bmatrix}$

13. (a) $\begin{vmatrix} 4 & 5 & 6 \\ 7 & 8 & 9 \\ 10 & 11 & 12 \end{vmatrix} = 0$

$\begin{vmatrix} 33 & 34 & 35 \\ 36 & 37 & 38 \\ 39 & 40 & 41 \end{vmatrix} = 0$ $\qquad$ $\begin{vmatrix} -5 & -4 & -3 \\ -2 & -1 & 0 \\ 1 & 2 & 3 \end{vmatrix} = 0$

$\begin{vmatrix} 19 & 20 & 21 & 22 \\ 23 & 24 & 25 & 26 \\ 27 & 28 & 29 & 30 \\ 31 & 32 & 33 & 34 \end{vmatrix} = 0$ $\qquad$ $\begin{vmatrix} 57 & 58 & 59 & 60 \\ 61 & 62 & 63 & 64 \\ 65 & 66 & 67 & 68 \\ 69 & 70 & 71 & 72 \end{vmatrix} = 0$

For an $n \times n$ matrix ($n > 2$) with consecutive integer entries, the determinant appears to be 0.

(b) For the 3×3 case we have the following:

$\begin{vmatrix} x & x + 1 & x + 2 \\ x + 3 & x + 4 & x + 5 \\ x + 6 & x + 7 & x + 8 \end{vmatrix} = x\begin{vmatrix} x + 4 & x + 5 \\ x + 7 & x + 8 \end{vmatrix} - (x + 1)\begin{vmatrix} x + 3 & x + 5 \\ x + 6 & x + 8 \end{vmatrix} + (x + 2)\begin{vmatrix} x + 3 & x + 4 \\ x + 6 & x + 7 \end{vmatrix}$

$= x[(x + 4)(x + 8) - (x + 7)(x + 5)] - (x + 1)[(x + 3)(x + 8)$

$- (x + 6)(x + 5)] + (x + 2)[(x + 3)(x + 7) - (x + 6)(x + 4)]$

$= x[(x^2 + 12x + 32) - (x^2 + 12x + 35)] - (x + 1)[(x^2 + 11x + 24)$

$- (x^2 + 11x + 30)] + (x + 2)[(x^2 + 10x + 21) - (x^2 + 10x + 24)]$

$= -3x - (x + 1)(-6) + (x + 2)(-3)$

$= -3x + 6x + 6 - 3x - 6 = 0$

15. There are a finite number of solutions.

    (a) If both equations are linear, then the maximum number of solutions to a finite system is *one*.

    (b) If one equation is linear and the other is quadratic, the the maximum number of solutions is *two*.

    (c) If both equations are quadratic, then the maximum number of solutions is *four*.

17. $6 = \pm\dfrac{1}{2}\begin{vmatrix} -2 & -3 & 1 \\ 1 & -1 & 1 \\ -8 & x & 1 \end{vmatrix}$

$\pm 12 = \begin{vmatrix} 1 & -1 \\ -8 & x \end{vmatrix} - \begin{vmatrix} -2 & -3 \\ -8 & x \end{vmatrix} + \begin{vmatrix} -2 & -3 \\ 1 & -1 \end{vmatrix}$

$\pm 12 = (x - 8) - (-2x - 24) + 5$

$\pm 12 = 3x + 21$

$x = \dfrac{-21 \pm 12}{3} = -7 \pm 4$

$x = -3$ or $x = -11$

19. Answers will vary. One example is:

$A = \begin{bmatrix} 1 & 0 \\ -1 & 0 \end{bmatrix}$

21. $(a - b)(b - c)(c - a)(a + b + c) = -a^3b + a^3c + ab^3 - ac^3 - b^3c + bc^3$

$\begin{vmatrix} 1 & 1 & 1 \\ a & b & c \\ a^3 & b^3 & c^3 \end{vmatrix} = \begin{vmatrix} b & c \\ b^3 & c^3 \end{vmatrix} - \begin{vmatrix} a & c \\ a^3 & c^3 \end{vmatrix} + \begin{vmatrix} a & b \\ a^3 & b^3 \end{vmatrix}$

$= bc^3 - b^3c - ac^3 + a^3c + ab^3 - a^3b$

Thus, $\begin{vmatrix} 1 & 1 & 1 \\ a & b & c \\ a^3 & b^3 & c^3 \end{vmatrix} = (a - b)(b - c)(c - a)(a + b + c).$

23. $\begin{vmatrix} x & 0 & 0 & d \\ -1 & x & 0 & c \\ 0 & -1 & x & b \\ 0 & 0 & -1 & a \end{vmatrix} = x\begin{vmatrix} x & 0 & c \\ -1 & x & b \\ 0 & -1 & a \end{vmatrix} - d\begin{vmatrix} -1 & x & 0 \\ 0 & -1 & x \\ 0 & 0 & -1 \end{vmatrix}$

$= \underbrace{x(ax^2 + bx + c)}_{\text{From Exercise 22}} - d\left(-\begin{vmatrix} -1 & x \\ 0 & -1 \end{vmatrix}\right)$

$= ax^3 + bx^2 + cx + d$